Low-Voltage/Low-Power Integrated Circuits and Systems

IEEE Press
445 Hoes Lane, P.O. Box 1331
Piscataway, NJ 08855-1331

IEEE Circuits and Systems Society, *Sponsor*
CAS-S Liaison to IEEE Press, Gordon Roberts
IEEE Solid-State Circuits Society, *Sponsor*
SSC-S Liaison to IEEE Press, Stuart K. Tewksbury

Cover design: William T. Donnelly, *WT Design*

Technical Reviewers

Prof. John Choma, Jr., *University of Southern California*
Frode Larsen, *Lucent Technologies*

Low-Voltage/Low-Power Integrated Circuits and Systems

Low-Voltage Mixed-Signal Circuits

Edited by

Edgar Sánchez-Sinencio
Texas A&M University
Analog and Mixed Signal Center

Andreas G. Andreou
Johns Hopkins University
Department of Electrical and Computer Engineering

IEEE Circuits and Systems Society, *Sponsor*

IEEE Solid-State Circuits Society, *Sponsor*

IEEE Press Series on Microelectronic Systems
Stuart K. Tewksbury, *Series Editor*

The Institute of Electrical and Electronics Engineers, Inc., New York

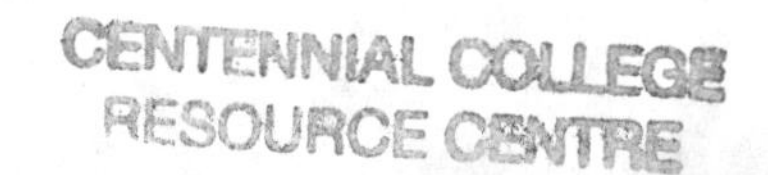

This book and other books may be purchased at a discount
from the publisher when ordered in bulk quantities. Contact:

IEEE Press Marketing
Attn: Special Sales
Piscataway, NJ 08855-1331
Fax: (732) 981-9334

For more information about IEEE PRESS products,
visit the IEEE Home Page: http://www.ieee.org/

Printed in the United States of America

10 9 8 7 6 5 4 3 2 1

ISBN 0-7803-3446-9
IEEE Order Number: PC4341

Library of Congress Cataloging-in-Publication Data

Low-voltage/low-power integrated circuits and systems: low-voltage mixed-signal circuits / edited by Edgar Sánchez-Sinencio, Andreas G. Andreou.
p. cm.—(IEEE Press series on microelectronic systems)
"IEEE Solid-State Circuits Society, sponsor"
"IEEE Circuits and Systems Society, sponsor"
Includes bibliographical references and index
ISBN 0-7803-3446-9
1. Low voltage integrated circuits. I. Sánchez-Sinencio, Edgar.
II. Andreou, Andreas G. III. IEEE Solid-State Circuits Society.
IV. Series.
TK7874.66.L69 1998
621.3815—dc21 98-36702
CIP

To
Rene, Nellyda, Feliciano,
and Maria Esther

and

To
Gregory

Contents

Foreword

The design of low-voltage and low-power circuits is a most timely subject in electronics. The need for such circuits is immense: Portable computers and communications devices are becoming so pervasive that hardly a day goes by without one using such a device. And, of course, more traditional applications such as medical devices are still with us and are increasing in number and scope.

The design of low-voltage and low-power circuits poses difficult challenges for all involved: process, device, circuit, and system designers. In the last few years, we have seen advances in process technologies, in device modeling for computer simulation, in circuit design techniques, and in approaches to system design—all aimed at the production of electronic systems that operate from very low supply voltages and that dissipate very low power. These electronic systems employ both analog and digital circuitry and many mix the two types in an optimal manner.

The editors of this book have assembled a distinguished group of contributors who address nearly every aspect of low-voltage and low-power design. They provide a snapshot of the state of the art in this rapidly evolving area. Their contributions range from device modeling to novel systems, with a great deal of emphasis placed on the constraints and trade-offs available in the design of such circuits. The book includes a wealth of design ideas for such fundamental building blocks as amplifiers, logic gates, filters, and data converters.

This volume should prove invaluable to engineers involved in this area of design and to researchers who want to find out the state of knowledge and the exciting problems that remain to be solved in low-voltage and low-power electronics.

Adel S. Sedra
Toronto, Canada

Preface

Modern analog circuits and systems of mixed-signal very large scale integrated (VLSI) chips for multimedia, perception, control, instrumentation, and telecommunications operate with much smaller supply voltages than the discrete circuits and integrated circuits of the last decade. This new environment and needs have forced designers to develop new design approaches more amenable to low-voltage and low-power integrated circuits. In this book, we are the first to attempt to bring together these new design methodologies together with a number of applications in an understandable manner. Furthermore, since the design of analog and mixed-signal circuits with low voltage requires suitable system architectures, we discuss proper circuit models, optimization, and innovative circuit design. In the past, a circuit designer required little or no knowledge of the overall system and where the circuit was going to be used; these days a complete notion of the system, applications, requirements, and limitations is essential for a circuit designer to optimize a sound circuit design.

This book deals with analog and digital circuit design. When combined, these circuits are known as *mixed-signal circuits*. A particular emphasis of this book is on low-power supply voltages and low-power consumption. The revival of analog circuits in the last few years has been significant. The optimal use of analog circuits versus digital circuits has been a controversial issue, which is application dependent. There is a trend to combine analog and digital circuits in an optimal way. In current applications, high-speed analog front ends impose high speed and power consumption requirements. This book has developed from the need to fill a vacuum of information about low-power and low-voltage integrated circuits. We cover fundamental aspects of metal-oxide-semiconductor (MOS) device modeling, analog and digital building blocks, analog filters, DC-to-DC conversion, as well as analog, digital, and mixed-signal systems, including several important applications.

Several unique features of this book are (1) the introduction of an all-region transistor model that allows optimization of the circuit design; (2) device performance limitations under low-voltage operation; (3) discussion of both continuous-time and switched-capacitor filters including high-frequency design techniques; (4) a key issue regarding the optimal design of DC-to-DC conversion for portable operation; (5) digital circuit design and new directions in low-power digital VLSI design; (6) a complete discussion of data conversion for low-voltage conditions; (7) a host of applications, from very low frequency (a few hertz) to telecom circuits, including a digital approach to solve a YUV-to-RGB converter. The advantages and drawbacks of digital and analog circuits are also discussed at different hierarchical levels. A quick overview of the book is given in Chapter 1. We have tried to present this book as a combination of a bottom-up and a top-down approach. Systems are covered as well as basic device transistor physics, models, and limitations to understand devices. Systems that can operate with less than 2 V are already demanded in industry; thus, it is a must for designers to quickly become acquainted with current low-voltage circuit design techniques. Even though the applications presented in this book are a reflection of present trends, we also cover a set of underlying concepts required to understand low-voltage circuit design.

This book is intended as a key reference for upper level courses as well as for graduate courses. It is also a resource for practicing professional engineers employed in circuit and system design involving both digital and analog circuits.

Edgar Sánchez-Sinencio
Andreas G. Andreou

Acknowledgments

We thank our colleagues who contributed to this book. The editors also thank the reviewers. We also thank Ella Gallagher for her valuable assistance during the preparation of the book and the former and present Director of the IEEE Press, Dudley Kay and Ken Moore, for their facilitation in producing the book. Foremost, we thank Assistant Editor Marilyn G. Catis, Production Editor Surendra Bhimani, and Copy Editor Beatrice Ruberto for their excellent work on this volume. Edgar Sánchez-Sinencio wants to acknowledge the financial support from the National Science Foundation and, in particular, from the Mixed-Signal Group from Texas Instruments. He gratefully acknowledges the constant encouragement, understanding, and love from his wife, Yolanda (Jamy), as well as the use of family time to accomplish this book. Andreas Andreou acknowledges support from the National Science Foundation and the ONR MURI on automated sensing and vision systems.

Edgar Sánchez-Sinencio
Andreas G. Andreou

Contributors

Andreas G. Andreou Department of Electrical and Computer Engineering, Johns Hopkins University, 105 Barton Hall/3400 North Charles Street, Baltimore, MD 21218-2686; email: andreou@jhu.edu

Jacob H. Botma MESA Research Institute, Twente University, P.O. Box 217, 7500AE Enschede, The Netherlands

Derek F. Bowers Analog Devices, Precision Monolithics, Inc., 1500 Space Park Drive, P.O. Box 58020, MS B4-40, Santa Clara, CA 95050; email: derek.bowers@analog.com

Robert W. Brodersen Department of Electrical Engineering and Computer Sciences, University of California, 211-106 Cory Hall #1772, Berkeley, CA 94720-1772; email: rwb@eecs.berkeley.edu

Navin Chaddha Computer Systems Laboratory, Standford University, CIS-132, MC 4070, Stanford, CA 94305-4070; email: navin@leland.standford.edu

Richard Coggins University of Sydney, System Engineering and Design Automation Laboratory, Department of Electrical Engineering, J03 NSW 2006, Australia; email: richardc@sedal.usyd.edu.au

J. Crols Departement Elektrotechniek Afdeling ESAT, Katholieke Universiteit Leuven, Kardinaal Mercierlaan 94, B-3001 Heverlee, Belgium; email: jan.crols@esat.kuleuven.ac.be

Ana Isabela Araújo Cunha Departamento de Engenharia Elétrica, Escola Politécnica, UFBA 40210-630 Salvador, BA, Brazil; email: aiac@ufba.br

Klaas-Jan de Langen Delft Institute of Microelectronics and Submicron Technology (DIMES), Delft University of Technology, Mekelweg 4, 2628 CD Delft, The Netherlands; email: kjlangen@et.tudelft.nl

Sherif H. K. Embabi Department of Electrical Engineering, Texas A&M University, College Station, TX 77843-3128; email: embabi@eesun2.tamu.edu, http://amesp02.tamu.edu

Rudy G. H. Eschauzier Delft Institute of Microelectronics and Submicron Technology (DIMES), Delft University of Technology, Mekelweg 4, 2628 CD Delft, The Netherlands; email: reschauzier@et.tudelft.nl

Paul M. Furth Department of Electrical and Computer Engineering, New Mexico State University, Las Cruces, New Mexico 88003; email:pfurth@nmsu.edu

Randall L. Geiger Department of Electrical and Computer Engineering, Iowa State University, 201 Coover Hall, Ames, IA 50011-3060; email: rlgeiger@iastate.edu

Sander L. J. Gierkink MESA Research Institute, Twente University, P.O. Box 217, 7500AE Enschede, The Netherlands; email: S.L.G.Gierkink@el.utwente.nl

S. Gogaert Departement Elektrotechniek Afdeling ESAT, Katholieke Universiteit Leuven, Kardinaal Mercierlaan 94, B-3001 Heverlee, Belgium; email: Stan.Gogaert@esat.kuleuven.ac.be

Benjamin M. Gordon Computer Systems Laboratory, Stanford University, CIS-132, MC 4070, Stanford, CA 94305-4070; email: bgordon@snooze.standford.edu

Ron Hogervorst Delft Institute of Microelectronics and Submicron Technology (DIMES), Delft University of Technology, Mekelweg 4, 2628 CD Delft, The Netherlands; email: rhogervorst@et.tudelft.nl

Johan H. Huijsing Delft Institute of Microelectronics and Submicron Technology (DIMES), Delft University of Technology, Mekelweg 4, 2628 CD Delft, The Netherlands; email: jhhuijsing@et.tudelft.nl

Marwan Jabri University of Sydney, System Engineering and Design Automation Laboratory, Department of Electrical Engineering, J03 NSW 2006, Australia; email: marwan@ee.su.oz.au

Vitit Kantabutra Department of Mathematics, Idaho State University, Pocatello, ID 83209; email: kantviti@fs.isu.edu

Teresa H. Meng Computer Systems Laboratory, Stanford University, CIS-132, MC 4070, Stanford, CA 94305-4070; email: teresa@tilden.standford.edu

Carlos Galup Montoro LINSE, Departamento de Engenharia Elétrica, UFSC, CP.476, 88040-900 Florianópolis, SC, Brazil; email: carlos@linse.ufsc.br

Marcel J. M. Pelgrom Philips Research, Nederlandse Philips Bedrijven B.V., Eindhoven, The Netherlands; email: pelgrom@natlab.research.philips.com

Jaime Ramírez-Angulo The Klipsch School of Electrical and Computer Engineering, New Mexico State University, Box 30001/Dept. 3-0, Las Cruces, NM 88003-0001; email: jramirez@nmsu.edu

Edgar Sánchez-Sinencio Department of Electrical Engineering, Texas A&M University, College Station, TX 77843-3128; email: e.sanchez@ieee.org, http://amesp02.tamu.edu

Seth R. Sanders Department of Electrical Engineering and Computer Sciences, University of California, 211-106 Cory Hall #1772, Berkeley, CA 94720-1772; email: sanders@eecs.berkeley.edu

Márcio Cherem Schneider LINSE, Departamento de Engenharia Elétrica, UFSC, CP.476, 88040-900 Florianópolis, SC, Brazil; email: marcio@linse.ufsc.br

Sterling L. Smith Texas Instruments, P.O. Box 655303, MS 8213, Dallas, TX 75243; email: slss@msg.ti.com

Michel Steyaert Departement Elektrotechniek Afdeling ESAT, Katholieke Universiteit Leuven, Kardinaal Mercierlaan 94, B-3001 Heverlee, Belgium; email: Michiel.Steyaert@esat.kuleuven.ac.be

Anthony J. Stratakos Department of Electrical Engineering and Computer Sciences, University of California, 211-106 Cory Hall #1772, Berkeley, CA 94720-1772; email: anthony@eecs.berkeley.edu

Charles R. Sullivan Department of Electrical Engineering and Computer Sciences, University of California, 211-106 Cory Hall #1772, Berkeley, CA 94720-1772; email: charless@cs.berkeley.edu

Paul Vanoostende Alcatel-Bell, Advanced CAD for VLSI, F. Wellelspein 1, B-2018 Antwerp, Belgium

Geert Van Wauwe Alcatel-Bell, Advanced CAD for VLSI, F. Wellelspein 1, B2018 Antwerp, Belgium

Roelof F. Wassenaar MESA Research Institute, Twente University, P.O. Box 217, 7500AE Enschede, The Netherlands; email: R.F. Wassenaar@el.utwente.nl

Remco J. Wiegerink MESA Research Institute, Twente University, P.O. Box 217, 7500AE Enschede, The Netherlands; email: remco@ice.el.utwente.nl

Chong-Gun Yu Department of Electronics Engineering, University of Inchon, 177, Tohwa-dong, Namgu, Inchon, Korea

Chapter 1

Introduction

Edgar Sánchez-Sinencio
Department of Electrical Engineering, Texas A&M University, College Station, Texas 77843-3128

1.1. BACKGROUND

Interest in and evolution of low-voltage supply and low-power circuits have grown rapidly from applications on watches and medical electronics (pacemakers, hearing aids, blood flow meters) to a host of other applications. This increased interest is mainly due to commercial implications of portable equipment, power reduction on non-battery-powered systems, and consumer electronics. Some of these examples are the laptop/notebook computers as well as workstations, PCs, electronic organizers, language translators, electronic dictionaries, implantable devices, portable radios and TV sets. One specially attractive device with growing importance in multimedia and wireless applications is the personal digital assitant, which together with cellular phones, pagers, and palmtops, among others, has revived interest on low-voltage supply and/or low-power consumption system design.

Low-voltage and low-power electronic systems have been pursued continuously since the transistor invention in the 1940s and the integrated circuit (IC) invention in the late 1950s. Until the late 1960s analog techniques were the dominant approaches. With the phenomenal development of microprocessors in the 1980s and the efficient digital computer-aided design (CAD) tools developed in the last years, the digital approach has become the favorite design methodology.

Simultaneously, a number of significant acquirements during the late 1980s using massively interconnected analog signal processors were accomplished. These systems, inspired by biological systems, have converted theory to practical implementations. Among the applications are the artificial cochlea, the artificial retina, and other useful image processors based on neural network concepts. This recent trend plus the fact that low-voltage supply, high-speed digital circuit design problems are becoming an analog problem has motivated researchers both in academia and industry to work in mixed-signal circuits and systems. This edited book deals with both low-voltage/low-power digital and analog circuits and systems. The need to design and fabricate mixed-mode processors in a unified approach is becoming more evident. System designers need to consider the hardware issues at the highest hierarchical design level. The design of low-voltage supply and/or low-power consumption systems requires proper system architecture, suitable circuit models, optimization, and often innovative circuit design.

It is interesting to note that early circuit design was based on bipolar technology; then digital techniques drifted into N-type metal-oxide-semiconductor (NMOS) technology. The dominant digital circuits forced analog circuits to have a common complementary MOS (CMOS) technology; however in recent years a combination of CMOS and bipolar (BiCMOS) has attracted the attention of many designers. This BiCMOS was developed for higher speed performance. Other expensive process technologies have been proposed also for high-frequency circuits. Currently CMOS is still the dominant process tecnology with some serious competition from the BiCMOS technology and bipolar technology. This last process technology is reviving due to a number of RF communication applications.

In the last four years there have been several books as well as special journal issues and special sessions and workshops in several international symposia on this growing and exciting field. The interest shared by industry, academia, and governmental institutions on this topic is reflected by the eager attendance at the increasing number of short courses, workshops, and conferences held recently.

1.2. OVERVIEW OF BOOK

The chapter components encompassed in this book are illustrated in Figure 1.1. The major areas presented in this book deal with six topics:

i. MOS device modeling, constraints, devices, and circuit design methodology
ii. Design and characterization of building blocks, analog and digital, using bipolar, CMOS, and BiCMOS techology
iii. Analog filters using continuous-time as well as switched-capacitor techniques
iv. High-efficiency low-voltage DC-DC conversion for portable applications
v. Low-power CMOS data conversion and new directions in low-power digital CMOS very large scale integrated (VLSI) design
vi. Applications:
    a. A low-power multiplierless YUV-to-RGB converter based on human vision perception
    b. Micropower systems for implantable defibrillators and pacemakers
    c. Bit-energy comparison of signal representations at the circuit level
    d. Low-power design in Telecom circuits

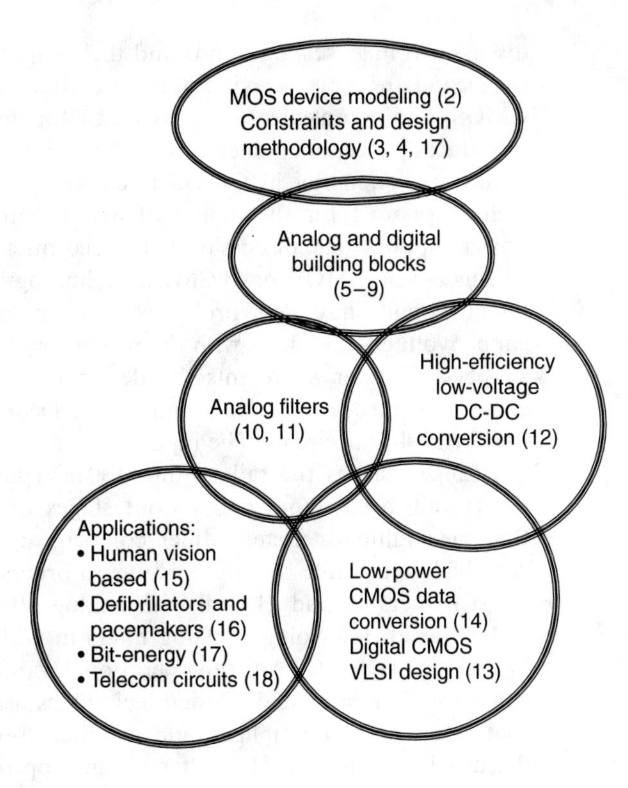

Figure 1.1 Areas of low-voltage, low-power integrated circuits and systems. Corresponding chapters are in parentheses.

A brief description of each chapter follows.

Chapter 2 provides a physically based continuous, all region charge-controlled MOS transistor model for circuit simulation and design. A major emphasis on the modeling is on low-power operation. Behavioral and simplified models are discussed. Noise modeling is presented. Then Chapter 3 discusses the practical limitations of integrated circuits operating under low-power considerations and/or low-voltage power supply. Even though the emphasis of the discussion is on analog circuits, the conclusions can be applied to mixed-signal processors. In Chapter 4, circuit and design methodology are the main focus. The basic device is the MOS transistor. The design constraints for high integration density and micropower operation are met by using current-mode techniques based on the translinear properties and operating the transistors in the subthreshold (weak-inversion) region.

Chapter 5 focuses on the promising characteristics of the floating-gate MOS transistors to tackle the low voltage circuit design problems. This approach uses standard process technology and does not require a variable threshold voltage (V_T) process. In fact, it is shown that the V_T of a floating-gate MOS can be reduced without device scaling and can be programmed and tuned precisely. As a consequence, it is possible to obtain practical low-voltage circuits.

Recent developments in the area of low-power CMOS digital circuits are presented in Chapter 6. This chapter includes two appendices. One of the appendices discusses voltage scaling trends and their impact on the delay of CMOS circuits. In the second appendix, a review of three CMOS logic families—static CMOS, dynamic CMOS, and complementary pass transistor logic—is presented. This chapter also introduces concepts on energy-delay products as a figure of merit and strategies to reduce power/energy in CMOS circuits and systems. Chapter 7 presents several techniques used for the design of analog circuits that operate with low-voltage power supplies. The discussion is focused on continuous-time circuits implemented in single-well CMOS or BiCMOS technology. It addresses the question of why digital circuits have less problems than analog circuits to operate with reduced supply voltages and deals with the problems of analog circuits with reduced-voltage supplies. Chapter 8 presents the design of a low-voltage two-state op-amp. The common-source output state employs a feedback circuit which also contains an MOS translinear circuit. The input range exceeds the supply rails, whereas the output range reaches the rails within 130 mV. Chapter 9 discusses low-voltage/low-power rail-to-rail input and output stages of operational amplifiers. It is shown that the multipath nested Miller compensation combines a very high bandwidth with high gain while being insensitive to process parameters.

Chapters 10 and 11 deal with analog filters. Chapter 10 discusses the design techniques of low-voltage analog filters in CMOS technologies. In particular, two filter design styles, OTA-C (an operational transconductance amplifier) and switched capacitors, are presented. Design techniques using low-V_T transistors, single device switches, voltage multipliers, and a switched-op-amp technique are analyzed and discussed. In Chapter 11 a filter design approach based on a pseudo-differential current-mode integrator is presented. This approach avoids the use of a conventional differential pair, eliminating the bias current of the differential pairs. Thus this approach is very suitable for low-voltage-supply operation. The key basic building blocks, filter architectures, and a practical test-chip example are introduced.

Chapter 12 is intended to describe the application-specific design techniques necessary to meet the rigorous size and efficiency requirements of portable systems based on a CMOS implementation of a low-output-voltage buck converter. In portable systems, high-efficiency low-voltage DC-DC conversion is required to efficiently generate each low-voltage supply from a single battery source.

Chapter 13 describes two methods of low-power CMOS VLSI design. The first involves eliminating races in asynchronous sequential circuits while keeping the circuit size reasonable. The second method of low-power design involves using complex gates and is applicable to general combination functions and to sequential circuits as well.

Chapter 14 deals with low power consumption in A/D and D/A converters. The converters are characterized on the level of application, integrated circuit, implementation, and physics. The main focus is on CMOS circuits intended for embedded operation.

Chapter 15 describes a converter for use in portable video applications. The proposed low-complexity architecture requires only fixed shift-and-adds to perform color conversion, eliminating the need for multiplication while maintaining image quality. It is shown that applying low-power circuit design techniques and perception-based distortion metrics to the design of color converters, the power consumption can be reduced by more than three orders of magnitude.

Chapter 16 introduces methods, circuits, and algorithms for the implementation of analog micropower neural networks to be used in signal processing and pattern recognition for intracardiac electrogram classification.

Chapter 17 proposes an outline of framework to formalize and quantify the notion of low-power information processing, low-power computation. Microelectronic systems are treated as communication channels transmitting messages in the presence of noise.

Chapter 18 shows how a synchronous gated-clock strategy can result in a reduction by 25% of the complete chip-power (including logic, random-access memory, Clock, input/output), while retaining the ease and safety of fully synchronous design.

References

[1] R. J. Widlar, "Low-Voltage Techniques," *IEEE J. Solid-State Circuits*, vol. SC-13, pp. 838–846, 1978.

[2] H. R. Camenzind and R. B. Kash, "A Low Voltage IC timer," *IEEE J. Solid-State Circuits*, vol. SC-13, pp. 847–852, 1978.

[3] C. A. Mead, *Analog VLSI and Neural Systems*, Addison-Wesley, Reading, MA, 1989.

[4] E. Vittoz, "Micropower Techniques," in *Design of VLSI Circuits for Telecommunication and Signal Processing*, J. Franca and Y. Tsivids (eds.), Prentice-Hall, Englewood Cliffs, NJ, 1993.

[5] M. Ismail and T. Fiez (eds.), *Analog VLSI Signal and Information Processing*, McGraw-Hill, New York, 1994.

[6] D. A. Johns and K. Martin, *Analog Integrated Circuit Design*, Wiley, New York, 1997.

[7] R. T. Howe and C. G. Sodini, *Microelectronics: An Integrated Approach*, Prentice-Hall, Upper Saddle River, NJ, 1997.

[8] R. C. Jaeger, *Microelectronic Circuit Design*, McGraw-Hill Co., New York, 1997.

[9] *IEEE Symposium on Low Power Electronics, Digest of Technical Papers*, October 10–12, San Diego, 1994.

[10] A. Bellaouar and M. I. Elmasry, *Low-Power Digital VLSI Design Circuits and Systems*, Kluwer Academic, Boston, 1995.

[11] A. P. Chandrakasan and R. W. Brodersen, *Low-Power Digital CMOS Design*, Kluwer Academic, Boston, 1995.

[12] W. Serdijn, A. C. van der Woerd, and J. C. Kuenen (eds.), Special Issue: Low-Power Analog Integrated Circuits, *Analog Integrated Circuits Signal Process.*, vol. 8, no. 1, pp. 5–123, 1995.

[13] A Rodríguez-Vázquez and E. Sánchez-Sinencio, "Special Issue on Low-voltage and Low Power Analog and Mixed-Signal Circuits and Systems," *IEEE Trans. Circuits Systems I*, vol. 42, no. 11, pp. 825–977, 1995.

[14] W. K. Chen (Ed.-in-Chief), *The Circuits and Filters Handbook*, CRC Press, Boca Raton, FL, 1995.
[15] L. M. Terman and R-H Yan, Editors, Special Issue on Low Power Electronics, *Proc. IEEE*, vol. 83, no. 4, 1995.

Chapter 2

A Current-Based MOSFET Model for Integrated Circuit Design

Carlos Galup Montoro
and Márcio Cherem Schneider
LINSE, Departamento de Engenharia Elétrica, UFSC, 88040–900 Florianópolis, SC, Brazil
Ana Isabela Araújo Cunha
Departamento de Engenharia Elétrica, Escola Politécnica, UFBA, 40210–630, Salvador, BA, Brazil

2.1. INTRODUCTION

The modeling of metal-oxide-semiconductor (MOS) transistors for integrated circuit (IC) design has been driven by the needs of digital circuit simulations for many years. The present trend toward mixed analog-digital chips makes necessary the development of a MOS field-effect transistor (MOSFET) model adapted to analog design as well [1–3].

Another technological trend is toward very large scale integrated (VLSI) low-voltage and high-speed circuits for portable equipment and wireless communication systems [2]. As a consequence of this tendency to shorter channel lengths and reduced supply voltages, MOS devices are expected to often operate in the moderate- and weak-inversion regions [4]. To satisfy the present requirements of IC designers, the desirable properties of a MOSFET model can be summarized [2, 3, 5] as follows:

i. The model should be single piece and continuous and present accurate expressions. Models that use different sets of equations in different regions of device operation often produce large errors or discontinuities in small-signal parameters such as conductances and capacitances. Moreover, the discontinuity in AC parameters can lead to numerical oscillations during DC or transient simulations.

ii. The model should preserve the intrinsic symmetry of the device. A MOSFET model that keeps its symmetry allows straightforward analysis and design of circuits containing MOSFETs acting as switches, voltage-controlled resistors, or current dividers [6].

iii. The model should conserve charge for the correct simulation of charge-sensitive circuits such as random-access memory (RAM) and switched capacitor (SC) and switched current (SI) filters.

iv. It must correctly represent not only the strong- and weak-inversion regions but also the moderate-inversion region, where the MOSFET often operates.

v. The model should have a minimum of independent parameters, all physically based, such that it can be applied to any technology and be useful for statistical analysis.

vi. Finally, analog IC designers need simple expressions to compute transistor dimensions for any current level.

Models based on the surface potential are inherently continuous; however, the MOSFET equations are numerically ill-conditioned when expressed in terms of the surface potential. In effect, in the subthreshold (weak-inversion) regime, the drain current and the channel charge can vary many orders of magnitude while the surface potential varies too modestly. To overcome this problem, the model of [7] uses the approximate linear relation between inversion charge density and surface potential to eliminate the latter from the MOSFET equations. As a consequence, the MOSFET equations are written as functions of the inversion charge density.

This chapter reviews a physically based model for the MOS transistor [7–9], suitable for analysis and design of integrated circuits, and extends it to include short-channel and non-quasi-static effects. This MOSFET model is useful for designing not only high-current circuits but also low-voltage-operated circuits because it accurately represents the moderate- and weak-inversion regions. All the static and dynamic characteristics of the MOSFET, described by single-piece functions with infinite order of continuity for all regions of operation, are expressed in terms of two components of the drain current. Therefore, hand calculation for circuit design is substantially simplified. The proposed model preserves the structural source-drain symmetry of the transistor and uses a reduced number of physical parameters. It is also charge conserving and has explicit equations for the MOSFET capacitances of the quasi-static model and admittances of the non-quasi-static model as well. Simple expressions for the transconductance-to-current ratio, the drain-to-source saturation voltage, and the cutoff frequency in terms of the drain current are given. Short-geometry effects are included by adapting results previously reported in the technical literature to our model. The MOSFET model presented in this chapter has been included in the SMASH [1] circuit simulator.

The basic principles to derive our MOSFET model and the expressions for current and charges are presented in Section 2.2. In Section 2.3, we show small-signal models including noise as well as a non-quasi-static model. Short-channel effects are considered in Section 2.4. Section 2.5 presents the application of our MOSFET model to the design of a common-source amplifier.

2.2. FUNDAMENTALS OF THE MOSFET MODEL

In this section, we present a current-based DC model of the long-channel MOSFET. The fundamental approximation to derive the MOSFET model is the linear relationship between inversion charge density and surface potential. As a consequence, the MOSFET drain current and charges are expressed as very simple functions of two components of the drain current, namely, the forward and reverse saturation currents. A very simple relation between these two components of the drain current and the applied voltages is shown.

2.2.1. Basic Concepts and Definitions

The expressions and discussions that follow are related to the long-channel N-type MOS (NMOS) transistor, illustrated in Figure 2.1. Uniform substrate doping and field-independent mobility have been assumed in our analysis. Unless stated otherwise, voltages are referred to the local substrate, allowing one to exploit the intrinsic symmetry of the device.

In the gradual channel approximation, the relationship between ϕ_S, the surface potential, and V_G, the gate voltage, obtained applying Gauss's law to the MOS structure [10] is

$$V_G - V_{\text{FB}} = \phi_S - \frac{Q'_B + Q'_I}{C'_{\text{ox}}} \tag{2.1}$$

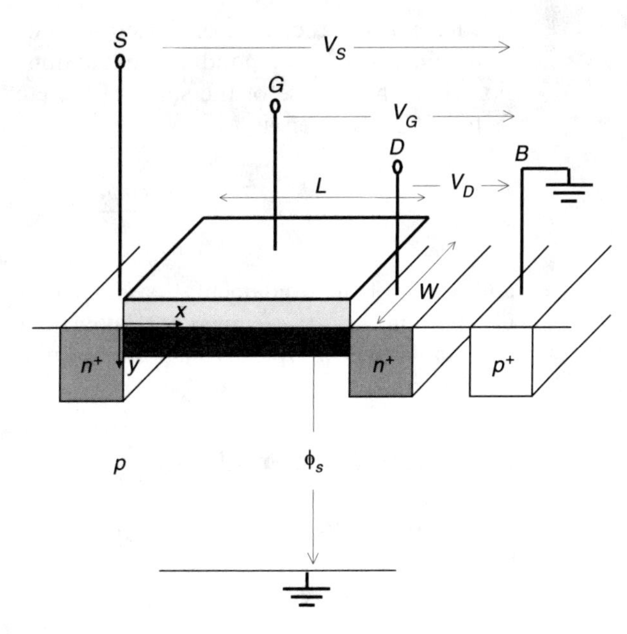

Figure 2.1 Idealized structure of NMOS transistor.

where V_{FB} is the flat-band voltage, C'_{ox} is the oxide capacitance per unit area, and Q'_B and Q'_I are the depletion and inversion charge densities, respectively.

The total semiconductor charge density, calculated by integrating Poisson's equation in the semiconductor [10], is

$$Q'_B + Q'_I = -\gamma C'_{ox}\sqrt{\phi_S + \phi_t e^{(\phi_S - 2\phi_F - V_C)/\phi_t}} \tag{2.2}$$

where ϕ_t is the thermal voltage, V_C is the channel potential, $\phi_F + V_C$ is the electron quasi-Fermi potential, and γ is the body effect factor. Equation (2.2) is valid for both the depletion and inversion regions of operation of the MOSFET.

According to the charge sheet approximation, the depletion charge density Q'_B [10] is given by

$$Q'_B = -\gamma C'_{ox}\sqrt{\phi_S} \tag{2.3}$$

If V_C tends toward infinity, the inversion charge density tends to zero, as readily noted from (2.2) and (2.3). The surface potential ϕ_{Sa} for which $Q'_I = 0$ is the solution of (2.4a):

$$V_G - V_{FB} = \phi_{Sa} + \gamma\sqrt{\phi_{Sa}} \tag{2.4a}$$

Solving (2.4a) [10], we get

$$\phi_{Sa} = \left(\sqrt{V_G - V_{FB} + \tfrac{1}{4}\gamma^2} - \tfrac{1}{2}\gamma\right)^2 \tag{2.4b}$$

Here, ϕ_{Sa} is the value of the surface potential disregarding the channel charge. Consequently, ϕ_{Sa} is a good approximation of the surface potential in weak inversion [10]. The inverse of the slope of the curve ϕ_{Sa} versus V_G, known as the slope factor [10], is written as

$$n = \left(\frac{d\phi_{Sa}}{dV_G}\right)^{-1} = 1 + \frac{\gamma}{2\sqrt{\phi_{Sa}}} \tag{2.5}$$

and is one of the fundamental parameters in the MOSFET model. Even though n depends on V_G, it is sometimes taken as constant. This approximation is quite acceptable since n typically changes not more than 30% for several decades of the drain current.

2.2.2. Basic Approximations

According to (2.1) and (2.3), Q'_I is expressed [10] as

$$Q'_I = -C'_{ox}(V_G - V_{FB} - \phi_S - \gamma\sqrt{\phi_S}) \tag{2.6}$$

The main approximation in this work has been to consider the inversion (Q_I') and depletion (Q_B') charge densities as incrementally linear functions of ϕ_S for a constant gate-to-bulk voltage.

Expanding (2.6) in power series about ϕ_{Sa} and disregarding second and higher order terms [7], we obtain, for constant V_{GB},

$$Q_I' \cong C_{ox}' n[\phi_S - \phi_{Sa}] \tag{2.7a}$$

The depletion charge density is approximated likewise, resulting in

$$Q_B' \cong -C_{ox}'(n-1)[\phi_S - \phi_{Sa}] + Q_{Ba}' = -\frac{n-1}{n} Q_I' + Q_{Ba}' \tag{2.7b}$$

where

$$Q_{Ba}' = -\gamma C_{ox}' \sqrt{\phi_{Sa}} \tag{2.7c}$$

is the depletion charge density as V_C tends toward infinity.

Let us now define the pinch-off voltage, an important transistor parameter in our model. According to (2.2), the inversion charge density never equals zero. The name pinch-off voltage is retained herein by historical reasons and means the channel potential corresponding to a small (but well-defined) amount of carriers in the channel. For reasons that will become apparent later, the pinch-off voltage is defined herein as the channel voltage for which the inversion charge density is equal to $-nC_{ox}'\phi_t$, that is

$$Q_I'(V_C = V_P) = Q_{IP}' = -nC_{ox}'\phi_t \tag{2.8}$$

Using approximation (2.7a) to calculate the surface potential ϕ_{SP} at the pinch-off condition gives

$$\phi_{SP} = \phi_{Sa} - \phi_t \tag{2.9a}$$

In order to calculate the pinch-off voltage, we use expressions (2.2) and (2.3), replacing ϕ_S, V_C, and Q_I' with ϕ_{SP}, V_P, and Q_{IP}', respectively. Thus

$$V_P = \phi_{Sa} - 2\phi_F - \phi_t\left[1 + \ln\left(\frac{n}{n-1}\right)\right] \tag{2.9b}$$

For the practical implementation of the MOSFET model, (2.9b) can be written as

$$V_P = \phi_{Sa} - \phi_{S0} \tag{2.9c}$$

where ϕ_{S0} can be considered as a fitting parameter.

Recalling that the threshold voltage in equilibrium ($V_C = 0$) is given by

$$V_{T0} = V_{FB} + 2\phi_F + \gamma\sqrt{2\phi_F} \tag{2.9d}$$

expression (2.9b) can be rewritten as

$$V_P = \left(\sqrt{V_G - V_{T0} + \left(\sqrt{2\phi_F} + \tfrac{1}{2}\gamma\right)^2 + \phi_t} - \tfrac{1}{2}\gamma\right)^2 - 2\phi_F - \phi_t\left[1 + \ln\left(\tfrac{n}{n-1}\right)\right] \tag{2.9e}$$

A slightly modified version of the definition of V_P in (2.9e), presented in [10] as V_{CBM} and in [11, 12] as V_P, is derived under the assumption that the inversion layer charge is negligible in the upper limit of weak inversion. If the thermal voltage is canceled out in (2.9e), the same simplified expression as in [3, 10–12] is readily obtained. In effect, the exclusion of the thermal voltage does not significantly affect the accuracy of (2.9e). Since the slope factor is almost independent of V_G, a useful approximation for the pinch-off voltage [3] is

$$V_P = \frac{V_G - V_{T0}}{n} \tag{2.9f}$$

where n can be assumed to be constant for hand calculations. Usually, $n = 1.2, \ldots, 1.6$ for typical values of the gate voltage.

Figure 2.2 shows the dependence of Q_I' on ϕ_S. At constant V_G, the linearization of Q_I' with respect to ϕ_S, according to Eq. (2.7a), fits very well the charge-sheet model. It should be pointed out that, unlike the conventional linearization about the source voltage, the linearization of Q_I' about ϕ_{Sa} preserves the symmetry of the model with respect to drain and source. The symmetry of the MOSFET implies that, no matter what its current level is or whatever its size is, the interchange of source and drain must result in identical electrical characteristics. For this reason, referring all voltages to the local substrate is preferable to referring them to the source. The substrate-referred MOSFET model exploits the functional symmetry of the MOSFET [3], allowing one to calculate very easily the properties of components like switches, MOSFETs operating as voltage-controlled resistors [13], or MOSFET-only current dividers [14].

The choice of $Q_I' = -nC_{ox}'\phi_t$ to define the pinch-off voltage is not occasional; rather, this point has been judiciously chosen since it represents the transition from weak inversion, where the transport mechanism is dominated by diffusion, to strong inversion, where the prevailing transport mechanism is drift.

The pinch-off voltage as well as the slope factor are plotted in Figure 2.3 against the gate voltage. The pinch-off voltage varies almost linearly with the gate voltage while the slope factor only changes a little for large gate voltage variations. For hand design, n can be assumed constant for several decades of current.

The fundamental approximations (2.7)–(2.9) have been applied throughout this work in order to obtain general expressions for the drain current and the total charges in terms of the inversion charge densities at the channel boundaries.

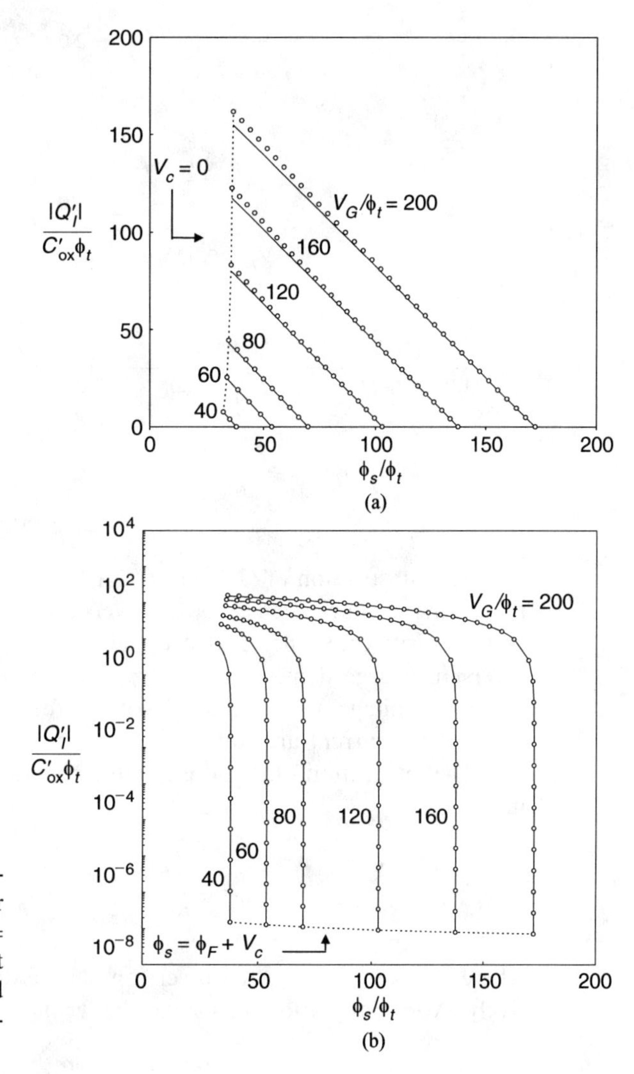

Figure 2.2 Inversion charge density calculated from: (a) incrementally linear approximation (2.7a) with $\phi_{Sa} = \phi_{S0} + V_P$; (b) classical charge-sheet expression (2.6), with ϕ_S calculated numerically. (—) Calculated; (○) experimental.

2.2.3. Drain Current

In this section we formulate the drain current I_D as a very simple function of the inversion charge densities at the channel ends. According to [10], I_D can be calculated from

$$I_D = \mu W \left(-Q_I' \frac{d\phi_S}{dx} + \phi_t \frac{dQ_I'}{dx} \right) \tag{2.10}$$

where the first and second terms account for drift and diffusion, respectively. From the approximate relationship between Q_I' and ϕ_S [Eq. (2.7a)], it follows that, for constant V_G,

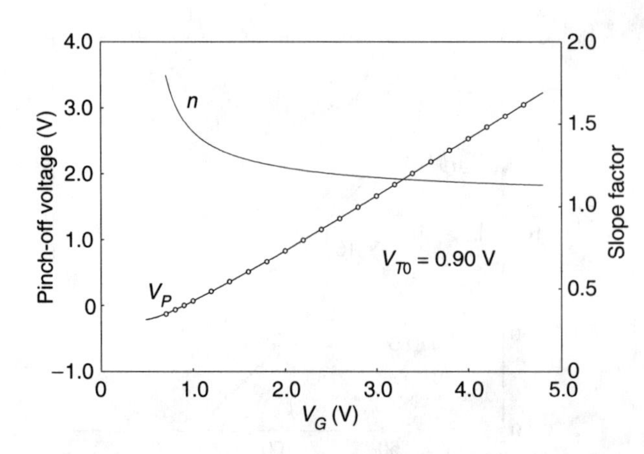

Figure 2.3 Pinch-off voltage and slope factor vs. gate-to-bulk voltage for NMOS transistor ($t_{ox} = 280$, $W = L = 25\,\mu\text{m}$): (○) measured values of V_P; (—) values of V_P and n calculated from their definitions in (2.9c) and (2.5).

$$dQ'_I = nC'_{ox}\, d\phi_S \tag{2.11}$$

The substitution of (2.11) into (2.10) allows writing the current as a function of the inversion charge density. Moreover, one can readily conclude that the diffusion and the drift components at a certain point of the channel are equal if the local inversion charge density is $-nC'_{ox}\phi_t$. As pointed out before, we have chosen the pinch-off voltage as the channel voltage for which the drift and diffusion components of the current are equal.

The integration of (2.10) along the channel length, together with (2.11), results in

$$I_D = -\frac{\mu}{nC'_{ox}} \frac{W}{L} \int_{Q'_{IS}}^{Q'_{ID}} (Q'_I - nC'_{ox}\phi_t)\, dQ'_I \tag{2.12}$$

where Q'_{IS} and Q'_{ID} are the inversion charge densities at source and drain, respectively. After integration, (2.12) can be written as

$$I_D = I_F - I_R = I(V_G, V_S) - I(V_G, V_D) \tag{2.13a}$$

with

$$I_{F(R)} = \mu n C'_{ox} \frac{W}{L} \frac{\phi_t^2}{2} \left[\left(\frac{Q'_{IS(D)}}{nC'_{ox}\phi_t} \right)^2 - 2\frac{Q'_{IS(D)}}{nC'_{ox}\phi_t} \right] \tag{2.13b}$$

where $I_{F(R)}$ is the forward (reverse) saturation current and $Q'_{IS(D)}$ is the inversion charge density evaluated at the source (drain) end [7, 9]. Note that, for a long-channel device, the forward (reverse) current depends on both the gate voltage and the source (drain) voltage, being independent of the drain (source) voltage.

Equations (2.13) emphasize the source-drain symmetry of the MOSFET. Now let us explain how to determine the forward and reverse components of the drain current from the transistor output characteristic, as shown in Figure 2.4 for a long-channel MOSFET. Note that there is a region, usually called the saturation region, where the drain current is almost independent of V_D. This means that, in saturation, $I(V_G, V_D) \ll I(V_G, V_S)$. Therefore, $I(V_G, V_S)$ can be interpreted as the drain current in forward saturation. Similarly, in reverse saturation, I_D is independent of the source voltage. Since the long-channel MOSFET is a symmetric device, the knowledge of the saturation current $I(V_G, V_S)$ for any V_G, V_S allows computing the drain current for any combination of source, drain, and gate voltages.

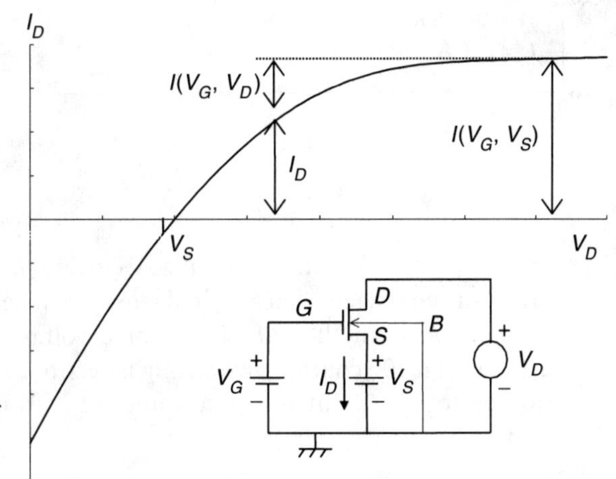

Figure 2.4 Output characteristics of long-channel NMOS transistor at constant V_S and V_G.

Expression (2.13b) can be rewritten in the form

$$-\frac{Q'_{IS(D)}}{nC'_{ox}\phi_t} = \sqrt{1 + i_{f(r)}} - 1 \tag{2.14a}$$

where

$$i_{f(r)} = \frac{I_{F(R)}}{I_S} = \frac{I(V_G, V_{S(D)})}{I_S} \tag{2.14b}$$

is the forward (reverse) normalized current [3] and

$$I_S = \mu n C'_{ox} \frac{\phi_t^2}{2} \frac{W}{L} \tag{2.14c}$$

is the normalization current, which is four times smaller than the corresponding factor presented in [3]. The factor $I_{S\square} = \frac{1}{2}\mu n C'_{ox}\phi_t^2$, herein denominated the sheet normalization current, is a technological parameter, slightly dependent on V_G, through μ and n.

Figure 2.5 depicts the saturation current of a long-channel MOS transistor versus the gate voltage. The saturation current variation around its average value is about ±30% for a gate voltage ranging from 0.6 to 5 V.

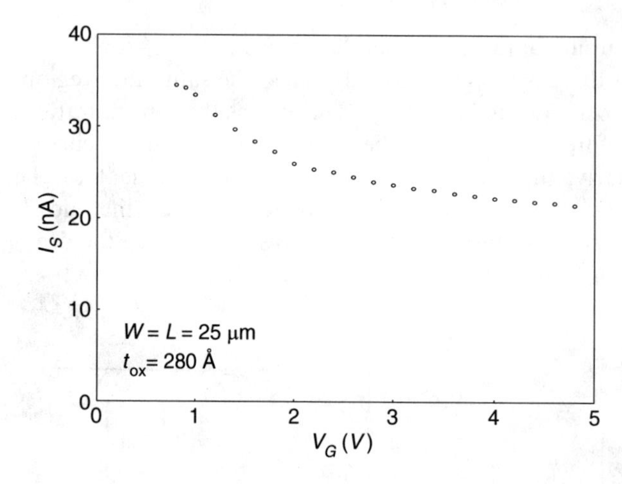

Figure 2.5 Normalization current of NMOS transistor ($t_{\mathrm{ox}} = 280$, $W = L = 25\,\mu\mathrm{m}$).

In [3] the forward normalized current i_f is also properly referred to as the inversion coefficient since it indicates the inversion level of the MOSFET, which depends on both the gate and source voltages. As a rule of thumb, values of i_f greater than 100 characterize strong inversion. The transistor operates in weak inversion up to $i_f = 1$. Intermediate values of i_f from 1 to 100, indicate moderate inversion.

2.2.4. Total Charges

The total inversion (Q_I) and bulk (Q_B) charges are defined [10] as

$$Q_I = W \int_0^L Q_I' \, dx \tag{2.15a}$$

$$Q_B = W \int_0^L Q_B' \, dx \tag{2.15b}$$

Substituting the basic approximation stated by (2.11) into expression (2.10) of the drain current, we have

$$dx = -\frac{\mu W}{nC_{\mathrm{ox}}' I_D} (Q_I' - nC_{\mathrm{ox}}' \phi_t) \, dQ_I' \tag{2.16}$$

which can be readily used to calculate the integrals in (2.15a) and (2.15b). Substituting (2.16) into (2.15a) yields

$$Q_I = -\frac{\mu W^2}{nC_{\mathrm{ox}}' I_D} \left[\int_{Q_{IS}'}^{Q_{ID}'} (Q_I'^2 - nC_{\mathrm{ox}}' \phi_t Q_I') \, dQ_I' \right] \tag{2.17a}$$

The integration of (2.17a) gives the total charge as a combination of both the inversion charge densities at source and drain:

$$Q_I = WL\left[\frac{2}{3}\frac{Q_F'^2 + Q_F'Q_R' + Q_R'^2}{Q_F' + Q_R'} + nC_{ox}'\phi_t\right] \tag{2.17b}$$

where

$$Q_{F(R)}' = Q_{IS(D)}' - nC_{ox}'\phi_t \tag{2.17c}$$

The total depletion charge is directly determined using (2.7b), resulting in

$$Q_B = -\frac{n-1}{n}Q_I + Q_{Ba}'WL \tag{2.18}$$

The gate charge can be calculated [10] from

$$Q_G = -Q_B - Q_I - Q_0 \tag{2.19}$$

where Q_0 is the effective interface charge, assumed to be independent of the terminal voltages.

Introducing (2.14a) into both (2.17b) and (2.18) allows the total charges to be written as functions of the normalized forward and reverse currents, as shown in Table 2.1. The derivation of the source (Q_S) and drain (Q_D) charges is shown in Appendix A.

Let us now interpret two very simple cases of transistor operation with the help of Table 2.1. Deep in strong inversion ($i_f \gg 1$) and saturation ($i_f \gg i_r$), the total inversion charge is proportional to the square root of the drain current while the source charge (Q_S) is equal to 60% of the total inversion charge. Deep in weak inversion ($i_f \ll 1$) and saturation ($i_f \gg i_r$), the total inversion charge is proportional to the current and the source charge is two-thirds of the total inversion charge.

2.2.5. Relationship between Inversion Charge Density and Terminal Voltages

The application of Boltzmann statistics to a long-channel MOSFET with a uniformly doped substrate leads to the expression

$$Q_I'(x) = -qn_0\int_{y_s}^{y_c} e^{[\phi(x,y)-V_C(x)]/\phi_t}\,dy \tag{2.20}$$

for the inversion charge density [10]. In (2.20), q is the electronic charge, n_0 is the free-electron concentration in equilibrium, y is the coordinate perpendicular to the oxide-semiconductor interface (see Fig. 2.1), and $\phi(x, y)$ is the electrostatic potential

TABLE 2.1 STATIC AND DYNAMIC MOSFET PARAMETERS

Variable	Expression
I_S	$(W/L)\mu n C'_{ox}(\frac{1}{2}\phi_t^2)$
I_D	$I_S(i_f - i_r)$
V_P	$(V_G - V_{T0})/n$
$V_P - V_{S(D)}$	$\phi_t\left[\sqrt{1+i_{f(r)}} - \sqrt{1+i_P} + \ln\left(\frac{\sqrt{1+i_{f(r)}}-1}{\sqrt{1+i_P}-1}\right)\right]$
Q_I	$-C_{ox}n\phi_t\left[\frac{2}{3}\left(\sqrt{1+i_f}+\sqrt{1+i_r}-\frac{\sqrt{1+i_f}\sqrt{1+i_r}}{\sqrt{1+i_f}+\sqrt{1+i_r}}\right)-1\right]$
Q_B	$-\frac{n-1}{n}Q_I - C_{ox}\frac{\gamma^2}{2(n-1)}$
Q_S	$-C_{ox}n\phi_t\left[\frac{2}{15}\left(\frac{3(\sqrt{1+i_f})^3+6(1+i_f)\sqrt{1+i_r}+4\sqrt{1+i_f}(1+i_r)+2(\sqrt{1+i_r})^3}{(\sqrt{1+i_f}+\sqrt{1+i_r})^2}\right)-\frac{1}{2}\right]$
Q_D	$Q_I - Q_S$
$g_{ms(d)}$	$(2I_S/\phi_t)(\sqrt{1+i_{f(r)}}-1)$
g_{mg}	$(g_{ms} - g_{md})/n$
$C_{gs(d)}$	$\frac{2}{3}C_{ox}(\sqrt{1+i_{f(r)}}-1)\frac{\sqrt{1+i_{f(r)}}+2\sqrt{1+i_{r(f)}}}{(\sqrt{1+i_f}+\sqrt{1+i_r})^2}$
C_{gb}	$[(n-1)/n](C_{ox} - C_{gs} - C_{gd})$
$C_{bs(d)}$	$(n-1)C_{gs(d)}$
$C_{ds(sd)}$	$-\frac{4}{15}nC_{ox}(\sqrt{1-i_{f(r)}}-1)\frac{1+i_f+3\sqrt{1+i_f}\sqrt{1+i_r}+1+i_r}{(\sqrt{1+i_f}+\sqrt{1+i_r})^3}$
C_m	$(C_{sd} - C_{ds})/n$

(equal to zero at infinity). The lower limit of the previous integral, y_S, corresponds to the interface between the oxide and the semiconductor. The upper limit y_C is an arbitrary depth in the bulk, where the electron concentration is negligible.

According to the charge sheet model, there is no significant voltage drop in the inversion layer [10]. Thus, $\phi(x, y)$ in (2.20) can be approximated to the surface potential $\phi_S(x)$, and taking the derivative of (2.20) gives

$$dQ'_I(x) = \frac{Q'_I(x)}{\phi_t}[d\phi_S(x) - dV_C(x)] \tag{2.21}$$

The substitution of the fundamental approximation (2.11) into (2.21) leads to the following relationship between Q_I' and V_C:

$$dQ_I'\left(\frac{1}{nC_{\text{ox}}'} - \frac{\phi_t}{Q_I'}\right) = dV_C \tag{2.22}$$

Integrating (2.22) from an arbitrary channel potential V_C to the pinch-off voltage V_P yields

$$V_P - V_C = \frac{Q_{\text{IP}}' - Q_I'}{nC_{\text{ox}}'} + \phi_t \ln\left(\frac{Q_I'}{Q_{\text{IP}}'}\right) \tag{2.23}$$

Expression (2.23) is almost identical to the unified charge control model (UCCM) [15, 16]. The similarity between (2.23) and the UCCM is readily verified in (2.24), which has been obtained by substituting the approximate expression of V_P, (2.9f), into (2.23):

$$Q_{\text{IP}}' - Q_I' + nC_{\text{ox}}'\phi_t \ln\left(\frac{Q_I'}{Q_{\text{IP}}'}\right) = C_{\text{ox}}'(V_G - V_{T0} - nV_C) \tag{2.24}$$

Even though the UCCM has been presented as an empirical result [15, 16], it can be deduced from Boltzmann statistics, the charge sheet approximation and the linear relationship between inversion charge density and surface potential, as previously shown.

Using (2.14a) and (2.23), we find the following relationship between current and voltage:

$$V_P - V_{S(D)} = \phi_t\left[\sqrt{1 + i_{f(r)}} - \sqrt{1 + i_P} + \ln\left(\frac{\sqrt{1 + i_{f(r)}} - 1}{\sqrt{1 + i_P} - 1}\right)\right] \tag{2.25}$$

where i_P is the value of the normalized current at pinch-off. As long as we have chosen $Q_{\text{IP}}' = -nC_{\text{ox}}'\phi_t$, $i_P = 3$ according to (2.14a). Expression (2.25) is a universal relationship for long-channel MOSFETs, valid for any technology, gate voltage, dimensions, and temperature. The accuracy of (2.25) has been verified in [9] for different gate voltages, technologies, and transistor dimensions.

Since (2.25) is not invertible in terms of elementary functions, simple approximations of this expression in which the current is an explicit function of the applied voltages are very useful for some applications such as, for example, digital circuits, usually driven by voltage signals. One example of such an approximation is provided in Appendix B.

Some widely known results can be derived from (2.25). For example, deep in weak inversion, both i_f and $i_r \ll 1$. Therefore, (2.25) can be approximated by

$$\frac{V_P - V_{S(D)}}{\phi_t} \cong -1 + \ln\left(\tfrac{1}{2} i_{f(r)}\right) \tag{2.26a}$$

$$I_D = I_S(i_f - i_r) \cong 2I_S \exp\left(\frac{V_P + \phi_t}{\phi_t}\right)\left[\exp\left(-\frac{V_S}{\phi_t}\right) - \exp\left(-\frac{V_D}{\phi_t}\right)\right] \tag{2.26b}$$

Equation (2.26b) is similar to the expression presented in [3] for the current in weak inversion. Now, assuming that the drain and source ends of the channel are strongly inverted, both i_r and $i_f \gg 1$. Thus, (2.25) can be approximated by

$$\frac{V_P - V_{S(D)}}{\phi_t} \cong \sqrt{i_{f(r)}} \tag{2.27a}$$

$$I_D = I_S(i_f - i_r) \cong I_S\left[\left(\frac{V_P - V_S}{\phi_t}\right)^2 - \left(\frac{V_P - V_D}{\phi_t}\right)^2\right] \tag{2.27b}$$

Equation (2.27b) is equal to the expression presented in [3] for the drain current in strong inversion.

Figure 2.6 shows plots of the drain current in saturation versus V_S at constant V_G while Figure 2.7 displays the drain current against V_G at constant V_S. The values of I_F, calculated by using (2.25), are compared to the measured values for a typical device. The experimental results agree very well with the theoretical results obtained from the model presented here.

The MOSFET output characteristics described by the universal relationship

$$\frac{V_{DS}}{\phi_t} = \sqrt{1 + i_f} - \sqrt{1 + i_r} + \ln\left(\frac{\sqrt{1 + i_f} - 1}{\sqrt{1 + i_r} - 1}\right) \tag{2.28}$$

are readily derived from (2.25). Expression (2.28) demonstrates that the normalized output characteristics of a long-channel MOSFET are independent of technology and transistor dimensions, corroborating again the universality and consistency of our model. In Figure 2.8 we compare the measured output characteristics for several gate voltages and the curves obtained by using (2.28).

In Figure 2.9 we present the theoretical drain-to-source saturation voltage $V_{DS,\mathrm{SAT}}$, defined here as the value of V_{DS} for which the ratio $Q'_{ID}/Q'_{IS} = \xi$, where ξ is an arbitrary number much smaller than 1. Note that $1 - \xi$ represents the saturation level of the MOSFET. From (2.14a) and (2.28) we have

$$V_{DS,\mathrm{SAT}} = \phi_t\left[\ln\left(\frac{1}{\xi}\right) + \sqrt{1 + i_f} - 1\right] \tag{2.29}$$

In (2.29), the term $1 - \xi$ that multiplies $\sqrt{1 + i_f} - 1$ has been approximated to 1. The definition in (2.29) is extremely powerful for circuit design since it gives the boundary between the triode and saturation regions in terms of the inversion level. Note that, in weak inversion, $V_{DS,\mathrm{SAT}}$ is independent of the inversion level while in strong inversion it is proportional to the square root of the inversion level. Our

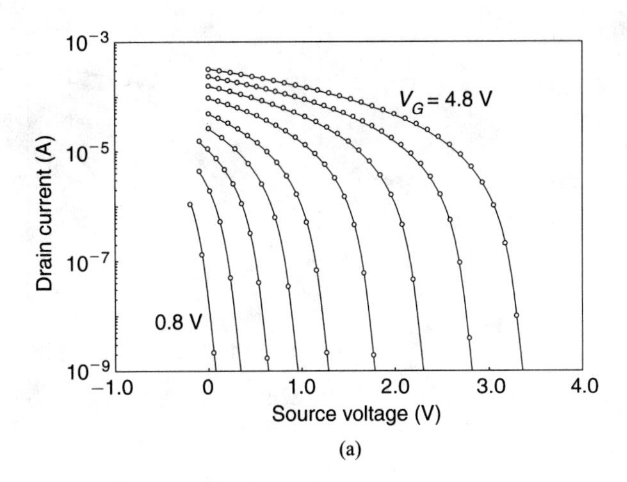

(a)

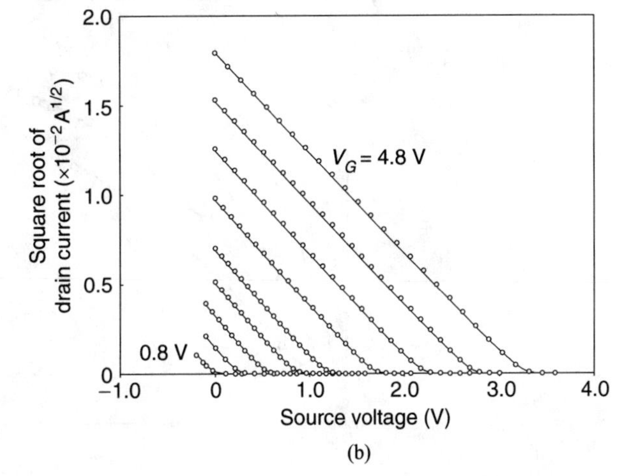

(b)

Figure 2.6 Common-gate characteristics of NMOS transistor ($t_{ox} = 280$, $W = L = 25\,\mu m$) in saturation ($V_G = 0.8$, 1.2, 1.6, 2.0, 2.4, 3.0, 3.6, 4.2, 4.8 V): (—) simulated curves calculated from Table 2.1; (○) measured data.

definition of saturation is arbitrary but gives designers a very good first-order approximation of the minimum V_{DS} required to keep the MOSFET in the "constant-current region".

2.3. SMALL-SIGNAL MOSFET MODEL

Small-signal equivalent circuits are linear circuits that represent a device when the terminal voltage variations are sufficiently small. We begin this section with the development of a small-signal model for the MOSFET that is valid when the voltage and current variations are so slow that charge storage effects can be neglected. We then derive a complete small-signal model valid up to medium frequencies, assuming quasi-static operation. Finally, a high-frequency model wherein the charges and

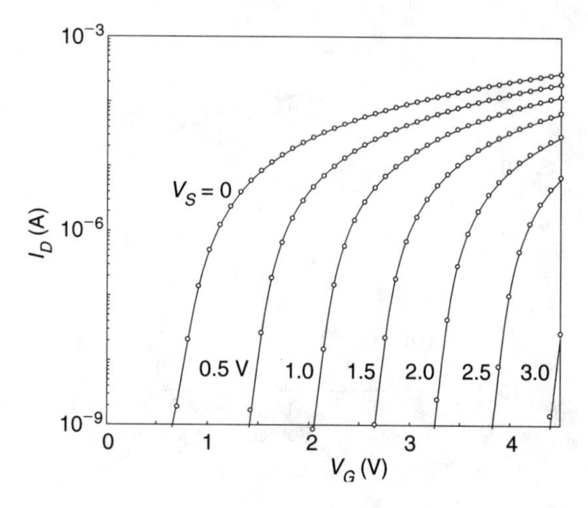

Figure 2.7 Common-source characteristics of NMOS transistor ($t_{ox} = 280\,\text{Å}$, $W = L = 25\,\mu\text{m}$) in saturation.

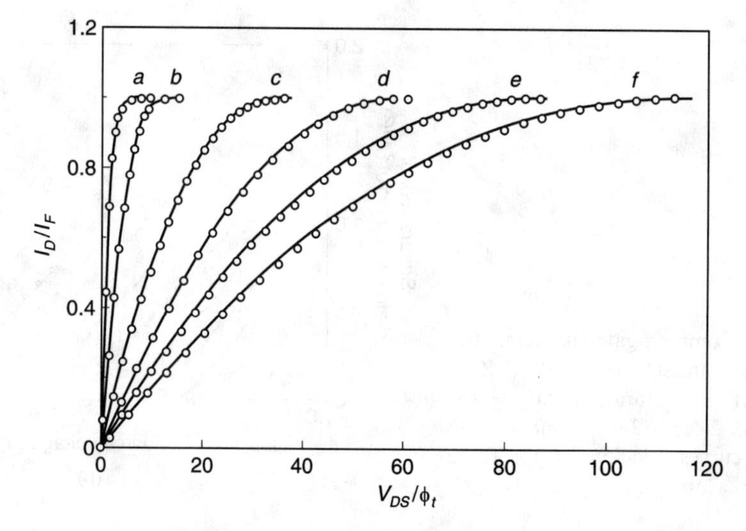

Figure 2.8 Normalized output characteristics of NMOS transistor ($t_{ox} = 280\,\text{Å}$, $W = L = 25\,\mu\text{m}$) for $V_S = 0$ [○, measured data; —, calculated from (2.28)]: (a) $i_f = 4.5 \times 10^{-2} (V_G = 0.7\,\text{V})$; (b) $i_f = 65\,(V_G = 1.2\,\text{V})$; (c) $i_f = 9.5 \times 10^2\,(V_G = 2.0\,\text{V})$; (d) $i_f = 3.1 \times 10^3\,(V_G = 2.8\,\text{V})$; (e) $i_f = 6.8 \times 10^3\,(V_G = 3.6\,\text{V})$; (f) $i_f = 1.2 \times 10^4\,(V_G = 4.4\,\text{V})$.

currents do not change instantaneously with the terminal voltages will be presented in Section 2.3.6.

2.3.1. Transconductances

At low frequencies, the variation of the drain current due to small variations of the gate, bulk, source, and drain voltages is

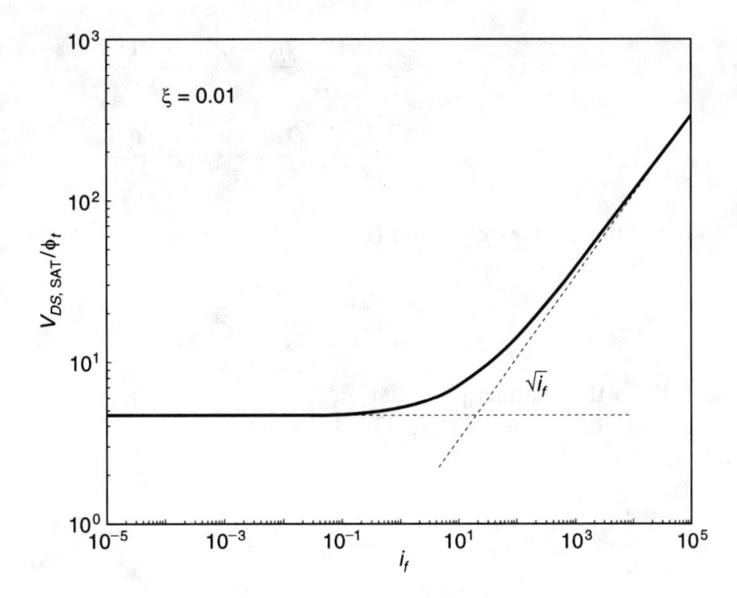

Figure 2.9 Drain-to-source saturation voltage vs. inversion coefficient, calculated for $\xi = 0.01$.

$$\Delta I_D = \left.\frac{\partial I_D}{\partial V_G}\right|_{V_S,V_D,V_B} \Delta V_G + \left.\frac{\partial I_D}{\partial V_S}\right|_{V_G,V_D,V_B} \Delta V_S + \left.\frac{\partial I_D}{\partial V_D}\right|_{V_G,V_S,V_B} \Delta V_D + \left.\frac{\partial I_D}{\partial V_B}\right|_{V_G,V_S,V_D} \Delta V_B \tag{2.30a}$$

and

$$g_{mg} = \left.\frac{\partial I_D}{\partial V_G}\right|_{V_S,V_D,V_B} \qquad g_{ms} = -\left.\frac{\partial I_D}{\partial V_S}\right|_{V_G,V_D,V_B}$$
$$g_{md} = \left.\frac{\partial I_D}{\partial V_D}\right|_{V_G,V_S,V_B} \qquad g_{mb} = \left.\frac{\partial I_D}{\partial V_B}\right|_{V_G,V_S,V_D} \tag{2.30b}$$

are the gate, source, drain, and bulk transconductances, respectively [3].

If the variation of the gate, source, drain, and bulk voltages is the same, $\Delta I_D = 0$. Therefore, we can conclude that

$$g_{mg} + g_{md} + g_{mb} = g_{ms} \tag{2.30c}$$

Thus, three transconductances are enough to characterize the low-frequency small-signal behavior of the MOSFET.

Applying the definition (2.30b) of source and drain transconductances to the equation of the drain current (2.13), the transconductances g_{ms} and g_{md} become

$$g_{ms} = -\frac{\mu W}{C'_{ox} L} \frac{Q'_{IS} - nC'_{ox}\phi_t}{n} \frac{\partial Q'_{IS}}{\partial V_S} \tag{2.31a}$$

$$g_{md} = -\frac{\mu W}{C'_{ox} L} \frac{Q'_{ID} - nC'_{ox}\phi_t}{n} \frac{\partial Q'_{ID}}{\partial V_D} \tag{2.31b}$$

From the UCCM equation (2.22), it follows that

$$\frac{\partial Q'_I}{\partial V_C} = \frac{nC'_{ox} Q'_I}{Q'_I - nC'_{ox}\phi_t} \tag{2.32}$$

where V_C is the channel potential.

Substituting expression (2.32) of the derivative of the inversion charge density into (2.31) yields

$$g_{ms} = -\mu \frac{W}{L} Q'_{IS} \tag{2.33a}$$

$$g_{md} = -\mu \frac{W}{L} Q'_{ID} \tag{2.33b}$$

These compact expressions for the transconductances can be derived directly from the quasi-Fermi potential formulation of the drain current [3, 7]. In the above derivation, the use of a specific relationship between charge and voltage (UCCM) was necessary to obtain (2.33). It can be concluded that the UCCM is the only voltage-to-charge relationship fully compatible with (2.11), the basic approximation of our model.

Using (2.14a), the relationsip between inversion charge and inversion level, and the definition of the normalization current I_S, (2.14c), the source and drain transconductances can be expressed as

$$g_{ms(d)} = \frac{2I_S}{\phi_t}\left(\sqrt{1 + i_{f(r)}} - 1\right) \tag{2.34}$$

The above expression for the transconductance is very useful for circuit design because it is very compact, is valid for any inversion level, and uses easily measurable parameters. Moreover, (2.34) is a universal relationship for MOSFETs. The only technology-dependent parameter in (2.34) is the normalization current, which depends on the transistor aspect ratio as well.

Applying the definition of gate transconductance in (2.30b) to the relationship between current and inversion charge, (2.13), and neglecting the variations of both the slope factor and the mobility with V_G, we obtain

$$g_{mg} = \frac{\mu W}{C'_{ox} L}\left(\frac{Q'_{IS} - nC'_{ox}\phi_t}{n} \frac{\partial Q'_{IS}}{\partial V_G} - \frac{Q'_{ID} - nC'_{ox}\phi_t}{n} \frac{\partial Q'_{ID}}{\partial V_G}\right) \tag{2.35}$$

From the UCCM (2.23), it follows that

$$\frac{\partial Q_I'}{\partial V_P} = \frac{nC_{ox}' Q_I'}{nC_{ox}'\phi_t - Q_I'} = -\frac{\partial Q_I'}{\partial V_C} \tag{2.36a}$$

Recalling that $dV_P/dV_G = 1/n$ [see (2.5) and (2.9c)] and applying the chain rule to calculate the derivative of the inversion charge density with respect to the gate potential yield

$$\frac{\partial Q_I'}{\partial V_G} = -\frac{1}{n}\frac{\partial Q_I'}{\partial V_C} \tag{2.36b}$$

The comparison of (2.35) and (2.36b) with (2.31) results in

$$g_{mg} = \frac{1}{n}(g_{ms} - g_{md}) \tag{2.37}$$

Equation (2.37) gives the conventional (gate) transconductance in terms of the source and drain transconductances. For a long-channel MOSFET in saturation, $i_r \ll i_f$; consequently, $g_{mg} \cong g_{ms}/n$.

Figure 2.10 compares measured and simulated values of both the source and gate transconductances in saturation, thus demonstrating the satisfactory precision of the proposed model.

It can be noticed that all the above-calculated transconductances approximate their well-known asymptotic values in weak and strong inversion [3]. For instance, deep in weak inversion, that is, for $i_f \ll 1$, $\sqrt{1+i_f}$ can be approximated by $1 + \frac{1}{2}i_f$. Therefore, g_{ms} tends to its expected value, I_F/ϕ_t. On the other hand, in very strong inversion, g_{ms} is proportional to $\sqrt{I_F}$ since i_f is much greater than 1.

2.3.2. Transconductance-to-Current Ratio

An important design parameter in analog circuits is the transconductance-to-current ratio [17], a measurement of speed per unit power consumed [18]. In the following we will demonstrate that this design parameter can be expressed in terms of a normalized saturation current.

The substitution of I_S by I_F/i_f in (2.34) allows one to write the ratio of the forward (reverse) saturation current to the source (drain) transconductance:

$$\frac{I_{F(R)}}{\phi_t g_{ms(d)}} = \frac{\sqrt{1+i_{f(r)}}+1}{2} \tag{2.38}$$

Equation (2.38) is a universal expression for MOS transistors, as is the transconductance-to-current ratio for bipolar transistors. Expression (2.38) is a very powerful tool for circuit design since it allows designers to compute the available transconductance-to-current ratio in terms of the inversion level i_f. Moreover, (2.38) provides a straightforward procedure for extracting the value of the normalization current, the most important parameter in our model. The value $\frac{3}{2}$ for the ratio $I_F/(g_{ms}\phi_t)$, for instance, corresponds to $i_f = 3$, that is, $I_S = \frac{1}{3}I_F$. This very simple procedure for extracting I_S is illustrated in Figure 2.11 for two different values ($\frac{3}{2}$ and 2) of the current-to-transconductance ratio.

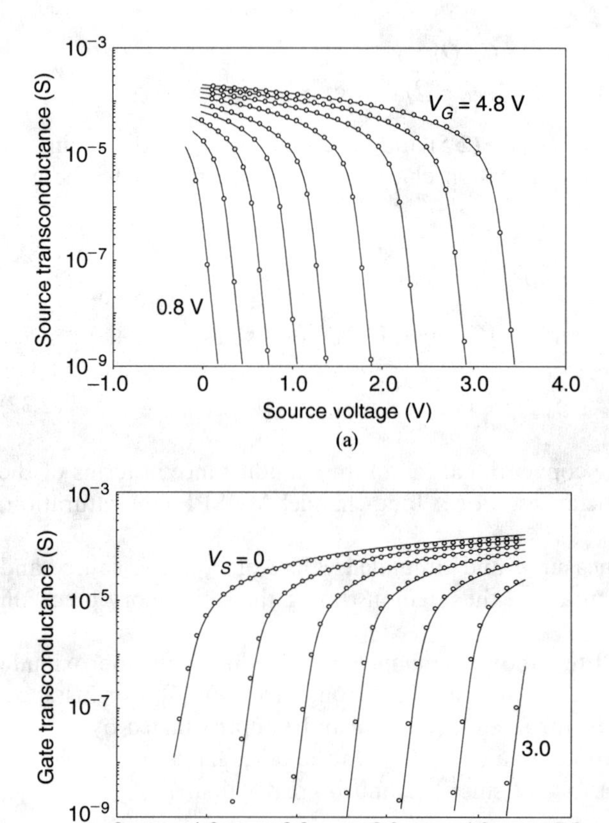

Figure 2.10 Measured and simulated values of source and gate transconductances of NMOS transistor (t_{ox} = 280, $W = L = 25\,\mu$m) in saturation.

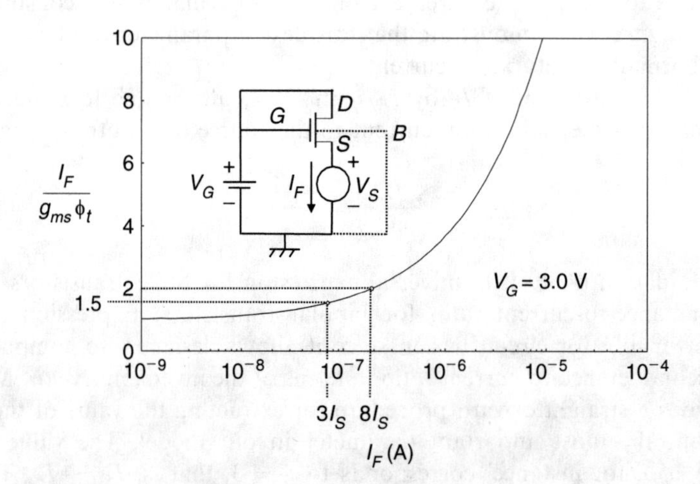

Figure 2.11 Determination of saturation current from current-to-transconductance ratio.

The universality of expression (2.38) is confirmed in Figure 2.12, where measured and simulated current-to-tranconductance ratios are plotted for different gate voltages, technologies, and channel lengths. The accuracy of (2.38) is excellent for any of these cases.

2.3.3. Intrinsic Capacitances

We will first consider the device part under the gate, between the source and drain. It is called the intrinsic part and is where transistor action takes place. The rest of the device constitutes the extrinsic part and can be modeled by a network of parasitic (two-terminal) capacitances and diodes. For low and moderate frequencies the quasi-static approximation [10] is used. The quasi-static approximation is valid when the charges stored in the transistor depend only on the instantaneous terminal voltages and not on their past variation. The MOSFET intrinsic capacitances for quasi-static operation are defined by the general expressions

$$C_{xy} = -\left.\frac{\partial Q_X}{\partial V_Y}\right|_0 \qquad X \neq Y \tag{2.39}$$

$$C_{xx} = \left.\frac{\partial Q_X}{\partial V_X}\right|_0 \tag{2.40}$$

where Q_x can be any of the charges Q_S, Q_D, Q_B, or Q_G and V_X and V_Y can be any of the voltages V_G, V_S, V_D, or V_B. The notation "0" indicates that the derivatives are evaluated at the bias point [10]. Because the MOSFET is an active device, the capacitances C_{xy} are nonreciprocal; that is, in general, $C_{xy} \neq C_{yx}$ [10]. Taking into account charge conservation, $Q_S + Q_D + Q_B + Q_G = 0$, and that only three voltage differences out of four can be chosen independently, it follows that the MOSFET is characterized by nine independent capacitances [10]. From the definitions of capacitance in (2.39) and (2.40) and the expressions of the MOSFET charges presented in the previous section, we can derive formulas for the small-signal capacitances (Appendix C). Capacitances C_{gs}, C_{gd}, C_{gb}, C_{bs}, and C_{bd} are widely used in AC modeling because together they accurately describe charge storage in MOSFETs up to moderate frequencies and can be calculated directly from the gate and bulk charges [3, 10]. It should be recalled, however, that a MOSFET model containing only these five capacitances does not conserve charge. Therefore, for the electrical simulation of charge-sensitive circuits such as switched-capacitor or switched-current filters, the complete quasi-static model must be taken into account [5, 10, 19]. Considering that in our model $C_{bg} = C_{gb}$ (Appendix C) only three more independent capacitances must be added to the basic MOSFET model in order to attain a model that conserves charge. To keep the symmetry of the device in the small-signal schematic, we have chosen C_{sd}, C_{ds}, and C_{dg} to complete the quasi-static MOSFET model. Small-signal analysis shows that the effects of C_{dg} and C_{gd} can be combined and replaced with C_{gd} and transcapacitance C_m (see Figure 2.13), where

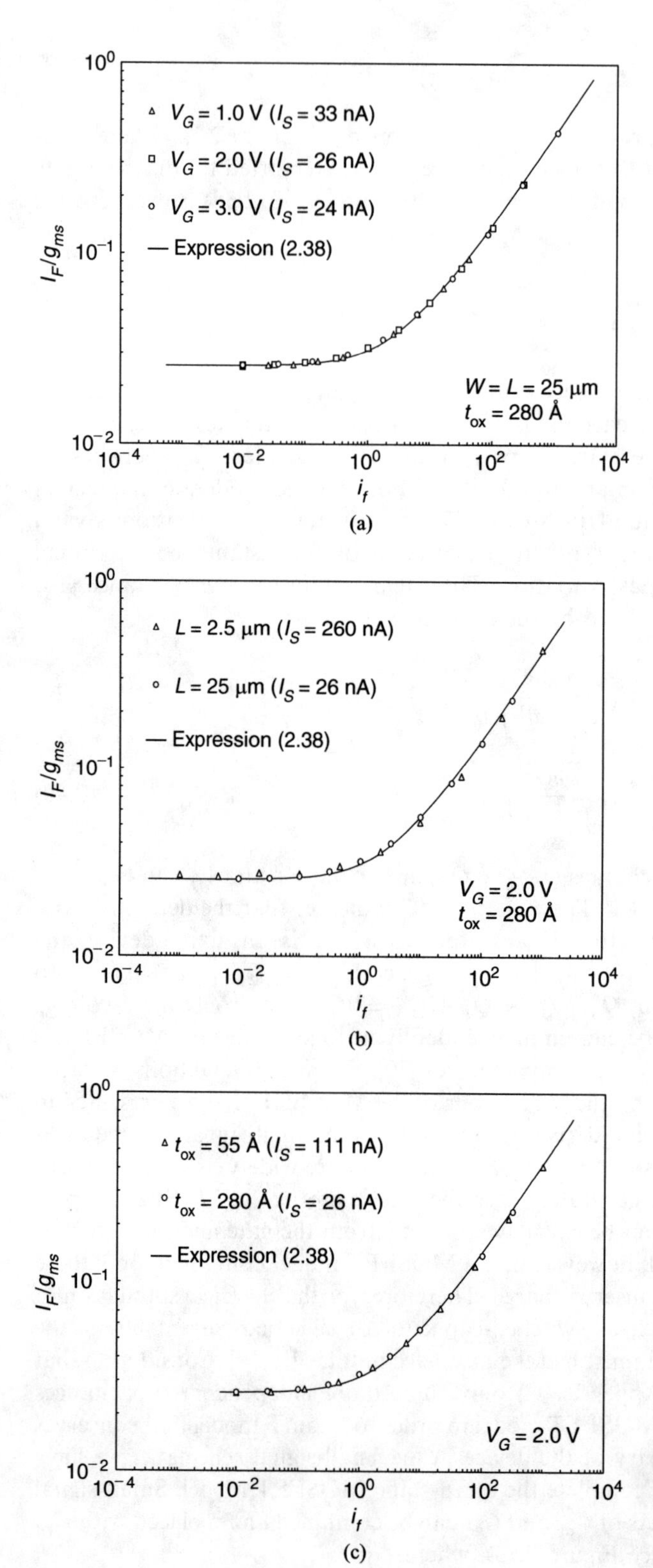

Figure 2.12 Forward current-to-transconductance ratio (I_F/g_{ms}) vs. inversion coefficient (i_f) of NMOS transistors: (a) biased at different gate voltages; (b) with different channel lengths; (c) from different technologies.

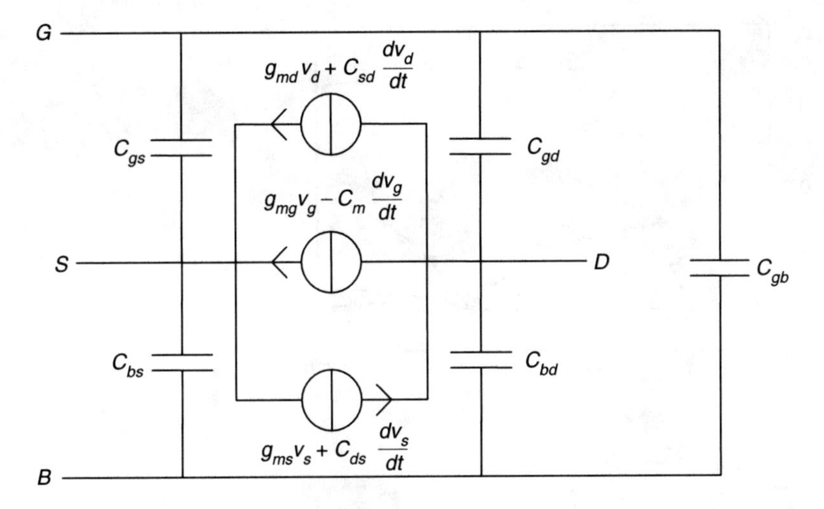

Figure 2.13 Small-signal MOSFET model for medium frequencies.

$$C_m = C_{dg} - C_{gd} = -(C_{sg} - C_{gs}) \tag{2.41}$$

From the expressions in Appendix C, it can be readily verified that $C_m = (C_{sd} - C_{ds})/n$. Consequently, the small-signal schematic of Figure 2.13, constituted of five capacitances, three transcapacitances, and three transconductances, preserves the inherent symmetry of the MOSFET. Figure 2.14 illustrates the capacitances as functions of the gate-to-bulk voltage. All the curves are continuous and vary smoothly. The five basic intrinsic capacitances of our model are in close agreement with those of the Enz-Krummenacher-Vittoz model [3], as can be seen in Figure 2.14.

An important figure of merit for a MOSFET is the unity-gain or intrinsic cutoff frequency, defined as the frequency value at which the short-circuit current gain in the common-source configuration drops to 1 [10]. The intrinsic cutoff frequency of a MOSFET in saturation [10] is given by

$$f_T = \frac{g_{mg}}{2\pi(C_{gs} + C_{gb})} = \frac{g_{ms}}{2\pi n(C_{gs} + C_{gb})} \tag{2.42a}$$

From the expressions of the source transconductance g_{ms} presented in Section 2.3.1 and C_{gs} and C_{gb} in Appendix C, f_T can be readily written in terms of the inversion coefficient i_f as

$$f_T = \frac{\mu n\phi_t}{2\pi L^2}\frac{i_f(\sqrt{1+i_f}+1)}{(n-1)(\sqrt{1+i_f}+1)^2 + \frac{2}{3}(i_f + \sqrt{1+i_f} - 1)} \tag{2.42b}$$

The first term on the right-hand side of (2.42b) shows the dependence of f_T on the channel length, slope factor, and mobility. The second term represents the

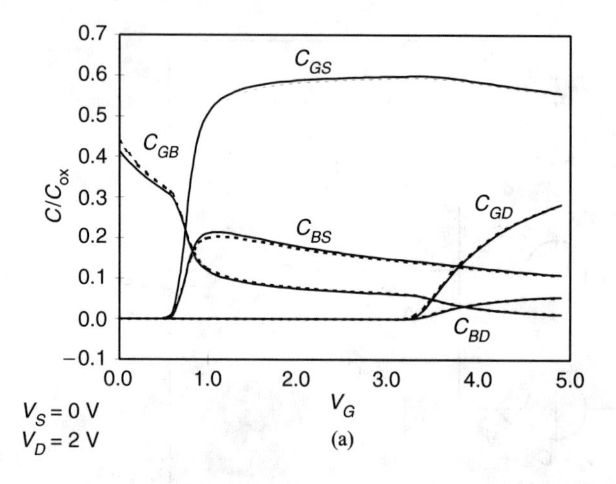

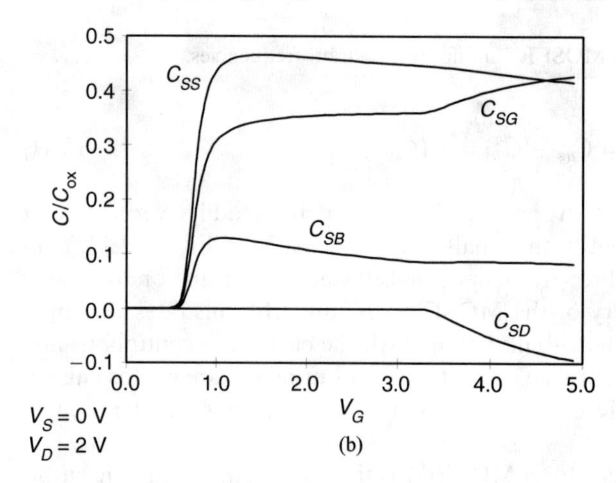

Figure 2.14 (a) Comparison between five capacitances available in EKV model [11] and those of our model. (b) Four capacitances in our model that are not modeled in EKV model. Simulations were run in SMASH.

dependence of the cutoff frequency on the inversion level. Due to the lack of adequate models, designers usually employ transistors whose f_T is much higher than that required for a specific application, thus leading to an unnecessary increase in power consumption.

Assuming the slope factor n in the denominator of (2.42b) to be equal to $\frac{4}{3}$, a typical value, f_T, can be roughly approximated to

$$f_T \cong \frac{\mu\phi_t}{2\pi L^2} 2(\sqrt{1+i_f} - 1) \tag{2.42c}$$

for any inversion level.

Figure 2.15 shows the intrinsic cutoff frequency calculated from (2.42b) for $n = \frac{4}{3}$ and $n = \frac{5}{3}$ and from (2.42c) for several decades of the inversion coefficient. Even though (2.42c) is not too accurate, it gives designers a first-order estimation of the unity-gain frequency.

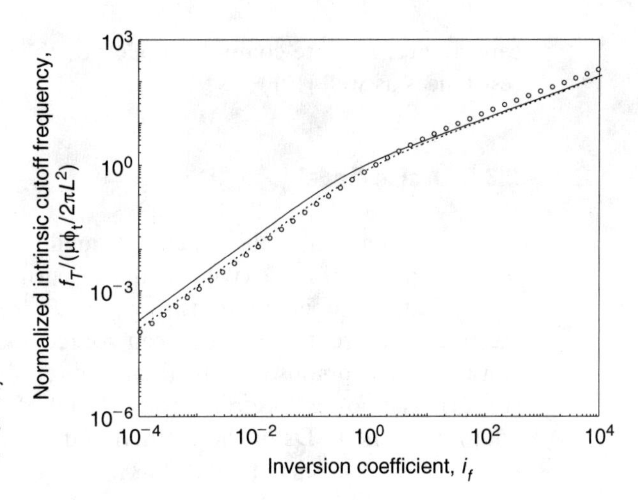

Figure 2.15 Normalized intrinsic cutoff frequency: (—) Eq. (2.42b) with $n = \frac{4}{3}$; (- - - -) Eq. (2.42b) with $n = \frac{5}{3}$; (○) Eq. (2.42c).

2.3.4. Extrinsic Element Modeling

The charge storage associated with the extrinsic part of the MOS transistor can be modeled by using up to six capacitances [10], one between each pair of terminals (see Figure 2.16). The unavoidable overlap between the gate and the source and drain diffusions gives rise to overlap capacitances. The substrate-source and the substrate-drain junctions must also be modeled by diode (nonlinear) capacitances. The very small drain-to-source proximity capacitance can usually be neglected and the extrinsic gate-to-bulk capacitance can be incorporated into the gate wiring

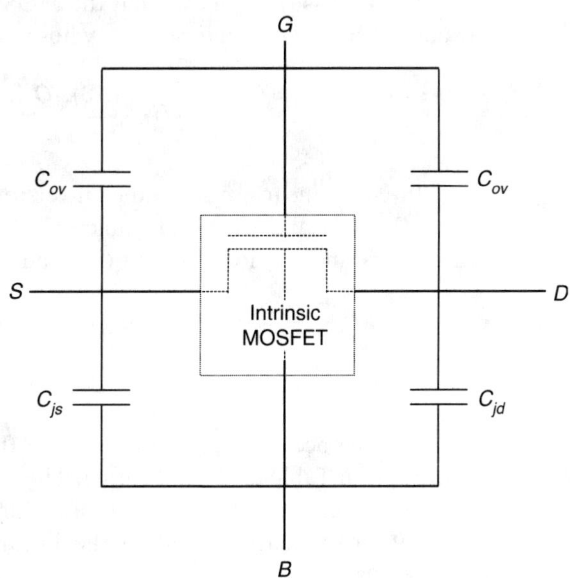

Figure 2.16 Extrinsic transistor components.

capacitance. A more complete model for the extrinsic part should include parasitic resistances as well [10].

2.3.5. Noise Model

Noise is an internally generated small signal in the device that can be modeled by the addition of noise (voltage and/or current) sources to the small-signal equivalent circuit of a noiseless transistor [10, 20]. MOSFET noise is usually modeled by including a current source between source and drain. The MOSFET noise can be considered as composed of thermal (white) noise and flicker (low-frequency) noise [10, 21]. The power spectral density (PSD) of the noise current source is simply the addition of the PSDs of the thermal and the flicker noise, because these two noise sources are uncorrelated [10]. Thus

$$S_{Id} = S_{\text{thermal}} + S_{\text{flicker}} \tag{2.43}$$

The expression of the total current, including both diffusion and drift components, together with Nyquist relationship [10, 22] allows calculation of the PSD of the thermal noise:

$$S_{\text{thermal}} = \frac{-4kT\mu Q_I}{L^2} \tag{2.44a}$$

where k is the Boltzmann constant, T the absolute temperature, and Q_I the total inversion charge. Equation (2.44a) is valid from weak to strong inversion and, as quoted in [10, 22, 23], includes the contribution of shot noise in weak inversion.

From (2.44a), it follows that the MOSFET thermal noise is the same as the one produced by a conductance $G_{N,\text{th}}$ whose value is

$$G_{N,\text{th}} = \frac{\mu|Q_I|}{L^2} = g_{ms}\frac{Q_I}{Q'_{IS}WL} \tag{2.44b}$$

In the linear region, the inversion charge density is almost uniform, $Q'_I \cong Q_I/WL$, and the conductance $G_{N,\text{th}}$ equals the source transconductance. In saturation, the relation between $G_{N,\text{th}}$ and g_{ms} becomes

$$G_{N,\text{th}} = \tfrac{1}{2}g_{ms} \quad \text{in weak inversion} \tag{2.44c}$$

$$G_{N,\text{th}} = \tfrac{2}{3}g_{ms} \quad \text{in strong inversion} \tag{2.44d}$$

For the accurate calculation of the thermal noise, the expression of the total charge Q_I in Table 2.1 must be used. This very compact expression, a function of the normalized currents i_f and i_r, is valid for any bias set of voltages. For both the calculation of thermal noise in the linear region or its estimation in saturation, one can use

$$S_{\text{thermal}} \cong 4kTg_{ms} \tag{2.44e}$$

A very common mistake in the calculation of the thermal noise is the substitution of g_{ms} with g_{mg}, the gate transconductance, in (2.44e). This substitution gives completely erroneous values for the thermal noise in the nonsaturation region [3] because $g_{mg} \ll g_{ms}$, particularly near the origin ($V_D = V_S$).

Flicker noise in MOS transistors is generally associated with charge fluctuation arising from trapping and detrapping of electrons by interface states [24]. As shown in [21], the PSD of the drain current is given by

$$S_{\text{flicker}} = \frac{\text{KF}\, g_{mg}^2}{WLC'_{\text{ox}}f} \tag{2.45}$$

where KF is the technology-dependent flicker noise constant. Note that the PSD of the input-referred voltage noise is independent of g_{mg} and, consequently, of bias current.

2.3.6. Non-quasi-static Model

The small-signal model of Figure 2.13 is valid for low- and medium-frequency analysis. A complete non-quasi-static model [10], suitable for high-frequency operation, can be derived by taking into account the continuity equation:

$$\frac{\partial i_I(x, t)}{\partial x} = W \frac{\partial q'_I(x, t)}{\partial t} \tag{2.46}$$

where $i_I(x, t)$ is the time-varying inversion channel current, no longer supposed to be constant along the channel length, and $q'_I(x, t)$ is the time-varying inversion charge density. The time-varying version of (2.10) together with approximation (2.11) leads to

$$i_I(x, t) = -\frac{\mu W}{nC'_{\text{ox}}} [q'_I(x, t) - nC'_{\text{ox}}\phi_t] \frac{\partial q'_I(x, t)}{\partial x} \tag{2.47}$$

Equations (2.46) and (2.47) can be solved either numerically or by an iterative procedure such as the one proposed in [25] and applied in [26]. The method of [26] allows deriving approximate expressions of the MOSFET transadmittances for any desired order (highest exponent of the frequency in the denominator). For instance, by applying the method described in [25] to solve (2.46) and (2.47), we obtain the nine independent transadmittances presented in Table 2.2.

The expressions in Table 2.2 are similar to the corresponding ones derived in [26]. However, the non-quasi-static parameters shown here are expressed in terms of the forward and reverse normalized currents while the parameters in [26] are functions of the surface potentials at source and drain.

TABLE 2.2 MOSFET TRANSADMITTANCES

$$-y_{gs} = j\omega C_{gs}\frac{1+j\omega\tau_2}{1+j\omega\tau_1} \qquad y_{bs} = (n-1)y_{gs}$$

$$-y_{gd} = j\omega C_{gd}\frac{1+j\omega\tau_3}{1+j\omega\tau_1} \qquad y_{bd} = (n-1)y_{gd}$$

$$-y_{gb} = -y_{bg} = j\omega C_{gb}\frac{1+j\omega\tau_4}{1+j\omega\tau_1}$$

$$-y_{ds} = \frac{g_{ms}}{1+j\omega\tau_1} \qquad -y_{sd} = \frac{g_{md}}{1+j\omega\tau_1} \qquad y_{dg} = \frac{g_{mg}}{1+j\omega\tau_1} + y_{gd}$$

$$\tau_1 = \frac{1}{\omega_0}\frac{4}{15}\frac{1+i_f+3\sqrt{1+i_f}\sqrt{1+i_r}+1+i_r}{(\sqrt{1+i_f}+\sqrt{1+i_r})^3}$$

$$\tau_{2(3)} = \frac{1}{\omega_0}\frac{1}{15}\frac{2(1+i_{f(r)})+8\sqrt{1+i_f}\sqrt{1+i_r}+5(1+i_{r(f)})}{(\sqrt{1+i_f}+\sqrt{1+i_r})^2(\sqrt{1+i_{f(r)}}+2\sqrt{1+i_{r(f)}})}$$

$$\tau_4 = \frac{C_{ox}\tau_1 - C_{gs}\tau_2 - C_{gd}\tau_3}{C_{ox} - C_{gs} - C_{gd}}$$

$$\omega_0 = \frac{\mu\phi_t}{L^2}$$

From the expressions in Table 2.2, it can be readily verified that

$$C_{sd} = -g_{md}\tau_1 \qquad C_{ds} = -g_{ms}\tau_1 \qquad C_m = g_{mg}\tau_1 \tag{2.48}$$

As shown in Appendix D, $-y_{gb}$ can be modeled by the constant capacitance C_{gb} and $\tau_{2(3)}$ is always of the order of $\frac{1}{2}\tau_1$ allowing the high-frequency model to be simplified to the schematic of Figure 2.17. Assuming $\omega\tau_1 \ll 1$, we can use the

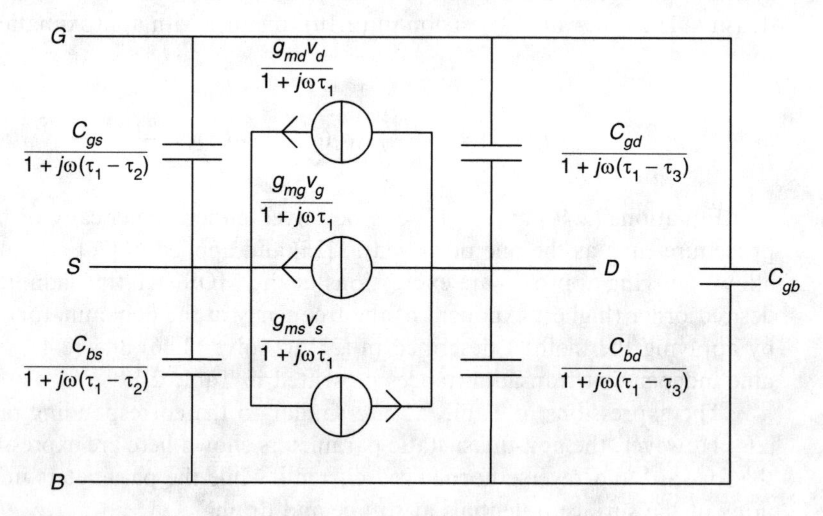

Figure 2.17 Simplified high-frequency small-signal MOSFET model.

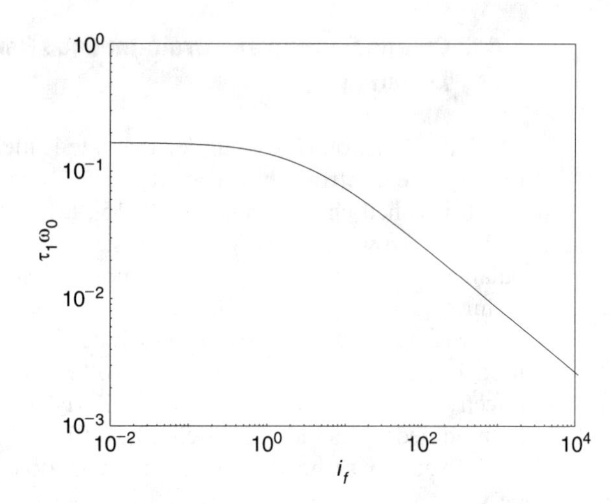

Figure 2.18 Normalized time constant of high-frequency model for operation in saturation.

approximation $1/(1 + j\omega\tau_1) \cong 1 - j\omega\tau_1$. From this approximation follows (2.48), the relation between time constant τ_1 and capacitances C_{sd} and C_{ds}. Moreover, for $\omega\tau_1 \ll 1$, the current sources of the non-quasi-static model connected between drain and source (Figure 2.17) reduce to the parallel combination of transconductances and negative transcapacitances of the complete quasi-static model of Figure 2.13 [10].

Figure 2.18 illustrates the variation of the time constant τ_1 in saturation with the forward normalized current. Since the intrinsic cutoff frequency is proportional to $\sqrt{1 + i_f} - 1$, the non-quasi-static correction of the transadmittances is only significant for moderate and strong inversion. In weak inversion, the quasi-static model predicts dynamic operation with satisfactory precision at frequencies up to the intrinsic cutoff frequency. In moderate and strong inversion the applicability of the quasi-static model should be restricted to frequency values up to one-third of the intrinsic cut-off frequency [10].

2.4. SECOND-ORDER EFFECTS ON MOSFET CHARACTERISTICS

In the previous sections, MOSFETs have been assumed to be long- and wide-channel devices. The mobility has been assumed to be independent of the transversal electric field. The effects of small geometry as well as the transverse field, here denominated as second-order effects, are examined in this section. The common approach to model second-order effects in MOSFETs, also employed in this section, is to decorate the long-channel model with a list of corrections related to these effects [27]. Presented is a discussion on small-geometry MOSFETs as well as some approximate expressions to calculate the influence of the geometry on the drain current. All the expressions are derived from the charge-based physical model of the long-channel MOSFET reported in [7–9] and are valid for any inversion level.

2.4.1. Charge Sharing and Drain-Induced Barrier Lowering

The depletion charge under the gate is identified not only with counter charges on the gate electrode but also in the N^+ regions of source and drain. This effect, which is called charge sharing [10, 15], affects decisively the performance of short- and/or narrow-channel MOS transistors. The depletion charge under the gate balanced by charges in the N^+ regions of source and drain depends not only on technological parameters but also on transistor dimensions and drain and source voltages. The usual approach to include these small-channel effects in the MOSFET model is to replace the threshold voltage by another quantity, denominated the effective threshold voltage [10]. Similarly, the long-channel pinch-off voltage can be modified to include small-channel effects.

If velocity saturation effects are negligible, the drain current of a MOSFET can be written as

$$I_D = f(V_P, V_S) - f(V_P, V_D) \tag{2.49}$$

For a long and wide transistor the pinch-off voltage is a function of V_G only, but the short- and narrow-channel devices V_P depends on V_S and V_D as well. To keep the symmetry of Eq. (2.49), V_P is modeled as

$$V_P(V_G, V_S, V_D) = V_{P0}(V_G) + \frac{\sigma}{n}(V_D + V_S) \tag{2.50}$$

Here, $V_{P0}(V_G)$ is the pinch-off voltage [3] at equilibrium ($V_D = V_S = 0$) and is given by

$$V_{P0} = \left(\sqrt{V_G - V_{T0} + 2\phi_F + \gamma\sqrt{2\phi_F} + \left(\tfrac{1}{2}\gamma'\right)^2 + \phi_t} - \tfrac{1}{2}\gamma'\right)^2 - \phi_{S0} \tag{2.51a}$$

where

$$\gamma' = \gamma - \frac{\varepsilon_0 \varepsilon_{Si}}{C'_{ox}}\left[\frac{2\eta_L}{L_{eff}} - \frac{3\eta_W}{W_{eff}}\right]\sqrt{\phi_{S0}} \tag{2.51b}$$

and ϕ_{S0} is about twice the Fermi potential ($2\phi_F$). The term γ is the body effect coefficient of a wide, long-channel device; γ' includes the short- and narrow-channel effects; η_L and η_W are fitting parameters; and L_{eff} and W_{eff} are the effective length and width, respectively. The parameter σ accounts for the drain-induced barrier lowering (DIBL) [15] and is roughly proportional to $1/L_{eff}^2$.

2.4.2. Mobility Reduction Due to Transversal Field

The electron mobility in the inversion layer depends on the transverse electric field [10]. The effective mobility is dependent on all terminal voltages [28] but is modeled here by

$$\mu = \frac{\mu_0}{1 + \theta\gamma\sqrt{V_{P0} + \phi_0}} \tag{2.52}$$

where μ_0 is the zero-bias mobility and θ is a fitting parameter. Equation (2.52) has been derived assuming that the transverse field is mainly determined by the average depletion charge, a quite reasonable assumption for low and moderate inversion levels.

2.4.3. Velocity Saturation

The effect of velocity saturation in our model is based on the expression [29]

$$\mu_S = \frac{\mu}{1 + (\mu/v_{\text{lim}})(d\phi_S/dx)} \tag{2.53}$$

where the mobility μ is a function of the gate-to-bulk potential V_G only and v_{lim} is the saturation velocity.

The substitution of both the fundamental relationship between inversion charge density and surface potential in (2.11) and (2.53) into the differential equation of the drain current leads, after integration along the channel [8, 29], to

$$I_D = \frac{\mu W_{\text{eff}}}{C'_{\text{ox}} L_{\text{eq}}} \frac{1}{1 + |Q'_{IS} - Q'_{ID}|/Q'_A} \frac{[(Q'^2_{IS} - Q'^2_{ID}) - 2nC'_{\text{ox}}\phi_t(Q'_{IS} - Q'_{ID})]}{2n} \tag{2.54a}$$

where

$$Q'_A = nC'_{\text{ox}} L_{\text{eq}}\,\text{UCRIT} \quad \text{where} \quad \text{UCRIT} = \frac{v_{\text{lim}}}{\mu} \tag{2.54b}$$

UCRIT being the critical electric field for which the carrier velocity is one-half of v_{lim}, the saturation velocity. In our analysis, the saturated MOSFET is divided into two parts: the channel, whose length is denominated L_{eq}, and a part closer to the drain where the carriers reach the saturation velocity. The latter part, the channel length modulation (ΔL), is modeled as in [6, 15]. Therefore, the channel length is given by $L_{\text{eq}} = L_{\text{eff}} - \Delta L$.

The charge-based expression (2.54a) of the drain current includes the effects of diffusion, drift, and carrier velocity saturation. Equation (2.54a) is a general expression that is valid from weak to strong inversion. The charge in the denominator models velocity saturation; the term $|Q'_{IS} - Q'_{ID}|$ correctly represents the channel potential in strong as well as in weak inversion.

The maximum current that can flow is limited by saturation velocity and the minimum amount of charge density (Q'_{ID}) at the drain end of the channel:

$$I_D = -Wv_{\text{lim}} Q'_{ID} \tag{2.55}$$

Equating (2.54) to (2.55) allows one to calculate $Q'_{ID,\mathrm{SAT}}$, the value of Q'_{ID} which corresponds to the onset of saturation, for any regime of operation:

$$Q'_{ID,\mathrm{SAT}} = Q'_{IS} - nC'_{\mathrm{ox}}\phi_t - Q'_A\left[1 - \sqrt{1 - \frac{2(Q'_{IS} - nC'_{\mathrm{ox}}\phi_t)}{Q'_A} + \frac{(nC'_{\mathrm{ox}}\phi_t)^2}{Q'^2_A}}\right] \tag{2.56}$$

Note that the inversion charge density at the onset of saturation is a function of the source and gate voltages, through Q'_{IS}, and of the channel length, through Q'_A.

From the charge-voltage relationship given by (2.24) one can readily calculate $V_{DS,\mathrm{SAT}}$, the drain-to-source voltage for which the charge density at the drain end corresponds to the onset of saturation:

$$V_{DS,\mathrm{SAT}} = \phi_t\left[\ln\left(1 + \frac{\sqrt{1 + i_{\mathrm{sat}}} - 1}{\frac{1}{2}\varepsilon i_{\mathrm{sat}}}\right) + \sqrt{1 + i_{\mathrm{sat}}} - 1\right] \tag{2.57a}$$

$$\varepsilon = \frac{\phi_t}{L_{\mathrm{eff}}\,\mathrm{UCRIT}} \tag{2.57b}$$

where i_{sat} is the normalized drain current at the onset of saturation. Typical values of ε, a "short-channel factor", range from 0.01 to 0.02.

Equations (2.56) and (2.57) are alternative descriptions of MOSFET saturation owing to the effect of carrier velocity saturation. Equations (2.56) and (2.57) are valid from weak to strong inversion, clearly showing the effect of current level on the saturation characteristics. The saturation voltage, as defined by (2.57), does not depend on the current for low current levels ($i_{\mathrm{sat}} \ll 1$) and is proportional to the square root of the current for high current levels ($i_{\mathrm{sat}} \gg 1$), coinciding with the value obtained from the long-channel theory. Note that (2.57a) is quite similar to (2.29), which describes the saturation voltage of a long-channel device. Therefore, for practical purposes, (2.29) can also be used to calculate the saturation voltage for short-channel transistors.

2.4.4. Channel Length Modulation

For a MOSFET operating in the saturation regime, the gradual channel approximation becomes less valid, specially in the vicinity of the drain junction, where the two-dimensional nature of the space charge region must be considered [10, 30]. Therefore, an analytical formulation of the saturated part of the conducting channel is not an easy task; as a consequence, many semiempirical models have been tried [28] to describe channel length modulation (CLM). The approach employed here divides the region between drain and source into two parts, the nonsaturated part of the channel, closer to the source, and the saturated part, closer to the drain. In the saturated part of the channel, the carrier velocity is assumed to be constant and equal to the saturation velocity. The channel length L_{eq}, that is, the nonsaturated part of the channel, is generally written as $L_{eq} = L - \Delta L$, where ΔL is the channel shrinkage due to CLM. Here we model the CLM as in [6, 15]

$$\Delta L = \lambda L_C \ln\left[1 + \frac{V_{DS} - V_{DS,\mathrm{SAT}}}{L_C\,\mathrm{UCRIT}}\right] \tag{2.58}$$

if $V_{DS} \geq V_{DS,\mathrm{SAT}}$ and $\Delta L = 0$ if $V_{DS} < V_{DS,\mathrm{SAT}}$. Here, λ and L_C can be considered as fitting parameters.

2.4.5. Transconductance and Output Conductance

In analog circuits, MOSFETs often operate in the saturation region, where the available transistor voltage gain, equal to the transconductance-to-output-conductance ratio, can be made large. We are going to show how to use the results from the previous section in order to calculate the MOSFET transconductance and output conductance.

The transconductance-to-current ratio determined in Section 2.3 for a long-channel device can be derived for the more general case of short-channel transistors. In the saturation regime, both g_{mg} and I_D are weakly dependent functions of V_D. Therefore, in order to compute the ratio g_{mg}/I_D in saturation, one can assume $I_D = I_{D,\mathrm{SAT}}$, where $I_{D,\mathrm{SAT}}$ is the drain current at the onset of saturation. From (2.55) one can write

$$\frac{g_{mg}}{I_{D,\mathrm{SAT}}} = \frac{1}{Q'_{ID,\mathrm{SAT}}}\frac{dQ'_{ID,\mathrm{SAT}}}{dV_G} \tag{2.59a}$$

Using (2.56), the relationship between charge and voltage given by (2.24) and $dV_P/dV_G = 1/n$, yields

$$\frac{g_{mg}}{I_{D,\mathrm{SAT}}} = \frac{2}{n\phi_t(1 + \sqrt{1 + i_{\mathrm{sat}}})}\left(1 - \frac{\varepsilon}{2}\frac{(\sqrt{1 + i_{\mathrm{sat}}} - 1)}{1 + \varepsilon\sqrt{1 + i_{\mathrm{sat}}}}\right)\left(1 - \frac{\varepsilon}{2}\frac{i_{\mathrm{sat}}}{\sqrt{1 + i_{\mathrm{sat}}} + \frac{1}{2}\varepsilon\, i_{\mathrm{sat}}}\right) \tag{2.59b}$$

where $i_{\mathrm{sat}} = I_{D,\mathrm{SAT}}/I_S$ is the normalized saturation current. The first term on the right-hand side of (2.59b) is the transconductance-to-current ratio of a long-channel MOS transistor. If $i_{\mathrm{sat}} \ll 1/\varepsilon^2$, (2.59b) can be simplified to the expression derived for long-channel devices. On the other hand, if $i_{\mathrm{sat}} \gg 1/\varepsilon^2$, the transconductance is given by the well-known expression $g_{mg} = WC'_{\mathrm{ox}}v_{\mathrm{lim}}$. Figure 2.19 shows the transconductance-to-current ratios of both long- and short-channel transistors. According to (2.59b), the transconductance-to-current ratio of a short-channel device is one-half the transconductance-to-current ratio of a long-channel device if the normalized current of both transistors is equal to $1/\varepsilon^2$, ε being the short-channel factor calculated for the short-channel transistor.

In the classical approach to model the saturation region, the output conductance is assumed to be proportional to the drain current and inversely proportional to the Early voltage V_A [10], a constant parameter in first-order models such as SPICE1. However, a constant Early voltage is inadequate to model the output conductance for the simulation of analog circuits. Other models [15, 28, 31] use smoothing functions to unify the linear and saturation regions but demand the extraction of parameters which neither have a simple physical interpretation nor are easy to extract. Another drawback of some models is that the effects of carrier velocity saturation that are only valid in strong inversion [3, 10, 31, 32] are extended to weak inversion, where the channel is almost equipotential [10]. Consequently, saturation is modeled with a physical background in strong inversion while, on the other hand, it is defined in an empirical manner [3, 33] in weak and moderate inversion.

Our model of the output conductance includes velocity saturation effects, CLM, and DIBL. Impact ionization and substrate current-induced body effects can be easily included in our model by adding a substrate current such as in [31].

From (2.55) one can write

$$g_o = \frac{dI_D}{dV_D} = -Wv_{\text{lim}} \frac{dQ'_{ID,\text{SAT}}}{dV_D} \tag{2.60a}$$

where g_o is the output conductance in saturation, previously defined as g_{md}. According to (2.56), the saturation charge $Q'_{ID,\text{SAT}}$ depends on the modulated channel length, L_{eq}, and on the inversion charge density at the source, Q'_{IS}. At this point we can include both the CLM and DIBL effects to calculate the small-signal output conductance.

Now, (2.60a), (2.56), and (2.55), together with (2.24), which gives us the relationship between voltage and charge, allow calculation of the MOSFET output conductance-to-current ratio:

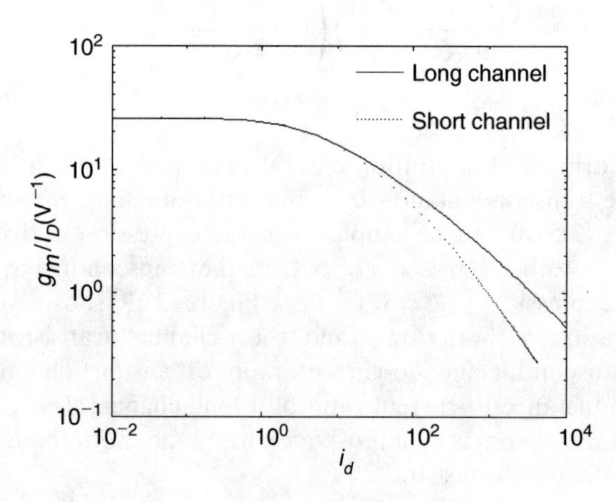

Figure 2.19 Transconductance-to-current ratio of long- and short-channel ($1/\varepsilon^2 = 800$) MOS transistors vs. normalized drain current.

$$\frac{g_o}{I_D} = \frac{1}{V_A} = \frac{1}{V_{A,\text{DIBL}}} + \frac{1}{V_{A,\text{CLM}}} \tag{2.60b}$$

where the two components of the Early voltage can be approximated by

$$\frac{V_{A,\text{DIBL}}}{\phi_t} \cong \frac{n}{2\sigma}\left(\sqrt{1+i_{\text{sat}}}+1\right) \tag{2.60c}$$

$$\frac{V_{A,\text{CLM}}}{\phi_t} \cong \frac{1}{\phi_t}\left(\frac{1}{L_{\text{eq}}}\frac{d\,\Delta L}{dV_D}\right)^{-1} \tag{2.60d}$$

for $i_{\text{sat}} < 1/\varepsilon^2$.

If ΔL is given by (2.58), then (2.60d) can be written as

$$\frac{V_{A,\text{CLM}}}{\phi_t} \cong \frac{L_{\text{eq}}\text{UCRIT}}{\lambda\phi_t}\left(1+\frac{V_{DS}-V_{DS,\text{SAT}}}{L_C\,\text{UCRIT}}\right) \tag{2.60e}$$

The set of equations (2.60) is a generalization, for any bias condition, of the MOSFET output conductance presented in [32]. In weak inversion $i_{\text{sat}} \ll 1$; therefore, the Early voltage V_A is independent of the current level. If $1 \ll i_{\text{sat}} < 1/\varepsilon^2$, the DIBL component of the Early voltage is proportional to the square root of the current while the CLM component does not depend on the current level. Moreover, the CLM component of V_A depends on the effective voltage drop across the shrunk part of the channel. Typically, the DIBL component of the Early voltage can be neglected for high inversion levels, CLM being the dominant factor in the output conductance.

Figure 2.20 illustrates the output conductance of a MOSFET, simulated according to our model. A smoothing function has been used to avoid discontinuous derivative of g_o with respect to the voltage in the transition from nonsaturation to saturation. The plot of the output conductance versus V_D is one of the benchmark tests proposed in [2] for evaluating MOSFET models.

In order to verify the consistency of the output conductance model, MOSFETs from a 0.75 μm technology with different channel lengths were measured for several bias conditions. Figure 2.21 displays the output characteristics of a MOSFET whose channel length $L = 1.25$ μm. The drain current in saturation is strongly dependent on the drain voltage. Figure 2.22 shows the variation of the Early voltage for several channel lengths in terms of the normalized drain current. One can conclude that the Early voltage increases with increasing channel lengths, is almost independent of the current level in weak inversion, and increases in moderate and strong inversion, as predicted by (2.60).

Figure 2.23 shows that the Early voltage is almost independent of V_D in weak inversion but increases slightly with the drain voltage in strong inversion. This difference can be explained with the help of expressions (2.60), where one can note that the influence of DIBL on the Early voltage is more important at low current densities.

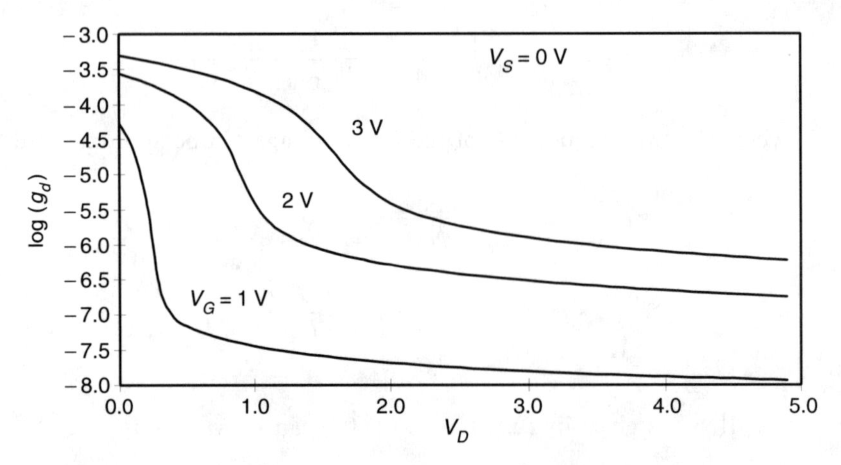

Figure 2.20 MOSFET output conductance, simulated according to our model.

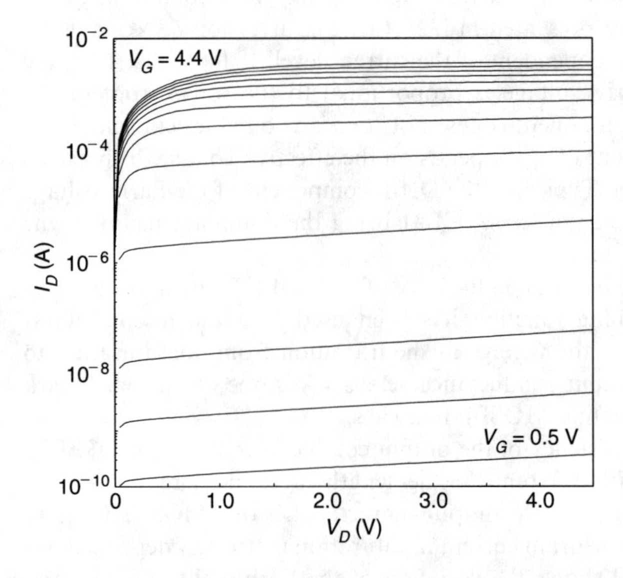

Figure 2.21 Output characteristics of NMOS transistor ($L = 1.25\,\mu m$) for $V_G = 0.5, 0.6, 0.7, 0.9, 1.2, 1.6, 2.0, 2.4, 2.8, 3.2, 3.6, 4.0, 4.4$ V ($V_S = 0$).

2.5. APPLICATION OF THE MOSFET MODEL TO THE DESIGN OF A COMMON-SOURCE AMPLIFIER

Many of the currently available strategies for designing CMOS amplifiers are based on MOSFET models developed for either weak inversion or strong inversion [34, 35]. However, the current trend toward low power often requires the transistor to be

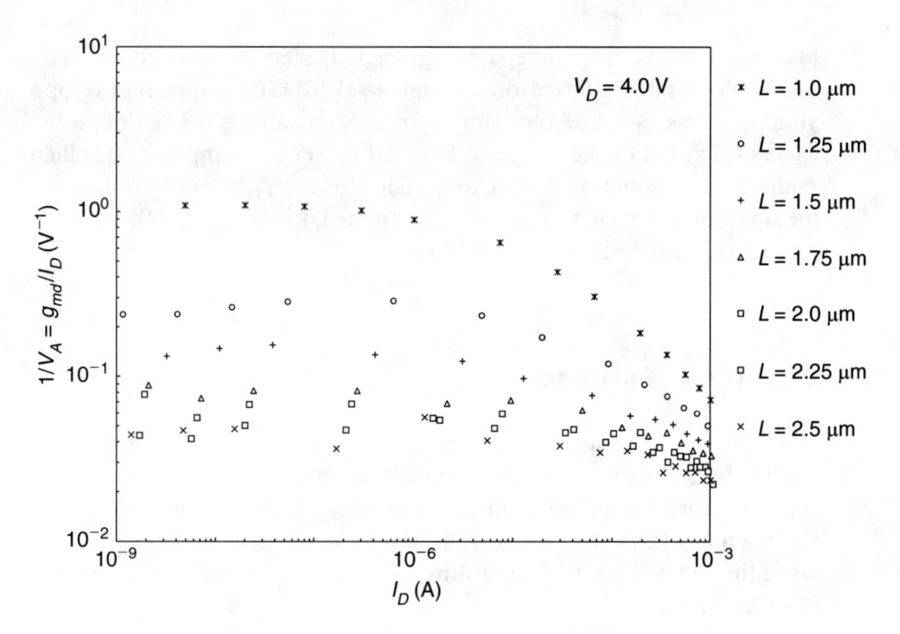

Figure 2.22 Output conductance-to-current ratio vs. drain current for MOSFETs whose channel lengths range from 1.0 to 2.5 μm.

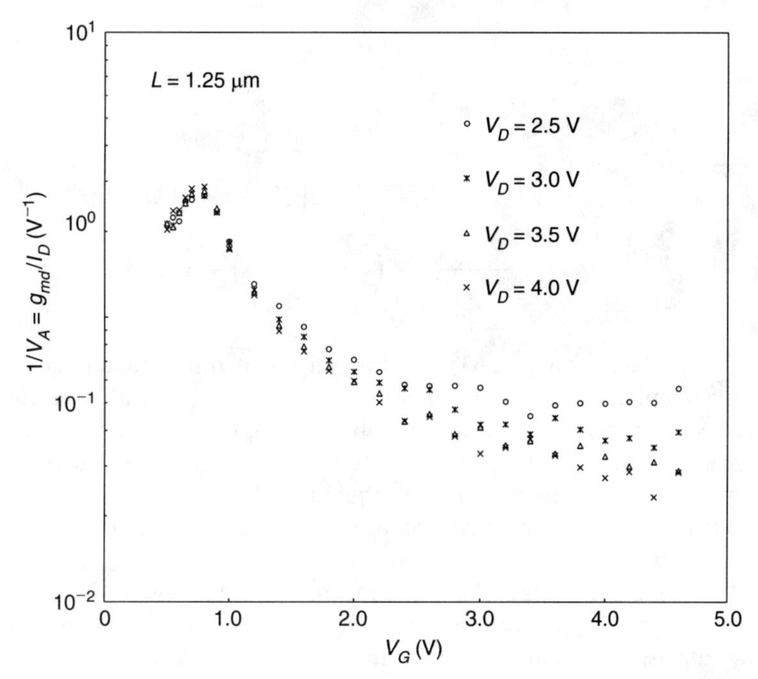

Figure 2.23 Output conductance-to-current ratio vs. gate voltage of NMOS transistor ($L = 1.25$ μm), with V_D varying from 2.5 to 4.0 V.

biased in the moderate-inversion region. This section presents a design methodology for MOS amplifiers based on the universal MOSFET model developed in the previous sections. A set of very simple expressions allows quick design by hand as well as an evaluation of the design in terms of power consumption and silicon real estate. A design of a common-source amplifier illustrates the applicability of the proposed methodology, which can be readily applied to more complex topologies such as differential and operational amplifiers.

2.5.1. Design Equations

Before writing some fundamental design equations, we would like to remind the reader that, in saturation, the reverse component of the drain current is much less than the forward component. Therefore, $I_D \cong I_F$ and $g_{mg} \cong g_{ms}/n$, according to (2.31a) and (2.37). Using previous results derived in Sections 2.2 and 2.3, we can write the following set of equations for designing circuits where MOSFETs operate in saturation:

$$\frac{\phi_t n g_{mg}}{I_D} = \frac{2}{1 + \sqrt{1 + i_d}} \tag{2.61a}$$

$$i_d = \frac{I_D}{I_S} \tag{2.61b}$$

$$I_S = \mu n C'_{\text{ox}} \frac{\phi_t^2}{2} \frac{W}{L} \tag{2.61c}$$

$$f_T \cong \frac{\mu \phi_t}{2\pi L^2} 2\left(\sqrt{1 + i_d} - 1\right) \tag{2.62}$$

$$\frac{V_{DS,\text{SAT}}}{\phi_t} \cong \left(\sqrt{1 + i_d} - 1\right) + 4 \tag{2.63}$$

Equations (2.61)–(2.63) constitute a set of fundamental expressions to design MOS amplifiers for any inversion level. Equation (2.61a) is readily derived from (2.37) and (2.38). Expressions (2.14b), (2.14c), (2.42c) and (2.29) have been rewritten in this section as (2.61b), (2.61c), (2.62), and (2.63), respectively. For design purposes, one can assume n to be equal to 1.3, a typical value. If higher accuracy is needed, n can be calculated according to expression (2.5). Recall that for bipolar transistors $\phi_t g_m / I_C = 1$. If g_m is defined in a bipolar design, so is I_C. However, in a MOSFET-based design, the specification of g_{mg} allows the designer to choose from a range of currents, according to (2.61a). Equation (2.62) is an approximation for the intrinsic cutoff frequency f_T in terms of i_d. An approximate formula for the source-to-drain saturation voltage as a function of the inversion level is shown in (2.63).

2.5.2. Common-Source Amplifier

The methodology presented in [17] to design amplifiers is based on either calculated [3] or measured curves of g_m/I_D. Up to this point, we have avoided using the notation g_m for the gate transconductance. Now we adopt the symbol g_m for the gate transconductance, instead of g_{mg}. According to our model, the g_m/I_D ratio, the saturation voltage, as well as the intrinsic cutoff frequency given by (2.61)–(2.63) are universal, being valid for any MOS technology. The only technology-dependent parameter used in our methodology is I_S, the normalization current.

In the ideal common-source amplifier shown in Figure 2.24, the transconductance required to achieve a gain-bandwidth product (GBW) for a load capacitance CL is $g_m = 2\pi$ GBW CL. Using this equation together with (2.61) allows writing the drain current and the aspect ratio as

$$I_D = 2\pi \,\mathrm{GBW}\ \mathrm{CL}\ n\phi_t\left[\tfrac{1}{2}(1+\sqrt{1+i_d})\right] \tag{2.64}$$

$$\frac{W}{L} = \frac{2\pi\ \mathrm{GBW}\ \mathrm{CL}}{\mu C'_{ox}\phi_t}\left(\frac{1}{\sqrt{1+i_d}-1}\right) \tag{2.65}$$

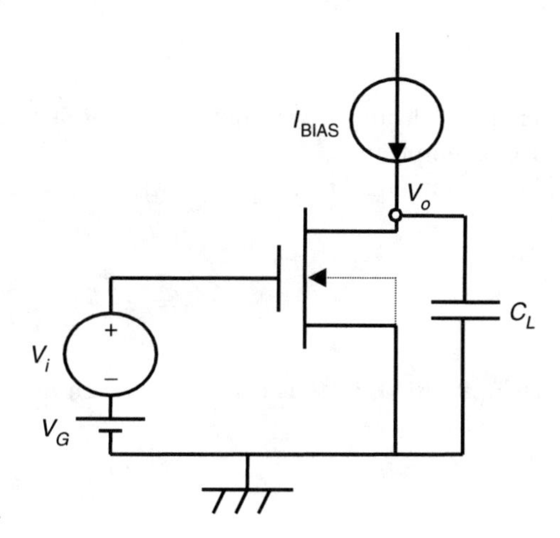

Figure 2.24 Common-source amplifier.

Equations (2.64) and (2.65) show that an infinite set of solutions are possible to meet the required GBW. Figure 2.25 shows plots of the bias current given by (2.64) and the aspect ratio in (2.65) as functions of i_d. Both curves have been normalized with respect to $i_d = 8$. The trade-off between area and power consumption can be reached by an appropriate choice of i_d. Power consumption is low but the aspect ratio is high for low inversion levels. The lowest current to meet the specified GBW is obtained in weak inversion, that is, for $i_d < 1$. A value of i_d close to 1 can be a good choice if low power is required. On the other hand, a normalized drain current much

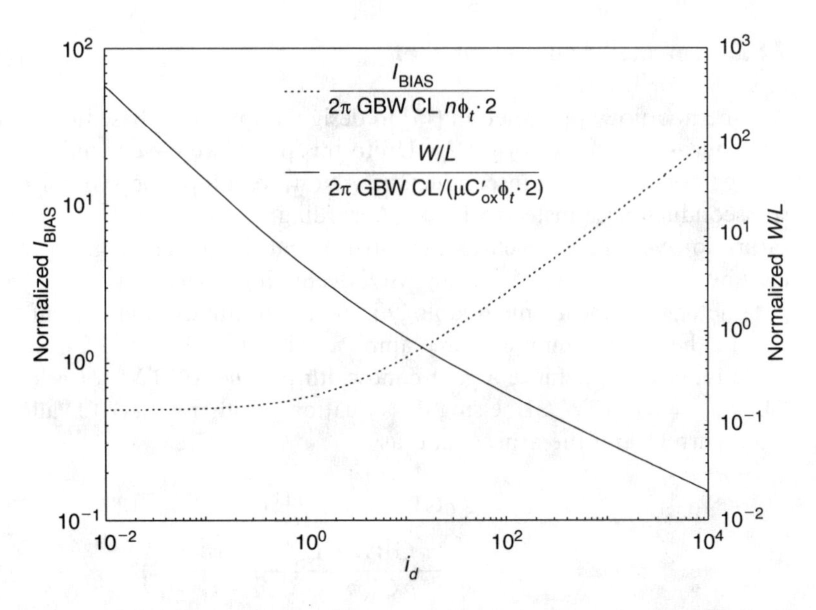

Figure 2.25 Bias current and aspect ratio vs. inversion coefficient. Both curves are normalized with respect to $i_d = 8$.

smaller than 1 leads to a prohibitively high aspect ratio for a negligible reduction in power consumption.

The following set of equations is used to determine the DC voltage gain [17] A_{V0}:

$$A_{V0} = -\frac{g_m}{g_o} = -\frac{g_m}{I_D} V_A \tag{2.66a}$$

Using (2.61a), (2.66a) can be rewritten as

$$A_{V0} = -\frac{V_A}{\phi_t}\left(\frac{2}{n(1+\sqrt{1+i_d})}\right) \tag{2.66b}$$

where V_A is dependent upon the channel length and the bias, as demonstrated in Section 2.4.

2.5.3. Design Methodology

Assuming that the GBW and the DC gain are the specifications to be met for a given load capacitance, the following design methodology for a common-source amplifier is suggested:

1. Given CL and GBW, use expressions (2.64) and (2.65) to determine I_D and W/L as functions of the inversion level. Plots of both the bias current and the aspect

ratio required to meet the GBW specification are shown in Figure 2.25. Choose a value for i_d and, consequently, a pair (I_D, W/L) that satisfies the design requirements for power consumption and area.

2. Select the dimensions W and L for the input transistor according to the aspect ratio determined in step 1. Verify if the unity-gain frequency of the input transistor is consistent with the GBW specified. Typically, f_T should be at least three times higher than the GBW in order to avoid the parasitic diffusion and overlap capacitances of the NMOSFET to be of the same order of magnitude as the load capacitance. From (2.62) and (2.65) one can write

$$W \cong 2 \frac{\text{CL}}{LC'_{\text{ox}}} \frac{\text{GBW}}{f_T} \tag{2.67}$$

The diffusion and overlap capacitances are both proportional to W; therefore, the parasitic capacitance of the NMOS transistor drain is proportional to GBW/f_T. If the ratio GBW/f_T is close to 1, the ratio of the parasitic capacitance to the load capacitance can be too large; thus, a new pair (I_D, W/L) should be chosen to compensate for the parasitic capacitance.

3. Verify if the gain specification is satisfied. If not, an increase in channel length and/or a reduction in i_d can be employed to increase the gain;
4. Design the current source. For a low-voltage design, the saturation voltage of the current source should be as small as possible. Therefore, inversion coefficients close to 1 can be used in order to achieve a good balance between area and saturation voltage.

2.5.4. Simulation Results

To illustrate the design methodology, a common-source amplifier whose specifications are GBW $= 10\,\text{MHz}$, CL $= 10\,\text{pF}$, and $A_{V0} \geq 40\,\text{dB}$ has been designed. The technological parameters are $\mu_N C'_{\text{ox}} = 107\,\mu\text{A/V}^2$ and n is supposed to be 1.3. The length $L = 5\,\mu\text{m}$ has been chosen for the specified gain. Table 2.3 illustrates the results obtained by hand design. The terms W/L and I_{BIAS} are calculated from (2.64) and (2.65). The simulated GBW has been obtained by using the EKV

TABLE 2.3 HAND DESIGN AND SIMULATION OF A COMMON-SOURCE AMPLIFIER

Design				Simulation	
W/L	I_{BIAS} (μA)	i_d	f_T/GBW	GBW (MHz)	A_{V0} (dB)
545	26	1	0.64	7.7	60
226	32	3	1.5	8.6	59
97	46	10	3.6	9.0	57
50	70	30	7.0	9.2	55
25	117	100	14.0	9.2	52
14	195	300	25.0	10.0	47

model from the SMASH simulator [36] using the ratio W/L and I_{BIAS} calculated from (2.64) and (2.65). The simulation shows that, for low inversion levels, W/L is large, and consequently, the parasitic capacitance in parallel with CL degrades the frequency response. More accurate results can be obtained by adding the parasitic capacitance to the load capacitance in order to calculate the bias current. For higher inversion levels, the GBW is within 10% of its specified value.

2.6. SUMMARY

In this chapter, we presented a very simple and compact MOSFET model suitable for the design of integrated circuits. All large- and small-signal characteristics of the MOSFET are expressed as functions of two components of the drain current, namely the forward and reverse saturation currents. The inclusion of a non-quasi-static model provides designers with an appropriate MOSFET model up to the transistor intrinsic cutoff frequency. Short-channel effects have been included in our model by properly modifying parameters of the long-channel transistor. The model presented in this chapter has been successfully applied to the design of a common-source amplifier.

Appendix A. Determination of Q_S and Q_D

Charge-conserving models must associate specific charges with each device terminal. Consequently, the channel charge must be divided into a source and a drain charge [10]. The integration of the continuity equation results [10] in

$$Q_S = W\int_0^L \left(1 - \frac{x}{L}\right) Q_I' \, dx \tag{A.1}$$

$$Q_D = Q_I - Q_S = W\int_0^L \frac{x}{L} Q_I' \, dx \tag{A.2}$$

In order to determine the drain and source charges, the coordinate x must be expressed in terms of the inversion charge density. The integration of (2.16) from the source to an arbitrary point of the channel leads to

$$x = \frac{\mu W}{2nC_{ox}' I_D}[(Q_{IS}' - nC_{ox}'\phi_t)^2 - (Q_I' - nC_{ox}'\phi_t)^2] \tag{A.3}$$

The substitution of (A.3) into (A.1) and (A.2) yields

$$Q_S = WL\left[\frac{6Q_F'^3 + 12Q_R'Q_F'^2 + 8Q_R'^2Q_F' + 4Q_R'^3}{15(Q_F' + Q_R')^2} + \frac{n}{2}C_{ox}'\phi_t\right] \tag{A.4}$$

$$Q_D = WL\left[\frac{6Q_R'^3 + 12Q_F'Q_R'^2 + 8Q_F'^2Q_R' + 4Q_F'^3}{15(Q_F' + Q_R')^2} + \frac{n}{2}C_{ox}'\phi_t\right] \tag{A.5}$$

Appendix B. Approximation of the Relationship between Current and Voltage

A first-order approximation to the inversion charge density can be derived by assuming it to be approximated to its value in weak inversion,

$$Q_I' \approx Q_{IP}' \exp\left(\frac{V_P - V_C}{\phi_t}\right) \tag{B.1}$$

and substituting (B.1) into (2.22) to obtain a more accurate approximation for any region of operation. Thus

$$Q_I' \cong -nC_{ox}'\phi_t \ln\left[1 + \exp\left(\frac{V_P - V_C}{\phi_t}\right)\right] \tag{B.2}$$

The relationship between current and voltage is readily obtained if we substitute (B.2) into (2.13b), resulting in

$$I_{F(R)} = I_S\left\{\left[1 + \ln\left(1 + \exp\left(\frac{V_P - V_{S(D)}}{\phi_t}\right)\right)\right]^2 - 1\right\} \tag{B.3}$$

A more precise (and cumbersome) approximation of the relationship between current and voltage in a MOSFET is given in [1].

Appendix C. Calculation of the Intrinsic Small-Signal Capacitances

The total gate charge is given by

$$Q_G = -Q_B - Q_I - Q_0 \tag{C.1}$$

where Q_0 is the effective interface charge, assumed to be independent of the terminal voltages. Substituting the expression for the bulk charge, (2.18), into the charge balance equation, (C.1), we obtain

$$Q_G = -\frac{Q_I}{n} - Q_{Ba}'WL - Q_0 \tag{C.2}$$

Capacitances C_{gs} and C_{gd} can be obtained as follows. Using definition (2.39) and deriving the gate charge (C.2) with respect to $V_{S(D)}$:

$$C_{gs} = \frac{1}{n}\frac{\partial Q_I}{\partial V_S} \tag{C.3a}$$

$$C_{gd} = \frac{1}{n}\frac{\partial Q_I}{\partial V_D} \tag{C.3b}$$

Substituting expression (2.17b) of the total inversion charge Q_I into (C.3) gives

$$C_{gs} = \frac{2WL}{3n}\left[\frac{1 - Q'^2{}_R}{(Q'_F + Q'_R)^2}\right]\frac{\partial Q'_{IS}}{\partial V_S} \tag{C.4a}$$

$$C_{gd} = \frac{2WL}{3n}\left[\frac{1 - Q'^2{}_F}{(Q'_F + Q'_R)^2}\right]\frac{\partial Q'_{ID}}{\partial V_D} \tag{C.4b}$$

From (2.32), the equation of the derivative of the inversion charge with respect to the channel voltage, and from (2.14a), the relationship between the inversion charge density and the inversion coefficient, it follows that

$$\frac{\partial Q'_{IS(D)}}{\partial V_{S(D)}} = nC'_{\rm ox}\frac{\sqrt{1 + i_{f(r)}} - 1}{\sqrt{1 + i_{f(r)}}} \tag{C.5}$$

The substitution of (2.14a) and (C.5) into (C.4) results in

$$C_{gs(d)} = \tfrac{2}{3}C_{\rm ox}(\sqrt{1 + i_{f(r)}} - 1)\frac{\sqrt{1 + i_{f(r)}} + 2\sqrt{1 + i_{r(f)}}}{(\sqrt{1 + i_f} + \sqrt{1 + i_r})^2} \tag{C.6}$$

We can calculate the source capacitance C_{ss} and the source-to-drain capacitance C_{sd} by directly calculating the partial derivatives of the total source charge, given by (A.4), with respect to V_S and V_D and by using (C.5) to express the derivatives of the inversion charges as functions of the inversion coefficients. Thus

$$C_{ss} = \tfrac{2}{15}nC_{\rm ox}(\sqrt{1 + i_f} - 1)\frac{3(1 + i_f) + 9\sqrt{1 + i_f}\sqrt{1 + i_r} + 8(1 + i_r)}{(\sqrt{1 + i_f} + \sqrt{1 + i_r})^3} \tag{C.7}$$

$$C_{sd} = -\tfrac{4}{15}nC_{\rm ox}(\sqrt{1 + i_r} - 1)\frac{1 + i_f + 3\sqrt{1 + i_f}\sqrt{1 + i_r} + 1 + i_r}{(\sqrt{1 + i_f} + \sqrt{1 + i_r})^3} \tag{C.8}$$

Using expression (2.18) for the bulk charge, the source-to-bulk capacitance C_{bs} and the drain-to-bulk capacitance C_{db} are given by

$$C_{bs} = \frac{n - 1}{n}\frac{\partial Q_I}{\partial V_S}\bigg|_{V_D V_G V_B} \tag{C.9a}$$

$$C_{bd} = \frac{n-1}{n} \frac{\partial Q_I}{\partial V_D}\bigg|_{V_S V_G V_B} \tag{C.9b}$$

Comparison of (C.9) with (C.3) allows writing

$$C_{bs} = (n-1)C_{gs} \tag{C.10a}$$

$$C_{bd} = (n-1)C_{gd} \tag{C.10b}$$

To obtain the gate-to-bulk capacitance C_{gb} and the bulk-to-gate capacitance C_{bg}, the derivatives of the depletion charge density Q'_{Ba}, Eq. (2.7c), must be calculated. Recalling that

$$\frac{\partial V_P}{\partial V_G}\bigg|_{V_B} = -\frac{\partial V_P}{\partial V_B}\bigg|_{V_G} = \frac{1}{n}$$

it follows that

$$\frac{\partial Q'_{Ba}}{\partial V_G}\bigg|_{V_B} = -\frac{\partial Q'_{Ba}}{\partial V_B}\bigg|_{V_G} = -\frac{n-1}{n} C'_{ox} \tag{C.11}$$

Using (C.11) and neglecting the variation of n with V_G, it follows from expressions (C.2) and (2.18) of charges Q_G and Q_B that

$$C_{gb} = \frac{1}{n} \frac{\partial Q_I}{\partial V_B}\bigg|_{V_S V_D V_G} + \frac{n-1}{n} C_{ox} \tag{C.12a}$$

$$C_{bg} = \frac{n-1}{n} \frac{\partial Q_I}{\partial V_G}\bigg|_{V_S V_D V_B} + \frac{n-1}{n} C_{ox} \tag{C.12b}$$

Noting that an increase in the bulk voltage with respect to a reference voltage is equivalent to a simultaneous decrease of the gate, source and drain voltages keeping the bulk voltage constant, it follows that

$$\frac{\partial Q'_{IS(D)}}{\partial V_B} = -\frac{\partial Q'_{IS(D)}}{\partial V_G} - \frac{\partial Q'_{IS(D)}}{\partial V_{S(D)}} \tag{C.13}$$

From (2.36) and (C.13) we obtain

$$\frac{\partial Q'_{IS(D)}}{\partial V_B} = -\frac{n-1}{n} \frac{\partial Q'_{IS(D)}}{\partial V_{S(D)}} \tag{C.14}$$

Applying (2.36) and (C.14) to calculate the partial derivatives of Q_I and comparing the results with expressions (C.3) for C_{gs} and C_{gd}, we can write

$$\frac{\partial Q_I}{\partial V_G} = \frac{\partial Q_I}{\partial Q'_F}\frac{\partial Q'_F}{\partial V_G} + \frac{\partial Q_I}{\partial Q'_R}\frac{\partial Q'_R}{\partial V_G} = -(C_{gs} + C_{gd}) \tag{C.15a}$$

$$\frac{\partial Q_I}{\partial V_B} = \frac{\partial Q_I}{\partial Q'_F}\frac{\partial Q'_F}{\partial V_B} + \frac{\partial Q_I}{\partial Q'_R}\frac{\partial Q'_R}{\partial V_B} = -(n-1)(C_{gs} + C_{gd}) \tag{C.15b}$$

Finally, from (C.12) and (C.15),

$$C_{gb} = C_{bg} = \frac{n-1}{n}(C_{ox} - C_{gs} - C_{gd}) \tag{C.16}$$

Calculating the partial derivative of Q_S with respect to V_G using (2.36b) and comparing the result with the expressions for C_{ss} and C_{sd} gives

$$C_{sg} = \frac{C_{ss} - C_{sd}}{n} \tag{C.17}$$

and using (2.36b) and (C.13) yields

$$C_{sb} = (n-1)C_{sg} \tag{C.18}$$

Appendix D. Discussion of the Non-quasi-static Model

From the expressions for time constants in Table 2.2 and from expression (C.6) for the capacitances, it follows that

$$\tau_4 = \frac{1}{\omega_0}\frac{2}{15}\frac{[4(1+i_f) + 7\sqrt{1+i_f}\sqrt{1+i_r} + 4(1+i_r)](\sqrt{1+i_f} + \sqrt{1+i_r})}{[(\sqrt{1+i_f} - \sqrt{1+i_r})^2 + 6(\sqrt{1+i_f} + \sqrt{1+i_r})](\sqrt{1+i_f} + \sqrt{1+i_r})^2}$$

$$\frac{+7(1+i_f) + 16\sqrt{1+i_f}\sqrt{1+i_r} + 7(1+i_r)}{[(\sqrt{1+i_f} - \sqrt{1+i_r})^2 + 6(\sqrt{1+i_f} + \sqrt{1+i_r})](\sqrt{1+i_f} + \sqrt{1+i_r})^2} \tag{D.1}$$

In strong inversion ($i_f \gg 1$) and assuming $i_f \neq i_r$, (D.1) can be approximated by

$$\tau_4 \cong \frac{1}{\omega_0}\frac{2}{15}\frac{4(1+i_f) + 7\sqrt{1+i_f}\sqrt{1+i_r} + 4(1+i_r)}{(\sqrt{1+i_f} - \sqrt{1+i_r})^2(\sqrt{1+i_f} + \sqrt{1+i_r})} \tag{D.2}$$

Here, (D.2) is similar to the expression presented in [10].

If $V_{DS} = 0$, then $i_f = i_r$. From the expression of τ_1 in Table 2.2 and from (D.1), we obtain

$$\frac{\tau_4}{\tau_1} = \frac{1}{2}(1 + \sqrt{1+i_f}) \tag{D.3}$$

Therefore, the singularity of the conventional approximation in strong inversion for τ_4 in the linear region is removed.

Finally, in weak inversion ($i_f, i_r \ll 1$), we calculate τ_4 from (D.1) and τ_1 from Table 2.2 and conclude that

$$\tau_4 \cong \tau_1 \cong \frac{1}{6\omega_0} \tag{D.4}$$

The term τ_4 is always greater than τ_1, but in weak inversion τ_4 and τ_1 are almost the same. Consequently, in weak inversion, the non-quasi-static effects in $C_{gb(bg)}$ are canceled out. According to (D.3), τ_4 is much greater than τ_1 in the linear region in strong inversion, but in that case y_{gb} can be neglected in comparison with the other gate admittances. As a consequence, the effects of the gate-to-bulk admittance can be modeled by a constant capacitance equal to C_{gb}.

References

[1] O. C. Gouveia-F., A. I. A. Cunha, M. C. Schneider, and C. Galup-Montoro, "The ACM Model for Circuit Simulation and Equations for SMASH," in Application Note in Home-page Dolphin, http://www.dolphin.fr.

[2] Y. P. Tsividis and K. Suyama, "MOSFET Modeling for Analog CAD: Problems and Prospects," *IEEE J. Solid-State Circuits*, vol. 29, no. 3, pp. 210–216, 1994.

[3] C. C. Enz, F. Krummenacher and E. A. Vittoz, "An Analytical MOS Transistor Model Valid in all Regions of Operation and Dedicated to Low-Voltage and Low-Current Applications," *Analog Integrated Circuits Signal Process.*, vol. 8, pp. 83–114, 1995.

[4] E. A. Vittoz, "Micropower Techniques," in *Design of MOS VLSI Circuits for Telecommunications*, J. Franca and Y. Tsividis (eds.), Prentice-Hall, Englewood Cliffs, NJ, 1993.

[5] K. S. Kundert, *The Designer's Guide to SPICE & SPECTRE*, Kluwer Academic, London, 1995.

[6] C. C. Enz and E. A. Vittoz, "Low-Power Analog CMOS Design," in *Emerging Technologies*, R. Cavin and W. Liu (eds.), IEEE, New York, 1996, Chap. 1.2.

[7] A. I. A. Cunha, M. C. Schneider, and C. Galup-Montoro, "An Explicit Physical Model for the Long-Channel MOS Transistor Including Small-Signal Parameters," *Solid-State Electron.*, vol. 38, no. 11, pp. 1945–1952, 1995.

[8] A. I. A. Cunha, "Um Modelo do Transistor MOS para Projeto de Circuitos Integrados," Ph.D. Thesis, Universidade Federal de Santa Catarina, December 1996 (in Portuguese).

[9] A. I. A. Cunha, O. C. Gouveia-F., M. C. Schneider, and C. Galup-Montoro, "A Current-Based Model for the MOS Transistor," *Proc. 1997 IEEE Int. Symp. Circuits Systems*, June 1997, Hong-Kong, pp. 1608–1611.

[10] Y. Tsividis, *Operation and Modeling of the MOS Transistor*, McGraw-Hill, New York, 1987.

[11] C. Enz, "High-Precision CMOS Micropower Amplifiers," Ph.D. Thesis, no. 802, Swiss Federal Institute of Technology, Lausanne, 1989.
[12] E. A. Vittoz, *Intensive Summer Course on CMOS VLSI Design—Analog & Digital*, Swiss Federal Institute of Technology, Lausanne, 1989.
[13] M. Banu and Y. Tsividis, "Fully Integrated Active RC Filters in MOS Technology," *IEEE J. Solid-State Circuits*, vol. 18, no. 6, pp. 644–651, 1983.
[14] K. Bult and G. J. G. M. Geelen, "An Inherent Linear and Compact MOST-Only Current-Division Technique," *IEEE J. Solid-State Circuits*, vol. 27, no. 6, pp. 1730–1735, 1992.
[15] K. Lee, M. Shur, T. A. Fjeldly, and T. Ytterdal, *Semiconductor Device Modeling for VLSI*, Prentice-Hall, Englewood Cliffs, NJ, 1993.
[16] C. K. Park, C. Y. Lee, K. Lee, B. J. Moon, Y. H. Byun, and M. Shur, "A Unified Current-Voltage Model for Long-Channel *n*MOSFET's," *IEEE Trans. Electron Devices*, vol. 38, no. 2, pp. 399–406, 1991.
[17] F. Silveira, D. Flandre, and P. G. A. Jespers, "A g_m/I_D Based Methodology for the Design of CMOS Analog Circuits and Its Application to the Synthesis of a Silicon-on-Insulator Micropower OTA," *IEEE J. Solid-State Circuits*, vol. 31, no. 9, pp. 1314–1319, 1996.
[18] A. G. Andreou and K. A. Boahen, "Neuromorphic Information Processing," in *Analog VLSI Signal and Information Processing*, M. Ismail and T. Fiez (eds.), McGraw-Hill, New York, 1994.
[19] P. Yang, B. D. Epler, and P. K. Chatterjee, "An Investigation of the Charge Conservation Problem for MOSFET Circuit Simulation," *IEEE J. Solid-State Circuits*, vol. 18, no. 1, pp. 128–138, 1983.
[20] P. R. Gray and R. G. Meyer, *Analysis and Design of Analog Integrated Circuits*, Wiley, New York, 1984.
[21] G. Ghibaudo, "A Simple Derivation of Reimbold's Drain Current Spectrum Formula for Flicker Noise in MOSFETs," *Solid-State Electron.*, vol. 30, no. 10, pp. 1037–1038, 1987.
[22] G. Reimbold and P. Gentil, "White Noise of MOS Transistors Operating in Weak Inversion," *IEEE Trans. Electron Devices*, vol. 29, no. 11, pp. 1722–1725, 1982.
[23] R. Sarpeshkar, T. Delbruck, and C. A. Mead, "White Noise in MOS Transistors and Resistors," *IEEE Circuit Devices Mag.*, vol. 9, no. 6, pp. 23–29, 1993.
[24] G. Reimbold, "Modified $1/f$ Trapping Noise Theory and Experiments in MOS Transistors Biased from Weak to Strong Inversion—Influence of Interface States," *IEEE Trans. Electron Devices*, vol. 31, no. 9, pp. 1190–1198, 1984.
[25] J. A. Van Nielen, "A Simple and Accurate Approximation to the High-Frequency Characteristics of Insulated-Gate Field-Effect Transistors," *Solid-State Electron.*, vol. 12, pp. 826–829, 1969.
[26] M. Bagheri and Y. Tsividis, "A Small Signal dc-to-High-Frequency Nonquasistatic Model for the Four-Terminal MOSFET Valid in All Regions of Operation," *IEEE Trans. Electron Devices*, vol. 32, no. 11, pp. 2383–2391, 1985.

[27] D. Foty, *MOSFET Modeling with SPICE: Principles and Practice*, Prentice-Hall, Upper Saddle River, NJ, 1997.

[28] A. Chatterjee, C. F. Machala III, and P. Yang, "A Submicron DC MOSFET Model for Simulation of Analog Circuits, *IEEE Trans. Computer-Aided Design Integrated Circuits Systems*, vol. 14, no. 10, pp. 1193–1207, 1995.

[29] M. A. Maher and C. A. Mead, "A Physical Charge-Controlled Model for MOS Transistors," in *Advanced Research in VLSI*, P. Losleben (ed.), MIT Press, Cambridge, MA, 1987.

[30] Y. A. El-Mansy and A. R. Boothroyd, "A Simple Two-Dimensional Model for IGFET Operation in the Saturation Region," *IEEE Trans. Electron Devices*, vol. 24, no. 3, pp. 254–262, 1977.

[31] Y. Cheng, M. Jeng, Z. Liu, J. Huang, M. Chan, K. Chen, P. Ko, and C. Hu, "A Physical and Scalable I-V Model in BSIM3v3 for Analog/Digital Simulation, *IEEE Trans. Electron Devices*, vol. 44, no. 2, pp. 277–287, 1997.

[32] K. Toh, P. Ko, and R. G. Meyer, "An Engineering Model for Short-Channel MOS Devices," *IEEE J. Solid-State Circuits*, vol. 23, no. 4, pp. 950–958, 1988.

[33] B. Iñiguez and E. G. Moreno, "A Physically Based $C\infty$ Continuous Model for Small-Geometry MOSFET's," *IEEE Trans. Electron Devices*, vol. 42, no. 2, pp. 283–287, 1995.

[34] D. A. Johns and K. Martin, *Analog Integrated Circuit Design*, Wiley, New York, 1997.

[35] K. Laker and W. Sansen, *Design of Analog Integrated Circuits and Systems*, McGraw-Hill, New York, 1994.

[36] *SMASH Users Manual*, Dolphin Integration, Meylan, France, 1995.

Chapter 3

Derek F. Bowers
Analog Devices, Precision Monolithics, Inc., Santa Clara, California 95050

A Review of the Performance of Available Integrated Circuit Components Under the Constraints of Low-Power Operation

3.1. INTRODUCTION

This chapter explores the limitations of practical integrated circuit (IC) devices when faced with the requirement for low-power operation. This is, however, not a treatise on semiconductor physics, since the necessary theory is well covered in the usual reference books. Neither is it a "designer's guide", since this subject is the material for most of the other chapters in this book. I have also chosen to keep the mathematics to a minimum (though some is obviously necessary). It should be noted that I am an analog IC designer, and as such my analyses are heavily biased in this direction: But many of my data converter and application-specific IC (ASIC) designs also include a significant logic content, and I think that most of my conclusions apply well to the digital realm as well. At least the fundamentals are correct and at a minimum should provoke useful thought.

3.2. WHAT EXACTLY IS LOW POWER?

Power, of course, is the product of voltage and current. Producing a low-power circuit, therefore, requires a minimization of both these parameters. Unfortunately, the trade-offs between these two parameters are highly application dependent. The user of an IC is unlikely to be upset when presented with a very low operating current but almost certainly has a (probably very) restricted operating voltage with which to work. Operating voltage also has severe ramifications for the

IC designer; but in this case low operating currents also pose very difficult design problems.

These trade-offs are usually at least somewhat calculable when designing custom circuits or ASICs but are extremely nebulous for the designer of general-purpose ICs. Additionally, many techniques for optimizing an integrated design for low-voltage operation (usually involving increased complexity) are incompatible with the goal of low current drain, further complicating the trade-off issue.

Although I am directly concerned with the design of circuits in both the aforementioned categories, I have decided to devote this chapter to more fundamental issues. Instead, then, this is basically a tutorial concerned with some of the limitations of semiconductor devices when presented with low-voltage or low-current requirements.

This focus neatly absolves me of the need to explore system aspects such as shutdown, "sleep-mode," or switched operation. What it does mean, however, is that I must take a highly generalized view of all technologies used in analog IC design. This would have been much simpler 20 years ago when the basic six-mask silicon bipolar IC process was in use for virtually all mainstream analog design. In recent years, however, the importance of analog metal-oxide-semiconductor (MOS) circuits, complementary bipolar processes, and even materials other than silicon make this a very complex topic.

3.3. SEMICONDUCTOR CHOICE

Silicon is not the only choice of semiconductor available for IC realization. Alternative semiconductors (almost always compound types) have been explored mostly to improve speed in digital ICs. In the light of reduced voltage and current operation, it is enlightening to reexplore the possibilities for analog design work.

Silicon is an element located in group 4 of the periodic table with a bandgap voltage of approximately 1.12 V. As a semiconductor, it offers some highly undeniable advantages, notably a highly stable thermal oxide and a melting point high enough to allow doping at normal pressures. Silicon is also cheap, easy to manufacture, and relatively mechanically stable. On the black side, it is an indirect gap semiconductor, meaning that electron mobility is poor compared to some more exotic alternatives (such as gallium arsenide). At the current levels used in micropower circuits, the low electron mobility is much less of a drawback than with circuits designed purely for optimum speed (particularly considering that the hole mobility is less than a factor of 3 slower, unlike some compound semiconductors). So what more could we want from a semiconductor for micropower applications?

The major consideration is the bandgap voltage. The V_{be}(on) of a bipolar transistor is proportional to the bandgap voltage (as well as other things), and this places a limit on the lowest operating voltage of a bipolar IC [or section in the case of a bipolar MOS (BiMOS) IC]. Even when this limit is not approached, a high V_{be} makes stacking of devices more impractical, creating additional circuit complexity, consequent increase in current consumption and die area, and probable performance sacrifices. There is a trade-off here, because low bandgap voltages lead to greatly increased leakage currents, which can place limits on the lower operating *current*. This is particularly bothersome at elevated temperatures, since such leakages tend to double for every 10 K or so of operating temperature.

To get some feel for these numbers, an NPN bipolar transistor on a "conventional" (8-μm-type rules) bipolar process might have a saturation current around 0.2 fA. At 1 μA collector current, this would yield a V_{be} of roughtly 580 mV at 25°C. At 85°C, the same transistor would have a collector-substrate leakage in the region of 60 pA, and at 125°C this would increase to about 1 nA.

These numbers tend to favor low current operation rather than low voltage operation, and if given a choice, reducing the bandgap voltage to 0.9 V or so has proven to be more optimum for the vast majority of bipolar micropower ciruits with which this author has been involved. Unfortunately, none of us get a vote on the bandgap voltage of silicon, and if we are to change things, we need to look to alternatives. The only alternative semiconductor even vaguely considerable in group 4 is germanium (diamond has a bandgap voltage exceeding 5 V and the bandgap voltage of tin is so low that it cannot be considered as a semiconductor at room temperature).

Despite the fact that no practical technology currently exists for the production of germanium ICs, the low bandgap voltage (≈ 0.66 V), while being close to ideal for low-voltage operation, yields devices some three orders of magnitude more leaky than that of silicon. It should be noted that if germanium were a practical IC technology, it probably would have already been exploited for its potentially high speed characteristics; germanium's electron mobility is about 2.6 times that of silicon, and its hole mobility some 4 times higher. Silicon-germanium heterojunction bipolar transistors (with F_t approaching 100 GHz) are looking promising for high-speed applications and indeed offer a reduced V_{be} over pure silicon (but only by 60 mV or so, not a great advantage for low-voltage operation). No other compound semiconductors have been developed with even the slightest promise for practical micropower circuits, and while this may change, they are not a realistic possibility at present.

The situation with MOS field-effect transistors (MOSFETs) is even more straightforward since, barring some technological breakthrough, they are virtually impossible to produce on anything but silicon (thermally grown gate oxide is impractical for most compound semiconductors, and germanium dioxide is unfortunately water soluble). Additionally, the bandgap voltage is not necessarily a limitation on the low-voltage application of MOSFETs since the threshold voltage is more directly related to process control. It is true that a reduced bandgap voltage should yield a reduced temperature coefficient for the threshold voltage, but again, none of us get to choose this. The inescapable conclusion is that we are stuck with silicon as our semiconductor for now. But things could be worse, and the remainder of this chapter will assume that we are resigned to this fact.

3.4. ACTIVE DEVICES

3.4.1. Bipolar Transistors

The history of the bipolar junction transistor (BJT) dates from around 1948 when Bell Laboratories announced an alternative amplifying element to the vacuum tube based upon the semiconductor, germanium. Schockley, Bardeen, and Brattain subsequently won a Nobel Prize for this invention, though it might more accurately be described as a discovery, since their initial research was to produce what we now call a MOSFET, a device postulated in the 1930s by Julius Lilienfield. The first silicon BJTs were produced by Texas Instruments around 1954, but the real start of the IC industry was the invention of the junction-isolated planar bipolar IC process by Fairchild in the late 1950s.

The original "planar" IC process produced vertically integrated (double-diffused) NPN transistors using an N-type epitaxial layer grown upon a P-type substrate. A deep P-type isolation diffusion enabled individual transistors to be electrically "isolated" from each other (provided that the P-type material was always reverse biased with respect to the collector), and this has subsequently become known as the *junction isolated bipolar process.*

One of the most comprehensive books on analog IC design, *Analysis and Design of Analog Integrated Circuits* by P. R. Gray and R. G. Meyer, is now in its (excellent) third edition [1]. It is interesting to note that page 109 of the first edition of this book (1977) had the following to say (concerning the bipolar process): "The basic fabrication process consisting of six or seven mask steps and four diffusions is used for the vast majority of the analog integrated circuits produced." Even in 1997, a very large number of commercially successful ICs are still being produced by the same basic process: and with some possible modifications, such a process is still highly suitable for the realization of very low power ICs. Technology has hardly been stagnant, however, and a veritable cornucopia of new (and considerably more advanced) bipolar processes are emerging. The major thrust in purely bipolar development has been in the very high speed transistor arena, with 20-GHz NPN transistors (for RF applications) and 5-GHz complementary NPN/PNP processes (for video-bandwidth analog functions) becoming a reality. Such processes are not directly intended for micropower operation, of course, but sooner or later will be used for such purposes as they become economically viable, if only because they can offer greatly improved packing density (and lower junction leakage) because of the smaller mask geometries generally used.

Another complicating factor is the trend for adding bipolar transistors (NPN type, PNP type, or both) to CMOS processes. Regardless of the original motivation for doing this, such BiCMOS processes, as they have been labeled, cannot be ignored when considering low-power operation.

In the light of the above, the conventional bipolar process will be considered first.

Low-Voltage Limitations of Conventional Bipolar Process. The conventional bipolar process consists essentially of double-diffused NPN transistors and lateral PNP transistors; both types are usually used to their full advantage in the design of low-power circuits. To quote from the late (and legendary) Bob Widlar: "The intrinsic operating voltage limit of bipolar I.C.'s is somewhat greater than the emitter-base voltage of the transistors" [2]. This limitation assumes that all sections of the circuitry have been given full optimization in terms of low operating headroom; Darlington pairs, Wilson current sources, and conventional cascode circuits are all definitely out of the question as this limit is approached. With careful design, the "somewhat greater" aspect comes down to a saturation voltage (V_{sat}), possibly some resistive degeneration to improve matching, and some additional safety margin to accommodate processing and temperature variations. Clearly, then, the V_{be} and V_{sat} of the bipolar transistor, and their likely variations, become very important as the operating supply voltage is reduced. The traditional literature represents the V_{be} of a bipolar transistor by

$$V_{be} = \frac{kT}{q} \ln\left(\frac{I_c}{I_s} + 1\right)$$

where k is Boltzmann's constant (1.38×10^{-23} J/K), T is absolute temperature (K), q is the charge on an electron (1.6×10^{-19} C), I_c is the collector current (A), and I_s is the theoretical reverse saturation current (A).

The term I_c/I_s is usually presumed to be much greater than unity, in which case the equation simplifies to

$$V_{be} = \frac{kT}{q} \ln\left(\frac{I_c}{I_s}\right)$$

Exceptions are to be found in such applications as wide-range logarithmic amplifiers, but for the most part (even in very low current applications), ignoring the unit adder to the I_c/I_s ratio is unlikely to disturb the mathematics to any practical degree.

A typical NPN transistor with 20 V or so of breakdown and a beta of 100 or so would have an I_s in the region of 10^{-18} A/μm^2 of emitter area. This implies that the V_{be} of a 12 × 12-μm emitter transistor at $I_c = 10$ μA would be about 650 mV at room temperature. The current I_s tends to double for each 8 K increase in temperature, and this makes the overall temperature coefficient of V_{be} negative. It can be shown that the V_{be} of all transistors approaches the extrapolated bandgap voltage (≈ 1.2 V) at absolute zero [3], enabling the temperature coefficient to be estimated from just one measurement. Thus, our example transistor would be expected to exhibit a temperature coefficient close to -1.83 mV/K, assuming the initial V_{be} was the value at 300 K. Clearly as far as base-emitter voltage is concerned, the major problems for low-voltage operation are at low temperatures.

Fortunately, V_{be} is only a weak function of process variations (unlike the threshold voltage of a MOSFET, for example) and the saturation current tends to track beta variations. So a 3 : 1 variation in beta, for example (most processes can maintain this or better), would produce a total deviation in V_{be} of only 29 mV (higher betas yielding lower V_{be}).

For a similar geometry, lateral PNP transistors have a V_{be} that is somewhat lower (typically by 30 mV or so) due to the lighter doping of the base-emitter junction. The V_{be} of lateral devices, if anything, is more predictable than that of vertical devices, since the overall base doping is not dependent on a thin "pinched" region. The beta of a lateral transistor is influenced mostly by recombination effects well away from the emitter injection region, and so unlike double-diffused devices, there is little correlation between V_{be} and beta. It should also be noted that the so-called lateral injection paths are not truly lateral in practice and involve a complicated series of carrier routes originating at the edges and under all parts of the emitter. Thus, the V_{be}(on) does not scale linearly with emitter area, or, as might at first be expected, with emitter periphery. In practice, it is somewhere between, and this has led to the technique of scaling lateral transistors using multiple emitters rather than experimenting with the unknown.

The high predictability of the V_{be}(on) of silicon transistors is clearly benign but has the disadvantage that nothing substantial can be done to reduce it. Large emitter areas obviously lower this value, but only logarithmically, which makes increased emitter area a very inefficient way of reducing the operating voltage of an IC. Lower operating currents also reduce the V_{be} in an identical fashion, and clearly reduced operating currents also assist in the goal of achieving low power. In general, though, the limits on operating current are set by constraints totally divorced from the meager benefits of a slightly reduced base-emitter voltage.

The situation with the saturation voltage is considerably more complicated than dealing with V_{be}. The first area of confusion concerns the terminology. A "saturated" bipolar transistor is generally regarded as a transistor in a state where the base-collector junction is sufficiently forward biased to remove base current from its normal function of increasing the collector current. This is in stark contrast to the use of the term in dealing with FET technology, which describes almost the exact opposite operating region. To add to the confusion, nothing actually "saturates" in a bipolar transistor operating in this region: Several texts describe the base as being saturated with carriers, but this is not so, since increased base current clearly produces additional carriers in the base region, even in so-called heavy saturation. But I am not proposing a change in the nomenclature at this point (it is already obfuscated sufficiently as it is), and for the present I will ignore this ambiguity.

As a starting point, it is instructive to look at the *intrinsic* saturation voltage predicted by the well-known Gummel-Poon model (used by the majority of circuit simulators, and a good general reference for such things is Ian Getreu's book, *Modeling the Bipolar Transistor*, available from Tektronix Inc. by, somewhat oddly, ordering part number 062-2841-00).

Analysis of the Gummel-Poon model yields the following expression for V_{sat} [4] (or, more correctly, $V_{CE(sat)}$):

$$V_{sat} = \frac{kT}{q} \ln\left(\frac{1/\alpha + (I_c/I_b)(1/\beta_r)}{1 - (I_c/I_b)(1/\beta_f)}\right)$$

where β_f and β_r are the forward and reverse betas (current gains), respectively, and α_r is the reverse alpha (collector-emitter current transfer ratio). For example, with $\beta_f = 100$, $\beta_r = 2$, and $I_c/I_b = 10$, the intrinsic saturation voltage would be 50 mV at room temperature.

In addition to the intrinsic saturation voltage, there are also voltage drops across the bulk resistances of the collector and emitter (the latter term is usually small enough to be ignored). To further complicate matters, due to effects such as current crowding and conductivity modulation, the collector resistance is not a simple value like most circuit simulators assume it to be. Fortunately, at the current levels usually used in low-power applications, these extrinsic terms are small enough that they do not need to be modeled particularly accurately.

Ignoring these terms, what we really need is to relate the saturation voltage to the headroom required to maintain a transistor in its forward active region.

Taking the transistor parameters of the previous example, let us assume that an error of 1% is acceptable in collector current. This sets I_c/I_b at $\beta_f/1.01$, or 99 in this case. Solving for V_{sat} gives

$$V_{\text{sat}} = \frac{kT}{q} \ln(5100)$$

which at room temperature would indicate that about 220 mV of headroom is required, increasing, of course, with temperature. In practice, even this may not be sufficient headroom to give acceptable performance, due to complicating factors in the saturation equations not predicted by the Gummel-Poon model. The first of these concerns reciprocity; the fundamental assumption of the model that the saturation current in the inverse region is the same as that in the forward region. This is theoretically true for a one-dimensional cross section, but of course a planar transistor can never be one dimensional in cross-section because each diffusion has to be totally overlapped by the previous one. Some parametric extraction systems tend to compensate for this by adjuting β_r to get a best fit in the saturation region, a compromise at best. A much more serious problem, however, arises from the fact that a double-diffused transistor on a junction isolated process has a parasitic substrate transistor (an NPN transistor would have a PNP parasitic) which causes most of the base current in the saturated region to ignore the emitter completely and become diverted to the substrate. The gain of this parasitic transistor can be substantial, 10–80 being common, and the effect of this is to bring the effective I_c/I_b ratio closer to β_f.

For a parasitic β of 20, the previous example would need to be recalculated for an I_c/I_b of $\beta_f/1.005$.

$$V_{\text{sat}} = \frac{kT}{q} \ln(103{,}000)$$

or about 300 mV at room temperature. In many cases, considerably more error than 1% may be tolerable in the collector current, in which case, of course, less headroom will be necessary.

In the case of a lateral PNP transistor, the same parastic PNP exists, but it now acts to divert current from the emitter to the substrate when the collector fails to collect it (some argue that the collector reemits current after collecting it, a viewpoint of which I fail to see the validity). In this case, the parasitic has no serious effect on the I_c/I_b ratio, and lateral transistors tend to have lower intrinsic saturation voltages than double-diffused types.

Another complication occurring in double-diffused transistors is the so-called quasi-saturation effect [5]. This term is used to describe the region of operation where the voltage drop across the internal collector resistance causes forward biasing of the internal base-collector metallurgical junction, even though the external base-collector terminal remains reverse biased. This produces a drastic reduction in current gain and a large drop in F_t (the transistor cutoff frequency). Part of this forward biasing occurs as a result purely of the ohmic drop due to the bulk-collector resistance, but in double-diffused transistors carrier velocity saturation in the epitaxial region produces an additional nonohmic term in the collector resistance, which can mean that the transistor is in quasi-saturation at surprisingly high collector-base voltages. Fortunately, even small-geometry transistors are relatively free from this effect at collector currents below a few hundred microamperes, and for most low-power circuits this effect causes no big problems. Sometimes, though, low-voltage ICs are operated at much higher current densities (particularly where bandwidth is the major concern) and the quasi-saturation phenomenon becomes a very important constraint. Several circuit simulators have now implemented the quasi-saturation model described in ref. [5], and hopefully a more universal adoption of this, and other improved simulation techniques, will alert designers to possible problems in this area.

It should be noted that lateral transistors have collector doping very much greater than that of the base and, despite their other quirky characteristics, are pretty much free from the nonohmic quasi-saturation effect. It should now be obvious that the business of precisely determining when a bipolar transistor is in saturation is somewhat application dependent. To add further complication, even when a transistor is deemed to be in the "forward-active" region, there is still the Early effect to be taken into account. This effect is commonly assumed to be relevant only with significant collector-base voltages and thus would not reasonably be expected to be an important consideration for low-voltage operation. Alas, this is not always the case. The Early effect is an increase in current gain (and consequent reduction in V_{be}) caused by modulation of the transistor basewidth by depletion effects of both the collector and emitter voltages. The Gummel-Poon model takes these into account by adding a multiplier to the effective saturation current [6]. This boils down to the following modifier:

$$I_{s(\text{effective})} = \frac{I_s}{1 + V_{be}/V_{ar} + V_{bc}/V_{af}}$$

where V_{be} and V_{bc} are the respective *forward* biases on the base-emitter and base-collector, respectively, and V_{ar} and V_{af} are the reverse and forward Early voltages. The V_{be}/V_{ar} term (sometimes referred to as the *late* effect, a play on James Early's name) is relatively constant in the forward-active region and at any rate is absorbed in the parameter extraction process. So, to avoid getting bogged down in modeling details, it is easier to write an expression for the modified collector current (with increasing collector voltage) as

$$I_{c(\text{effective})} = \frac{I_c}{1 - V_{cb}/V_{af}}$$

where V_{cb} now represents *reverse* bias on the collector-base junction.

When $V_{cb} = V_{af}$, this equation becomes a singularity, which physically represents punch-through in the transistor. Because singularities tend to cause overflow errors in simulators, and because other breakdown effects are normally more prevalent than punch-through, most circuit simulators use a linear approximation for this equation which actually amounts to the first variable term in a standard binomial expansion:

$$I_{c(\text{effective})} = I_c\left(1 + \frac{V_{cb}}{V_{af}}\right)$$

This, of course, is the traditional simplified expression for the Early effect.

Such a linear approximation to what is (theoretically) a reciprocal effect is not particularly important for low-voltage operation: what is more relevant is the fact that neither accurately represents the true behavior of actual depletion regions under reverse-bias conditions, particularly at low collector-base voltages (the effect is similar to the behavior of depletion capacitance and is worse for step junctions than for graded ones). Measurements of actual devices show a marked reduction in effective Early voltage at low collector-base voltages, even when saturation effects are not present. This effect is distinct from quasi-saturation because it occurs even at very low collector currents. Double-diffused transistors (with their highly graded base-collector junctions) typically show Early voltages at zero collector-base voltage some 3–10 times lower than the projected large-signal value. Lateral devices (with an approximate step junction in the collector-base region) are even worse; and indeed, it is often difficult to detect the onset of saturation just by looking at a typical set of I_c curves. Circuit simulators do not generally take this into account, and the ramifications of this include amplifiers with very much lower gain than simulation might predict.

Hopefully, better modeling will at least enable the designer to account for this in the future (this, and other effects, are reasons why low-power analog designers still often resort to breadboarding as a verification tool).

Low-Current Limitations of Conventional Bipolar Process. As previously mentioned, the other "dimension" to reducing overall power is reduced operating current. Actually, the bipolar transistor is quite well behaved at surprisingly low currents, and furthermore simple modeling techiques can hold up well in such situations. Speed is always a sacrifice at low currents, and this subject will be addressed later; but there are other compromises which need to be considered.

One of the more heavily discussed topics concerning bipolar transistors is the reduction of forward currrent gain, β_f, with reduced collector current. While it is well known that the exponential relationship of V_{be} with collector current is extremely well behaved over many decades, the base current is not necessarily so predictable.

Measurements on typical small NPN transistors indicate that β_f peaks at roughly 1 mA of collector current and at 100 pA is on the order of one-third of the peak value. This is very adequate for the vast majority of micropower applications.

Many IC texts quote the lateral PNP transistor as being a very low gain device in general but with particular problems at low currents. This is characteristic of a device with high surface recombination, usually a result of poor surface oxide annealing. Modern lateral PNPs are actually quite good in both these respects; on light (3–6 Ω-cm) epitaxial regions peak gains of 50–200 are now common, with the peak usually occurring in the 1–50 μA range. On devices produced at the author's company (Analog Devices), we still record gains of above 20 at 10 pA of collector current (it should be noted that lateral PNP transistors have been used in input bias current compensation circuits for many years, where good performance at low currents is mandatory).

In practice the gain reduction at low currents is caused by three additional mechanisms [7]:

1. Recombination of carriers at the surface
2. Recombination of carriers in the emitter-base space charge layer
3. Formation of emitter-base surface channels

All three mechanisms have a similar varation with base-emitter voltage V_{be} [8]:

$$I_{b(\text{surface})} = I_{s(\text{surface})}\left[\exp\left(\frac{qV_{be}}{2kT}\right) - 1\right]$$

$$I_{b(\text{EBscl})} = I_{s(\text{EBscl})}\left[\exp\left(\frac{qV_{be}}{2kT}\right) - 1\right]$$

$$I_{b(\text{channel})} = I_{s(\text{channel})}\left[\exp\left(\frac{qV_{be}}{4kT}\right) - 1\right]$$

In SPICE, all three of these are lumped into one equation, which is added to the base current:

$$I_{b(\text{composite})} = I_{se}\left[\exp\left(\frac{qV_{be}}{n_e kT}\right) - 1\right]$$

The parameter I_{se} is the base-emitter leakage saturation current, and n_e is the leakage emission coefficient, which lies between 1 and 4. Two similar parameters, I_{sc} and n_c, are added to model the leakages in the reverse mode of operation.

Another consideration for low-power operation is the effect of junction leakage.

In a junction isolated process, the biggest junction is the collector-substrate junction (which becomes the base-substrate junction in the case of a lateral PNP transistor). Measurements suggest that at room temperature this leakage saturates out at around $0.03\,fA/\mu m^2$. Like most semiconductor leakages, this tends to double for every 10 K increase in temperature, so even at 125°C, a small NPN transistor with around $800\,\mu m^2$ of collector area would have a collector-to-substrate leakage of only 240 pA. Actually, sidewall leakage will add to this for a small device, but even so the resuling high-temperature leakage will almost certainly be less than 500 pA. This corresponds to an 85°C value of 30 pA and a 25°C value of 0.5 pA. The corresponding collector-base leakage has been measured at $0.08\,fA/\mu m^2$, and while this is higher, it is usually less important since the junction area is much smaller.

One area of concern when operating at low currents is the effect of minor defects in junctions, which can induce excess and random leakages which might not be important at higher operating currents, resulting in yield loss. Such defects can be particularly troublesome in the emitter region, because heavily doped areas are more susceptible in this regard. Even harder to deal with are *latent* defects, which only show up at temperature extremes. These can be a real headache, because they can cause a circuit, which tests good at room temperature, to fail at some other temperature.

At high operating voltages, the field threshold also can become a major issue when stage currents become very low. Traditionally, bipolar ICs have been produced using silicon with a surface orientated on the ⟨111⟩ crystal plane, since this yields a high surface charge and a corresponding high field threshold. Nowadays, light surface implants are often used to achieve the same effect, and this is particularly important for BiCMOS circuits which are normally produced on ⟨100⟩ silicon. The catch is that field threshold is measured (and controlled) on some test structure at a current easy to handle in production. Often a process with a claimed threshold of, say 30 V, at low currents can run into trouble well below this value. Careful layout, guard rings, and the like can circumvent this problem, but if possible, it is better to keep the field threshold well above the operating voltage.

A final point concerns the nature of offset voltage in bipolar transistors. Because the logarithmic nature of V_{be}, offset voltage and emitter area mismatch can be considered equivalent (theoretically, the "offset voltage" between any two bipolar transistors should vary as kT/q, or in other words be proportional to absolute temperature; to a first order, in practice, this has indeed been verified to be the case). Theoretically, then, the offset voltage of, say, a differential pair should be independent of collector current over a wide range. Current mirrors, one of the more fundamental building blocks for analog circuits, have a gain error resulting from offset voltage, but once more this term is theoretically independent of operating current. This "current scaling" behavior enables many bipolar building blocks to be used at varying currents without any drastic design modifications and is a major strength of the bipolar transistor in micropower applications.

More Recent Bipolar Process Trends. Recent years have seen many variations on the basic bipolar process. Complementary processes have appeared using radically different steps from the older approach; some even turn the process upside down using an N-type substrate and P-type epitaxial layer. Low-voltage processes are also displacing the older 36 V-type processes in a quest for higher speed and packing density. New isolation techniques such as trench isolation are becoming prevalent as the technology and equipment needed becomes standardized. BiCMOS processes exist in many forms; usually these consist of a CMOS process modified to include either an NPN or PNP bipolar transistor, but occasionally both types are available simultaneously (this has been termed CBCMOS by some).

In general, the trend toward smaller transistors and improved isolation techniques tends to lower leakage, which can only be good news for low-current operation. As far as low voltage is concerned, obviously reducing the emitter size will increase V_{be}, but also the higher base doping used on lower voltage processes tends also to raise V_{be} also. This effect is small, however, and the reduced stray capacitances and leakages of such processes often enable lower operating currents to be used for a given level of performance, tending to offset the V_{be} disadvantage in many applications.

On BiCMOS processes, shortcuts are often taken to avoid the use of buried layers and expitaxial deposition. This can result in transistors with significant collector resistance (often several kilohms). Clearly, this must be taken into account when considering low-voltage operation.

On the whole, though, bipolar transistors remain bipolar transistors, and the fundamentals considered so far hold for a wide variety of modern processes.

3.4.2. MOSFETs

In recent years, CMOS technology has become very important for the integration of analog (as well as, of course, digital) functions. The prevalence of very large scale integrated (VLSI) circuits with large amounts of both digital and analog circuitry (the so-called mixed-signal technology) has made the quest for low power very important. Additionally, some of the newer submicrometer CMOS (and BiCMOS) processes cannot be operated above 3 V, forcing the design to be a low voltage one. Virtually all low-power MOS circuitry is complementary, and designing for low power in a purely NMOS or PMOS technology is mostly obsolete. For the purposes of this chapter only CMOS technology will therefore be considered. The depletion mode MOSFET, common as an active load in earlier NMOS processes, has also largely disappeared from the scene. The depletion device is not especially useful in low-power design (except possibly as a substitute for high-value resistors) and anyway has characteristics similar to the junction field-effect transistor (JFET), discussed in a later section. The enhancement mode device is therefore assumed to be the prevalent one in low-power CMOS design.

Low-Voltage Limitations of MOSFETs. Like bipolar transistors, similar constraints limit the analog operation of MOSFET circuitry. The V_{be} of the former is replaced by the V_{gs} of the latter. The term "saturation voltage" is not used for MOSFETs; the equivalent term is the drain-to-source voltage at which the device leaves the saturated region and enters the linear one. This is a much softer effect than bipolar saturation, since the V_{ds} can theoretically approach zero at low currents. Even so, it is often necessary to ensure that MOSFETs are maintained in the saturation region for correct circuit performance.

The V_{gs} of a MOSFET consists of two terms: the effective threshold voltage and the enhancement voltage. The enhancement voltage is directly under the control of the designer, the "textbook" square-root equation governed by the W/L of the device and the drain current. A simplistic expression for V_{gs} might therefore be written as

$$V_{gs} = V_t + \sqrt{\frac{I_d}{K'}\frac{L}{W}}$$

where L and W are the *effective* dimensions of the gate length and width, respectively; I_d is the drain current; and K' is the transconductance factor for the MOSFET. The enhancement (second) term can be made arbitrarily small by making the ratio W/L large compared to the ratio I_d/K' but there are definite limits to this; even if I_d is low, large W/L ratios can result from attempts to minimize the enhancement term with a consequent die area penalty. Even if this is acceptable, it may not be practical from a performance standpoint due to noise susceptibility and subthreshold effects (considered later).

The value of K' is theoretically given by

$$K' = \tfrac{1}{2}\mu C_{ox}$$

where μ is the surface carrier mobility: about 250 cm^2/V-s for holes (P-channel devices) and about 700 cm^2/V-s for electrons (N-channel devices). Note that the KP term in SPICE (and most other simulation programs) is actually μC_{ox} rather than $\frac{1}{2}\mu C_{ox}$, adding confusion to the situation. The oxide capacitance per unit area C_{ox} is further given by

$$C_{ox} = \frac{\varepsilon_{ox}}{t_{ox}}$$

where t_{ox} is the oxide thickness and ε_{ox} is the permittivity of silicon dioxide, which is approximately 3.45×10^{-11} F/m.

Oxide thicknesses (at the time of writing) range from 15 nm for submicrometer processes up to 150 nm or so for processes intended for operation above 30 V. Other dielectrics, notably silicon nitride, have sometimes been used to replace part of the gate oxide, since the higher dielectric constant gives a higher value of K'. This practice is virtually obsolete, and among its problems are a severe threshold hysteresis effect, a particularly odious characteristic for low-voltage operation.

To get a feel for the numbers involved, consider a PMOS device on a 2 μm process operating at 10 μA drain current. Remembering that the enhancement voltage V_e is equal to $V_{gs} - V_t$, solving for W/L yields

$$\frac{W}{L} = \frac{I_d}{K'(V_e)^2}$$

A typical K' value might be around 8 $\mu A/V^2$, so designing for a V_e of 80 mV (say) would in this case require a W/L of 200 or so. It is obvious that the square-law dependence of V_{gs} has a more severe effect than the corresponding logarithmic term governing the V_{be} of a bipolar transistor.

The onset of saturation (remember, this is essentially the *reverse* of the terminology used with a bipolar device) occurs when the drain-to-gate voltage becomes less than the threshold voltage, and enhancement no longer occurs at the drain end of the channel. Thus, the onset of this region is described as

$$V_{dg} \geq -V_t \quad \text{or } V_{ds} \geq (V_{gs} - V_t) \quad \text{or } V_{ds} \geq V_e$$

Again, the enhancement voltage is a very important quantity when dealing with MOS devices in low-voltage operation. The inherent relationships linking V_e with the onset of saturation and the minimum gate-source voltage facilitate a number of design possibilities not inherently possible in bipolar technology. One point to remember, however, is that in CMOS design the mobility differences between NMOS and PMOS transistors give very different values for V_e for a given current and device size, and even if scaling has been performed to compensate for the differences (PMOS devices usually have a K' 2.5–3 times smaller than NMOS ones), they cannot be expected to track perfectly.

Assuming that the circuit can be designed with devices large enough (or currents small enough) that the enhancement term is small, the fundamental limitation becomes the device threshold voltage V_t. The term V_t is generally extrapolated from large-signal measurements, and various effects (such as subthreshold operation) make it less of a constant than simple theory would predict. However, no matter how involved the theory becomes, there is still some V_t-related voltage which places a lower limit on the operating voltage of an MOS circuit. In theory, the V_t can be targeted considerably lower than the V_{be} of a bipolar transistor, potentially allowing lower voltage operation than in the case of bipolar circuit. Practially, however, this is not a robust option.

First, the threshold voltage depends on quantities (particularly surface charge) which are far more difficult to control than the variables which determine the V_{be} of a bipolar transistor. So while the latter may be controllable to ±15 mV or so, it is not uncommon for threshold to deviate from its nominal value by ±250 mV or more.

Second, the threshold voltage exhibits a negative temperature coefficient (usually 1–3 mV/K) which causes devices with low thresholds to become depletion-mode devices at elevated temperatures. Furthermore, even if the threshold remains technically positive, subthreshold leakage can ruin the performance of a low-current circuit unless the V_t is targeted to provide an adequate safety margin. In practice, this usually means a room temperature minimum somewhat above 0.5 V, with a realistic nominal of 0.8 V or so. Clearly this is not an inherently advantageous situation over a bipolar transistor.

It should be noted that there is little inherent tendency for the thresholds of PMOS and NMOS devices to track each other in magnitude. In fact, variation in surface charge theoretically causes the threshold magnitudes of the respective devices to move in different directions, causing the sum of the magnitudes to be more constant than the threshold of a single device. It is conceivable that advantage could be taken of this phenomenon, but usually process engineers are reluctant to guarantee anything but absolute values, and attempting to design outside these constraints is treading on very dangerous ground.

As a final note, the threshold voltage also is increased by reverse bias on the back gate, the so-called body effect. The "native device" (PMOS on a *p*-well process and NMOS on an *n*-well process) is the one least affected by this phenomenon, but the "nonnative device" can have the body effect effectively removed by tying the source and body together, at some penalty in source capacitance and die area. On most processes, therefore, the body effect on the native device represents an unremovable component, but fortunately, as the supply voltage is lowered, this effect becomes small. It is not usually a serious constraint on the operating voltage of a practical circuit.

Low-Current Limitations of MOSFETs. The limitations on operating current due to junction leakages are similar in the case of a MOSFET to that of a bipolar transistor, though generally the effects are somewhat reduced. This is due to the fact that MOSFETs can often be made smaller than their bipolar counterparts (for a given set of photolithography limitations) due to removal of nested diffusions and a reduction in mask tolerance accumulations. Additionally, MOSFETs are self-isolating, removing the need for (and associated leakage of) an isolation pocket. Of much more importance, however, is that the highly beneficial "current scaling" characteristic of a bipolar transistor is not duplicated in the case of a MOSFET.

Offset voltage among MOSFETs intended to have matched V_{gs} characteristics consists of two terms: one due to device scaling mismatch, which obviously can be minimized by increasing the area of the transistor; the other due to random fluctuations in the threshold votlage. The latter term is caused by local variations of oxide thickness, surface charge, and other complex effects, but this term tends to act as a voltage independent of operating current and thus cannot be represented as an area mismatch. The $1/f$ noise, discussed further in Section 3.6.4, also behaves as a time-variant term in the threshold voltage and therefore creates similar problems.

At low currents, the g_m reduces as the square root of the drain current, which in turn means that the effect of the threshold mismatch term on drain current matching (of a current mirror, for instance) effectively increases (in a square-root fashion) as the current is lowered. The net result of all this is that MOSFETs need to be scaled correctly for correct current matching as the operating current is changed. In general, a suitable gate area needs to be chosen to minimize area mismatches, and then a suitable W/L ratio needs to be calculated to optimize the device g_m such that the expected threshold mismatch (a highly process-dependent term) does not dominate the overall matching term.

To present an example, assume that a current mirror must be designed with less than 10% error running at 1 µA. Being very conservative, any reasonable MOS process should be able to yield better than 3% area mismatch with a gate area of 200 μm^2 or so. This leaves 7% for threshold mismatch effects. At this stage, some estimate must be made concerning the threshold mismatch. This itself is highly process dependent and also depends heavily on layout aspects. Large gate areas theoretically average out threshold mismatch terms but also make devices more prone to gradient effects. Cross coupling of critical device pairs can reduce the effects of the latter, but for now assume that no (area-consuming) layout optimization is being used and use a (reasonable) worst case of 30 mV. So for 7% of 1 µA (70 nA) this leaves us with a desired g_m of 70 nA/30 mV, or 2.3×10^{-6} (siemens, mhos, reciprocal ohms, or whatever; I'll use siemens henceforth). This incidentally corresponds to an enhancement voltage of about 850 mV, a figure which actually may be too high for some low-voltage applications, and a compromise may have to be made.

Using the classic g_m equation for a MOSFET,

$$g_m = 2\sqrt{K'I_d \frac{W}{L}}$$

and rearranging for W/L,

$$\frac{W}{L} = \frac{g_m^2}{4K'I_d}$$

gives the "optimized" ratio W/L.

For a 3 µm CMOS process, K' might be 15 $\mu A/V^2$ for an N-channel device, in this case fixing the L/W ratio at 11. For our 200 μm^2 gate area, this would correspond to a MOSFET with a width of 4.3 µm and a length of 47 µm, a device at least practically realizable (these are effective widths and lengths; allowance has to be made for side diffusion, "bird's beak" and all the other MOS processing anomalies).

At this point, it is instructive to compare this with a similar bipolar situation. A 10% mismatch would correspond to 2.5 mV of offset voltage, and almost any geometry bipolar transistor should achieve this type of matching. At 1 µA of collector current, the bipolar transistor would have a (room temperature) g_m of 38 µS, while the MOSFET is operating at a g_m of only 2.33 µS. This amply illustrates the effect of loss of "current scaling".

Many complete ICs are now being designed to work with *total* currents of less than a few microamperes, so this might be regarded as a rather optimistic example. For a drain current of 50 nA, for example, to keep the same (850 mV) enhancement voltage would require the L/W to be increased to 220 in this case, a rather large value. Smaller size MOSFETs *are* often used at such currents, though, because they operate in the *weak-inversion* region.

The *weak-inversion*, or *subthreshold*, region occurs when the gate voltage is very close to V_t. Under these conditions the classic square-root nature of V_{gs} versus I_d breaks down, and the relationship becomes logarithmic [9] in nature, somewhat like a bipolar transistor:

$$\Delta V_{gs} = \frac{nkT}{q} \ln\left(\frac{I_{d1}}{I_{d2}}\right)$$

where n is a process-related variable. Experience has shown, however, that a value for n of about 1.85 is quite common. This would yield a subthreshold g_m of

$$g_m = \frac{qI_d}{1.85kT}$$

which is independent of the W/L and a little more than half that of a bipolar transistor operating at the same current.

The subthreshold effect is useful in that it limits the maximum transconductance and therefore the maximum error caused by offset voltage at low currents. It is dangerous to carry the comparison with the bipolar transistor too far, because it must be remembered that threshold mismatch terms (and $1/f$ noise) are much higher than the corresponding terms in a typical bipolar transistor.

For example, with fairly large devices and layout tricks such as common-centroid quad geometries, it should be possible to achieve less than 10 mV of mismatch between two devices. Factoring in the value for n, this would cause the same mismatch in transfer currents as a 10 mV/n or 5.4 mV mismatch in a bipolar pair. The latter figure is of course much worse than might be expected from even minimum-geometry bipolar devices.

3.4.3. JFETs

The JFET was proposed by Shockley in 1952 [10] and later demonstrated by Dacey and Ross [11] and found occasional use on early IC processes, mostly as a high-input-impedance device for operational amplifiers. The first integrated JFETs were formed using the NPN emitter and collector as the gate and substituting a lighter P-type diffusion for the normal NPN base, which acted as the channel, forming a P-channel JFET. These JFETs had very poor pinch-off (V_p) and I_{dss} control, but in 1972 National Semiconductor Corporation developed an ion-implanted P-channel JFET process which was completely compatible with linear bipolar processing. This has been subsequently trade-marked (by National) as the "Bi-FET" process. Almost all current JFET processes are variations on the National idea. The advantage of the JFET over the MOSFET lies in its below-the-surface nature, giving superior noise and matching performance. The main disadvantage is that only depletion devices are really practical (a JFET *can* be enhanced slightly, by forward biasing the gate-channel junction, but this is a somewhat dubious practice, especially where leakage is a concern).

The P-channel device is the most popular, because it can be conveniently fabricated by two extra mask steps in the same isolation pocket used by a standard NPN transistor, but N-channel processes also exist (to my knowledge, no complementary JFET processes are in use, but they *are* technically feasible).

As far as low-current operation is concerned, JFETs have all the same junction leakage considerations as bipolar transistors and MOSFETs but otherwise are highly suited for such applications. For example, a typical P-channel device might have a V_p of 1.5 V and a transconductance parameter (β) of 7 μA/V^2. A semiempirical expression for the I_{dss} of such a device has been found to be

$$I_{dss} = \frac{W}{L} \beta V_p^h$$

where h is approximately constant for a given process and is usually in the range 1.5–2.0.

So a device with an L/W of 10 might exhibit an I_{dss} of as low as 1.4 μA. Two such devices could be used to make a self-biased unity-gain follower with good DC accuracy.

I have not seen any good mathematical treatment of subthreshold effects on JFETs, but experience with integrated JFETs on our processes indicates that there is a marked deviation from the simple "square-law" theory at operating currents less than a few percent of I_{dss}. At very low currents, the V_{gs}-versus-I_d characteristic becomes logarithmic (like the MOSFET) but with values for "n" in the kT/nq term of 1.2. Again, like the MOSFET, this can result in considerably lower values of transconductance than simple theory (or simplistic simulation models) would predict.

Unfortunately, JFETs are not so well suited to low-voltage operation since it is not generally practical to target the pinch-off voltage to less than about 1 V. Below this figure, the channel becomes very lightly doped, and small variations in the gate doping can have drastic effects on I_{dss} control (remember, I_{dss} varies as a power of the pinch-off voltage). Even above 1 V, pinch-off control is not often much better than MOSFET threshold control, and it becomes hard to maintain a JFET in saturation with supply voltages less than 2 V or so.

3.5. PASSIVE DEVICES

3.5.1. Resistors

Resistors of the diffused, junction-isolated type have always been available on integrated circuits. Because such components come essentially for "free", manufacturers have typically been reluctant to add additional processing steps (and masks) to produce a more ideal component.

The basic drawbacks of diffused resistors include limited value range, high temperature coefficient, and value sensitivity to operating voltage and current.

Typical P-type diffused resistors have temperature coefficients around +1300 ppm/K and voltage coefficients around −350 ppm/V. In general, such resistors present no unusual drawbacks for low-voltage operation but become very unwieldly components for low-current circuits.

The base diffusion of a typical bipolar process has a sheet resistance in the range of 100–300 Ω/square, making resistors above a few kilohms costly in terms of die area. Additionally, even if the area penalty is acceptable, large diffused resistors can have significant leakage and stray capacitance, a particular concern at low operating currents. The drain-source diffusions in a CMOS process can also be used to form resistors, but these usually have an even lower sheet resistance; values of the order of 20 Ω/square for N-type and 40 Ω/square for P-type are usual.

Because long-channel MOSFETs can be made self-degenerating, and because bipolar scaling remains accurate at low currents, much low current analog design can sometimes be realized using almost no resistors at all. Nevertheless, a suitable higher valued resistor is a great asset to a process intended for high-performance low-power circuit integration. Epitaxial resistors, pinch resistors, and CMOS well resistors, while offering the potential for integrating a large amount of resistance, will be considered in the next section for they are crude (but often useful) resistors in most respects. Perhaps this treatment is a little harsh, for almost any conducting element can be used as a "resistor" under the right conditions; but my criterion for a true resistive element is that it should at least display a reasonable approximation to Ohm's law, and many of these realistically do not.

The easiest way of adding a high-sheet-resistance layer is to use a deliberate ion-implant step of opposite doping to the surface layer (the latter being N-type in the case of a standard bipolar and P-well CMOS process, P-type in the case of N-well CMOS). Ion-implanted resistors can achieve wide value ranges, but above 1–2 Ω/square they tend to have very poor temperature and voltage coefficients. Additionally, the relatively light doping of such resistors can also create leakage problems (especially at elevated temperatures) for low-current operation.

To remove the leakage and voltage coefficient and also greatly alleviate stray capacitance, resistors deposited on the dielectric layer above the silicon are highly attractive. The major requirement of the type of material used to form deposited resistors is that it be compatible with normal silicon processing, particularly the metallization system. Common resistive materials include polycrystalline silicon (polysilicon), silicon-chrome (sometimes referred to as cermet), and nickel-chrome (nichrome), though this by no means exhausts the possibilities.

Polysilicon deposition is already part of many IC processes, notably silicon gate CMOS (and BiCMOS) and polysilicon emitter bipolar. Unfortunately, the 15–80 Ω/square achieved with the high phosphorus doping inherent in these processes is not much use for the generation of high resistance values. An additional mask and implant step is often used to generate higher sheet resistance values, and extremes can be obtained; static random-access memories (SRAMs) have traditionally used an undoped polysilicon resistor which can attain sheet resistances of over a gigohm per square. Such "resistors", of course, would certainly be relegated to the next section because their near-intrinsic characteristics give rise to temperature coefficients that are nearly exponential, but there is an opportunity for compromise here.

Heavily doped polysilicon has a *positive* temperature coefficient, usually 20 Ω/square gate material exhibiting about 1000 ppm/K, while lightly doped polysilicon has a *negative* one. Therefore, implanting normal 500-nm-thick polysilicon to a value of 500–2000 Ω/square can yield resistors with very low temperature coefficients. I am aware of some processes with polysilicon resistors in the 40 kΩ/square range, and these yield negative temperature coefficients of the order of −8000 ppm/K. While this is high, such processes do make practical the integration of resistors of several tens of megohms in a reasonable die area. Resistors of the thin-film type offer the best performance in terms of overall stability and can give almost zero voltage and current coefficients at the cost of considerable processing complexity and, with careful control, temperature coefficients less than 100 ppm/K.

Nichrome and silicon-chrome are the classic film materials, though the former is hard to deposit at a sheet resistance greater than 200 Ω/square, a severe disadvantage for low-power circuits. The traditional sheet resistances for silicon-chrome are in the 500–4000 Ω/square range, and these are the easiest to deposit with low temperature coefficients. It is, however, possible to formulate silicon-chrome compounds capable of yielding much higher sheet resistances if higher temperature coefficients can be tolerated.

One big advantage of thin-film resistors is that they can be trimmed to precision values by means of a laser. Even with today's design techniques of Electrically Alterable Read Only Memory (EAROM) trimming and autocalibration, laser trimming is still used to great effect in the production of precision analog circuits.

3.5.2. Other Biasing Elements

Even if the majority of current generation in a particular design is performed by active elements, some "start-up" circuitry is required to provide an initial current. Often, it is arranged that such a current is completely removed from circuit operation when nominal bias conditions have been achieved, or sometimes such a current is processed by heavily nonlinear regulator circuitry to reduce sensitivity to the value of the current. The latter technique is generally to be preferred, since it automatically checks for the integrity of the start-up current (often power supply transients or noise will start up a circuit during test, but deficiencies in the activation circuitry will show up in normal operation as a failure to bias correctly).

It is possible to devise schemes where junction leakage alone can provide the necessary start-up condition, but the small values of current involved (especially at low temperatures) can cause even minor defects to prevent proper circuit operation.

Some early micropower circuits used an external high-value resistor as a start-up device; but even where this is acceptable, external effects such as printed circuit board leakage or moisture can make this an undesirable option.

Circuitry operating exclusively at low voltage simplifies the task of generating a suitable internal current, but even at a few volts of supply voltage, generating hundreds of nanoamperes of reliable current requires effective resistances of tens of megohms, values totally impractical with the sheet resistances normally available.

Even if the process has a deliberate high-value resistor available, it still may not be a suitable candidate for the start-up function, especially if the circuit is not a low-voltage design. What is needed under these circumstances is a very high value resistor to perform this task, whose value, nonlinearity temperature coefficient, and so on are somewhat unimportant. Such a function can be realized in many ways.

The ion-implanted JFET is probably the best behaved element for such a function; above pinch-off it behaves almost as a true current source, normally has a breakdown equal to the full operating voltage of the process, can produce small currents without drastic die area implications, and has a reasonably well controlled I_{dss} (modern implantation techniques can control the latter over a range of better than 3:1). Depletion-mode MOSFETs also fall into this category, but, alas, often neither of these devices are available. Enhancement-mode MOSFETs are also frequently pressed into service as resistors, and if any significant voltage swing is present, this usually takes the form of a weak transmission gate in a permanently "on" condition. Again, the voltage coefficient of this arrangement is extreme, to say the least, particularly at low supply voltages. Enhancement-mode MOSFETs used as biasing elements have the unfortunate property of generating a current proportional to the square of the supply voltage, but this may be tolerable where the latter is reasonably well defined.

On a standard bipolar process, two techniques are currently used to obtain noncritical high-value resistors: transistor base "pinch" resistors and epitaxial resistors. The pinched base resistance under the emitter of a bipolar transistor typically has a sheet resistance of something like 5–20 kΩ/square, but unfortunately it has a maximum voltage limitation equal to the breakdown voltage (BV_{ebo}) of the process (it is not usually the actual resistor which breaks down, but the region where the emitter mask inevitably terminates to allow a surface connection to the base region). Since the latter is usually at least several volts, this may not matter much for low-voltage circuits, but higher voltage applications mandate that several of these be connected in series. This actually consumes significant die area, since each pinch resistor must be contained within its own isolation pocket. The resistance thus obtained has a (positive) temperature coefficient of several thousand ppm per Kelvin, and allowance must be made for this. The tolerance (at least in theory) correlates with the transistor forward current gain, but this usually constitutes a spread of 3:1 or more. Even when the breakdown voltage is not exceeded, such pinch resistors exhibit very high voltage coefficients and are usually best modeled as JFETs with pinch-off voltages in the neighborhood of a few volts.

An epitaxial reistor uses an elongated isolated pocket as a resistor, and several kilohms per square can be achieved in this way. The minimum width of such a resistor is, however, determined by side diffusion during the isolation process, which makes such a resistor most costly in terms of die area than may first be apparent. Although this resistor usually has a breakdown as great as anything else on the process, it is plagued by poor tolerance due to epitaxial thickness and doping variations and also variations of the isolation side diffusion and substrate up diffusion. Such resistors also exhibit high positive temperature coefficients and severe voltage coefficients.

The sheet resistance of an epitaxial resistor can be considerably raised by overlaying it with the base diffusion of the NPN transistor. This produces a "buried" resistor with a sheet resistance of 3–6 times the native epitaxial resistance but also greatly increases the voltage coefficient. Like pinch resistors, it is not uncommon for such devices to enter the pinch-off (saturation) region well before their breakdown voltage (measurements indicate that 3 Ω/cm epi material at 12 μm thickness with normal diffusion cycles pinches off at somewhere around 11 V). As such, these devices are commonly referred to as EPI-FETs, although I_{dss} typically varies by 6:1 or more with process variations.

The classic "noncritical" resistor on a CMOS process is the well resistor (obviously N-type for an N-well process and P-type for a P-well process), and this again gives a resistor with several kilohms per square of sheet resistance. Typical temperature coefficients are high (9000 ppm/K not being uncommon), and once more the voltage coefficient is definitely not negligible, although like epitaxial resistors the full process breakdown voltage is usable. Side diffusion of the well limits the spacing of a meander resistor, and so the area of a very large resistor occupies more area than might at first be expected, but still much less than the epitaxial resistor of a typical bipolar process.

The resistance of a well resistor (like an EPI-FET) can be increased by overlaying it with the drain-source diffusion of the oppositely doped MOSFET. This can increase the effective resistance several-fold, depending on the type of process.

3.5.3. Capacitors

Capacitors of the P-N junction type have always been available on ICs in some form or other. Such a capacitor is inherently polarized (of course) and delivers a severe voltage dependence which is in simple theory given by [1, p. 6 all editions]

$$Cj_{(\text{effective})} = \frac{Cj_{(\text{nom})}}{\sqrt[M_j]{1 + V_{\text{bias}}/V_j}}$$

where $C_{j(\text{nom})}$ is the measured capacitance at zero bias, V_{bias} is the *reverse* bias on the junction, V_j is the built-in voltage for the junction, and M_j is the junction grading coefficient that is theoretically equal to 2 for a step junction and 3 for a fully graded junction. Thus, the capacitance obeys a square-root law for an abrupt junction and a cube-root law for a graded one; practical results, as might be expected, are somewhere in between. If V_{bias} becomes negative, corresponding to a forward-biased junction, the equation displays a singularity at $V_{\text{bias}} = -V_j$. Simple theory breaks down in this region, and several papers have been written on the subject [12–14], but these are somewhat academic for the purposes of this discussion since the junction would be carrying enough forward current at this point to render it useless as a purely capacitive element.

Because the depletion capacitance is at a maximum at low voltages, it would seem a good choice as a component for low-voltage operation. This may well be the case, but the severe voltage dependence and polarization make the junction capacitor awkward to use, at best. And if the operating currents are also low, leakage once again becomes a major concern. Zero-bias values for the junctions found on a standard bipolar process are typically 0.15 fF/μm^2 for collector-base capacitance and 0.8 $fF/\mu m^2$ for base-emitter capacitance. While the latter boasts high value, it is limited in voltage to the BV_{ebo} of the NPN transistor, typically 5–8 V.

The collector-base junction exhibits the same leakage as a corresponding bipolar transistor, of course, typically 0.08 $fA/\mu m^2$ for 5 V reverse bias. Thus a 10 pF junction capacitor could leak some 8 pA at 25°C, 500 pA at 85°C, and 8 nA at 125°C. The room temperature value is probably negligible in most applications, but clearly the leakage at elevated temperatures must be taken into account.

Base-emitter capacitance can be far more leaky, depending on the applied voltage. Room temperature values of less than 0.15 $fA/\mu m^2$ for voltages below 2 V are common, but this increases abruptly as the reverse bias is increased.

Another capacitor sometimes used where a lot of capacitance to the negative supply rail is needed is the emitter-to-isolation capacitor. On a bipolar process, running emitter diffusion in the isolation walls between transistors can generate a lot of capacitance ($\cong$ 1.5 $fF/\mu m^2$) in an area that would otherwise be wasted. Unfortunately, this tends to be a defect-ridden area, especially if implanted isolation is used. Consequently, such capacitors can be very leaky indeed, even to the point of being a yield hazard. As with bipolar transistors, latent defects can sometimes show up at different temperatures, further complicating yield and test issues.

To alleviate these problems, most analog processes these days include a quality capacitor of some kind. The dielectrics used are almost always silicon dioxide or silicon nitride (more correctly, trisilicon tetranitride). Dielectric constants vary with growth and deposition techniques, but for 10 nm oxide the resulting capacitance is usually around 0.3 $fF/\mu m^2$. Silicon nitride gives roughly double this figure, which is why it is preferred by some manufacturers. The leakages of these dielectrics are truly negligible (except near breakdown) and cause no problems for low-power operation. Of more concern, however, are the parasitics associated with the structure.

The classical way to make an oxide-type capacitor is to use one of the diffusions as the back plate, grow oxide (or deposit nitride) on top, and use metallization as the top plate. The back plate of such a capacitor has exactly the same type of parasitic capacitance and leakage as the corresponding junction capacitor, and similar care needs to be taken to prevent these from disturbing circuit operation. In this case, however, the task is easier because the parasitics only affect one side, and usually the circuit can be arranged such that this side is insensitive to such parasitics.

The CMOS processes often feature double polysilicon layers, and these are used to create capacitors with both plates fabricated from polysilicon. Such capacitors are essentially leakage free, and though there is still some residual parasitic capacitance from the bottom plate to substrate, it is much less than that of a P-N junction.

On processes featuring dual-level metallization, the possibility exists of depositing a thin dielectric between the levels to create a metal-oxide-metal (MOM) capacitor. Such capacitors have similar parasitic characteristics to the double polysilicon types.

In passing, it should be noted that the silicon dioxide dielectric is usually thermally grown, except for the MOM capacitors where it has to be deposited. Silicon nitride is always deposited. Deposited dielectrics have poor dielectric absorption characteristics compared to grown types, and this can dictate the type of capacitor used for circuits especially sensitive to dielectric absorption, such as sample-hold and switched-capacitor circuits.

3.5.4. Inductors

Microwave ICs use spiral metallization techniques (often coupled with airbridge isolation) to fabricate inductors in the nanohenry range. For the purposes of this book, such techniques, and the values of inductance obtained, are impractical.

The function of an inductor can, of course, be theoretically synthesized with circuit design techniques using active transconductance elements and capacitors. At the bandwidths with which low-power operation is usually concerned, this is indeed a very practical option, one not available to ultra-high-frequency designers. But for the record, there is currently no *direct* way of fabricating a useful inductor on a low-power IC.

3.6. BANDWIDTH AND NOISE CONSIDERATIONS

Perhaps the most easily anticipated sacrifice when reducing the *current* consumption of an IC is loss of speed. Voltage noise also tends to increase as the square root of circuit impedances and so would also be expected to be a major compromise with reduced currents. With some qualification, both these presumptions are quite true.

When reducing operating *voltage*, however, the compromises are less obvious. From the device point of view, the speed penalty is relatively small and the noise penalty negligible. However, very low supply voltages demand unusual and often complicated design techniques, which can seriously impact the performance of a practical circuit in both these respects. Additionally, as far as noise is concerned, signal swings must by definition be reduced at low supply voltages, so even if the device noise is not unduly compromised, the signal-to-noise ratio almost certainly will be. Additionally, good noise performance depends upon keeping the internal noise currents as low as possible, and this involves generous amounts of emitter degeneration (in the case of a bipolar device) or large L/W ratios (in the case of a MOSFET) for active loads and bias lines. In the case of a low-voltage device, there may not be sufficient headroom for such degeneration; and at low currents the required die area resulting from large resistors or MOSFETs may simply be impractical.

This whole point is raised to illustrate the fact that device considerations alone cannot predict the speed and noise trade-offs of a low-power circuit. Even so, some conclusions can be drawn from the examination of the relevant device physics.

3.6.1. Speed Restrictions at Low Voltages

Commercially available linear ICs frequently lose both bandwidth and slew rate when operated from reduced supplies, but in many cases this can be largely traced to deficiencies in regulator circuits, with correspondingly reduced stage currents at low voltages. In a circuit specifically designed for low-voltage operation, the major speed penalty comes from the fact that junction capacitors approach their maximum values as their reverse bias is reduced. At 0 V, parasitic capacitances are typically between 1.5 and 2.5 times their values at 10 V reverse bias.

For a bipolar transistor, F_t is also at a minimum at low voltages due mostly to the fact that the base width is at its widest under nondepleted conditions. However, the F_t is not often a speed-limiting mechanism in a practical low-power circuit.

On the other hand, low-voltage operation enables the adoption of small-geometry processes, with reduced stray capacitances and higher F_t. Thus the speed penalty may be somewhat ameliorated.

3.6.2. Speed Restrictions at Low Currents

Since reduced currents lead to low transconductance values but not to reduced stray capacitances, there is a marked loss of speed in low-current operation. For a bipolar circuit, g_m is basically proportional to the supply current, and to a first order it thus might be expected that bandwidth and slew rate would behave similarly, and in general they do. The MOSFET has a transconductance normally proportional to the square root of the operating current and thus would be expected to behave differently. However, when device scaling and subthreshold operation are taken into account, the overall effect remains much the same as the bipolar case.

One capacitance that is not independent of device current is the bipolar diffusion capacitance, the product of transit time $(1/2\pi F_t)$ and transistor g_m. Since in theory this capacitance also scales with operating current, it becomes small compared to parasitic capacitances (particularly base-emitter depletion capacitance) at low currents. This is tantamount to saying that F_t, a common "figure of merit" in assessing the speed of a process, may actually have little relevance on a low-power circuit, since it is almost always quoted at a peak (high-current) value. It is interesting to note that the lateral PNP transistor is notorious for its speed characteristics as a result of a low F_t; at low collector currents, however, it is not at a very great disadvantage to double-diffused transistors.

There is a good point to be made here: The classic equation for voltage dependence of a depletion capacitance,

$$Cj_{(\text{effective})} = \frac{Cj_{(\text{nom})}}{\sqrt[M_j]{1 + V_{\text{bias}}/V_j}}$$

exhibits a singularity if the junction is *forward* biased by an amount equal to V_j (usually 0.5–0.6 V). In the case of the emitter-base junction, this is almost exactly where the junction will be operated at low currents. More accurate expressions [13, 14] show that the capacitance actually stops increasing at about V_j and starts reducing again as forward bias increases. These improved models, however, have yet to be implemented in most circuit simulators.

In SPICE (and most other simulators) the singularity is removed by replacing this equation with a linear one above a forward-bias point called the function corner. This corner is defined by the parameter FC, which represents the fraction of V_j at which the equation reverts to a linear slope (equal to the derivative of change at the function corner). The FC defaults to 0.5 in SPICE.

This clearly does not represent the true behavior of the capacitance under these conditions, but the justification has been made that the *diffusion* capacitance swamps the depletion term in this region. As discussed here, this may not be the case at low currents, and one way to verify this is to experiment with the effects of adjusting the FC (which has no physical significance whatsoever). If simulation results prove sensitive to this parameter, they should be viewed with some skepticism.

3.6.3. Noise Restrictions at Low Voltages

Device noise is only a weak function of operating voltage. In a bipolar transistor, the base resistance and current gain increase with collector voltage due to base depletion (Early effect), and this causes base current noise to increase slightly and voltage noise to reduce slightly at low voltages. The MOS devices that are not fully in saturation can also appear noisy due to a reduced g_m, but this is really more of a design consideration.

3.6.4. Noise Restrictions at Low Currents

There is a considerable penalty to be paid in noise performance as an attempt is made to reduce overall operating current. Bipolar transistor circuits and MOSFET circuits definitely have differing trade-offs as far as this is concerned and will be considered separately. There are also two differing noise mechanisms to consider: one is broadband (white) noise, which is well understood; the other is flicker noise (often termed $1/f$ noise because it is generally assumed to have a squared component inversely proportional to frequency). A third noise mechanism, burst noise (sometimes called popcorn noise), will not be discussed in detail here because it is more of a process problem (and, in any case, the physics are poorly understood). Suffice it to say, however, that there is evidence that the latter *is* more troublesome at low currents and that it is *not* a phenomenon totally peculiar to bipolar transistors; I have definitely observed it in MOSFETs.

I have assumed that attention to design optimization has been made for operation at the particular current in use; there are far more severe penalties for designs merely "starved down" to low-current levels.

Low-Current Noise in Bipolar Transistors. In a bipolar circuit, broadband voltage noise mainly comes from the Johnson noise of resistors (intentional or parasitic) and Schottky noise effects on the V_{be} of the transistors. The latter can be shown to be equivalent to the Johnson noise of a resistor equal to $1/2g_m$ of the particular device in question. The Johnson noise of a resistor (R) in a bandwidth (Δf) can be easily calculated from the expression

$$V_{\text{noise}}(\text{RMS}) = \sqrt{4kTR(\Delta f)}$$

Since the g_m of a bipolar transistor is proportional to operating current, and if all other resistors are correctly scaled, the noise voltage will be inversely proportional to the square root of operating current. Of course, it may be impractical to include current source degeneration resistors large enough to mimic the degeneration (100 mV or more) used to reduce noise of circuits without severe low-power requirements, in which case the noise penalty may be more severe than the square-root law would imply. To partially compensate for this, some parasitic resistances do not increase as the power is reduced (the base resistance of a bipolar transistor, one of the largest parasitic noise sources, does reduce somewhat with increased current, but it is less than a linear relationship, and this ameliorates the noise penalty in some cases).

Flicker noise in bipolar transistors is not as serious as in the case of a MOSFET, and voltage noise "corners" are often well below 100 Hz with a good process technology. The analysis of such noise is complicated by the fact that it is not truly $1/f$ in behavior, tending to follow the general relationship

$$V_{n(\text{flicker})} \propto \sqrt{\frac{1}{f^n}}$$

where n can be considerably greater than unity (values of 2 or more having been observed). Even so, the tendency in a bipolar transistor is for the flicker noise to vary with current in similar proportion to the broadband noise.

Base current noise (Schottky noise) will naturally reduce as the square root of the base current, but this is of no overall benefit when the impedances driving these currents are scaling in inverse proportion to operating currents.

Low-Current Noise in MOSFETs. Like a bipolar transistor, an expression can be derived for the Johnson noise equivalent of V_{gs} noise for a MOSFET (or JFET) in terms of its operating transconductance. In this case, though, the equivalent noise is theoretically that of a resistor having a value of $2/3g_m$. If W/L scaling is performed for a constant enhancement voltage, the broadband voltage noise should follow the same square-root law with increasing current as for a bipolar device. Without scaling, the broadband noise in simple theory is dependent on the fourth root of the operating current, since g_m varies as the square root of this current (apparently very good news for low-current designers). Unfortunately, as discussed earlier under Low-Current Limitations of MOSFETs, subthreshold effects can easily invalidate such a simple theory, and in general the overall broadband noise performance of an MOS circuit becomes highly sensitive to circuit and biasing details.

Flicker noise in MOS devices is of considerably more importance than in bipolar devices, and typical noise corners are in the 1–100 kHz range. For what it is worth, the flicker noise in MOSFETs is found to quite accurately behave as a true $1/f$ characteristic (or more correctly, the square root of $1/f$). The physical mechanisms giving rise to $1/f$ noise in MOSFETs have received extensive study [15]. The precise dependence of flicker noise on bias conditions and device geometry is highly process dependent. Practical measurements show that the flicker noise component adds to the threshold mismatch term in a manner essentially independent of bias current. The only way to design a device for low flicker noise seems to be to increase the total gate area, which tends to reduce the noise in a square-root fashion by purely averaging the effect [1, 1993, p. 746]. It may be fortuitous that the large L/W ratios required for the scaling mentioned earlier tend to reduce the flicker noise component also, since in conjunction with minimum channel width design rules they tend to force a larger gate area. The converse of this happens when attempting to minimize device size; not only does this result in larger flicker noise, but it also forces devices into subthreshold operation, where the transconductance per unit current is at a maximum and flicker noise is able to produce a large amount of drain current fluctuation.

3.7. CONCLUSIONS

Modern silicon IC processes are reasonably well suited to the design of low-power analog circuitry, even when not specifically intended to do so.

The practical limit on operating voltage is probably in the region of 0.8 V or so, but simpler (and lower operating current) designs can result from raising this somewhat.

The lower current limit is more nebulous since it strongly depends on the circuit complexity and operating temperature range, but suffice it to say that surprisingly involved systems can currently be designed with a supply current of less than a microampere or so.

References

[1] P. R. Gray and R. G. Meyer, *Analysis and Design of Analog Integrated Circuits*, Wiley, New York, 1977, 1984, 1993, three editions.

[2] R. J. Widlar, "Low Voltage Techniques," *ISSCC Dig. Tech. Papers*, Feb. 1978, pp. 238–239.

[3] J. S. Brugler, "Silicon Transistor Biasing for Linear Collector Current Temperature Dependence," *IEEE J. Solid-State Circuits*, vol. SC-2, pp. 57–58, 1967.

[4] D. A. Hodges and H. G. Jackson, *Analysis and Design of Digital Integrated Circuits*, McGraw-Hill, New York, 1983.

[5] G. M. Kull, L. W. Nagel, P. Lloyd, E. J. Prendergast, and H. Dirk, "A Unified Circuit Model for Bipolar Transistors Including Quasi-Saturation Effects," *IEEE Trans. Electron Devices*, vol. ED-32, no. 6, pp. 1103–1113, 1985.

[6] I. Getreu, *Modeling the Bipolar Transistor*, Tektronix Inc., 1979 (third printing).

[7] C. A. Bittman, G. H. Wilson, R. J. Whittier, and R. K. Waits, "Technology for the Design of Low-Power Circuits," *IEEE J. Solid-State Circuits*, vol. SC-5, pp. 29–37, 1970.

[8] C. T. Sah, "Effect of Surface Recombination and Channel on *p-n* Junction and Transistor Characteristics," *IRE Trans. Electron Devices*, vol. ED-9, pp. 94–108, 1962.

[9] R. S. Muller and T. I. Kamins, *Device Electronics for Integrated Circuits*, Wiley, New York, 1977, p. 377.

[10] W. Shockley, "A Unified Field-Effect Transistor," *Proc. IRE*, vol. 40, p. 1365, 1952.

[11] G. C. Dacey and I. M. Ross, "Unipolar Field-Effect Transistor," *Proc. IRE*, vol. 41, p. 970, 1953.

[12] H. C. Poon and H. K. Gummel, "Modeling of Emitter Capacitance," *Proc. IEEE*, vol. 57, pp. 2181–2182, 1969.

[13] B. R. Chawla and H. K. Gummel, "Transistor Region Capacitance of Diffused *p-n* Junctions," *IEEE Trans. Electron Devices*, vol. ED-18, pp. 178–195, 1971.

[14] P. Van Halen, "A New Semiconductor Junction Diode Space Charge Layer Capacitance Model," *IEEE BCTM Proc.*, 1988, pp. 168–171.

[15] M. B. Das and J. M. Moore, "Measurements and Interpretation of Low-Frequency Noise in FETs," *IEEE Trans. Electron Devices*, vol. ED-21, pp. 247–257, 1974.

Chapter 4

Andreas G. Andreou
Department of Electrical and Computer Engineering, Johns Hopkins University, Baltimore, Maryland 21218

Exploiting Device Physics in Circuit Design for Efficient Computational Functions in Analog VLSI

4.1. INTRODUCTION

Computation as performed by "real" systems is an irreversible physical process and as such it is associated with an inevitable amount of energy dissipation [1, 2]. This is true both for human engineered very large scale integrated (VLSI) systems (Chapter 9 in [3]) and for nature's machinery, biological systems.

Biological organisms excel at solving hard problems in sensory communication and motor control—speech and vision—by sustaining high computational throughput while keeping energy dissipation at a minimal level. The total power consumed by the awaken human brain is about 10 W [4, 5]. The existence of such remarkably efficient computational structures is a result of an evolutionary necessity toward systems that are *truly autonomous* and thus are constrained to operate under strict bounds of size and weight and at temperatures where favorable conditions exist for the development of life as we "know" it ($T = 300\,\text{K}$).

Neural systems operate from power supplies of approximately 100 mV, only a few kT/q, while the bandwidth limitation at the *macroscopic* scale of constituent components is only a few hundred kilohertz, which is comparable and sometimes even lower than the bandwidth of the signals that have to be processed. Typical current values in biology are in the pico- to nanoampere range [7]. Neural "wetware" is characterized by components that are inherently highly variable and noisy. Nevertheless, natural sensors and natural computing systems are capable of remarkable performance that our best technology is unable to match. Consider the mammalian retina, a thin structure (approximately 500 μm) at the back of the eyeball and an outpost of the brain. It serves as both sensor and preprocessor for the visual cortex. Nobel laureate David Hubel writes [6, p. 33]:

> The eye has often been compared to a camera. It would be more appropriate to compare it to a TV camera attached to an automatically tracking tripod—a machine that is self-focusing, adjusts automatically for light intensity, has a self cleaning lense, and feeds into a computer with parallel processing capabilities so advanced that engineers are only just starting to consider similar strategies for the hardware they design. . . . No human inventions, included computer assisted cameras can begin to rival the eye.

The effectiveness and efficiency of biological systems stem partly from exploiting *prior* knowledge about the problems that they encounter [9]. Such information, in the form of *internal models*, reflects the statistical properties of the natural environments in which they function. The exploitation of such prior knowledge plays an important role in both the evolutionary development of neural structures and the adaptation mechanisms during system function. Exploitation of prior knowledge is also the key to successful algorithm design for human engineered speech and vision systems (see discussion in [8, 18]).

Perhaps equally important is the way *function* emerges from *form*. Systems of remarkable complexity, not just in the number of elements but also in composition, have evolved into a hierarchy of structures with remarkable computational capabilities that clearly the best digital technologies cannot match (Fig. 4.1).

4.1.1. From Device Physics to Analog VLSI Systems

At the most basic level, analog VLSI technology offers the possibility of exploring experimentally computation by truly complex, *real systems* which lie beyond digital computing and the symbolic processing paradigm.

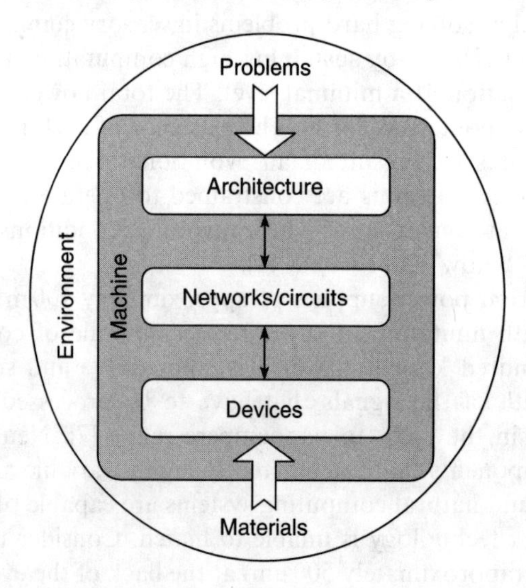

Figure 4.1 A physics-of-computation view for the synthesis of complex, highly integrated microsystems. This methodology optimizes the design at and between all levels of a system: from devices to circuits and network levels and all the way to the architecture.

It is appropriate at this point to ask the question: What kind of computational primitives does one have? In complementary metal-oxide-semiconductor (CMOS) silicon these are continuous functions (analog) of *time*, *space*, *voltage*, *current*, and *charge*. To help manage the complexity in VLSI systems, these functions will be considered at three hierarchical levels: the *device level*, the *circuit level*, and the *architectural level*.

Device Level. At the lowest level, gain is provided by MOS transistors operating in the subthreshold region [11, 13, 17]. In this regime device physics yield the following functional form for the drain current in terms of the voltages at its four terminals:

$$I = I_0 S \mathcal{G}(\kappa V_{GB})[\mathcal{H}(V_{SB}) - \mathcal{H}(V_{DB})] \tag{4.1}$$

where $\mathcal{G}$ and $\mathcal{H}$ are growing and decaying exponential functions, respectively. The terminal voltages V_{GB}, V_{SB}, V_{DB} are referenced to the substrate and are normalized to the thermal voltage (kT/q). The constant I_0 depends on mobility (μ) and other silicon physical properties; S is a geometry factor and is equal to the ratio of width W and length L of the device. The Pauli exclusion principle dictates that the prefactors that normalize the voltages to thermal voltages in the exponentials be less than or at best equal to unity. The MOS transistor has excellent circuit properties as a voltage input, current output device (transconductance amplifier) with good fan-out capabilities (high transconductance $\mathcal{G}$) and good fan-in capability (almost zero conductance at the input).

The exponential functions of voltage in the square brackets of Eq. (4.1) correspond to Boltzmann distributed charges at the source and drain:

$$I \propto [\mathcal{Q}_S - \mathcal{Q}_D] \tag{4.2}$$

The charge-based representation depicted in Eq. (4.2) suggests that the MOS transistor in the subthreshold region is a highly linear device, a property that finds many uses in analog circuit design. This property was first observed by Kwabena Boahen and is discussed in [24], where the concept of a *diffusor* is introduced. The view of an MOS transistor in the subthreshold as a basic diffusive element allows for the effective implementation of systems that exploit properties of elliptic partial differential equations.

The transfer characteristics of MOS transistors are plotted in Figure 4.2 for both the above and subthreshold regime. The transconductance per unit current increases as the current decreases—throughout the above-threshold and transition regions—and reaches a maximum in the subthreshold region. In highly integrated VLSI systems, small-geometry devices must be used to achieve high densities. Small device geometries and high transconductance per unit current make the drain current strongly dependent on variations of the process-dependent parameters, in particular I_0, which is the source for the variability observed in the drain currents of Figure 4.2. The apparent improvement in device matching for higher values of gate-source

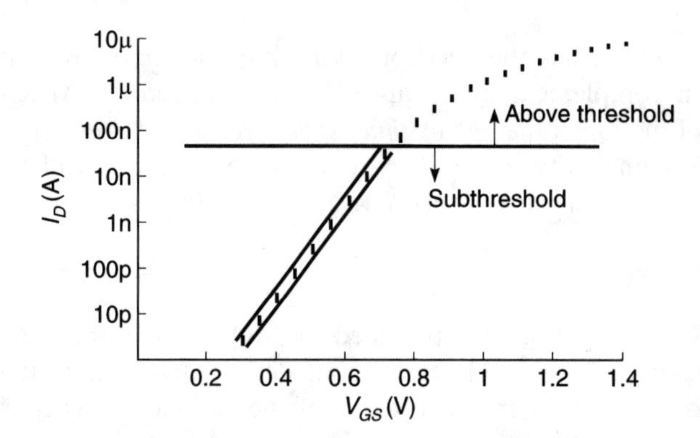

Figure 4.2 Measured drain current I_d versus gate-source voltage V_{GS} for 32 small-geometry transistors ($4 \times 4\,\mu m$) fabricated in a 2-μm N-well CMOS process; drain-source voltage $V_{DS} = 1.5\,V$. The fuzziness in the current (mismatch between devices) is constant in the subthreshold [on a log(I) scale] and decreases as the device enters the transition and above-threshold regime (data from [15]).

voltage is simply a manifestation of reduced transconductance per unit current as the device enters the above threshold regime.

Our preference for subthreshold operation (despite what seems to be a worse matching characteristics) is based on the observation that *active devices should be used in the region where their transconductance per unit current is maximized.* In this way one can minimize the energy per operation and maximize the speed per unit power consumed, i.e., minimize the power-delay product:

$$\frac{\text{Speed}}{\text{Power}} = \frac{1/\tau}{I\,\Delta V} = \frac{g_m/C}{I^2/g_m} = \frac{1}{C}\left(\frac{g_m}{I}\right)^2 \tag{4.3}$$

A squared factor is obtained because both voltage swing (ΔV) and propagation delay (τ) are inversely proportional to the transconductance g_m for a given current level. However, only a linear factor is realized if the power supply voltage is not reduced to match the voltage swing $\Delta V \sim I/g_m$. When the device is operated in the subthreshold, the drain–source conductance saturates at a few kT/q [see Eq. (4.1)]. Power supplies of a few kT/q are also possible and thus power supplies can theoretically match the voltage swing levels. The capacitance C is analogous to an inevitable "mass" of the switching node. When physical structures are miniaturized, this capacitance is reduced and the power-delay product improves. This simple scaling "law" has been one of the driving forces toward high levels of system integration and miniaturization in the microelectronics industry.

The maximum useful frequency of operation possible with an MOS transistor when operating in the subthreshold is determined by its transition frequency f_T, which has an upper limit $f_{T,\max}$ of

$$f_{T,\max} < \frac{\mu\,(kT/q)}{\pi L^2} \tag{4.4}$$

where μ is the effective carrier mobility and L is the device channel length. The transition frequency of a device is essentially the frequency where its gain-bandwidth product (as determined by the internal gain and parasitic capacitances of the transistor) is unity.

Circuit Level. It is at this level that the synthesis of computational structures begins and manifests itself as the emergence of *networks*. Conservation laws, that is, conservation of charge (Kirchhoff's current law), $\sum_i I_i = 0$, and conservation of energy (Kirchhoff's voltage law), $\sum_i V_i = 0$, are used to realize simple constraint equations. The important concept of *negative feedback* is also exploited to trade the gain in the active elements for precision and speed in the circuits.

Aside from the benefits of a device with a large gain, the exponential relationships between the controlling voltages and the current depicted in Eq. (4.1) endow the MOS transistor with some interesting circuit properties. There exists a powerful synthesis (and analysis) procedure which can be used to generate a wide variety of circuits that perform linear and nonlinear operations in the current domain and it relies on the exponential form of current-voltage nonlinearities. This procedure is based on what is known as the *translinear principle* [21], originally used in the context of bipolar transistors. The synthesized circuits are called *translinear* and may involve operations of one or more variables, such as products, quotients, power terms with fixed exponents, as well as scalar normalization of a vector quantity.

The application of the translinear principle to circuits implemented with MOS devices operating in subthreshold saturation and an extension to the subthreshold ohmic regime are discussed in the next section. One fascinating aspect of translinear circuits is that while the currents in their constitutive elements (the transistors) are exponentially dependent on temperature, the overall input/output relationship is insensitive to isothermal temperature variations. The effect of small local variations in fabrication parameters can also be shown to be temperature independent.

To demonstrate how computational primitives emerge at the network level from the device physics of the underlying technology, let us consider an example of a summing operation, *local aggregation*. Such linear addition of signals over a confined region of space occurs throughout the nervous system. Aggregation was discussed in Chapter 6 of [11] (also in [78]), and it is the basis for many neuromorphic silicon VLSI systems described therein. Here we take a close look at *diffusion*, the physical process that underlies local aggregation in the nervous system, contrast it with the process of diffusion in MOS transistors, and come up with a novel network design technique.

The first network uses voltages and currents [Figure 4.3(a)]. Its node equation is

$$\frac{dV_n}{dt} = \frac{4G}{C}[\tfrac{1}{4}(V_j + V_k + V_l + V_m) - V_n] \tag{4.5}$$

which is homologous to the diffusion equation since the term in brackets is a first-order approximation to the Laplacian. However, this solution is not amenable to VLSI integration because transconductances (G) with a large linear range consume large amounts of area and power.

The second network uses charges (positive) and currents [Fig. 4.3(b)]. Its node equation is

$$\frac{dQ_n}{dt} = 4D[\tfrac{1}{4}(Q_j + Q_k + Q_l + Q_m) - Q_n] \tag{4.6}$$

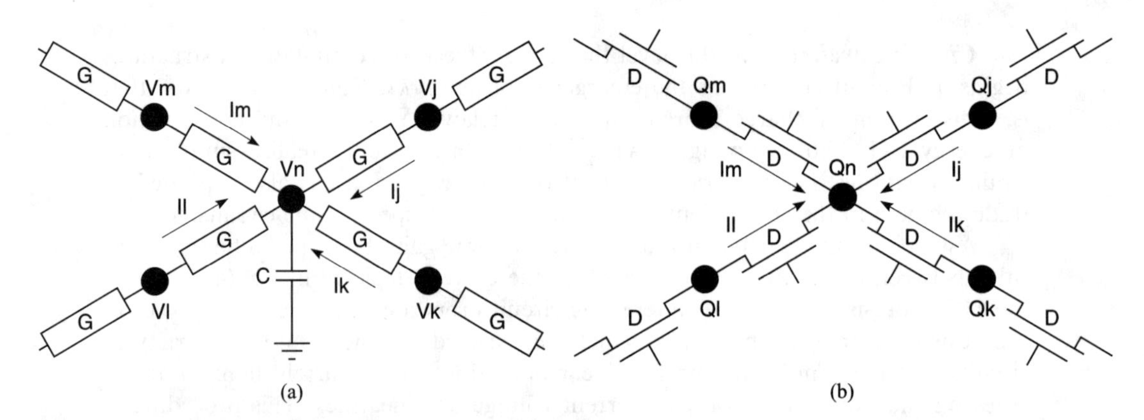

Figure 4.3 Simulating diffusion with (a) conductances and voltage/current variables or (b) diffusors and charge/current variables.

Note that dQ_n/dt is the same as the current supplied to node n by the network. This solution is easily realized by exploiting diffusion in subthreshold MOS transistors. As shown in the device section, the current is linearly proportional to the charge difference across the channel [see Eq. (4.2)]. Therefore, the diffusion process may be modeled using devices with identical geometry S and identical gate voltages. The former guarantees they have the same diffusivity and the latter guarantees that the charge concentrations at all the source/drains connected to node n are the same and equal to Q_n.

In both of these networks, the boundary conditions may be set up by injecting current into the appropriate nodes. In the voltage-mode network, the solution is the node voltages. They are easily read without disturbing the network. However, the network depicted by Eq. (4.6) represents the solution by charge concentrations Q_S and Q_D at source/drains, not the charge on the node capacitance. The source/drain charge cannot be measured directly without disturbing the network. It may be inferred from the node voltage.

Architectural Level. At this level, differential equations from mathematical physics will be employed to implement useful signal-processing functions, still in the form of constraint equations. For example, the *biharmonic* equation

$$\lambda\nabla^2\nabla^2\Phi + \Phi = \Phi_{\text{in}} \tag{4.7}$$

where $\nabla^2 \equiv \partial^2/\partial x^2 + \partial^2/\partial y^2$ is the Laplacian operator, constrains the sum of the fourth derivative of Φ and Φ itself to be equal to a fixed input Φ_{in}. From a *statistical* signal-processing viewpoint, solutions to this equation could represent an *optimal estimation* Φ of the underlying smooth continuous function, given a set of noisy, spatially sampled observations Φ_{in}. The solution is optimal in the sense that it simultaneously minimizes the squared error and the energy in the second derivative; the parameter λ is the relative cost associated with the derivative term. A large value for λ favors smooth solutions while a small value favors a closer fit.

We have already seen how a diffusive grid can be used to compute a discrete approximation of the Laplacian. In Section 4.5 we show how a model of early visual processing is related to the biharmonic equation and can be realized using diffusive networks.

In this chapter we begin at the basic technology device level and proceed all the way to the system level demonstrating how a neuromorphic approach and a physics computation view can yield fruitful results in analog VLSI.

The chapter is divided into six sections. Section 4.2 begins with an introduction to the physics of MOS transistors, including floating-gate MOS transistors (FGMOS). A basic review of subthreshold MOS and bipolar transistor device physics contrasting their operation is provided since the large-signal properties of the devices are *key* to the subject matter. In particular, we emphasize the translinear properties of bipolar transistors and those of MOS transistors in the subthreshold. The current-mode design methodology is introduced in Section 4.3, where the concepts of a current conveyor, a canonical block that is widely employed in minimal complexity circuits, is presented. Circuit techniques based on the translinear principle that employ MOS transistors in subthreshold saturation and ohmic regimes are introduced in Section 4.4. In the same section, we discuss both translinear loops (TLs) composed of generalized diodes and current sources and translinear networks (TNs) that include voltage sources as well. In Section 4.5 we discuss an analog VLSI system for early vision, a silicon retina that has evolved from the design methodology presented in the earlier sections. A general discussion in Section 4.6 concludes the chapter.

4.2. TECHNOLOGY AND DEVICES

Complementary MOS technology, in particular subthreshold MOS operation, has long been recognized as the technology for implementing digital VLSI and analog LSI circuits that are constrained by power dissipation requirements [13, 14]. Complementary MOS has the highest integration density attainable today, making it especially attractive for systems that demand high overall computational throughput. Moreover, the physical properties of silicon and its native oxides, together with recent advances in micromachining of electromechanical elements [31], make silicon-based technologies the prime candidate for highly integrated systems that are complex not only in the strict sense of a large number of gates but also in the design of the individual components.

In the last few years we have seen analog computation based on physical properties of silicon devices emerge as an alternative to abstract mathematical algorithms implemented on digital hardware [11]. Much like biological systems, high processing throughput is attained through massive parallelism of slower but energy-efficient analog circuits operating from low-power supplies and operating in subthreshold CMOS [11, 13, 14] where the devices have the highest possible gain and the lowest noise. Our work [15–19] has followed a similar line of thinking.

At the lowest level, gain is provided by MOS transistors operating in the subthreshold region [11, 13, 17]. Since the adopted design methodology depends critically on an in-depth understanding of the underlying device physics, a charge-based transistor model [11, 17] that preserves the symmetry between the source/drain terminals of an MOS transistor is briefly discussed here. Which terminal of the device actually serves as the source or the drain is determined by the circuit, the bias conditions, and even the input signals. This symmetric view of an MOS transistor enabled us to extend the translinear principle to operation in the subthreshold ohmic regime [17]. (See Fig. 4.4.)

The MOS device has a very simple current-charge relationship because diffusion and drift are both proportional to the concentration gradient. As shown in Appendix A of [11] and in [17], as well as in Chapter 2 of this book, this yields a quadratic expression for the current that consists of two independent opposing current components: the source-driven current I_{Q_s} and the drain-driven current I_{Q_d}. These currents are related to charge densities at the source $Q_s^\diamond$ and at the drain $Q_d^\diamond$ of the device.

The device drain–source current can thus be written as

$$I \equiv I_{Q_s} - I_{Q_d} = \mu \frac{W}{L + \left(\frac{\mu}{v_o}\frac{Q_s^{\diamond 2}}{C_{\text{ox}}^\diamond + C_{\text{dep}}^\diamond} + \frac{kT}{q}Q_s^\diamond\right)} \left[\left(\frac{1}{2}\frac{Q_s^{\diamond 2}}{C_{\text{ox}}^\diamond + C_{\text{dep}}^\diamond} + \frac{kT}{q}Q_s^\diamond\right) - \left(\frac{1}{2}\frac{Q_d^{\diamond 2}}{C_{\text{ox}}^\diamond + C_{\text{dep}}^\diamond} + \frac{kT}{q}Q_d^\diamond\right)\right] \tag{4.8}$$

where W is the width, L is the length of the channel, μ is the effective channel mobility, and v_o is the saturation velocity of the carriers. The capacitances $C_{\text{ox}}^\diamond$ and $C_{\text{dep}}^\diamond$ are the gate oxide and depletion area capacitances of the channel. A key property of the MOS device that makes this possible is lossless channel conduction. Unlike a bipolar transistor, the controlling charge on the gate is isolated from the charge in transport by the almost infinite gate-oxide resistance. Therefore, there is no recombination between the current-carrying charge in the channel and the current-modulating charge on the gate.

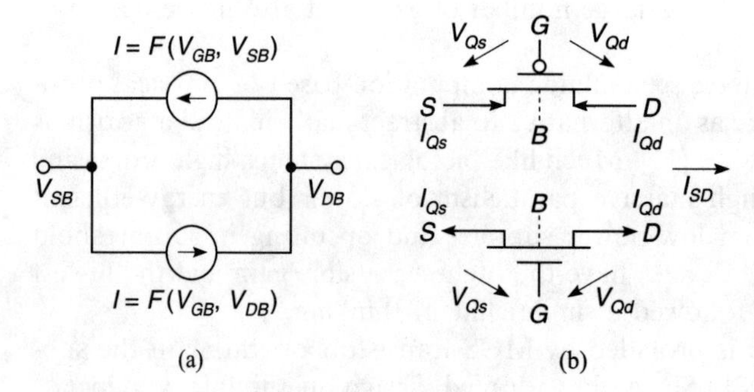

Figure 4.4 Large-signal model for an NMOS transistor (a). Adopted conventions for current decomposition in PMOS and NMOS devices (b).

The familiar ohmic/saturation dichotomy introduced in voltage-mode design can be reformulated in terms of the opposing drain- and source-driven current components. In the saturation region, $|I_{Q_d}| \ll |I_{Q_s}|$; $I \approx I_{Q_s}$; therefore the current is independent of the drain voltage. In the ohmic region, $I_{Q_d} \simeq I_{Q_s}$ and $I = I_{Q_s} - I_{Q_d}$, and therefore the current depends on the drain voltage as well as the source and gate voltages. The functional dependence of the current components on the terminal voltages is fixed and remains the same throughout the ohmic and saturation regions.

The charge densities at the source and drain terminals can be related to the terminal voltages. In general, the charge-voltage relationship is much more complicated than the current-charge one because both the mobile charge and the depletion charge are involved in the electrostatics. The device current in Eq. (4.8) can thus be written as a function $\mathcal{F}$ of the terminal voltages with a general functional form for the current-voltage relationship valid for all the regions of operation given by

$$I_{SD} \propto \mathcal{F}(V_{GB}, V_{SB}) - \mathcal{F}(V_{GB}, V_{DB}) \qquad (4.9)$$

This functional form was first introduced by [34] for above-threshold operation and is also discussed in [35]. For an N-type device, $\mathcal{F}$ is a nonpositive, monotonically increasing function of V_{GB} and a monotonically decreasing function of V_{SB}.

In the subthreshold region, the following factorization of $\mathcal{F}$ is also possible [11, 14]:

$$I \propto \mathcal{G}(V_{GB})[\mathcal{H}(V_{SB}) - \mathcal{H}(V_{DB})] \qquad (4.10)$$

where $\mathcal{G}$ and $\mathcal{H}$ are exponential functions. This shows that the source-driven and drain-driven components are controlled independently by V_{SB} and V_{DB}. However, V_{GB}, acting through the surface potential, also controls both components in a symmetric and multiplicative fashion. In this mode of operation the MOS transistor has been called a *diffusor* [24], in analogy with the variable-conductance electrical junctions in biological systems.

An expression for the current in an NMOS transistor operating in the subthreshold can thus be written [11, 14] as

$$I_D \equiv I_{DS} = I_{n0} S \exp\left(\frac{\kappa_n V_{GB}}{V_t}\right)\left[\exp\left(-\frac{V_{SB}}{V_t}\right) - \exp\left(-\frac{V_{DB}}{V_t}\right)\right] \qquad (4.11)$$

and for a PMOS transistor

$$I_D \equiv I_{SD} = I_{p0} S \exp\left(-\frac{\kappa_p V_{GB}}{V_t}\right)\left[\exp\left(\frac{V_{SB}}{V_t}\right) - \exp\left(\frac{V_{DB}}{V_t}\right)\right] \qquad (4.12)$$

The terminal voltages V_{GB}, V_{SB}, and V_{DB} are referenced to the substrate ("bulk"). The constant I_0 depends on mobility (μ) and other silicon physical properties; S is the width W to length L ratio (W/L) of the device. A plot for the transfer characteristics of MOS transistors in the subthreshold and of a bipolar transistor is shown in Figure 4.5.

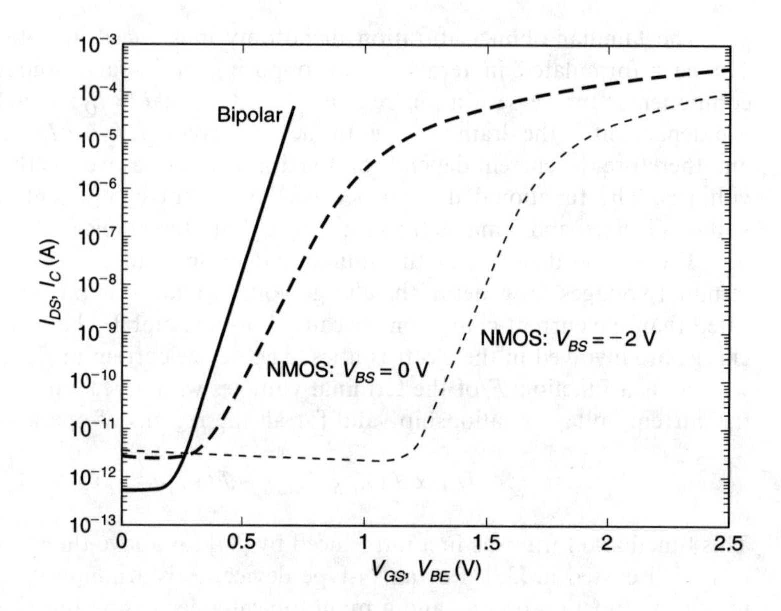

Figure 4.5 Measured current I_{DS} and I_C versus controlling voltage V_{GS} and V_{BE} respectively. The MOS transistor has dimensions of $16 \times 16\,\mu m^2$ and is fabricated in a 1.2-μm N-well CMOS process and biased at a drain-source voltage $V_{DS} = 1.5\,V$. The current is measured at two different substrate voltage bias conditions. The bipolar transistor is a vertical device with an emitter area of $16 \times 16\,\mu m^2$ fabricated in a 2-μm N-well CMOS process and biased with $V_{CE} = 1.5\,V$, $T = 301.5\,K$.

For devices that are biased with $V_{DS} \geq 4\,V_t$ (saturation) (Fig. 4.6), the drain current for an NMOS transistor is reduced to

$$I_{DS} = SI_{n0} \exp\left(\frac{(1-\kappa_n)V_{BS}}{V_t}\right) \exp\left(\frac{\kappa_n V_{GS}}{V_t}\right) + g_d V_{DS} \tag{4.13}$$

and for a PMOS transistor

$$I_D \equiv I_{SD} = SI_{p0} \exp\left[-\left(1 - \frac{\kappa_p V_{BS}}{V_t}\right)\right] \exp\left(-\frac{\kappa_p V_{GS}}{V_t}\right) + g_d V_{DS} \tag{4.14}$$

This shows explicitly the dependence on V_{BS} and the role of the bulk as a *back gate* that underlies this. This equation, having only the dependence on V_{GS} and V_{BS}, is used for circuit designs where devices operate in saturation as transconductance amplifiers. When the device is operating as a controlled current source (saturation), its characteristics are not ideal. The variation of the current with drain voltage results in an output conductance. For a regular MOS transistor

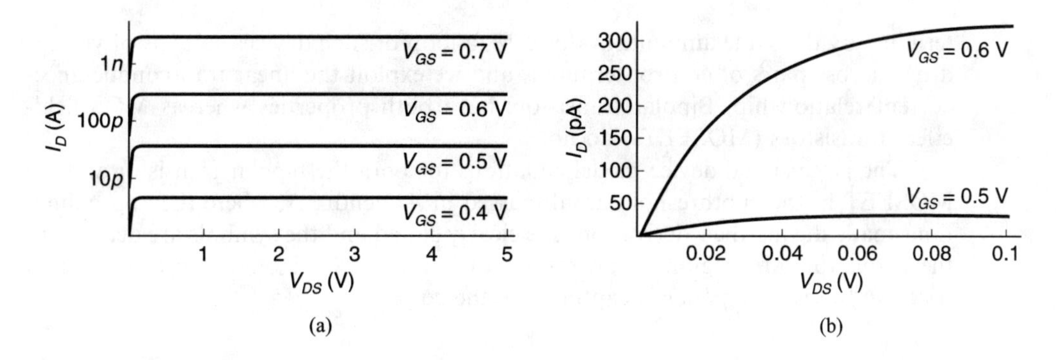

Figure 4.6 Measured output characteristics of NMOS transistors in subthreshold. Device dimensions are $W = L = 16\,\mu\text{m}$. The two sets of data emphasize the good saturation characteristics (a), and the nonlinear ohmic behavior at low drain source voltages (b).

$$g_d \equiv \frac{\partial I_{ds}}{\partial V_d} = \frac{\partial I_{ds}}{\partial L}\frac{\partial L}{\partial V_d} = \left(-\frac{I_{ds}}{L}\right)\left(-\sqrt{\frac{q\varepsilon_{Si}}{2N_A}}\frac{1}{\sqrt{V_{DB}}}\right) \approx \frac{I_{ds}}{V_0} \tag{4.15}$$

The inverse square-root dependence of the depletion layer width is replaced by a nominal value V_0 which has units of volts and is proportional to L. This voltage is analogous to the Early voltage in bipolars and is determined by experiment.

The parameter κ is defined as

$$\kappa = \frac{C_{\text{ox}}^{\diamond}}{C_{\text{ox}}^{\diamond} + C_{\text{dep}}^{\diamond}} \tag{4.16}$$

The physical significance of κ is apparent if the observation is made that the oxide capacitance $C_{\text{ox}}^{\diamond}$ and depletion capacitance $C_{\text{dep}}^{\diamond}$ form a capacitive divider between the gate and bulk terminals that determines the surface potential [11]. Lighter doping reduces $C_{\text{dep}}^{\diamond}$ and pushes the divider ratio closer to unity. A larger surface potential also reduces $C_{\text{dep}}^{\diamond}$. The parameter κ takes values between 0.6 and 0.9.

4.2.1. Translinear Devices

A translinear element is a physical device whose transconductance and current through the device are linearly related, that is, the current is exponentially dependent on the controlling voltage. A two terminal P-N junction (diode), with its exponential I-V characteristics, is a translinear element and used often as an example in circuits [46].

Three-terminal devices are termed "translinear" if the relationship between the current and the controlling voltage is exponential *and* the two terminals across which the controlling voltage is applied exhibit true diodelike behavior, i.e., increasing the voltage on one terminal is exactly equivalent to decreasing the voltage on the other terminal by the same amount. In this case, a loop of such devices consists of voltage drops across pairs of control terminals and we exploit the linear transconductance-current relationship. Bipolar transistors have both properties whereas MOS field-effect transistors (MOSFETs) do not.

The large-signal device model equations for both the bipolar transistor and the MOSFET in the subthreshold are discussed in Appendix A, where the approximations made during their derivations are clearly stated and the symbols are defined. In the active-forward region of operation, the function of a bipolar transistor as a *transconductance amplifier* is captured by the equation

$$I_C = I_S\, e^{(V_B - V_E)/V_t} \tag{4.17}$$

where $V_t = kT/q$ and I_S is defined in Appendix A.

The magnitude of the transconductance from the base is identical to the magnitude of the transconductance from the emitter:

$$g_m \equiv \frac{\partial I_C}{\partial V_{BE}} = -\left.\frac{\partial I_C}{\partial V_E}\right|_{V_B=c} = \left.\frac{\partial I_C}{\partial V_B}\right|_{V_E=c} = \frac{I_C}{V_t} \tag{4.18}$$

We now contrast the operation of a bipolar transistor as a translinear element with that of an MOS transistor operating in the subthreshold. Much like a bipolar transistor, the MOSFET in the subthreshold has exponential voltage-current characteristics (see Fig. 4.5). There are, however, two fundamental differences between MOSFET and bipolar devices that have implications in the design of translinear circuits:

1. Unlike a bipolar transistor, the current in a MOSFET is controlled by the surface potential, which is capacitively coupled to the gate (front-gate) and bulk (back-gate) terminals.
2. The MOSFET is symmetric with respect to the source and drain terminals while a bipolar is not.

In summary, the MOS transistor is a *four*-terminal device with symmetric drain and source terminals, a result of lossless channel conduction, and an isolated control potential set by one or more control gates. As we will see in subsequent sections, the latter property of the MOS transistor is a mixed blessing when the design of translinear circuits is considered.

It should be pointed out that the voltage difference that controls the current in a MOSFET to yield the translinear behavior is the potential difference between the channel surface potential ψ_s and the potential at the source V_S and/or drain V_D so that the current between the drain and source for an NMOS is given by

$$I_{DS} \sim \left[\exp\frac{(\psi_s - V_S)}{V_t} - \exp\frac{(\psi_s - V_D)}{V_t}\right] \tag{4.19}$$

Since the MOS transistor has two "gates," the relationship between $\psi_s(V_B, V_G)$ and the bulk or gate terminal voltages V_B and V_G can be obtained using the simple capacitive divider model depicted in Figure 4.7. The introduction of the parameter $\kappa \equiv C^{\diamond}_{\text{ox}}/(C^{\diamond}_{\text{ox}} + C^{\diamond}_{\text{dep}})$ is convenient for modeling the effect of the two gates. Note that κ is a function of the surface potential ψ_S as $C^{\diamond}_{\text{dep}}$ is a function of the applied gate and substrate voltages.

In saturation (i.e., when $V_{DS} \geq 4V_t$) and when current-controlling voltages are referenced to source, Eq. (4.19) can be rewritten as

$$
\begin{aligned}
I_{DS} &= SI_{n0} \exp\left(\frac{C^{\diamond}_{\text{dep}}}{C^{\diamond}_{\text{ox}} + C^{\diamond}_{\text{dep}}} \frac{V_{BS}}{V_t}\right) \exp\left(\frac{C^{\diamond}_{\text{ox}}}{C^{\diamond}_{\text{ox}} + C^{\diamond}_{\text{dep}}} \frac{V_{GS}}{V_t}\right) \\
&= SI_{n0} \exp\left(\frac{(1-\kappa_n)V_{BS}}{V_t}\right) \exp\left(\frac{\kappa_n V_{GS}}{V_t}\right)
\end{aligned}
\tag{4.20}
$$

Equation (4.20) can be rewritten as a function of dimensionless current quantities i_G and i_B. Each of these currents would correspond to the device current if the surface potential ψ_S could assume the voltage at the gate or bulk terminal. In essence these currents correspond to ideal diode junctions between the source and surface potential weighted by the appropriate capacitive divider ratio. Therefore, the equation for the drain current can be written as

$$
I_{DS} \equiv SI_{no}\, i_G^{\kappa}\, i_B^{(1-\kappa)} \tag{4.21}
$$

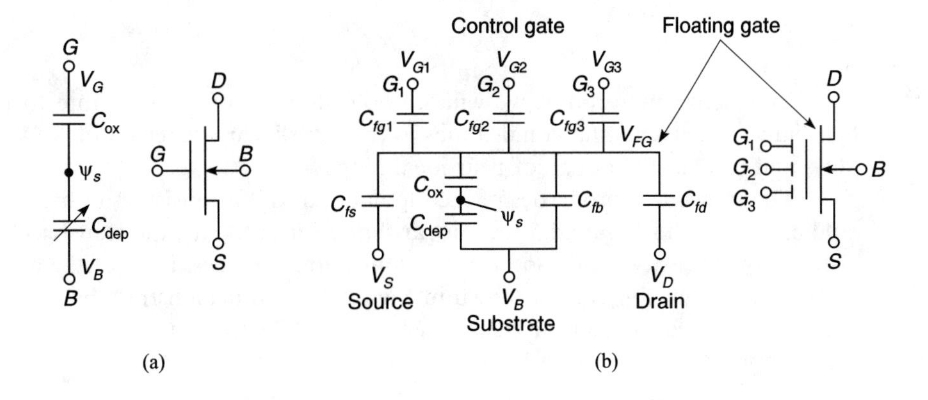

Figure 4.7 (a) Symbol and simple capacitance model for an NMOS transistor. The current between the drain and source is controlled by the difference between the surface potential ψ_s and the potential at the source terminal. The surface potential ψ_s is set by the potential at the gate and bulk terminals through the capacitive divider between C_{ox} and C_{dep}. (b) Symbol and capacitance model for a floating-gate MOS (FGMOS) transistor. The device depicted in this figure has three control gates: G_1, G_2, and G_3.

In subsequent sections, we will see how the latter formulation facilitates the analysis of MOS translinear circuits and an extension of it will be used to analyze FGMOS translinear loops. Since the dimensionless quantities are related to the surface potential, they will be called ψ-currents or *psi-currents*.

The transconductance from the gate is given by

$$g_m \equiv \left.\frac{\partial I_{DS}}{\partial V_G}\right|_{V_b, V_S=c} = \frac{\kappa I_{DS}}{V_t} \tag{4.22}$$

and from the local substrate (back-gate) terminal

$$g_{mb} \equiv \left.\frac{\partial I_{DS}}{\partial V_B}\right|_{V_G, V_S=c} = \frac{(1-\kappa) I_{DS}}{V_t} \tag{4.23}$$

The conductance g_s at the source is given by

$$g_s \equiv \left.\frac{\partial I_{DS}}{\partial V_S}\right|_{V_G, V_B=c} = \frac{I_{DS}}{V_t} \tag{4.24}$$

The transconductances depicted in Eqs. (4.22)–(4.24) are linear functions of the current—to a first order—and hence each MOS transistor in saturation has the equivalent of three different translinear elements. Note that the source transconductance is equal to the gate transconductance by shorting the local substrate and the source of the transistor ($V_B = V_S$), in which case Eq. (4.20) for the current becomes

$$I_{DS} = SI_{n0}\, e^{\kappa_n (V_G - V_S)/V_t} \tag{4.25}$$

In subsequent sections, we will see how shorting of the substrate to the bulk partially circumvents the nonidealities in the translinear properties of MOS transistors and enables the design of near-ideal loops.

The translinear properties of the bipolar and MOS transistors in the subthreshold are evident in Figure 4.5. In a logarithmic current scale, the transfer characteristics show linearity with respect to the controlling voltages. Plots of the normalized transconductance (g_m/I) are shown in Figure 4.8 and demonstrate how the bipolar device is an ideal translinear element while the MOS transistor in the subthreshold only approximates it over a limited range.

4.2.2. Floating-Gate MOS Transistor

In this section the MOS device model is extended to describe floating-gate MOS transistors (FGMOSs). The FGMOS is an important element in the design of micropower neuromorphic systems. It can be used for long-term nonvolatile storage of neural network model parameters [36–39] and for compensating parametric variations in device characteristics [40–42]. The floating gate can also be employed as a

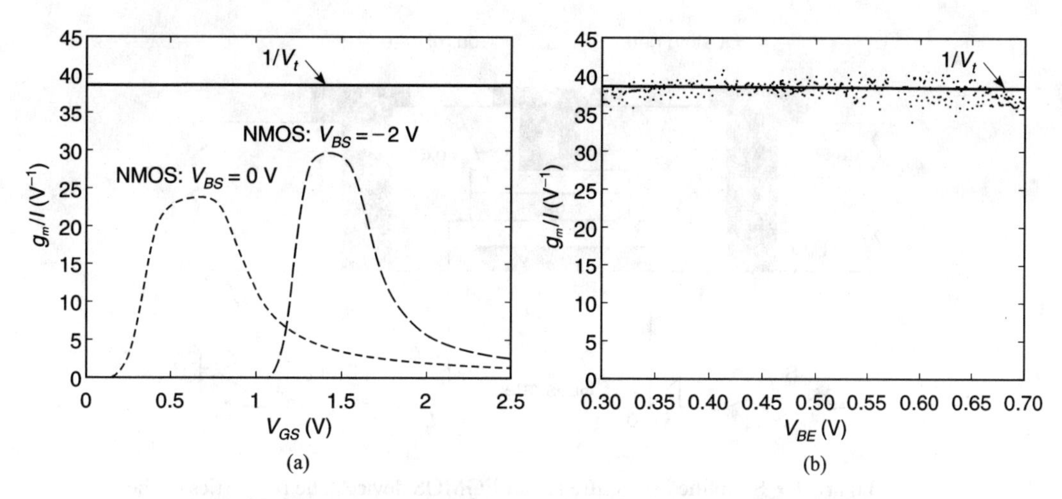

Figure 4.8 Normalized transconductance curves. The transconductance is computed through numerical differentiation of the data in Figure 4.2 and subsequent smoothing: (a) for the MOS transistor; (b) for the bipolar transistor. The dramatic decrease of the transconductance at low gate-source voltages is attributed to the leakage current.

summing node to perform mathematical operations in the charge domain with high linearity [43].

In an FGMOS transistor fabricated in a standard double-polysilicon CMOS process, the floating gate is first polysilicon and the control gate is second polysilicon (see Fig. 4.9). The floating gate controls the current in the channel beneath it and is capacitively coupled to the control gate above it. The voltage on the floating gate is determined by the amount of charge deposited on it as well as the voltages on the control gate, drain, source, and substrate.

Several methods have been reported in the literature for metering charge onto or off a floating gate. For a literature survey and reference to key papers in the field, refer to [17].

The best way to incorporate a floating-gate transistor as part of a system is in a *closed-loop* configuration. The relaxation phenomena in the different dielectrics [44] introduce hysteresis in the device characteristics. The time constants depend on the details of the manufacturing process. Thus open-loop control of the charge on the floating gate may be difficult.

4.2.3. Large-Signal FGMOS Model

The large-signal model of a floating-gate MOS transistor can be derived by first obtaining the voltage on the floating gate as a function of the voltages on the nodes that are capacitively coupled to it. Then, the large-signal FGMOS equation is obtained by substituting this voltage for V_{GB} in the equation for the regular MOS transistor. The resulting model is not much different except for two extra terms

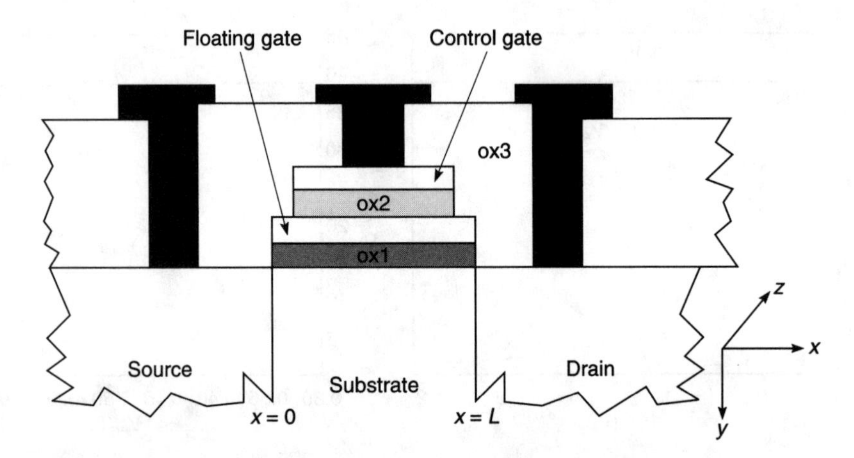

Figure 4.9 Simplified structure for an FGMOS device. The properties of the different dielectric layers (ox1, ox2, and ox3) that surround the two polysilicon gates depend on the particulars of the manufacturing process. Injecting charge on the floating gate involves current flow through these dielectric layers, and thus programming the device is highly process dependent.

arising from capacitive coupling to the source and drain. This effect is significant and cannot be ignored in the design of analog circuits using FGMOS transistors.

The voltage on the floating gate of the FGMOS can be calculated with the aid of the simple circuit model in Figure 4.7(b). If transient oxide interface states are ignored, the voltage on the floating gate is given by

$$V_{FGB} = \frac{C_{fcg}V_{CGB} + C_{fs}V_{SB} + C_{fd}V_{DB} + C_{\text{ox}}\psi_s + Q_{fg}}{C_{\text{sum}}} \tag{4.26}$$

where

$$C_{\text{sum}} = C_{fcg} + C_{fs} + C_{fd} + C_{fb} + C_{\text{ox}}$$

Q_{fg} is the charge on the floating gate, C_{fcg} is the capacitance between the control gate and the floating gate, C_{fs} and C_{fd} are the capacitances between the floating gate and source and drain, respectively, C_{ox} is the oxide capacitance between the floating gate and the channel, and C_{fb} is the capacitance between the floating gate and the substrate along the edge of the channel. Here, V_{CGB} is the control gate voltage, V_{FGB} is the floating-gate voltage, V_{SB} is the source voltage, V_{DB} is the drain voltage, and ψ_s is the channel potential, all referenced to the bulk.

The surface potential ψ_s is eliminated using the kappa approximation (see Chapters 2 and 3) and an expression for V_{FGB} is obtained in terms of the capacitances and the terminal voltages,

$$V_{FGB} = \frac{C_{fcg}V_{CGB} + C_{fs}V_{SB} + C_{fd}V_{DB} + C_{ox}(1.5\phi_F + V_{SB} - \kappa V_{GB}^*) + Q_{fg}}{C_{sum}^*} \tag{4.27}$$

where $C_{sum}^* = C_{sum} - \kappa C_{ox}$.

Equation (4.27) yields the following equation for the current as a function of the terminal voltages:

$$I = SI_{fn0}\, e^{\zeta_C V_{CGB}/(kT/q)}\, e^{(\zeta_S V_{SB} + \zeta_D V_{DB})/(kT/q)}\, (e^{-V_{SB}/(kT/q)} - e^{-V_{DB}/(kT/q)}) \tag{4.28}$$

where I_{fn0} is given by

$$I_{fn0} = I_{n0} \exp\left(\frac{\kappa Q_{fg} + \kappa C_{ox}(1.5\phi_F + V_{SB} - \kappa V_{GB}^*)}{C_{sum}^* kT/q}\right) \tag{4.29}$$

and $\zeta_C \equiv \kappa C_{fcg}/C_{sum}^*$, $\zeta_S \equiv \kappa C_{fs}/C_{sum}^*$, and $\zeta_D \equiv \kappa C_{fd}/C_{sum}^*$. The capacitive divider between the control gate and the floating gate reduces its loglinear slope coefficient ζ_C. Note that the current is also exponentially dependent on the drain and source voltages with loglinear slope coefficients of ζ_S and ζ_D due to capacitive coupling to the fringes of the floating gate.

Ideally, we would like to have $\zeta_C = 1$, $\zeta_S = \zeta_D = 0$. This may be achieved by adding some extra capacitance between the control and floating gates. The floating-gate charge, Q_{fg}, is the only variable that can be changed after the chip is fabricated; it changes I_{fn0} and shifts the I-V curve.

The output conductance of an FGMOS transistor has an additional term due to capacitive coupling between the drain and the floating gate. This term is exponentially related to the drain voltage, with slope coefficient ζ_D. The output conductance is therefore given by

$$g_d^* = g_d + \zeta_D \frac{I_{ds}}{kT/q} \tag{4.30}$$

The extra term increases the output conductance significantly, and therefore, cascoding the FGMOS is strongly advised.

4.3. CURRENT-MODE APPROACH

The VLSI system considerations and requirements for a high degree of integration make the subthreshold MOS transistor the device of choice for complex neuromorphic systems. Subthreshold operation offers the highest processing rates per unit power. Current-mode (CM) operation yields large dynamic range, simple and elegant implementations of both linear and nonlinear computations, and low-power dissipation without sacrificing speed.

By taking advantage of the high subthreshold transconductance per unit current, voltage swings are kept to a few thermal voltages, and reasonable processing bandwidths are achieved. Dynamic power dissipation and supply noise are also reduced as a result of the smaller voltage swings. Smaller voltage swings eliminate the current that is wasted in charging and discharging parasitic capacitances, thereby allowing us to use smaller current signals and cut quiescent power dissipation as well. Thus, this approach yields relatively fast analog circuits with power dissipation levels compatible with future trends in system integration. Fast digital circuits can also be designed using source-coupled logic gates and current steering (ECL-like circuits).

The essence of CM VLSI signal processing is signal representation and normalization in terms of a unit current. Since signals are represented by currents, the CM approach enables the design of systems that function over a wide dynamic range. The low end is limited by leakage currents in the junctions and by noise in the system; the high end is limited by degrading transconductance per unit current above threshold. Large dynamic range is essential in neuromorphic systems that receive inputs from real-world environments (e.g., silicon retinas). The value of the unit current ultimately determines the overall power dissipation and controls the temporal response of the system. Thus, power consumption can be managed and *adaptively* controlled to satisfy the temporal response requirements of the system.

We now introduce the *current conveyor* concept to facilitate the analysis and synthesis of minimal complexity CM circuits.

4.3.1. Current Conveyor

Sedra and Smith originated the notion of a current conveyor [67]—a hybrid voltage/current three-port device (Fig. 4.10). It is a versatile building block for analog signal-processing applications. The current conveyor is characterized by small-signal relationships between the voltages and currents at different nodes.

The design methodology presented here employs the concept of the current conveyor as a canonical element with minimal components that can be employed to aid the synthesis of analog VLSI systems. In analogy with the operational amplifier, there is a virtual short for voltage from node Y to node X and a virtual short for current from node X to node Z. Therefore, node X can transmit a current signal to node Y while simultaneously receiving a voltage signal from node Y. This original current conveyor implementation (CCI) has been exploited effectively in linear analog LSI circuit design [71]. The high functionality embodied in this very simple, yet elegant circuit makes it a good candidate for analog VLSI systems where it satisfies

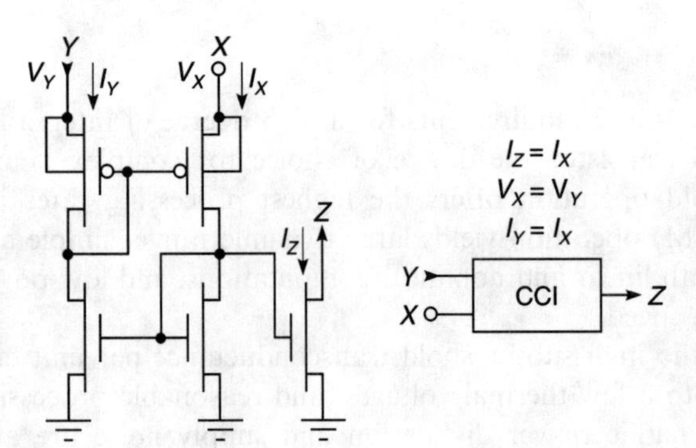

Figure 4.10 Original Sedra and Smith five-transistor (5T) current-conveyor circuit (CCI) implemented using MOS transistors. In this conveyor, Y is a hybrid voltage input/current output node, X is a hybrid voltage output/current input node, and Z is a current output node.

the need for large fan-in and large fan-out. In the context of the CM systems described here, node X is used as a *communication* node, node Y as a *control* node, and node Z as an *output* node.

Although the original implementation of the current conveyor concept used five transistors, it can in fact be realized with just one (the 1T form). An MOS transistor, in saturation, can transfer a current from its high-conductance source terminal to its low-conductance drain terminal or a voltage from its high-impedance gate terminal to its low-impedance source terminal. These two actions can be exploited to simultaneously achieve a voltage short from gate to source and a current short from source to drain [see Fig. 4.11(a)]. This dual role, obtained with a single device, captures the essence of the current conveyor. As such, the gate is the control voltage (Y) node, the drain is the output current (Z) node, and the source is the hybrid input current/output voltage communication (X) node. In this minimalist implementation, we forsake the redundant current output at node Y and introduce a nominally constant voltage offset between nodes X and Y.

The large-signal behavior for the 1T current conveyor is described by the following set of equations:

$$I_Z = I_X \qquad V_X = \kappa V_Y - \frac{kT}{q}\ln\left(\frac{I_X}{SI_0}\right) \tag{4.31}$$

assuming the drain conductance is zero. Note that, unlike Sedra's CCI circuit, the 1T conveyor is highly nonlinear. However, the compressive log function does not seriously degrade the voltage-following property. The small-signal behavior of the 1T conveyor is due to the combined action of a single-transistor voltage follower and a single-transistor current buffer, realized by the same device. That is, if the current source at node X has an output conductance g_i, then the output conductance at node Z is approximately g_i/A_{v_0} and the input conductance at node X is g_m, where A_{v_0} is the intrinsic gain of the device and g_m is its transconductance.

A two-transistor (2T) current-controlled current conveyor is shown in Figure 4.11(b). This circuit has a hybrid current input/voltage output communication node X, a current input control node Y, and a current output node Z. The voltage V_X at the communication node X is determined by the current supplied to node Y, instead

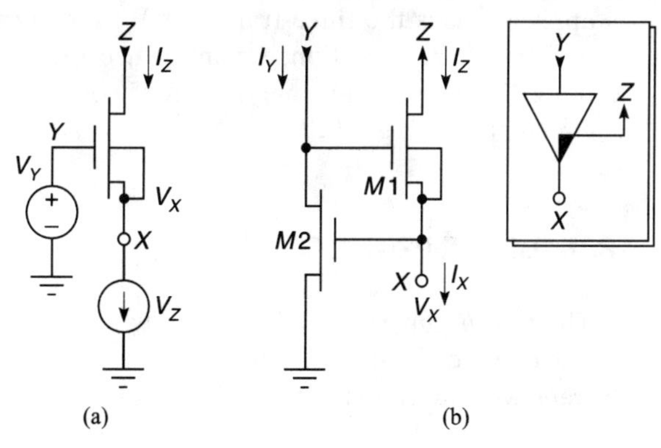

Figure 4.11 (a) One-transistor (1T) current conveyor and (b) two-transistor (2T) current-controlled current conveyor circuit and conveyor symbol. The white triangle represents the voltage buffering action from Y to Z while the in-laid black triangle represents the current buffering action from X to Z.

of using a control voltage. The current-controlled conveyor's large-signal behavior is described by the equations

$$
\begin{aligned}
I_Z &= I_X \\
V_X &= \frac{kT}{q}\frac{1}{\kappa_2}\ln\left(\frac{I_Y}{S_2 I_0}\right) \\
V_Y &= \frac{kT}{q}\frac{1}{\kappa_2\kappa_1}\ln\left(\frac{I_Y}{S_2 I_0}\right) + \frac{kT}{q}\frac{1}{\kappa_1}\ln\left(\frac{I_X}{S_1 I_0}\right)
\end{aligned}
\tag{4.32}
$$

The voltage buffering between nodes Y and X has been replaced with a generalized transimpedance (M_2) that generates a voltage V_X proportional to $\log(I_Y)$. As will be seen in a later section, in a current-mode circuit design style where all signals are implicitly represented by "$\log I$" voltages, this nonlinearity is transparent. The logging action is achieved by diode-connecting M_2 using the voltage-following action of M_1 from node Y to node X. Here, M_1 sets V_X so as to make M_2's current equal to I_Y; M_2 inverts and amplifies small changes in V_X to generate the requisite voltage at node Y.

Compared to the 1T conveyor, the small-signal conductance at the current output node (Z) is lower by a factor of A_{v_0} and the small-signal conductance at the communication node (X) is increased by the same factor. More specifically, the conductance seen at node Z is approximately $g_i/(A_{v_1}A_{v_2})$; i.e., the buffering action of M_1 is improved by the gain A_{v_2} of M_2. The conductance seen at node X is approximately $A_{v_2}g_{m_1}$, i.e., the source conductance of M_1 times M_2's gain.

The 2T current-controlled conveyor was used in a two-way communication scheme for a current-mode, clamped bit-line, associative memory design [15, 69]. In that application, the dual role of the 2T current-controlled current conveyor was fully exploited to simultaneously fan in currents and fan out voltage over the same wire. Säckinger independently proposed this arrangement to increase the output impedance of an MOS transistor current source, as we have shown here. He used this circuit in a high-performance op-amp design and dubbed it the *regulated cascode* [70]. Indeed, the negative-feedback arrangement used in the 2T conveyor is not new and is well known to veteran analog circuit designers. One of its most elegant applications is the three-transistor Wilson current mirror [72].

In the next section, we introduce the translinear principle, which provides a powerful design and analysis framework for linear and nonlinear computational circuits in analog VLSI.

4.4. TRANSLINEAR PRINCIPLE

The *translinear principle* [21] exploits the exponential current-voltage nonlinearity in semiconductor devices and offers a powerful circuit analysis and synthesis [45] framework. Originally formulated for bipolar transistors [21], this principle enables the design of analog circuits that perform complex computations in the current domain, including products, quotients, and power terms with fixed exponents [21, 45]. Translinear circuits perform these computations without using differential voltage signals and are amenable to device-level circuit design methodology.

In this section, we provide an overview of nanopower translinear circuit design using MOS transistors operating in the subthreshold region. We contrast the bipolar and MOS subthreshold characteristics and extend the translinear principle to the subthreshold MOS ohmic region through a drain/source current decomposition. A front/back-gate current decomposition is adopted; this facilitates the analysis of translinear loops, including multiple-input FGMOS transistors. Circuit examples drawn from working systems designed and fabricated in standard digital CMOS oriented processes are used as vehicles to illustrate key design considerations, systematic analysis procedures, and limitations imposed by the structure and physics of MOS transistors. This performs phototransduction and amplification, edge enhancement, and local gain control at the pixel level.

Most of the work on translinear circuits to date use bipolar transistors and the emphasis is on high precision and high speed. One fascinating aspect of translinear circuits is their insensitivity to isothermal temperature variations, though the currents in their constitutive elements (the transistors) are exponentially dependent on temperature. The effect of small local variations in fabrication parameters can also be shown to be temperature independent. An excellent up-to-date overview of translinear current-mode analog circuits using bipolar transistors can be found in [46].

The increased commercial interest in analog CMOS LSI and VLSI has renewed interest in the translinear principle for MOS circuit design. A generalized form of the translinear principle was recently proposed for MOSs operating above threshold [47]; this extension, however, does not follow the original definition of a translinear circuit [21]. This extension is simply a design principle that exploits conservation of energy [Kirchhoff's voltage law (KVL)] around circuit loops which have specific topological properties. A novel class of translinear circuits that employs multiple-input gates with FGMOS transistors in the subthreshold, has been recently proposed and experimentally demonstrated [49]. Another exciting research area that emerged in the last few years is the synthesis of analog VLSI for sensory information processing, which is the focus of discussion in this section.

In the next two sections we discuss translinear circuits that employ translinear elements, both MOSs operating in the subthreshold and bipolar transistors. We follow the convention proposed by Barrie Gilbert in [46] and make a distinction between a translinear loop (TL) and a translinear network (TN). We begin with translinear loops.

4.4.1. Translinear Loops

In "strictly" TLs the translinear principle [21] can be stated as follows:

In a closed loop containing an equal number of oppositely connected translinear elements, the product of the current densities in the elements connected in the

ClockWise (CW) direction is equal to the corresponding product for elements connected in the Counter ClockWise (CCW) direction.

As an example let us consider the circuit of Figure 4.12(a) consisting of four *ideal* diodes in the loop *X-Z-Y-W-X*. Following the translinear principle, we can write

$$\frac{\Pi_{\mathrm{CCW}} J}{\Pi_{\mathrm{CW}} J} = 1 \quad \text{or} \quad J_2 J_4 = J_1 J_3 \tag{4.33}$$

Note that the translinear principle is derived by beginning with KVL or the principle of conservation of energy, so that

$$\sum_{i=1}^{i=N/2} V_{D(2i-1)} - \sum_{i=1}^{i=N/2} V_{D(2i)} = 0 \tag{4.34}$$

Equation (4.33) follows from Eq. (4.34) if the voltages are summed around loops of translinear devices.

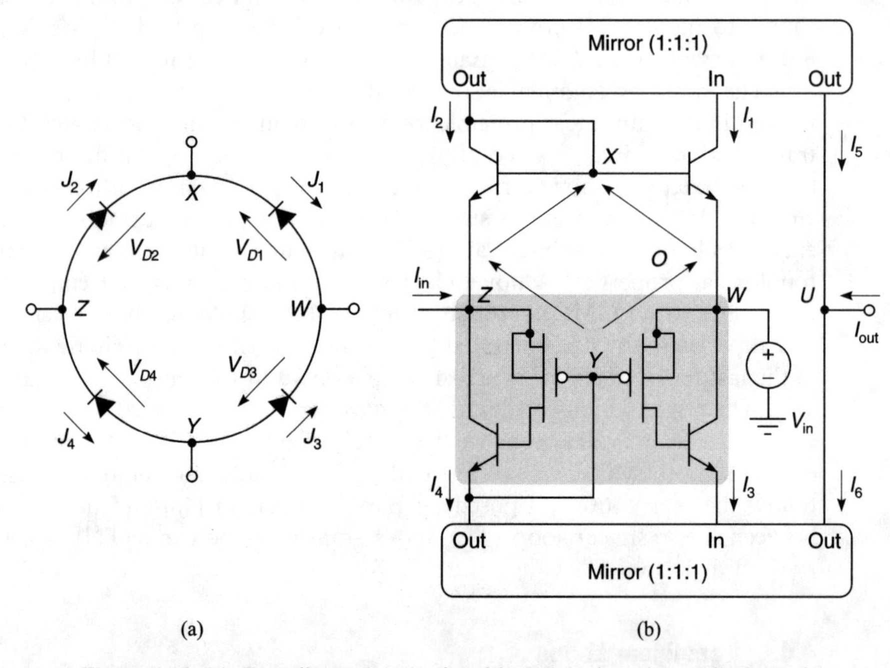

Figure 4.12 (a) Translinear loop using ideal P-N junctions (diodes). (b) Translinear loop of the circuit (left) implemented using composite bipolar and subthreshold MOS transistors. The loop is employed in a current conveyor configuration where the bidirectional output current I_{out} equals to the bidirectional current I_{in}.

In a circuit graph composed of two terminal elements such as ideal diodes [see Fig. 4.12(a)], there is a direct relationship between the voltage difference among each pair of nodes transversed by the translinear device and the current in the arc that joins the nodes. This is a consequence of having the voltage nodes that control the current be the same as the current output nodes of the device. In practical systems, the ideal diodes in Figure 4.12(a) would correspond to base-emitter junctions of bipolar transistors with shorted collector-base terminals.

Analogous behavior can be obtained using translinear three-terminal devices such as bipolar transistors, MOS transistors in the subthreshold, or any other device that yields diodelike characteristics. However, in three-terminal devices, the diode control nodes in the circuit need not correspond to the current path. In bipolar transistors the diode control nodes are available and thus can be used explicitly in constraint equations such as Eq. (4.34). This is not true for MOS transistors! As we have seen already, one of the diode control nodes (namely the node corresponding to the surface potential ψ_s) is not *directly* accessible. The situation becomes even more complex in MOS transistors with a floating gate (FGMOS) (see Fig. 4.9) coupled to one or more controlling gates (see [48, 49] and references therein). At first sight, the floating gate appears to make the situation worse, but actually it opens the possibility for a new class of translinear circuits proposed and experimentally demonstrated recently by Minch et al. [49].

Essentially the physical structure of FGMOS transistors offers an extra degree of freedom which can be exploited systematically through another set of constraint equations of the form

$$V_{FGi} = \frac{Q_{FGi}}{C_{Ti}} + \sum_{j=1}^{j=N} \Lambda_{ij} V_{Gij} \tag{4.35}$$

where V_{FGi}, Q_{FGi} are the floating-gate voltage and charge on the ith transistor and V_{Gij} is the voltage of the jth control gate. The total capacitance seen in the floating gate is C_{Ti} and Λ_{ij} is a design parameter that depends on the ratio of the control gate to floating-gate capacitance, i.e., $\Lambda_{ij} \equiv C_{fgi}/C_{Ti}$. The details of a systematic analysis procedure for FGMOS translinear circuits can be found in [49].

Analysis of Translinear Circuits with MOS Transistors in Saturation. The current mirror is a trivial example of a translinear circuit. It has a single loop with two translinear elements, one CCW and the other CW.

Two current mirrors implemented with complementary devices and connected back to back yield the circuit shown in Figure 4.12(b). This loop includes four three-terminal devices and corresponds to the ideal diode example of Figure 4.12(a). The circuit can be readily recognized as a BiCMOS implementation of an AB stage in a digital oriented CMOS process where only one type (NPN) of bipolar transistor is available [52]. A composite structure made of an MOS in the subthreshold and an NPN bipolar yields a pseudo-PNP device with good driving capabilities. Translinear loops using both PNP and NPN bipolar transistors were first studied by Fabre [50].

Applying the translinear principle to the loop X-Z-Y-W-X of Figure 4.12(b) yields the following constraint equation for the currents in the circuit:

$$I_2 I_4 = I_1 I_3 \tag{4.36}$$

This classical four-junction loop can be combined with two current mirrors to implement a current conveyor [51] where $I_{out} = I_{in}$ and $V_Z = V_{in}$.

Our second example is the MOS transistor one-quadrant multiply-divide circuit shown in Figure 4.13(a). A large number of these CMOS multipliers have been employed in the implementation of a correlation-based motion-sensitive silicon retina [53].

Applying the translinear principle to the loop GND-A-B-C-GND, we find a total of four equivalent diode junctions and obtain

$$I_1 I_2 = I_3 I_4 \quad \text{or} \quad I_4 = \frac{I_1 I_2}{I_3} \tag{4.37}$$

The above relationship can also be derived by summing the voltages around the loop (conservation of energy):

$$V_1 + V_2 - V_3 - V_4 = 0$$

Replacing the gate-source voltages for M_1, M_2, M_3, M_4 with their respective drain-source currents using Eq. (4.25) (assuming all devices are in saturation, have

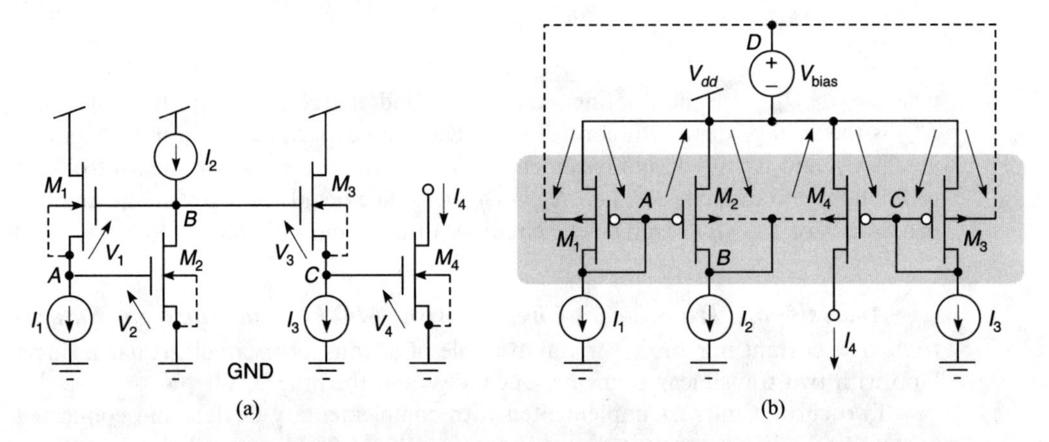

Figure 4.13 (a) Translinear circuit that performs one-quadrant normalized multiplication; I_1, I_2, and I_3 are the inputs and I_4 is the output. (b) Four-transistor translinear circuit that performs a one-quadrant normalized multiplication and exploits the back gate in an MOS transistor. Device pairs M_1, M_3 and M_2, M_4 share local substrate terminals (in this case N-wells); I_1, I_2, and I_3 are the inputs and I_4 is the output.

$V_{SB} = 0$, have negligible drain conductance, have identical κ, and have identical I_0 and geometry S), we obtain

$$\frac{kT}{\kappa q}\left[\ln\left(\frac{I_1}{S I_0}\right) + \ln\left(\frac{I_2}{S I_0}\right) - \ln\left(\frac{I_3}{S I_0}\right) - \ln\left(\frac{I_4}{S I_0}\right)\right] = 0$$

or

$$\ln\left(\frac{I_1 I_2}{S I_0}\right) = \ln\left(\frac{I_4 I_3}{S I_0}\right)$$

from which Eq. (4.37) readily follows. Note that the assumption of identical κ holds true to a first order because $V_{SB} = 0$ and the gate voltages of all transistors are within a few hundred millivolts from each other.

Yet another way of viewing the function of this circuit is that of a *log-antilog* block. Transistors M_1 and M_2 do the *log-ing*, M_4 does the *antilog-ing*, and M_3 is a level shifter.

Another single-quadrant multiplier is shown in Figure 4.13(b). This circuit was proposed and its function experimentally demonstrated in [58]. The operation of the circuit can be understood by noting that a single transistor (M_4) can perform a single-quadrant multiplication because the voltages on the gate and bulk control the current in a multiplicative fashion [see Eq. (4.20)]. Since in the subthreshold the transistors saturate at only a few V_t of drain–source voltage, the bulk terminal of the device can be connected to the drain without turning on the bulk-source junction.

An expression for the output current I_4 can be obtained by applying the translinear principle around the four loops (V_{dd}-A-V_{dd}), (V_{dd}-B-V_{dd}), (V_{dd}-C-V_{dd}), (V_{dd}-D-V_{dd}) to obtain the following equations for the *psi-currents* introduced in Eq. (4.21):

$$i_{G1} = i_{G2} \qquad i_{B2} = i_{B4} \qquad i_{G4} = i_{G3} \qquad i_{B3} = i_{B1} \tag{4.38}$$

The actual currents in the four MOSFETs M_1, M_2, M_3, M_4 can be written as a function of the psi-currents

$$\begin{aligned} I_1 &= I_{DS1} = SI_{n0}\, i_{G1}^{\kappa_1}\, i_{B1}^{(1-\kappa_1)} \\ I_2 &= I_{DS2} = SI_{n0}\, i_{G2}^{\kappa_2}\, i_{B2}^{(1-\kappa_2)} \\ I_3 &= I_{DS3} = SI_{n0}\, i_{G3}^{\kappa_3}\, i_{B3}^{(1-\kappa_3)} \\ I_4 &= I_{DS4} = SI_{n0}\, i_{G4}^{\kappa_4}\, i_{B4}^{(1-\kappa_4)} \end{aligned} \tag{4.39}$$

where the devices have been assumed to have the same S and I_{n0}. If now the assumption is made that $\kappa_1 = \kappa_2 = \kappa_3 = \kappa_4$, Eqs. (4.38) and (4.39) yield the following expression for the output current I_4 in terms of the input currents I_1, I_2, and I_3:

$$I_1 I_4 = I_2 I_3 \quad \text{or} \quad I_4 = \frac{I_2 I_3}{I_1} \tag{4.40}$$

In the original implementation [58], it was suggested that, to improve accuracy, the voltage on the local substrate (N-well) of devices M_3 and M_1 should be set at a value close to that of node B, the local substrate of devices M_2 and M_4. This is indeed necessary, as the bulk voltage determines κ of all transistors, which was assumed to be the same for all transistors. Another implicit assumption here is that the gate voltage is approximately the same for all transistors.

Our next example addresses the problem of converting a bidirectional current to two unidirectional currents, which is equivalent to a current-mode half-wave rectification. A translinear circuit that computes this nonlinear function is shown in Figure 4.14(a). The bidirectional current I_{BD} is steered through transistor M_3 when $I_{BD} > 0$ and through transistors $M_{4,5}$ when $I_{BD} < 0$. Concentrating on transistors $M_{1,2,3,4}$, we identify a loop (V_{DD}-A-B-C-V_{DD}) and apply the translinear principle to yield

$$I_B I_B = I_1(I_1 - I_{BD})$$

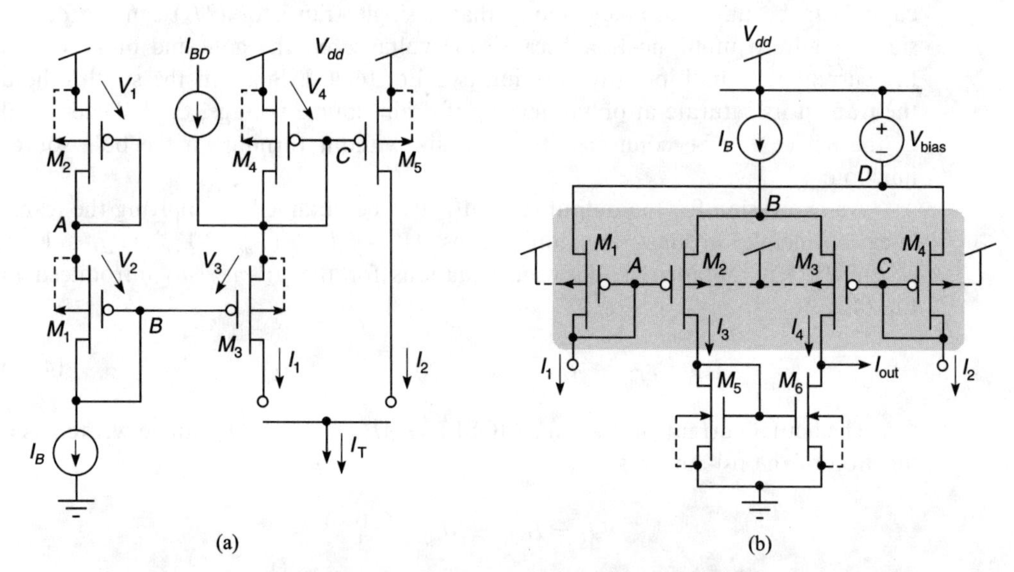

Figure 4.14 (a) Circuit that converts a bidirectional current on a single wire into two unidirectional currents on separate wires. This is a current-mode absolute-value circuit. The sign of the bidirectional input current is assumed to be positive when it adds positive charge to node C. The bidirectional current I_{BD} is the input, the unidirectional currents I_1 and I_2 are the outputs, and I_B sets the operating point of the circuit. (b) A translinear circuit that computes the normalized difference of two current signals; I_1 and I_2 are the inputs and the bidirectional current I_{out} is the output normalized to $I_1 + I_2$.

A second equation is obtained from the trivial loop V_{DD}-C-V_{DD}:

$$I_2 = I_1 - I_{BD}$$

which can also be obtained by observing that transistors M_4 and M_5 form a simple current mirror.

These equations may be solved for I_1 and I_2 in terms of I_B and I_{BD}:

$$I_1 = \tfrac{1}{2}\left(I_{BD} + \sqrt{4I_B^2 + I_{BD}^2}\right) \qquad I_2 = \tfrac{1}{2}\left(-I_{BD} + \sqrt{4I_B^2 + I_{BD}^2}\right) \tag{4.41}$$

which shows that I_1 is equal to the rectified value of I_{BD} and $I_2 \simeq 0$ when $I_{BD} \gg I_B$ and vice versa. The absolute value is obtained by connecting the two output wires together, in which case

$$I_T \equiv I_1 + I_2 \simeq |I_{BD}| \quad \text{if} \quad I_{BD} \gg I_B$$

This circuit has been employed in a CMOS integration of an autoadaptive linear recursive network for the separation of sources [23].

The next translinear circuit performs a current ratio computation. This functional block is part of the readout amplifier in an analog VLSI system that integrates monolithically a one-dimensional array of photodiodes and selective polarization film to form a *polarization contrast* retina [54].

The simple translinear circuit in Figure 4.14(b) is excellent for rescaling differential current signals and thus computing the contrast. Currents I_1 and I_2 represent those from two selected photodiodes. The heart of the computation circuit will be recognized as a Gilbert gain cell [46] implemented in subthreshold MOS.

The analysis of this circuit is typical for translinear circuits that involve differential current signals. Application of the translinear principle around the loop A-B-C-D-A yields

$$I_1 I_4 = I_3 I_2 \Rightarrow \frac{I_3}{I_1} = \frac{I_4}{I_2}$$

Using basic algebra,

$$\frac{I_3 - I_4}{I_1 - I_2} = \frac{I_3 + I_4}{I_1 + I_2}$$

and

$$\Delta I^* = \frac{I_B}{I_{\text{in}}}\,\Delta I$$

where $\Delta I \equiv I_1 - I_2$, $I_{\text{in}} \equiv \text{I}_1 + \text{I}_2$ and similarly for $\Delta I^* \equiv I_3 - I_4$, $I_B \equiv I_3 + I_4$. The differential output current ΔI^* is a scaled version of the differential input current ΔI. The voltage between node B and V_{dd} should be such that the current source I_B stays in saturation.

The mirror composed of transistors M_5 and M_6 converts the unidirectional differential signal ΔI to the bidirectional signal I_{out} so that

$$I_{\text{out}} \equiv -I_B \frac{\Delta I}{I_{\text{in}}}$$

Analysis of Circuits with MOS Transistors in Ohmic Regime. In this section, we extend the translinear principle to subthreshold MOS transistors operating in the ohmic region. In Appendix A (see Fig. 4.4), we show how the source–drain current of a MOS transistor can be decomposed into a source component I_{Q_s} and a drain component I_{Q_d} and that these components superimpose linearly to yield the actual current $I_{SD} \equiv I_{Q_s} - I_{Q_d}$.

In the ohmic region, these components are comparable. Decomposition and linear superposition may be used to exploit the intrinsic translinearity of the gate-source and gate-drain "junctions." This is the basis for extending the translinear principle to the ohmic region. However, in the saturation region, we can exploit the translinearity of the gate-source "junction" directly because the drain component is essentially zero and decomposition is of no consequence.

The translinear circuits based on subthreshold ohmic operation are only possible because of the symmetry between drain and source operation of an MOS transistor. One could argue that decomposition is also possible with bipolar devices, it is only an *approximation* for bipolars, due to the fact that the forward and reverse current gains of the device never reach unity [79, 80] [see Eqs. (4.62) in Appendix A]. This distinction is a fundamental and important difference between MOS and bipolar transistors due to recombination in the base. It is possible to use CMOS-compatible lateral bipolar transistors as symmetric devices [55], but at the expense of a large base current that increases the power dissipation in the system.

To demonstrate the application of the translinear principle to circuits that include MOS transistors in the ohmic regime, consider the one-quadrant current-correlator circuit in Figure 4.15. Transistor M_2 operates in the ohmic region. Proper circuit operation requires that the output voltage is high enough to keep M_3 in saturation. This circuit was first introduced by Delbrück [56] and later incorporated in a larger circuit that implements the nonlinear Hebbian learning rule in an auto-adaptive network [57, 59] and in a micropower autocorrelation system [60].

An expression relating the output current I_3 to the input currents I_1 and I_4 can be derived by treating the source-gate and the drain-gate "junctions" of the ohmic device as separate translinear elements and applying the translinear principle. For the two loops formed by nodes GND-*A*-GND and GND-*A*-*B*-*C*-GND in Fig. 4.15, we obtain

$$I_1 = I_{Q_s} \qquad I_3 I_{Q_s} = I_4 I_{Q_d} \tag{4.42}$$

In writing Eq. (4.42), we have tacitly assumed that the source–drain current of the MOS transistor can be decomposed into a source component I_{Q_s} and a drain component I_{Q_d}—controlled by their respective "junction" voltages V_{Q_s} and V_{Q_d}. These opposing components superimpose linearly to give the actual current passed by M_2, i.e.,

$$I_{DS2} = I_{Q_s} - I_{Q_d}$$

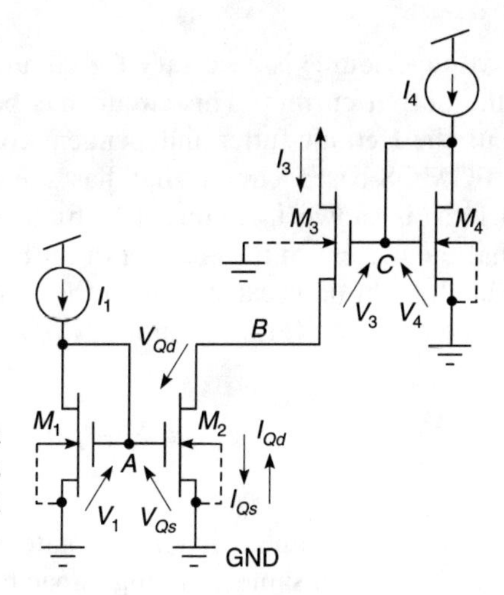

Figure 4.15 Translinear circuit that employs subthreshold MOS transistors in saturation and ohmic regime and computes the product of two input currents I_1 and I_2 normalized to $I_1 + I_2$.

Furthermore, both M_2 and M_3 pass the same current, and therefore

$$I_3 = I_{Q_s} - I_{Q_d} \tag{4.43}$$

Substituting for I_{Q_s} and I_{Q_d} from Eqs. (4.42) and (4.43), the output current is given by

$$I_3 = \frac{I_1 I_4}{I_1 + I_4} \tag{4.44}$$

4.4.2. Translinear Networks

In the previous section we discussed "strictly" translinear loops. Translinear networks [46] differ from translinear loops in that they contain independent voltage sources, and the following equation can be employed in their analysis:

$$\frac{\Pi_{\text{CCW}} J}{\Pi_{\text{CW}} J} = \mathcal{G} e^{E/V_t} \quad \text{or} \quad J_2 J_4 \cdots J_N / J_1 J_3 \cdots J_{N-1} = \mathcal{G} e^{E/V_t} \tag{4.45}$$

where E is the independent voltage source and $\mathcal{G}$ is a constant coefficient that lumps device design and fabrication parameters. The above extension to the translinear principle was proposed by Hart [61].

We begin the discussion of TNs using a simple circuit that has the topology of a current mirror and incorporates a voltage source in the loop [see Fig. 4.16(a)]. If the input current is I_1, the output current is I_2, and V_R is a constant voltage source, application of the translinear principle around the loop GND-A-B-C-D-GND yields

$$I_2 = \frac{I_1^3}{\mathcal{G} e^{V_r/V_t}} \tag{4.46}$$

The voltage source is necessary for circuit operation and it appropriately normalizes the output current. This circuit has been employed in a small system that implements the Herault-Jutten independent component analyzer [23].

An FGMOS-based circuit that has the same functionality as the circuit in Figure 4.16(a) is shown in Figure 4.16(b). We begin the analysis of the circuit by noting that the current in the channel of an FGMOS is controlled by multiple gates that can be thought of as extensions to the front gate of the transistor. As such, Eq. (4.21) can be rewritten for an N-input NMOS transistor as

$$I_{DS} \equiv SS_Q I_{n0}\, i_B^{(1-\kappa)} \prod_{j=1}^{j=N} i_{Gj}^{\kappa/N} \tag{4.47}$$

where it has been assumed that all N gates of the ith transistor have the same strength, i.e., have the same coupling capacitance to the floating gate. The charge Q_{FG} on the floating gate is incorporated through a geometry-related multiplicative constant S_Q so that when the charge Q_{FG} is zero, $S_Q = 1$.

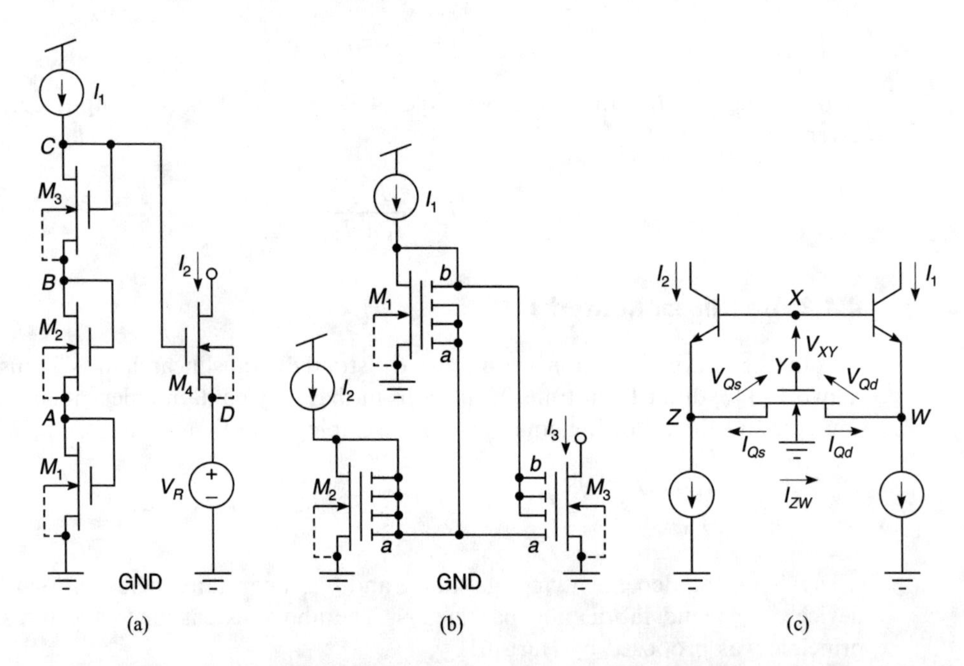

Figure 4.16 Translinear networks in subthreshold MOS. (a) Current-mode circuit that implements a normalized cubic nonlinearity; I_1 is the input, I_2 is the output, and the voltage source V_R normalizes the result. (b) Another translinear circuit using FGMOS transistors to compute the ratio of a cubic to square functions; I_1 and I_2 are the inputs and I_3 is the output. (c) BiCMOS translinear network that exploits the MOS subthreshold ohmic characteristics.

An expression for the output current I_3 can be obtained by applying the translinear principle around three loops that include the floating gates and source of transistor M_3 together with the floating-gate nodes and sources of the other transistors. When the three loops are traversed, the following equations for psi-currents are obtained:

$$i_{3a} = i_{2a} \qquad i_{3a} = i_{1a} \qquad i_{3b} = i_{1b} \tag{4.48}$$

We have adopted a notation where, for example, i_{3a} denotes the psi-current in device 3 controlled by the voltage on its gate a. Using Eq. (4.27), the current at the source of M_1, M_2, M_3, can be expressed as functions of psi-currents, so that

$$\begin{aligned} I_1 &= I_{DS1} = SS_{Q1} I_{n0}\, i_{1a}^{3\kappa_1/4}\, i_{1b}^{\kappa_1/4} \\ I_2 &= I_{DS2} = SS_{Q2} I_{n0}\, i_{2a}^{\kappa_2} \\ I_3 &= I_{DS3} = SS_{Q3} I_{n0}\, i_{3a}^{\kappa_3/4}\, k_{3b}^{3\kappa_3/4} \end{aligned} \tag{4.49}$$

where the devices are assumed to have the same S and I_{n0}; the back-gate contribution to the current in each device is eliminated as all transistors have the source shorted to the substrate. Now, by making the assumption that $\kappa_1 = \kappa_2 = \kappa_3 = \kappa_4$ and with no charge on the floating gates, Eqs. (4.48) and (4.49) yield the following expression for the output current I_3 in terms of the input currents I_1 and I_2:

$$I_3 = \frac{I_1^3}{I_2^2} \tag{4.50}$$

The assumption of equal κ is reasonable as long as the voltage on the floating gate is such that all devices stay in the subthreshold. An alternative way of obtaining a functional description of the circuit can be found in the paper by Minch et al. [49].

A TN that incorporates bipolar transistors, an MOS in subthreshold, and an independent voltage source V_{XY} is shown in Figure 4.16(c). We will use Equation (4.45) to derive the relationship between I_1, I_2, I_{ZW}, and V_{XY}. The MOS transistor is assumed to be ideal with $\kappa = 1$. A discussion of networks with nonideal devices will be done in the next section.

A ratio relationship between I_{ZW} and $I_1 - I_2$ can be derived by employing Eq. (4.45) applied to the two loops X-Z-Y-X and X-Y-W-X to yield the equations

$$I_{Q_s} = I_2 \mathcal{G} e^{-V_{XY}/V_t} \qquad I_{Q_d} = I_1 \mathcal{G} e^{-V_{XY}/V_t} \tag{4.51}$$

Using the adopted conventions for current decomposition in NMOS transistors (see Fig. 4.4), I_{Q_s} and I_{Q_d},

$$I_{ZW} = I_{Q_d} - I_{Q_s}$$

together with Eq. (4.51), we obtain the following expression that relates the currents in the circuit:

$$I_{ZW} = (I_1 - I_2)\mathcal{G}e^{-V_{XY}/V_t} \tag{4.52}$$

It is immediately apparent that the current ratio $I_{ZW}/(I_1 - I_2)$ can be controlled both by a fixed parameter ($\mathcal{G}$) that is designed prior to the fabrication of the circuit and a variable quantity (e^{-V_{XY}/V_t}) that can be programmed (postfabrication) during circuit/system operation. This property will be utilized in the design of *linear* MOS transistor-only spatial averaging networks.

Translinear Spatial Averaging Networks. Often, models of neural computation necessitate the realization of spatial averaging networks [11]. To demonstrate the analogies between linear and translinear networks as well as their subtle and important differences, we begin with networks that employ linear conductances, voltages, and currents and contrast them with translinear current-mode [17] networks.

A voltage-mode circuit model for a loaded network is shown in Figure 4.17(a) for which

$$I_{PQ} = \frac{G_1}{G_2}(I_Q - I_P)$$

This is a lumped-parameter model where G_1 and G_2 correspond to resistances. The voltages on nodes P and Q, referenced to ground, represent the state of the network and can be read out using a differential amplifier with the negative input grounded.

The equivalent circuit using idealized nonlinear conductances is shown in Figure 4.17(b). The difference in currents through the diodes D_1 and D_2 are linearly related to the current through the *diffusor* MOS transistor.[1] This relationship can be derived from Eq. (4.11) describing subthreshold conduction and the ideal diode characteristics where $I_D = I_S \exp[V_D/V_t]$. An expression can be derived for the current I_{PQ} in

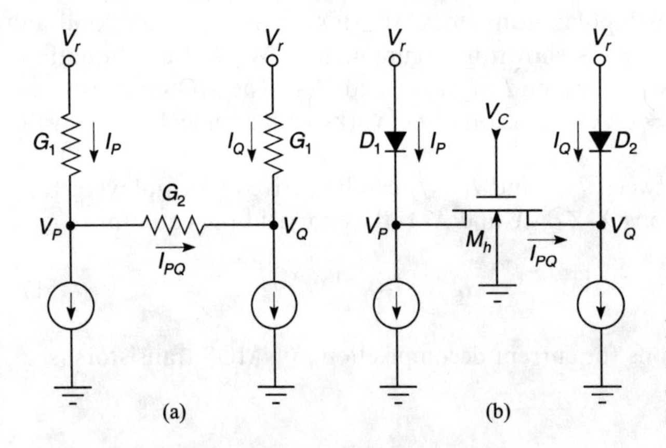

Figure 4.17 Building blocks for linear loaded networks. Using segments that employ ideal (a) linear and (b) nonlinear elements.

[1]The term diffusor was adopted in [24] to describe the exploitation of diffusion transport in MOS transistors to spread signals in a manner analogous to gap junctions between neural cells.

terms of the currents I_p and I_Q, the reference voltage V_r, and the bias voltage V_C when diodes are replaced by transistors:

$$I_{PQ} = \left(\frac{SI_{n0}}{I_S}\right) \exp\left[\frac{\kappa_n V_C - V_r}{V_t}\right](I_Q - I_P) \tag{4.53}$$

The current I_{n0} and S are the zero intercept current and geometry factor, respectively, for the diffusor transistor M_h; I_S is the reverse saturation current for the diode that is assumed to be ideal. The currents in these circuits are identical if

$$\frac{G_1}{G_2} = \left(\frac{SI_{n0}}{I_S}\right) \exp\left[\frac{\kappa_n V_C - V_r}{V_t}\right]$$

Increasing V_C or reducing V_r has the same effect as increasing G_1 or reducing G_2. The state of this network is represented by the charge at the nodes P and Q. Since the anode of a diode is the reference level (zero negative charge), the currents I_P and I_Q represent the result. Unfortunately, the anode of a diode or a diode connected transistor is not a good current source.

When diodes are not explicitly available in the process, diode-connected PMOS or NMOS transistors can be used as shown in Figure 4.18. When the loads are PMOS, the current I_{PQ} is given in terms of voltages normalized to kT/q:

$$I_{PQ} = \left(\frac{S_h I_{n0}}{S_v I_{p0}}\right) \exp(\kappa_n V_C - \kappa_p V_r)(I_Q^{1/\kappa_p} - I_P^{1/\kappa_p}) \tag{4.54}$$

When NMOS transistors are used as loads, there is the additional benefit, that of exploiting the current-conveying properties of a single transistor [17], to obtain the current outputs I_P and I_Q, on nodes that are low conductance (the drain terminals are now excellent outputs for the currents). Using Eqs. 8.45 in [17], the current I_{PQ} is given as

$$I_{PQ} = \left(\frac{S_h}{S_v}\right) \exp(\kappa_n V_C - \kappa_n V_r)(I_Q - I_P) \tag{4.55}$$

where S_h and S_v are geometry parameters for transistors M_h and M_v, respectively.

The one-dimensional MOS transistor-only network corresponding to the Helmholtz equation shown in Figure 4.19 can model the averaging that occurs at the horizontal cell layer of the outer retina. This equation is the basis of the well-known silicon retina architecture proposed by Maholwald and Mead [11, 62].

Summing the currents at node j, we get

$$I_j^* = I_{ij} - I_{jk} + I_j \tag{4.56}$$

Using the results from the previous section for the currents I_{ij} and I_{jk} given by Eq. (4.55) substituted in Eq. (4.56) yields

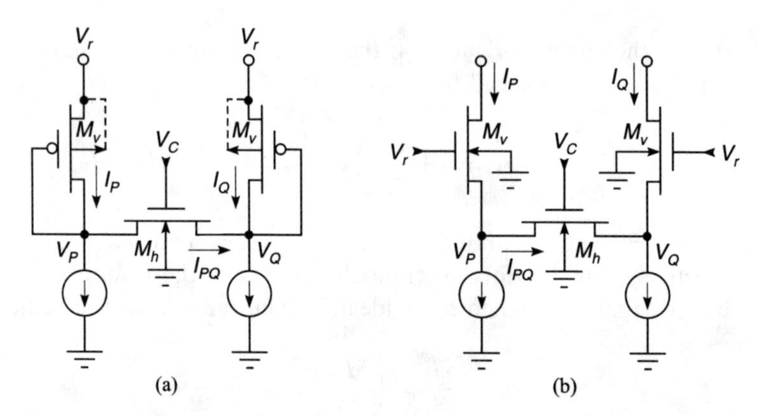

Figure 4.18 Current-mode building blocks for linear loaded networks using (a) PMOS transistor implementation and (b) NMOS single transistor current-conveyor implementation.

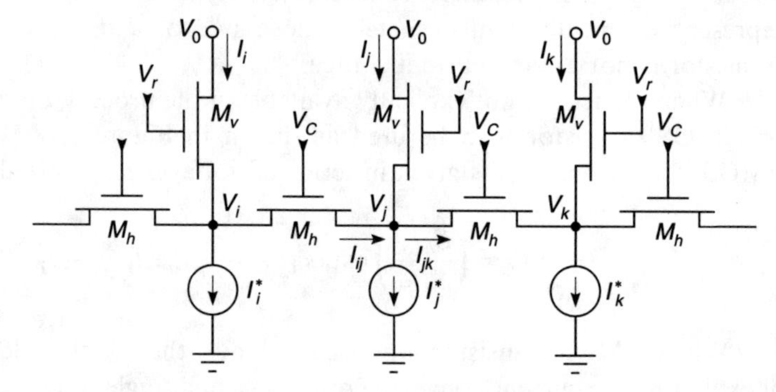

Figure 4.19 One-dimensional MOS translinear network to perform local aggregation—spatial averaging. The back-gate terminals of all devices are connected to the substrate.

$$I_j^* = I_j + \left(\frac{S_h}{S_v}\right)\exp(\kappa_n v_C - \kappa_n v_r)(2I_j - I_i - I_k) \tag{4.57}$$

Normalizing internode distances to unity, the above equation can be written on the continuum as

$$I^*(x) = I(x) + \lambda\,\frac{d^2 I(x)}{dx^2}$$

This equation yields the solution to the following optimization problem: find the smooth function $I(x)$ that best fits the data $I^*(x)$ with the minimum energy in its first derivative. Input is the current $I^*(x)$ and output the currents $I(x)$.

The parameter $\lambda \equiv (S_h/S_v)\exp(\kappa_n v_C - \kappa_n v_r)$ is the cost associated with the derivative energy relative to the squared error of the fit.

The diffusive network in Figure 4.19 was recently described in terms of "pseudo-conductances" [63]. Here we use the charge/current-based formulation first proposed in [24] to explain its behavior. This current-mode approach relies on an intuitive understanding of the device physics and yielded the insight which enabled us to extend the translinear principle to subthreshold MOS transistors in the ohmic region as well as the decomposition of the current into dimensionless components corresponding to an ideal junction. We now have a comprehensive *current-mode* approach for analyzing subthreshold MOS circuits. The essence of this approach is the representation of variables and parameters by charge, current, and diffusivity. Voltages and conductances are not used explicitly.

Bult and Geelen proposed an identical network for linear current division above threshold and used it in a digitally controlled attenuator [64]; they also analyzed its subthreshold behavior. However, they stipulate that all gate voltages must be identical and control the division by manipulating the geometric factor W/L of the devices. We have shown here, and previously in [24], that this constraint may be relaxed in the subthreshold without disrupting linear operation. This is a real bonus because it allows us to modify the divider ratio or space constant of the network after the chip is fabricated by varying $V_c - V_r$. Tartagni et al. have demonstrated a current-mode centroid network [65] using subthreshold MOS devices whose operation is described by the current division principle.

4.4.3. General Result for MOS Translinear Loops

Three of the circuits discussed in the previous section, namely the translinear multiplier of Figure 4.13(b), the MOS implementation of the Gilbert gain stage in Figure 4.14(b), and the current correlator (Fig. 4.15), have been experimentally shown to exhibit near-"exact" translinear behavior even though they are built from MOS transistors and do not have their source connected to the local substrate.

A recent result by Vittox [30] can be employed to partially explain this rather surprising behavior. He considers translinear loops constructed from MOS transistors in subthreshold saturation with common substrate connection (similar to the one shown in Fig. 4.20). If the pairing of transistors in the CW and CCW direction is such that they have their gates connected to gates and sources connected to sources and are alternated (much like even- and odd-numbered devices in Fig. 4.20), Vittoz shows that the translinear loop does not suffer from the MOS transistor nonideal translinear behavior. He notes also that loops containing transistors in the ohmic regime can also be included in this formulation as they can be decomposed as two parallel connected saturated devices sharing a common gate and a common substrate (see Fig. 4.4).

However, to account for the near-exact operation of the multiplier in Figure 4.13(b), Vittoz's argument must be extended to include loops that go through the back gate of the MOS transistors, as illustrated in Figure 4.20. The common substrate restriction can thus be removed and replaced by a local substrate connection, and the result still holds true. In a standard CMOS process, this will of course be possible only for one type of device.

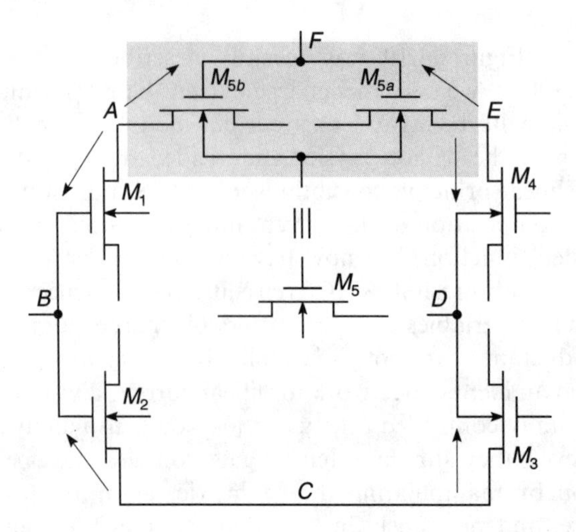

Figure 4.20 Translinear loop composed of five MOS transistors in the subthreshold. All devices are in saturation except device M_5, which is in the ohmic regime and therefore can be decomposed as two devices in saturation, back to back sharing the same gate and substrate.

Now, we will reexamine the operation of the circuits in Figures 4.14(b), 4.13(b), and 4.15.

Consider the largest loop (*A-B-C-D-A*) in Figure 4.14(b). Devices M_1 and M_2 have common gate and common bulk and so do devices M_3 and M_4. When adjacent devices are paired in different ways, we observe that M_3 and M_2 share the same source and bulk, which is the case also for M_1 and M_4.

In the largest loop (*A-B-C-D-A*) of the translinear multiplier circuit of Figure 4.13(b), we can verify that transistors M_1 and M_2 as well as M_3 and M_4 share common gate and source. The alternative pairing finds M_4 and M_2 sharing the same bulk and source, which is also the case for M_3 and M_1.

When devices in the loop are operating in the ohmic regime, such as M_2 in the circuit of Figure 4.15, we can verify that the loop GND-*B*-*C*-GND incorporates two adjacent sets of devices. Note that M_3 and M_2 share the same bulk and source/drain while M_3 and M_4 share bulk and gate; the bulk in this circuit is the same for all devices.

4.4.4. Translinear Circuit Dynamics

The dynamics of translinear circuits and systems have not been discussed in this chapter. However, it was pointed out in [66] that, in networks with nonlinear conductances without complementary nonlinear reactances, the state equations that describe the dynamics of the system are nonlinear. Given an architecture and a particular network, a method was outlined to test for stability [66].

4.5 CONTRAST-SENSITIVE SILICON RETINA

The analog silicon system is modeled after neurocircuitry in the distal part of the vertebrate retina, called the outer plexiform layer. Figure 4.21 illustrates interactions

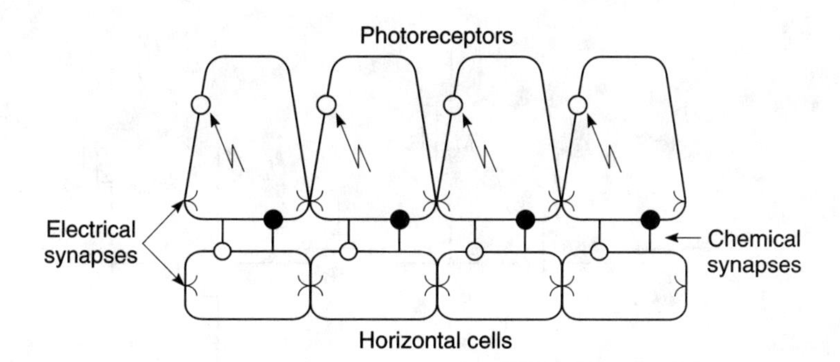

Figure 4.21 One-dimensional model of neurons and synapses in the outer plexiform layer. Based on the red-cone system in the turtle retina.

between cells in this layer [76]. The well-known center/surround receptive field emerges from this simple structure, consisting of just two types of neurons. Unlike the ganglion cells in the inner retina and the majority of neurons in the nervous system, the neurons that we model here have graded responses (they do not spike); thus this system is well suited to analog VLSI.

The photoreceptors are activated by light; they produce activity in the horizontal cells through excitatory chemical synapses. The horizontal cells, in turn, suppress the activity of the receptors through inhibitory chemical synapses. The receptors and horizontal cells are electrically coupled to their neighbors by electrical synapses. These allow ionic currents to flow from one cell to another and are characterized by a certain conductance per unit area.

In a biological system, contrast sensitivity—the normalized output that is proportional to a local measure of contrast—is obtained by shunting inhibition. The horizontal cells compute the local average intensity and modulate a conductance in the cone membrane proportionately. Since the current supplied by the cone outer segment is divided by this conductance to produce the membrane voltage, the cone's reponse will be proportional to the ratio between its photo input and the local average, i.e., to contrast. This is a very simplified abstraction of the complex ion channel dynamics involved. The advantage of performing this complex operation at the focal plane is that the dynamic range is extended (local automatic gain control).

The basic analog MOS circuitry for a one-dimensional pixel with two-neighbor connectivity is shown in Figure 4.22. The analysis of the system can be found in [17, 24]; here we present an outline and approximations to the main results.

We begin with the nonlinear aspects of system operation, its *contrast sensitivity*. The nonlinear operation that leads to a local gain-control mechanism in the silicon system is achieved through a mechanism that is qualitatively similar to the biological counterpart but quantitatively different (see discussion in [24]). Referring to Figure 4.22, the output current $I_c(x_m, y_n)$ at each pixel can be given (approximately) in terms of the input photocurrent $I(x_m, y_n)$ and a local average of this photocurrent in a pixel

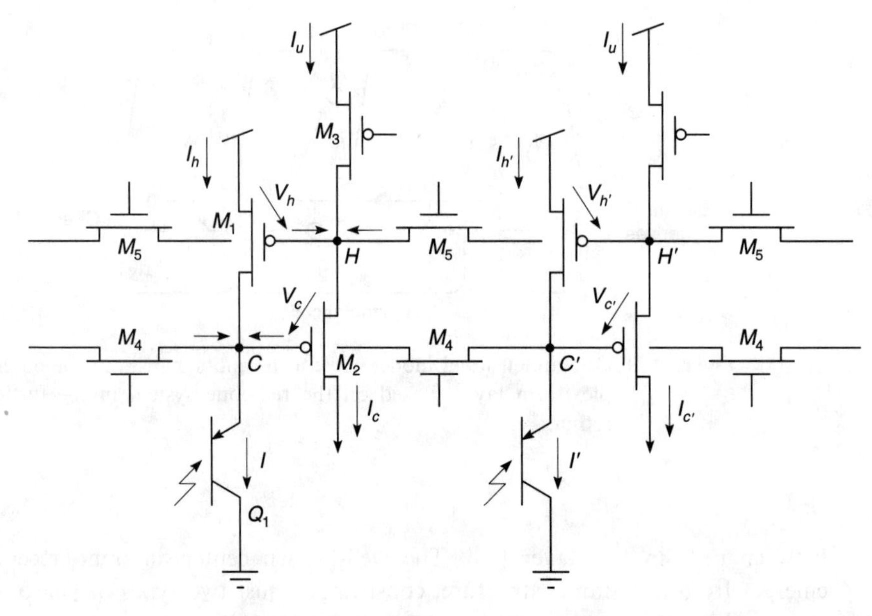

Figure 4.22 One-dimensional implementation of outer plexiform retinal processing. There are two diffusive networks implemented by transistors M_4 and M_5, which model electrical synapses. These are coupled together by controlled current sources (devices M_1 and M_2) that model chemical synapses. Nodes H in the upper layer correspond to horizontal cells while those in the lower layer (C) correspond to cones. The bipolar phototransistor Q_1 models the outer segment of the cone and M_3 models a leak in the horizontal cell membrane. Note that the actual system has six-neighbor connectivity.

neighborhood (M, N). This region may extend beyond the nearest neighbor. The fixed current I_u supplied by transistor M_3 normalizes the result:

$$I_c(x_m, y_n) = I_u \frac{I(x_m, y_n)}{\left(I(x_m, y_n) + \sum_{M,N} I(x_i y_j) \right)} \tag{4.58}$$

At any particular intensity level, the outer plexiform behaves like a linear system that realizes a powerful second-order regularization algorithm for edge detection. This can be seen by performing an analysis of the circuit about a fixed operating point. To simplify the equations, we first assume that $\hat{g} = \langle I_h \rangle g$, where $\langle I_h \rangle$ is the local average. Now we treat the diffusors (devices M_4) betweens nodes C and C' as if they had a fixed diffusivity $\hat{g}$. The diffusivity of the devices M_5 between nodes H and H' in the horizontal network is denoted by h. Then the simplified equations describing the full two-dimensional circuit on a square grid are

$$I_h(x_m, y_n) = I(x_m, y_n) + \hat{g} \sum_{\substack{i=m\pm1 \\ j=n\pm1}} \{I_c(x_i, y_j) - I_c(x_m, y_n)\}$$

$$I_c(x_m, y_n) = I_u + h \sum_{\substack{i=m\pm1 \\ j=n\pm1}} \{I_h(x_m, y_n) - I_h(x_i, y_j)\}$$

Using the second-difference approximation for the Laplacian, we obtain the continuous versions of these equations,

$$I_h(x, y) = I(x, y) + \hat{g}\nabla^2 I_c(x, y) \tag{4.59}$$

$$I_c(x, y) = I_u - h\nabla^2 I_h(x, y) \tag{4.60}$$

with the internode distance normalized to unity. Solving for $I_h(x, y)$, we find

$$\hat{g}h\nabla^2\nabla^2 I_h(x, y) + I_h(x, y) = I(x_i, y_j) \tag{4.61}$$

This is the *biharmonic* equation used in computer vision to find an optimally smooth interpolating function $I_h(x, y)$ for the noisy, spatially sampled data $I(x_i, y_j)$; it yields the function with minimum energy in its second derivative [77]. The coefficient $\lambda = \hat{g}h$ is called the regularizing parameter; it determines the trade-off between smoothing and fitting the data.

A one-dimensional solution to this equation can be obtained using Green's functions, valid for vanishing boundary conditions at plus and minus infinity:

$$I_h(x, \lambda) = \frac{1}{2\lambda^{1/4}} \exp(-|x|/\sqrt{2}\lambda^{1/4}) \cos\left(\frac{|x|}{\sqrt{2}\lambda^{1/4}} - \frac{\pi}{4}\right)$$

In the original work [24], the chip was fabricated with 90×92 pixels on a 6.8×6.9-mm die in a 2-μm N-well double-metal, double-polysilicon, garden variety digital oriented CMOS technology and was fully functional. More recently the same system has been fabricated with 230×210 pixels on a 1×1-cm die in a 1.2-μm N-well double-metal, double-polysilicon, digital oriented CMOS technology. The chip incorporates 590,000 transistors and 48,000 pixels operating in subthreshold/transition region with power dissipation on the order of a few milliwatts when powered from a 5-V power supply. Temporal response is in the order of a few microseconds.

To find the energetic efficiency of this system, we assume that a total of 18 low-precision operations (OP) are performed per pixel. Six operations are necessary for the convolution with bandpass kernel of Figure 4.23, six for the Laplacian operator [Eq. (4.60)] and six for the local gain control computation [Eq. (4.58)]. If the system is biased so that at the pixel level the frequency response is 100 kHz, approximately 1×10^{12} low-precision calculations per second are performed in the 210×230 pixels. The power dissipation under the above biasing conditions is about 50 mV when operating from 5-V power supplies. This is equivalent to 0.05 pW/OP. This performance is a result of an optimization done at the system level, by mapping the

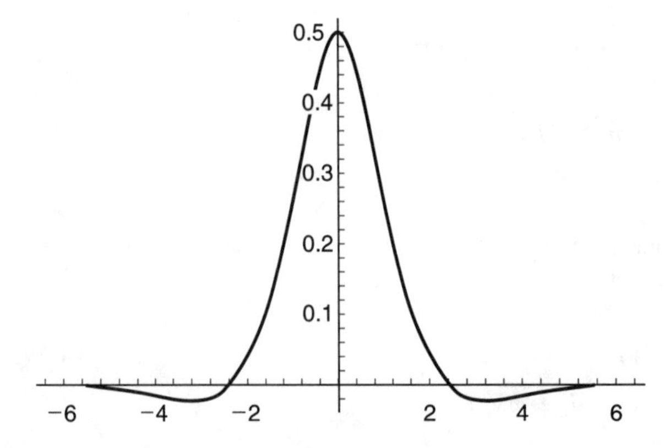

Figure 4.23 Plot for one-dimensional solution of the biharmonic equation; $\lambda = 1$.

problem on an effective physical computational model, rather than trying to optimize the energetic efficiency of an individual gate.

An image captured through the silicon retina is shown in Figure 4.24. Note the edge enhancement properties of the system and the absence of dynamic range (flat image).

Figure 4.24 Image of the author as captured by the silicon system. There is a large gradient in illumination from left to right. A regular CCD camera does a poor job in imaging both high- and low-illumination regions.

4.6. DISCUSSION

The exponential characteristics of a subthreshold MOS device offer the strongest nonlinearity relating a voltage and a current in solid-state devices [74] (within the constant κ). When plotted on a logarithmic axis, it manifests itself as a linear function with a constant slope (see Fig. 4.5). In this chapter we have seen how a bottom-up design methodology can yield energy-efficient circuits for computing and signal processing that exploit the translinear property and zero gate conductance of the MOS transistors.

The importance of this limiting steepness, and associated with it the highest possible amplification, has long been recognized by engineers involved in the design of analog linear integrated circuits, and in their literature it is referred to as the "Boltzmann limited" slope. Mead often points out the striking similarity between the electrical properties of excitable membranes and the MOS subthreshold characteristics (see Fig. 1 in [12]), as both exhibit the Boltzmann limited behavior. Furthermore, he cites this similarity as one motivation for pursuing the synthetic approach in analog VLSI using subthreshold MOS devices. Having pursued such an approach, we are tempted to ask a question that has to do with differences rather than similarities. What is fundamentally different at this level of description that could have implications at the *system* level?

A careful examination of the slopes in Figure 1 of [12] (also Fig. 4.6 in [11]) reveals that in biological structures the prefactor in the exponent of the controlling node is larger than unity! That is, the transconductance per unit current is not limited to a value equal to or greater than kT/q millivolts per e-fold of current change. Ions in biological membranes are not limited by the Pauli exclusion principle; the conductance dependence is steeper in excitable membranes because of *correlated* charge control of the current (see discussion on p. 55 of Hille [7]). In subthreshold MOS operation, the slope can only asymptotically achieve the minimum value of kT/q millivolts per e-fold of current change. The minimum value can however be seen in bipolar transistors and in junction field effect transistors when operating in the subthreshold.

The ramifications of this fundamental difference can be appreciated if one attempts to realize physically an information-processing system that operates in the neighborhood of 300 K from power supplies that are only $4 \times (kT/q) \approx 100\,\text{mV}$ (biological hardware operates under these conditions). The advantage of reduced power supplies is reduced power dissipation and thus an improved figure for the power-delay product [see Eq. (4.3)]. Since the adopted figure of merit is quadratically related to the transconductance per unit current, a device with exponential voltage-to-current characteristics is always better. Bipolar transistors, field effect transistors operating in the subthreshold, or any other barrier-controlled device capable of power gain with the "Boltzmann limited" steepness is "optimum" in this sense.

We now consider a very simple operation at this reduced power supply, the quantization of a scalar signal for reliable communication. This could be an inverter circuit in VLSI or the generation of an action potential in biology. The effects of thermal agitation in the system make reliable operation of the quantizer possible only when the energy barriers that separate the two states are more than a few kT electron-volts apart. This has been discussed extensively in the literature (see, e.g., [1, 2]). The problem becomes more serious in large, complex information systems such as VLSI with millions of computational elements and where structural variability (i.e., "noise" in the individual components) has to be taken into account (see transistor data in Figure 4.2). The problem of component variability in complex VLSI systems has been addressed by Mead and Conway in Chapter 9 of [3] and by Keyes in Chapter 4 of [75].

So, how is it possible for an information-processing system that has the complexity of biological systems to operate reliably with power supplies of the order of a few kT/q volts?

The issue of structural "noise" in biological systems can be addressed at the architectural level through robust algorithms and representations much like it was done for our silicon retina or through local *adaptation* learning mechanisms [41]. The problem of noise in a thermodynamic sense is a more difficult one. It can perhaps be addressed by the *fine* details of signal amplification mechanisms that are found in biological systems. For example, the biophysics of excitable membranes allow polyvalent charged entities of charge z to respond as a *unit* rather than independently to an applied potential energy differential. This is a cooperative phenomenon that produces Boltzmann-limited, nonlinear effects that are stronger than those possible in solid-state devices. This would correspond to an effective "cooling" of the system to a temperature T/z!

It is unlikely that the question posed in the previous paragraph has a simple answer and, therefore, our explanations must be inadequate. They do, however, point to some intriguing possibilities worth further consideration as better understanding of the chemistry and physics of computation in neural systems may contribute to long-term fundamental improvements in solid-state electronics and, in particular, in the fields of low-energy, low-voltage integrated microsystems.

Appendix A. Bipolar Transistor Model

The Ebers-Moll model [79, 80] for an NPN bipolar transistor is given as

$$I_E = -I_F + \alpha_R I_R \qquad I_C = \alpha_F I_F - I_R \tag{4.62}$$

$$I_F = I_{ES}(e^{qV_{BE}/kT} - 1) \qquad I_R = I_{CS}(e^{qV_{BC}/kT} - 1) \tag{4.63}$$

$$\alpha_F I_{ES} = \alpha_R I_{CS} \tag{4.64}$$

where I_C and I_E are the collector and emitter currents, respectively, V_{BE} is the base-to-emitter voltage, V_{BC} is the base-to-collector voltage, I_{ES} is the saturation current of emitter junction with zero collector current, I_{CS} is the saturation current of collector junction with zero emitter current, α_F is the common-base current gain, and α_R is the common-base current gain in inverted mode, i.e., with the collector functioning as an emitter and the emitter functioning as a collector.

By convention, the currents for bipolar transistors are positive when flowing into its terminals.

Combining Eqs. (4.62)–(4.64), the collector current can be expressed as

$$I_C = \alpha_F I_{ES}\left[(e^{qV_{BE}/kT} - 1) - \frac{1}{\alpha_R}(e^{qV_{BC}/kT} - 1)\right] \tag{4.65}$$

For an ideal device with common-base current gain α_F and common-base current gain α_R in the inverted mode, very close to unity, the above equation becomes

$$I_C = I_{ES}(e^{qV_{BE}/kT} - e^{qV_{BC}/kT}) \tag{4.66}$$

However, regular bipolar transistors do not have both α_F and α_R near unity.

When the collector-to-base voltage equals zero or the collector is reverse biased with respect to the base, the above equation simplifies to the familiar

$$I_C = A_E J_{ES}(e^{qV_{BE}/kT} - 1) \approx A_E J_{ES} e^{qV_{BE}/kT} = I_S e^{V_{BE}/V_t} \tag{4.67}$$

where A_E is a design parameter, the area of the emitter junction. The terms J_{ES} and I_S are the saturation current density and current for the emitter, respectively. In this case, $I_R \ll I_F$ and the equations above give $I_C = -\alpha_F I_E$. Using the relation $I_E + I_C + I_B = 0$ (Kirchhoff's current law), we get the familiar result

$$I_C = \frac{\alpha_F}{1 - \alpha_F} I_B \equiv \beta_F I_B \tag{4.68}$$

where β_F is the common-emitter current gain.

In a standard N-well CMOS process, a vertical PNP transistor is available for circuit design, but only in the common-collector configuration since it has the P-substrate as the collector—the N-well forms the base and a P-diffusion the emitter. This device is useful as a light sensor; the smallest possible phototransistor permitted by the 2-μm design rules has base dimensions of 16 × 16 μm when the emitter has an area $A_E = 6 \times 6$ μm. The dark current in these minimum-size phototransistors is approximately 100 fA. These sensors show a linear response over at least eight orders of magnitude in light intensity. Experimentally determined responsivity of a device with $A_E = 100 \times 100$ μm was 73.8 A/W at $\lambda = 632.8$ nm and 118.5 A/W at $\lambda = 834$ nm. The β is approximately 200 with Early voltage $V_o = 48$ V. The frequency response is limited to a few hundred kilohertz by the large base-collector capacitance.

In some processes, an NPN transistor is offered through an extra implant in the N-well to form the base. Typical forward β for these devices is 60 for emitter area $A_E = 8 \times 8$ μm. The performance of such bipolar devices is limited by the collector resistance r_c, which is in the kilohm range if there is no buried collector implant. The Early voltage of these devices is approximately 45 V. At high collector currents (high injection conditions) the characteristics of bipolars deviate from exponential and their current gain β is also reduced. For the NPN vertical bipolars, with a minimum emitter area, this high current effect becomes important at current levels above a few hundred microamperes. At low collector currents β is also limited by recombination in the base. Typical betas range from 20 at current levels of a few nanoamperes to their maximum at current levels of a few hundred microamperes.

References

[1] R. Landauer, "Irreversibility and Heat Generation in the Computing Process," *IBM Res. Dev.*, pp. 183–191, July 1961.
[2] R. Landauer, "Information Is Physical," *Proceedings of the 1992 Physics of Computation Workshop*, Dallas, TX, 1992, pp. 1–4.
[3] C. A. Mead and L. Conway, *Introduction to VLSI Systems*, Addison-Wesley, Reading, MA, 1980.
[4] P. Wheeler and L. C. Aiello, "The Expensive-Tissue Hypothesis," *Curr. Anthropol.*, vol. 36, no. 2, pp. 199–201, 1995.
[5] M. Jones, J. Allman, A. Hakeem, and G. R. Sandoval, *Brain and Life Span in Primates*, 4th ed., Academic, 1996, Chapter 5.
[6] D. Hubel, *Eye, Brain and Vision*, Scientific American Library, 1988.
[7] B. Hille, *Ionic Channels of Excitable Membranes*, Sunderland, MA, Sinauer Associates, 1984.
[8] A. G. Andreou, "On Physical Models of Neural Computation and Their Analog VLSI Implementation," in *Proceedings of the 1994 Workshop on Physics and Computation*, Los Alamitos CA, IEEE Computer Society Press, 1994, pp. 255–264.
[9] H. B. Barlow, "Unsupervised Learning," *Neural Computation*, vol. 1, no. 3, pp. 295–311, 1989.
[10] S. Haykin, *Neural Networks; A Comprehensive Foundation*, McMillan College Publishing, New York, 1994.
[11] C. A. Mead, *Analog VLSI and Neural Systems*, Addison-Wesley, Reading, MA, 1989.
[12] C. A. Mead, "Neuromorphic Electronic Systems," *Proceedings IEEE*, vol. 78, no. 10, pp. 1629–1636, 1990.
[13] E. A. Vittoz and J. Fellrath, "CMOS Analog Integrated Circuits Based on Weak Inversion Operation," *IEEE J. Solid-State Circuits*, vol. SC-12, no. 3, pp. 224–231, 1977.
[14] E. A. Vittoz, "Micropower Techniques," in *VLSI Circuits for Telecommunications*, Y. P. Tsividis and P. Antognetti (eds.), Prentice-Hall, Englewood Cliffs, NJ, 1985.
[15] A. G. Andreou and K. A. Boahen, "Synthetic Neural Circuits Using Current Domain-Signal Representations," *Neural Computat.*, vol. 1, pp. 489–501, 1989.
[16] A. G. Andreou, K. A. Boahen, P. O. Pouliquen, A. Pavasović, R. E. Jenkins, and K. Strohbehn, "Current-Mode Subthreshold MOS Circuits for Analog VLSI Neural Systems," *IEEE Trans. Neural Networks*, vol. 2, no. 2, pp. 205–213, 1991.

[17] A. G. Andreou and K. A. Boahen, "Neural Information Processing (II)," in *Analog VLSI Signal and Information Processing*, M. Ismail and T. Fiez (eds.), McGraw-Hill, New York, 1994, Chapter 8.

[18] A. G. Andreou, "Low Power Analog VLSI Systems for Sensory Information Processing," in *Microsystems Technologies for Multimedia Applications: An Introduction*, B. Sheu, Edgar Sánchez-Sinencio, and M. Ismail (eds.), IEEE Press, Los Alamitos, CA, 1995.

[19] K. A. Boahen, "Retinomorphic Vision Systems," in *Proceedings of MicroNeuro-96*, IEEE Computer Society Press, Los Alamitos, CA, 1996.

[20] R. Douglas, M. Mahowald, and C. A. Mead, "Neuromorphic Analog VLSI," *Annu. Rev. Neurosci.*, vol. 18, pp. 255–281, 1995.

[21] B. Gilbert, "Translinear Circuits: A Proposed Classification," *Electron. Lett.*, vol. 11, no. 1, pp. 14–16, 1975; errata vol. 11, no. 6, p. 136.

[22] A. G. Andreou, "Synthetic Neural Systems Using Current-Mode Circuits," in *Proceedings of the IEEE 1990 International Symposium on Circuits and Systems*, New Orleans, May 1990, pp. 2428–2432.

[23] M. Cohen and A. G. Andreou, "Current-Mode Subthreshold MOS Implementation of the Herault-Jutten Autoadaptive Network," *IEEE J. Solid-State Circuits*, vol. 27, no. 5, pp. 714–727, 1992.

[24] K. A. Boahen and A. G. Andreou, "A Contrast Sensitive Silicon Retina with Reciprocal Synapses," in *Advances in Neural Information Processing Systems*, vol. 4, J. E. Moody, S. J. Hanson, and R. P. Lippmann (eds.), Morgan Kaufmann, San Mateo, CA, 1992.

[25] A. G. Andreou and K. A. Boahen, "A 48,000 Pixel, 590,000 Transistor Silicon Retina in Current-Mode Subthreshold CMOS," in *Proceedings of the 37th Midwest Symposium on Circuits and Systems*, Lafayette, Aug. 1994, pp. 97–102.

[26] A. G. Andreou and K. A. Boahen, "A 590,000 Transistor, 48,000 Pixel Contrast Sensitive, Edge Enhancing CMOS Imager-Silicon Retina," in *Proceedings of the 16th Conference on Advanced Research in VLSI*, Chapel Hill, NC, Mar. 1995, pp. 225–240.

[27] E. Vittoz, E. Dijkstra, and D. Shiels, *Low-Power Design: A Collection of CSEM Papers*, edited volume, Electronic Design Magazine, Hasbrouck Heights, NJ, 1995.

[28] Papers by Eric Vittoz at ISSCC96, the logitech mouse and the sun tracker.

[29] E. Vittoz, "Analog VLSI Signal Processing: Why, Where and How?" *J. Analog Integrated Circuits Signal Process.*, vol. 6, pp. 27–44, June 1994.

[30] E. A. Vittoz, "Analog VLSI Implementation of Neural Networks," in *Handbook of Neural Computation*, Institute of Physics Publishing and Oxford University Press.

[31] R. S. Muller, R. T. Howe, S. D. Senturia, R. L. Smith, and R. M. White (eds.), *Microsensors*, IEEE Press, Piscataway, NJ, 1991.

[32] E. A. Vittoz, "The Design of High-Performance Analog Circuits on Digital CMOS Chips, *IEEE J. Solid-State Circuits*, vol. SC-20, no. 3, pp. 657–665, 1985.

[33] Y. P. Tsividis, "Analog MOS Integrated Circuits—Certain New Ideas, Trends and Obstacles," *IEEE J. Solid-State Circuits*, vol. SC-22, pp. 317–321, 1987.

[34] J. E. Meyer, "MOS Models and Circuit Simulation," *RCA Rev.*, vol. 32, pp. 42–63, 1971.
[35] Y. Tsividis, *Operation and Modeling of the MOS Transistor*, McGraw-Hill, New York, 1987.
[36] C.-K. Sin, A. Kramer, V. Hu, R. R. Chu, and P. K. Ko, "EEPROM as an Analog Storage Device, with Particular Application in Neural Networks," *IEEE Trans. Electron Devices*, vol. ED-39, no. 6, pp. 1410–1419, 1992.
[37] M. Holler, S. Tam, H. Castro, and R. Benson, "An Electrically Trainable Artificial Neural Network ETANN with 10420 'Floating Gate' Synapses," in *Proceedings of the International Joint Conference on Neural Networks*, vol. II, Washington DC, June 1989, pp. 191–196.
[38] T. H. Borgstrom, M. Ismail, and S. B. Bibyk, "Programmable Current-Mode Neural Network for Implementation in Analog MOS VLSI," *IEE Proc.*, vol. 137, pt. G, no. 2, pp. 175–184, 1990.
[39] E. Vittoz, H. Oguey, M. A. Maher, O. Nys, E. Dijkstra, and M. Chevroulet, "Analog Storage of Adjustable Synaptic Weights," in *VLSI Design of Neural Networks*, U. Ramacher and U. R'uckert (eds.), Kluwer Academic, Boston, MA, 1991.
[40] E. Sackinger and W. Guggenbuhl, "An Analog Trimming Circuit Based on a Floating Gate Device," *IEEE J. Solid-State Circuits*, vol. SC-23, no. 6, pp. 1437–1440, 1988.
[41] C. A. Mead, "Adaptive Retina," in *Analog VLSI Implementation of Neural Systems*, C. A. Mead and M. Ismail (eds.), Kluwer Academic, Boston, MA, pp. 239–246, 1989.
[42] L. R. Carley, "Trimming Analog Circuits Using Floating-Gate Analog Memory," *IEEE J. Solid-State Circuits*, vol. SC-24, pp. 1569–1575, 1989.
[43] K. Yang and A. G. Andreou, "Multiple Input Floating-Gate MOS Differential Amplifier and Applications for Analog Computation," in *Proceedings of the 36th Midwest Symposium on Circuits and Systems*, Detroit, MI, Aug. 1993.
[44] G. Cauwenberghs, C. F. Neugebauer, and A. Yariv, "Analysis and Verification of an Analog VLSI Incremental Outer-Product Learning Systems," *IEEE Trans. Neural Networks*, vol. 3, no. 3, May 1992.
[45] E. Seevinck, "Analysis and Synthesis of Translinear Integrated Circuits," in *Studies in Electrical and Electronic Engineering*, vol. 31, Elsevier, Amsterdam, 1988.
[46] B. Gilbert, "Current-Mode Circuits from a Translinear Viewpoint: A Tutorial," in *Analogue IC Design: The Current-Mode Approach*, C. Toumazou, F. J. Lidgey, and D. G. Haigh (eds.), IEEE Circuits and Systems Series, vol. 2, Peter Peregrinus, London, 1990.
[47] E. Seevinck and R. J. Wiegerink, "Generalized Translinear Circuit Principle," *IEEE J. Solid-State Circuits*, vol. SC-26, no. 8, pp. 1198–1102, 1991.
[48] K. Yang and A. G. Andreou, "A Multiple Input Differential Amplifier Based on Charge Sharing on a Floating Gate MOSFET," *J. Analog Integrated Circuits Signal Process.*, vol. 6, pp. 197–208, 1994.

[49] B. A. Minch, C. Diorio, P. Hasler, and C. A. Mead, "Translinear Circuits Using Subthreshold Floating-Gate MOS Transistors," *J. Analog Integrated Circuits Signal Process.*
[50] A. Fabre, "Dual Translinear Voltage/Current Convertor," *Electron. Lett.*, vol. 19, no. 24, pp. 1030–1031, 1983.
[51] A. Fabre, "Wideband Translinear Current Convertor," *Electron. Lett.*, vol. 20, pp. 241–242, 1984.
[52] P. M. Furth and A. G. Andreou, "A High-Drive Low-Power BiCMOS Buffer Using Compound PMOS/NPN Transistors," in *Proceedings of the 36th Midwest Symposium on Circuits and Systems*, Detroit, MI, Aug. 1993, pp. 1369–1372.
[53] A. G. Andreou, K. Strohbehn, and R. Jenkins, "A Silicon Retina for Motion Computation," in *Proceedings of the IEEE 1991 International Symposium on Circuits and Systems*, Singapore, June 1991, pp. 1373–1376.
[54] Z. Kalayjian, A. G. Andreou, L. Wolff, and J. Williams "Integrated 1-D Polarization Imagers," in *Proceedings of the 29th Annual Conference on Information Sciences and Systems*, Baltimore, MD, March 1995.
[55] X. Arreguit, E. A. Vittoz, F. A. van Schaik, and A. Mortara, "Analog Implementation of Low-Level Vision Systems," in *Proceedings of European Conference on Circuit Theory and Design*, Elsevier, 1993, pp. 275–280.
[56] T. Delbrück, " 'Bump' Circuits for Computing Similarity and Dissimilarity of Analog Voltages," in *Proceedings of the International Joint Conference on Neural Networks*, Seattle, WA, 1991, pp. 475–479.
[57] M. Cohen and A. G. Andreou, "MOS Circuit for Nonlinear Hebbian Learning," *Electron. Lett.*, vol. 28, no. 9, pp. 809–810, 1992.
[58] M. van der Gevel and J. C. Kuenen, "Square-Root X Circuit Based on a Novel, Back-Gate-Using Multiplier," *Electron. Lett.*, vol. 30, no. 3, pp. 183–184, 1994.
[59] M. H. Cohen and A. G. Andreou, "Analog CMOS Integration and Experimentation with an Autoadaptive Independent Component Analyzer," *IEEE Trans. Circuits Systems: Part II: Analog Digital Signal Proc.*, vol. 42, no. 2, pp. 65–77, 1995.
[60] T. S. Lande, J. A. Nesheim, and Y. Berg, "Auto Correlation in Micropower Analog CMOS," *J. Analog Integrated Circuits Signal Process.*, vol. 7, pp. 61–68, 1995.
[61] B. L. Hart, "Translinear Circuit Principle: A Reformulation," *Electron. Lett.*, vol. 15, no. 24, pp. 801–803, 1979.
[62] C. A. Mead and M. A. Mahowald, "A Silicon Model of Early Visual Processing," *Neural Networks*, vol. 1, pp. 91–97, 1988.
[63] E. Vittoz and X. Arreguit, "Linear Networks Based on Transistors," *Electron. Lett.*, vol. 29, pp. 297–299, 1993.
[64] K. Bult and G. J. G. M. Geelen, "An Inherently Linear and Compact MOST-Only Current Division Technique," *IEEE J. Solid-State Circuit*, vol. SC-27, no. 12, pp. 1730–1735, 1992.
[65] M. Tartagni and P. Perona, "Computing Centroids in Current-Mode Technique," *Electron. Lett.*, vol. 21, 1993.

[66] F. J. Pineda and A. G. Andreou, "An Analog Neural Network Inspired by Fractal Block Coding," in *Advances in Neural Information Processing Systems*, vol. 7, Morgan Kaufmann, San Mateo, 1995.
[67] K. C. Smith and A. S. Sedra, "The Current Conveyor—A New Circuit Building Block," *IEEE Proc.*, vol. 56, pp. 1368–1369, 1968.
[68] K. A. Boahen, P. O. Pouliquen, A. G. Andreou, and R. E. Jenkins, "A Heteroassociative Memory Using Current-Mode MOS Analog VLSI Circuits," *IEEE Trans. Circuits Sys.*, vol. CAS-36, no. 5, pp. 747–755, 1989.
[69] K. A. Boahen, A. G. Andreou, P. O. Pouliquen, and A. Pavasovič, "Architectures for Associative Memories Using Current-Mode Analog MOS Circuits," in *Proceedings of the Decennial Caltech Conference on VLSI*, C. Seitz (ed.), MIT Press, Cambridge, MA, 1989.
[70] E. Säckinger and W. Guggenbühl, "A High-Swing, High Impedance MOS Cascode Circuit," *IEEE J. Solid-State Circuits*, vol. 25, no. 1, pp. 289–298, 1990.
[71] C. Toumazou, F. J. Lidgey, and D. G. Haigh (eds.), *Analogue IC Design: The Current-Mode Approach*, IEEE Circuits and Systems Series, vol. 2, Peter Peregrinus, London, 1990.
[72] G. R. Wilson, "A Monolithic Junction FET-npn Operational Amplifier," *IEEE J. Solid-State Circuits*, vol. SC-3, no. 4, pp. 341–348, 1968.
[73] A. Pavasović, A. G. Andreou, and C. R. Westgate, "Characterization of Subthreshold MOS Mismatch in Transistors for VLSI systems," *J. Analog Integrated Circuits Signal Process.*, vol. 6, pp. 75–84, 1994.
[74] W. Shockley, *Electrons and Holes in Semiconductors*, Van Nostrand, Princeton, NJ, 1963, p. 90; J. B. Gunn, "Thermodynamics of Nonlinearity and Noise in Diodes," *J. Appl. Phys.*, vol. 39, no. 12, pp. 5357–5361, 1968.
[75] R. W. Keyes, *The Physics of VLSI Systems*, Addison-Wesley, Wokingham, England, 1987.
[76] J. E. Dowling, *The Retina: An Approachable Part of the Brain*, Belknap, Harvard University, Cambridge, MA, 1987.
[77] T. Poggio, V. Torre, and C. Koch, "Computational Vision and Regularization Theory," *Nature*, vol. 317, pp. 314–319, 1985.
[78] C. Koch, "Seeing Chips: Analog VLSI Circuits for Computer Vision," *Neural Computat.*, vols 1, 2, pp. 184–200, 1989.
[79] J. J. Ebers and J. L. Moll, "Large-Signal Behavior of Junction Transistors," *Proc. IRE*, vol. 42, pp. 1761–1772, 1954.
[80] R. S. Muller and T. I. Kamins, "Device Electronics for Integrated Circuits," Wiley, New York, 1977.

Chapter 5

Low-Voltage Circuit Techniques Using Floating-Gate Transistors

Chong-Gun Yu
Department of Electronics Engineering, University of Inchon, 177, Tohwa-dong, Namgu, Inchon, Korea
Randall L. Geiger
Department of Electrical and Computer Engineering, Iowa State University, Ames, Iowa 50011

5.1. INTRODUCTION

With an emergence of an increasing number of battery-operated applications, great interest has been aroused in low-voltage circuit techniques. The research efforts for low-supply-voltage-based operation have been focused mainly on digital circuits [1], especially on high-density memory circuits such as dynamic random-access memory (DRAM) and static random-access memory (SRAM) [2, 3]. The present technology trends for low-voltage operation are paralleling the scaling of device feature sizes and threshold voltages. By scaling down the device sizes, power supply voltages (V_{DD}) and threshold voltages (V_T), low-voltage circuits are possible with lower power dissipation while maintaining good operating speed.

The V_{DD} scaling is required for submicrometer processes due to power dissipation and reliability limitations (e.g., hot-carrier effects). As a device is scaled down, the delay time decreases due to decreased loading capacity and increased transconductance. However, the current drive capability cannot be scaled that much because the fringing capacitances of interconnects do not scale [4]. Moreover, the fabrication variations do not scale with dimensions. This will cause noise margin problems if voltages are scaled down with dimensions. The threshold voltage should also be scaled down when scaling down the supply voltages and device sizes because without V_T scaling, the delay will increase.

The V_T scaling, however, has some limitations. The V_T components due to the built-in junction potential and the bulk charge in the channel do not scale as desired. General V_T scaling requires a decreased impurity concentration in the channel, which leads to increased depletion width and thus an increased punch-through current. This is the point where classical scaling breaks down. To scale down V_T without punch-through, more sophisticated *substrate engineering* is required such as reducing the surface doping while increasing average substrate doping. A low-impurity layer lowers the threshold voltage and the highly doped wells prevent punch-through. The maximum doping levels may be limited by junction breakdown. There also exists a lower limit in scaling down the threshold voltage because the scaled-down devices experience problems such as increased leakage currents, short channel effects, and parasitic effects which are much more severe than in large-feature-size devices [5].

One severe problem with device scaling to obtain low-voltage circuits is the increased threshold voltage variation [6–10]. The threshold voltage depends on both device and process parameters such as channel length and width, oxide thickness, junction depth, and substrate doping concentration. It has been shown through statistical analyses [10] that as the channel length decreases, the probability density curve of V_T becomes more broadened, and the standard deviation of the root-mean-square error of V_T increases. This is because the effects of the variation are much higher for short-channel devices. It has also been shown in [6] that the threshold voltage variation increases 200% when the channel width is reduced from 1 to 0.5 μm. For 10% deviations of parameters affecting V_T, the calculated $3\sigma_{V_T}$ is 140 mV for a 1-μm process [11]. The increased V_T variation of scaled devices will have fatal effects on the integrated circuit performance and thus yield. It will lead to poor component matching for analog circuits and poor noise immunity for digital circuits. Although the threshold voltage variation can be minimized by complicated and expensive processes for structural improvements [7] and can be tolerated for digital circuits operating with moderately low supply voltages (>1.5 V), it cannot be tolerated for high-precision analog circuits and circuits operating with very low supply voltages (<1.0 V). These problems can be circumvented if a device called an FGT (floating-gate MOS Transistor) is used as a basic circuit element [12]. The threshold voltage of an FGT can be easily reduced without device scaling and without any substrate engineering, and further, it can be programmed/tuned precisely [13].

Motivated by the unique and promising characteristics of the floating-gate MOS transistors, a new method to obtain low-voltage circuits is discussed in this chapter. The main idea of the new technique is to use FGTs as basic circuit elements for precise V_T control and to reduce V_T in proportion to the supply voltage scaling without scaling the device size significantly. By doing this, the short-channel effects can be relieved and, thus, much lower voltage operation can be attained at the expense of moderately degraded speed. The speed degradation will not be severe. It has been shown that without device size scaling the speed can be improved by choosing optimum supply and threshold voltages [11]. Moreover, several architectural techniques such as parallelism, pipelining, and parallelism-pipelining can be used to compensate for the reduced speed of digital circuits [1]. In Section 5.2, characteristics of the FGTs and the concept of V_T programming are presented. Performance of low-voltage FGT circuits is evaluated in Section 5.3. Low-voltage circuit design issues and conclusions are given in Sections 5.4 and 5.5, respectively.

5.2. CONCEPT OF V_T PROGRAMMING

Floating-gate transistors have been used primarily as data storage devices in erasable programmable read-only memory (EPROM) and electronically erasable programmable read-only memory (EEPROM) circuits [14–18]. Recently, however, the device has started to attract considerable interest as a nonvolatile analog storage device and as a precision analog trim element [19–24] because it has threshold voltage programmability with nearly infinite resolution as well as long-term charge retention. Experimental results have demonstrated that the threshold voltage of a test FGT can be adjusted in submillivolt range increments with a charge loss of less than 2% in 10 years at room temperature [13]. To utilize these salient features of the FGTs in obtaining low-voltage circuits without requiring major architectural changes in the basic circuit structure, floating-gate MOS transistors should have characteristics similar to those of conventional MOS transistors. Thus, the characteristics of FGTs are discussed first. Then methods for adjusting the threshold voltage of an FGT and an on-chip scheme for programming the threshold voltages of FGTs within integrated circuits will be discussed.

5.2.1. Characteristics of Floating-Gate MOS Transistors

A cross section of a typical floating-gate MOS transistor is shown in Figure 5.1. The FGT differs from a conventional MOS transistor in that it has one more gate called the floating gate which is completely isolated within the oxide. The floating gate is capacitively coupled with the control gate, source, substrate, and drain. Their capacitances are denoted as C_g, C_{gs}, C_{gb}, and C_{gd}, respectively. The drain current equation for the FGT can be readily obtained by modifying the equations for the conventional MOS transistor equations. A detailed formulation can be found in [25].

The equation is expressed in the ohmic region as

$$I_{DS}^{0} = \mu C_{\mathrm{ox}} \frac{W}{L}\left(\frac{C_g}{C_T}\right)\left[\left(V_{CGS} - V_{TC0} + \frac{C_{gd}}{C_g} V_{DS} + \frac{C_{gb}}{C_g} V_{BS} + \frac{Q_{FG}}{C_g}\right) \times V_{DS} - \frac{1}{2}\frac{C_T}{C_g} V_{DS}^2\right] \tag{5.1}$$

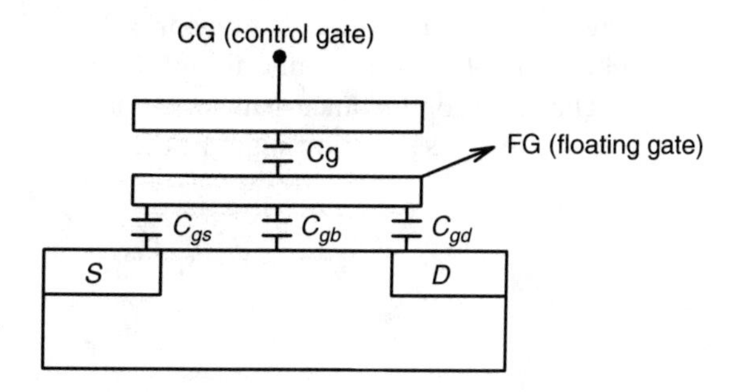

Figure 5.1 Cross section of typical floating-gate MOS transistor.

and in the saturation region as

$$I_{DS}^{s} = \frac{\mu C_{\text{ox}}}{2}\frac{W}{L}\left(\frac{C_g}{C_T}\right)^2\left(V_{CGS} - V_{TC0} + \frac{C_{gd}}{C_g}V_{DS} + \frac{C_{gb}}{C_g}V_{BS} + \frac{Q_{FG}}{C_g}\right)^2 \tag{5.2}$$

where V_{CGS} is the control gate-to-source voltage, Q_{FG} is the amount of charge stored on the floating gate, V_{TC0} is the threshold voltage from the control gate with $Q_{FG} = 0$, and C_T is the total capacitance on the floating gate,

$$C_T = C_g + C_{gs} + C_{gb} + C_{gd}$$

It can be seen from (5.2) that if the drain terminal is strongly capacitively coupled to the floating gate, then the drain current in the saturation region is not independent of the drain voltage any more but will increase continuously with the drain voltage and will not saturate. However, if the capacitive coupling between the drain and the floating gate is small, then the *I-V* characteristics of the FGT become similar to those of a conventional MOS transistor. In fact, the saturation current equation (5.2) reduces to the well-known square-law equation if C_{gd} and C_{gb} are negligible compared to C_g. The capacitance between the floating gate and the control gate, C_g, can be readily made to be much greater than the others by using layout techniques for the floating-gate shape such as dog-bone [14], multifinger [20], and a single large coupling area [21].

Assuming $C_g \gg C_{gd}, C_{gb}$, Eqs. (5.1) and (5.2) can be approximated as

$$I_{DS}^{0} \simeq \beta_{\text{eff}}^{0}\left[(V_{CGS} - V_{T,\text{eff}})V_{DS} - \frac{1}{2}\frac{C_T}{C_g}V_{DS}^2\right]$$

$$I_{DS}^{s} \simeq \beta_{\text{eff}}^{s}(V_{CGS} - V_{T,\text{eff}})^2$$

where

$$\beta_{\text{eff}}^{0} = \mu C_{\text{ox}}\frac{W}{L}\left(\frac{C_g}{C_T}\right) \qquad \beta_{\text{eff}}^{s} = \mu C_{\text{ox}}\frac{W}{L}\left(\frac{C_g}{C_T}\right)^2 \qquad V_{T,\text{eff}} = V_{TC0} - \frac{Q_{FG}}{C_g} \tag{5.3}$$

By intentionally making C_g large, the floating-gate MOS transistor has characteristics similar to those of conventional MOS transistors while its effective threshold voltage, $V_{T,\text{eff}}$, can be adjusted by controlling the charge on the floating gate Q_{FG}, as can be seen in (5.3). The similarity has been validated by experiments in [25] and [20].

The transconductance gain and the output conductance of the floating gate transistor in the saturation region become

$$g_m \simeq \sqrt{2\beta_{\text{eff}}^{s} I_{DS}} = \frac{C_g}{C_T}g_m^c$$

$$g_d \simeq \lambda I_{DS} + \frac{C_{gd}}{C_g}g_m = g_d^c + \frac{C_{gd}}{C_g}g_m$$

where λ is the channel length modulation parameter and g_m^c and g_d^c are the transconductance and the output conductance of a conventional MOS transistor. These equations indicate that the FGT has a smaller transconductance and a larger output conductance than a conventional MOS transistor.

5.2.2. Methods for V_T Adjustment

It has been shown in the previous section that the threshold voltage of a floating-gate MOS transistor can be adjusted by changing the amount of the charge on the floating gate. According to Eq. (5.3), the operation transferring electrons onto the floating gate (programming) will increase the threshold voltage while the operation transferring electrons from the floating gate (erasing) will decrease it.

Two programming mechanisms, channel hot-electron injection and Fowler-Nordheim (FN) tunneling, have been widely used for digital EEPROM circuits. Programming using channel hot-electron injection has better control of the program-disturb problem and thus has been preferably used in flash EEPROM circuits [16–18], whereas FN tunneling is used for the erasing operation. Although FN tunneling needs higher voltages than hot-electron injection, it requries very little current, which makes generation of on-chip high voltages much easier. For analog applications, the primary interests are in accurate control of the floating-gate charge and good charge retention. Thus, FN tunneling for both operations are preferred for analog applications because of the exponential relation between programming voltage and current which allows accurate control of the floating-gate charge [20–24].

One popular way to adjust V_T based on FN tunneling is to use EEPROM technologies with a very thin gate oxide for tunneling by applying a high voltage (V_{pp}) at the control gate for programming and at the drain terminal for erasing, as shown in Figure 5.2. When a large enough field is present across the gate oxide, in most existing FGTs, FN electron tunneling allows charge to be transferred to or from the floating gate, depending on the polarity of the field. The charge amount to be transferred depends on the magnitude and duration of the programming pulse that is needed to produce a large enough electric field in the tunnel oxide. Since charge transfer to or from the floating gate affects the threshold voltage of the FGT, three variables, the magnitude, polarity, and duration of the programming pulse, can be used to control the threshold voltage.

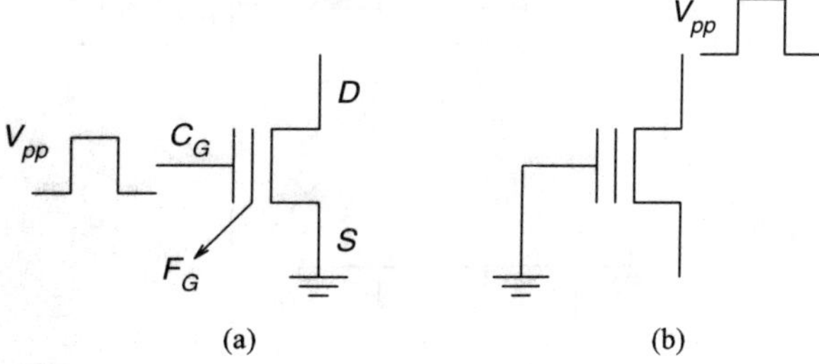

Figure 5.2 Typical V_T adjustment of FGTs using EEPROM technologies: (a) programming (increasing V_T); (b) erasing (decreasing V_T).

Recently, FGTs fabricated with conventional standard CMOS technologies have been reported by Carley [23], Thomsen [21], and Yang [20], where any special processing associated with expensive EEPROM technologies such as ultra-thin oxide and textured polysilicon can be avoided. The three FGTs are depicted in Figure 5.3. The FGT proposed by Carley, which uses field enhancement at the edge of the polysilicon over the diffusion region, has different characteristics between programming and erasing operations resulting in difficult bidirectional control. On the other hand, the FGTs proposed by Thomsen and Yang use the

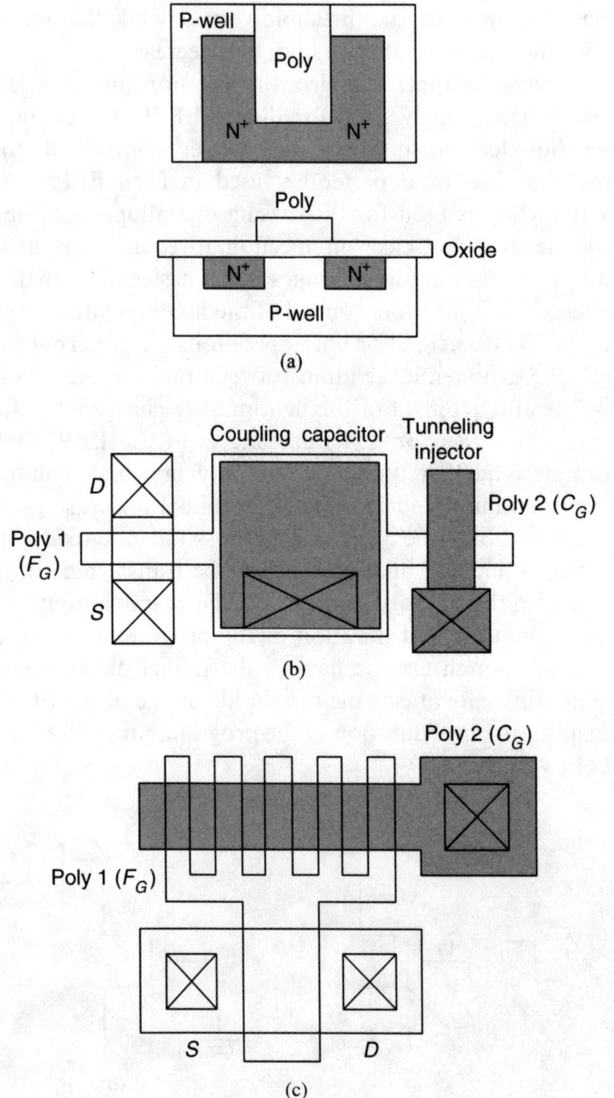

Figure 5.3 FGTs that can be fabricated with standard CMOS processes proposed by (a) Carley [23], (b) Thomsen [21], and (c) Yang [20].

polysilicon oxide as a tunneling area instead of the gate oxide. The sharp edge of the bottom polysilicon, which is used as the floating gate, causes the field enhancement and the thinning of the oxide which can reduce the operating voltages. This scheme is promising because it can be implemented with standard complementary MOS (CMOS) double-polysilicon processes. It is also possible to apply both the programming and the erasing voltages at the control gate only, avoiding use of the drain terminal. However, this will require two different on-chip high-voltage generation circuits.

5.2.3. A V_T Programming/Tuning Scheme

To utilize the FGTs as circuit elements, their threshold voltages must be programmed and tuned using on-chip automatic techniques. A threshold voltage tunable circuit structure that can be operated with a very low power supply (e.g., 0.5 V) is thus presented in this section [12]. Although this scheme can be implemented with both V_T adjustment methods associated with EEPROM and standard CMOS technologies, the method using EEPROM technologies shown in Figure 5.2 is used here to avoid the requirement of two different on-chip high voltages. This scheme can be applicable for both analog and digital circuits. Good matching, which is of great concern for analog circuits, can also be achieved by tuning the threshold voltages.

Threshold Voltage Tunable Circuit Structure. One method of tuning the threshold voltage entails placing the FGTs in an array, as shown in Figure 5.4. Each cell terminology is shown in Figure 5.5. Each cell consists of six transistors: an FGT, three select transistors (S_D, S_G, and S_S) and two switch transistors (S_1 and S_2). The select MOS transistors are required to tune the threshold voltage of the selected FGT only, and thus, the other FGTs that are not selected will not be affected by the tuning process. The switch transistors are used to connect or disconnect the FGT with other FGTs. A switch transistor at the source terminal of the FGT is not required since high voltages are not applied at the source terminal during the threshold voltage tuning. Although each cell has two switch transistors, as shown in Figure 5.4, two switch transistors are not always required for all cells. The number of switch transistors can be reduced and depends upon the circuit topology.

The threshold voltage (V_T) tunable circuit has two operating modes: a V_T tuning mode and a normal mode. In a V_T tuning mode the cells are disconnected from the main circuit and sequentially selected through the row and column selection lines so that the threshold voltages of the FGTs can be tuned. In a normal mode the cells are connected to each other according to the circuit topology by turning on the switch transistors S_1 and S_2 in Figure 5.5. The signal C_E is used to connect or disconnect the cells from the circuit.

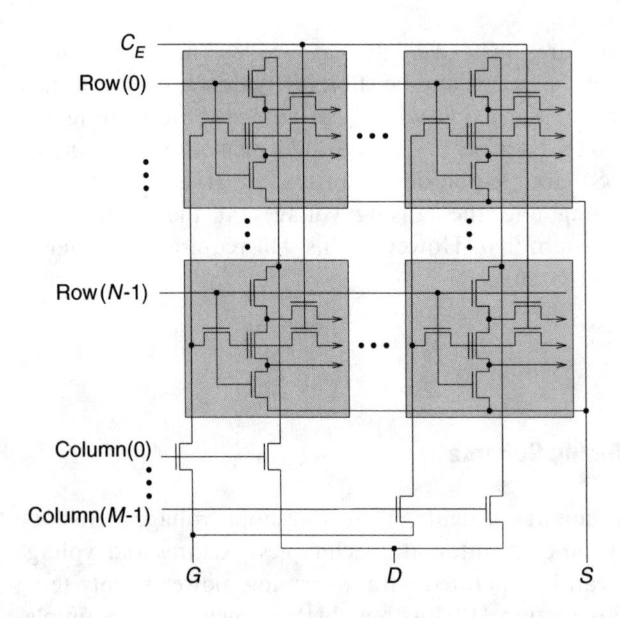

Figure 5.4 Floating-gate MOS transistor array.

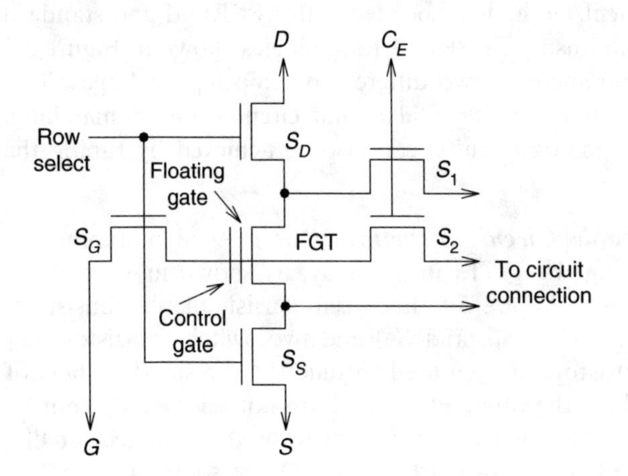

Figure 5.5 Floating-gate MOS transistor cell terminology.

V_T Tuning Strategy. One V_T tuning strategy is discussed here. The simplified entire block diagram of the V_T tunable low-voltage circuit is shown in Figure 5.6. A counter, a row decoder, and a column decoder can be used to sequentially select the cells of the FGT array. The V_T tuning is performed in two steps: a coarse tuning and a fine tuning. The coarse tuning is a preliminary step to provide an environment where the on-chip charge pump and the main circuit are capable of operating with a low voltage. The fine tuning is for providing good matching properties and a desired operating point.

In the coarse tuning, all the FGTs that are the elements of the FGT array are approximately programmed in a one-tuning cycle to have a very low threshold

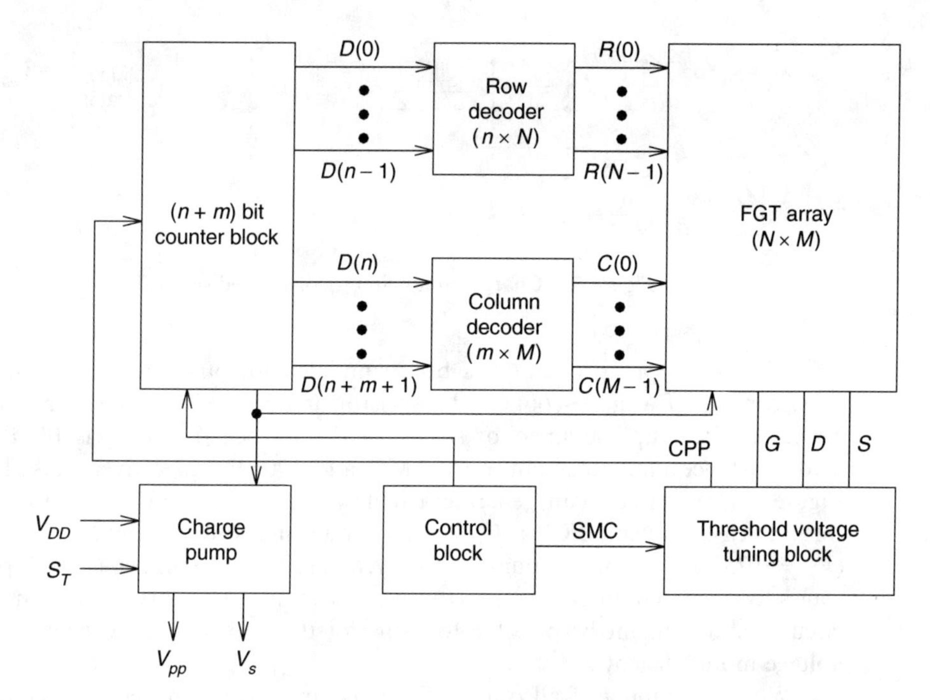

Figure 5.6 Simplified block diagram of V_{th} tunable low-voltage circuit (CCP = counter clock pulse; SMC = select measurement control).

voltage (e.g., 100 mV) using an external high voltage (e.g., 20 V). The coarse tuning can be performed using either a closed-loop mode or an open-loop mode. This action is performed only once. After the coarse tuning the entire circuit operates with a very small power supply (e.g., $V_{DD} = 0.5$ V). The fine tuning is performed under the external control signal S_T. Whenever the signal S_T is detected, the fine tuning is performed, and the circuit automatically returns to a normal mode when the fine tuning is finished. The fine tuning need not be a frequent event because of the long-term charge retention characteristics of the FGTs. This can afford the possibility of near continuous-time operation.

In a fine-tuning mode a high voltage is also required to adjust the threshold voltages of the FGTs. The high voltage V_{pp} can be developed from V_{DD} with an on-chip charge pump, and thus, no external high voltages are required. A charge pump circuit is shown in Figure 5.7, which consists of an oscillator, diode-connected floating-gate MOS transistors, capacitors, and a voltage regulator. The detailed operation principles of the charge pump can be seen in [26]. To make the charge pump operate with a low power supply, the oscillator is also constituted of FGTs. The threshold voltages of the FGTs in the charge pump circuit are also adjusted to a very low value (e.g., 100 mV) during the coarse tuning step. After that the charge pump can generate a high voltage V_{pp} from V_{DD}, and the internally generated V_{pp} is used for the fine tuning.

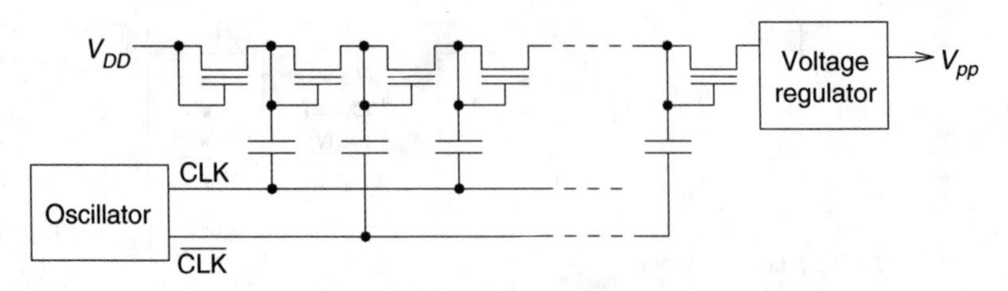

Figure 5.7 Charge pump circuit constituted of FGTs.

The charge pump circuit has been simulated for 50 stages. A multiplication stage consists of a diode-connected transistor and a coupling capacitor. The capacitance of the coupling capacitor is 2 pF, and the clock frequency is 10 MHz. The load resistance and capacitance are 1 MΩ and 100 pF, respectively. As shown in Figure 5.8, the circuit can generate a high voltage (over 20 V) from $V_{DD} = 0.5\,V$ for both cases of using FGTs ($V_T = 0.1$ V) and using conventional MOS transistors ($V_T = 0.75$ V). The charge pump circuit requires a large number of multiplication stages to generate a high voltage. However, there are various types of charge pump circuits, and it might be possible for some of them to be able to generate a high voltage more efficiently.

When the tuning of all cells in the FGT array is completed, the circuit automatically returns to a normal mode, and the oscillator of the charge pump circuit is disabled. Thus, the charge pump does not generate the high voltage V_{pp} during the normal mode and is left in a state where it awaits another S_T signal. In the normal mode the switch transistors must be turned on for the circuit to function correctly. Since the switch transistors are conventional MOS transistors, they will not be turned on by the very low supply voltage. Hence, another charge pump circuit is

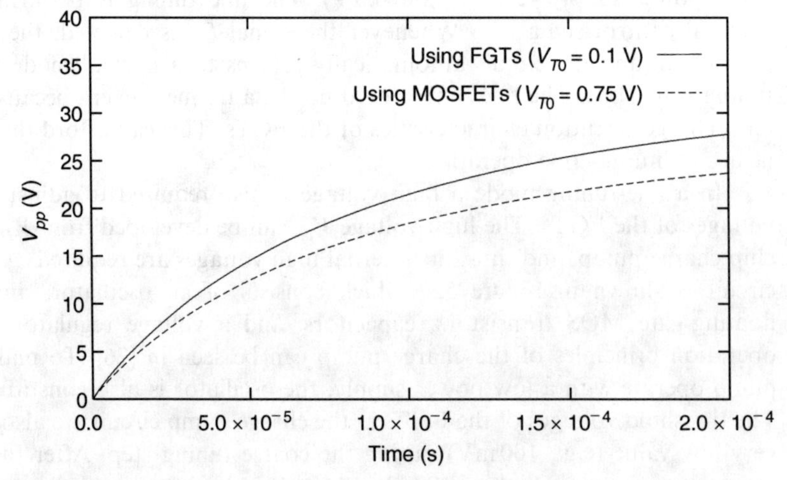

Figure 5.8 Simulation results of charge pump circuit ($V_{DD} = 0.5$ V, $f_{CLK} = 1$ MHz, $R_L = 1\,M\Omega$, $C_L = 100$ pF).

required which can generate a voltage that is high enough to turn the switch transistors on in the normal mode. The power dissipation by the generated voltage will be very small since it is applied only at the gate terminals of the MOS transistors and possibly to one or two small drain or source diffusions. The frequency of the charge pump oscillator for the switch transistors does not have to be high because the load resistance is very high.

There can be many fine-tuning methods to obtain a good matching property and a desired operation of the coarsely tuned circuit. Two fine-tuning methods are discussed here. The first method is to adjust the threshold voltages of all FGTs to a predetermined value. The second method is to adjust the intermediate node voltages of the circuit to preassigned values by adjusting the threshold voltages. The first method will be simpler and more generally applicable to all kinds of circuit structures than the second method. However, the second method will give better matching results because it can provide node voltage matching and hence compensation for other parameter variations as well while the first method can provide only threshold voltage matching of the presumably matched transistors.

The block diagram of the fine-tuning block for threshold voltage tuning is shown in Figure 5.9. During a V_T tuning mode this block alternately measures the threshold voltage of a selected FGT and adjusts it to a desired threshold voltage. It is controlled by the control signal SMC generated from the control block. The V_T control block shown in Figure 5.10 compares the measured V_T with the desired V_T and then determines the adjustment direction and also determines through a zero crossing detector whether to keep adjusting or to finish it. If further adjusting is required, the unit pulse generator generates a unit programming pulse for V_T adjustment. The resolution of the V_T tuning depends on the magnitude and width of the unit programming pulse.

The accuracy of the tuning results also depends on the performance of the V_T measurement block. A simple scheme can be used to measure the threshold voltage. For example, $V_T \approx V_{GS}$ may be assumed at low I_{DS} [27]. This method is very simple,

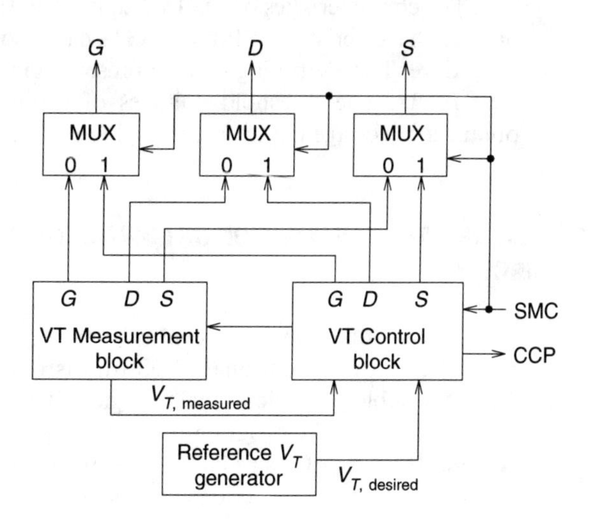

Figure 5.9 Block diagram of fine-tuning block.

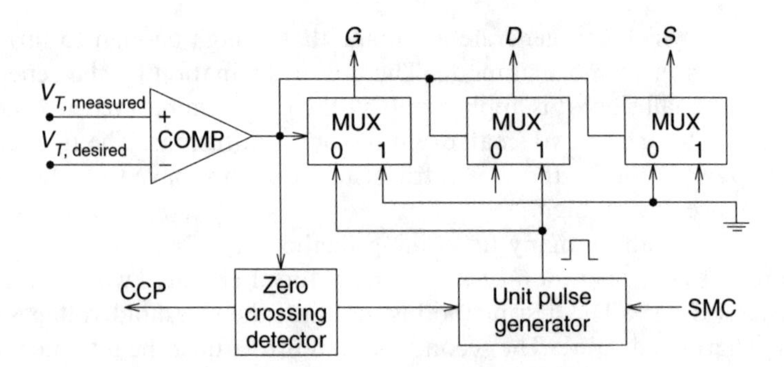

Figure 5.10 The V_T control block.

but the measurement error is somewhat large. To obtain more accurate tuning, a more complicated measurement circuit [28] is required at the expense of much large area. This topic will be discussed in a later section.

To obtain a better transistor pair matching of the circuit, the internal node voltages can be adjusted to preassigned values by adjusting the threshold voltages. For node voltage fine tuning, a sequential tuning method for a circuit structure is required because the adjustment of one node voltage may affect the other node voltages. The order of the FGTs to be tuned must be carefully determined, and the FGTs must be placed in the FGT array according to the order. To use the node voltage fine-tuning method, the circuit should be changed a little. Switch transistors for the nodes to be tuned are additionally required in the FGT array for comparison with preassigned values. No measurement circuit is, however, required because the node voltages are directly compared with the preassigned values, as defined by a reference voltage generator which can generate the same number of reference voltages as the number of nodes to be tuned. The reference voltages can be selected by the same counter and decoding circuits that are used for the FGT array.

The characteristics of FGTs discussed in this section are for N-type FGTs only, but the characteristics of P-type FGTs can also be easily investigated using the same procedure. The V_T tuning scheme presented in this section can also be applied for P-type FGTs. The threshold voltages of P-type FGTs also need to be trimmed to obtain low-voltage complementary FGT circuits.

5.3. PERFORMANCE EVALUATION OF LOW-VOLTAGE FGT CIRCUITS

It has been shown in the previous sections that FGTs can have very similar characteristics with conventional MOS transistors and can be fabricated using standard CMOS double-polysilicon technologies. It has also shown that the threshold voltages of FGTs can be programmed/tuned by an on-chip automatic technique, making it possible to use FGTs as basic elements for low-voltage circuits. Now, the effects of voltage scaling on the circuit performance will be investigated. As mentioned in

Section 5.1, the supply voltage and the threshold voltage are scaled down without scaling the device size significantly in order to reduce the problems associated with small-size transistors.

Three scaling laws have been used, namely constant-electric-field (CE), constant-voltage (CV), and quasi-constant-voltage (QCV) scaling laws. The CE scaling proposed by Dennard [29] improves circuit performance by reducing voltage swing and capacitances while it ensures high reliability by keeping the electric fields constant. However, the performance improvement brought by the CE scaling is not as good as expected by the first-order law because of second-order effects such as nonscalability of process parameter variations and logarithmic scaling of interconnection capacitances due to fringing effects [30]. The CV scaling has been widely used to achieve higher circuit performance and transistor-transistor-logic (TTL) compatibility. However, the CV scaling will cause reliability problems for scaled-down devices, and it is not suitable for low-voltage and low-power circuit applications. Taking into account the trade-off between circuit performance and reliability, the QCV scaling law was proposed by Chatterjee [30]. It has been shown in [30] that the QCV scaling can provide optimum current drive capability in digital circuits.

Studies on the performance of scaled circuits have been concentrated mainly on digital or memory circuits [1, 5, 11, 29–34]. Effects of the scaling laws on the performance of analog circuits have been investigated by only a few researchers [35–37]. Wong [35] and Enomoto [36] have investigated the first-order scaling laws and have shown that the first-order scaling theories agree well with simulation and experimental results for scaled devices with channel length down to 2 μm. They also have shown that the QCV scaling law is optimum and gives the best overall analog performance, resulting in increase of density and unity-gain bandwidth with moderate degradation of gain and signal-to-noise ratio. However, Sano [37] has shown that the short-channel effects such as mobility degradation and drain-induced barrier lowering reduce the performance improvement expected by the first-order theories for further shrinking devices to submicrometer channel lengths.

Now, we will investigate new scaling laws (CD, QCD1, and QCD2) for obtaining low-voltage circuits by mainly reducing the operating voltages. These scaling laws are shown in Table 5.1 along with the conventional scaling laws. In the constant-dimension (CD) scaling, only the voltages are scaled while the device dimensions remain unchanged. In the quasi-constant-dimension scaling laws (QCD1 and QCD2), the dimensions are also scaled, but slowly, which can somewhat compensate for the performance degradation expected by the CD scaling.

TABLE 5.1 SCALING LAWS

Parameter	CE	QCV	CV	CD	QCD1	QCD2
Voltages (V)	s^{-1}	$s^{-1/2}$	1	s^{-1}	s^{-1}	s^{-1}
Lateral dimensions (L, W)	s^{-1}	s^{-1}	s^{-1}	1	1	$s^{-1/2}$
Vertical dimensions (T_{ox})	s^{-1}	s^{-1}	$s^{-1/2}$	1	$s^{-1/2}$	$s^{-1/2}$
Substrate doping (N_B)	s	s	s	1	1	$s^{1/2}$

5.3.1. First-Order Scaling

Since the FGT and the MOS transistors have similar characteristics, the conventional MOS field-effect transistor (MOSFET) models will be used for modeling the FGTs to test the FGT circuit performance. Given the scaling laws in Table 5.1, the performance of the basic MOS parameters have been evaluated and are shown in Table 5.2. The MOS parameters that will be investigated are modeled by the following first-order model equations:

- Drain current in the saturation region:

$$I_{DS} = \frac{\mu C_{\text{ox}}}{2} \frac{W}{L} (V_{GS} - V_T)^2 \propto \left(\frac{V^2}{t_{\text{ox}}}\right)$$

- Transconductance:

$$g_m = \mu C_{\text{ox}} \frac{W}{L} (V_{GS} - V_T) \propto \left(\frac{V}{t_{\text{ox}}}\right)$$

- Output conductance:

$$g_d = \frac{I_{D,\text{sat}}\sqrt{2\varepsilon_{si}/qNB}}{2L\sqrt{V_{DS} - V_{D,\text{sat}}}} \propto \frac{I}{L}\left(\frac{1}{VN_B}\right)^{1/2}$$

- Subthreshold current ($V_{DS} \gg V_t$ and $V_{BS} = 0$):

$$I_D \simeq \frac{W}{L} I_{D0} \exp\left(\frac{V_{GS} - V_{\text{on}}}{nV_t}\right)$$

TABLE 5.2 SCALED DEVICE PARAMETERS

Parameter	CE	QCV	CV	CD	QCD1	QCD2
I_{DS}	s^{-1}	1	$s^{1/2}$	s^{-2}	$s^{-3/2}$	$s^{-3/2}$
g_m	1	$s^{1/2}$	$s^{1/2}$	s^{-1}	$s^{-1/2}$	$s^{-1/2}$
g_d	1	$s^{3/4}$	s	$s^{-3/2}$	s^{-1}	$s^{-3/4}$
$C_D/C_{\text{ox}}, nVt, S\,(V_{BS} \neq 0)$	1	$s^{-1/4}$	1	$s^{1/2}$	1	$s^{1/4}$
$C_D/C_{\text{ox}}, nVt, S\,(V_{BS} = 0)$	$s^{-1/2}$	$s^{-1/2}$	1	1	$s^{-1/2}$	$s^{-1/4}$
v_{nf}	1	1	$s^{1/2}$	1	$s^{-1/2}$	1
$v_{nT}/\sqrt{\Delta f}$	1	$s^{-1/4}$	$s^{-1/4}$	$s^{1/2}$	$s^{1/4}$	$s^{1/4}$
τ	s^{-1}	$s^{-3/2}$	s^{-2}	s	s	1
P	s^{-2}	$s^{-1/2}$	$s^{1/2}$	s^{-3}	$s^{-5/2}$	$s^{-5/2}$
P/A	1	$s^{3/2}$	$s^{5/2}$	s^{-3}	$s^{-5/2}$	$s^{-3/2}$

where

$$V_t = \frac{kT}{q}$$
$$V_{\text{on}} = V_T + nV_t$$
$$n = 1 + \frac{q\,\text{NFS}}{C_{\text{ox}}} + \frac{C_D}{C_{\text{ox}}}$$
$$C_D = \left[\frac{q\varepsilon_{si}N_B}{2(2\phi_F - V_{BS})}\right]^{1/2}$$
$$\frac{C_D}{C_{\text{ox}}} \propto t_{\text{ox}}\left(\frac{N_B}{V}\right)^{1/2}$$

- Difference between V_{on} and V_T:

$$\begin{aligned} nV_t &= V_{\text{on}} - V_T \\ &= 1 + \frac{q\,\text{NFS}}{C_{\text{ox}}} + \frac{C_D}{C_{\text{ox}}} \\ &\propto t_{\text{ox}}\left(\frac{N_B}{V}\right)^{1/2} \end{aligned} \tag{5.4}$$

- Gate voltage swing needed to reduce the drain current a decade in subthreshold:

$$\begin{aligned} S &= \left[\frac{\partial(\log I_{DS})}{\partial V_{GS}}\right]^{-1} \\ &\simeq \frac{kT}{q}(\ln 10)\left(1 + \frac{C_D}{C_{\text{ox}}}\right) \\ &\propto t_{\text{ox}}\left(\frac{N_B}{V}\right)^{1/2} \end{aligned}$$

- Root-mean-square flicker noise voltage:

$$v_{nf} = \left(\frac{a_n}{WL}\frac{q^2}{C_{\text{ox}}^2}\frac{\Delta f}{f}\right)^{1/2} \propto \left(\frac{t_{\text{ox}}}{\sqrt{WL}}\right)$$

- Root-mean-square thermal noise voltage:

$$v_{nT} = \left(\frac{8}{3}\frac{kT}{g_m}\Delta f\right)^{1/2}$$
$$\frac{v_{nt}}{\sqrt{\Delta f}} \propto \left(\frac{1}{g_m}\right)^{1/2}$$

- Intrinsic gate delay:

$$\tau = \frac{V_{DD}}{I_{DS}} C_{\text{ox}} WL$$

- Power dissipation:

$$P = I_{DS} V_{DS}$$

- Power dissipation density:

$$\frac{P}{A} = \frac{I_{DS} V_{DS}}{LW}$$

The term NFS is the subthreshold region model fitting parameter.

Compared to the conventional scaling laws (CE, QCV, and CV), CD scaling gives performance degradation in current drive capability, transconductance gain, thermal noise density, and delay but results in the lowest power dissipation and power dissipation density and also improved output conductance. Thus, the CD scaling law can be used for very low power (or very low voltage) circuits with limited circuit performances. The performance degradation by the CD scaling can be partially compensated by using the QCD scaling laws (QCD1 and QCD2) at the expense of increased power dissipation, as can be seen in Table 5.2. To evaluate the accuracy of the first-order CD scaling theory it is compared with SPICE simulation results. Figure 5.11 shows that the basic MOS parameters scaled by the first-order theory agrees well with the SPICE simulation results for two different geometries.

Figure 5.12 shows the effects of the V_T scaling on V_{T0}, V_{on}, and nV_t, where nV_t is the difference between V_{on} and V_T as defined in (5.4). The differently denoted threshold voltages are defined as follows:

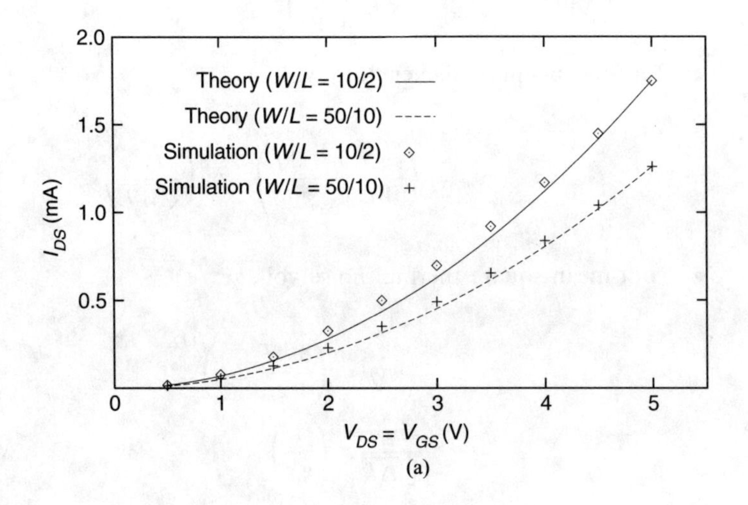

(a)

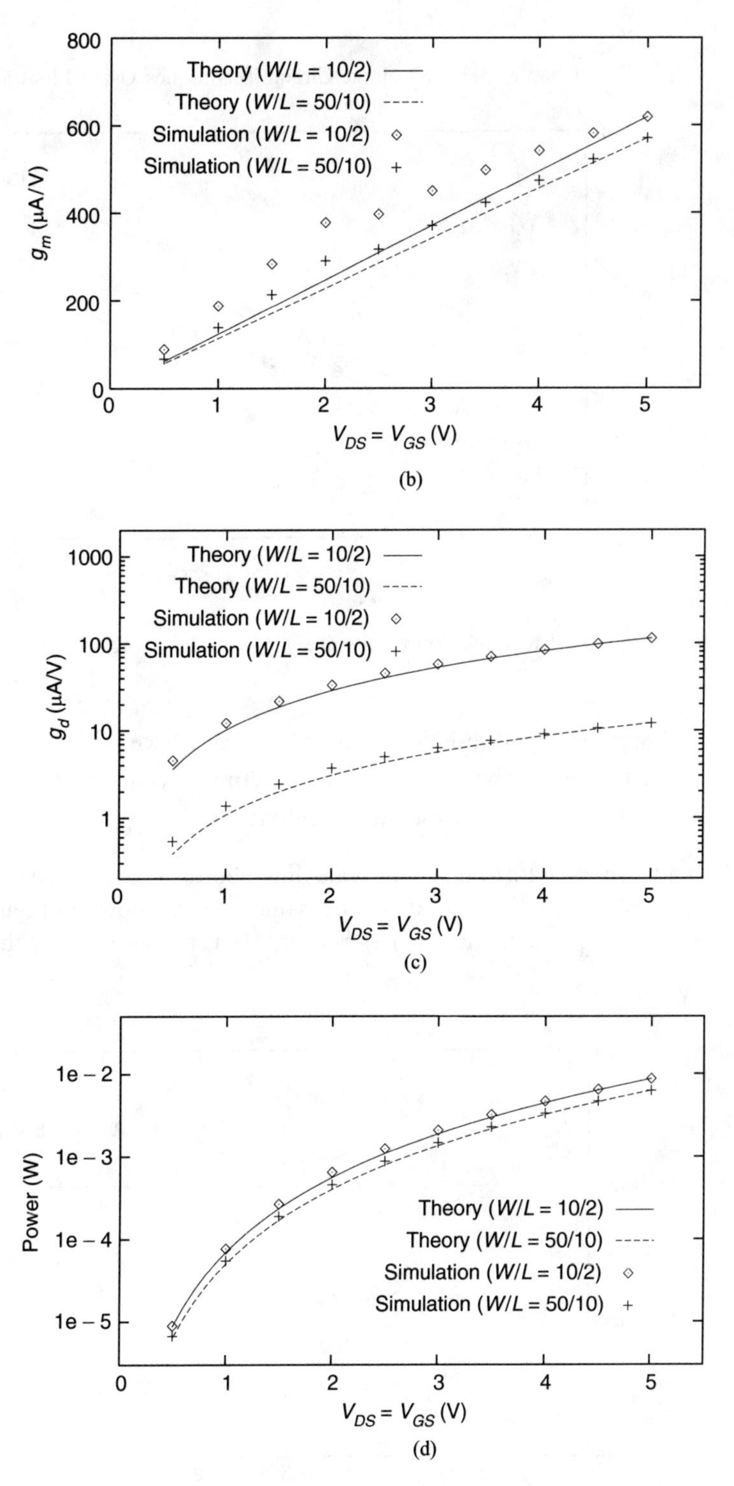

Figure 5.11 Comparison of the first-order CD scaling theory with simulation results ($V_{\text{th}} = 0.2V_{DS}$): (a) drain current I_{DS}; (b) transconductance g_m; (c) output conductance g_d; (d) power ($I_{DS} \times V_{DS}$).

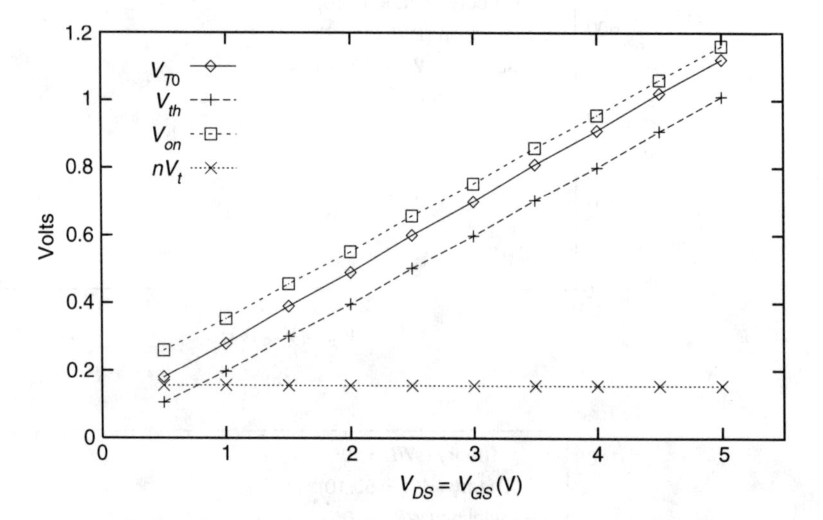

Figure 5.12 Simulation results of scaled threshold voltages ($V_T = 0.2V_{DS}$).

V_{T0}: Zero-bias threshold voltage of a large device
V_T: Including device size effects and terminal voltage effects
V_{on}: Including subthreshold current effects

It can be seen that nV_t does not scale down but is almost constant. This is because V_{on} scales down with V_T at almost the same rate. As shown in Figure 5.13, the effect of V_{SB} on nV_t is small, and if V_{SB} is zero, nV_t is not affected by the voltage scaling.

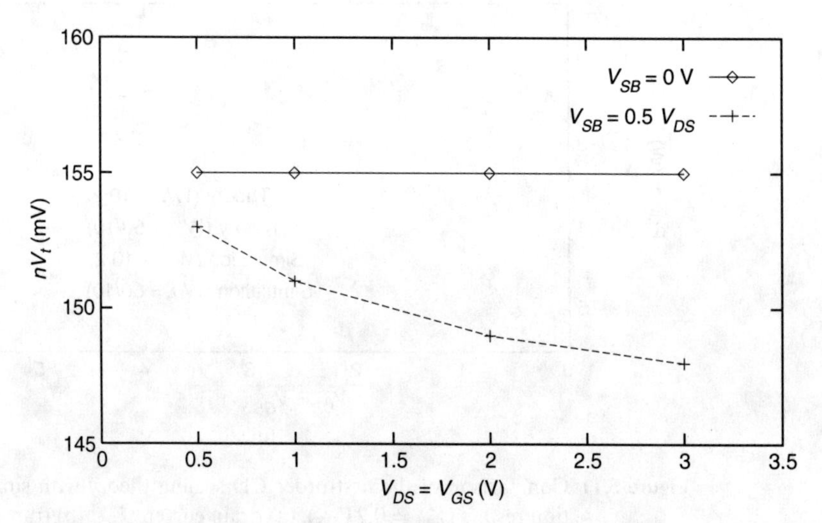

Figure 5.13 Simulation results showing effects of V_{SB} on scaling of $nV_t (V_T - 0.2V_{DS})$.

This should be taken into account when adjusting and/or programming the threshold voltage (V_T) of a FGT in order not to drive it into the subthreshold region.

5.3.2. Effects of Voltage Scaling on Analog Circuits

It is very important to evaluate the effects of voltage scaling on circuit performance. Although very low voltage/power circuits are obtainable using FGTs and scaling them based upon the CD or QCD laws, other performance degradation caused by the scaling should lie in an acceptable range. To investigate this, the CMOS differential stage shown in Figure 5.14 is selected. The basic performance parameters to be evaluated and their model equations that are well known are as follows, where i, l, and o subscripts denote the input transistors (M_1 and M_2), the load transistors (M_3 and M_4), and the transistor (M_5), respectively.

- Voltage gain:

$$A_v \simeq \frac{g_{mi}}{g_{di} + g_{dl}} \propto \left(\frac{g_m}{g_d}\right)$$

- Common-mode rejection ratio:

$$\text{CMRR} \simeq \frac{2g_{mi}g_{ml}}{g_o g_{di}} \propto \left(\frac{g_m}{g_d}\right)^2$$

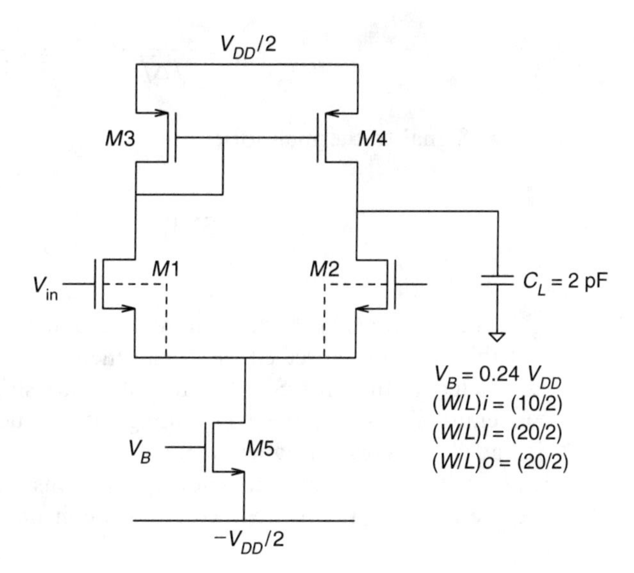

Figure 5.14 CMOS differential stage.

- Load capacitance:

$$C_L \propto AC_{\text{ox}} \propto \left(\frac{WL}{t_{\text{ox}}}\right)$$

- Bandwidth (−3 dB frequency):

$$\text{BW} \simeq \frac{g_{di} + g_{dl}}{C_L} \propto \left(\frac{g_d}{C_L}\right)$$

- Unity-gain frequency:

$$\text{UGF} \simeq A_v \text{BW} = \frac{g_{mi}}{C_L} \propto \left(\frac{g_m}{C_L}\right)$$

- Slew rate:

$$\frac{\text{SR}}{V_{DD}} \simeq \frac{I_D}{V_{DD} C_L} \propto \left(\frac{1}{VC_L}\right)$$

- Thermal noise density:

$$\frac{v_{nT}}{\sqrt{\Delta f}} \simeq \left[\frac{16kT}{3}\frac{1}{g_{mi}}\left(1 + \frac{g_{ml}}{g_{mi}}\right)\right]^{1/2} \propto \left(\frac{1}{g_m}\right)^{1/2}$$

- Root-mean-square thermal noise voltage:

$$v_{nT} = \frac{v_{nT}}{\sqrt{\Delta f}}\sqrt{\text{BW}} \propto \left(\frac{g_d}{C_{Lg_m}}\right)^{1/2}$$

- Signal-to-thermal-noise ratio:

$$\text{SNR}_T \propto V_s \left(\frac{C_L g_m}{g_d}\right)^{1/2}$$

These parameters have been evaluated according to the first-order scaling laws in Table 5.1 and the scaled MOS parameters in Table 5.2. The scaled performance parameters of the CMOS differential stage are shown in Table 5.3. Compared to the conventional scaling, the CD scaling gives much degraded bandwidth and thus degraded unity-gain frequency and slew rate, which suggests that the CD scaling is not suitable for high-frequency applications, as expected. The CD scaling, however, results in not only the lowest power consumption but also improved open-loop

TABLE 5.3 SCALED PERFORMANCE PARAMETERS OF CMOS DIFFERENTIAL STAGE

Parameter	CE	QCV	CV	CD	QCD1	QCD2
A_{V_0}	1	$s^{-1/4}$	$s^{-1/2}$	$s^{1/2}$	$s^{1/2}$	$s^{-1/4}$
CMRR	1	$s^{-1/2}$	s^{-1}	s	s	$s^{-1/2}$
C_L	s^{-1}	s^{-1}	$s^{3/2}$	1	$s^{1/2}$	$s^{-1/2}$
BW	s	$s^{7/4}$	$s^{5/2}$	$s^{-3/2}$	$s^{-3/2}$	$s^{1/4}$
UGF	s	$s^{3/2}$	s^2	s^{-1}	s^{-1}	1
SR/V_{DD}	s	$s^{3/2}$	s^2	s^{-1}	s^{-1}	1
$v_{nT}/\sqrt{\Delta f}$	1	$s^{-1/4}$	$s^{-1/4}$	$s^{1/2}$	$s^{1/4}$	$s^{1/4}$
SNR_T	$s^{-3/2}$	$s^{-9/8}$	s^{-1}	$s^{-3/4}$	$s^{-1/2}$	$s^{-11/8}$
Power	s^{-2}	$s^{-1/2}$	$s^{1/2}$	s^{-3}	$s^{-5/2}$	$s^{-5/2}$

voltage gain and thus better CMRR owing to the improved output conductance indicated in Table 5.2. Despite the increased noise density, the signal-to-noise ratio of the CD scaling is less degraded due to reduced bandwidth. Consequently, the FGT circuit scaled by the CD law can be well applicable to very low voltage/power circuits but limited speed applications.

It must be questioned how low-voltage operation is possible and how low-voltage circuits scaled by the CD law can maintain the scaled performance parameters to follow the results expected by the first-order law as shown in Table 5.3. To investigate this, the CMOS differential stage has been simulated using SPICE, where the transistor sizes are shown in Figure 5.14. The effects of the voltage scaling according to the CD law on the circuit performance parameters such as open-loop voltage gain, bandwidth, unity-gain frequency, power consumption, and signal-to-noise ratio have been simulated and depicted in Figures 5.15(b), (c), (d), (e), and (f), respectively. In this simulation the power supply voltage V_{DD} has been scaled from 5 to 0.5 V, and the threshold voltage scaling has been performed based on two conditions as follows:

Condition 1: $V_T = 0.1 V_{DD}$
Condition 2: $V_{\text{on}} = 0.1 V_{DD}$

The threshold voltages V_T and V_{on} have been defined in the previous section. As expected in Table 5.3, the open-loop gain increases and the bandwidth and unity-gain frequency decrease as V_{DD} is scaled down.

It can be seen from Figure 5.15(a) that the input transistors already step into the subthreshold region at $V_{DD} = 1V$, that is V_{GS} is less than V_{on}, when the threshold voltages are scaled down according to condition 1. Thus, the performance parameters start to deviate from the expected values, which would result in much degraded circuit performance with further scaled-down V_{DD}. In contrast, with the threshold voltages scaled down according to condition 2, the input transistors are close to the subthreshold region but still in the strong-inversion region when V_{DD} are

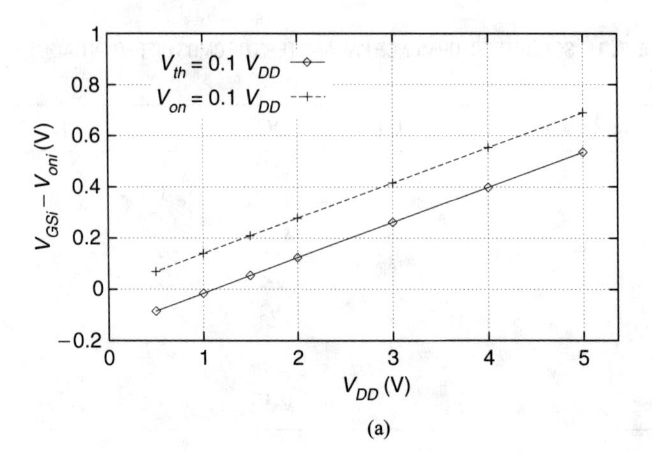
V_{th} = 0.1 V_{DD}
V_{on} = 0.1 V_{DD}
$V_{GSi} - V_{oni}$ (V)
V_{DD} (V)
(a)

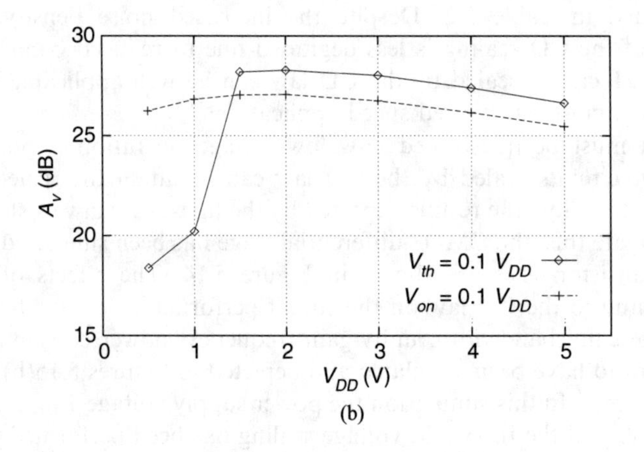
A_v (dB)
V_{th} = 0.1 V_{DD}
V_{on} = 0.1 V_{DD}
V_{DD} (V)
(b)

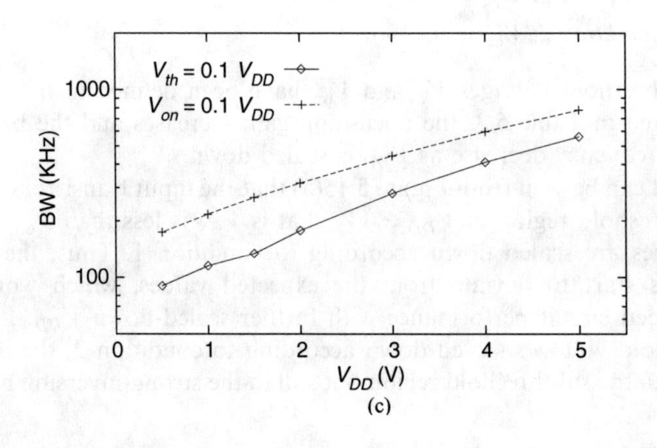
V_{th} = 0.1 V_{DD}
V_{on} = 0.1 V_{DD}
BW (KHz)
V_{DD} (V)
(c)

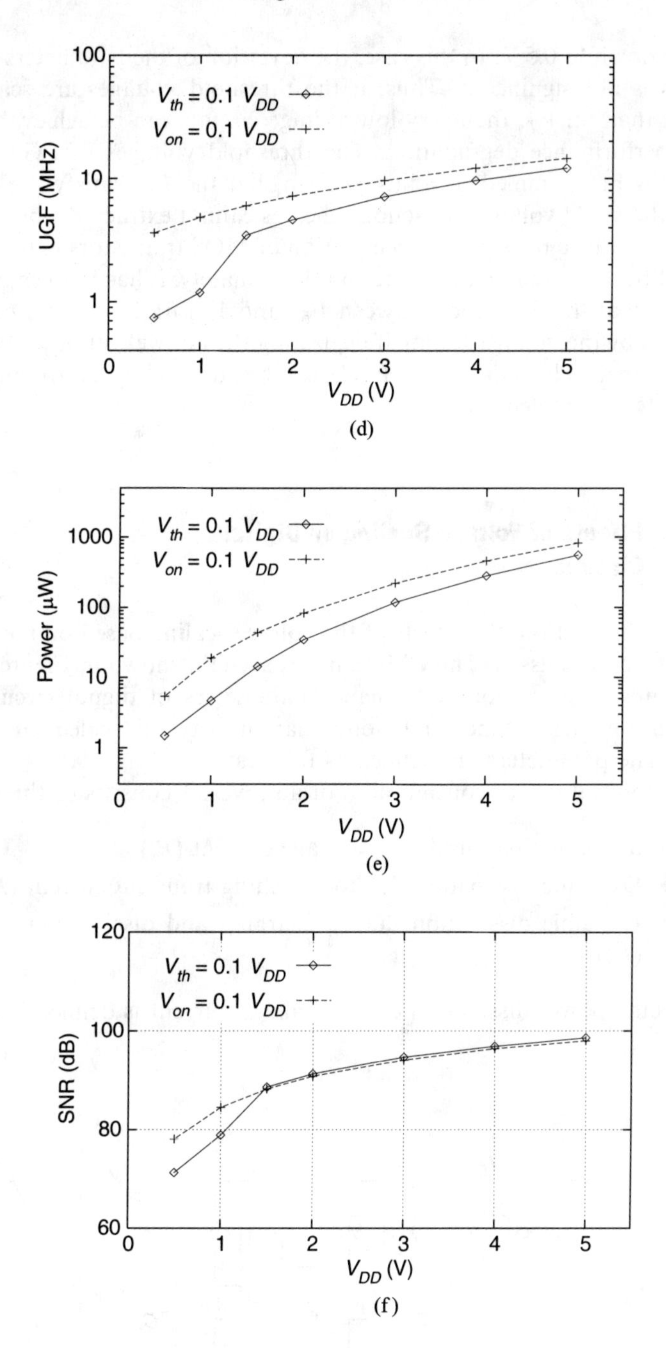

Figure 5.15 Simulation results of first-order CD scaling of CMOS differential stage: (a) difference between gate-to-source voltage V_{GSi} and threshold voltage $V_{\mathrm{on},i}$ of input transistor; (b) open-loop voltage gain A_v; (c) Bandwith ($-3\,\mathrm{dB}$ frequency); (d) unity-gain frequency; (e) power; (f) signal-to-noise ratio (SNR).

scaled down to 0.5 V. In this case, the deviation of the parameters from the expected values is not significant. Thus, if the threshold voltages are scaled based on V_{on} rather than on V_T, then very low voltage circuits can be achievable without significant performance degradation. The threshold voltages of FGTs should, thus, be carefully programmed or adjusted such that the V_{on} satisfy the condition because most threshold voltage extraction schemes cannot extract V_{on} but V_T due to the fact that they utilize the square-law equation of MOS transistors in the saturation region, as will be discussed in a later section. Fortunately, it has been shown in the previous section that the difference between V_{on} and V_T, nV_t, is almost constant and is not affected by the voltage scaling if neglecting the body effect of which the effect on nV_t is very small. Thus, the V_{on} of FGTs can be adjusted by controlling the V_T based on accurately extracted V_T values.

5.3.3. Effects of Voltage Scaling on Digital Circuits

In this section, the effects of the voltage scaling based on the CD law on digital circuits are discussed. The CMOS inverter chain shown in Figure 5.16(a) is selected to compare the major performance parameters of digital circuits such as power dissipation, delay time, and noise margin between scaled circuits and unscaled ones. The parameters are defined as follows:

The total power consumption of an inverter consists of three parts [38]:

- Static dissipation due to leakage current (P_s)
- Dynamic dissipation due to switching transient current (P_{sc})
- Dynamic dissipation due to charging and discharging of load capacitance (P_d)

The static power dissipation due to leakage current is defined here as

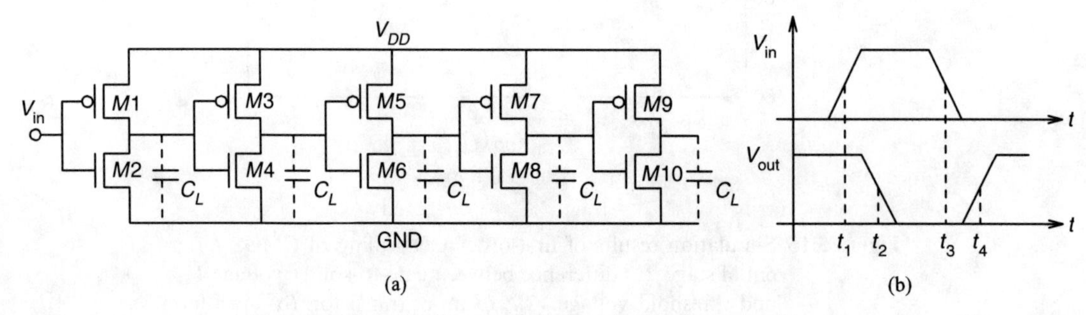

Figure 5.16 CMOS inverter chain. (a) Transistor-level implementation. (b) Input-output delay characteristics.

$$P_s = \tfrac{1}{2}[I_{Dsn}(V_{\text{in}} = 0) + I_{DSp}(V_{\text{in}} = V_{DD})] \times V_{DD} \tag{5.5}$$

The average dynamic power dissipated through the load capacitor when it is charged and discharged during switching of a square-wave input with frequency $f_p = 1/T_p$ is defined as

$$P_d = C_L V_{DD}^2 f_p \tag{5.6}$$

where

$$\begin{aligned} C_L &= C_{db1} + C_{db2} + C_{g3} + C_{g4} \\ &= D_{db1} + C_{db2} + C_{gs3} + C_{gd3} + C_{ch3} + C_{gs4} + C_{gd4} + C_{ch4} \end{aligned} \tag{5.7}$$

During transitions between 0 and 1, both the N- and P-transistors are on for a short period of time. This results in a short current pulse from V_{DD} to V_{SS}. The switching transient current pulse results in a short-circuit power dissipation which is dependent on the inverter design. The short-circuit dissipation P_{sc} of an inverter without load is given by [39]

$$P_{\text{sc}} = \frac{\beta}{12}(V_{DD} - 2V_T)^3 \frac{\tau}{T_p} \tag{5.8}$$

where τ is the rise and fall times (assumed equal) of the input signal. Here, P_d is usually more dominant than P_{sc}, and P_{sc} can be neglected under the assumption that $\tau \ll T$. In this simulation P_{sc} is neglected. The total power of an inverter is thus the sum of the three terms:

$$P_t = P_s + P_{\text{sc}} + P_d \tag{5.9}$$

The delay time is defined here as

$$T_d = \tfrac{1}{2}[(t_2 - t_1) + (t_4 - t_3)] \tag{5.10}$$

where t_1, t_2, t_3, t_4 are shown in Figure 5.16(b). In the delay time calculation, the fourth inverter of the inverter chain is used to avoid the effects of the input signal rise time.

Noise margin describes the allowable noise voltage on the input of a gate such that the output will not be affected. The specification most commonly used to specify noise margin is in terms of two parameters: the low-level noise margin NM_L,and the High-level noise margin NM_H [38]. These are depicted in Figure 5.17(a):

$$\text{NM}_L = V_{IL} - V_{OL} \qquad \text{NM}_H = V_{OH} - V_{IH}$$

where the parameters are defined as shown in Figure 5.17(b). Note that NM_H and NM_L increase as threshold voltages are increased. If either NM_H or NM_L for a gate

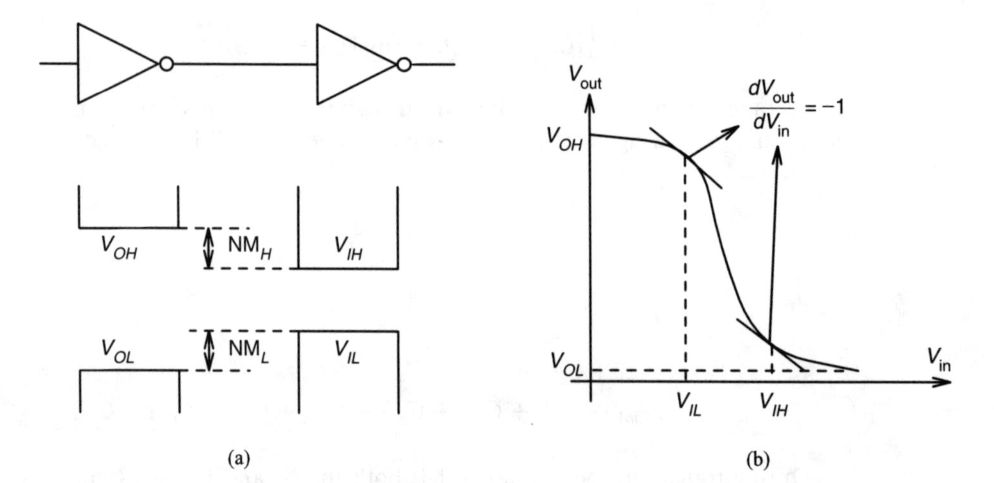

Figure 5.17 (a) NM_H and NM_L definition of noise margin. (b) Illustration of noise margin.

is reduced to around $0.1V_{DD}$, then the gate may be susceptible to switching noise. The total noise immunity can be defined as [33]

$$\text{NI} = \frac{\text{NM}_L + \text{NM}_H}{V_{DD}} \tag{5.11}$$

The delay time simulated with SPICE and calculated using (5.10) is depicted in Figure 5.18(a). When the threshold voltage V_{T0} is fixed at 1.0 V, the delay time T_d increases very quickly with scaling down of V_{DD}. Scaling the threshold voltage according to $V_{T0} = 0.2V_{DD}$ significantly relieves the increase of the delay time, as shown in the figure. However, the delay abruptly increases at very low V_{DD} values due to the subthreshold region operation. Therefore, digital circuits scaled based on the CD law will suffer from the increased delay when operated with very low power supplies. The noise immunity of the inverter calculated from (5.11) is shown in Figure 5.18(b). The noise immunity slightly increases with scaling V_{DD} down but begins to decrease rapidly at around $V_{DD} = 0.5$ V due to the subthreshold operation and increased leakage currents. The badly degraded performance in the delay and noise immunity becomes an obstacle to further scaling to lower voltage (<0.5 V) operation of digital circuits.

The effects of voltage scaling on the power dissipation of the CMOS inverter are also simulated and shown in Figure 5.19. As can be seen in Figure 5.18, a possible lowest power supply voltage is 0.5 V, where the delay and noise immunity degradation is not significant. Thus, three scaled inverters which have the same scaled power supply voltage of 0.5 V but different threshold voltages V_{T0} are compared with the unscaled inverter, which has V_{DD} of 5 V. It can be seen from Figure 5.19(a) that the unscaled inverter shows the lowest static power dissipation, and the scaled inverters

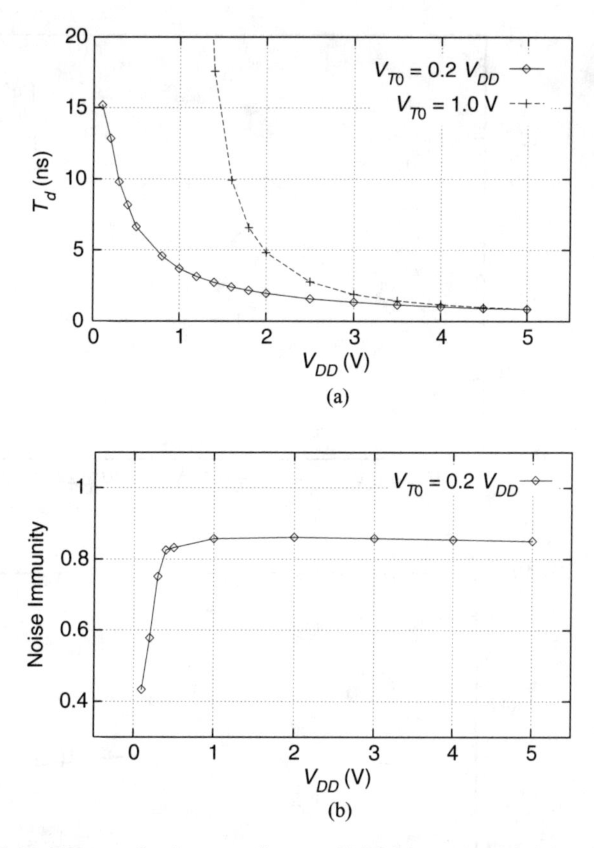

Figure 5.18 Effects of voltage scaling on CMOS inverter: (a) delay time; (b) noise immunity.

show increased static power dissipation as the threshold voltages are reduced. This is because the threshold voltage scaled inverters have increased leakage currents. However, the dynamic power dissipation of the unscaled inverter is much larger than those of the scaled inverters. It can also be seen that the dynamic power dissipations depend on the switching frequency. Figure 5.19(b) shows that the total power dissipation of the unscaled inverter is much larger than those of the scaled inverters at a switching frequency of 1 MHz. Since the dynamic power dissipation due to the short circuit currents is neglected in this calculation, the actual dynamic power dissipation of the unscaled inverter will be larger as the switching frequency is increased. Figure 5.19(b) also shows that high threshold voltages reduce power dissipation when the supply voltage is fixed. Increased threshold voltage will, however, increase the delay time, as indicated in Figure 5.18(a). This means that the power and the speed are in a trade-off relation, and thus, an optimum threshold voltage at a fixed V_{DD} should be selected for improved overall performance or for satisfying a given specification.

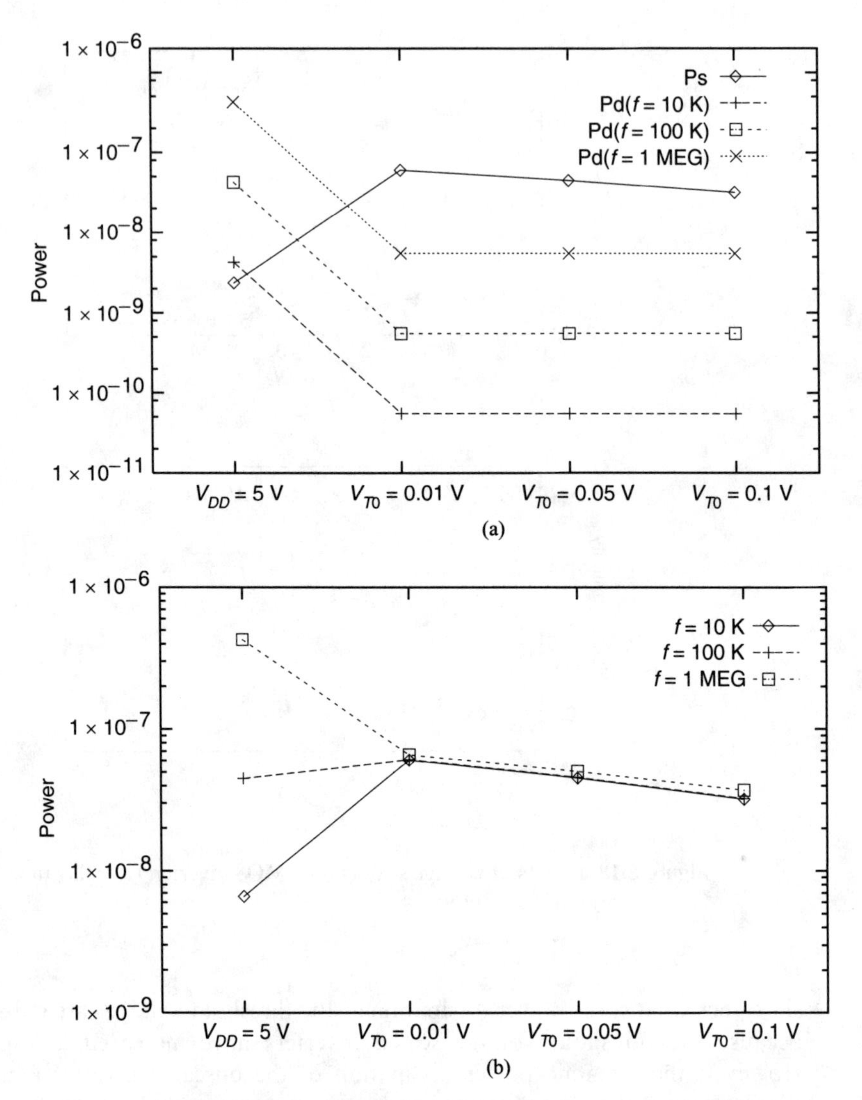

Figure 5.19 Power dissipation (Pd) of CMOS inverter: (a) static and dynamic power dissipation; (b) total power dissipation.

5.4. DESIGN ISSUES

In this section some of the important issues in designing low-voltage circuits using FGTs are discussed.

5.4.1. V_T Extraction Schemes

It has been pointed out that threshold voltages play an important role in the performance of low-voltage and low-power circuits. Threshold voltages must be adjusted with high accuracy to achieve very low voltage analog circuits while avoiding subthreshold region operation and to obtain optimum performance between speed and power in digital circuits, as indicated in Section 5.3. In order to precisely control threshold voltages, there must be an accurate threshold voltage extraction scheme. One is presented here [28]. Compared to other methods recently reported in the literature, the proposed scheme does not require matched replicas of the transistor under test and thus can be applied more effectively and accurately to real-time on-chip applications where threshold voltage measurements are required for many transistors with various geometries and bias conditions.

Comparison of V_T Extraction Schemes. One popular V_T extraction scheme [40] uses numerical techniques in which a linear regression on measurements of I_{DS} at many V_{GS} values is used to extract threshold voltages accurately. The numerical techique is, however, computationally intense and not well-suited for real-time on-chip applications. A real-time on-chip V_T extraction scheme is necessary for V_T programming of the low-voltage V_T tunable circuits, as indicated in Section 5.2.3.

Recently, several real-time V_T extraction methods based on circuit implementations have been proposed for overcoming the above disadvantages [27, 41–43]. These methods are very fast and have ample applications, although the accuracy is degraded compared to that attainable by the numerical methods. Most methods [41–43] require matched devices to extract V_T for one test device of a fixed geometry. Their accuracy thus depends on the matching between the devices under test. These methods are inefficient when extraction of V_T is required for many transistors with various geometries, including small size, since the matching of small-size transistors is poorer than that of large transistors. These methods also require other component matching in their extraction circuits, such as current mirror transistors and resistors. Their mismatches will also degrade the accuracy of the extracted V_T values. Moreover, the methods [41–43] are not applicable for transistors with different bias conditions, that is, nonzero substrate-to-source voltages ($V_{BS} \neq 0$), since they need a cascode configuration of matched test transistors or transistor arrays. The method discussed in [27] uses only one test device and thus does not require device matching. The method is very simple but produces relatively large errors (about 100 mV) due to the uncertainty of choosing the proper threshold current used to measure V_T.

We will discuss an accurate matching-free VT extraction scheme which does not require any replica of the device under test and is applicable for transistors with different geometries and different substrate bias conditions. The features of the proposed scheme are comparatively summarized in Table 5.4 with other extraction schemes mentioned above.

TABLE 5.4 FEATURE COMPARISON OF V_T EXTRACTION SCHEMES

Scheme	Required Number of Matched Test Transistors	Required Matched Components	Applicability at Different Geometries	Applicability at Different Substrate Bias Conditions	Comments
Numerical [40]	None	None	Efficient	Yes	Accurate but not suitable for real time
Wang [41]	9	Current mirror transistors	Inefficient	No	Using a transistor array
Tsividis [42]	3	Current mirror transistors	Inefficient	No	Using a transistor string
Alini [43]	2	Resistor and current mirror transistors	Inefficient	Yes	BiCMOS implementation
Lee [27]	None	None	Efficient	Yes	Simple but poor accuracy
Proposed	None	None	Efficient	Yes	Dynamic implementation

Principle of Matching-Free V_T Extractor. A conceptual schematic of the matching-free V_T extraction scheme is depicted in Figure 5.20. Applying the outputs of a current mirror I_{D1} (with S_1 closed and S_2 open) and I_{D2} (with S_1 open and S_2 closed) to a test transistor which operates in the saturation region and assuming that the transistor has square-law characteristics, we obtain respectively

$$K(V_{GS1} - V_T)^2 = I_{D1} \tag{5.12}$$

$$K(V_{GS2} - V_T)^2 = I_{D2} \tag{5.13}$$

where

$$I_{D2} = nI_{D1} \qquad K = \frac{\mu C_{ox}}{2} \frac{W}{L}$$

Equations (5.12) and (5.13) have the same K and V_T because only one test transistor is used, as contrasted with other extraction schemes. Solving these equations, we readily obtain the threshold voltage V_T:

$$V_T = \frac{1}{\sqrt{n} - 1} (\sqrt{n} V_{GS1} - V_{GS2}) \tag{5.14}$$

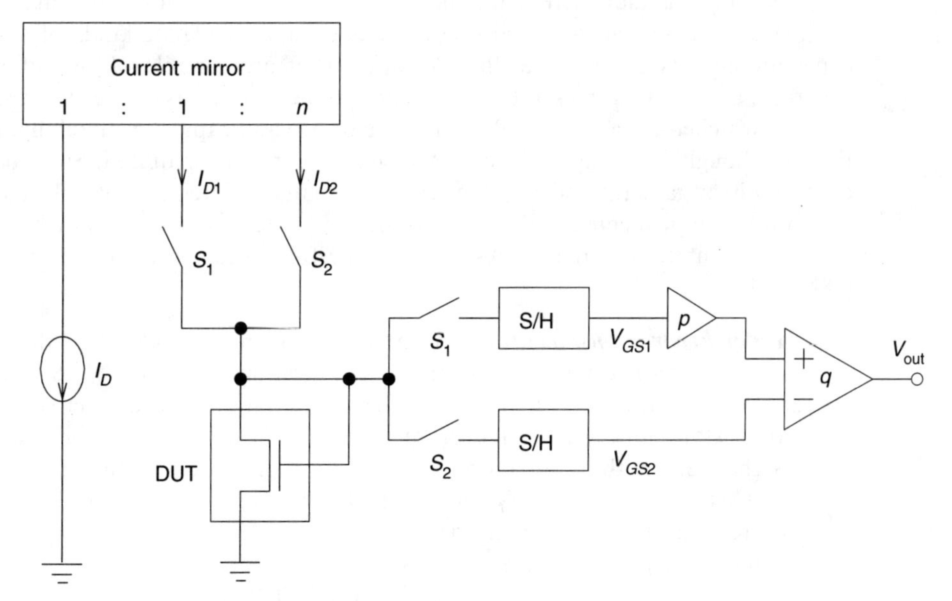

Figure 5.20 Conceptual schematic of proposed V_T extraction scheme (S/H = sample and hold; DUT = device under test).

Assume S_1 and S_2 are driven by a complementary nonoverlapping clock. When S_1 is closed, V_{GS1} is sampled and multiplied by p. When S_2 is closed, V_{GS2} is sampled and subtracted from pV_{GS1}. The result is then multiplied by q. The output voltage V_{out} of Figure 5.1 is then

$$V_{\text{out}} = q(pV_{GS1} - V_{GS2}) \tag{5.15}$$

Comparing (5.14) with (5.15), it can be seen that the output voltage will be

$$V_{\text{out}} = mV_T$$

if

$$p = \sqrt{n}, \qquad q = \frac{m}{\sqrt{n} - 1}$$

Thus, an integer multiple of V_T can be readily obtained by choosing an integer m. The easiest way to obtain V_T is to choose $n = 4$ and $m = 1$, resulting in $p = 2$ and $q = 1$, and thus,

$$V_{\text{out}} = 2V_{GS1} - V_{GS2} = V_T$$

where V_{GS1} and V_{GS2} are the gate-to-source voltages of the test transistor when the drain currents are I_D and $4I_D$, respectively.

The simple analog arithmetic operation $2V_{GS1} - V_{GS2}$ can be accurately implemented using a switched-capacitor subtracting amplifier. These kinds of switched-capacitor circuits for basic arithmetic operations are capable of providing high accuracy, as indicated in [44]. Most common implementations of the current mirror and the switched-capacitor amplifier require device and capacitor matching, respectively, although matching with the device under test is not required. Since both the current mirror gain and the amplifier gain are small and integral, both blocks can be dynamically implemented without requiring any matching of devices or capacitors. The implementation of the blocks and their performance can be found in detail in [28] and [45].

Model-Error Consideration. As are most other methods [41–43], the proposed V_T extraction scheme is also based on the assumption that MOS transistors operating in the saturation region obey the square law. The characteristics of real MOS transistors, however, deviate from the square law due to the nonideal effects such as channel length modulation and mobility degradation, resulting in a discrepancy between the extracted V_T and a real V_T. Including the nonideal effects, the drain current can more accurately be described by

$$I_{DS} = \left[\frac{\mu_0}{1 + \theta(V_{GS} - V_T)}\right]\left[\frac{1}{L(1 - \lambda V_{DS})}\right]\frac{C_{\text{ox}}W}{2}(V_{GS} - V_T)^2 \tag{5.16}$$

where λ is the channel length modulation parameter, θ is the mobility degradation parameter, and μ_0 is the zero-field mobility of carriers.

The error voltage due to the λ and θ effects can be readily derived using Eq. (5.16) and neglecting the second-order effects, resulting in

$$\begin{aligned} V_{T,\mathrm{err}} &\simeq \tfrac{1}{2}(\lambda - \theta) V_{\mathrm{ex1}} V_{\mathrm{ex2}} \\ &\simeq \tfrac{1}{4}(\lambda - \theta) V_{\mathrm{ex2}}^2 \end{aligned} \tag{5.17}$$

where $V_{\mathrm{ex}i}$ for $i = 1, 2$ is the excess voltage $V_{GSi} - V_T$. Small excess voltages will help reduce the error voltage, which is an expected result because the λ and θ effects increase with V_{DS} and V_{GS}, respectively, and the test device in our extractor is diode connected to guarantee its saturation region operation, resulting in $V_{DS} = V_{GS}$. It is interesting to note that the two parameters in (5.17) are in a relation of canceling each other, and fortunately, both parameters are inversely proportional to the channel length. Therefore, the variance of $\lambda - \theta$ and thus the error voltage will not increase substantially with the channel length reduction. For example, if the maximum-difference value of the two parameters is $0.1\,\mathrm{V}^{-1}$, $V_T = 0.8\,\mathrm{V}$, and $V_{\mathrm{ex2}} = 0.4\,\mathrm{V}$, then the error voltage will be less than 0.5%.

The proposed scheme has been simulated for two test devices which have different geometries using SPICE with level 2 MOS models ($V_{T0} = 0.924\,\mathrm{V}$). In this simulation, no error associated with the current mirror and the analog arithmetic block was assumed to examine the pure model error. With the assumption that the V_T computed by SPICE is the actual threshold voltage, the error voltage $V_{T,\mathrm{err}}\,(V_{T,\mathrm{ext}} - V_T)$ is plotted in Figure 5.21(a) as a function of bias current I_D. As expected, the error voltages for the long-channel device ($W/L = 200\,\mu\mathrm{m}/40\,\mu\mathrm{m}$) are smaller at all I_D values than those for the short-channel device ($W/L = 20\,\mu\mathrm{m}/4\,\mu\mathrm{m}$). It can be seen that the error increases with I_D since large I_D increases the excess voltage, as shown in Figure 5.21(b), where the error voltages are plotted as a function of the excess voltage $V_{\mathrm{ex2}}\,(V_{GS2} - V_T)$. The figure exhibits that the error variance of the proposed scheme to the excess voltage V_{ex2} is comparable with the variance to the device size. It can also be seen that the slope of the curves in Figure 5.21 changes substantially at a small I_D or a small V_{ex2} that corresponds to the transition point between the strong-inversion region and the weak-inversion region. Therefore, the bias current I_D should be selected carefully such that the excess voltages V_{ex1} and V_{ex2} are greater than the transition point but not too big for small model error. It can be seen in Figure 5.21(b) that if $V_{\mathrm{ex2}} \leq 0.4\,\mathrm{V}$, then the error voltage due to the model error will be less than 5 mV even with the short-channel device ($L = 4\,\mu\mathrm{m}$).

The proposed scheme has also been compared with the linear regression (LR) method [40] in Table 5.5. In the LR method, I_{DS} values are collected at 20 V_{GS} values using SPICE, so no measurement error is assumed. For consistency in excess voltages, the V_{GS} values are selected such that the highest sample value V_{GSh} is V_{GS2} and the lowest sample value G_{GSl} is V_{GS1}. Since threshold voltages are functions of device terminal voltages, and their variance increases as the device size decreases, the actual threshold voltage V_T of the short-channel device ($L = 4\,\mu\mathrm{m}$) computed by SPICE

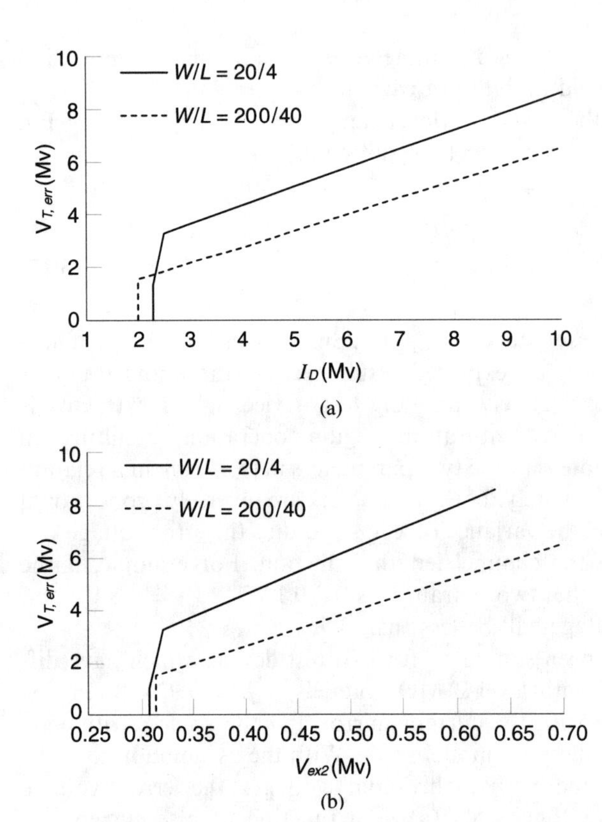

Figure 5.21 Error voltage of extracted voltage $V_{T,\text{ext}}$ from actual threshold voltage V_{th} computed by SPICE: (a) as function of bias current I_D; (b) as function of excess voltage V_{ex2}.

varies with V_{GS} or V_{DS}, as shown in the table. At $V_{GS} = 1.066\,\text{V}$, $V_T = 0.889\,\text{V}$, while at $V_{GS} = 1.525\,\text{V}$, $V_T = 0.887\,\text{V}$. Thus, the V_T variation is about 2 mV when the V_{GS} change is 0.46 V. This variation will be significant for shorter channel devices. The V_T variation of the long-channel device ($L = 40\,\mu\text{m}$) is almost negligible. In the proposed scheme the variation is due to the two different V_{GS} values, V_{GS1} and V_{GS2}, and in the LR method the variation is also due to the different V_{GS} values used to grab the I_D data. Thus, the average values

$$V_{T,av} = \begin{cases} \frac{1}{2}[V_T(V_{GS1}) + V_T(V_{GS2})] & \text{for proposed scheme} \\ \frac{1}{2}[V_T(V_{GSl}) + V_T(V_{GSh})] & \text{for LR method} \end{cases}$$

were used to calculate the error of extracted threshold voltages. It can be seen from the table that the accuracy of the LR method is similar to that of the proposed scheme, and the LR method also gives large error when the samples are taken from large V_{GS} values.

The proposed scheme can be accurately implemented in a matching-free way using a ratio-independent switched-capacitor subtracting amplifier and a dynamic current mirror. Taking into account unexpected process variations, the error voltage associated with the designed circuits are in a few millivolt range [28, 45]. This error is smaller compared with the model error. To make the V_T extractor be applicable for

TABLE 5.5 ACCURACY COMPARISON BETWEEN PROPOSED SCHEME AND LR METHOD

Test Device Size (W/L)	Number of samples	V_{GS1} (V)	V_T (V)	V_{GSh} (V)	V_T (V)	Extracted $V_{T,\mathrm{ext}}$ (V)	Error $V_{T,\mathrm{ext}}$-$V_{T,\mathrm{av}}$ (mV)
		LR Method [40]					
20 μm/4 μm	20	1.066	0.888	1.241	0.888	0.8919	3.43
		1.211	0.888	1.525	0.887	0.8971	9.58
200 μm/40 μm	20	1.116	0.921	1.309	0.921	0.9230	2.03
		1.275	0.921	1.623	0.921	0.9281	7.09
Test Device Size (W/L)	**Bias Current I_D**	**V_{GS1} (V)**	**V_T (V)**	**V_{GSh} (V)**	**V_T (V)**	**Extracted $V_{T,\mathrm{ext}}$ (V)**	**Error $V_{T,\mathrm{ext}}$-$V_{T,\mathrm{av}}$ (mV)**
		The Proposed Scheme					
20 μm/4 μm	3 μA	1.066	0.888	1.241	0.888	0.8921	3.61
		1.211	0.888	1.525	0.887	0.8961	8.58
200 μm/40 μm	10 μA	1.116	0.921	1.309	0.921	0.9233	2.26
		1.275	0.921	1.623	0.921	0.9275	6.47

various transistors which have different geometries and different bias conditions and to achieve a high accuracy, the model error should always be kept small. The scheme is applicable to various applications where many V_T measurements are required. Thus, it can be well applied for implementation of the low-voltage floating-gate MOSFET circuits where V_T measurement of many floating-gate MOSFETs with different geometries are essential for V_T tuning.

5.4.2. Limitations and Design Challenges

There may exist many challenges in designing very low voltage and low power circuits, and there also exist several factors limiting the lowest level of the power supply voltage where the circuit performance cannot be acceptable with further reduction of the supply voltage. Among them, the most severe problem is the degradation of subthreshold characteristics when V_T is scaled down to very low levels. This is due to the fact that the subthreshold turn-off does not scale because it is dominated by the exponential dependence on potential of the channel carrier density. The degraded subthreshold characteristics increase the leakage current at zero gate bias. This will increase the static power dissipation in digital circuits, and analog circuits employing some transistors as switches will suffer from signal information leakage. This gives the limit for V_T scaling down and the lower bound for obtainable low-voltage circuits. Thus, the problem of the subthreshold leakage current and their possible solutions will be discussed in this section.

The leakage current is governed by a bipolarlike mechanism, because when the surface is not inverted, minority carriers and diffusion currents are dominant just like in bipolar device [4]. To keep the off-current low at zero gate bias, the following condition should be satisfied [5, 32–34]:

$$V_T \geq S(\log R) + V_{\text{off}}$$

where the on/off current ratio R is defined as the ratio of the on current at threshold ($V_{GS} = V_T$) to the off current flowing in the off state. Here, $V_{GS} = V_{\text{off}} \geq 0\,\text{V}$, and S is the subthreshold slope, that is, the gate voltage swing needed to reduce the drain current by a decade in subthreshold:

$$\begin{aligned} S = \left[\frac{\partial(\log I_{DS})}{\partial V_{GS}}\right]^{-1} &\simeq \frac{kT}{q}\ln 10\left(1 + \frac{C_D}{C_{\text{ox}}}\right) \\ &= \frac{kT}{q}\ln 10\left(1 + \frac{\Omega T_{\text{ox}}}{d}\right) \end{aligned} \tag{5.18}$$

where $\Omega = \varepsilon_{si}/\varepsilon_{\text{ox}}$ and d is the channel depletion width. Typical values for S are between 60 and 90 mV/decade current at room temperature, with 60 mV/decade being the lower limit, which means that for an ideal PN-junction, every 60-mV increase in the barrier height will reduce the current by a factor of 10. The finite rate of change of the current with the gate potential gives a fundamental limit on how much the threshold potential can be scaled down. If V_T is reduced too much, the potential difference between 0 V and V_T may be insufficient to turn off the transistor completely.

This leakage current criterion gives constraints on the design that are more severe for submicrometer devices than for larger devices because of larger short-channel effects. Although V_T is designed to minimize the leakage current in the off state, excess subthreshold current results from the DIBL (drain-induced barrier lowering) effect as the channel length is reduced. The barrier height at the source end is lowered as drain potential is raised. A drop in the potential barrier height exponentially increases the subthreshold leakage current. Other short-channel effects include channel length modulation, velocity saturation, mobility degradation, source/drain resistance, and dependence of V_T on device geometry.

The CMOS inverters have been simulated to investigate the amount of the subthreshold leakage current. Simulation has been performed for large-size devices ($L = W = 40\,\mu\text{m}$) and small-size devices ($L = 2\,\mu\text{m}$ and $W = 4\,\mu\text{m}$) when the power supply voltage (V_{DD}) is scaled down with differently scaled threshold voltages. The simulation results are shown in Tables 5.6 and 5.7. The different threshold voltage notations in the tables have been defined in Section 3.1. For a large-size device inverted with V_{DD} of 5 V and unscaled threshold voltages, the turn-off leakage current of the NMOS transistor was 474 pA, and the turn-off leakage current of the PMOS transistor was 7.0 pA. The nonsymmetry between the two turn-off currents is due to different mobilities (μ) and threshold voltages (V_{on}).

From the tables, it can be seen that the turn-off leakage currents significantly increase with V_{DD} and V_T scaling. The reduced V_T increases the leakage currents, as expected. It can also be seen that device size scaling down with constant V_T also increases the leakage current. This is because a number of short-channel effects contribute to the increase of the turn-off leakage currents. For example, when the power supply voltages are scaled down to 0.5 V and the threshold voltages V_{T0} are

TABLE 5.6 TURN-OFF LEAKAGE CURRENTS OF LARGE-SIZE NMOS TRANSISTOR

V_{TO} (V)	0.1	0.05	0.01
V_T (V)	0.0992	0.0492	0.00925
V_{on} (V)	0.257	0.207	0.167
$V_{on} - V_T$ (V)	0.1578	0.1578	0.1578
$I_{DS} \mid_{V_{GS}=0}$ (nA)	79.99	105.7	136.3

Note: $L = W = 40\ \mu\text{m},\ V_{SB} = 0\ \text{V},\ V_{DS} = 0.5\ \text{V}.$

TABLE 5.7 TURN-OFF LEAKAGE CURRENTS OF SMALL-SIZE NMOS TRANSISTOR

V_{TO} (V)	0.1	0.05	0.01
V_T (V)	0.0397	−0.0103	−0.0503
V_{on} (V)	0.196	0.146	0.106
$V_{on} - V_T$ (V)	0.1563	0.1563	0.1563
$I_{DS} \mid_{V_{GS}=0}$ (nA)	293.2	404.2	522.5

Note: $L = 2\ \mu\text{m},\ W = 4\ \mu\text{m},\ V_{SB} = 0\ \text{V},\ V_{DS} = 0.5\ \text{V}.$

scaled down to 0.01 V, the NMOS transistor's turn-off leakage current of a large-size inverter (Table 5.6) is 136 nA, which can be compared with 523 nA of a small-size inverter (Table 5.7). The difference can be reduced if V_T or V_{on} instead of V_{T0} are scaled to a certain value. However, the small-size inverter will still have much larger leakage currents compared to the large-size inverter, which can be readily observed from Tables 5.6 and 5.7. This can be one advantage of the new technique based on the CD scaling law over the conventional device size scaling techniques. The new technique may be preferred to the device size scaling techniques for very low voltage circuits because the subthreshold leakage current increase due to DIBL is much higher in short-channel devices than in long-channel devices. The device size scaling can also cause reliability problems such as hot carriers, oxide and junction breakdown, electrostatic discharge, electrical overstress, and so on [4].

The subthreshold leakage current problem can be somewhat relieved by using the low-voltage circuit technique which employs FGTs for easy V_T scaling and uses the CD scaling law. It is, however, still the major obstacle to obtaining ultralow voltage circuits and thus should be overcome. There are several possible solutions to reduce the leakage current. One is to use low operating temperature. As can be seen in (5.18), the gate voltage swing S required to reduce the subthreshold current a decade can be decreased by reducing the temperature T. Reducing the operation temperature to liquid nitrogen levels can be a good solution but may not be the cost-effective one [4].

Another possible solution is the switched-source-impedance (SSI) technique recently reported in the literature [46]. By inserting a switched impedance at the source of the NMOS transistor or at the source of the PMOS transistor according to the standby condition, as shown in Figure 5.22, the subthreshold leakage current can be significantly reduced. During the active mode, the switch S_s it turned on for high-speed operation. During the standby mode, the switch is turned off to reduce the standby leakage current. For the first case [Figure 5.22(a)], during the standby

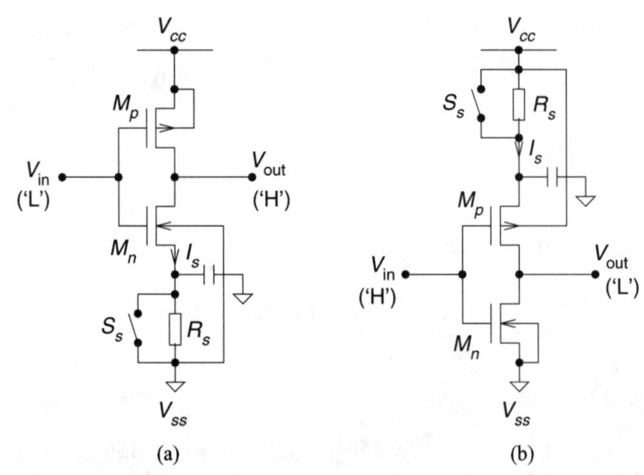

Figure 5.22 Concept of SSI circuits. (a) SSI applied to the NMOS transistor. (b) SSI applied to the PMOS transistor.

mode, the threshold voltage of the NMOS transistor increases, and the gate-to-source voltage of the NMOS transistor becomes negative due to the source potential increase to R_sI_s, where I_s is the subthreshold current. These two factors, increased threshold voltage and negative V_{GS}, contribute to the reduction of the leakage current. Using this technique, the subthreshold leakage current can be reduced by three to four orders of magnitude. The current variation due to threshold voltage and temperature fluctuations can also be reduced by this technique. Moreover, a switched impedance can be shared by many logic gates, which results in minimum area penalty. However, one major disadvantage of this technique is that it is applicable to only the logic circuits for which the standby node voltages are predictable.

5.5. CONCLUSIONS

The low-voltage circuit techniques based upon device size scaling with V_T scaling can offer reduced delay and power dissipation of the devices and improved packing density. They, however, require more sophisticated substrate engineering to scale down V_T and suffer from the short-channel effects which become more severe as the device size is reduced. Moreover, the increased subthreshold leakage current, which is more severe than for large devices, will give a limitation on the techniques.

The new technique for low-voltage circuits which uses FGTs as basic circuit elements for precise V_T control can provide much higher accuracy for low-voltage analog circuits. The V_T programming and adjusting of the FGTs are relatively easy, and the complicated substrate engineering can thus be avoided. By scaling the power supply voltage and threshold voltages based upon the CD scaling law, the new technique can provide advantages over the conventional device size scaling techniques in that the former does not suffer from the short-channel effects as much as the latter, and thus its subthreshold characteristics are better than the conventional counterparts. This means that the new technique can offer more favorability to very low voltage circuits at the expense of increased delay. The increased delay time can be partially compensated by optimum selections of the threshold voltage and power supply level.

References

[1] A. P. Chandrakasan, S. Sheng, and R. W. Brodersen, "Low-Power CMOS Digital Design," *IEEE J. Solid-State Circuits*, vol. SC-27, pp. 473–484, 1992.

[2] Y. Nakagome, H. Tanakar, K. Takeuchi, E. Kume, Y. Watanabe, T. Kaga, Y. Kawamoto, F. Murai, R. Izawa, D. Hishamoto, T. Kisu, N. Nishida, E. Takeda, and K. Itoh, "An Experimental 1.5-V 64-Mb DRAM," *IEEE J. Solid-State Circuits*, vol. 26, no. 4, pp. 465–470, 1991.

[3] A. Sekiyama, T. Seki, S. Nagai, A. Iwase, N. Suzuki, and M. Hayasaka, "A 1-V Operating 256-kb Full-CMOS SRAM," *IEEE J. Solid-State Circuits*, vol. 27, no. 5, pp. 776–782, 1992.

[4] H. B. Bakoglu, *Circuits, Interconnections, and Packaging for VLSI*, Addison-Wesley, Reading, MA, 1990.

[5] M. Nagata, "Limitations, Innovations, and Challenges of Circuits and Devices into a Half Micrometer and Beyond," *IEEE J. Solid-State Circuits*, vol. SC-27, pp. 465–472, 1992.

[6] E. H. Li and H. C. Ng, "Parameter Sensitivity of Narrow-Channel MOSFET's," *IEEE Electron Device Lett.*, vol. EDL-12, pp. 608–610, 1991.

[7] K. Yokoyama, A. Yoshii, and S. Horiguchi, "Threshold Sensitivity Minimization of Short-Channel MOSFET's by Computer Simulation," *IEEE Trans. Electron Devices*, vol. ED-27, pp. 1509–1514, 1980.

[8] S. Selberherr, A. Schutz, and H. Potzi, "Investigation of Parameter Sensitivity of Short Channel MOSFETs," *Solid-State Electron.*, vol. 25, pp. 85–90, 1982.

[9] E. H. Li, K. M. Hong, Y. C. Cheng, and K. Y. Chan, "The Narrow-Channel Effect in MOSFET's with Semi-Recessed Oxide Structures," *IEEE Trans. Electron Devices*, vol. ED-37, pp. 692–701, 1990.

[10] M. Conti, C. Turchetti, and G. Masetti, "A New Analytical and Statistical-Oriented Approach for the Two-Dimensional Threshold Analysis of Short-Channel MOSFETs," *Solid-State Electron.*, vol. 32, pp. 739–747, 1989.

[11] D. Liu and C. Svensson, "Trading Speed for Low Power by Choice of Supply and Threshold Voltages," *IEEE J. Solid-State Circuits*, vol. SC-28, pp. 10–17, 1993.

[12] C. G. Yu and R. L. Geiger, "Very Low Voltage Operational Amplifiers Using Floating Gate MOS Transistors," *IEEE Symp. Circuits Sys.*, vol. 2, pp. 1152–1155, 1993.

[13] J. Sweeney and R. L. Geiger, "Very High Precision Analog Trimming Using Floating Gate MOSFETs," *Proceedings of European Conference on Circuit Theory and Design (ECCTD)*, Brighton, United Kingdom, Sep. 1989, pp. 652–655.

[14] C. Kuo, J. R. Yeargain, W. J. Downey III, K. A. Ilgenstein, J. R. Jorvig, S. L. Smith, and A. R. Bormann, "An 80ns 32K EEPROM Using the FETMOS Cell," *IEEE J. Solid-State Circuits*, vol. SC-17, no. 5, pp. 821–827, 1982.

[15] *M68HC11 Reference Manual*, Motorola, 1991.

[16] B. Gerber, J. C. Martin, and J. Fellrath, "A 1.5V Single-Supply One-Transistor CMOS EEPROM," *IEEE J. Solid-State Circuits*, vol. SC-16, no. 3, pp. 195–199, 1981.

[17] C. Kuo, M. Weidner, T. Toms, H. Choe, K.-M. Chang, A. Harwood, J. Jelemenski, and P. Smith, "A 512-kb flash EEPROM Embedded in a 32-b Microcontroller," *IEEE J. Solid-State Circuits*, vol. SC-27, no. 4, pp. 574–582, 1992.

[18] Y. Miyawaki, T. Nakayama, S. Kobayashi, N. Ajika, M. Ohi, Y. Terada, H. Arima, and T. Yoshihara, "A New Erasing and Row Decoding Scheme for Low Supply Voltage Operation 16-Mb/64-Mb Flash Memories," *IEEE J. Solid-State Circuits*, vol. SC-27, no. 4, pp. 583–588, 1992.

[19] K. Yang and A. G. Andreou, "Multiple Input Floating-Gate MOS Differential Amplifiers and Applications for Analog Computation," paper presented at the Thirty-Sixth IEEE Midwest Symposium on Circuits and Systems, pp. 1212–1216, Detroit, Aug. 1993.

[20] H. Yang, B. J. Sheu, and J. C. Lee, "A Nonvolatile Analog Neural Memory Using Floating-Gate MOS Transistors," *Analog Integrated Circuits Signal Process.*, vol. 2, pp. 19–25, 1992.

[21] A. Thomsen and M. A. Brooke, "A Floating-Gate MOSFET with Tunneling Injector Fabricated Using a Standard Double-Polysilicon CMOS Process," *IEEE Electron Device Lett.*, vol. EDL-12, pp. 111–113, 1991.

[22] A. Thomsen and M. A. Brooke, "A Programmable Piecewise Linear Large-Signal CMOS Amplifier," *IEEE J. Solid-State Circuits*, vol. SC-28, pp. 84–89, 1993.

[23] L. R. Carley, "Trimming Analog Circuits Using Floating-Gate Analog MOS-Memory," *IEEE J. Solid-State Circuits*, vol. SC-24, pp. 1569–1575, 1989.

[24] E. Sackinger and W. Guggenbuhl, "An Analog Trimming Circuit Based on a Floating Gate Device," *IEEE J. Solid-State Circuits*, vol. SC-23, pp. 1437–1440, 1988.

[25] S. T. Wang, "On the I-V Characteristics of Floating-Gate MOS Transistors," *IEEE Trans. Electron Devices*, vol. ED-26, pp. 346–348, 1979.

[26] J. F. Dickson, "On-Chip High-Voltage Generation in MNOS Integrated Circuits Using an Improved Voltage Multiplier Technique," *IEEE J. Solid-State Circuits*, vol. SC-11, No. 3, pp. 374–378, 1976.

[27] H. G. Lee, S. Y. Oh, and G. Fuller, "A Simple and Accurate Method to Measure the Threshold Voltage of an Enhancement-Mode MOSFET," *IEEE Trans. Electron Devices*, vol. ED-29, no. 2, pp. 346–348, 1982.

[28] C. G. Yu and R. L Geiger, "An Accurate and Matching-Free Threshold Voltage Extraction Scheme for MOS Transistors," *IEEE Int. Symp. Circuits Sys.*, 1994.

[29] R. H. Dennard, F. H. Gaenssler, H.-N. Yu, E. Bassous, A. R. LeBlanc, and V. L. Rideout, "Design of Ion Implanted MOSFET's with Very Small Physical Dimensions," *IEEE J. Solid-State Circuits*, vol. SC-9, pp. 256–266, 1974.

[30] P. K. Chatterjee, W. R. Hunter, T. C. Holloway, and Y. T. Lin, "The Impact of Scaling Laws on the Choice of *n*-channel or *p*-Channel for MOS VLSI," *IEEE Electron Device Lett.*, vol. EDL-1, pp. 220–223, 1980.

[31] M. Kakumu and M. Kinugawa, "Power-Supply Voltage Impact on Circuit Performance for Half and Lower Submicrometer CMOS LSI," *IEEE Trans. Electron Devices*, vol. ED-37, pp. 1902–1908, 1990.

[32] J. R. Brews, "Subthreshold Behavior of Uniformly and Nonuniformly Doped Long-Channel MOSFET," *IEEE Trans. Electron Devices*, vol. ED-26, pp. 1282–1291, 1979.

[33] J. R. Pfiester, J. D. Shott, and J. D Meindl, "Performance Limits of CMOS ULSI," *IEEE J. Solid-State Circuits*, vol. SC-20, pp. 253–263, 1985.

[34] R. K. Watts, *Submicron Integrated Circuits*, Wiley, New York, 1989.

[35] S. Wong and C. A. T. Salama, "Impact of Scaling on MOS Analog Performance," *IEEE J. Solid-State Circuits*, vol. SC-18, pp. 106–114, 1983.

[36] T. Enomoto, T. Ishihara, M.-A. Yasumoto, and T. Aizawa, "Design, Fabrication, and Performance of Scaled Analog IC's," *IEEE J. Solid-State Circuits*, vol. SC-18, pp. 395–402, 1983.

[37] E. Sano, T. Tsukahara, and A. Iwata, "Performance Limits of Mixed Analog/Digital Circuits with Scaled MOSFET's," *IEEE J. Solid-State Circuits*, vol. SC-23, pp. 942–949, 1988.

[38] N. Weste and K. Eshraghian, *Principles of CMOS VLSI Design—A Systems Perspective*, 2nd ed., Addison-Wesley, Reading, MA, 1993.

[39] H. J. M. Veendrick, "Short-Circuit Dissipation of Static CMOS Circuitry and Its Impact on the Design of Buffer Circuits," *IEEE J. Solid-State Circuits*, vol. SC-19, pp. 468–473, 1984.

[40] P. E. Allen and D. R. Holberg, *CMOS Analog Circuit Design*, Holt, Rinehard and Winston, New York, 1987.

[41] Z. Wang, "Automatic V_T Extractors Based on an $n \times n^2$ MOS Transistor Array and Their Application," *IEEE J. Solid-State Circuits*, vol. 27, pp. 1057–1066, 1992.

[42] Y. P. Tsividis and R. W. Ulmer, "Threshold Voltage Generation and Supply-Independent Biasing in CMOS Integrated Circuits," *Electron. CAS*, vol. 3, pp. 1–4, 1979.

[43] R. Alini, A. Baschirotto, R. Castello, and F. Montecchi, "Accurate MOS Threshold Voltage Detector for Bias Circuitry," *Proc. IEEE Int. Symp. Circuits Systems*, May 1992, pp. 1280–1283.

[44] R. Unbehauen and A. Cichocki, *MOS Switched-Capacitor and Continuous-Time Integrated Circuits and Systems*, Springer-Verlag, New York, 1989.

[45] C. G. Yu, "A Digital Tuning Scheme for Digitally Programmable Integrated Continuous-Time filters and Techniques for High-Precision Monolithic Linear Circuit Design and Implementation," Ph.D. Dissertation, Iowa State University, Ames, Dec. 1993.

[46] M. Horiguchi, T. Sakata, and K. Itoh, "Switched-Source-Impedance CMOS Circuit for Low Standby Subthreshold Current Giga-Scale LSI's," *IEEE J. Solid-State Circuits*, vol. 28, pp. 1131–1135, 1993.

Chapter 6

Low-Power CMOS Digital Circuits

Sherif H. K. Embabi
Department of Electrical Engineering, Texas A&M University, College Station, Texas 77843-3128

6.1. INTRODUCTION

In the United States, computer equipment is consuming 5–10% of electrical power consumption [1]. This number will increase with the tremendous rise in computer applications even in the household. At the same time, the demand for portable computers and portable communication devices is on the rise. The battery life time for such products is crucial. Both desktop computers and portable electronic products need to operate with minimum energy. System and circuit designers have to use low-energy design strategies.

This chapter is an overview of the recent developments in the area of low-power complementary metal-oxide-semiconductor (CMOS) digital circuits. The chapter starts with definitions of power and energy. Section 6.3 includes a discussion on the static and dynamic power dissipation in digital CMOS circuits. The use of the energy-delay product as a metric for measuring the performance of CMOS circuits is introduced. Section 6.4 touches on the switching activity in static and dynamic CMOS logic. This is an important aspect for power estimation. In Section 6.5, various approaches and strategies to reduce power/energy in CMOS circuits and systems are reviewed. Adiabatic CMOS, a new class of circuits, which can overcome the CV^2 barrier, is introduced in Section 6.6. The chapter includes two appendices. The first is a discussion on the voltage scaling trends and their impact on the delay of CMOS circuits. In the second appendix, a review of three CMOS logic families, static CMOS, dynamic CMOS, and complementary pass transistor logic (CPL), is presented.

6.2. POWER AND ENERGY DEFINITIONS

It is important, at this point, to distinguish between energy and power. The power consumed by a device is, by definition, the energy consumed per unit time. In other words, the energy (E) required for a given operation is the integral of the power (P) consumed over the operation time (T_{op}), hence,

$$E = \int_0^{T_{op}} P(t)\, dt \tag{6.1}$$

Here, the power of digital CMOS circuits[1] is given by

$$P = CV_{dd}V_s f \tag{6.2}$$

where C is the capacitance being recharged during a transition, V_{dd} is the supply voltage, V_s is the voltage swing of the signal, and f is the clock frequency. If we assume that an operation requires n clock cycles, T_{op} can be expressed as n/f. Hence, Equation (6.1) can be rewritten as

$$E = nCV_{dd}V_s \tag{6.3}$$

It is important to note that the energy per operation is independent of the clock frequency. Reducing the frequency will lower the power consumption *but will not change the energy required to perform a given operation.* Since the energy consumption is what determines the battery life, it is imperative to reduce the energy rather than just the power. It is, however, important to notice that the power is critical for heat dissipation considerations.

6.3. POWER AND ENERGY CONSUMPTION IN DIGITAL CIRCUITS

It is more convenient to talk about power consumption of digital circuits at this point. Although power depends greatly on the circuit style, it can be divided, in general, into static and dynamic power. The static power is generated due to the DC bias current, as is the case in transistor-transistor-logic (TTL), emitter-coupled logic (ECL), and N-type MOS (NMOS) logic families, or due to leakage currents. In all of the logic families except for the push-pull types such as CMOS, the static power tends to dominate. That is the reason why CMOS is the most suitable circuit style for very large scale integration (VLSI). In this chapter, we will focus on low-power CMOS circuits only.

[1]The power expression will be derived in Section (6.3.1).

6.3.1. Power Consumption in CMOS Digital Circuits

Static Power Dissipation. The static power in CMOS circuits is mainly due to two sources of leakage currents. The first is the reverse leakage current of the parasitic drain-substrate and source-substrate diodes. This current is in the order of a few femtoamperes per diode, which translates into a few microwatts of power for a million transistors. The second source is the subthreshold current of the MOS field-effect transistor (MOSFETs), which is in the order of a few nanoamperes. For a million transistors, the total subthreshold leakage current results in a few milliwatts of power. The subthreshold current is sensitive to temperature and increases dramatically with the increase in temperature. For further readings on the leakage currents in CMOS the reader is advised to consult ref. [2]. Leakage current increases with temperature.

Dynamic Power Dissipation. During the low-to-high output transition, the path from the V_{dd} rail to the output node is conducting, and the capacitance (C_o) at the output node is charging. Let us assume that the output voltage (V_o) is charging from an initial level V to $V + V_s$. The energy provided by the supply source is

$$E = \int_0^{\infty} V_{dd}\, I(t)\, dt \tag{6.4}$$

where $I(t)$ is the current drawn from the supply and is given by

$$I(t) = \left(\frac{V_s}{R}\right) e^{-t/RC_o} \tag{6.5}$$

where R is the resistance of the path between the V_{dd} rail and the output node. Substituting from (6.5) in (6.4) and integrating, we get

$$E = C_o V_{dd} V_s \tag{6.6}$$

during the high-to-low transition, no energy is supplied by the source and, hence, the total energy provided by the power supply during one full clock cycle is that given in Eq. (6.6). The average power consumed during one clock cycle is therefore

$$P = \frac{E_{\text{(per cycle)}}}{T} = C_o V_{dd} V_s f \tag{6.7}$$

For conventional CMOS gates, $V_s = V_{dd}$. Equations (6.6) and (6.7) reduce to

$$E = C_o V_{dd}^2 \tag{6.8}$$

$$P = C_o V_{dd}^2 f \tag{6.9}$$

Another source for dynamic power dissipation is the short-circuit current, which is drawn from the supply to ground through the P-logic and the N-logic when they are both conducting during the low-to-high or high-to-low input transitions. The power due to short-circuit current is typically much smaller than the CV^2f component. In the rest of this chapter, we will assume that the power consumption in CMOS is dominated by the dynamic power resulting from the charging process. For more details on the power dissipation in CMOS, the reader is referred to [2].

6.3.2. Energy-Delay Product: A Metric for Low-Energy Design

The scaling of V_{dd} is beneficial from the energy point of view but may have serious side effects on the delay. This implies that using the energy as the metric is not sufficient. Horowitz et al. [3] have proposed an alternative metric which accounts for both energy and delay by using the product of the *energy per operation* and the *delay per operation*. This metric can be used as the basis for design optimization and comparison between different systems.

To minimize the energy-delay product, we need to consider the trends of CMOS scaling and its implications on the delay (see Appendix A). From the discussion in Appendix A, we may conclude that the delay of CMOS circuits will most probably increase as the supply voltage increases. This is illustrated in Figure 6.1. The delay times in Figure 6.1 were calculated based on Eq. (A.3), which accounts for short-channel effects. In the calculations we have assumed that the threshold voltage will scale as $0.7–0.05(5 - V_{dd})$.[2] Figure 6.1 also shows the energy as a function of V_{dd}.

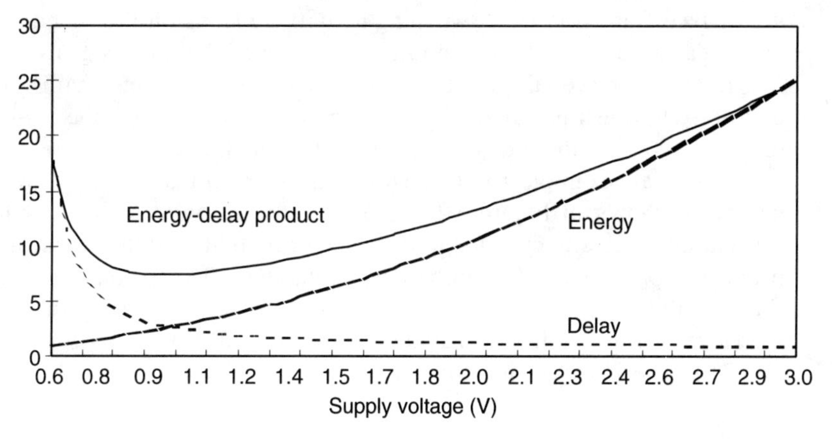

Figure 6.1 Normalized delay, energy, and energy-delay product vs. supply voltage. Delay is normalized to delay value at largest supply voltage (3 V), and energy is normalized to energy value at smallest supply voltage (0.6 V).

[2]This means that $V_T = 0.5\,\text{V}$ at $V_{dd} = 1\,\text{V}$, which is a realistic assumption.

The product of the energy and the delay, which is also shown in the same figure, demonstrates the trade-off between the delay and the energy. For low supply voltages, the energy is minimum but the delay is not. Increasing the supply voltage may improve the speed but at the expense of the energy. The energy-delay product is a metric that accounts for both, the energy and the delay, and can be used to compare different processes. The closer the minimum of the energy-delay curve to the 1-V supply, the better the process is. The optimum supply voltage can also be determined from the energy-delay product.

6.4. SWITCHING ACTIVITY IN CMOS DIGITAL SYSTEMS

Equations (6.8) and (6.9) can only be used to estimate the energy or the power of a single gate if its output exhibits a low-to-high transition. To estimate the power and energy consumed by a system at any given point in time, it is imperative to account for the actual number of gates switching at that point in time and also any undesired glitches that may result due to time skew of the input signals. To account for that, Eqs. (6.8) and (6.9) can be rewritten as

$$E = p_t C_L V_{dd}^2 \tag{6.10}$$

$$P = p_t C_L V_{dd}^2 f \tag{6.11}$$

where p_t is the probability that an energy-consuming transition occurs (0-to-1 transition for static CMOS). This is also referred to as the activity factor. The term C_L is the total loading capacitance. The product $C_L p_t$ is known as the effective switching capacitance C_{eff}, which represents the actual chip capacitance being recharged during a given transition. The switching activity depends on the functions of the individual gates, the logic family, and the statistics of the input signals.

If we assume equal probabilities for the input transitions, that is, the occurrence of a $0 \to 1$ transition is equal to that of a $1 \to 0$, we find the probability p_t for different types of logic gates can be easily calculated. Let us use a static CMOS inverter as an example to start with. The state transition diagram of a CMOS inverter is shown in Figure 6.2. The $0 \to 1$ transition may only result if the input was initially high and changed to low. The probability for this to occur is, hence, the product of p_0, which is the probability of having a 0 at the output, and p_1, which is the probability of having a 1 at the output. For an inverter both p_0 and p_1 are $\frac{1}{2}$; therefore, the probability of having a $0 \to 1$ output transition is $\frac{1}{4}$.

The general expression for p_t is

$$p_t = p_0 p_1 \tag{6.12}$$

Both p_0 and p_1 can be expressed as

$$p_0 = \frac{n_0}{2^n} \tag{6.13}$$

$$p_1 = \frac{n_1}{2^n} = \frac{2^n - n_0}{2^n} = 1 - p_0 \tag{6.14}$$

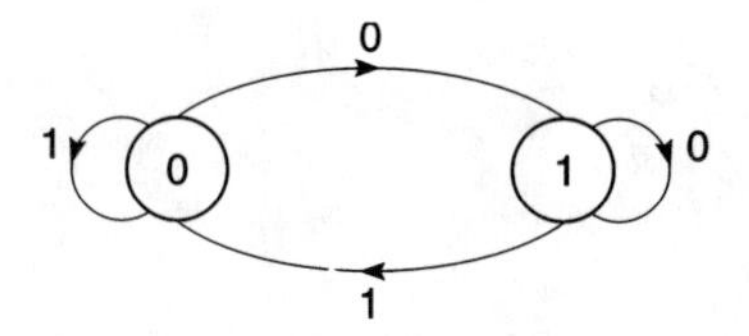

Figure 6.2 State transition diagram of an inverter.

where n_0 is the number of 0's in the output column of the truth table, n_1 is the number of 1's, and n is the number of inputs.

Equations (6.12)–(6.14) can be used to compute the probabilities for energy-consuming transitions for any static (and CPL) CMOS gate. Table 6.1 shows these probabilities for a variety of gates. For dynamic CMOS the $0 \rightarrow 1$ output transition occurs if the output is 0 by the end of the evaluation phase (regardless of the state of the input); therefore, the probability p_t becomes

$$p_t = p_0 = \frac{n_0}{2^n} \tag{6.15}$$

The switching activity of dynamic logic is higher than that of static logic. This is demonstrated by a few examples for different gate types shown in Table 6.1. For all types except for the NAND gates, the probability of a $0 \rightarrow 1$ transition in dynamic circuits is worse. This is expected since using the precharge phase the output of a dynamic circuit must be pulled high whether the input changes or not.

If the inputs are not equiprobable, p_t becomes a function of the probabilities of the input transitions. For further reading on this topic the reader is advised to consult other references [4–7].

So far we have considered static transitions, which result from the static behavior of the gates when their delays are neglected. This is usually referred to as the zero-delay model. Dynamic transitions occur in CMOS circuits frequently and have to be considered in the analysis of switching activity of any system. Dynamic transi-

TABLE 6.1 PROBABILITIES FOR A $0 \rightarrow 1$ TRANSITION FOR DIFFERENT TYPES OF STATIC AND DYNAMIC CMOS GATES

Gate Type	Static	Dynamic
Inverter	$\frac{1}{4}$	$\frac{1}{2}$
Two-input NAND	$\frac{3}{16}$	$\frac{1}{4}$
Two-input NOR	$\frac{3}{16}$	$\frac{3}{4}$
Two-input XOR	$\frac{1}{4}$	$\frac{1}{2}$
Three-input NAND	$\frac{7}{64}$	$\frac{1}{8}$
Three-input NOR	$\frac{7}{64}$	$\frac{7}{8}$
Three-input XOR	$\frac{1}{4}$	$\frac{1}{2}$

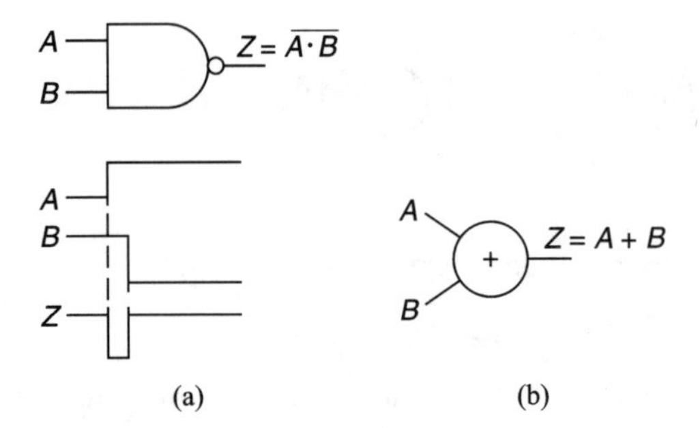

Figure 6.3 Spurious transitions (glitches) in (a) a NAND Gate and (b) an adder block.

tions result due to the imbalance in the paths of the input signals of a single gate or logic block in general. For example, consider the two-input NAND gate with inputs A and B shown in Figure 6.3(a). If B arrives after A as illustrated in the figure, the output would exhibit a "spurious" transition, which is known as a "glitch." This would consume "useless" energy since the output ought to stay unchanged. The same concept applies for more complex logic blocks. If the operand A to the adder shown in Figure 6.3(b) arrives before B, the sum (Z) will experience intermediate transitions before settling to its final state.

6.5. REDUCTION OF ENERGY IN CMOS DIGITAL CIRCUITS

So far we have discussed the energy consumption in digital CMOS circuits. In this section, we would like to look at the means of reducing the energy/power in digital CMOS systems.

6.5.1. Power Supply Reduction

Equations (6.8) and (6.9) show that the energy and power consumed by CMOS digital circuits are sensitive to the power supply voltage. Reducing the power supply voltage is an efficient approach to lower the energy and power. For example, scaling the supply voltage from 5 to 3 V reduces the energy/power by 64%, while scaling it to 1 V results in 96% energy/power saving. The power supply voltage is actually the most crucial factor in reducing energy/power. This will, however, be at the expense of the delay of the circuits. Using the energy-delay product as a metric one can derive the optimum supply voltage that would yield minimum energy-delay product. To simplify the analysis, we will assume that the saturation current of deep submicrometer MOSFETs is proportional to $(V_{GS} - V_T)^\alpha$ [8].[3] Assuming that $V_{GS} = V_{dd}$ (for maximum current) and using the delay expression in (A.3), it can be shown that the delay becomes $KV_{dd}/(V_{dd} - V_T)^\alpha$, where K is a constant independent of V_{dd}. The energy-delay product can hence be expressed as

[3]The exponent α is between 1 and 2. It tends to be closer to 1 for deep submicrometer MOS transistors (where carrier velocity saturation may occur) and increases toward 2 for longer channel transistors.

$$E \times t_d = \text{const} \frac{V_{dd}^3}{(V_{dd} - V_T)^{\alpha}} \tag{6.16}$$

The optimum supply voltage (for minimum energy-delay product) can be found from Eq. (6.16) and is given by

$$V_{dd(\text{opt})} = \frac{3V_T}{3 - \alpha} \tag{6.17}$$

The above expression is valid for long-channel and deep submicrometer devices. For long-channel transistors ($\alpha = 2$) the optimum supply voltage is equal to $3V_T$, which agrees with the result of the analysis presented in [5]. For deep submicrometer devices with α closer to unity the optimum voltage is expected to be less than $3V_T$. For example, if $\alpha = 1.5$, then $V_{dd(\text{opt})} = 2V_T$. At any rate, the optimum value for V_{dd} is proportional to the threshold voltage. In Figure 6.1, the minimum of the energy-delay product is at $V_{dd} = 0.95\,\text{V}$, which is twice the value of V_T (0.5 V).

So the conclusion is that the supply voltage must be reduced to minimize the energy-delay-product. Scaling the supply voltage below the point of minimum energy-delay-product will cause severe degradation in the delay. The second point is that the optimum supply voltage is related to the threshold voltage.

In certain types of applications the architecture can be designed to allow for scaling the voltage supply without sacrificing the system's performance. Good examples are digital processing systems, where the throughput is more critical than the speed. The feature can be exploited to reduce the voltage at the expense of speed without throughput degradation. This has been demonstrated by Chandrakasan et al. [9]. We briefly review their concept of trading area and hardware for lower power to preserve throughput. This can be achieved by using parallelism and/or pipelining. Both techniques will be discussed next.

Let us first consider the data path of the generic uniprocessor shown in Figure 6.4. This is made up of two latches which synchronize the data flow in and out of a logic function. The switched capacitance is C, the clock frequency is f, and the supply voltage is V_{dd}. The energy consumed by this implementation is

$$E = CV_{dd}^2 \tag{6.18}$$

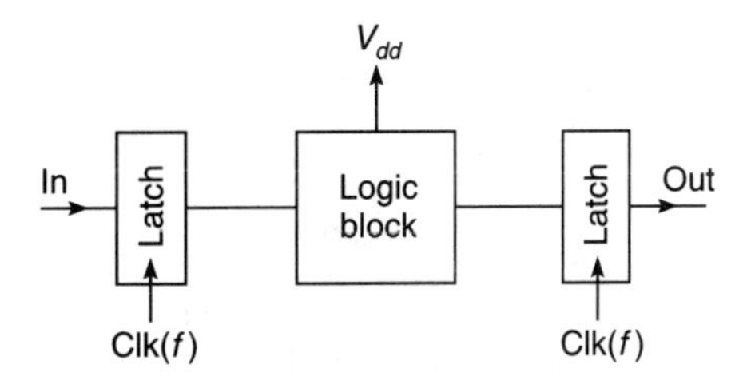

Figure 6.4 Uniprocessor architecture operating at clock frequency f and with supply voltage V_{dd} [5].

If the logic is duplicated n times using n parallel blocks and the input is fed to each logic block at a frequency f/n, this implies that the latches to each block are clocked at one nth the frequency of the latch in uniprocessor implementation. The output of the parallel blocks is sent to the output latch through a multiplexer. The parallel implementation is illustrated in Figure 6.5. Since each logic block operates at f/n, the delay of these blocks can be relaxed by a factor n. Note that the throughput will be similar to that of the uniprocessor. By allowing for increased delay, the supply voltage can be scaled down until the block delays equal n times the delay of the uniprocessor implementation. Hence, the delay in the parallel architecture is given by

$$T_p = k\frac{V_{dd(P)}}{(V_{dd(P)} - V_T)^{\alpha}} = nT = nk\frac{V_{dd}}{(V_{dd} - V_T)^{\alpha}} \tag{6.19}$$

where T_P and T are the delay times of the logic blocks in the parallel and the uniprocessor architectures, respectively. Here, $V_{dd(P)}$ and V_{dd} are the supply voltages in the parallel and the uniprocessor architectures, respectively. The supply voltage required for the supply voltage can be obtained from Eq. (6.19). It has been shown that if a square-law model ($\alpha = 2$) is assumed and if the threshold voltage is neglected, the supply voltage for the parallel architecture is given by [5]

$$V_{dd(P)} = \frac{V_{dd}}{n} \tag{6.20}$$

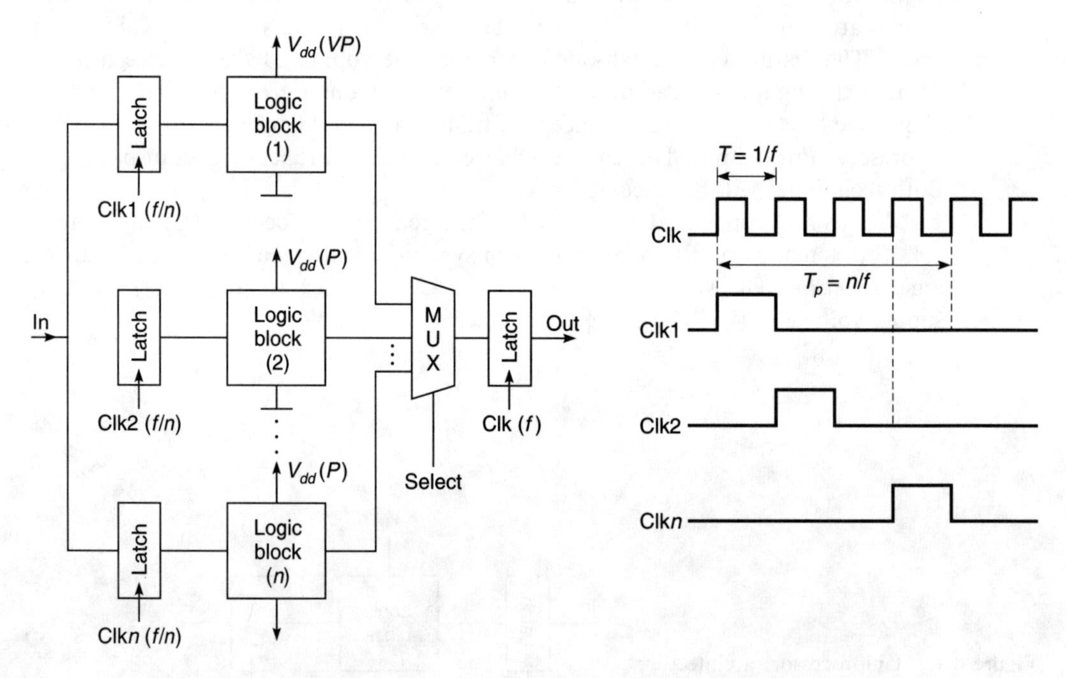

Figure 6.5 Parallel processor architecture operating at clock frequency f/n and with reduced supply voltage V_{dd} [5].

For all values of V_T it was demonstrated that as the number of parallel slices increases (i.e., n increases), the supply voltage can be scaled down up to a certain point where the supply voltage saturates. For submicrometer transistors with carrier velocity saturation the analysis will become more complex. To obtain a closed-form expression for $V_{dd(P)}$, we will assume that $\alpha = 1$. In this case,

$$V_{dd(P)} = \frac{nV_T}{n - V_T/V_{dd} - 1} \tag{6.21}$$

Equation (6.21) can be used to show that for deep submicrometer CMOS the supply voltage may scale at a slightly faster rate compared to longer channel CMOS as n increases.

If the power overhead due to the multiplexer is ignored, then the energy of the parallel architecture can be expressed as

$$E_P = CV_{dd(P)}^2 = E\,\frac{V_{dd(P)}^2}{V_{dd}^2} \tag{6.22}$$

The above equation indicates that the energy saving by going to parallelism is proportional to the square of the voltage scaling factor. So by choosing the proper value of n and depending on the threshold voltage, the new supply voltage will be determined. This will lead to a corresponding energy scaling without sacrificing the throughput. Yet this energy reduction will come at the expense of area.

If the overhead is considered, there will be a value for n which corresponds to the minimum energy consumption. This yields the optimum supply voltage of the parallel architecture. For further reading on this issue refer to [5].

Another approach to relax the speed requirement without degrading the throughput is to pipeline the data path. If the logic block of Figure 6.4 is broken down in n levels and latches are inserted between them, as shown in Figure 6.6, the delay of each level or subblock is reduced. This allows for reducing the supply voltage.

An example of a simple adder-comparator data path has been used to demonstrate the power savings achieved through parallelism and pipelining [5]. The result of the comparison is summarized in Table 6.2. Note that the pipeline approach offers comparable power savings to the parallel architecture but with less area overhead.

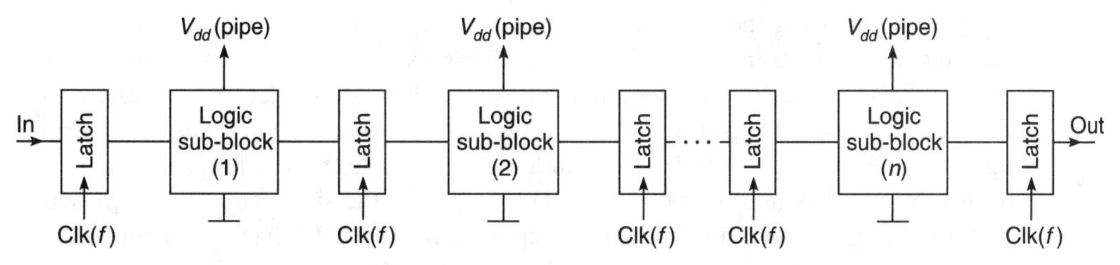

Figure 6.6 Pipelined processor architecture operating at a clock frequency and with reduced supply voltage V_{dd} [5].

TABLE 6.2 RESULTS OF ARCHITECTURE-BASED VOLTAGE SCALING [5]

Architecture	Voltage (V)	Area (Normalized)	Power (Normalized)
Simple	5	1	1
Parallel	2.9	0.75	0.36
Pipelined	2.9	1.3	0.39
Pipelined-parallel	2.0	3.7	0.2

6.5.2. Switching Activity Reduction

In Section 6.4 we have demonstrated the importance of the impact of switching activity on the power consumption in CMOS digital circuits. Switching activity can be reduced by algorithmic optimization, architecture optimization, logic topology, and circuit optimization. Each of these aspects will be discussed briefly in the following sections.

Algorithmic Optimization. Algorithmic optimization depends heavily on the application and on the characteristics of the data, such as the dynamic range, the correlation, the statistics of the data transmission, and so on. Some of the techniques apply only to applications such as Digital Signal Processing (DSP) and cannot be used for general-purpose processing. An example for selecting the vector quantization algorithm with the minimum switching activity can be found in [5]. Three algorithms, the full search, the tree search, and the differential tree search, have been compared. The differential tree search uses the least number of operations of all three. For example, the number of memory accesses, the number of multiplications, and the number of additions reduce by a factor of 30 if the differential tree search is used rather than the full search.

The data representation may have a significant impact on the switching activity. Two examples have been used in [5] to demonstrate that. The first is the use of a gray code versus the binary code. In applications where the data are sequential, the use of gray code leads to reduced number of transitions. The address bits to access instructions is a good example. Since the address will most probably change sequentially (except if there is a jump), only one bit in the address word will change (if the gray code is used). In the case of binary code the number of transitions will be 2 on the average. Also, using the sign magnitude instead of the conventional 2's complement may result in less transitions if the data change sign frequently. A change in sign causes transitions of the most significant bits of the 2's complement representation. In the case of sign magnitude, only the sign bit will change. For example, reversing the sign of the number 2 in the 2's complement representation requires six transitions if an 8-bit word is used (00000010 $\rightarrow$ 11111110). In the sign magnitude representation, only the sign bit, namely the most significant bit (MSB), has to be changed. The reduction of the switching activity of the sign magnitude will diminish if the amplitude of the signal that is being processed is close to the full dynamic range.

Architecture Optimization

DELAY BALANCING. Several architectural techniques have been proposed to reduce the switching activity, such as ordering of input signals [5] and delay path balancing to remove glitching. Delay path balancing can be achieved by using balanced tree topologies rather than cascaded or chain topologies. Figure 6.7 shows two architectures for the same operation $A+B+C+D$. The tree yields less glitches. In some cases, extra delay may be added to the shortest delay path for balancing purposes.

It is interesting to note that both static and dynamic transitions are to be considered when comparing between different structures. Structures with high static switching activity may have very low dynamic switching and vice versa.

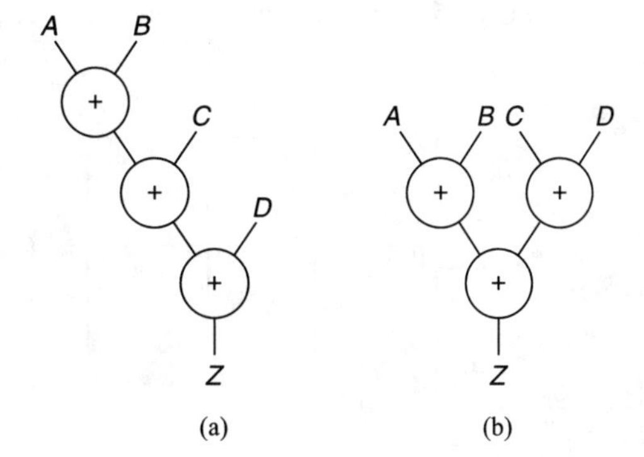

Figure 6.7 Implementations of summation of $A, B, C,$ and D using (a) a chain structure and (b) a tree structure.

PRECOMPUTATION-BASED LOGIC. The concept of precomputation has been used to predict the output signal one clock cycle ahead using minimum circuit overhead [10]. Consider the circuit shown in Figure 6.8. A combinational logic block A which implements a Boolean function f is separated by registers R_1 and R_2. The inputs to A can be partitioned into two sets. One set is transmitted through R_1 and the other through R_2. The inputs to R_1 are also fed to two combination blocks G_1 and G_2. These blocks implement the Boolean predictor functions g_1 and g_2 which have to satisfy the conditions

$$g_1 = 1 \Rightarrow f = 1 \qquad g_2 = 1 \Rightarrow f = 0$$

Note that g_1 and g_2 depend on a small subset of input signals. If any of them evaluates to 1 during clock cycle n, R_2 is disabled so that the input set transmitted through it is blocked [see Fig. 6.8(b)]. Only the small input set corresponding to R_1 passes over to the combinational block A. This is enough for the correct evaluation of f (according to the above condition). Since only a small set of the inputs has changed, the switching activity in A will be reduced. This is achieved if G_1 and G_2 are significantly less complex than A. The selection of the input signal partitioning and the synthesis of G_1 and G_2 are discussed in [10]. An application of this concept for a comparator is shown in Figure 6.9 [10].

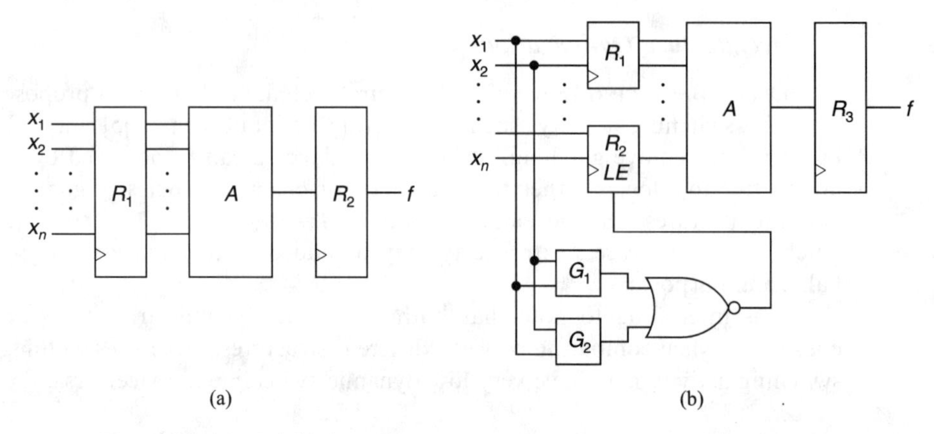

Figure 6.8 (a) Original architecture and (b) architecture with precomputation [10].

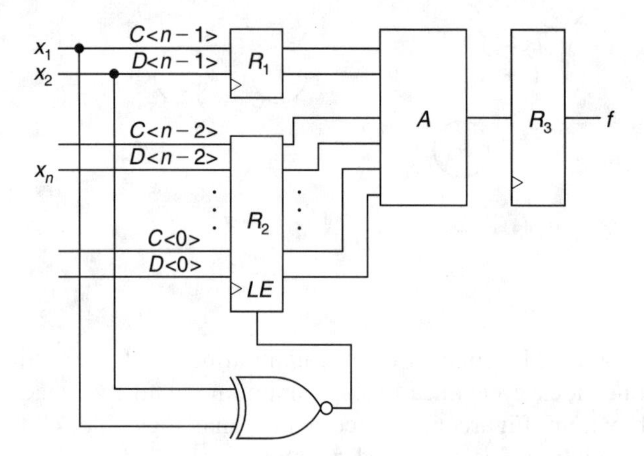

Figure 6.9 Comparator with precomputation [10].

Power Management. The technique that is by far the most effective is the power-down approach which is used to put the circuits in a *sleep mode* when they are idle. This can be used for different levels of hierarchy. It can be applied at the chip level as well as at the printed circuit board (PCB) level. Consider the conceptual chip shown in Figure 6.10. The chip is divided in several blocks. The clock to each block is gated so that the clock to a given block can be disabled upon demand. For example, powering down a floating-point unit in a processor when integers are processed may result in a 20% power reduction, as demonstrated in [11].

Circuit Optimization. In Appendix B, we discuss the difference between static and dynamic CMOS logic families. We see that the dynamic CMOS circuits have higher probabilities of $0 \rightarrow 1$ transition compared to static CMOS. However, dynamic logic seems to have less tendency to exhibit glitches. This makes the choice between static and dynamic logic difficult. In the cases where it is possible to balance the delay of the paths, static CMOS may be the best choice.

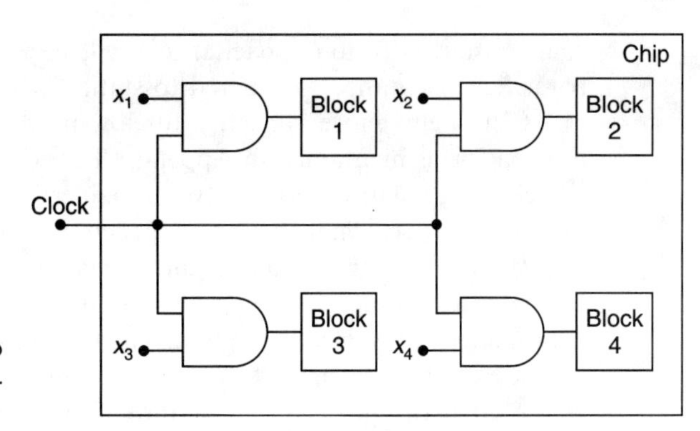

Figure 6.10 Using gating signals X_i to enable clock to certain blocks on demand.

The switching activity may depend heavily on the topology. A good example is the adder, which can be implemented using ripple carry, carry lookahead, carry skip, carry select, or conditional sum. Callaway and Swartzlander [12] have compared the average number of transitions per addition for different topologies (see Table 6.3). The choice of the adder topology should not be based only on the power consumption but also on the delay. The ripple carry adder exhibits the least number of transitions yet is the slowest of all. The comparison may rather be based on a metric which accounts for both delay and power.

6.5.3. Reduction of Switching Capacitance

Energy consumption is proportional to switching capacitance, as shown by Eq. (6.8). The switching capacitance can be broken down into two categories, the capacitance in dense logic (which includes the transistor parasitic and wire capacitances at the output of the gates) and the capacitances of busses and a clock network (which is mainly wire capacitance). In some systems, the capacitance of busses and a clock network may comprise close to 50% of the overall chip capacitance. An example of such a system is the Alpha chip [13].

A comparison between the three logic families, static, dynamic, and CPL in Appendix B shows that the CPL has the least transistor count for a given Boolean function implementation. This implies that the parasitic capacitances (gate oxide and

TABLE 6.3 AVERAGE NUMBER OF GATE TRANSITIONS PER ADDITION

Adder Type	16 bits	32 bits	64 bits
Ripple carry	90	182	366
Carry lookahead	100	202	405
Carry skip	108	220	437
Carry select	161	344	711
Conditional sum	218	543	1323

source/drain diffusion capacitances) will be reduced. The dynamic gates have less parasitic capacitances compared to static CMOS.

CPL seems more attractive for low power applications. However, it has some drawbacks, as mentioned in Appendix B, because of the V_T swing loss. This results in delay degradation, especially for reduced supply voltages. Lowering V_T would improve the performance of the CPL circuits in terms of speed.

It is also important to note that the layout be optimized so that internal nodes of the gate are minimized. The fact that the capacitances of the internal nodes may consume power is usually ignored. The two-input NOR gate shown in Figure 6.11 is a good example to demonstrate that. Assume that input A was initially high and that B was low; the internal capacitance C_i will be discharged as well as the output capacitance, as shown by the waveforms in Figure 6.11. If both inputs switch (assuming that B does not switch after A), the internal node X would experience a $0 \rightarrow 1$ energy-consuming transition while the state of the output node remains unchanged.

Transistor sizing is imperative for speed and power. Conventional design strategies focused on speed. The delay specifications were satisfied by sizing the transistors. For low-power design, the rule is to size the transistors on the *critical paths* only such that the speed requirements are met. The transistors on the *noncritical paths* should be kept at the least possible dimensions. The name of the game for low-power design is to *use minimum-size transistors as much as possible.*

Layout optimization is another area where power can be saved by choosing the layout styles that will minimize the diffusion capacitances. The layout determines the interconnect length, which in turn determines the capacitive loading. The capacitance of global interconnects such as the data or address busses and the clock nets have to be optimized. This could lead to significant power savings. It, however, requires smart computer-aided design (CAD) tools that can do placement and routine to minimize the length of the wires which carry the signals with the maximum switching activity.

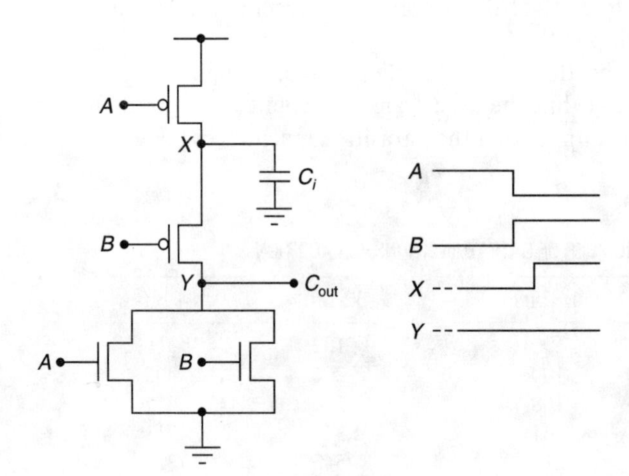

Figure 6.11 Example of internal node (X) that can switch from 0 to 1 while voltage of output node remains unchanged.

6.5.4. Reduction of Voltage Swing

As indicated by Eq. (6.3), the energy is proportional to the voltage swing of the signal. For most of the CMOS logic families the swing is typically rail to rail. It is difficult to reduce the swing in the logic because of the speed degradation, as is the case in CPL. It, however, seems more appealing to reduce the signal swing along the long busses and clock nets. This would result in reasonable power savings because of the large capacitance associated with the busses, as mentioned above. The reduction of the voltage swing requires special drivers and receivers. Figure 6.12 illustrates the architecture of a bus with reduced swing [14]. The driver attenuates the swing by a factor n, while the receiver amplifies the swing by the same factor to restore the original swing. Several structures for such drivers and receivers have been reported recently [14–16]. Some of these will be discussed next.

The swing can be reduced by shifting the rail voltages of the driver, as shown in Figure 6.13(a) [14]. Note that the driver is a simple CMOS inverter. The upper supply voltage has been lowered to V_{CL} and the lower rail has been raised to V_{SL}. This causes the output of the driver to swing between these two levels, while its input is driven by full swing signals. The drawback of this approach is that it requires internally generated supply voltage levels (V_{CL} and V_{SL}). This can be achieved using the generator shown in Figure 6.13(b), which has the serious disadvantage of consuming standby power. To restore the rail-to-rail signal swing at the receiving end, the symmetric-level converter shown in Figure 6.14 [14] can be used. The circuit consists of two symmetric-level converters: one, which consists of MN_3, MP_4, and MP_5, for converting V_{CL} to ground, and the other, which consists of MP_3, MN_4, and MN_5, for converting V_{SL} to V_{dd}. When the input goes to V_{CL}, node B is pulled up towards V_{CL} to turn MN_3 on, which in turn causes MN_5 to turn on. The output is pulled down to ground and MP_4 turns on to raise the voltage of node A to V_{dd}. As a result, MP_5 is cut off and prohibits any DC current from flowing from V_{dd} to ground. The overall transmission delay, including the delays of two inverters, the driver, the interconnect (2 pF), and the receiver, was estimated through simulations for different supply voltages. The comparison of the transmission delay of the reduced swing bus architecture versus conventional full swing bus is illustrated in Figure 6.15 [14]. The reduced swing bus shows better speed performance for low supply voltage.

Improvements to the bus architecture in [14] can be achieved by using driver architectures which produce a reduced output swing without internally generated supply voltages. Such structures are shown in Figure 6.16 [15].

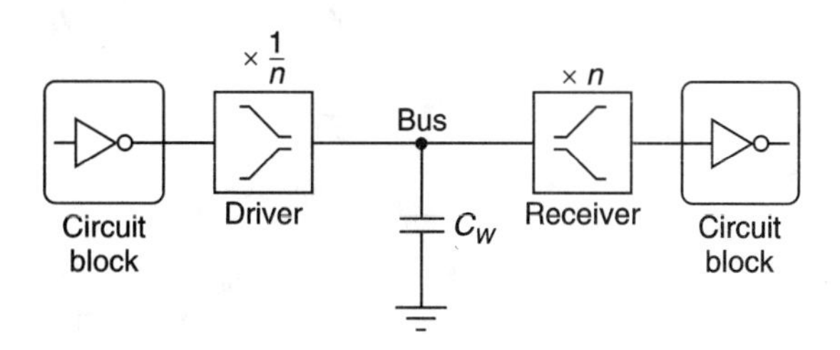

Figure 6.12 Bus architecture with reduced internal swing [14].

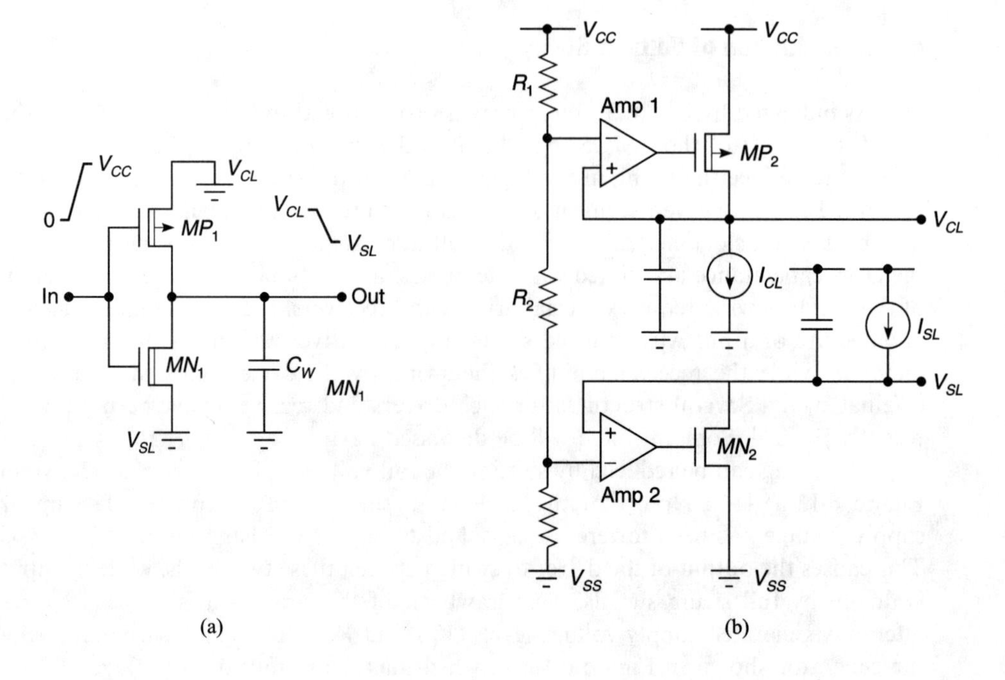

Figure 6.13 (a) Reduced swing bus driver and (b) generator of supply voltage for bus driver [14].

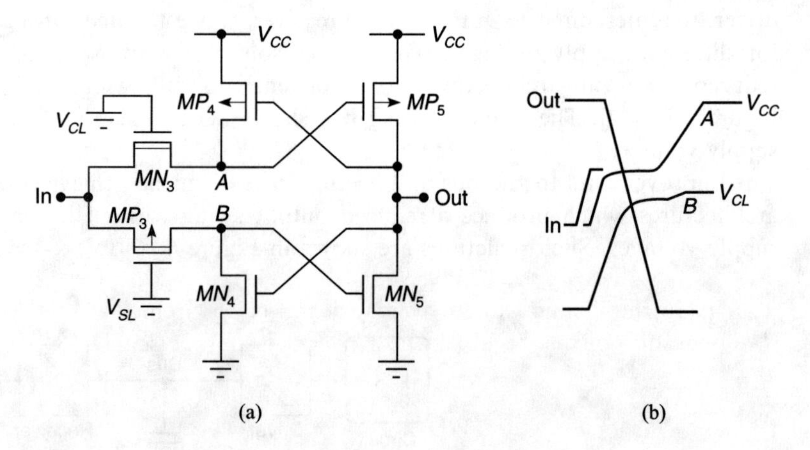

Figure 6.14 (a) Circuit schematic of symmetric level converter; (b) voltage waveforms at input, output, and nodes A and B [14].

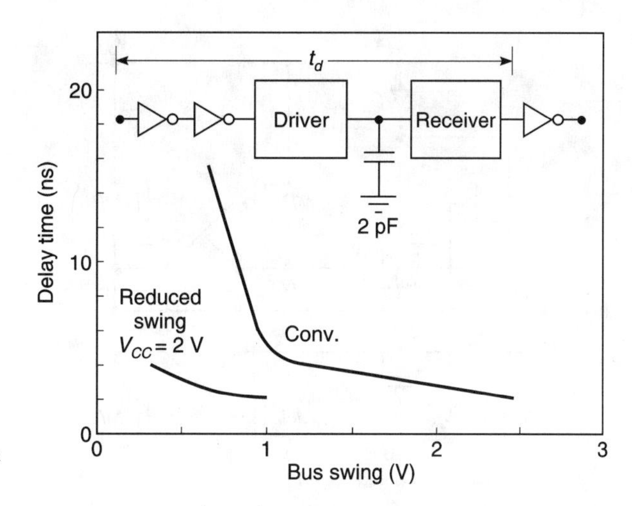

Figure 6.15 Simulated bus transmission delay vs. bus signal swing [14].

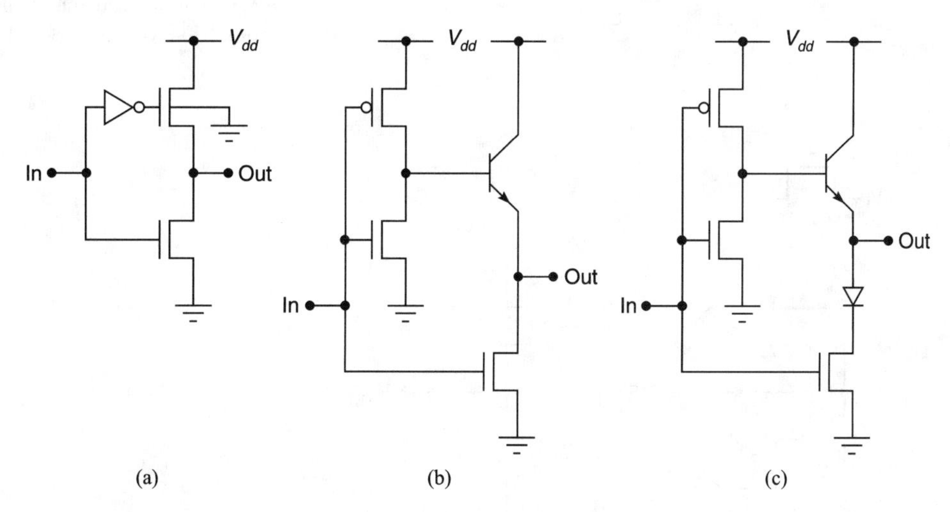

Figure 6.16 Circuit schematics of (a) CMOS driver with NMOS pull-up, (b) bipolar NMOS (BiNMOS) driver, and (c) BiNMOS with a diode in the pull-down [15].

The symmetric-level converter can also be modified to avoid the use of V_{CL} and V_{SL}, as shown in Figure 6.17 [15].

In [16] a dynamic reduced swing driver/receiver architecture has been proposed. The operation of the driver can be explained using the schematic of Figure 6.18(a). The driver consists of a CMOS driver (Q_1 and Q_2), a voltage sense translator (VST), and controlling logic. Assume that the output of the VST is initially at a logic low. If the input goes low (0 V), Q_1 turns on, and the output node charges up until it reaches V_{SH}, which is the high level of the reduced output swing. This level will cause the VST to generate a logic high. The output of the OR gate is pulled to V_{dd} to turn Q_1

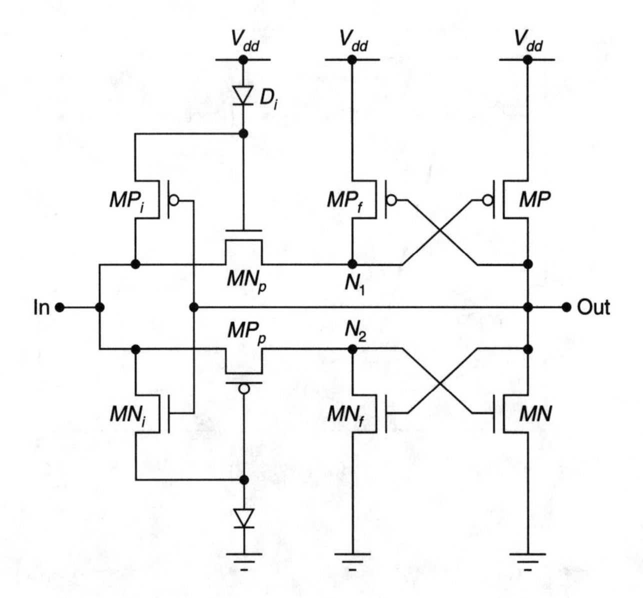

Figure 6.17 Circuit schematic of modified symmetric-level converter [15].

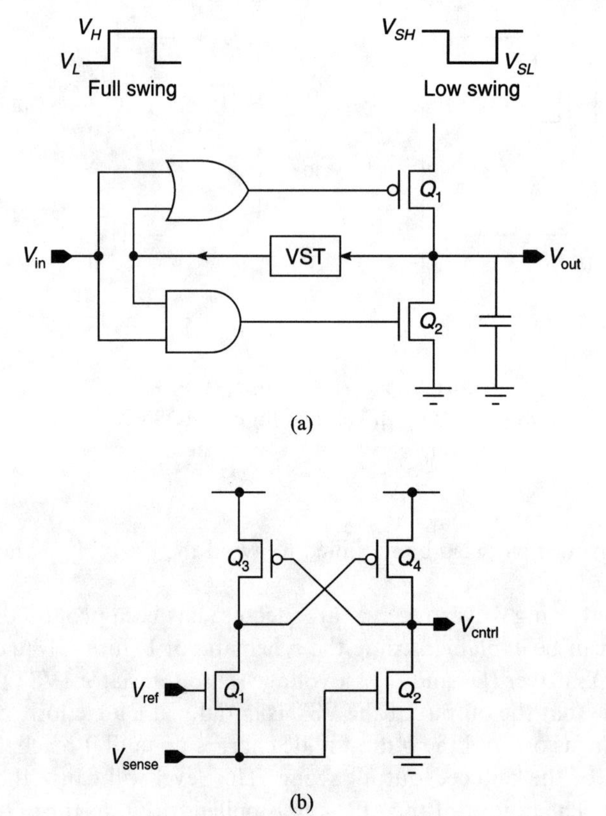

Figure 6.18 (a) Diagram of reduced swing driver using VST and (b) circuit schematic of VST [16].

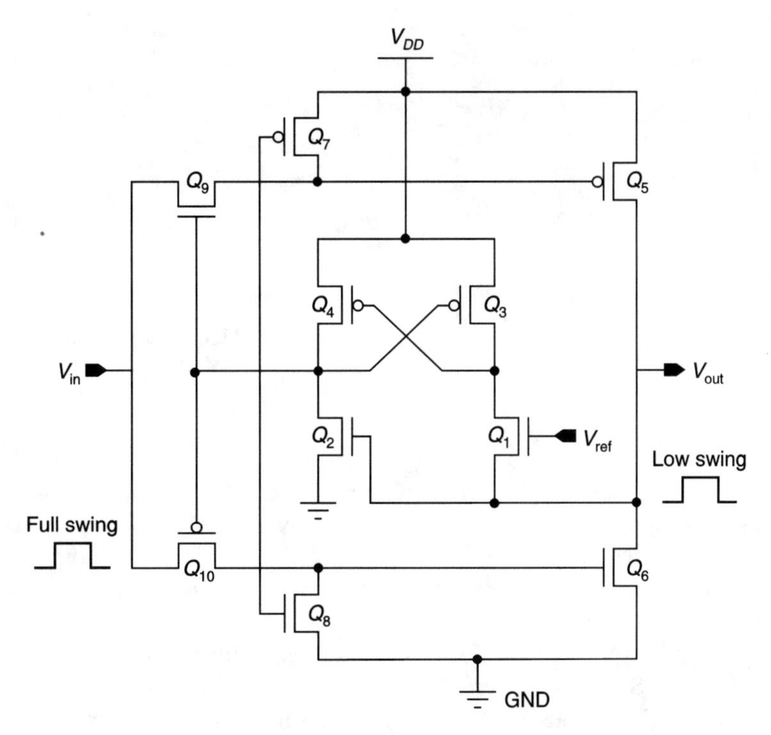

Figure 6.19 Full schematic of reduced swing driver [16].

off. The output of the driver ceases to charge beyond V_{SH}. If the input is high, the output discharges to V_{SL} when the VST generates a logic low. Here, Q_2 turns off and the output remains at V_{SL}. Figure 6.18(b) shows the schematic of the VST. When V_{sense} reaches V_{SH}, which is equal to $V_{\text{ref}} - V_{TN}$, Q_1 turns off. At the same time, Q_2 is on and pulls the gate of Q_3 down to turn it on. Then Q_4 cuts off. Thus the output of the VST (V_{cntrl}) will be low (0 V). Similarly, it can be shown that when V_{sense} is equal to V_{SL} ($=V_{TN}$), V_{ntrl} goes high ($=V_{dd}$). Figure 6.19 shows the complete driver structure with the VST implementation. The VST can be used at the receiving end for amplifying the levels to full swing.

6.6. ADIABATIC COMPUTING

The energy CV_{dd}^2, which is consumed in conventional CMOS circuits, is unavoidable since the charge is transferred from the supply and returned to ground [1]. The current drawn from the supply during a $0 \rightarrow 1$ transition is relatively large because of the large drain-source voltage. If, however, the supply voltage can be varied in a manner that would reduce the drain current, the energy will be signficantly reduced. This can be achieved by using adiabatic circuits. Consider the circuit shown in Figure 6.20. This circuit is sometimes referred to as a pulsed power supply CMOS (or PPS

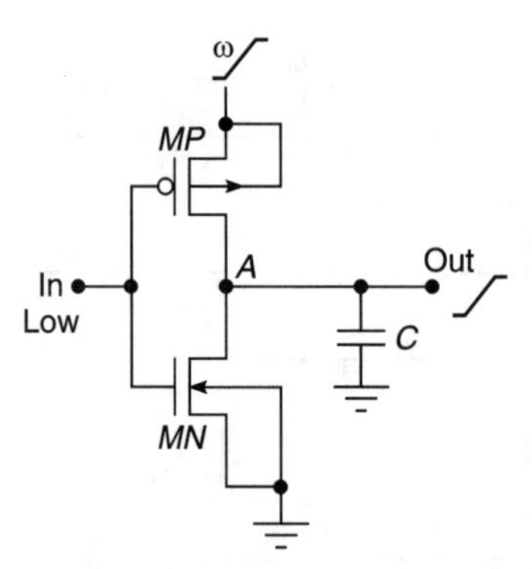

Figure 6.20 Schematic of (adiabatic) PPS CMOS inverter [17].

CMOS) [17]. Its topology is very similar to that of the conventional CMOS inverter except that its supply is driven with a pulsed supply waveform, ω. Assume that the input is low and that the output was initially low. With the supply voltage being low, the drain current is zero. As the supply voltage ramps up, the output follows the supply voltage. The drain-source voltage is always small and so is the current drawn from the supply.

To derive the energy for this type of adiabatic circuit, we will use a simple model for the PPS CMOS inverter when the input is low [see Fig. 6.21(a)]. The PMOS transistor is modeled by a resistor R. Assume that the supply is increasing in steps from 0 to V_{dd}, as shown in Figure 6.21. Let us first derive the energy per step. Between the ith step and the next one, the supply voltage changes from V_i to V_{i+1}. The drain current is given by

$$I = C\frac{dV_o}{dt} = \frac{V_{i+1} - V_o}{R} \tag{6.23}$$

Solving this differential equation from $t = t_i$ (when the supply switches to V_{i+1}) to any time $t < t_{i+1}$, we get the following expression for the output voltage as a function of time:

$$V_o = V_{i+1} - \frac{V_{dd}}{n}\, e^{-t/RC} \tag{6.24}$$

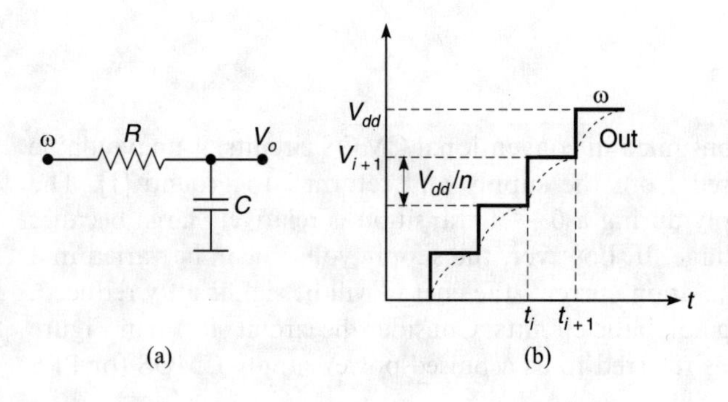

Figure 6.21 (a) the *RC* model of PPS CMOS inverter and (b) supply and output voltage waveforms.

where n is the number of steps of the supply voltage [see Fig. 6.21(b)]. Substituting from (6.24) into (6.23), we obtain the current expression which is then used for the derivation of the energy consumed per step:[4]

$$E_{\text{step}} = \int_0^\infty I^2 R\, dt = \int_0^\infty \left(\frac{V_{dd}}{nR} e^{-t/RC} \right)^2 R\, dt = \frac{1}{n^2} \frac{CV_{dd}^2}{2} \tag{6.25}$$

The energy consumed for one operation is $n \times E_{\text{step}}$. Theoretically, if n is infinite (i.e., the supply voltage is a slow ramp), the energy goes to zero.

The PPS CMOS can be used for complex Boolean function implementations. Figure 6.22 shows an example of a carry lookahead circuit [17].

Figure 6.23 shows a circuit which generates the supply voltage ramp. The circuit uses a low power oscillator, which employs a tank circuit and a cross coupled MOS transistor [17]. A stepwise supply, which uses switched supplies as shown in Figure 6.24, is another alternative [18].

This type of adiabatic circuit has, however, a major disadvantage. Such circuits can only be used if the delay is not critical. In addition, it is important to mention that in the above discussion we have ignored the energy overhead required for generating the supply ramp or the stepwise waveforms. In both schemes there is energy consumed by the switches used for the generation of the V_{dd} waveform (e.g., the generator in Figure 6.24). The drivers of the ramping or stepping supply voltage will also dissipate significant energy. It has been demonstrated that the

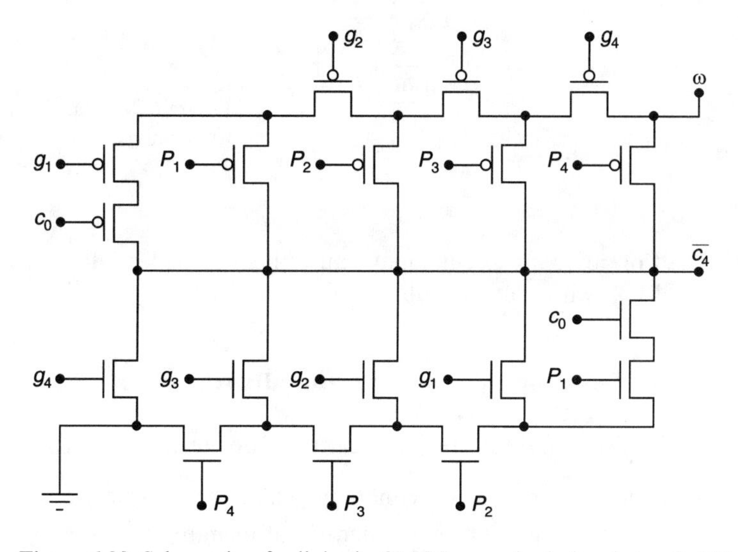

Figure 6.22 Schematic of adiabatic CMOS carry lookahead circuit [17].

[4]It is assumed that the duration of each step is much greater than the RC product.

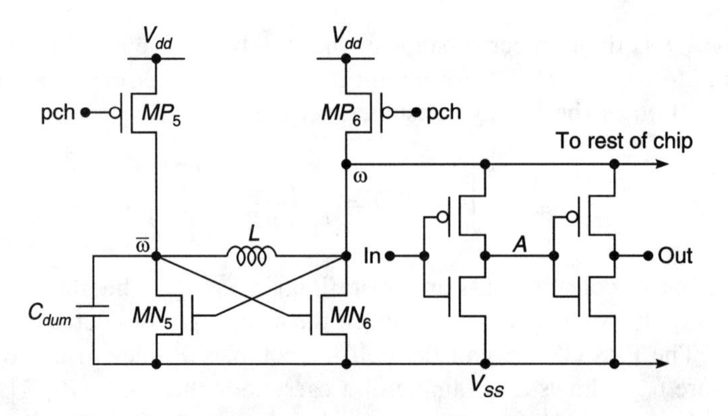

Figure 6.23 Tank circuit with cross-coupled NMOS transistors (MN_5 and MN_6) generate pulsed supply voltage. Frequency of oscillation is determined by voltage *pch* [17].

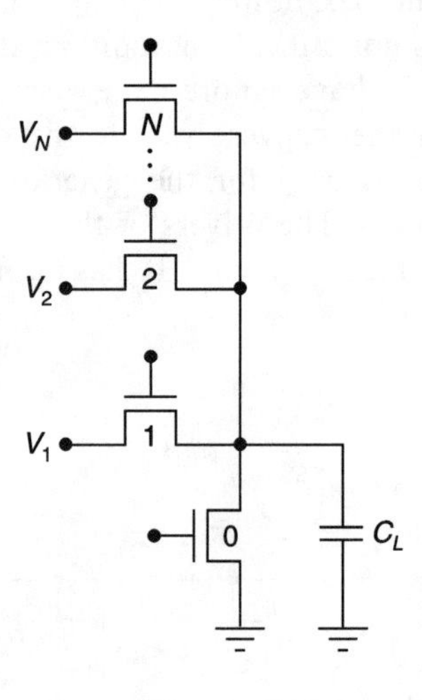

Figure 6.24 Transistors $1, \ldots, N$ switch uniformly distributed supplies $V_1, \ldots, V_N$ successively [18].

overhead of adiabatic circuits may cause them to be less efficient than the conventional CMOS with voltage scaling [19].

Appendix A

Hu [20] has summarized the historical trends of technology scaling as follows:

- Every 3 years a new technology generation was developed.
- With every generation the density of memory increased by a factor of 4 and that of logic by a factor of 2–3.
- With every two generations, the device feature size decreased by a factor of 2, while the current density, the circuit speed, and the chip size increased by 2.

Figure 6.25 shows the trend of MOS technology scaling from the end of the 1970s to the beginning of the 1990s (the illustrated data was extracted from [20]). Note that while the device dimensions were scaled, the supply voltage was maintained at 5 V.

Predictions for future scaling trends can be based on theoretical models. Next, we will derive an expression for the delay of a CMOS inverter based on the analyses in [20, 21].

The MOS current can be modeled by the expressions [21]

$$I_{DS,\text{sat}} = KWv_{\text{sat}}C_{\text{ox}}(V_{GS} - V_T) \tag{A.1}$$

and

$$K = \frac{1}{1 + E_c L_e/(V_{GS} - V_T)} \tag{A.2}$$

where

$$E_c = \frac{2v_{\text{sat}}}{\mu_{\text{eff}}}$$

and

$$L_e = L_{\text{eff}} - X_d$$

where W is the channel width, $V_{DS,\text{sat}}$ is the drain saturation voltage, v_{sat} is the carrier saturation velocity, L_{eff} is the effective channel length, and E_c is the velocity-saturation field. The full definitions of X_d and μ_{eff} can be found in [21].

The 50% delay time of a CMOS inverter driving a capacitance C can be written as [20]

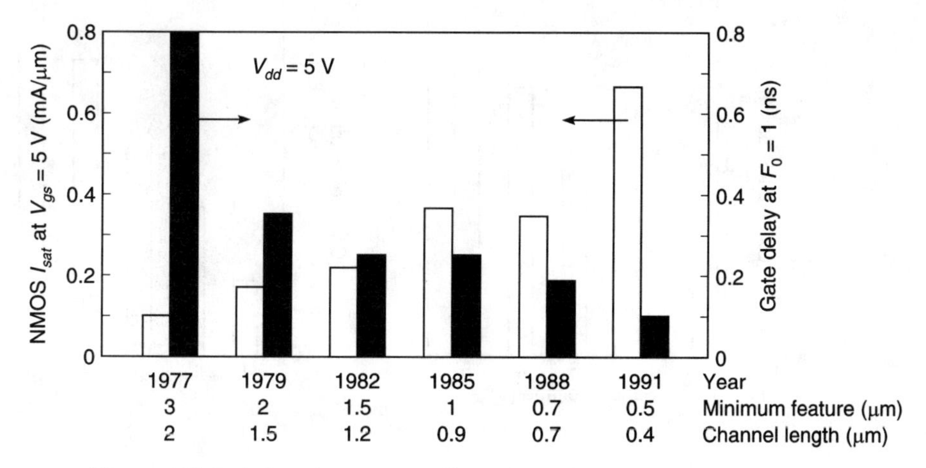

Figure 6.25 Delay and current scaling of MOS transistors for fixed 5 V supply voltage.

$$t_d = \frac{CV_{dd}}{4}\left(\frac{1}{I_{DSn,\text{sat}}} + \frac{1}{I_{DSp,\text{sat}}}\right) \tag{A.3}$$

where $I_{DSn,\text{sat}}$ and $I_{DSp,\text{sat}}$ are the saturation currents of the NMOS and PMOS transistors, respectively. Equation (A.3) shows that the delay time can be improved by increasing the MOS current drive. It is interesting to note that the channel length scaling has little impact on the improvement of the MOS current. Reducing the oxide thickness is, however, more effective in increasing the current and hence reducing the delay.

The aggressive scaling of the oxide thickness and channel length coupled with the demand for low energy has led and will continue to lead to the scaling of the supply voltage. Although it is difficult to accurately predict the impact of voltage scaling on the delay of CMOS circuits, we will use the analytical expression given by (A.3) to show the trend of the delay of the CMOS for different scaling scenarios. It is assumed that $V_{dd}/t_{ox} = 3\,\text{MV/cm}$. This is enough to satisfy the 20-year lifetime at 125°C [20]. It is also assumed that $V_{dd}/L_{\text{eff}} = 10\,\text{MV/m}$, which agrees with the state-of-the-art scaling trend. Figure 6.26 depicts the oxide thickness and channel length scaling versus supply voltage. The junction is scaled with the channel length ($x_j = \frac{1}{3}L_{\text{eff}}$). Figure 6.27(a) shows the delay of a CMOS inverter with a fan-out of 3. The widths of the NMOS and PMOS transistors of the CMOS inverter have not been scaled ($W_n = 1\,\mu\text{m}$ and $W_p = 2.3\,\mu\text{m}$). The fan-out transistors are assumed to be of the same size as the test inverter. The results displayed in the figure shows that the delay will drop with the supply voltage reduction only if the device threshold voltage is also scaled. Figure 6.27(b) depicts the delay as a function of the supply voltage when the load includes a fixed nonscalable interconnection capacitance (C_{int}). The shown normalized delay is for $C_{\text{int}} = 0.1\,\text{pF}$.

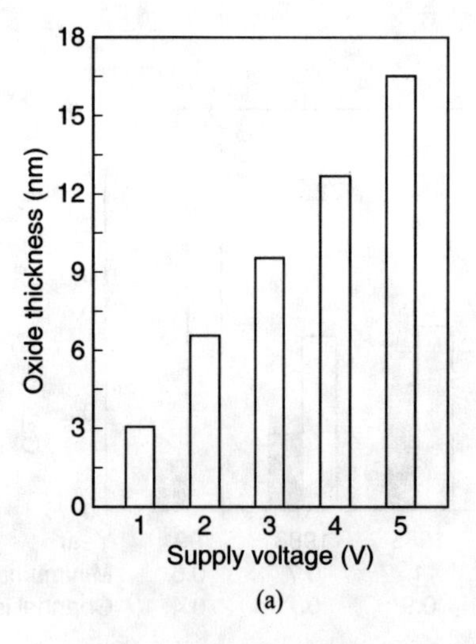

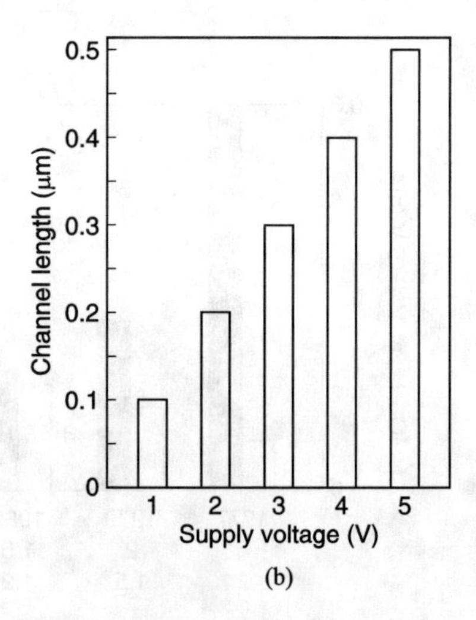

Figure 6.26 Scaling scenarios for (a) gate oxide thickness and (b) channel length vs. supply voltage.

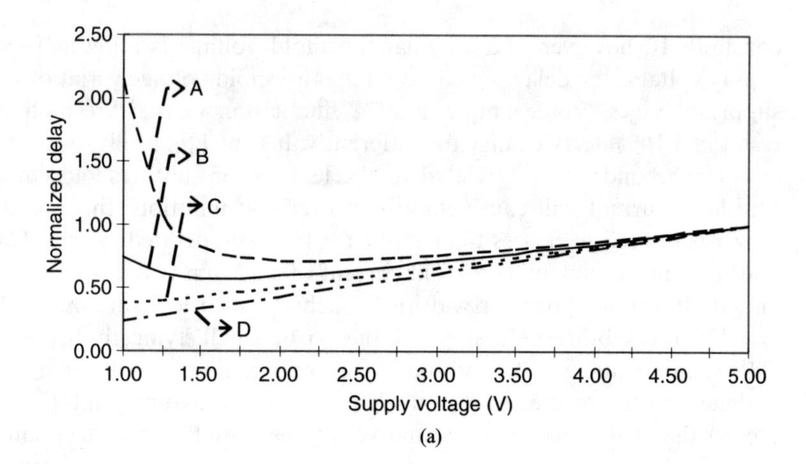

(a)

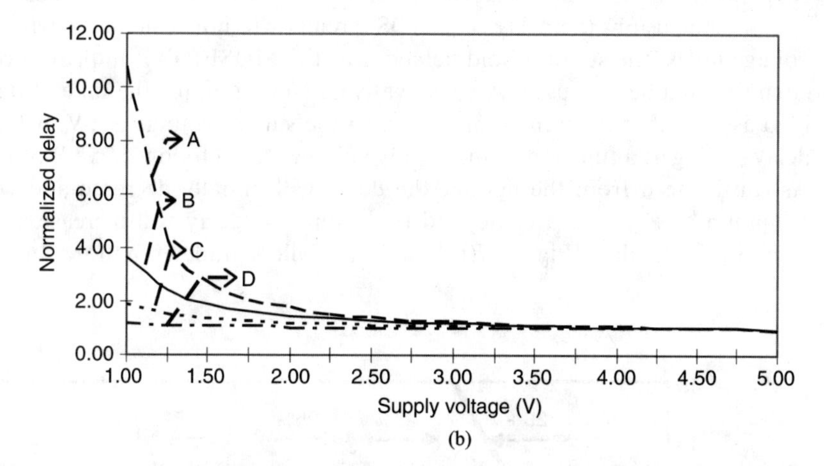

(b)

Figure 6.27 Normalized delay vs. supply voltage for different V_T scaling scenarios with (a) scaleable fan-out capacitance and (b) scaleable fan-out and nonscalable wire capacitances: case A, $V_T = 0.75$; case B, $V_T = 0.75 - 0.05(5 - V_{dd})$; case C, $V_T = 0.75 - 0.1(5 - V_{dd})$; case D, $V_T = 0.75 - 0.15(5 - V_{dd})$.

The future trend of CMOS speed will depend heavily on the feasibility of scaling V_T. It is, however, important to note that the threshold voltage scaling has two main drawbacks. The first is related to the sensitivity of the delay to the threshold voltage fluctuation. The second is the increase of the off-state leakage current as the threshold voltage is scaled down. These two points will be discussed next.

Kobayashi and Sakurai [23] have shown that the threshold voltage fluctuation due to process variations, which is around ± 0.15 V, would cause significant delay deviations from its nominal value for low V_{dd}. In their analysis, they compared the delay for different supply voltages assuming that the nominal[5] threshold voltage is

[5]The nominal threshold voltage is the threshold voltage which corresponds to the normalized delay for each supply voltage.

constant. If, however, the nominal threshold voltage is allowed to scale with the supply voltage, the delay sensitivity to the threshold voltage variations is equal for all supply voltages. For example, a 25% fluctuation in the threshold voltage would result in a 10% delay change for different values of V_{dd}, as illustrated in Figure 6.28.

The second point is related to the leakage or subthreshold current. The subthreshold current will cause standby power consumption. In general, the standby power is desired to be less than 1 mW for battery-operated systems [24]. This translates to a maximum subthreshold current of 1 nA/μm (assuming 1 V supply voltage and 10^6 μm total transistor width). To achieve such low subthreshold current with low V_T, the subthreshold slope[6] S has to be small, typically below 70 mV/decade [24]. Such a value is not easy to achieve and remains a challenging goal for process designers in the future. At the present it is safe to assume that V_T should be kept greater than 0.4 V. Another alternative is to use low V_T and use circuit techniques to suppress the subthreshold current when the MOSFET is off as proposed in [25].

In conclusion, the delay of CMOS circuits will not scale down with the supply voltage unless the subthreshold behavior of the MOSFETs is improved or if special circuit techniques are used. A conservative assumption, for the sake of our analysis, is to assume that the minimum V_T, when the supply voltage is 1 V, is 0.5 V. Hence delay scaling as a function of the supply voltage would follow curve B of Figure 6.27. As can be seen from the figures, the delay will initially decrease and will reach a minimum at $V_{dd} = 1.75$ V. Beyond that point the delay will increase again. In the case of fixed loading [Fig. 6.27(b)], the delay will degrade even more. In general, it is

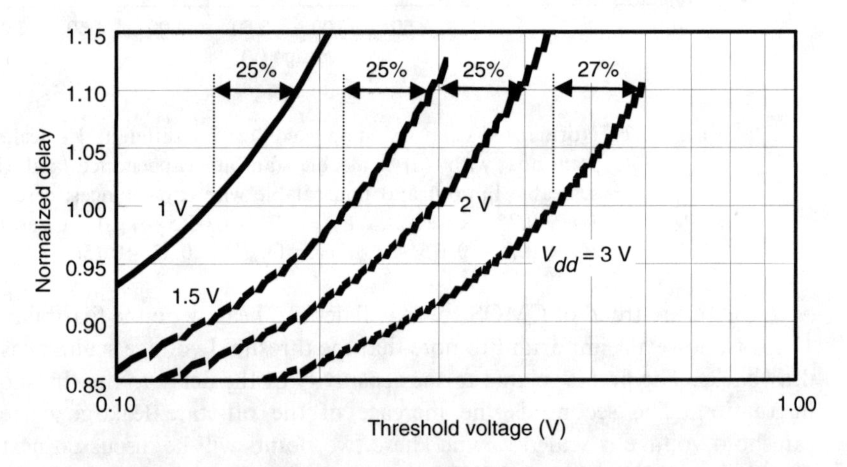

Figure 6.28 Sensitivity of delay to variation in V_T when it is scaled with V_{dd}.

[6]The subthreshold current changes by one decade for each change of the gate voltage by S millivolts.

reasonable to assume that the speed of CMOS circuits is not expected to improve under conventional scaling schemes; it may even decrease.

Appendix B

This appendix is a brief introduction to three digital circuit styles: static CMOS, complementary pass transistor logic (CPL), and dynamic CMOS. These are considered mainstream digital circuit styles. for more rigorous treatment of this topic the reader is referred to [2].

B.1. STATIC CMOS LOGIC

Static CMOS gates consist of two logic blocks: an N-logic block in the pull-down path and a P-logic block in the pull-up path, as shown in Figure 6.29. A static CMOS inverter, a NAND gate, and NOR gate are shown in Figure 6.30. The P- and N-logic

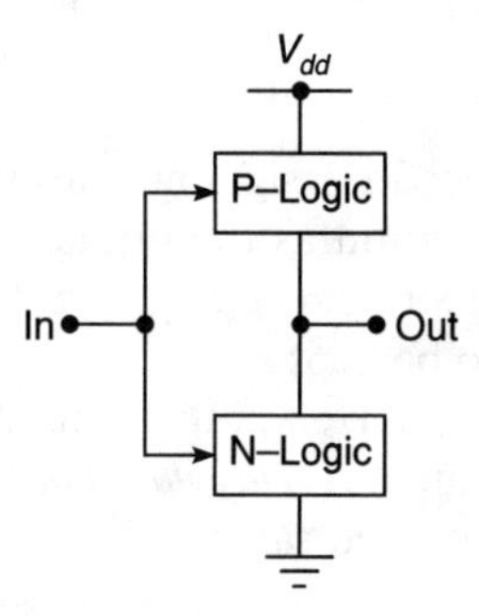

Figure 6.29 Block diagram of static CMOS gate.

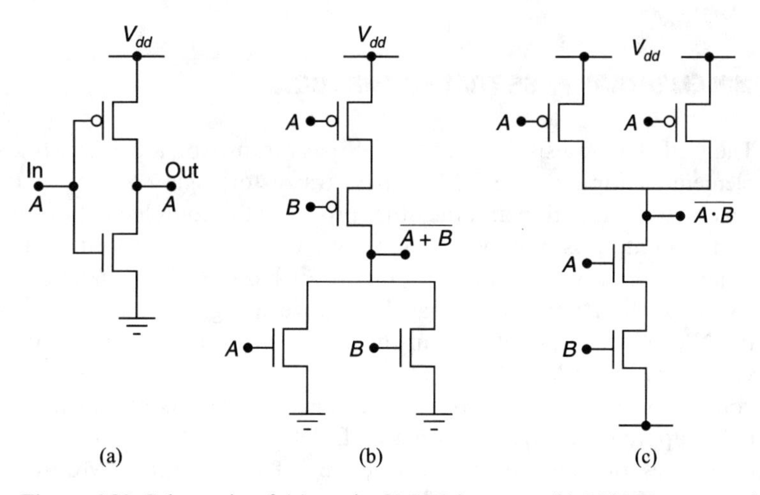

Figure 6.30 Schematic of (a) static CMOS inverter, (b) NAND gate, and (c) NOR gate.

are operating in a complementary manner so that only one block is conducting at any given time depending on the status of the input signals. If the N-block is conducting, the output will be discharged to ground, and if the P-block is conducting, the output is charged to V_{dd}. Since the steady-state output voltage levels are V_{dd} and ground, the noise immunity is high and the standby power is zero. These are among the most important features of static CMOS.

The rise and fall times of a static CMOS gate are determined, to first order, by the resistance of the conducting path to V_{dd} or ground and the capacitance at the output node. The on resistance of N- and P-logic blocks is a function of the sizes (aspect ratio) of the NMOS and PMOS transistor, respectively. To match the rise and fall times of a CMOS inverter, the on resistance of the PMOS should be equal to that of the NMOS, which may require in some cases that the size of the PMOS transistors be increased. This causes the capacitances of the internal nodes and the output node to increase. Sizing the transistors of a static CMOS gate is not a trivial task. It depends on several factors:

1. Required delay
2. Load (external capacitance) or fan-out
3. Fan-in

A common practice in designing gates for a cell library is to size the transistors with the delay minimization as the primary objective. For low-power design, however, a different sizing strategy is required. If the power is to be minimized, then the transistor sizes have to be kept as small as possible to minimize the intermediate and output node capacitances. Hence, the transistors should be sized large enough to just meet the delay requirements (*delay does not have to be minimum*). Even the fan-in can be determined based on the same objective (minimum power for a given required speed).

B.2. COMPLEMENTARY PASS TRANSISTOR LOGIC

The CPL [26] consists of an NMOS pass transistor logic network driven by complementary inputs. The NMOS pass transistor logic consists of two subnets: one generates the function and the other produces the complement of that function. Each subnet's output is buffered using a static CMOS inverter to drive load capacitances. Figure 6.31 shows the implementation for basic Boolean functions using CPL.

It is important to note that the voltage swing of the signal propagating through the NMOS tree is smaller than the supply voltage by the amount of the threshold voltage of the NMOSFET. This will result in reducing the available gate-source voltage for NMOS transistors, which may increase its propagation delay, especially if V_{dd} approaches V_T. Measured CPL delays for a 0.5-μm CMOS technology, with 0.4 V threshold voltage, are shown to be faster than static CMOS down to 2 V supply voltage [26]. The reduction of the swing and the input capacitance of CPL result in significant dynamic power savings (30%) compared to static CMOS [26].

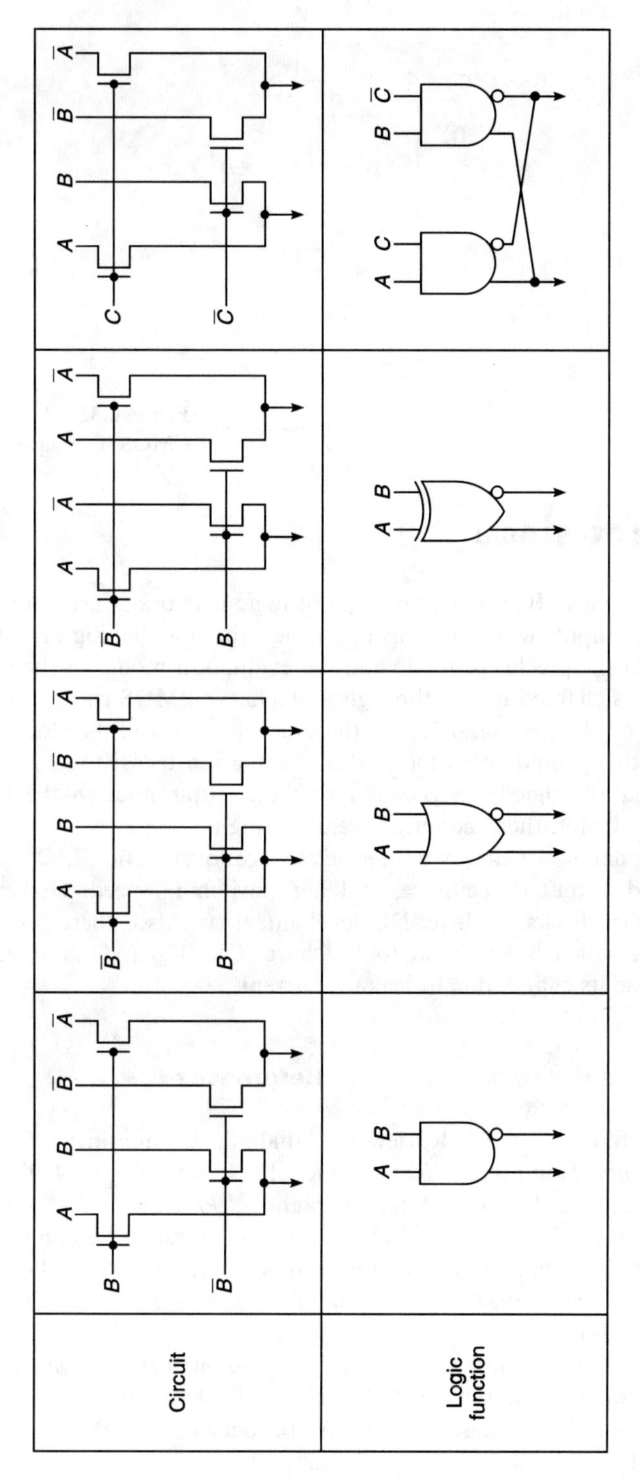

Figure 6.31 Sample CLP gates [26].

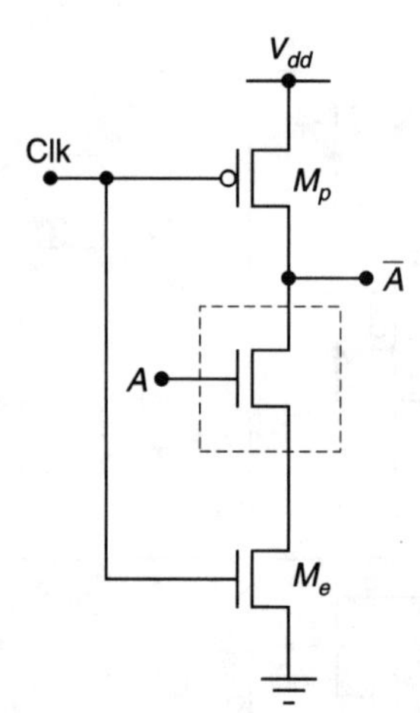

Figure 6.32 Schematic of dynamic CMOS inverter.

B.3. DYNAMIC CMOS LOGIC

In dynamic CMOS logic, one type of logic network is used, an NMOS or a PMOS net. The inputs will cause this net to be either conducting or not. The gate switches between the precharge mode and the evaluation mode. In the precharge phase, the output is charged to V_{dd} through a precharge PMOS transistor M_p which is driven by the clock (see Fig. 6.32). In the evaluation phase, the clock turns the precharge transistor off and turns the NMOS evaluation transistor M_e on. If the inputs are such that the logic block is conducting, the output node will be discharged to ground via M_e. If not, then the output remains high.

Dynamic circuits have some advantages over static CMOS, such as smaller area, reduced output capacitance, and high switching speed. However, they also have some drawbacks, such as DC level integrity. Also, there are instances when the output, which is supposed to be charged to V_{dd} (M_p and the N-block are off), may lose its charge due to leakage current.

References

[1] J. S. Denker, "A Review of Adiabatic Computing," *Tech. Dig. 1994 IEEE Symp. Low Power Electron.*, Oct. 1994, San Diego, 94–97.

[2] N. H. E. Weste and K. Eshraghian, *Principles of CMOS VLSI Design (A Systems Perspective)*, 2nd ed., Addison-Wesley, Reading, MA, 1993.

[3] M. Horowitz, T. Indennaur, and R. Gonzalez, "Low-Power Digital Design," *Tech. Dig. 1994 IEEE Symp. Low Power Electron.*, Oct. 1994, San Diego, pp. 8–11.

[4] A. Bellaouar and M. I. Elmasry, *Low-Power Digital Design VLSI (Circuits and Systems)*, Kluwer Academic, Norwell, MA, 1995.

[5] A. P. Chandrakasan and R. W. Brodersen, *Low-Power Digital CMOS Design*, Kluwer Academic, Norwell, MA, 1995.

[6] F. N. Najm, "A Survey of Power Estimation Techniques in VLSI Circuits," *IEEE Trans. Very Large Scale Integration (VLSI) Sys.*, vol. 2, no. 4, pp. 446–455, 1995.

[7] F. N. Najm, "Transition Density: A New Measure of Activity in Digital Circuits," *IEEE Trans. Computer-Aided Design*, vol. 12, no. 2, pp. 310–323, 1993.

[8] T. Sakurai and A. R. Newton, "Alpha-Power Law MOSFET Model and Its Applications to CMOS Inverter Delay and Other Formulas," *IEEE J. Solid-State Circuits*, vol. 25, no. 2, pp. 584–594, 1990.

[9] A. P. Chandrakasan, S. Sheng, and R. W. Brodersen, "Low-Power CMOS Digital Design," *IEEE. J. Solid-State Circuits*, vol. 27, no. 4, pp. 473–484, 1992.

[10] M. Alidina, J. Monteiro, S. Devadas, A. Ghosh, and M. Papaefthymiou, "Precomputation-Based Sequential Logic Optimization for Low-Power," *IEEE Trans. Very Large Scale Integration (VLSI) Sys.*, vol. 2, no. 4, pp. 426–436, 1994.

[11] J. Schutz, "A 3.3 V 0.6 μm BiCMOS Superscalar Microprocessor," *Tech. Dig. IEEE Int. Solid-State Circuits Conf.*, pp. 202–203, 1994.

[12] T. Callaway and E. Swartzlander, Jr., "Optimizing Arithmetic Elements for Signal Processing," *VLSI Signal Process.*, vol. 5, pp. 91–100, 1992.

[13] D. W. Dobberpuhl, R. T. Witek, R. Allmon, R. Anglin, D. Bertucci, S. Britton, L. Chao, R. A. Conrad, D. E. Dever, B. Gieseke, S. M. N. Hassoun, G. W. Hoeppner, K. Kuchler, M. Ladd, B. M. Leary, L. Madden, E. J. McLellan, D. R. Meyer, J. Montanaro, D. A. Priore, V. Rajagopalan, S. Samudrala, and S. Santhanam, "A 200 MHz 64-b Dual-Issue CMOS Microprocessor," *IEEE. J. Solid-State Circuits*, vol. 27, no. 11, pp. 1555–1567, 1992.

[14] Y. Nakagome et al., "Sub 1-V Swing Internal Bus Architecture for Future Low-Power ULSI's," *IEEE. J. Solid-State Circuits*, vol. 28, no. 4, pp. 414–419, 1993.

[15] A. Bellaouar, I-S. Abu-Khater, and M. I. Elmasry, "Low-Power CMOS/BiCMOS Drivers and Receivers for On-Chip Interconnects," *IEEE. J. Solid-State Circuits*, vol. 30, no. 6, pp. 696–700, 1995.

[16] R. Golshan and B. Haroun, "A Novel Reduced Swing CMOS Bus Interface Circuit for High Speed Low Power VLSI Systems," *Techn. Dig. Int. Symp. Circuits Systems*, pp. 351–354, 1994.

[17] T. Gabara, "Pulsed Power Supply CMOS–PPS CMOS," *Tech. Dig 1994 IEEE Symp. Low Power Electron.*, Oct. 1994, San Diego, pp. 98–99.

[18] L. "J." Svenson and J. G. Koller, "Driving a Capacitive Load without Dissipating fCV^2," *Tech. Dig. 1994 IEEE Symp. Low Power Electron.*, Oct. 1994, San Diego, pp. 100–101.

[19] T. Indermauer and M. Horowitz, "Evaluation of Charge Recovery Circuits and Adiabatic Switching for Low Power CMOS Design," *Tech. Dig. 1994 IEEE Symp. Low Power Electron.*, Oct. 1994, San Diego, pp. 102–103.

[20] C. Hu, "Future CMOS Scaling and Reliability," *Proc. IEEE*, vol. 81, no. 5, pp. 682–689, 1993.

[21] K.-Y. Toh, P.-K. Ko, and R. G. Meyer, "An Engineering Model for Short-Channel MOS Devices," *IEEE. J. Solid-State Circuits*, vol. 23, pp. 950–958, 1988.

[22] H. Oyamatsu, M. Kinugawa, and M. Kakumu, "Design Methodology of Deep Submicron CMOS Devices for IV Operation," *Tech. Dig. Symp. VLSI Technol.*, pp. 89–90, 1993.
[23] T. Kobayashi and T. Sakurai, "Self-Adjusting Threshold-Voltage Scheme (SATS) for Low-Voltage High-Speed Operation," paper presented at 1994 IEEE Custom Integrated Circuits Conference, pp. 271–274.
[24] A. Hori, A. Hiroki, H. Nakaoka, M. Segawa, and T. Hori, "Quarter-Micrometer SPI (Self-Aligned Pocket Implementation) MOSFET's and Its Application for Low Supply Voltage Operation," *IEEE Trans. Electron Devices*, vol. 24, no. 1, pp. 79–85, 1995.
[25] M. Horiguchi, T. Sakata, and K. Itoh, "Switched-Source-Impedance CMOS circuit for Low Standby Subthreshold Current Giga-Scale LSI's," *IEEE. J. Solid-State Circuits*, vol. 28, no. 11, pp. 1131–1135, 1993.
[26] K. Yano, T. Yamanaka, T. Nishida, M. Saito, K. Shimohigashi, and A. Shimizu, "A 3.8ns CMOS 16 × 16 Multiplier Using Complementary Pass Transistor Logic," *IEEE. J. Solid-State Circuits*, vol. 25, no. 2, pp. 388–395, 1990.

Chapter 7

Low-Voltage Analog BiCMOS Circuit Building Blocks

Jaime Ramírez-Angulo
The Klipsch School of Electrical and Computer Engineering, New Mexico State University, Las Cruces, New Mexico 88003

7.1. INTRODUCTION

In this chapter, some of the techniques used for the design of analog circuits that operate with low supply voltages are discussed. By low supply voltage, a total supply voltage range (V_{DD} to V_{SS} or ground) of 3.3 V or less is assumed. We constrain our discussion to continuous-time circuits implemented in single-well complementary or bipolar metal-oxide-semiconductor (CMOS or BiCMOS) technology. The term BiCMOS is used here in a very broad sense. It applies to circuits in BiCMOS technology as well as to circuits in conventional CMOS technology that make use of lateral and substrate bipolar transistors. Complementary MOS technology with features sizes down to 1 μm allow the implementation of lateral bipolar transistors that are relatively fast ($f_t > 100$ MHz), have relatively large common emitter current gain factors ($\beta > 100$), and offer relatively large current-handling capabilities (collector currents up to 50 μA using minimum-size lateral devices) [1]. Whenever we refer to the utilization of lateral bipolar transistors, we have to keep in mind their limitations: low Early voltages ($V_A < 20$ V), large collector resistances ($R_c > 500\,\Omega$), and large base resistances ($R_b > 2\,\text{k}\Omega$). We also have to take into account the fact that a lateral bipolar transistor always activates a vertical substrate transistor that absorbs a significant fraction of the emitter current; this effect worsens as the device sizes are reduced (from >30% for 2-μm CMOS technology up to >80% for 0.35-μm CMOS technology) [2, 3]. Although some of the techniques to be discussed here have been known for a long time, in the integrated circuit (IC) world where bipolar and MOS transistors coexist and must satisfy requirements of low-voltage operation, these techniques are finding renewed interest.

To begin with, we need to remember the main motivation for low-voltage operation: *compatibility with the digital CMOS world* that, without doubt, dominates the scene in mixed-mode very large scale integrated (VLSI) systems [4]. Typically, more than 80% of the circuitry in a mixed-mode VLSI system is digital. Digital CMOS circuits have much less of a problem operating in a low-voltage environment than analog circuits. CMOS digital circuits have (ideally) zero static power dissipation and their dynamic power dissipation increases as a function of the square of the supply voltage (V_{DD}^2) [5]. For power density to remain constant as technologies scale down, the best approach (and maybe the only one possible now) is the so-called constant field scaling approach. This approach requires supply voltages to scale down with feature sizes [5]. At the time of writing, the transition from 5 V to 3.3 V supplies in CMOS VLSI systems has occurred [6], and from the above discussion we can expect supply voltages to continue decreasing as feature sizes scale down. An important point that facilitates the transition to low-supply-voltage operation is the fact that the threshold voltage of MOS transistors also scales down, though not with the same scaling factor as for features sizes. In any case, even without the requirement for constant field in finer feature size technologies, and by constraining our discussion to a nonshrinking technology, voltage supply reduction is a natural option to reduce power consumption in analog and particularly in digital circuits. A host of applications ranging from biomedical implantable devices to cellular telephones, laptop computers and many other portable systems benefit from the utilization of reduced supply voltages. This reduction is due to the fact that it allows longer battery life and reduced equipment weight. Even in applications where reduced power dissipation does not seem to be of paramount importance, the reduction in power dissipation with voltage supply reduction has no adverse effects and leads to longer lifetime and increased system reliability as long as circuit performance is not seriously degraded.

Why do digital circuits present less problems than analog circuits when operating at reduced supply voltages? One of the most important reasons for the popularity of CMOS logic circuits is precisely the fact that they operate in class AB (or more precisely class B) with almost zero static currents. In these circuits, in static conditions, only one-half of the circuit (either the P or the N tree) is on at a given time. The transistors which are on operate in the triode mode. By stacking N and P transistors, the minimum supply voltage range (denoted here by $V_{SR} = V_{DD} - V_{SS}$) required for static operation corresponds to that of the half-circuit that is on and is equal basically to an MOS transistor's threshold voltage. In dynamic operation (during transitions from V_{DD} to V_{SS} and from V_{SS} to V_{DD}) only one P and/or one N transistor is required to operate in the saturated mode, while all other transistors operate in either the triode or the cutoff mode. Transistors operating in the triode mode have the maximum possible voltage applied, $V_{SR} = V_{DD} - V_{SS}$, acting as gate-to-source or source-to-gate voltage. For this reason they have the smallest possible drain-source equivalent resistance. This results in small drain-source voltages with large currents typically observed in the transitions from V_{DD} to V_{SS}. Under these conditions more transistors can be stacked between V_{SS} and V_{DD} in a digital circuit than in an analog circuit. A digital circuit *can still be functional* even with strongly reduced supply voltages as long as there is enough voltage on the input terminals to turn on at least one N- and one P-channel. The minimum supply range is approximately the sum of the threshold voltages of an N-channel and a P-channel transistor. Furthermore, even in the case that the supply voltage range is less than this minimum requirement, the circuit can still operate because $V_{DD} - V_{SS}$ transitions can take place (very slowly) with transistors operating in the subthreshold region. As more transistors are stacked, the penalty paid is longer time constants (rise-fall times) and smaller currents (maybe even subthreshold-level currents). The main point is that, in spite of the fact that by lowering their supply voltages CMOS digital circuits can become fairly slow, *they will in general still be functional.* Another important performance parameter that degrades with supply voltage reductions is noise margin. In effect, the noise margin of a digital circuit is reduced as supply voltages are reduced; however, digital circuits are more robust against noise than analog circuits in general.

Problems of analog circuits with reduced voltage supplies: Most conventional analog circuits operate in class A (nonzero static or quiescent currents). This means that all transistors connected between V_{SS} and V_{DD} must be on at the same time. On the input side, between the gate and source (or source and gate), the terminals of every transistor require a minimum voltage equal to the transistor threshold voltage; but in practice, at least 0.2 V above the threshold voltage is used as the quiescent voltage. Since most transistors (with the exception of switches and some circuits to be discussed later) operate in the saturated mode, their drain-to-source voltage requirements are relatively large: $V_{DS} > V_{GS} - V_T$. This limits the minimum supply voltage, especially on the output side. The body effect that affects at least one of the transistor types in a single-well CMOS technology increases the threshold voltage of stacked transistors and, with this, it increases the supply requirements on the input side. Performance improvement in analog circuits usually requires stacked transistors (a well-known example is the utilization of cascode or Wilson current mirrors to improve the linearity, accuracy, and output impedance of a mirror), and this again increases the voltage supply requirements. If high performance is required, then many factors seem to be against reducing the supply voltage of an analog circuit. Most conventional analog circuit structures operating with reduced supplies are slower and have a reduced dynamic range, as discussed later. But due to the fact that analog circuitry is usually only a small fraction of a VLSI system, additional area or power dissipation can be afforded if this is the price to pay for operation with lower supply voltages. The real problem is that most conventional *analog circuits become nonfunctional in a reduced-supply environment*. Consider, for example, the case of an N-transistor cascode current source feeding a P cascode current mirror. This circuit becomes nonfunctional if a single 3.3-V supply is used (nonfunctional at least in the saturated mode). With dual 5-V supplies available, the analog IC designer had far fewer problems in designing high-performance circuits with relatively large signal swings. The panorama changed completely as a single 3.3-V supply became the only supply available. Many circuits which had been commonplace became simply nonoperational and new circuit structures had to be introduced.

Dependence of some fundamental analog circuit performance parameters on supply voltage and motivation for the utilization of bipolar transistors: Lowering the supply voltage of (class A) analog circuits forces in general the design of circuits with reduced drain-source saturation voltages, $V_{DS,\text{sat}} = V_{GS} - V_t$, so that circuits can still be functional. Consider the following important relations for MOS transistors in the saturated mode (parameters μ, C_{ox}, W/L, and V_t have the usual meaning of mobility, oxide capacitance density, transistor aspect ratio, and MOS threshold voltage, respectively):

The saturation mode equation for the drain current is

$$I_D = \frac{\mu C_{\text{ox}}}{2} \frac{W}{L} (V_{GS} - V_t)^2 (1 + \lambda V_{DS}) \approx \beta V_{DS,\text{sat}}{}^2 \tag{7.1}$$

The small-signal transconductance gain is

$$g_m = \mu C_{\text{ox}} \frac{W}{L} V_{DS,\text{sat}} = \frac{2I_D}{V_{DS,\text{sat}}} = \sqrt{2\mu C_{\text{ox}} \frac{W}{L} I_D} \tag{7.2}$$

The maximum gain-bandwidth product is

$$\text{GB}^{\max} = \frac{g_m}{C_L^{\min}} = \frac{\mu V_{DS,\text{sat}}}{L^2} \tag{7.3}$$

The thermal noise generated over a frequency band Δf is

$$v_n = \sqrt{\frac{8}{3} \frac{kT}{g_m} \Delta f} \tag{7.4}$$

The slew rate is

$$\text{SR} = \frac{I_D}{C_L} \tag{7.5}$$

The load capacitance $C_L = C_L^{\min} + C_L'$ has two components: the nominal load capacitance C_L' and the parasitic load capacitance $C_L^{\min}$.

The input range and offset of a differential pair and the accuracy, input range, and bandwidth of a current mirror scale directly with $V_{DS,\text{sat}}$. Signal swings have to be reduced in order to keep circuits operational with reductions in $V_{DS,\text{sat}}$. In order to reduce $V_{DS,\text{sat}}$ there are two options: (i) to increase the transistor aspect ratio W/L while keeping the drain current I_D constant (in this case, parasitic load capacitances $C_L^{\min}$ are increased) and (ii) to decrease I_D maintaining W/L constant. In both cases, as can be seen from (7.3) to (7.5), reduced gain-bandwidth products and slew rate result. That means slower circuits from a small- and large-signal point of view. On the other hand, since, with the reduction of $V_{DS,\text{sat}}$, voltage swings have to be accordingly reduced, then, in order to keep distortion low and to avoid saturation of differential pairs and current mirrors, signal swings have to be reduced. On the upper end, maximum signals are decreased; on the lower end, minimum signals have to be increased because noise increases. This means that the dynamic range of a CMOS circuit is degraded if the supply voltage is reduced (not a surprising conclusion). It is precisely in this situation that the bipolar transistor can help to overcome some of the problems mentioned above. The bipolar junction transistor (BJT) has some very attractive features that allow reduction of supply voltage requirements. With comparable currents and geometry size as those of an MOS transistor, a BJT offers higher transconductance gain ($g_m = I_C/V_{\text{th}}$), low and approximately constant input voltages (base-emitter voltages of about 0.6 V), and minimum output voltages ($V_{CE,\text{sat}}$ of approximately 0.2 V), lower thermal noise, negligible popcorn and flicker ($1/f$) noise, and higher current-handling capabilities. The exponential nature of the bipolar transistor leads to very small voltage swings for large current swings (approximately 60 mV/decade of current change). If voltage signals are transformed into current signals, then voltage swings in bipolar devices remain very small. This allows reduction of supply voltage requirements without sacrificing bandwidth, dynamic range, slew rate, and so on. In fact, it is precisely the higher slew rate (or current-handling capability) of the bipolar devices that motivated originally the introduction of BiCMOS technology for digital applications such as buffers or line drivers for large capacitive loads. Another important feature of the BJT in the context of low-voltage applications is the fact that the BJT does not suffer from body effect. All these advantages have recently created interest in the design of high-performance low-voltage analog BiCMOS circuits [7–27].

7.2. TECHNIQUES TO REDUCE VOLTAGE SUPPLY REQUIREMENTS

In this section, some techniques to reduce voltage supply requirements without significant performance degradation and by utilization of BiCMOS circuits are discussed and illustrated with examples. In the discussion to follow, we refer to a gate-to-source (source-to-gate) drop and a minimum drain-to-source (source-to-drain) drop in terms of $V_{GS} = V_{DS,\text{sat}} + V_t = V_{GS}^Q + V_S$ and $V_{DS,\text{sat}} = V_{DS,\text{sat}}^Q + V_S$, respectively. Here, V_S represents a signal superimposed on a quiescent voltage. High-impedance DC current sources are assumed to require a minimum supply of $2V_{DS,\text{sat}}$ or $2V_{CE,\text{sat}}$. We will assume a minimum value $V_{DS,\text{sat}}^Q = 0.2$ V to keep MOS transistors safely in the saturated mode and "worst-case" threshold voltages $V_t = 0.9$ V which are typical worst case for 0.8-μm CMOS technology or for P transistors in lower feature size technologies. A superscript Q (i.e., $V_{DS,\text{sat}}^Q$ or V_{GS}^Q) will refer to a quiescent or constant drop (associated, e.g., with a DC current source). If no superscript Q is used, the voltage drop is increased by inclusion of the signal component V_S. To perform calculations of minimum voltage supply requirements in the circuits to be discussed, we will assume possibly open loop architectures with relatively large signal amplitudes $V_S = 1$ V. The signal might include a signal dependent common mode component on the same order of magnitude as the signal itself. These assumptions lead to nominal values $V_{GS}^Q = 1.1$ V, $V_{GS} = 2.1$ V, $V_{DS,\text{sat}}^Q = 0.2$ V, and $V_{DS,\text{sat}} = 1.2$ V. In regard to BJTs, we will assume that in the technology used

(CMOS or BiCMOS) there is at least one type of bipolar transistor available: either a "good" vertical bipolar in BiCMOS technology and/or a "not so good" lateral bipolar in conventional CMOS and BiCMOS technology. Nominal and constant values $V_{CE,\text{sat}} = 0.2\,\text{V}$ and $V_{BE} = 0.6\,\text{V}$, both for quiescent operation or including a signal, will be used for our calculations. The good bipolar devices can be used without restrictions in the signal path; the lateral bipolar transistors should be restricted to the implementation of DC current sources and voltage followers. For relatively high collector currents ($I_C > 100\,\mu\text{A}$), the nominal collector resistance R_c has to be taken into account in order to calculate the minimum collector-emitter voltage according to $V_{CE,\text{sat}} = 0.2\,\text{V} + I_c R_c$.

7.2.1. Folding Techniques

Utilization of folding techniques allows the designer to use only one type of transistor in the signal path. The complementary type is used for implementation of DC current sources with a constant-voltage requirement of $V^Q_{DS,\text{sat}}$ or $2V^Q_{DS,sat}$. This replaces a signal-dependent V_{GS} drop (2.1 V) by a constant-current source drop $V^Q_{DS,\text{sat}}$, so that supply requirements are reduced by at least 1 V. Figure 7.1 illustrates this techique for the case of a fixed-gain highly linear CMOS amplifier [28] consisting of a differential pair (M_1, M_2) loaded by diode-connected transistors (M_{1L}, M_{2L}). This circuit has a minimum supply voltage range requirement (denoted V_{SR} in what follows) given by $V_{SR} = 2V_{DS,\text{sat}} + V_{DS,\text{sat}} + V_{GS} = 3.5$ V (inputs to the differential pair can be DC level shifted so that they acquire an appropriate common-mode component). The load section requires, for isolation purposes, two NPN bipolar cascoding transistors (Q_{1C} and Q_{2C}). The maximum supply requirements (V_{SR}) of the circuit are determined by the load section and are $V_{SR} = 2V^Q_{DS,\text{sat}} + V_{CE,\text{sat}} + V_{GS} = 2.7\,\text{V}$. We will see later that by making use of DC level shifting techniques, reductions in V_{GS} by at least 0.6 V can be achieved. An additional advantage of the folding technique is that it improves matching between input and load transistors because they are of the same type and allow higher swing on the input and output nodes, since they are now isolated by the cascode transistors Q_{1C}, Q_{2C}. Another example where this technique can be applied, combined with DC level shifting techniques, is the well-known folded-cascode operational amplifier [29]. An example of a folded-cascode opamp is discussed in Section 7.3. This circuit can be augmented by including transistor M_R operating in the triode mode. This transistor permits control of the DC operating point at the output terminals, so that large output signal swings are possible [28]. The source-drain drop of M_R determines the output quiescent point (or output common-mode component) and this drop is controlled by the voltage V_c generated by the replica bias circuit formed by M_{1B}, M_{2B}, and M_{RB}. The voltage V_c causes the quiescent voltages at the output nodes to follow the value V_O applied to the gate of M_{1B}. If all MOS transistors (with the exception of M_R and M_{RB}) in this circuit are matched, then the circuit implements a highly linear differential-input, differential-output voltage follower characterized approximately by the single-pole transfer function $V_o/V_i = 1/(1 + s/\omega_p)$, where $\omega_p = g_m/C$. Here g_m and C are the MOS transistor small-signal transconductance gain and the parasitic capacitance at the output nodes of the circuit, respectively.

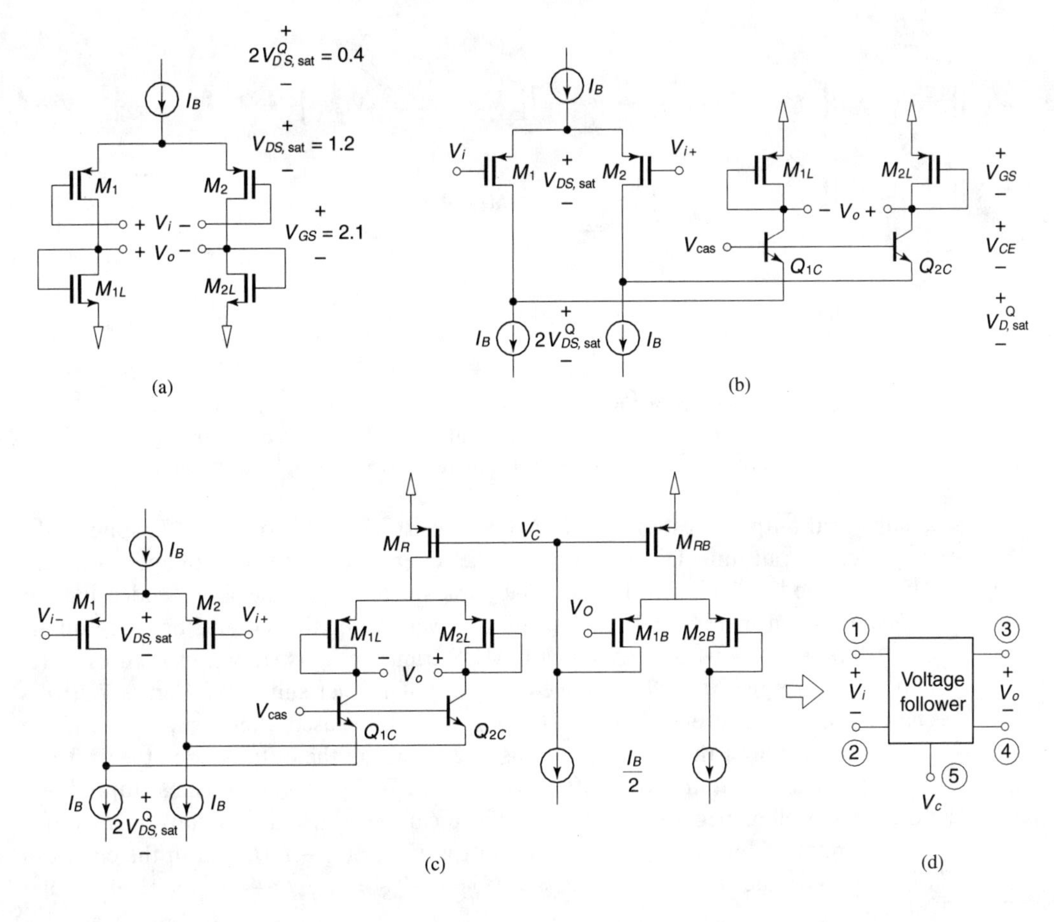

Figure 7.1 Illustration of folding technique: (a) fixed-gain amplifier; (b) folded BiCMOS version of fixed-gain amplifier; (c) inclusion of circuit to control output DC operating point; (d) symbol for differential voltage follower.

7.2.2. DC Level Shifting Techniques

This technique uses voltage followers implemented with bipolar (or MOS transistors) as DC level shifts [30, 31]. To reduce supply requirements, level shifting transistors have to be of the opposite type to that of transistors used in the signal path. The DC level shifters have a small effect on signal performance. To illustrate this technique, consider a P-transistor cascode current mirror [Fig. 7.2(a)] which requires a $2V_{SG}$ (4.2-V) drop on the input side and a $V_{GS} + V_{DS,\text{sat}}$ (3.3-V) drop on the output side. Inclusion of DC level shifters, V_{LS}, as shown in Figure 7.2(b), allows reduction by $2V_{LS}$ on the input and by V_{LS} on the output sides. Figure 7.2(c) shows the same technique applied to the regulated cascode current mirror, which is known to have a very high output impedance $R_{\text{out}} = r_o(g_m r_o)^2 \sim 100\,\text{M}\Omega$ (r_o is the

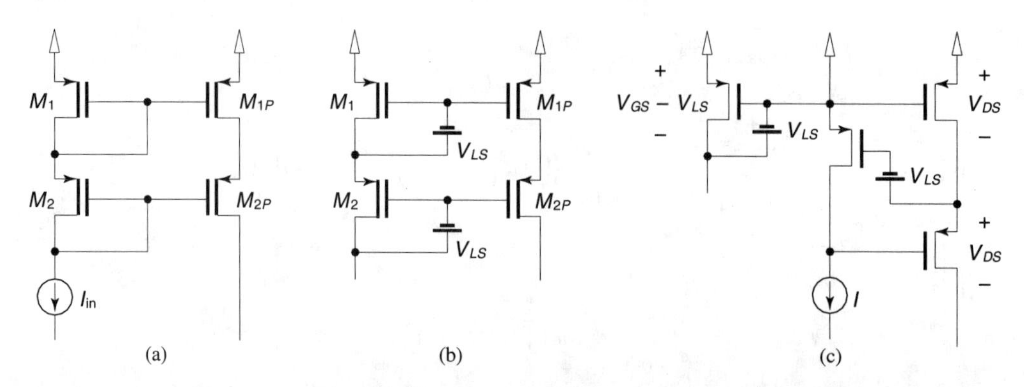

Figure 7.2 Utilization of DC level shifters to reduce supply requirements: (a) conventional cascode current mirror; (b) cascode current mirror DC level shifters; (c) regulated cascode with level shifters.

small signal output resistance of an MOS transistor). It is shown that in spite of the very high output impedance, this circuit can operate with output voltages as low as $2V_{DS,\text{sat}}$. The level shifters, V_{LS}, can be implemented using the single-ended bipolar follower shown in Figure 7.3(b), an NMOS version of this circuit, or using differential level shifters with NPN or with MOS transistors, as shown in Figures 7.3(c) and 7.3(d), respectively. The single-ended bipolar level shifter of Figure 7.3(b) is characterized by a level shift $V_{LS} = 0.6 + I_B R$. If this is used, then a supply reduction on the input side of the mirror is at least 1.2 V and on the output side at least 0.6 V. The use of differential level shifters with MOS or with bipolar transistors allows positive as well as negative level shifts which can be adjusted to a convenient value. In the circuit of Figure 7.3(c) the level shift is $V_{LS} = (I_2 - I_1)R$ and in the circuit of Figure 7.3(d) the level shift is $V_{LS} = V_{GS_2} - V_{GS_1} = \sqrt{I_1/\beta} - \sqrt{I_2/\beta}$. The bipolar

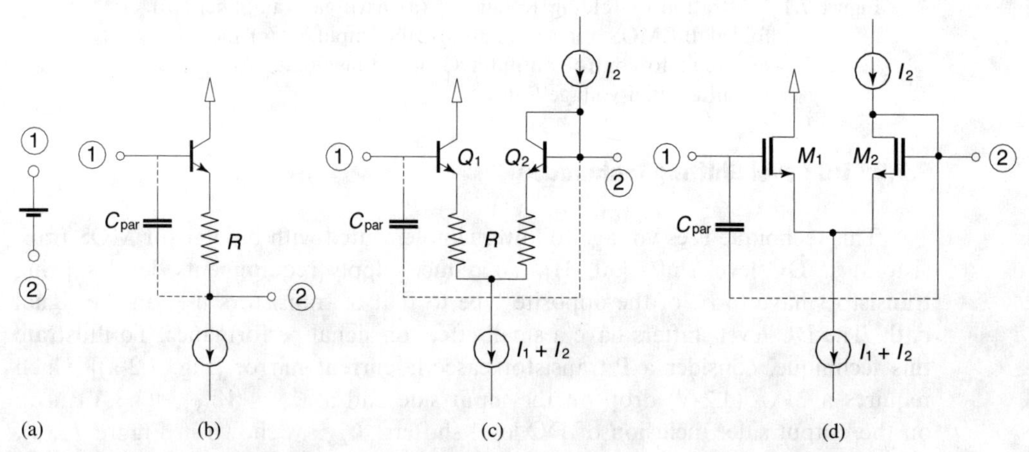

Figure 7.3 Implementation of DC level shifters: (a) power supply (b) single-ended BJT implementation; (c) differential bipolar level shifter; (d) differential MOS level shifter.

shifter introduces a small DC offset error due to the finite current gain factor β. To avoid performance degradation at high frequencies, the level shifters can be bypassed by means of capacitors C_{BP}. A word of caution: The low-impedance terminal of the DC shifter (terminals 2 in Figure 7.3) must be tied to high-impedance terminals of the circuits where they are used.

7.2.3. Floating-Gate Techniques

Multiple-input floating-gate (MIFG) transistors [Fig. 7.4(a)] have the interesting and useful characteristics that they perform weighted addition of voltage signals applied to their input terminals [32]. This is based on charge redistribution, since each input terminal can induce charge on the common floating gate. This fact can be used to advantage in the design of low-voltage systems. The basic idea is to use one of the input terminals for biasing purposes by connecting it to a large DC voltage source (V_{bias}) and use the other terminals for signal-processing purposes. With this, signal-processing input terminals are not required to include a biasing component. Figure 7.4(a) shows the layout of a two-input floating-gate transistor. Figure 7.4(b) shows a low-voltage current mirror using two-input floating-gate transistors [33]. In this circuit one of the input terminals of each transistor is connected to a DC bias

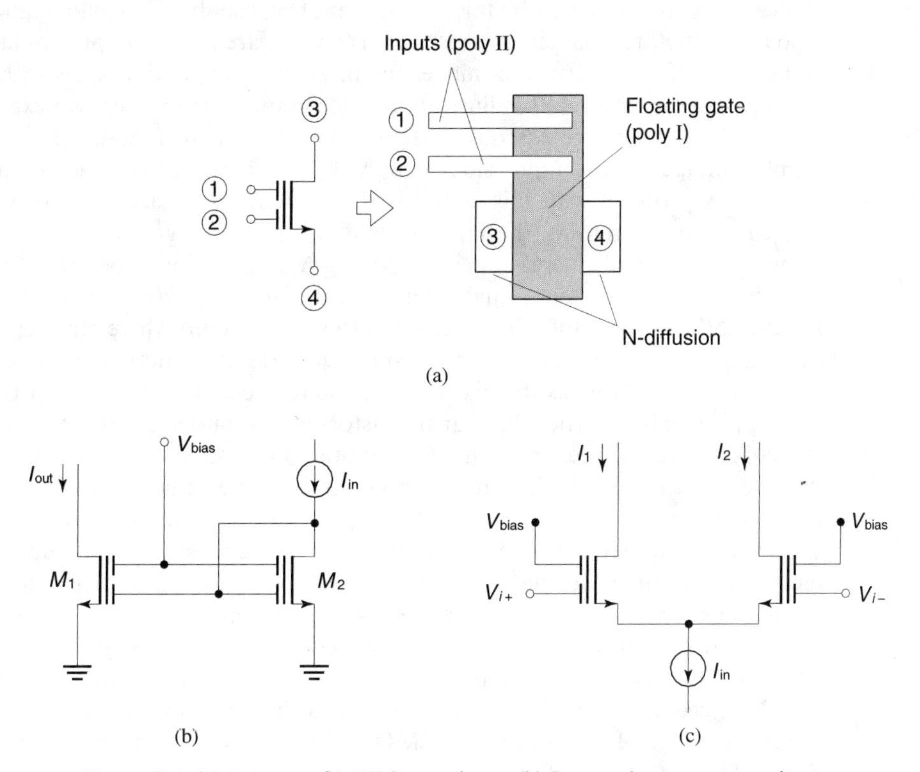

Figure 7.4 (a) Layout of MIFG transistor. (b) Low-voltage current mirror using MIFG transistor. (c) Low-voltage differential pair using MIFG transistor.

voltage, V_{bias}, while the other terminals are used as conventional input and output current mirror terminals.

Supply requirements on the input and output sides are reduced to only $V_S + V_{DS,\text{sat}}^Q = V_{DS,\text{sat}} = 1.2\,\text{V}$. Figure 7.3(c) shows a low-voltage differential pair using two-input floating-gate transistors that are biased using a floating voltage source, V_{bias}, connected between one biasing transistor input and the common source. Other low-voltage differential pairs using MIFG transistors are discussed in [33]. Any of the circuits shown in Figure 7.3 can be used to implement V_{bias}. Notice that the entire supply voltage range ($V_{DD} - V_{SS}$) is available for the biasing terminal of a floating-gate transistor.

7.2.4. MOS by Bipolar Replacement Techniques

These are based on the utilization of bipolar transistors in the signal path to replace MOS transistors. This allows the designer to replace variable $V_{DS,\text{sat}}$ and V_{GS} drops, which include a signal component V_S, by approximately constant and smaller V_{CE} and V_{BE} drops. Supply requirements are also reduced because the BJT does not suffer from body effect. The approximately constant V_{BE} and $V_{CE,\text{sat}}$ of the BJT is due to its exponential nature. From the small-signal point of view, circuit performance is usually improved by this replacement because the transconductance gain of bipolar transistors is larger, thermal and $1/f$ noise are lower, output conductance is higher and offset voltages are smaller for BJTs than for MOS devices with comparable or even larger sizes. To illustrate this technique we consider two examples.

1. The low-voltage cascode current mirror shown in Figure 7.5(a) requires a supply voltage on the input side of only V_{GS} (2.1 V) and on the output side of $2V_{DS,\text{sat}}$. A problem with this circuit is that it has a very small signal swing. This is because, by increasing the input current I_{in}, the voltage V_{GS} of the cascoding transistor (M1,M1P) increases, easily driving M2 into triode operation. This limits the swing of input current signals to a small fraction (say, 20%) of the bias current. Figure 7.5(b) shows a BiCMOS version of this same circuit where replacement of the cascoding MOS transistors by bipolar transistors results in input current swings that can be almost as large as the bias current due to the constant $V_{CE,\text{sat}}$ of the bipolar (this applies only if vertical bipolar transistors with a buried collector, and thus with low collector resistance, are used). The input and output small-signal impedances of the circuit of Figure 7.5(b) remain approximately the same. A small gain error is introduced by the finite β of the BJT. Figure 7.5(c) shows a version of the same circuit where an MOS DC level shifter is used to reduce the input supply requirements of the mirror so that it becomes a high-output-impedance circuit, with low input and output voltage requirements (signal swing is reduced in this case). Figure 7.5(d) shows an all-bipolar version of the cascode mirror of Figure 7.5(a) that has V_{BE} and $2V_{CE}$ voltage requirements on the input and output sides, respectively.

2. A second example to illustrate this techique is the low-voltage BiCMOS implementation of the class AB CMOS differential amplifiers shown in Figure 7.6(a). This circuit uses floating batteries, V_B, which are implemented using "composite CMOS transistors" (a series connection of an N- and P-channel transistor), as shown in Figure 7.6(c). A composite CMOS transistor requires an equivalent gate-

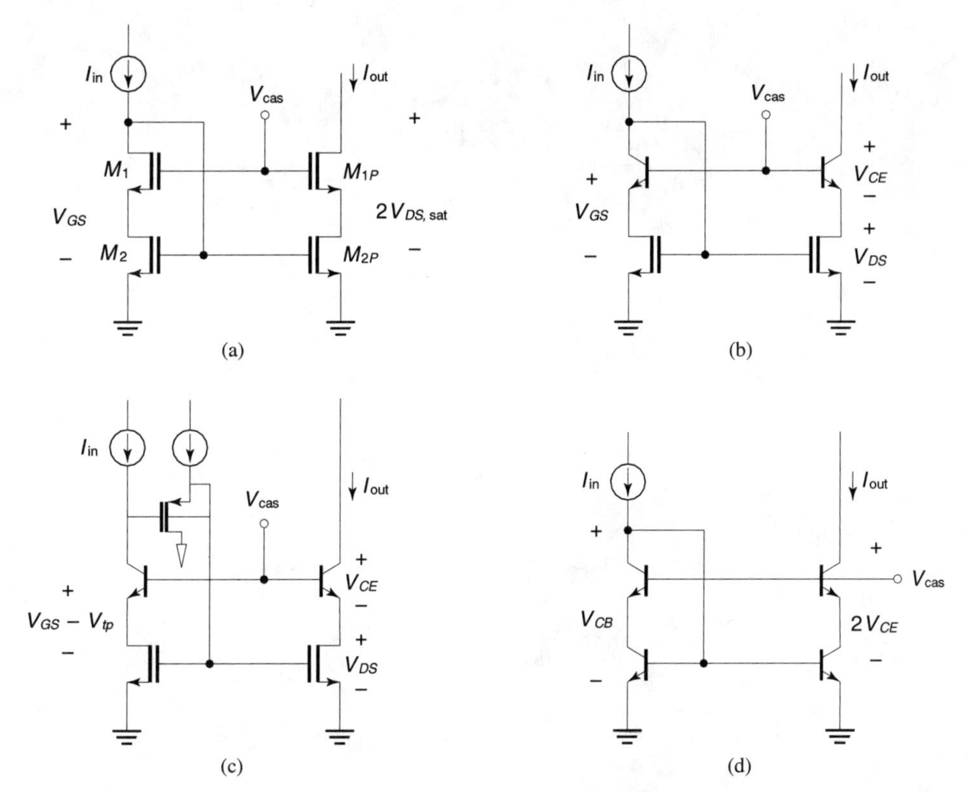

Figure 7.5 (a) Low-voltage CMOS cascode current mirror. (b) Low-voltage BiCMOS cascode mirror. (c) Very low voltage BiCMOS cascode mirror with DC level shifter. (d) Bipolar low-voltage cascode mirror.

source voltage $V_{GS}^* = V_{GSn} + V_{SGp}$ and drain-source voltage $V_{DS,\text{sat}}^* = V_{DSn,\text{sat}} + V_{SDp,\text{sat}}$. This corresponds to the sum of the supply requirements of the N and the P transistors. Since one of the transistors is subject to body effect, V_{GS}^* and V_{DS}^* can take large values on the order of 5 V. Class AB differential amplifiers find application in high-slew-rate opamps or/and in linear transconductors. Composite BiCMOS transistors or folded composite BiCMOS transistors can be used instead of composite CMOS transistors [34]. This is done by replacing an MOS transistor by a bipolar transistor, as shown in Figures 7.6(c) and 7.6(d). This replacement allows essential reduction of the supply requirements of the circuit to $V_{GS}^* = V_{SGn} + V_{BE}$ or to $V_{GSn} - V_{BE}$ for the folded version. Utilization of composite BiCMOS transistors leads also to improved small-signal performance: lower noise, higher effective transconductance, lower offset, and higher bandwidth.

7.2.5. Subthreshold Techniques

These techniques are based on the fact that transistors operating in the subthreshold mode use gate-to-source voltages which are lower than their nominal

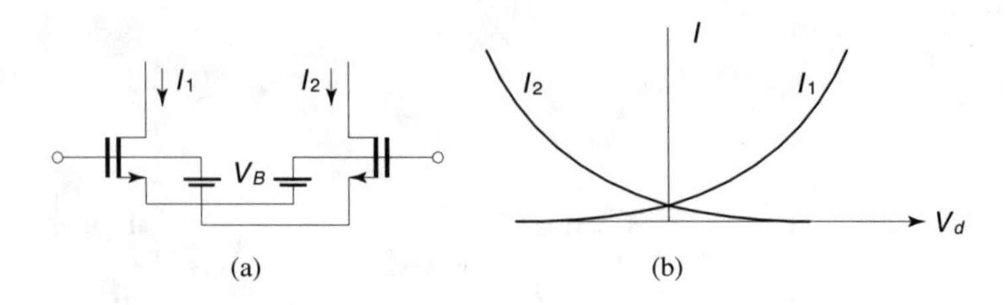

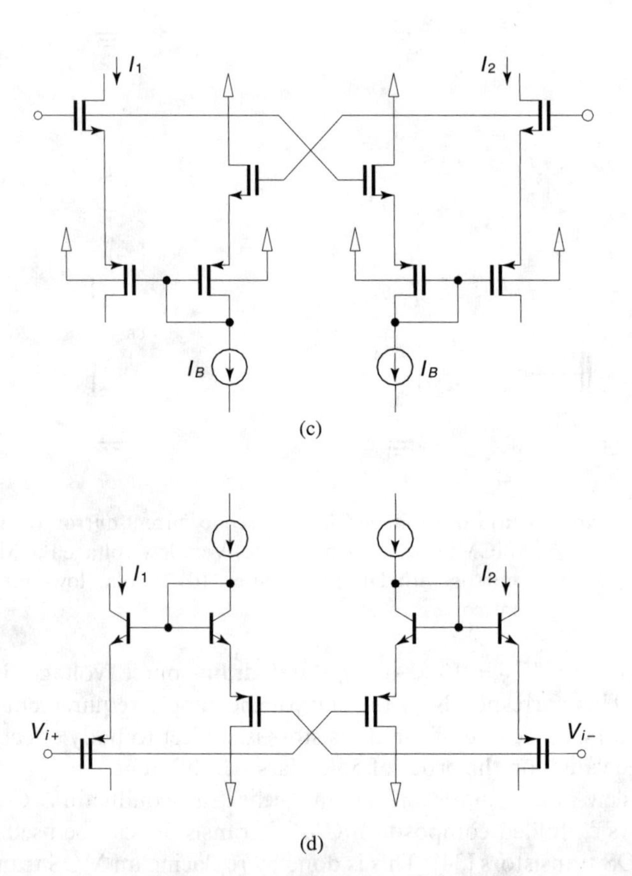

Figure 7.6 (a) CMOS class AB differential pair. (b) Transfer characteristic I_1, I_2 vs. V_d. (c) Implementation using composite CMOS transistors. (d) Implementation using composite BiCMOS transistors.

threshold voltage and their drain-to-source voltage requirements are reduced to just a few thermal voltages ($\sim 5V_{\text{TH}}$, where $V_{\text{TH}} = kT/q \approx 26\,\text{mV}$) [35]. This allows the designer to stack more transistors between V_{DD} and V_{SS}. In this mode of operation low supply voltages can be used, but currents, input signal voltage ranges, and small-signal transconductances are very small. Other disadvantages of circuits with small-sized MOS transistors operating in the subthreshold are less accurate current mirrors, larger offset voltages, lower bandwidth, higher noise spectral density, and lower slew rates. In spite of this, their low-voltage/low-power consumption and their high gain makes them attractive for a host of low-speed/low-power applications such as in biomedical implantable devices and for neural networks and neuromorphic systems [36] where massive parallel processing can compensate for low-speed. The lower limit on the gate-to-source voltages of MOS transistors operating in the subthreshold is determined by the value at which subthreshold currents become comparable to PN-junction leakage currents. This takes place at about one-half of the threshold voltage. One important problem is inaccuracy of MOS mirrors in the subthreshold if small-geometry MOS transistors are used. This obstacle can be overcome by replacing MOS transistors in current mirrors by bipolar transistors, as shown in Figure 7.7. This also results in improved speed and lower noise as compared to the MOS version. A problem of BJTs operating with subthreshold currents is the degradation of the current gain factor β for collector currents down to the nanoampere range.

7.2.6. MOS Transistors Operating in Triode Mode

Bipolar transistors can be used to force the operation of MOS transistors into the triode mode. This leads to circuits with very interesting characteristics and reduced voltage supply requirements. Consider, for example, the BiCMOS differential pair shown in Figure 7.8 [37, 38]. In this circuit the bipolar transistors are used to equalize the drain-to-source voltages of the MOS transistors to the approximately constant value $V_{DS}^{Q} = I_R R$. This value is such that M_1, M_2 satisfy the triode condition $V_{DS}^{Q} < V_{GS} - V_t$. The circuit performs as a linear differential transconductor. The following relations apply: $I_{\text{out}} = I_1 - I_2 = (2\mu C_{\text{ox}} W/L) V_{DS}(V_1 - V_2) = g_m V_d$. The transconductance gain is expressed by $g_m = (2\mu C_{\text{ox}} W/L) I_s R$ and can be adjusted over a wide range of values with the current I_s. An important feature of this circuit is that it has independent adjustable saturation levels (with I_B) and transconductance

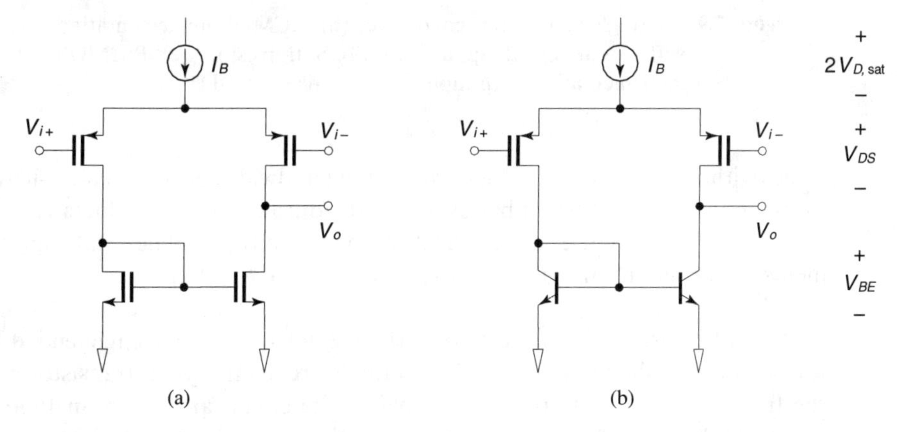

Figure 7.7 (a) CMOS actively loaded differential amplifier. (b) BiCMOS version for improved offset and bandwidth.

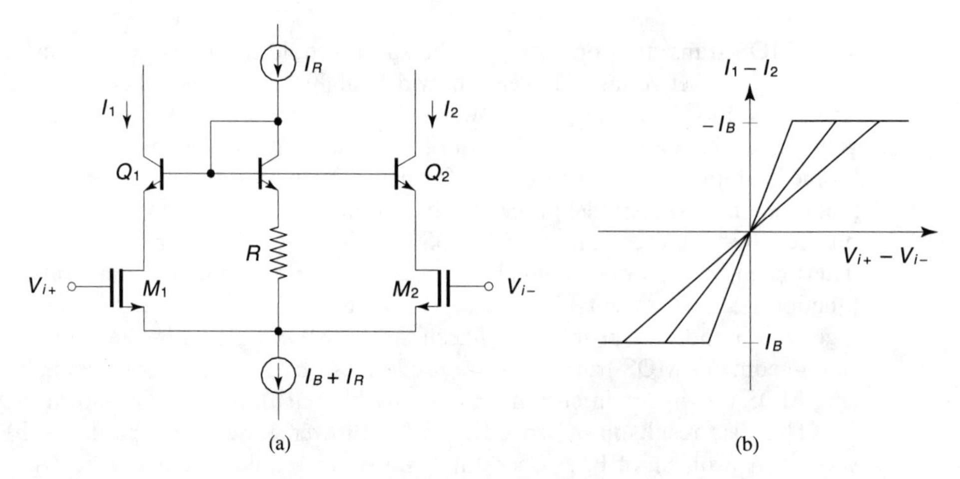

Figure 7.8 (a) Linear BiCMOS differential pair with MOS transistor in triode mode. (b) Transconductance characteristic.

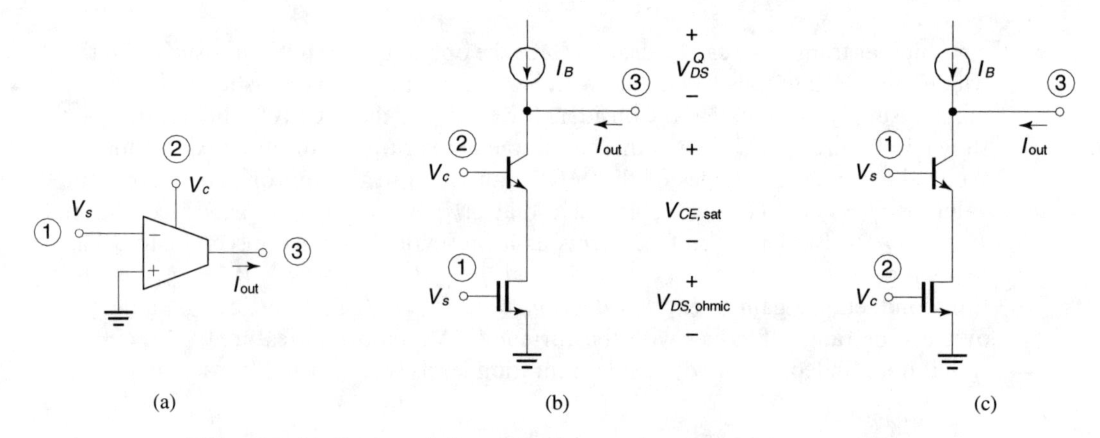

Figure 7.9 (a) Single-ended transconductor. (b) BiCMOS implementation with input signal applied to MOS transistor. (c) BiCMOS implementation with input signal applied to BJT.

slope (with I_s). The current I_B determines the bandwidth of the circuit. Therefore, the transconductor has constant bandwidth with adjustable transconductance slope. This characteristic is very useful for a host of applications [21]. The total supply requirements of the circuit are given as $V_{SR} = V_{DS,\text{sat}}^Q + I_s R + V_{BE} + V_{DS,\text{sat}}^Q$ (~1.5 V if we take $I_s R = 0.5\,\text{V}$).

Another circuit closely related to the previous one is the single-ended BiCMOS transconductor shown in Figure 7.9. In these circuits the MOS transistor operates in the triode mode [39]. Two ways to operate the circuit are shown in Figure 7.9. In Figure 7.9(b), a DC control voltage V_c is applied to the base of the bipolar transistor. The drain-to-source voltage is constant, $V_{DS} = V_b - 0.6\,\text{V}$, and it has a value that keeps the MOS transistor in the triode mode. The input voltage signal is applied to the gate of the MOS transistor. The output current I_o is linearly related to the input voltage. The control voltage V_c determines the transconductance gain. A different way to operate the same circuit is to apply the input signal to the base of the bipolar transistor [Fig. 7.9(c)], in which case it appears as a drain-source voltage in the MOS transistor. A constant voltage V_c is applied to the gate of the MOS transistor. In this case the MOS transistor works as an emitter degeneration resistor with an equivalent resistance R_{MOS} which is dependent on the control voltage V_c. Both circuits operate as linear transconductors. A constant offset current can be canceled using current replication techniques or by means of fully differential structures (see later). These circuits have very low voltage supply requirements (as low as 1.2 V) and can be used as building blocks of low-voltage current-mode and voltage-mode systems.

7.2.7. Class AB Operation and Complementary Techniques

This family of circuits, similar to digital circuits, use two circuit sections in which only one is active at a time. This reduces supply voltage requirements to those of the active section. Quiescent power supply requirements are also reduced. We can classify circuits into current-mode and voltage-mode circuits. The first ones split the current signal into an N-path for signals of one polarity and a P-path for signals of the opposite polarity. To illustrate this, consider the class AB current buffer (or current conveyor) shown in Figure 7.10 [40]. In this circuit, the

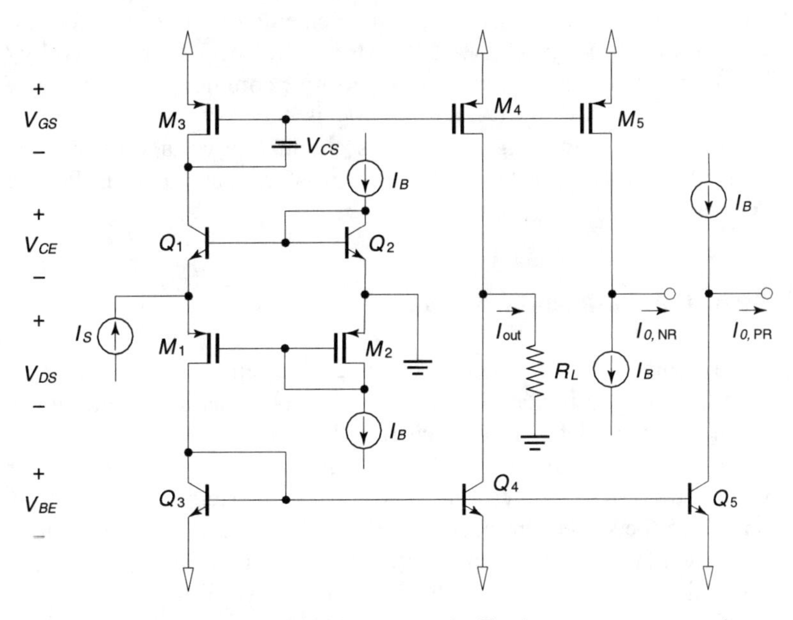

Figure 7.10 Class AB BiCMOS current buffer (or current conveyor).

BiCMOS version of a well-known four-transistor cell formed by M_1, Q_1, M_2, Q_2 [41] and a small bias current I_B are used to provide class AB operation. The four-transistor cell causes positive input currents to flow mainly through the bottom current mirror (Q_3, Q_4) and negative input currents to flow through the top current mirror (M_3, M_4); the two components add at the output node (R_L). Crossover distortion caused by unequal delays can be avoided by passing the current through an additional class AB stage so that each current signal goes through a top and bottom section. Since the circuit separates positive from negative signal components, it can be used as a precision positive/negative rectifier. Negative rectification is obtained by sensing only the current of the top mirror ($I_{o,\mathrm{NR}}$) with M_5, positive rectification by sensing only the current in the bottom mirror ($I_{o,\mathrm{NP}}$) with Q_5. Precision rectification is a fundamental operation in nonlinear systems.

In "voltage-domain" circuits the voltage signal is split into two complementary circuit sections, one of which is mainly active for positive signals and the other for negative signals. A good example of these types of structures are opamps with rail-to-rail common-mode input ranges. These opamps use complementary differential input stages, one with N-channel transistors and one with P-channel transistors [42]. For zero or small common-mode input signal components both differential circuits are active. For large positive or large negative common-mode signals only one of the differential pairs is active. For example, consider the high-slew-rate BiCMOS opamp with rail-to-rail common-mode input range shown in Figure 7.11 [43]. It is based on the class AB BiCMOS differential pair of Figure 7.6(d). In this circuit, advantage is taken of the fact that currents I_1, I_2, I_3, and I_4 are available both on top and bottom of the circuit; therefore the current generated by both differential pairs can be sensed using the same current mirrors CM_1 and CM_2. A standard three-mirror structure can be used to obtain the difference $I_o = I_a - I_b$ at a high-swing, high-impedance output node (these mirrors can be implemented with any of the low-voltage structures of Fig. 7.5). The class AB BiCMOS differential pair formed by Q_1, Q_2, M_1, M_2 is mainly active for medium to large positive common-mode signals while the differential pair formed by Q_3, Q_4, M_3, M_4 is the only one active for large negative common-mode input signals. In this circuit low-voltage operation is achieved by utilization of composite BiCMOS transistors but speed is limited by the MOS devices.

7.2.8. Current-Mode Techniques

In current-mode systems, variables are represented by current signals rather than by voltage signals. System operators (scaling, signal multiplication, addition, integration) are defined in the current domain [44–46]. If input signals are available as voltages and output signals are required in the form of voltages, then linear voltage-to-current and current-to-voltage converters are required at the input and output of the system, respectively. Basic advantages of current-mode signal processing are that current addition requires only a physical circuit node and that all nodes have low impedance. This leads to very good high-frequency performance. If bipolar transistors are used in the signal path of a current-mode circuit, the high

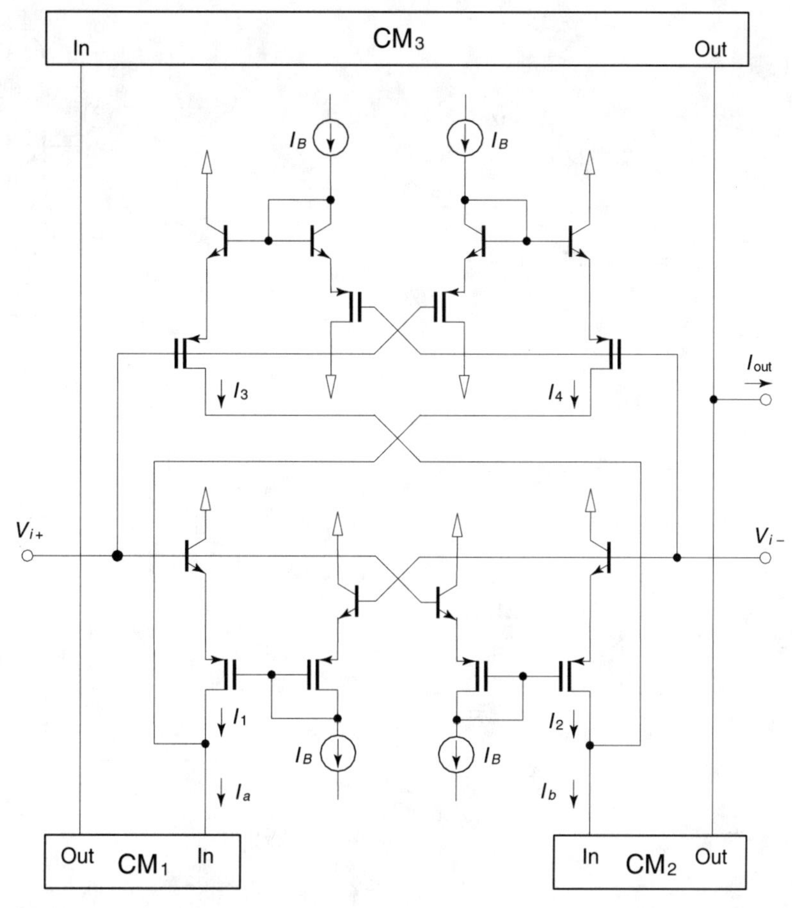

Figure 7.11 High-slew-rate low-voltage BiCMOS opamp with rail-to-rail common-mode input voltage range.

transconductance gain and small voltage swings that accompany large current swings result in circuits with reduced voltage supply requirements and improved performance. Consider Figure 7.12, in which the BiCMOS implementation of the basic current-mode operators is shown: integration [Fig. 7.12(a)], addition [Fig. 7.12(b)], scaling, and signal multiplication. The Gilbert translinear multiplier cell of Figure 7.12(d) is a one-quadrant current multiplier [47] that, if required, can be easily transformed into a two- or four-quadrant multiplier [48]. We can also use this cell for a linear gain programmable current mirror where the output current is given by $I_o = I_{\text{in}} I_{\text{in2}}/I_{\text{in3}} = AI_{\text{in}}$ and $A = I_{\text{in2}}/I_{\text{in3}}$ is the programmable gain. These building blocks plus the precision current rectifier of Figure 7.10 can be used to build linear and nonlinear low-voltage BiCMOS current-mode circuits. These systems can have fixed or programmable characteristics. Some examples will be given in Section 7.3.

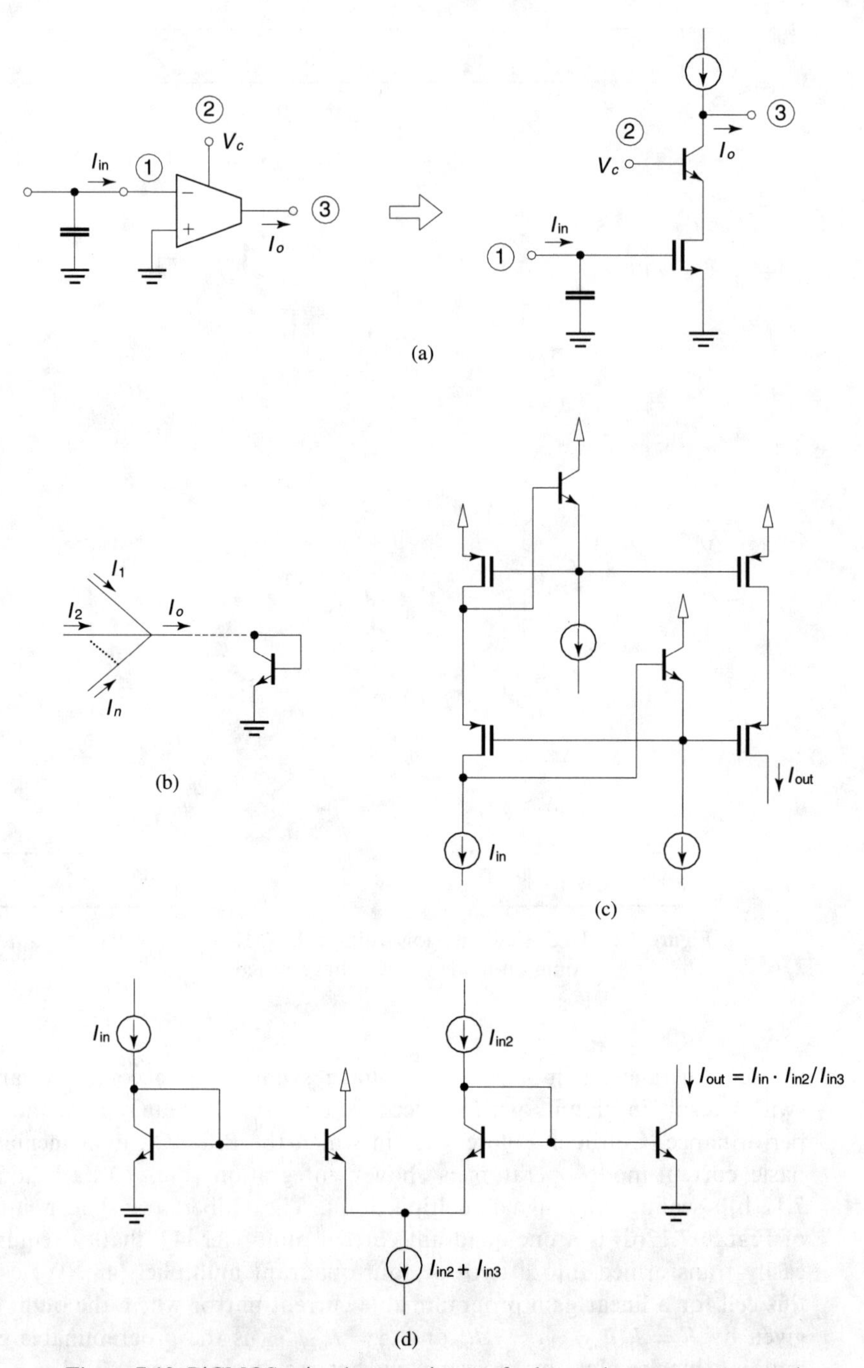

Figure 7.12 BiCMOS implementation of low-voltage current-mode operators: (a) integrator; (b) adder; (c) scaler; (d) multiplier/divider.

7.2.9. Fully Differential Structures

In fully differential circuits all current and voltage signals are available in complementary form. For each voltage node or branch current in the circuit there is node and a branch that has the same signal but with opposite polarity. Fully differential structures require duplication of circuit area and inclusion of a common-mode feedback network to stabilize the DC operating points. In spite of this, fully differential structures have become almost standard for mixed-mode applications due to the many advantages offered by these structures: (1) signal swings are doubled with respect to single-ended circuits, which increases dynamic range by 6 dB, (2) power supply and common-mode rejection are greatly improved (this is especially important in mixed-mode VLSI systems where analog circuitry is highly susceptible to dynamic range degradation due to digital noise induced through the common substrate, and (3) operation with lower supply voltages is possible. Other advantages arising from differential operation are possibility to increase output impedance (and therefore gain) using simple conductance cancellation schemes, cancellation of even-order nonlinear distortion terms generated by MOS devices, partial cancellation of systematic or common-mode errors using layout techniques, and elimination of the requirement for inverting elements. The delay associated with inverting elements in single-ended circuits is an important limitation for high-frequency operation. Inverting elements are not required in fully differential systems since signal inversion is simply achieved by injecting signals from the positive path into the negative path (i.e., crossing wires). To illustrate these concepts, consider the differential BiCMOS transconductor of Figure 7.13. This circuit performs as a current integrator by adding capacitors on the input side or as a voltage integrator by adding capacitors on the output

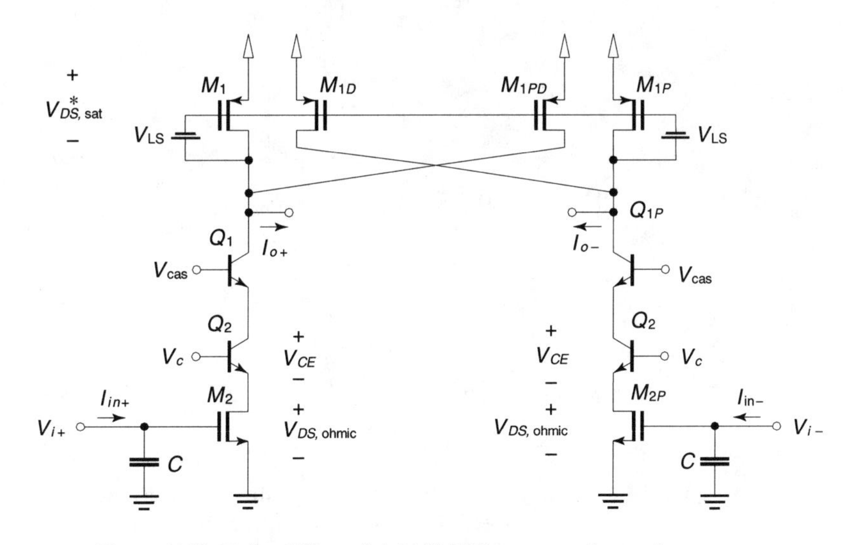

Figure 7.13 Fully differential BiCMOS transconductor integrator.

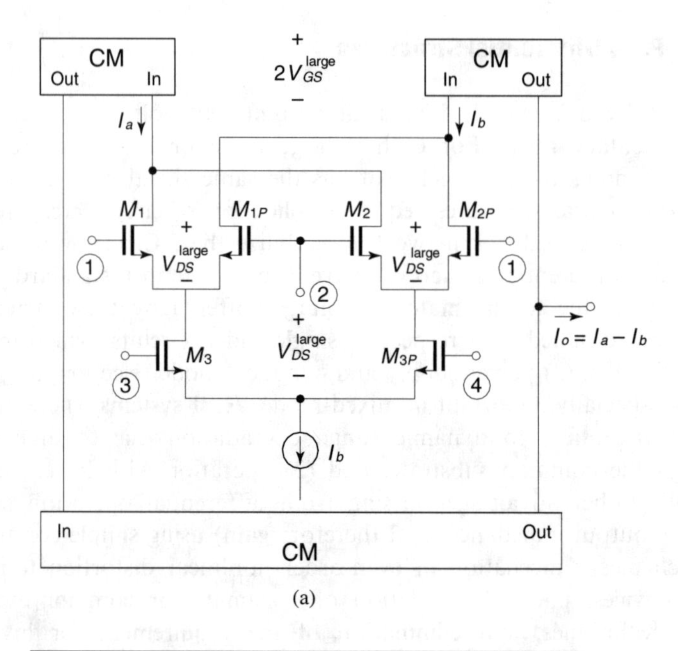

(a)

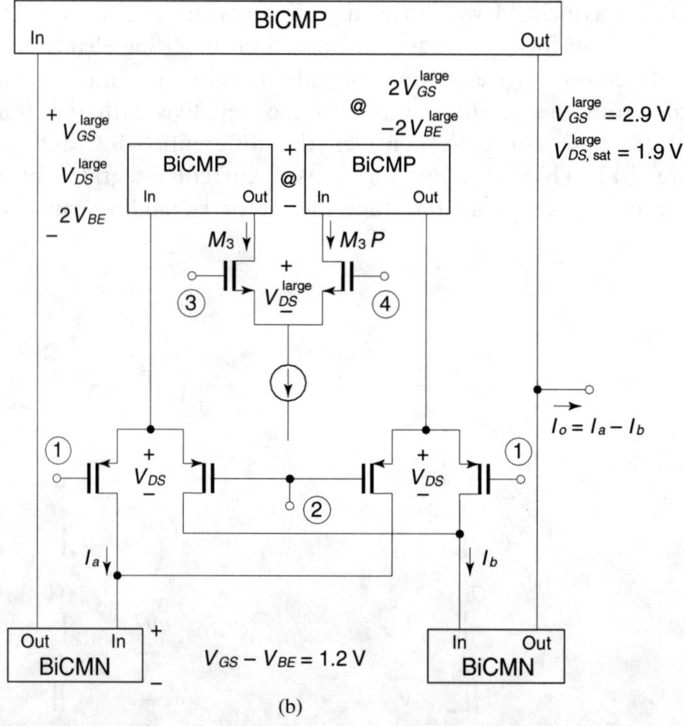

(b)

Figure 7.14 (a) CMOS version of four-quadrant Gilbert multiplier. (b) Low-voltage BiCMOS version of the circuit in (a).

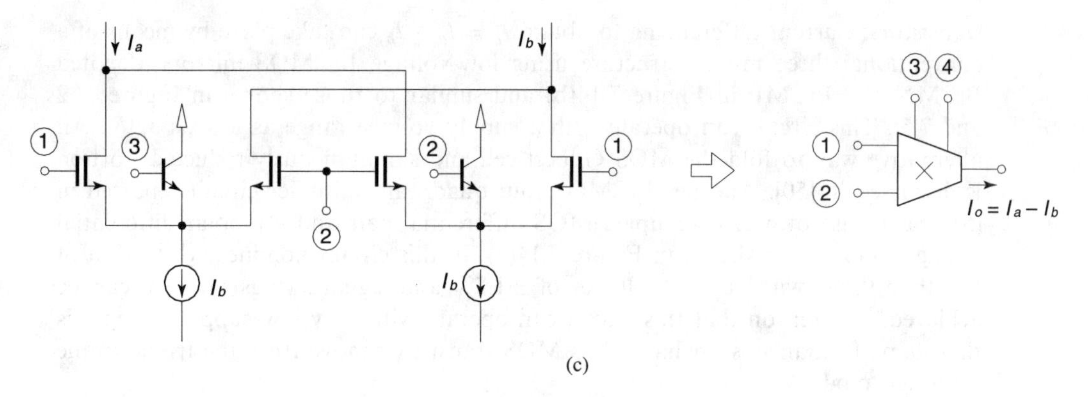

Figure 7.14 (*continued*). (c) Low-distortion four-quadrant BiCMOS transconductance multiplier cell and symbol.

side. Cross-coupled loads allow cancellation of the output conductance of M_1, M_{1P} for differential signals as discussed in [49]. The differential output resistance is determined mainly by the bottom part of the circuit and can be fairly high due to cascoding transistors Q_1, Q_{1P} and to the transconductance controlling transistors Q_2, Q_{2P}. Another advantage of the fully differential operation of this circuit is that nonlinear distortion terms associated with the MOS transistor are canceled. It can be easily shown that the differential output current $I_{o+} - I_{o-}$ is linearly related to the differential input voltage $V_{I+} - V_{i-}$. Operation of this circuit with a supply voltage range $V_{SR} = V_{DS,\text{sat}}^{Q} + V_{CE} + V_{CE} + V_{DS,\text{ohmic}} < 1.2\,\text{V}$ is possible.

7.2.10. Flip-Over Techniques

The last technique discussed here consists of the alternation of P and N (or NPN devices) in the signal path, denoted here as the flip-over technique. It prevents the designer from stacking more than three transistors between V_{DD} and V_{SS}. It can lead to significant reduction of the voltage supply requirements if low-voltage BiCMOS current mirrors are used for this purpose. To illustrate the technique, consider the CMOS counterpart to the well-known four-quadrant Gilbert multiplier shown in Figure 7.14(a). In order to keep distortion low in this circuit, relatively large V_{GS}^{Q} and $V_{DS,\text{sat}}^{Q}$ have to be used compared to the swing of the input differential signals $V_{d1} = V_1 - V_2$ and $V_{d2} = V_3 - V_4$. By stacking the cross-coupled differential amplifiers M_1, M_{1P} and M_2, M_{2P} on top of the differential amplifier formed by M_3, M_{3P} and by using conventional current mirrors (e.g., cascode mirrors) to obtain $I_o = I_a - I_b$, a large voltage range $V_{SR} = V_{DS}^{*} + V_{DS,\text{sat}} + V_{DS,\text{sat}} + 2V_{GS}$ results. This can be over 6 V for signal swings as small as ± 1 V. In order to reduce the supply requirements, we use low-voltage BiCMOS current mirrors to invert the direction of the currents generated by M_3, M_{3P}. This allows us to flip over the cross-coupled differential pair so that it can be implemented with P-channel transistors. Current differencing to obtain $I_o = I_a - I_b$ can take place by means of a conventional three mirror structure using low-voltage BiCMOS mirrors [denoted BiCMN and BiCMP in Figure 7.14(b) and similar to those shown in Figures 7.2 and 7.5]. This circuit can operate with a supply voltage range as low as 3.3 V. An alternative way to fold the MOS Gilbert cell and simultaneously reduce distortion is discussed in [50]. Another BiCMOS four-quadrant multiplier circuit reported in [51] makes use of a cross-coupled MOS differential pair and a bipolar differential voltage follower, as shown in Figure 7.14(c). In this circuit nonlinear distortion of less than 0.1% with supply voltages of ± 1.5 V and signal swings of ± 1 V can be achieved. The reason that this circuit can operate with very low supply voltages is that its performance is unchanged as MOS transistors move from the triode to the saturated mode.

7.3. EXAMPLES OF LOW-VOLTAGE BICMOS SUBSYSTEMS

In this section, examples of subsystems that use at least one of the techniques described in the previous section are given.

7.3.1. Programmable FIR and IIR Analog Filters

The voltage follower of Figure 7.1(c) together with the BiCMOS transconductance multiplier of Figure 7.14(c) can be used as building blocks of low-voltage continuous and discrete-time programmable analog filters of the programmable finite impulse response (FIR) and infinite impulse response (IIR) analog filters type [52–53]. Figure 7.15(a) shows the architecture of a discrete-time FIR analog filter that uses the voltage follower of Figure 7.1(c) and MOS switches to implement delay elements and transconductance multipliers to implement voltage programmable weighting coefficients. Additions are performed on current signals by simply connecting all currents to the output node. The circuit of Figure 7.15 is characterized by the z-domain transconductance transfer function $H(z) = I_{\text{out}}/V_i = a_0 + a_1 z^{-1} + \cdots + a_{n-1} z^{n-1} + a_n z^n$. The continuous-time counterpart is shown in Figure 7.16. Its s-domain transfer function is given by $H(s) = I_{\text{out}}/V_i = a_0 + a_1/(1 + s/\omega_p)^2 + \cdots + a_{n-1}/(1 + s/\omega_p)^{n-1} + an/(1 + s/\omega_p)^n$. This can be programmed with the weighting coefficients $a_0, a_1, \ldots, a_{n-1}, a_n$. The IIR discrete-time analog filters using voltage followers and switches as delay elements, transconductance multipliers as programmable weighting coefficients, and additions in the current mode can also be developed in a similar way. Figure 7.17 shows the block diagram of an nth order continuous-time programmable IIR filter derived from a canonical control structure. In this circuit, input and output signals are currents. Integrators in the canonical control structure are replaced by voltage followers which perform as lossy integrators. Additions are performed in the current domain and transconductance multipliers implement, as before, programmable coefficients.

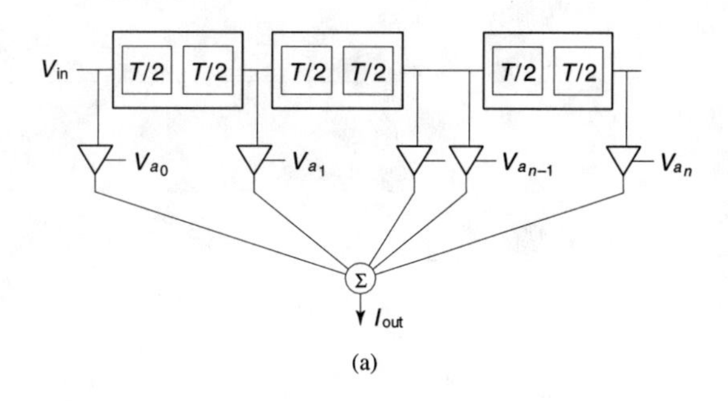

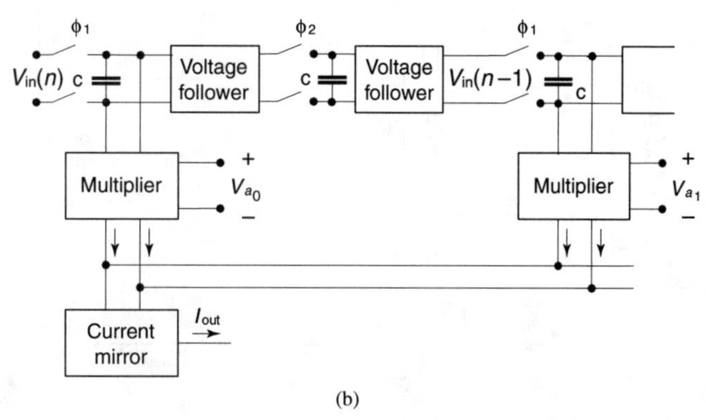

Figure 7.15 (a) Architecture of a discrete-time programmable analog FIR using voltage followers and analog multipliers. (b) Detailed circuit implementation of one FIR section.

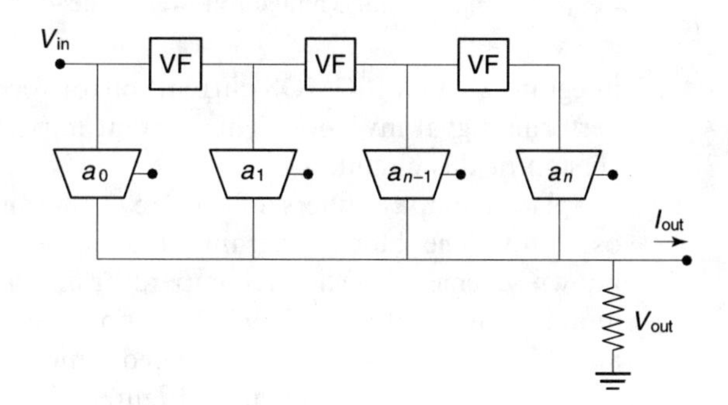

Figure 7.16 Architecture of a continuous-time analog FIR using voltage followers (VFs) and analog multipliers.

7.3.2. Current-Mode Continuous-Time Filter and Oscillators

The BiCMOS transconductors of the circuits of Figures 7.9(b) and 7.9(c) can be used as building blocks of voltage- or current-mode continuous-time systems. Figure 7.18(a) shows a BiCMOS current integrator; Figure 7.18(b) shows the block diagram of a current-mode bandpass filter with center frequency ω_0 and selectivity factor Q [54]. Figure 7.18(c) shows the circuit implementation using BiCMOS current

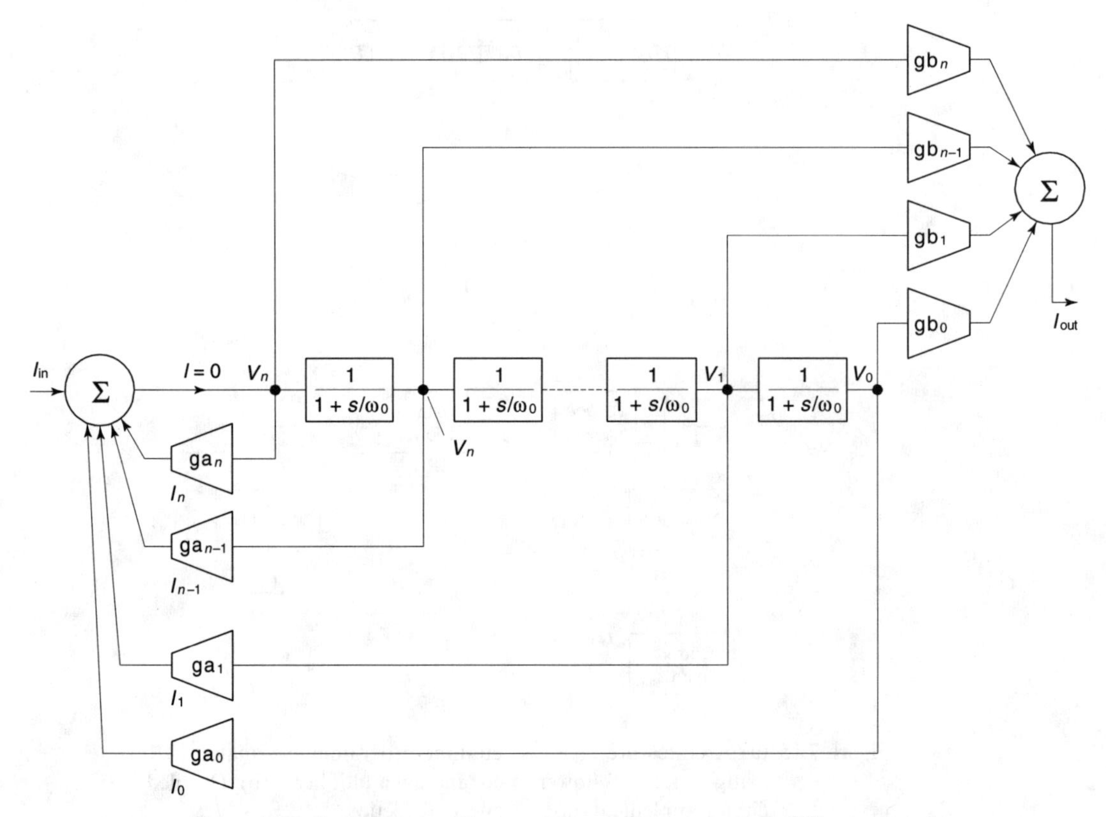

Figure 7.17 Architecture of a continuous-time canonical IIR structure using voltage followers instead of integrators.

integrators and a BiCMOS current mirror formed by Q_4, Q_5, MR_4, and MR_5 which performs signal inversion. This current mirror is not required in a fully differential version of this circuit.

The bandpass filters of Figure 7.18 can be used to implement a sinusoidal oscillator. The block diagram of a current-mode oscillator based on the well-known scheme of placing a comparator and a bandpass filter in a positive-feedback loop is shown in Figure 7.19 [54]. A novel feature of this scheme is the utilization of an additional "out-of-loop" matched bandpass filter to reduce distortion. The current starved inverter shown in Figure 7.20 is used in this oscillator as a current comparator with programmable saturation levels. This is convenient with respect to supply noise and power consumption.

7.3.3. Wide-Range Gain-Programmable OTA with Constant Bandwidth and Input Range

Figure 7.21(a) shows a linear BiCMOS operational transconductance amplifier (OTA) [55]. This has a folded-cascode architecture and uses as input stage the

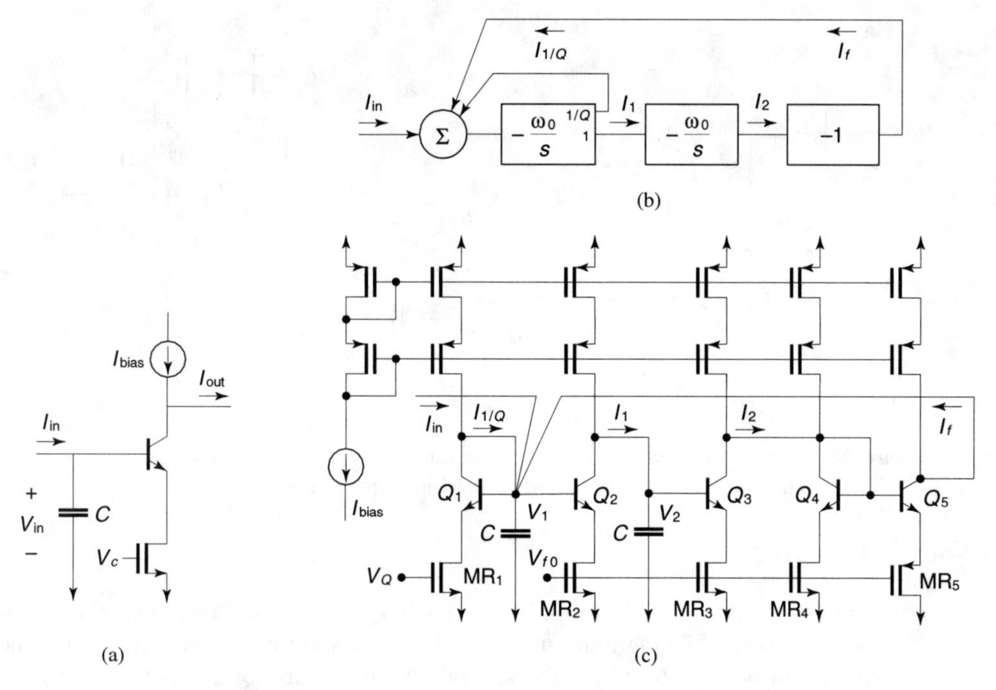

Figure 7.18 (a) BiCMOS current-mode integrator. (b) Block diagram of current-mode bandpass filter. (c) Single-ended circuit transistor-level implementation.

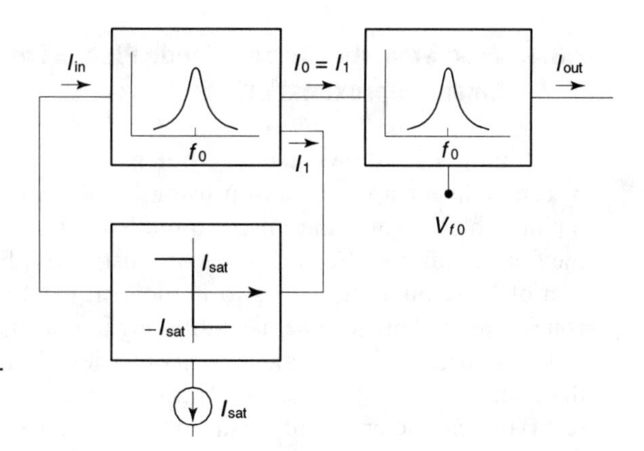

Figure 7.19 Architecture of a current-mode oscillator with out-of-loop bandpass filter.

BiCMOS linear differential amplifier of Figure 7.8 and a three-mirror current-differencing scheme. This circuit has two Gilbert translinear cells [Fig. 7.12(d)] which are used as linear gain programmable current mirrors and a BiCMOS low-voltage current mirror of Figure 7.2(b). The transconductance gain of this circuit is given by $g_m = 2\mu_n C_{ox}(W/L)I_s R(I_2/I_1)$. It can be adjusted over a very wide range of values with the current I_2 (or I_1) by keeping a constant input range and approximately

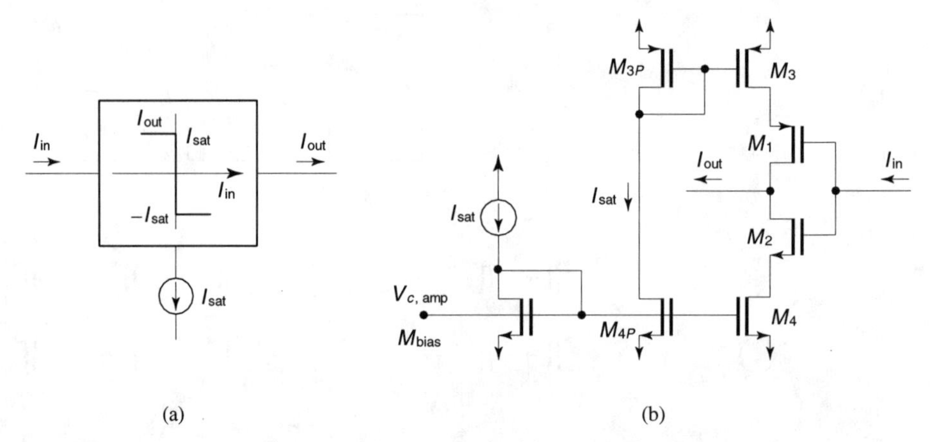

Figure 7.20 Current comparator with programmable saturation levels. (a) Input-output characteristics. (b) A transistor-level implementation.

constant bandwidth [Fig. 7.21(b)]. The current I_s can be used to adjust the transconductance gain while maintaining bandwidth and saturation levels constant. The current I_B can be used to adjust the bandwidth and saturation levels with constant transconductance gain. These characteristics are very useful for continuous-time OTA-C filter applications [56].

7.3.4. Programmable Current-Mode Piecewise Linear Approximation

Figure 7.22 shows a basic (but not very practical) scheme for current-mode piecewise linear approximation using bipolar current mirrors as rectifiers [57]. The building blocks shown are used to implement rectification with DC shifts in each of the four quadrants. For purposes of illustration, Figure 7.23 shows the interconnection of these building blocks to implement (a) a full-wave rectifier and (b) a "dead zone" circuit. This scheme has the following problems: (1) the transfer characteristic is determined solely by transistor geometries, (2) the high-frequency performance of the circuits is very poor since at times some transistors are turned off, and (3) the rectifying characteristic of a current mirror does not have a sharp corner and thus does not correspond to a precision rectifier. These last two problems can be solved by using the class AB current buffer rectifier of Figure 7.10. This circuit has very good high-frequency performance and precision rectifying characteristics. Slope programmability can be achieved if this circuit is followed by a Gilbert translinear multiplier of Figure 7.12(d) which is used as a linear gain programmable current mirror (LGPCM). The cell formed by these two circuits can be used along with fixed-gain low-voltage current mirrors to implement positive and negative rectifying characteristics with programmable slopes, as shown in Figure 7.24.

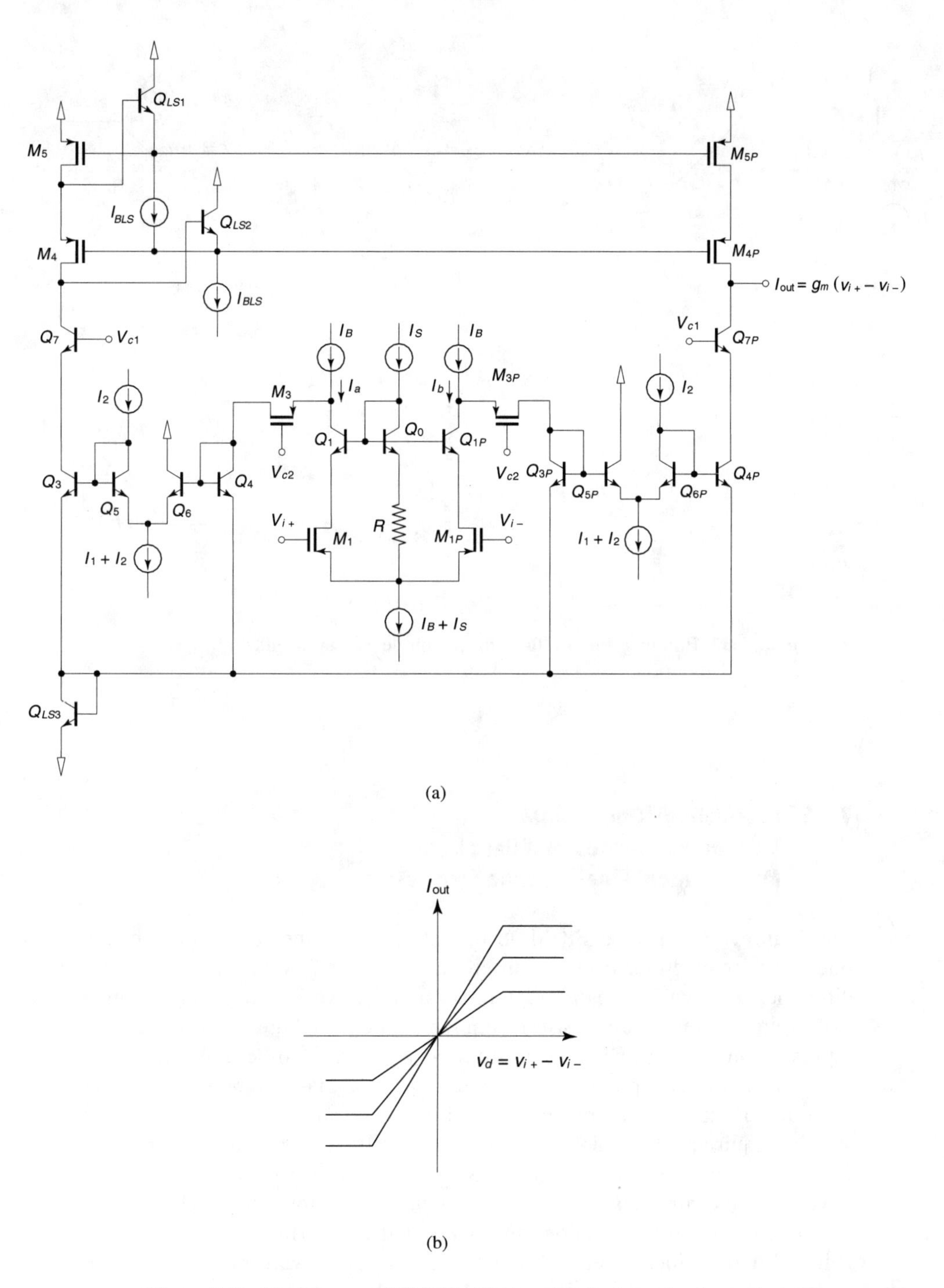

Figure 7.21 (a) Linear BiCMOS OTA with independently adjustable transconductance gain and bandwidth. (b) OTA transconductance characteristic with varying I_2.

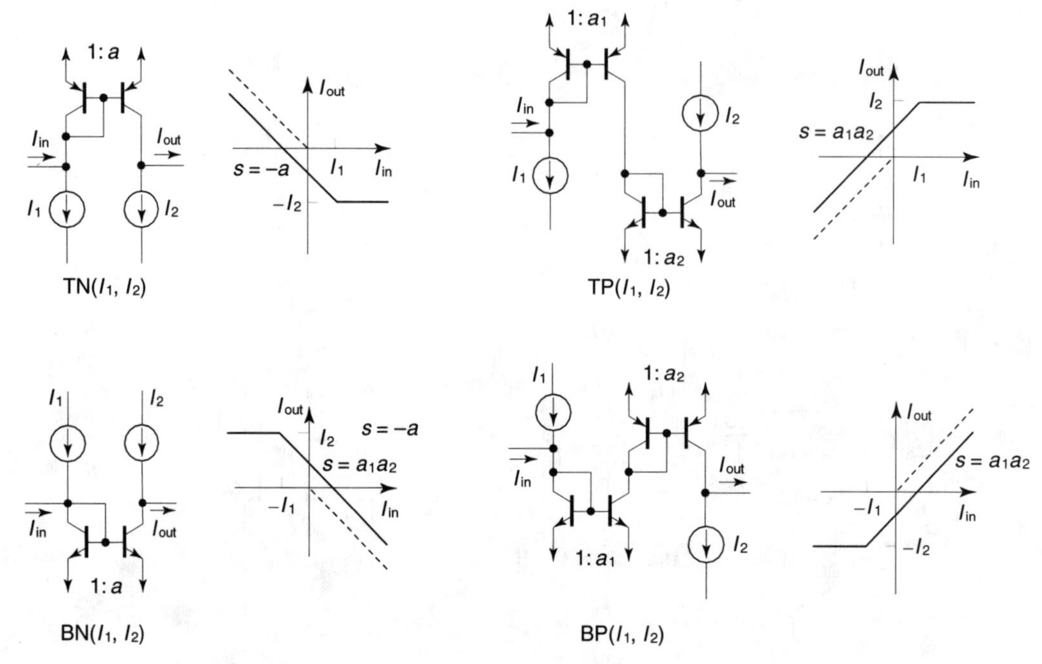

Figure 7.22 Building blocks for current-mode piecewise linear approximation.

7.3.5. Low-Voltage Operational Transconductance Amplifiers Using Multiple Input Floating-Gate Transistors

Figure 7.25 shows a BiCMOS operational amplifier based on a conventional one-stage architecture. It uses a differential input stage and a three-mirror current-differencing network. The differential pair and the two P-channel current mirrors use two-input floating-gate transistors similar to those of Figure 7.4, where one of the inputs is connected to DC biasing sources (V_B) and the other input terminal is used as a conventional input that does not require a bias component. A low-voltage cascode current mirror implemented with NPN transistors helps to minimize the supply requirements. This circuit operates with a single supply of 2 V or less. Low-distortion low-voltage analog BiCMOS multipliers using MIFG transistors have been recently reported [58–60]. Figure 7.26 shows the architecture of very compact low-distortion analog multipliers that use MIFG transistors [58]. The circuit of Figure 7.26(b) operates on a single 1-V supply with rail-to-rail input signals and a total harmonic distortion of less than 0.1% for 1-V peak-to-peak input signals. The circuit has only six transistors and silicon area of $100 \times 100\,\mu m^2$ in 2-μm CMOS technology.

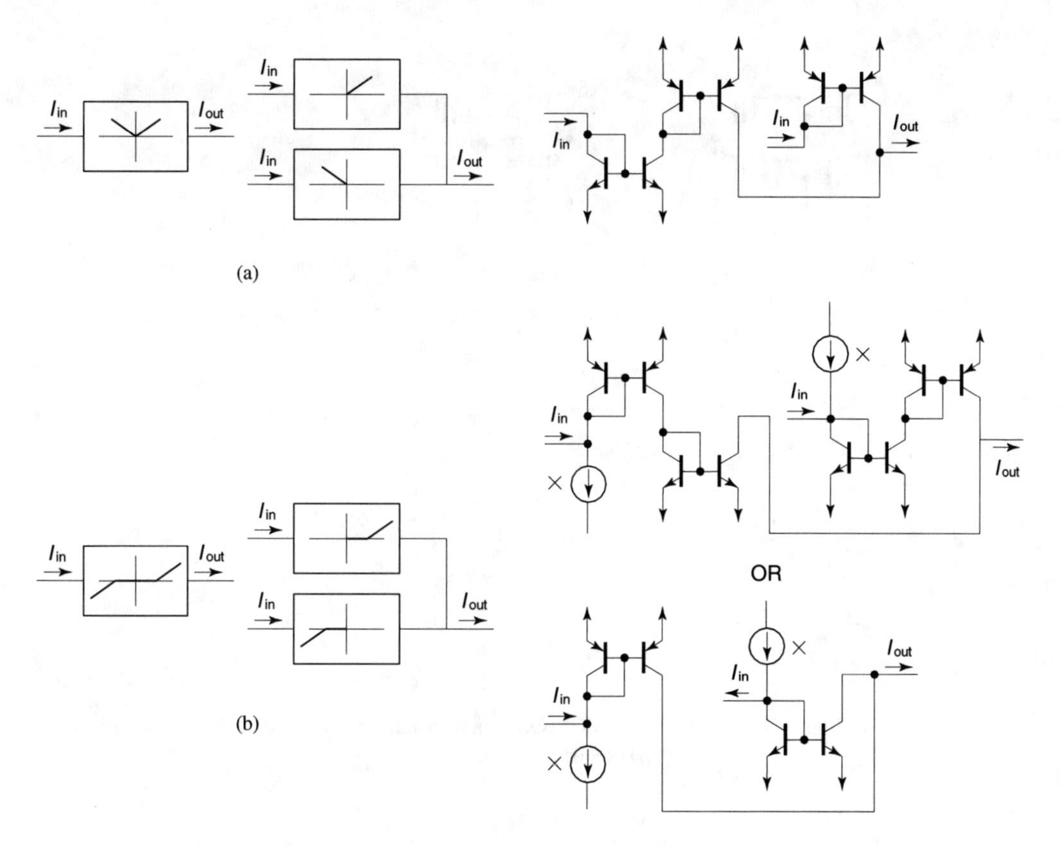

Figure 7.23 Examples of current-mode piecewise linear circuits: (a) full-wave rectifier and (b) dead zone circuit.

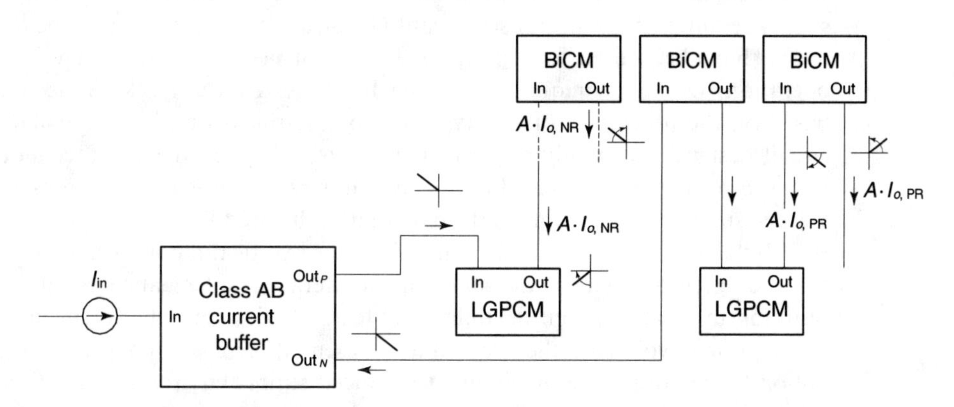

Figure 7.24 Scheme for obtaining programmable rectifying characteristics using class AB BiCMOS current buffers and gain programmable current mirrors.

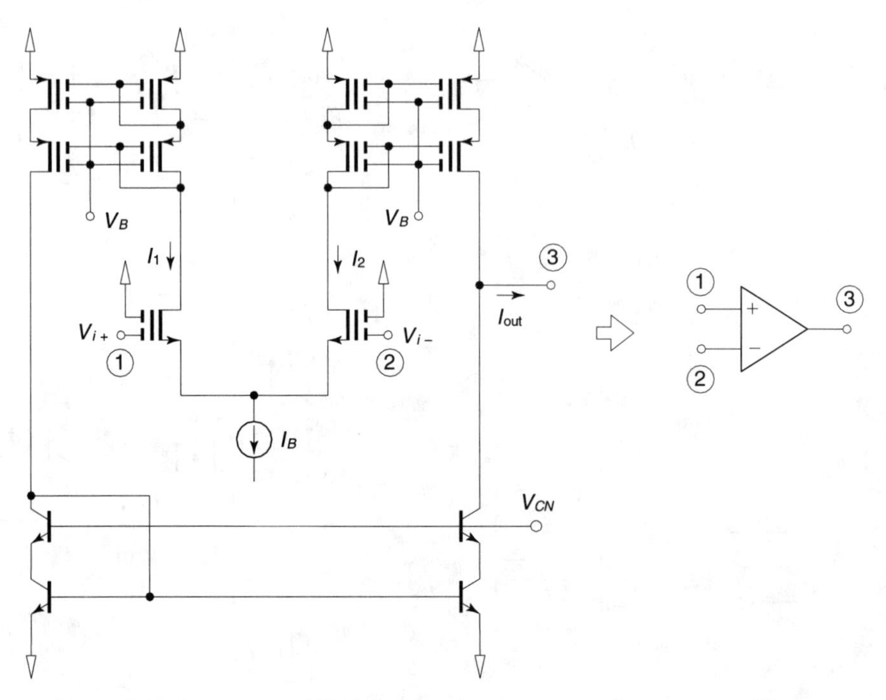

Figure 7.25 One-stage BiCMOS operational transconductance amplifier using MIFG transistors.

7.4. CONCLUSIONS

In this chapter we have discussed some techniques that allow reduction of voltage supply requirements without significant performance degradation. Some of the techniques have been used for a long time, but not necessarily to achieve low-voltage operation. Other techniques have been introduced recently. These techniques are based on the utilization of one type of bipolar transistor which is available to the circuit designer as a medium-performance lateral bipolar device in technology. The lateral bipolar transistors and/or vertical substrate transistors can be used to lower voltage supply requirements in the case that high speed is not of utmost importance. BiCMOS technology, besides having a dedicated vertical bipolar transistor that leads to faster circuits, offers at the same time a lateral complementary bipolar transistor resulting in greater design flexibility. As technologies continue to shrink, requiring lower supply voltages, the techniques presented here should offer an attractive option for adapting analog circuits to reduced supply environments without significantly sacrificing a circuit's performance.

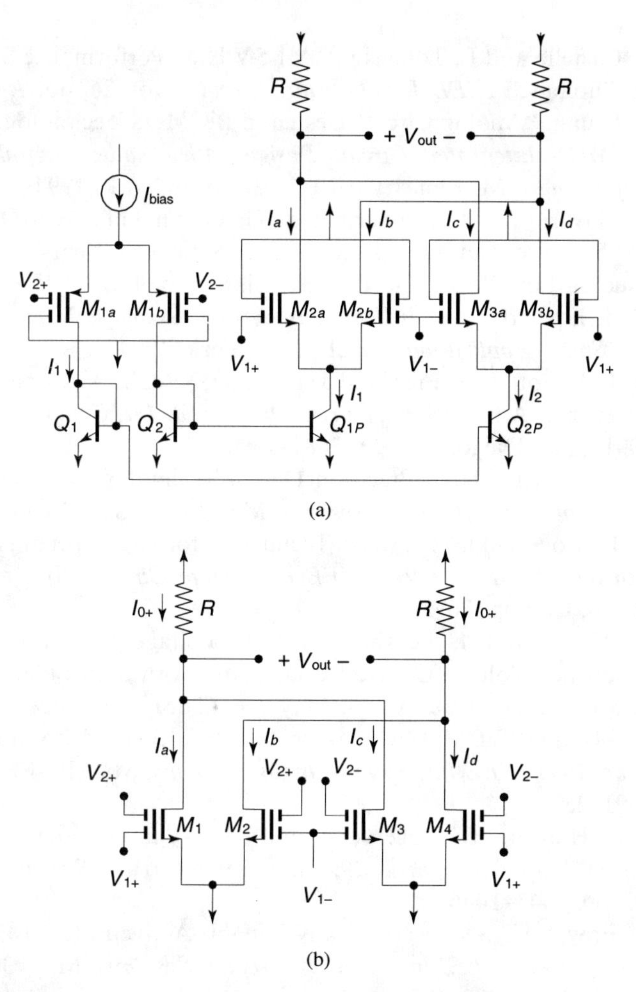

Figure 7.26 Low-voltage analog four-quadrant multipliers using MIFG transistors: (a) BiCMOS Gilbert cell; (b) voltage-biased cell operating from a single supply.

References

[1] E. A. Vittoz, "MOS Transistors Operated in the Lateral Bipolar Mode and Their Application in CMOS Technology," *IEEE J. Solid-State Circuits*, vol. SC-18, pp. 273–279, 1983.

[2] T. W. Pan and A. A. Abidi, "A 50 dB Variable Gain Amplifier Using Parasitic Bipolar Transistors in CMOS," *IEEE J. Solid-State Circuits*, vol. 24, no. 4, pp. 951–961, 1989.

[3] R. Castello and L. Tomasini, "A 1.5 V High Performance SC Filter in BiCMOS Technology," *IEEE J. Solid-State Circuits*, vol. 26, no. 7, pp. 930–936, 1991.

[4] B. Leung, "Analog Circuit Design in BiCMOS Technology: An Overview," in *BiCMOS Integrated Circuit Design with Analog Digital and Smart Power Applications*, M. Elmasry (ed.), IEEE, New York, 1994, pp. 367–374.

[5] B. Davari, R. H. Dennard, and G. G. Shahidi, "CMOS Scaling for High Performance and Low Power—The Next Ten Years," *Proc. IEEE*, special issue on Low Power Electronics, vol. 83, no. 4, pp. 595–606, 1995.

[6] M. I. Elmasry (ed.), *BiCMOS Integrated Systems Design with Analog, Digital and Smart Applications*, IEEE, New York, 1994.

[7] D. J. Allstot, T. S. Fiez, and J. Rahman, "BiCMOS Circuits for Analog Signal Processing Applications," *Proc. 1992 IEEE Int. Symp. Circuits Systems*, May 10–13, San Diego, CA, pp. 2687–1690.

[8] G. A. Miller, "High Precision Design Technologies in BiCMOS," *Proc. 1992 IEEE Int. Symp. Circuits Systems*, May 10–13, San Diego, CA, pp. 2703–2705.

[9] D. H. Robertson, "BiCMOS Techniques for High Speed, High Resolution A/D Converters," *Proc. 1992 IEEE Int. Symp. Circuits Systems*, May 10–13, San Diego, CA, pp. 2695–2698.

[10] H. Eaton and J. Leasho, "A Low Voltage Low Power BiCMOS Audio Frequency Voltage Controlled Oscillator with Sinusoidal Output," *Proc. 1993 Int. Symp. Circuits Systems*, May 3–6, Chicago, IL, pp. 1463–1466.

[11] K. Martin, "BiCMOS Components for Video Rate Continuous Time Filters," *Proc. 1994 IEEE Int. Symp. Circuits Systems*, May 10–13, San Diego, CA, pp. 2691–2694.

[12] J. H. Huijsing, R. Hogervorst, and Klass-Van de Langen, "Low Voltage Low Power Amplifiers," *Proc. 1993 Int. Symp. Circuits Systems*, May 3–6, Chicago, IL, pp. 1443–1446.

[13] D. Frey, "A 3.3 V Electronically Tunable Active Filter Usable Beyond 1 GHz," *Proc. 1994 IEEE Int. Symp. Circuits Systems*, May 30–June 2, London, England, pp. 493–496.

[14] K. Koli and K. Halonen, "A BiCMOS Current-Feedback Operational Amplifier with 60 dB Constant Bandwidth Range," *Proc. 1994 IEEE Int. Symp. Circuits Systems*, May 30–June 2, London, England, pp. 525–528.

[15] A. C. Van der Woerd, J. Davidse, A. H. M. van Roermind, and W. A. Serjin, "Low Voltage, Low Power Fully-Integratable Automatic Gain Control," *Proc. 1994 IEEE Int. Symp. Circuits Systems*, May 30–June 2, London, England, pp. 505–508.

[16] S. R. Zarabadi, "Design of Analog VLSI Circuits in BiCMOS/CMOS Technology," Ph.D. Dissertation, Ohio State University, June 1992.

[17] M. Sasaki, Y. Ogata, K. Taniguchi, F. Ueno, and T. Inoue, "BiCMOS Current-Mode Circuits with 1 V Supply Voltage and Their Application to ADC," *Proc. 1992 IEEE Int. Symp. Circuits Systems*, May 10–13, San Diego, CA, pp. 1273–1276.

[18] B. A. Wooley, "BiCMOS Analog Circuit Techniques," *Proc. 1990 IEEE Int. Symp. Circuits and Systems*, May 1–3, New Orleans, pp. 1983–1986.

[19] M. Ismail and T. Fiez, *Analog VLSI, Signal and Information Processing*, McGraw-Hill, New York, 1994.

[20] J. Ramirez-Angulo, "A Low-Voltage Vertical BiCMOS OTA," *IEE Proc. Part G*, vol. 139, no. 4, pp. 553–556, 1992.

[21] J. Ramirez-Angulo, "A BiCMOS Universal Membership Function Circuit with Fully Independent Adjustable Parameters," *Proc. 1995 IEEE Int. Symp. Circuits Systems*, April 29–May 3, Seattle, WA.

[22] A. De Luca, G. Gonzalez-Altamirano, J. Ramirez-Angulo, and F. Perez-Ramirez, "Low Voltage BiCMOS Comparator for Charge Threshold Detection," *Thirty-Seventh Midwest Symposium on Circuits and Systems*, August 3–5, Layfayette, LA, 1994, pp. 1137–1140.

[23] J. Ramírez-Angulo and K. H. Treece, "A Second Generation Linear Low-voltage BiCMOS OTA," *Proc. 1993 IEEE Int. Conf. Circuits Systems*, May 5–7, Chicago, IL, pp. 1172–1175.

[24] J. Ramírez-Angulo, "Analog VLSI Implementations of Adaptive Membership Function for Fuzzy Controllers," *IEEE Int. Symp. Circuits Systems*, May 30–June 2, 1994, London, England, pp. 965–968.

[25] J. Ramírez-Angulo, "A BiCMOS Universal Membership Function Circuit," *Proc. 1995 IEEE Int. Symp. Circuits Systems*, April 29–May 1, Seattle, WA.

[26] J. Ramírez-Angulo, K. Treece, P. Andrews, and S. C. Choi, "Current-Mode and Voltage Mode VLSI Fuzzy Processor Architectures," *Proc. 1995 IEEE Int. Symp. Circuits Systems*, April 29–May 1, Seattle, WA, pp. 1156–1159.

[27] J. Ramírez-Angulo, "Building Blocks for Fuzzy Processors: Voltage-Mode and Current-Mode Min–Max Circuits in CMOS Operate from a 3.3 V Supply," *IEEE Circuits Devices Mag.*, vol. 10, no. 4, pp. 48–50, 1994.

[28] J. Ramírez-Angulo, R. Sadkowski, and E. Sanchez-Sinencio, "Linearity Accuracy and Bandwidth Considerations in Wideband CMOS Voltage Amplifiers," *Proc. 1993 IEEE Int. Conf. Circuits Systems*, May 5–7, Chicago, IL, pp. 1251–1254.

[29] K. R. Laker and W. M. C. Sansen, *Design of Analog Integrated Circuits and Systems*, McGraw-Hill, New York, 1994.

[30] J. Ramírez-Angulo, "Current Mirrors with Low Input Voltage Requirements for Built in Current Sensors," *1994 IEEE Int. Symp. Circuits Systems*, May 30–June 2, London, England, pp. 529–532.

[31] J. Ramírez-Angulo, "Current Mirrors with Low Input and Output Voltage Requirements," *Thirty-Seventh Midwest Symposium on Circuits and Systems*, August 3–5, 1994, Lafayette, LA, pp. 107–110.

[32] T. Shibata and T. Ohmi, "A Functional MOS Transistor Featuring Gate-Level Weighted Sum and Threshold Operations," *IEEE Trans. Electron Devices*, vol. 39, no. 6, pp. 1444–1455, 1992.

[33] J. Ramírez-Angulo, S. C. Choi, and G. Gonzalez-Altamirano, "Low-Voltage OTA Architectures Using Multiple Input Floating Gate Transistors," *IEEE Trans. Circuits Systems*, vol. 42, no. 12, pp. 971–974, 1995.

[34] J. Ramírez-Angulo, M. Deyong, and W. J. Adams, "Applications of Composite BiCMOS Transistors," *Electron. Lett.*, vol. 27, no. 24, pp. 2236–2238, 1991.

[35] Y. Tsividis and P. Antogneti, *Design of MOS VLSI Circuits for Telecommunications*, Prentice-Hall, Englewood Cliffs, NJ, 1985.

[36] C. Koch and B. Mathur, "Neuromorphic Vision Chips," *IEEE Spectrum*, pp. 38–46, May 1996.

[37] R. Alini, A. Baschiroto, and R. Castello, "Tunable BiCMOS Continuous Time Filter for High Frequency Applications," *IEEE J. Solid-State Circuits*, vol. 27, no. 12, pp. 1905, 1915, 1992.

[38] M. Ali, M. Howe, E. Sánchez-Sinencio, and J. Ramírez-Angulo, "BiCMOS Low Distortion Tunable Transconductor for Continuous-Time Filters," *IEEE Trans. Circuits Systems*, vol. 40, no. 1, pp. 43–48, 1993.

[39] J. Ramírez-Angulo and E. Sánchez-Sinencio, "Programmable BiCMOS Transconductor for C-TA Filters," *Electron. Lett.*, vol. 28, no. 13, pp. 1185–1187, 1992.

[40] J. Ramírez-Angulo, "A BiCMOS Current Buffer Rectifier," *IEEE Trans. Circuits Systems*, vol. 39, no. 10, pp. 849–851, 1992.

[41] R. L. Geiger, P. E. Allen, and N. R. Strader, *VLSI Design Techniques for Analog and Digital Circuits*, McGraw-Hill, New York, 1990.

[42] J. N. Babenezhad, "Rail to Rail CMOS Op-amp," *IEEE J. Solid-State Circuits*, vol. 23, no. 6, pp. 1414–1417, 1988.

[43] J. Ramírez-Angulo, "Low Voltage, High Slew Rate Bipolar and BiCMOS Op-amp Architectures with Rail to Rail Common Mode Input and Output Swing Capabilities," *1994 IEEE Int. Symp. Circuits Systems*, May 30–June 2, London, England, pp. 743–746.

[44] J. Ramírez-Angulo, E. Sánchez-Sinencio, and M. Robinson, "Current Mode Continuous Time Filters: Two Design Approaches," *IEEE Trans. Circuits Systems*, vol. 39, no. 6, pp. 337–341, 1992.

[45] S. S. Lee, R. H. Zele, D. J. Allstot, and G. Liang, "A Continuous-Time Current-Mode Integrator," *IEEE Trans. Circuits Systems*, vol. 38, pp. 1236–1238, 1991.

[46] J. Ramírez-Angulo and E. Sánchez-Sinencio, "Two Approaches for VLSI Current-Mode Filters Based on Canonical Control Structures," *1994 IEEE Int. Symp. Circuits Systems*, May 30–June 2, London, England, pp. 669–672.

[47] B. Gilbert, "Translinear Circuits: A Proposed Classification," *Electron. Lett.*, vol. 11, pp. 14–16, 1975.

[48] J. Ramírez-Angulo, "Wide Range Programmable Class AB Current Mirrors for Low Supply Operation," *IEEE Trans. Circuits Systems*, vol. 41, no. 9, pp. 631–634, 1994.

[49] J. T. Nabitch, E. Sánchez-Sincencio, and J. Ramírez-Angulo, "A Programmable 1.8–18 MHz, High Q Fully Differential Continuous Time Filter with 1.5 V–2 V Power Supply," *1994 IEEE Int. Symp. Circuits Systems*, May 30–June 2, London, England, 653–657.

[50] J. Ramírez-Angulo and Sun Ming-Chen, "The Folded Gilbert Cell: A Low-Voltage, High Performance CMOS Multiplier," in *Proceedings of the Thirty-Fifth Midwest Symposium on Circuits and Systems*, M. Polis (ed.), IEEE, Piscataway, NJ, 1992, pp. 20–23.

[51] J. Ramírez-Angulo, "Highly Linear Four Quadrant BiCMOS Analogue Multiplier," *Electron. Lett.*, vol. 28, no. 19, pp. 1783–178, 1992.

[52] J. Ramírez-Angulo and A. Diaz-Sánchez, "Low Voltage Programmable Analog FIR Filters Using Voltage Followers and Multipliers," *Proc. 1993 IEEE Int. Conf. Circuits Systems*, May 5–7, 1993, Chicago, IL, pp. 1408–1411.

[53] J. Ramírez-Angulo, A. de Luca, and H. Pérez-Meana, "Continuous Time High Frequency Programmable Nonrecursive Filters," in *Proceedings of the Thirty-Sixth Midwest Symposium on Circuits and Systems*, R. W. Newcomb, B. Gerber, and M. E. Zaghloul (eds.), IEEE, Piscataway, NJ, pp. 153–156, 1993.

[54] J. Ramírez-Angulo and E. Sánchez-Sinencio, "BiCMOS High Frequency Linear Bandpass-Based Current Mode Oscillator for 3 V Supply Oscillator," paper presented at the 1993 European Conference on Circuit Theory and Design (EECTD '93), August 30–September 3, 1993, Davos, Switzerland, pp. 1119–1132.

[55] S. Dommaraju, "Low Voltage Highly Linear BiCMOS OTA Architectures," M.Sc. Thesis, Department of Electrical and Computer Engineering, New Mexico State University, March 1995.

[56] E. Sanchez-Sinencio, R. L. Geiger, and H. Nevarez-Lozano, "Generation of Continuous-Time Two Integrator Loop OTA Filter Structures," *IEEE Trans. Circuits Systems*, vol. 35, pp. 936–946, 1988.

[57] J. Ramírez-Angulo, E. Sánchez-Sinencio, and A. Rodriguez-Vazquez, "A Piecewise Linear Function Approximation Using Current Mode Circuits," *Proc. 1992 IEEE Int. Conf. Circuits Systems*, May 10–13, San Diego, CA, pp. 2021–2024.

[58] J. Ramírez-Angulo, "A ±0.75 V BiCMOS Multiplier with Rail–Rail Input Signal Swing," *1996 IEEE Int. Symp. Circuits Systems*, May 8–12, 1996, Atlanta, GA, pp. 242–245.

[59] J. Ramírez-Angulo, "Ultracompact Low-Voltage Analog CMOS Multiplier Using Multiple Input Gate Transistors," *1996 European Solid State Circuits Conference*, September 17–19, 1996, Neuchâtel, Switzerland.

[60] J. Ramírez-Angulo, G. Gonzalez-Altamirano, and S. C. Choi, "Low Voltage OTA Architectures Using Multiple Input Floating-Gate Transistors," *IEEE Trans. Circuits Systems*, vol. 4, no. 12, pp. 971–974, 1995.

Roelof F. Wassenaar,
Sander L. J. Gierkink,
Remco J. Wiegerink,
and Jacob H. Botma
*MESA Research Institute,
Twente University, 7500AE
Enschede, The Netherlands*

Chapter 8

Low-Voltage CMOS Operational Amplifiers

8.1. INTRODUCTION

To obtain the maximum dynamic range at low supply voltages, it is often necessary to have a rail-to-rail output voltage range. For opamps connected in a unity-gain configuration this implies that the input stage should have a rail-to-rail common-mode input voltage range; otherwise the maximum output swing will be restricted by the input stage.

In this chapter several rail-to-rail input and output stages will be discussed. The input stages have a transconductance which is independent of the common-mode input voltage. This is an important property, because the input transconductance is related not only to the unity-gain frequency of the amplifier but also to the position of the second pole (for a given capacitive load and required phase margin): The maximum value of the unity-gain frequency defines the minimum value of the second pole [1]. For example, if the input transconductance is not constant but has a maximum value which is twice the minimum value, the second pole also has to have twice the value that would otherwise be required. This strongly deteriorates the ratio between the (minimum) unity-gain frequency and the power consumption of the opamp, because in a two-stage opamp the second pole is usually defined by the output stage, which also has the largest power consumption.

The output stages presented in this chapter are based on a common-source configuration in order to obtain a rail-to-rail output voltage range. The output stages have class AB controlled output transistors to obtain a low quiescent power consumption. Two types of class AB control are discussed. In the first type, class AB operation is realized by using a local feedback loop. This type of output stage operates at very low supply voltages; however, due to the feedback loop, it is less suitable for high-speed applications. Furthermore, if a rail-to-rail input stage is used, the minimum required supply voltage will usually be defined by this stage and not by the output stage. Therefore, in the second type of output stage the feedback loop is omitted at the expense of a slightly higher supply voltage.

In the next section the input stages will be discussed. First, in Section 8.2.1, some input stages operating in weak inversion will be presented. Next, in Section 8.2.2, input stages operating in strong inversion will be considered. The output stages are discussed in Section 8.3. Finally, in Section 8.4 it will be shown how an input stage and output stage can be combined to form a complete low-voltage operational amplifier.

8.2. RAIL-TO-RAIL CONSTANT-g_m INPUT STAGES

A well-known technique for realizing an input stage with a rail-to-rail common-mode input range is to place an N- and P-type differential pair in parallel, as indicated in Figure 8.1(a) [1–5]. In this circuit transistor M_{10}, which provides the tail current for the N-type differential pair, requires a minimum drain-source voltage to conduct current. This determines the lower boundary of the operation range of the N-type metal-oxide-semiconductor (NMOS) differential pair. The upper boundary is determined by the supply voltage and the fact that the gate voltage of a transistor may exceed the drain voltage by only a threshold voltage before it brings the transistor out of saturation. Therefore, the N-type differential pair can operate

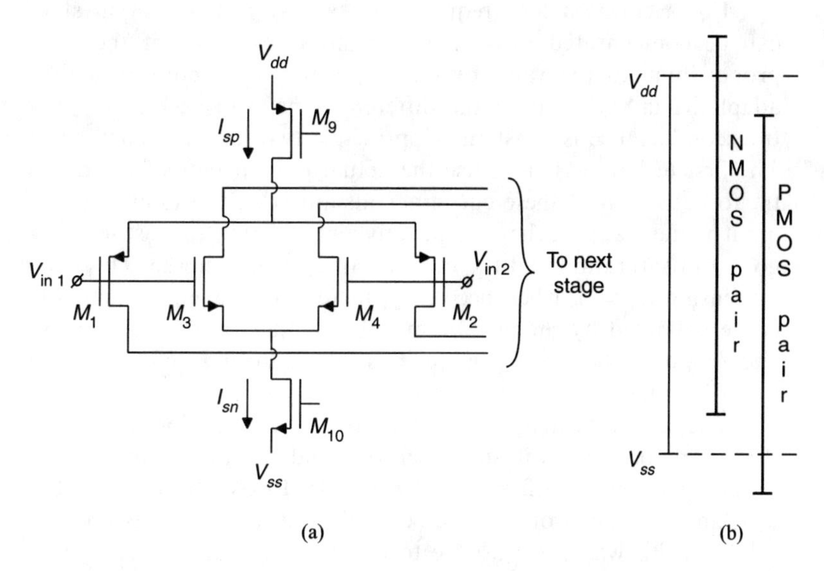

Figure 8.1 (a) NMOS and PMOS differential pair driven in parallel and (b) voltage ranges over which differential pairs operate.

from a certain common-mode input voltage above the negative supply rail up to a certain common-mode input voltage above the positive supply rail, as indicated in Figure 8.1(b). For the PMOS differential pair the opposite is true. Thus, applying the two complementary input pairs in parallel results in an input stage which has a common-mode input range that can even exceed the supply rails, provided that adequate circuitry is available for combining the output signals of the individual differential pairs.

Three different operation areas can be distinguished:

- The common-mode input voltage is near the negative supply rail; signal transfer will take place only by the P-type differential pair.
- The common-mode input voltage is in a region somewhere in the middle between the supply voltages; both the NMOS and the PMOS differential pairs will be active.
- The common-mode input voltage is near the positive supply rail; signal transfer will take place only by the N-type differential pair.

It is clear that without precautions the transconductance g_m of the combination strongly depends on the common-mode input level because only in a region somewhere in the middle of the supply voltages will both differential pairs be active. A nonconstant (i.e., a common-mode input-voltage-dependent) transconductance g_m is undesirable, because a variation in g_m causes a variation in the unity-gain bandwidth of the amplifier and prevents an optimal frequency compensation. This reduces the feasible unity-gain bandwidth of the compensated amplifier. Furthermore, a common-mode input-voltage-dependent transconductance causes extra distortion.

For a constant low-frequency voltage gain of the input stage, a variable g_m can be compensated by scaling the input resistance of the next stage inversely proportional to the value of g_m [6]. However, a more attractive approach is to adapt the tail currents of the differential pairs in such a way that the combined transconductance is constant. In principle two techniques exist to accomplish this. The first technique is to sense the actual tail currents of the differential pairs and try to keep each of these currents constant [7]. For example, if the common-mode input voltage approaches the negative supply rail, transistor M_{10} [Fig. 8.1(a)] will go out of saturation and the tail current I_{sn} will decrease. This can be compensated by sensing I_{sn} and, when necessary, increasing the gate voltage of M_{10}. This technique is limited by the maximum voltages that can be applied to the gates of M_9 and M_{10} and, therefore, it is not possible to keep g_m constant over the entire rail-to-rail common-mode input range.

The second technique is to accept the fact that close to the supply rails one of the differential pairs will stop operating and in that case increase the tail current of the complementary differential pair in order to keep the combined transconductance constant. In other words, increase I_{sn} if I_{sp} drops below its nominal value and vice versa. In this way it is possible to obtain a constant g_m over the entire rail-to-rail input range. A problem is the accuracy that can be reached in forcing the correct relation between I_{sn} and I_{sp}. Furthermore, this relation is dependent on the mode of operation of the differential input pairs. The weak-inversion mode is attractive when minimal offset voltage is important, while the strong-inversion mode must be used when high speed is required. These two cases will be considered separately in Sections 8.2.1 and 8.2.2, respectively.

8.2.1. Rail-to-Rail Constant-g_m Input Stages Operating in the Weak-Inversion Domain

In the weak-inversion mode, the transconductance of a single differential pair is proportional to its tail current [8]:

$$g_m = \frac{I_s}{2nU_T} \tag{8.1}$$

where I_s is the tail current, $n = (C_{\mathrm{ox}} + C_{\mathrm{depl}})/C_{\mathrm{ox}}$ (the weak-inversion slope factor [8]), and $U_T = kT/q$.

Thus, a constant combined transconductance is obtained if

$$\frac{I_{sn}}{n_n} + \frac{I_{sp}}{n_p} = \text{const} \tag{8.2}$$

Assuming equal slope factors for N- and P-type devices, this reduces to

$$I_{sn} + I_{sp} = \text{const} \tag{8.3}$$

Thus, the sum of the transconductances of the N- and P-type differential pairs will be constant if the sum of the tail currents is kept constant. A decrease in one of the tail currents has to be compensated by an equal increase of the other tail current.

Rail-to-Rail Constant-g_m Input Stage Using a Simple Current Mirror. Condition (8.3) is also valid for input stages employing bipolar transistors. Therefore, existing bipolar implementations can also be used in complementary MOS (CMOS). Figure 8.2 shows such an input stage, directly translated from a well-known bipolar version [1]. In this circuit the current $2I_0$ through M_0 is divided between the P-type differential pair and transistor M_5. The part of $2I_0$ flowing through M_5 is mirrored by M_6 and M_7 and used as tail current for the N-type differential pair. In this way the sum of the tail currents is always equal to $2I_0$. The gate of transistor M_5 is connected to a constant bias voltage. Therefore, at low common-mode input voltages, $2I_0$ will flow completely through the P-type differential pair. Transistor M_5 and the N-type differential pair will be turned off. When the common-mode input voltage increases, an increasing part of the current will flow through M_5 and the P-type differential pair will be gradually turned off while the N-type pair is turned on. For common-mode input voltages close to the

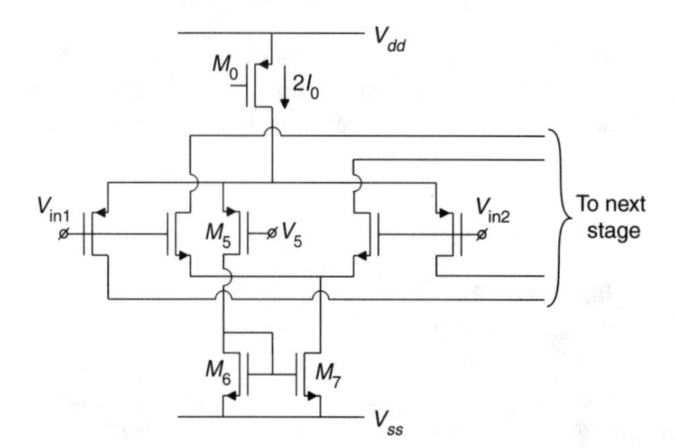

Figure 8.2 Rail-to-rail input stage operating in weak inversion and having constant transconductance.

positive supply rail $2I_0$ will flow completely through M_5 and only the N-type differential pair will operate.

A problem which occurs in all input stages consisting of complementary differential pairs connected in parallel is the non-constant-offset voltage. In the take-over range, the common-mode input voltage range, in which one differential pair takes over the transconductance from the other pair, the offset of the amplifier changes from the offset of one differential pair to the offset of the other differential pair. This significantly reduces the common-mode rejection ratio (CMRR), because the changing offset is in fact a common-mode-dependent differential input signal. To minimize the reduction of the CMRR, the take-over range between the differential pairs should be as large as possible, limited by the (low) supply voltage. In the circuit of Figure 8.2, the take-over range can be increased by giving M_5 a relatively small aspect ratio W/L. It is also possible to use a resistor in series with M_5 [4].

A difference in the slope factors of the NMOS and PMOS transistors can be compensated by adapting the current transfer ratio of current mirror M_6, M_7.

Rail-to-Rail Constant-g_m Input Stage Using Improved Wilson Current Mirrors. Another possibility to realize a rail-to-rail input stage having a constant g_m is shown in Figure 8.3 [5]. Again, a single current source $2I_0$ is divided between the two differential pairs and the sum of the tail currents always remains equal to $2I_0$. The circuit is based on the strong feedback that occurs in a Wilson current mirror when the voltage at the output drives the output transistor out of its normal operating range. In that case also the input current of the current mirror decreases and it remains the same as the output current.

When all current mirrors in Figure 8.3 are operating normally, the current $2I_0$ is divided equally between current mirrors CM-I and CM-II and both differential pairs receive a tail current equal to I_0. For common-mode input voltages approaching the positive supply rail, current mirror CM-I will not be able to provide enough current for the P-type differential pair. However, the input current of CM-I will decrease by the same amount, leaving more current for the N-type differential pair. In the same way, for common-mode input voltages close to the negative supply rail, current

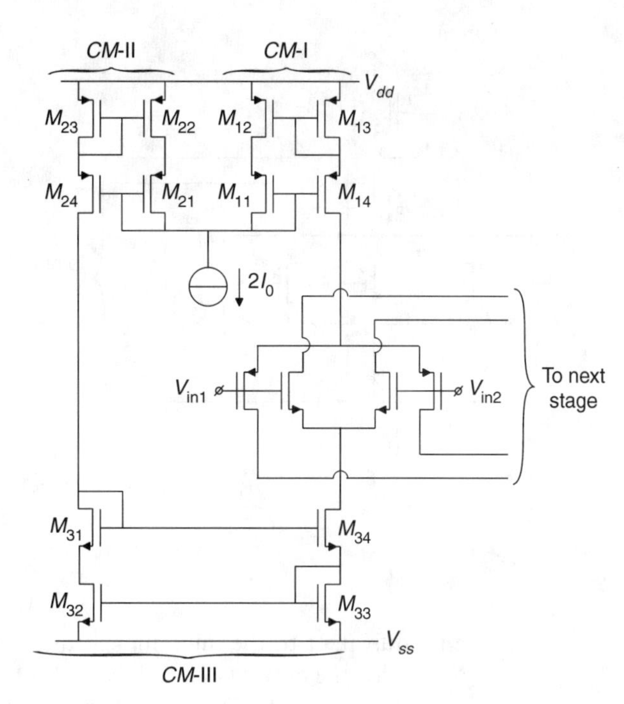

Figure 8.3 Rail-to-rail input stage operating in weak inversion and having a constant transconductance using Wilson current mirrors.

mirror CM-III will provide less current to the N-type differential pair and more of $2I_0$ will flow through the P-type differential pair.

Major drawbacks of the circuit are the large number of additional transistors that are needed and the fact that the minimum required supply voltage is relatively large. The supply voltage must be at least large enough to allow for a common-mode input voltage range where both input differential pairs operate at a tail current I_0. Compared to the circuit in Figure 8.2 the minimum required supply voltage has increased by two gate-source voltages necessary for M_{13} and M_{33}. A simpler solution which requires less supply voltage is discussed in the next section.

Simple Rail-to-Rail Constant-g_m Input Stage Using a Minimum-Current Circuit. The input stages discussed above need one or more current mirrors to force the sum of the tail currents to be constant. In this section, a more direct means of adapting the tail currents is presented which results in an improved high-frequency behavior. Furthermore, this approach does not affect the minimum required supply voltage.

Figure 8.4 shows the principle of the input stage. The sum of the tail currents is kept constant by means of a single additional current source I_x connected between the common source nodes of the two differential pairs. The value of I_x should be equal to half the minimum of the currents I_N and I_P (and is therefore dependent on the common-mode input voltage):

$$I_x = \tfrac{1}{2}\min(I_N, I_P) \tag{8.4}$$

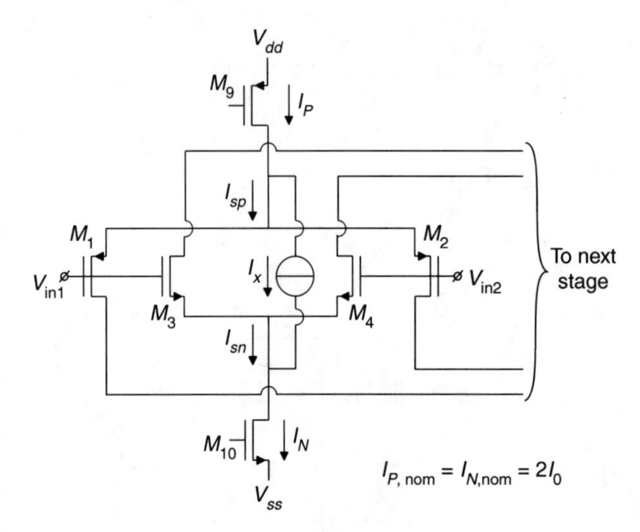

Figure 8.4 Scheme of simple rail-to-rail input stage operating in weak inversion and having constant transconductance.

which is equivalent to the minimum of the available tail currents I_{sn} and I_{sp}.

Normally, the currents I_N and I_P are both constant and equal to $2I_0$, resulting in a value of I_x equal to I_0. The tail currents of the input differential pairs are then also equal to I_0. When the common-mode input voltage approaches one of the supply rails, either M_9 or M_{10} will go out of saturation and, consequently, provide less current.

The tail currents of the input differential pairs are given by

$$I_{sn} = I_N - I_x = I_N - \tfrac{1}{2}\min(I_N, I_P) \tag{8.5}$$

and

$$I_{sp} = I_P - I_x = I_P - \tfrac{1}{2}\min(I_N, I_P) \tag{8.6}$$

Thus, the sum of the available tail currents is equal to

$$I_{sn} + I_{sp} = I_N + I_P - \min(I_N, I_P) = \max(I_N, I_P) = 2I_0 \tag{8.7}$$

which is constant. Therefore, a decrease in one of the tail currents due to M_9 or M_{10} going out of saturation does not influence the sum of the tail currents $I_{sn} + I_{sp}$. A decrease of I_{sn} due to a decrease of I_N is exactly compensated by the same decrease of I_x, which causes I_{sp} to increase by the same amount. Vice versa, a decrease of I_{sp} is compensated by an increase of I_{sn}.

As mentioned before, the current I_x must be equal to half the minimum of the currents I_N and I_P or, equivalently, equal to the minimum of the available tail currents I_{sn} and I_{sp}. To obtain this minimum current, we use the principle indicated in Figure 8.5. If a CMOS current sink and current source are connected in series, the

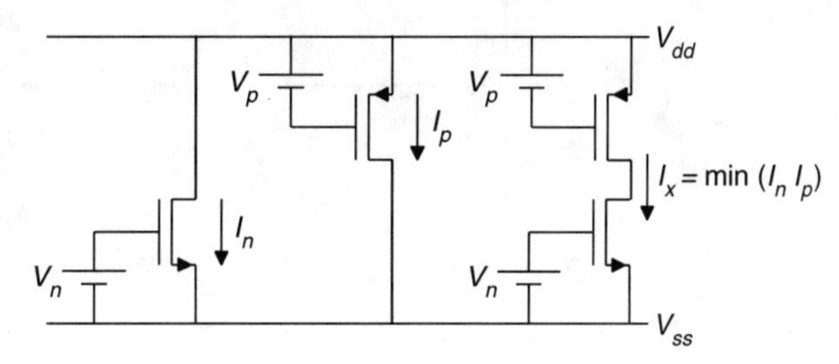

Figure 8.5 Principle of minimum current selection. If a CMOS current sink and current source are connected in series, the smallest current dominates, because the transistor which would normally carry the largest current is forced into the nonsaturated region.

smallest current will dominate. The transistor which would normally carry the largest current is forced out of saturation.

The complete constant-g_m input stage is shown in Figure 8.6. In this circuit both differential pairs are realized twofold. The combination M_5/M_6 senses the tail current I_{sp} of the P-Type differential pair, and transistors M_7/M_8 sense the tail current I_{sn} of the N-type differential pair. As in Figure 8.5 the current source M_5/M_6 and the current sink M_7/M_8 are connected in series. The result is that only the smallest current can flow. The combination M_5, M_6, M_7, and M_8 therefore acts as the current source I_x in Figure 8.4. Note that the minimum current principle indicated in Figure 8.5 does not apply to bipolar transistors and, therefore, this input stage cannot be implemented by bipolar transistors.

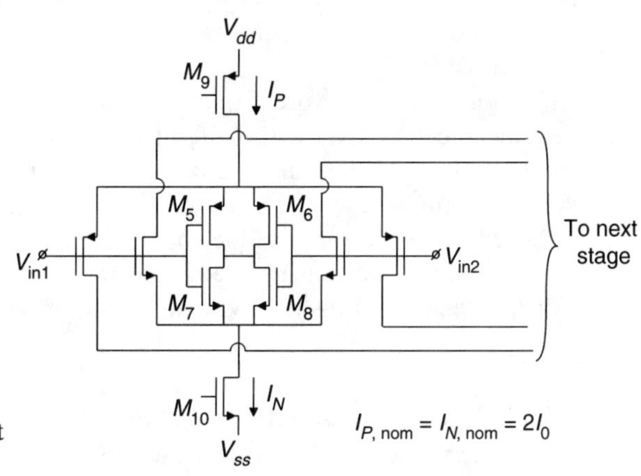

Figure 8.6 Complete constant-g_m input state operating in weak inversion.

The circuit in Figure 8.6 has been simulated with PSpice using MOS transistors with the following threshold voltages: $V_{Tp} = -0.77\,\text{V}$ and $V_{Tn} = 0.63\,\text{V}$. All transistors were 10 µm long. The PMOS transistors had a channel width of 100 µm and the NMOS transistors had a width of 50 µm. The simulations indicate a minimum required supply voltage of 1.4 V. Figure 8.7 shows the simulated tail currents I_{sn} and I_{sp} as well as the sum of these currents as a function of the common-mode input

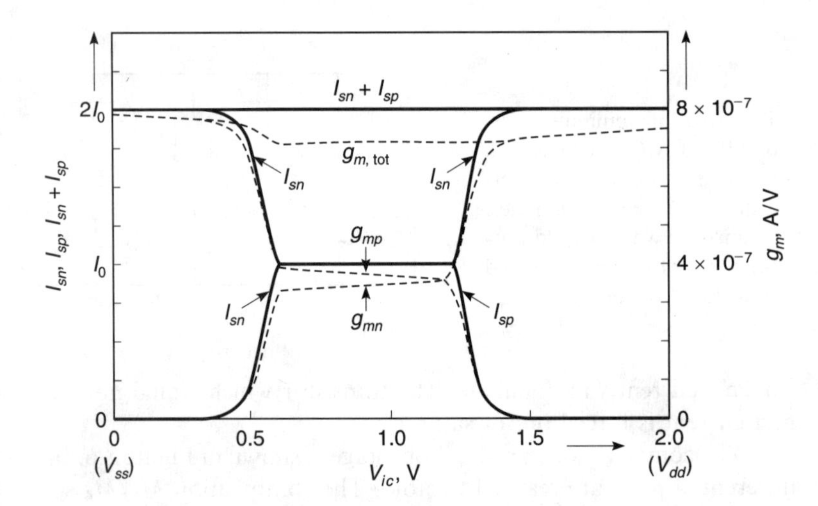

Figure 8.7 Simulated effective tail currents I_{sn} and I_{sp} and their sum for circuit in Figure 8.6 as function of common-mode input voltage. Dashed lines indicate resulting small-signal transconductances. Supply voltage was 2 V and I_0 was 25 nA.

voltage V_{ic}. The maximum variation of the sum current is less than 0.25%. The resulting combined transconductance is also shown in Figure 8.7 (dashed line) and has a maximum deviation from its nominal value of 5%. This relatively large deviation is due to the different slope factors [Eq. (8.2)] for N- and P-type transistors. Furthermore, the slope factor is also dependent on the source-bulk biasing voltage of the transistors.

As mentioned before, offset differences between the two input differential pairs degrade the CMRR. In the circuit of Figure 8.6, there is a common-mode range where both input pairs have a tail current equal to I_0 and when the common-mode voltage approaches either one of the supply rails one of the differential pairs is switched off quite rapidly due to M_9 or M_{10} leaving saturation. The CMRR may be improved by decreasing the tail currents of the differential pairs more gradually. This can be accomplished by simply adding two transistors, as indicated in Figure 8.8.

8.2.2. Rail-to-Rail Constant-g_m Input Stages Operating in the Strong-Inversion Domain

In the strong-inversion mode, the transconductance of a single differential pair is proportional to the square root of its tail current:

$$g_m = \sqrt{KI_s} \tag{8.8}$$

where I_s is the tail current and $K = \mu C_{ox}(W/L)$.

Thus, a constant combined tranconductance is obtained if

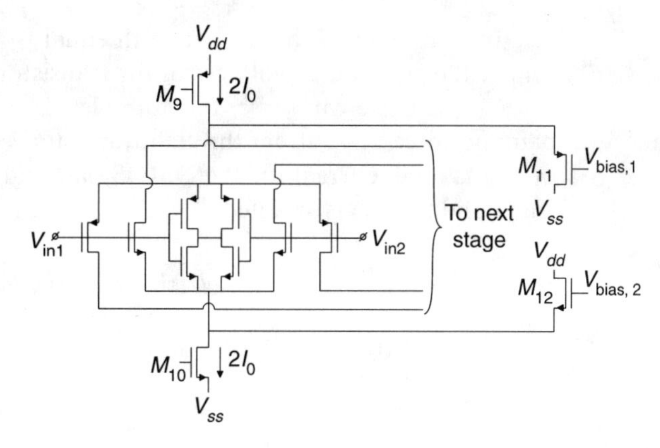

Figure 8.8 CMRR degradation due to offset differences between input differential pairs can be reduced by turning off input pairs more gradually when common-mode input voltage approaches a supply rail. In this circuit this is realized by adding M_{11} and M_{12}, which gradually reduce the available tail currents.

$$\sqrt{K_n I_{sn}} + \sqrt{K_p I_{sp}} = \text{const} \tag{8.9}$$

which is dependent on the factors K_n and K_p. Implementing (8.9) in an input stage results in complex circuits [9, 10], because of the necessity to create currents proportional to the ratio K_n/K_p. Furthermore, the accuracy obtained by these circuits is not very good because of relatively large errors due to mismatch. Therefore, we prefer to scale the input differential pair to have $K_n \approx K_p$. Then, condition (8.9) reduces to

$$\sqrt{I_{sn}} + \sqrt{I_{sp}} = \text{const} \tag{8.10}$$

which is much easier to realize.

Figure 8.9 shows a simple circuit that realizes relation (8.10). Transistors M_1, M_2, M_3, and M_4 are connected in a so-called MOS translinear loop [11, 12]. Applying Kirchhoff's voltage law, the following relation between the drain currents can be derived (assuming equal values of K for all transistors and neglecting the body effect):

$$\sqrt{I_{M_1}} + \sqrt{I_{M_2}} = \sqrt{I_{M_3}} + \sqrt{I_{M_4}} \tag{8.11}$$

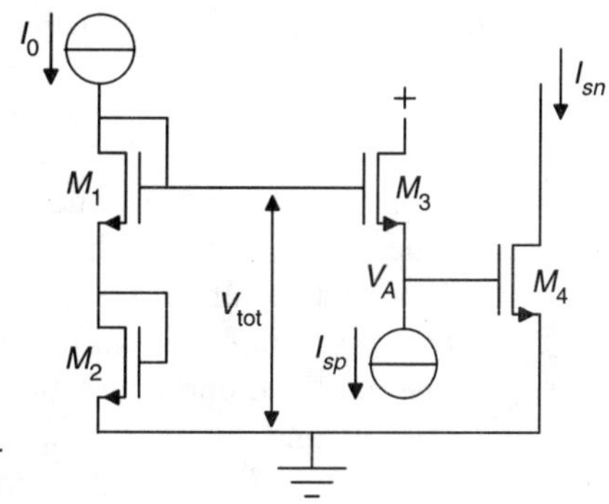

Figure 8.9 Simple circuit implementating relation (8.10).

In Figure 8.9, I_1 and I_2 are both equal to a constant current I_0. Therefore the sum of the gate-source voltages of the transistors M_3 and M_4 is constant and will be called V_{tot}. If we can sense, for example, the tail current of the PMOS differential pair and force this current through transistor M_3, then we can use the drain current of M_4 as a tail current for the NMOS pair and the condition for a constant overall transconductance is satisifed:

$$\sqrt{I_{sn}} + \sqrt{I_{sp}} = 2\sqrt{I_0} = \text{const} \tag{8.12}$$

Figure 8.10 shows the circuit of such an implementation.

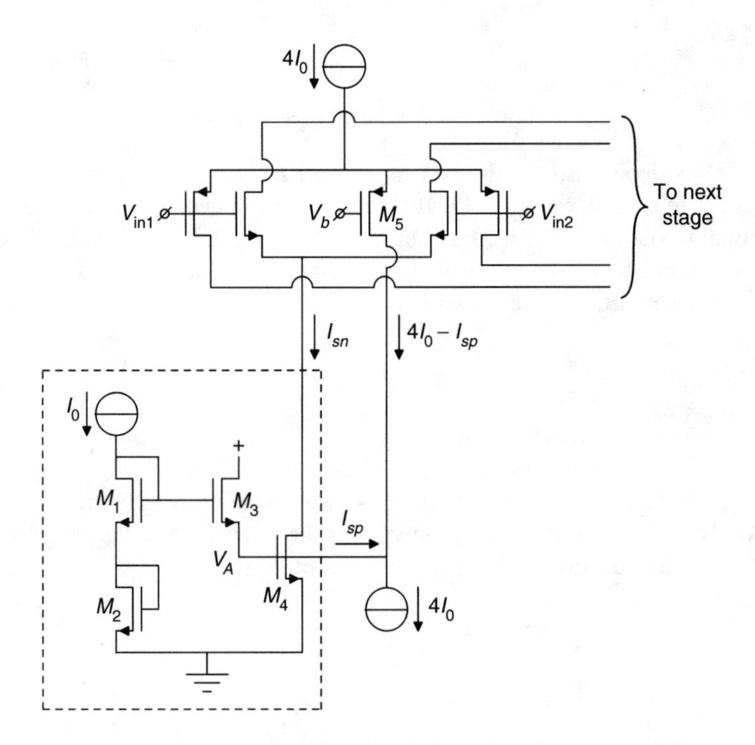

Figure 8.10 Constant-g_m input stage based on the circuit in Figure 8.9.

The relation between the currents I_{sn} and I_{sp} resulting from the circuit in Figure 8.9 can also be explained graphically, as indicated in Figure 8.11(a). In this figure, the drain currents of M_3 and M_4 are both plotted as a function of the voltage V_A at the gate of M_4. The line indicating I_{sn} is simply the transfer characteristic of M_4. The line indicating I_{sp} is the mirrored characteristic of M_3, because the sum of the gate-source voltages is constant and equal to V_{tot}. We can now derive the value of I_{sn} resulting from a certain value of I_{sp} as indicated by the dashed line.

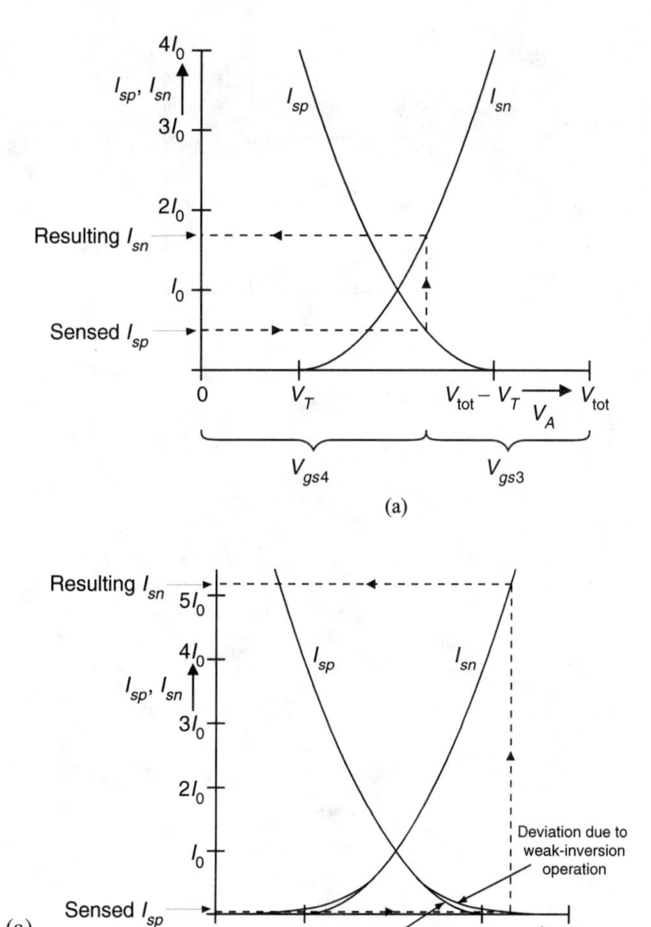

Figure 8.11 Graphical analysis of (a) circuit in Figure 8.9 and (b) deviation due to weak-inversion operation of transistor M_3 at small values of I_{sp}.

As long as M_3 and M_4 have an ideal square-law characteristic, there is no problem. However, a problem arises when the tail current I_{sp} becomes very small. Then, transistor M_3 will leave the strong-inversion domain and start to operate in weak inversion, resulting in a much lower gate-source voltage than expected. Because the sum of the gate-source voltages of M_3 and M_4 is constant, the gate-source voltage of M_4 will now become too large. As a result the current I_{sn} and also the overall transconductance of the complete input stage will be too large. This is illustrated in Figure 8.11(b).

A better solution has been found by using the current difference $I_{sn} - I_{sp}$ as the input current of the MOS translinear circuit instead of the current I_{sp} and at the same time limiting the absolute value of $I_{sn} - I_{sp}$ to $4I_0$, which can easily be achieved. Figure 8.12(a) shows how the translinear circuit in Figure 8.9 can be adapted for

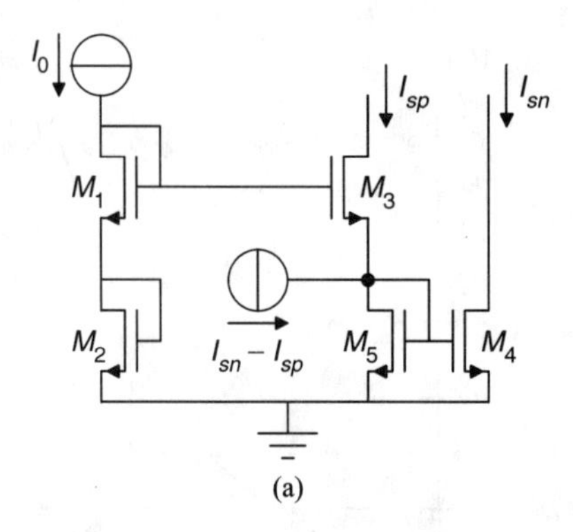

(a)

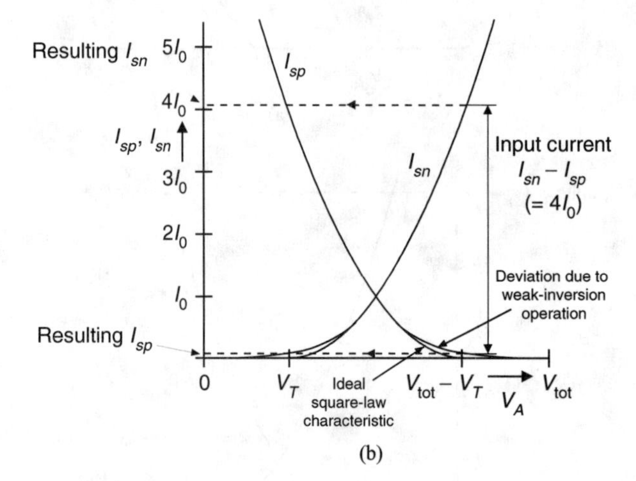

(b)

Figure 8.12 (a) Improved translinear circuit which is less sensitive to weak-inversion operation. (b) Graphical analysis shows that for small values of I_{sp}, I_{sn} is only slightly larger than $4I_0$.

receiving input current $I_{sn} - I_{sp}$. Both I_{sn} and I_{sp} are now output currents of the translinear circuit and are used as tail currents for the input differential pairs. The error introduced by weak-inversion operation is reduced significantly because the difference $I_{sn} - I_{sp}$ cannot become larger than $4I_0$, as illustrated graphically in Figure 8.12(b).

A drawback of the circuit in Figure 8.12(a) is the influence of the body effect due to the different source voltages of the transistors in the translinear loop. Therefore, the equivalent circuit shown in Figure 8.13(a) is used. The influence of the body effect is reduced significantly because the sources of the transistors in the translinear loop $M_1, \ldots, M_4$ are now pairwise at the same potential [12, 13]. The current source forcing the difference between I_{sn} and I_{sp} can easily be implemented by a differential pair M_5, M_6, as indicated in Figure 8.13(b). Limiting the absolute value of the difference $I_{sn} - I_{sp}$ to $4I_0$ is now easily accomplished by making the tail current of M_5, M_6 equal to $4I_0$.

Figure 8.14 shows the complete input stage with the additional differential pair M_5, M_6 providing the input current for the translinear loop. This additional differential pair is driven single ended by the common-mode input voltage of the input

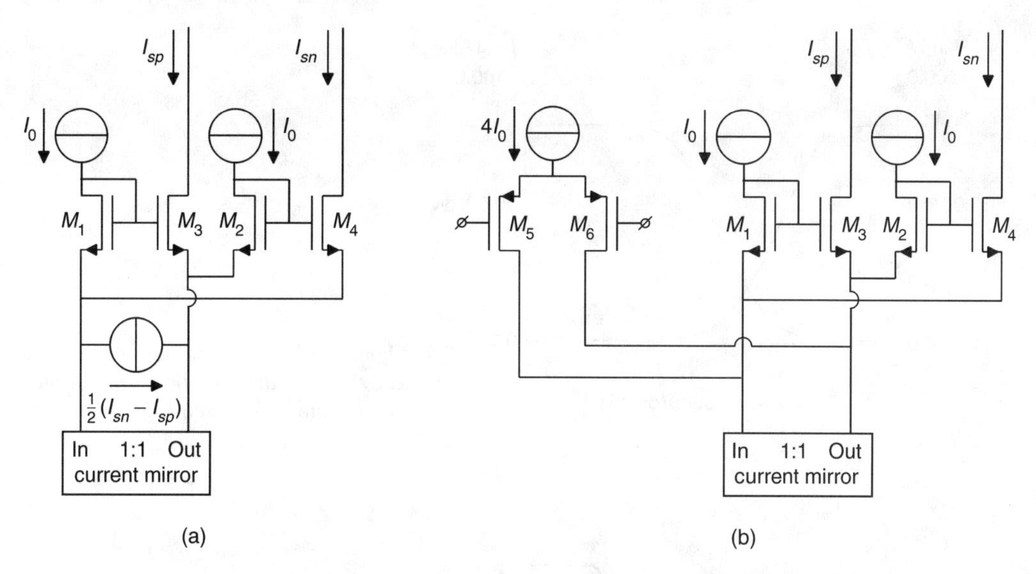

Figure 8.13 (a) Translinear circuit equivalent to Figure 8.12(a) but less sensitive to body effect because transistors in translinear loop have their sources pairwise connected to same potential. (b) Current source forcing difference between I_{sn} and I_{sp} can easily be implemented by a differential pair.

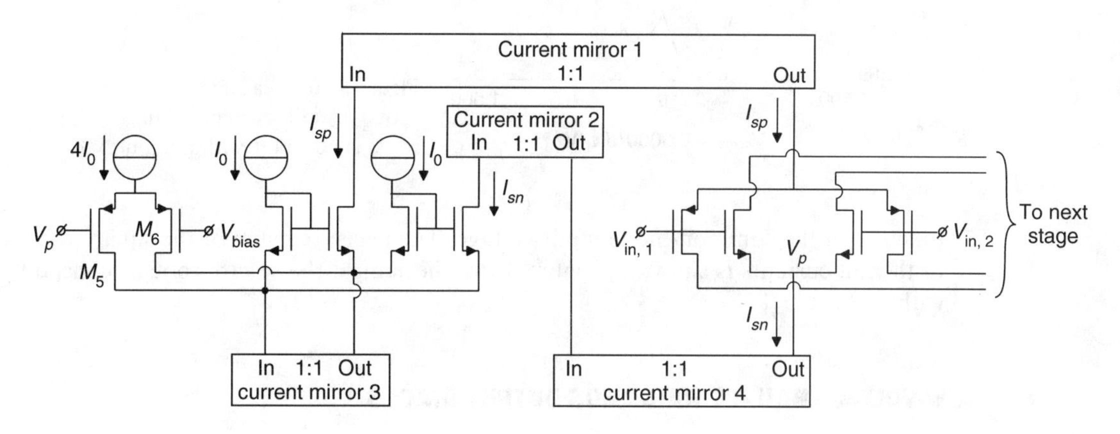

Figure 8.14 Complete constant-g_m input stage operating in strong inversion.

stage by connecting the gate of M_5 to the common-source node of the NMOS input differential pair. The output currents of the MOS translinear loop ($M_1, \ldots, M_4$) are fed to the input differential pairs with the help of some current mirrors. The combination of current mirrors 2 and 4 is not strictly necessary: They are only inserted to reduce the minimum required supply voltage.

The complete input stage, as shown in Figure 8.14, was integrated in a semi-custom CMOS process. Figure 8.15 shows the measured tail currents I_{sn} and I_{sp} as a

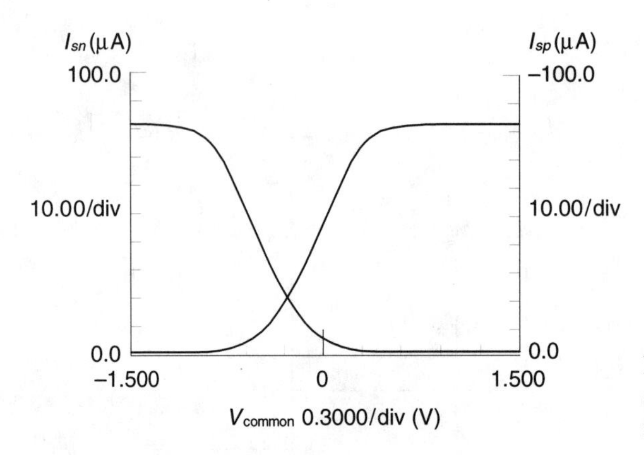

Figure 8.15 Measured tail currents I_{sn} and I_{sp} as function of common-mode input voltage for circuit of Figure 8.14.

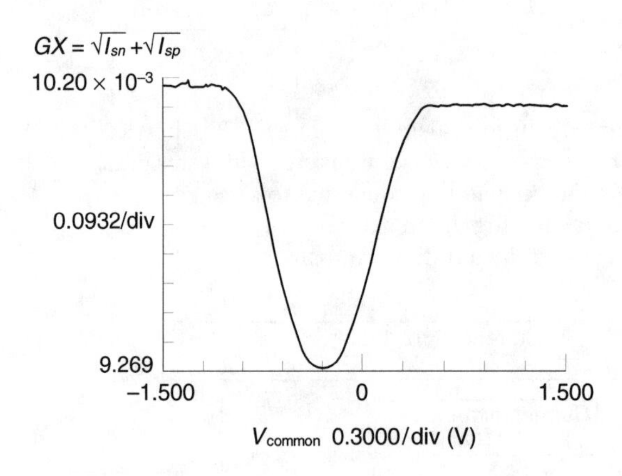

Figure 8.16 Measured sum of square roots of tail currents. Sum of square roots is constant within about 5%.

function of the common-mode input voltage. The measured sum of the square roots of the tail currents is shown in Figure 8.16. The sum of the square roots is constant within about 5%.

8.3. LOW-VOLTAGE RAIL-TO-RAIL CMOS OUTPUT STAGES

The output stage is an important part of the operational amplifier. It must be able to drive the load impedance of the opamp without disturbing the unloaded performance and without introducing unnecessary distortion. The output stage is usually the most power-consuming stage of the amplifier, and therefore special care should be taken to make it operate as power efficiently as possible.

Basically, there are two transistor configurations suitable to be used as an output stage. First we have the source-follower topology shown in Figure 8.17. The gain of an ideal source-follower stage approaches a value of 1. However, in practice it is somewhat smaller, mainly due to the body effect and the presence of a load resistance. The circuit features a low output impedance, but it has a relatively small

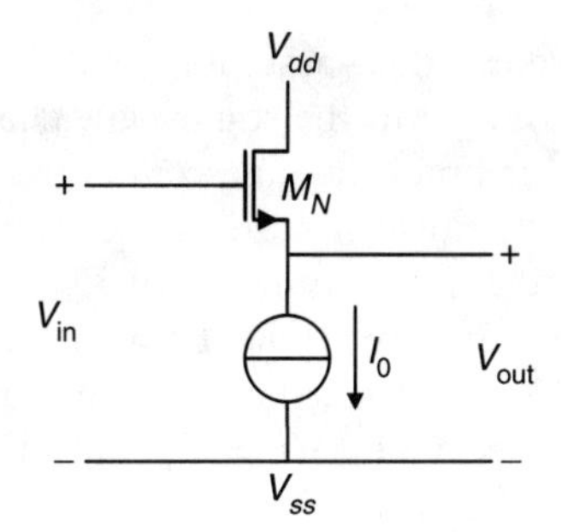

Figure 8.17 Simple source-follower stage.

output voltage swing capability. The latter is due to the fact that usually V_{in} cannot exceed the supply voltage V_{dd} and therefore V_{out} is limited to V_{dd} minus the gate-source voltage of the transistor. This limitation will be worsened by the body effect, which causes the gate-source voltage to increase with increasing V_{in}. Due to the low output swing, the source-follower is unsuitable for low-voltage applications.

A much larger output swing is offered by the second type of output stage: the common-source topology indicated in Figure 8.18(a). Contrary to the source-follower, it has a high output impedance and can provide some voltage gain, depending on the transconductance and the load impedance. The voltage gain decreases with smaller load impedances. Its large output swing capability is what makes this configuration attractive, especially at low supply voltages. For this reason, in this chapter we will only consider the common-source configuration.

The maximum output current of the common-source stage in Figure 8.18(a) is limited and equal to the bias current I_0. Therefore, in this basic circuit I_0 has to be relatively large, resulting in a large quiescent power consumption. For this reason it is better to use the complementary circuit of Figure 8.18(b). In this circuit both transistors are driven (in phase) by the input voltage. When the effective gate-source voltage $V_{gs} - V_t$ of M_N reaches zero, the effective gate-source voltage of M_P will be

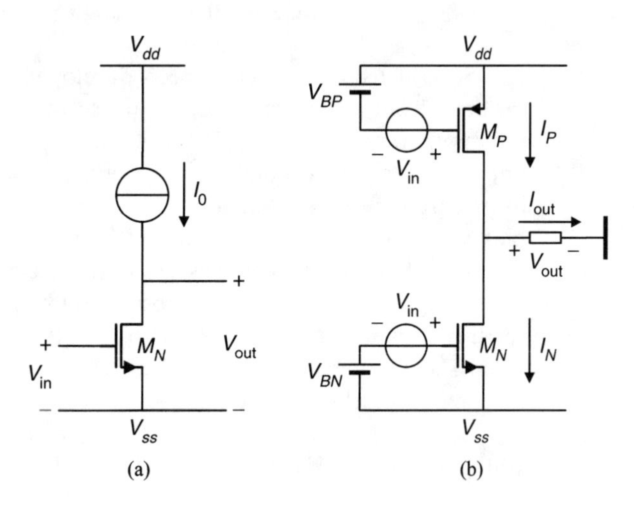

Figure 8.18 (a) Common-source output configuration and (b) complementary version.

twice the quiescent value, assuming that the transistors have equal transconductances ($K_N = K_P$). Due to the square-law character of the devices, a maximum output current of four times the quiescent current of M_P and M_N can now be obtained while both output transistors are still conducting currents. Figure 8.19(a) shows a plot of I_N, I_P, and I_{out} as a function of V_{in}. Figure 8.19(b) shows I_P as a function of I_N. Note that the relation between I_N and I_P can be described by $\sqrt{I_N} + \sqrt{I_P} = \sqrt{2K}V_0$ and that, in the indicated interval in Figure 8.19(a), $I_{out} = -2KV_0V_{in}$, assuming $K_N = K_P$. When the input voltage V_{in} exceeds the interval indicated in Figure 8.19(a), one of the transistors M_N or M_P will turn off, but the current through the other transistor will continue to increase.

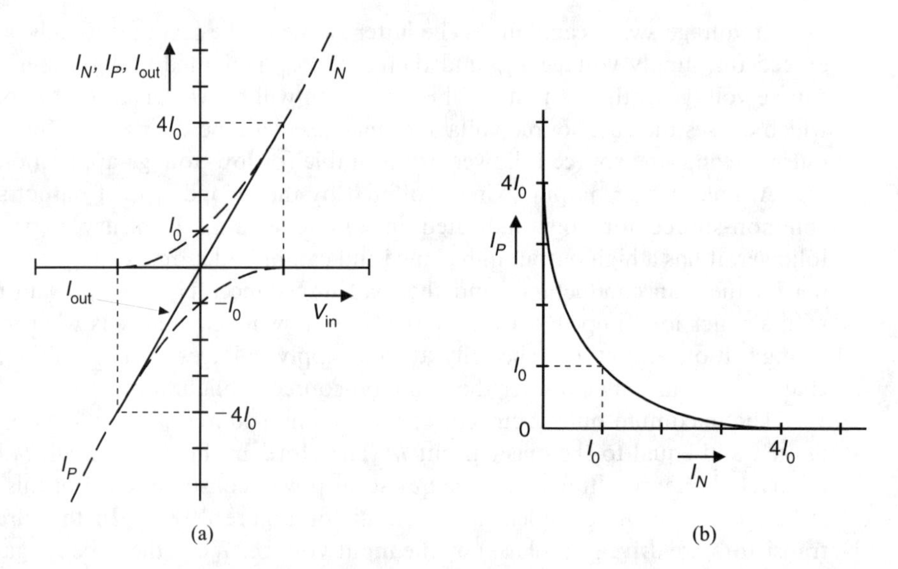

Figure 8.19 I_N, I_P, and I_{out} in the complementary common-source stage of Figure 8.18(b) as a function of (a) V_{in} and (b) I_P plotted as a function of I_N.

Several techniques have been developed for increasing the ratio between the maximum output current and the quiescent bias current and still keeping both output transistors conducting [13–17]. The latter is not essential; however, it slightly improves the step response and sometimes eases the design of the stage driving the output transistors (see Section 8.3.2). A way to accomplish this is by controlling the bias voltages V_{BN} and V_{BP} in such a way that the effective gate-source driving voltages of the output transistors never become smaller than a certain minimum value. Figure 8.20(a) shows a possible plot of V_{BN} and V_{BP} as a function of V_{in} for this situation. Note that these bias voltages have a differential-mode character: when V_{BN} is increasing, V_{BP} is decreasing and vice versa. Figure 8.20(b) shows the corresponding class AB relation between I_N and I_P.

In the next sections, two methods to implement the variation of the biasing voltages V_{BN} and V_{BP} in a circuit will be discussed. In Section 8.3.1 a circuit will be

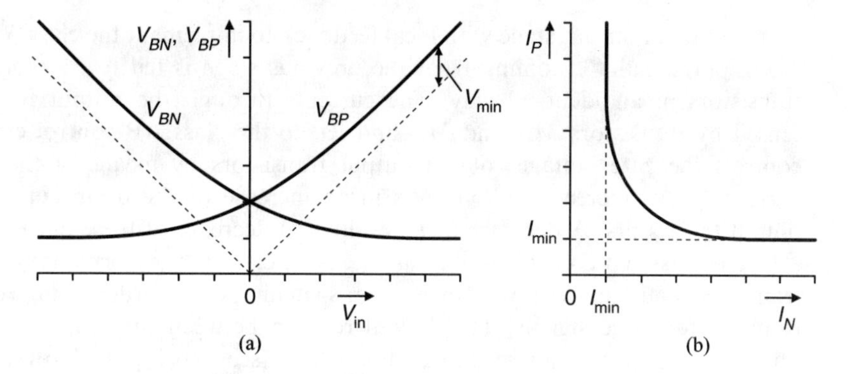

Figure 8.20 (a) Bias voltages V_{BN} and V_{BP} can be adapted as a function of input signal V_{in} (b) to obtain class AB current control.

presented where class AB operation is obtained with the help of a local feedback circuit. This feedback circuit senses the values of I_N and I_P and a deviation from the curve in Figure 8.20(b) is corrected by a feedback current. This circuit features very low voltage operation. However, the feedback loop limits the bandwidth of the amplifier. If a large bandwidth is important, the circuit discussed in Section 8.3.2 is more suitable. This circuit does not use a feedback loop to obtain class AB behavior; however, it requires a slightly higher supply voltage.

8.3.1. Low-Voltage Rail-to-Rail Output Stage with Feedback Class AB Control

As mentioned previously, the two output transistors in a complementary common-source output stage have to be driven in phase by the input signal, while the class AB control signal has a differential character. Figure 8.21 shows the basic

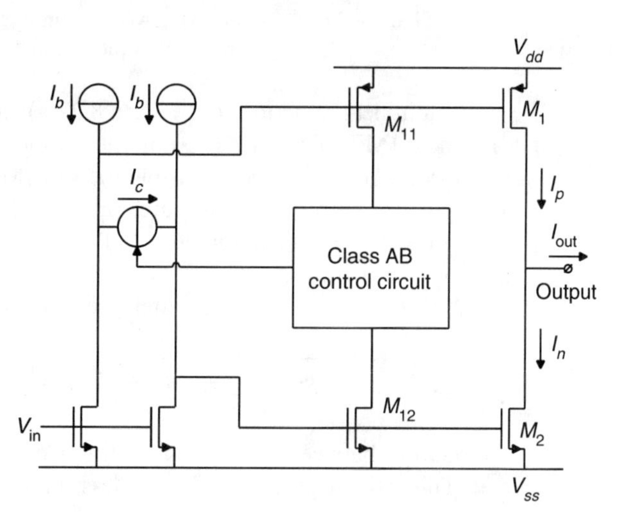

Figure 8.21 Basic rail-to-rail output stage with feedback class AB control circuit.

structure of an output stage with local feedback to implement the class AB behavior. The input signal V_{in} coming from the previous stage is fed to each of the output transistors in an identical way. The currents through the output transistors are sensed by transistors M_{11} and M_{12} and fed to the class AB control circuit, which controls the gate voltages of the output transistors by means of the differential current I_c. An increase of I_c results in an increase of the drain currents of both output transistors. A decrease of I_c results in a decrease of these currents.

The class AB control circuit has to ensure class AB performance of the output transistors and has to prevent them from switching off in order to improve the step response for large signals. The desired relation between the drain currents of the output transistors is indicated in Figure 8.22(a) [see also Fig. 8.20(b)]. If one of the drain currents becomes very large, the other transistor still conducts a minimum current I_{min}. Figure 8.22(b) shows the same relation, but now plotted as a function of I_{out} $(=I_p - I_n)$. An important property of the output stage is the ratio I_{bias}/I_{min}, in which I_{bias} represents the quiescent drain current of the output transistors. This ratio was chosen to be approximately equal to 2.

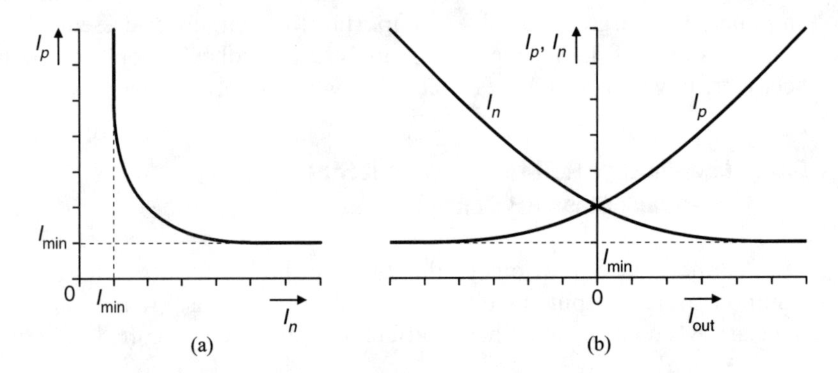

Figure 8.22 (a) Relation between drain currents of output transistors and (b) drain currents plotted as function of output current.

A nonlinear function like Figure 8.22(a) can easily be implemented by an MOS translinear (MTL) circuit. The nonlinear function must then be written in a sum-of-square-roots form to be compatible with the MTL loop equations [12, 13]. This was done with the help of the computer program MTLPLOT [12], which was developed as an interactive design tool for MTL circuits. Note that the exact shape of the function is not important.

With the help of the computer program the following equation is easily found:

$$\sqrt{I_n - \tfrac{1}{2}I_{min}} + \sqrt{I_p - \tfrac{1}{2}I_{min}} = \sqrt{I_n + I_p - I_{min}} + \sqrt{\tfrac{1}{2}I_{min}} \tag{8.13}$$

The relation between I_n and I_p resulting from (8.13) is plotted in Figure 8.23 (solid line). The ratio I_{bias}/I_{min}, is slightly less than 2.

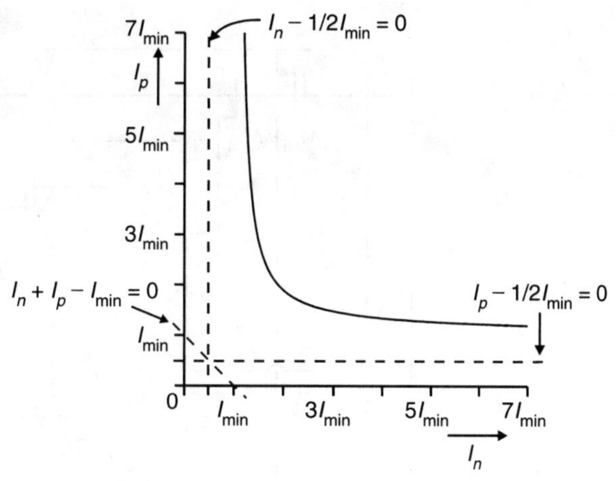

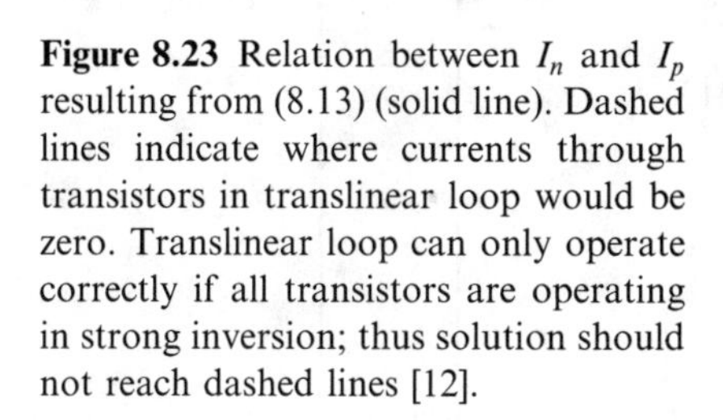
Figure 8.23 Relation between I_n and I_p resulting from (8.13) (solid line). Dashed lines indicate where currents through transistors in translinear loop would be zero. Translinear loop can only operate correctly if all transistors are operating in strong inversion; thus solution should not reach dashed lines [12].

Equation (8.13) can now be implemented by an MTL circuit. A possible implementation is shown in Figure 8.24. In this circuit, transistors M_{13} and M_{14} sense the value of I_p and M_9 senses the value of I_n. The translinear loop is formed by M_3, M_4, M_7, and M_8. Transistors M_3 and M_7 correspond to the left-hand side of (8.13). The drain current of M_7 is made equal to $I_n - \frac{1}{2}I_{\min}$ and the drain current of M_3 is equal to $I_p - \frac{1}{2}I_{\min}$. The drain currents of the transistors connected in the opposite direction in the loop (M_4 and M_8) are forced equal to $I_n + I_p - I_{\min}$ and $\frac{1}{2}I_{\min}$.

If (8.13) is satisfied, the translinear loop M_3, M_4, M_7, M_8 will be in equilibrium and there will be no voltage difference between the inputs of the differential pair M_5, M_6. If the equilibrium is disturbed, this will result in a differential input voltage across this differential pair. This, in turn, results in a differential current between the gates of the output transistors M_1 and M_2 and the drain currents of these transistors will be adjusted until the equilibrium state defined by (8.13) is reached. Due to the high loop gain, the differential input voltage of differential pair M_5, M_6 will always be approximately zero. The value of the bias voltage V_b should be approximately 0.5 V to ensure that all current sources operate correctly.

The output stage of Figure 8.24 was realized as part of a complete operational amplifier in a semicustom CMOS process [13]. Since the minimum available transistor channel length was 10 μm, the unity-gain bandwidth of this opamp was very low (approximately 550 Khz). However, simulations indicate that a unity-gain bandwidth of several megahertz is feasible if channel lengths of 2.5 μm are used. Correct operation of the class AB control circuit was verified by measuring the drain currents of the output transistors as a function of the opamp's output current. For this measurement the amplifier was connected in a unity-gain configuration, as indicated in Figure 8.25. The input was connected to a constant bias voltage equal to half the supply voltage. As a result, due to the unity-gain feedback, the output voltage will also remain fixed at half the supply voltage and both output transistors will operate in saturation. The output current can be adjusted by changing I_L. The sources of the output transistors were connected to the supply externally in order to

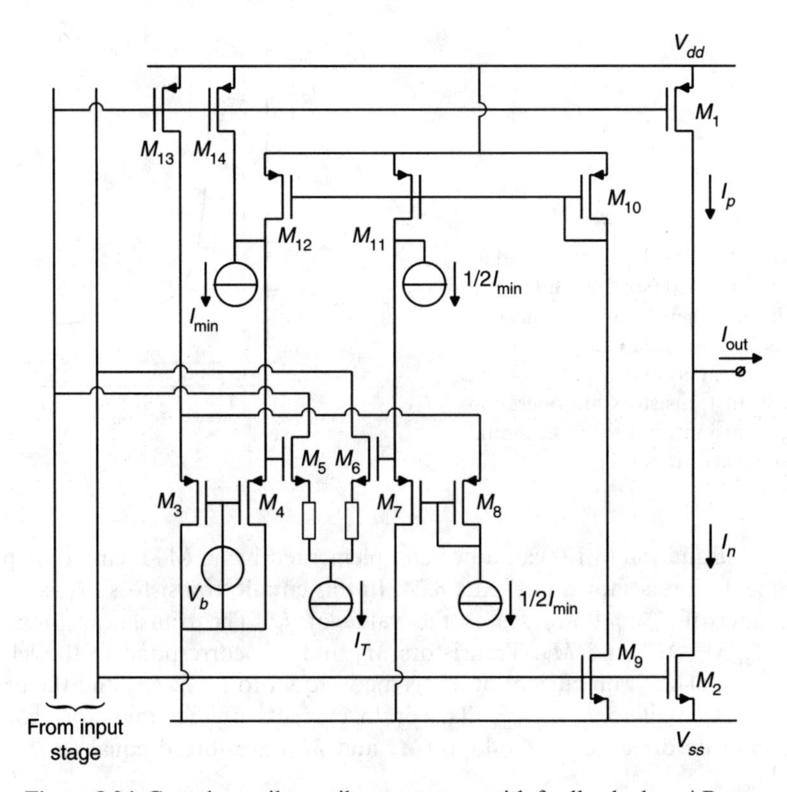

Figure 8.24 Complete rail-to-rail output stage with feedback class AB control.

be able to measure the currents I_n and I_p. Figure 8.26 shows the measured relation between these currents. This relation corresponds almost exactly to the theoretical relation given by (8.13) (see Figure 8.23).

A drawback of class AB control based on a local feedback loop is that additional phase shift introduced by the feedback loop reduces the phase margin of the complete amplifier. Furthermore, the class AB control circuit should have at least the

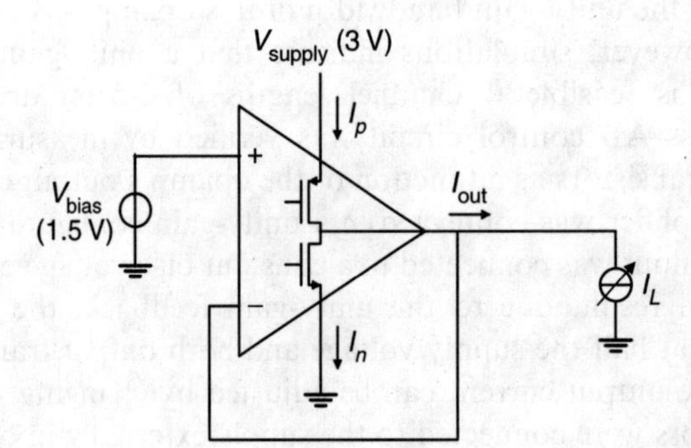

Figure 8.25 Measurement setup for measuring the relation between I_n and I_p.

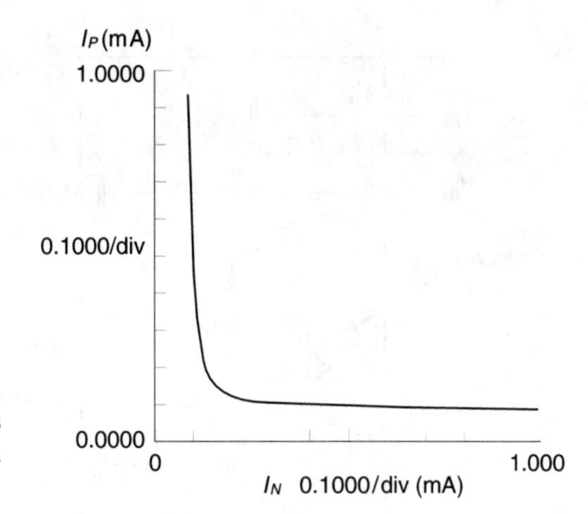

Figure 8.26 Measured relation between I_n and I_p. Measured relation corresponds almost exactly to theoretical relation resulting from (8.13).

same bandwidth as the complete amplifier. This large bandwidth combined with the large class AB loop gain (due to the high impedance level at the gates of the output transistors) could cause stability problems in the class AB control loop itself. This is why the resistors in series with the sources of M_5 and M_6 were added: They reduce the loop gain in order to keep the loop stable at the expense of a reduced accuracy of the class AB relationship. In the next section a class AB output stage is described which does not need a local feedback loop. This output stage might be preferred if a large bandwidth is required and when it is not necessary to operate at an extremely low supply voltage.

8.3.2. Low-Voltage Rail-to-Rail Class AB Output Stage without Local Feedback

In the output stage described in the previous section, each output transistor is driven by a separate cascode stage. This is illustrated in Figure 8.27(a). The class AB control current is indicated by the dashed circle. In this section, an output stage is discussed where both output transistors are driven by the same cascode stage, as shown in Figure 8.27(b). Now the dashed circle indicates a variable voltage source to obtain class AB behavior. In Figure 8.27(a), the class AB control has to provide a differential current between the two cascode stages to obtain the desired low quiescent current. In the previous section, it was shown that it is in fact possible to design the class AB circuit in such a way that neither of the output transistors ever cuts off completely. This possibility, which was optional in Figure 8.27(a), is indispensable in Figure 8.27(b). If in Figure 8.27(b) one of the output transistors is allowed to turn off, this will force the corresponding half of the cascode stage to operate in the triode region. This will in turn severely reduce the gain which is obtained in the cascode stage and thus reduce the gain of the entire amplifier. This effect does not occur in Figure 8.27(a), because only the cascode stage connected to the output transistor that is turned off has a reduced gain.

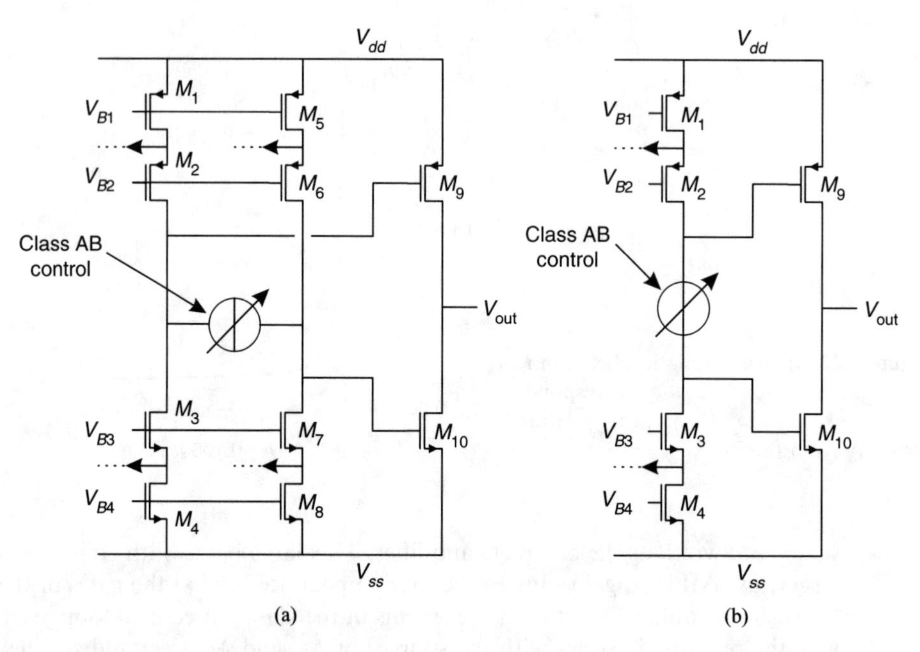

Figure 8.27 (a) Separate cascode stage for each output transistor and (b) both output transistors driven by same stage.

Figure 8.28 shows a simple output stage of the type indicated in Figure 8.27(b) which prevents the output transistors from cutting off [16]. In the quiescent situation ($I_{out} = 0$), the drain currents of the output transistors M_4 and M_8 are both equal to nI_b. This is accomplished by giving M_1 and M_5 twice the aspect ratio W/L of M_3 and M_7, respectively. In that case, in the quiescent situation the gate-source voltages of M_1 and M_3 and of M_5 and M_7 will be equal and, thus, the gate-source voltages of the output transistors are equal to the gate-source voltages of M_2 and M_6.

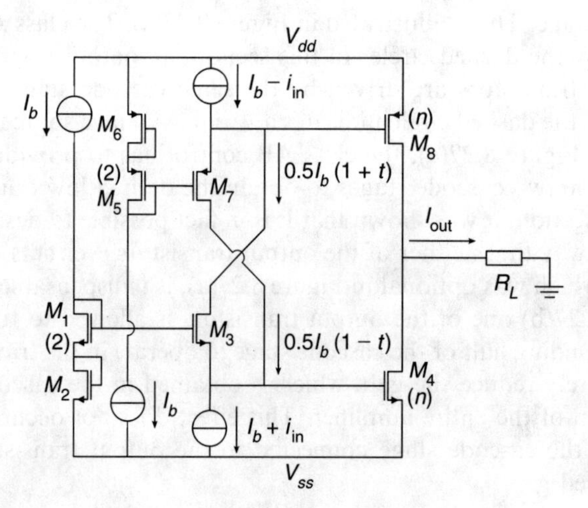

Figure 8.28 Simple output stage based on Figure 8.27(b).

The relation between I_n and I_p for $I_{out} \neq 0$ can be derived as follows. First the relation between the effective gate-source voltages of the output transistors $V_x - V_{gs4} - V_t$ and $V_y = V_{gs8} - V_t$ will be examined. For the translinear loops $M_1, \ldots, M_4$ and $M_5, \ldots, M_8$, we can write, respectively,

$$V_x - m\sqrt{\frac{I_b}{K}} = -\sqrt{\frac{I_b}{K}}\sqrt{1+t} \tag{8.14}$$

$$V_y - m\sqrt{\frac{I_b}{K}} = -\sqrt{\frac{I_b}{K}}\sqrt{1-t} \tag{8.15}$$

with $m = 1 + \sqrt{2}$, which results after squaring and subtracting in the following circle equation between V_x and V_y:

$$\left(V_x - m\sqrt{\frac{I_b}{K}}\right)^2 + \left(V_y - m\sqrt{\frac{I_b}{K}}\right)^2 = \frac{I_b}{K} \tag{8.16}$$

Figure 8.29(a) shows a plot of this relation. Since $V_x = \sqrt{2I_n/nK}$ and $V_y = \sqrt{2I_p/nK}$, the relation between I_n and I_p can now be constructed. Figure 8.29(b) shows the result.

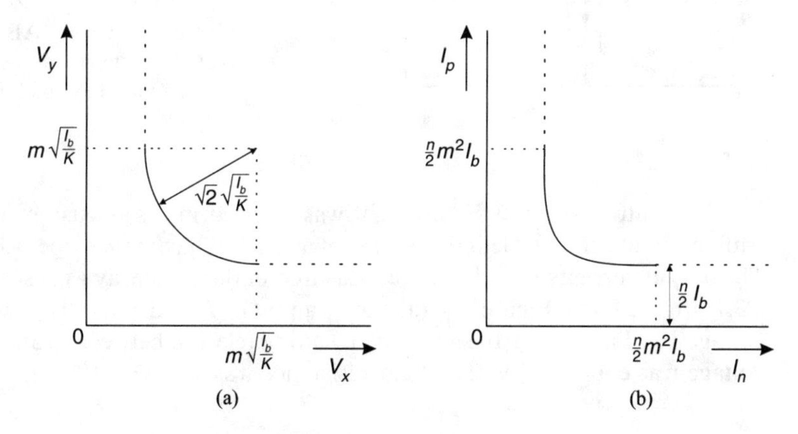

Figure 8.29 Relation between (a) effective gate-source voltages of output transistors and between (b) output currents I_n and I_p.

This relation is only valid under quasi-static conditions, when the input signal i_{in} from the previous stage is small. Figure 8.30 shows how i_{in} can be inserted in a practical circuit. In this figure M_9 and M_{12} provide constant currents I_b. The signal current i_{in} is inserted at the sources of the cascode transistors M_{10} and M_{11}. Usually, the value I_b is chosen as small as possible to minimize its contribution to the offset and noise of the amplifier. Therefore, I_b is chosen equal to the maximum possible value of i_{in} which is necessary to ensure that the currents through M_{10} and M_{11} always remain positive.

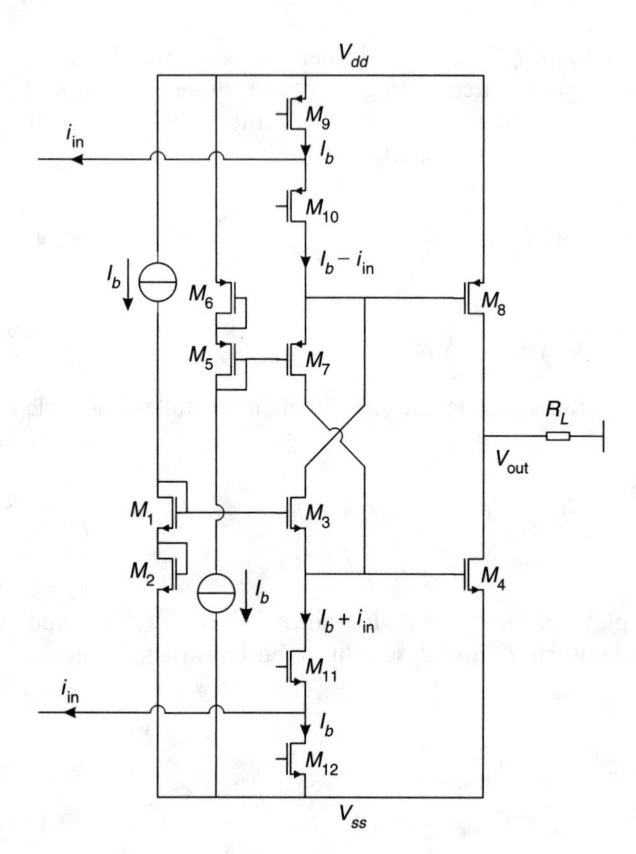

Figure 8.30 Class AB output stage. Note that under quasi-static conditions, only an infinitely small input signal i_{in} is required.

The output stage of Figure 8.30 was realized in a semicustom CMOS process with minimum channel lengths of 2.5 μm as part of a complete operational amplifier. The output currents I_n and I_p were measured in the same way as described in Section 8.3.1 (Figure 8.25). Figure 8.31(a) shows a plot of I_n and I_p as a function of the input voltage V_{in}. Figure 8.31(b) shows the resulting relation between I_n and I_p. The supply voltage was equal to 3 V. The load resistance was 1.1 kΩ.

8.4. COMPLETE RAIL-TO-RAIL INPUT AND OUTPUT OPERATIONAL AMPLIFIER

To obtain a complete rail-to-rail operational amplifier, we have to combine a rail-to-rail input stage with a rail-to-rail output stage. The first problem is then how to obtain the differential output current of the input stage without disturbing the rail-to-rail input range. A simple current mirror as indicated in Figure 8.32(a) cannot be used in series with the input differential pairs. The relatively large input voltage of the current mirror V_m imposes a severe restriction on the maximum value of the input voltage V_{in}. Therefore, for a maximum common-mode input voltage range a

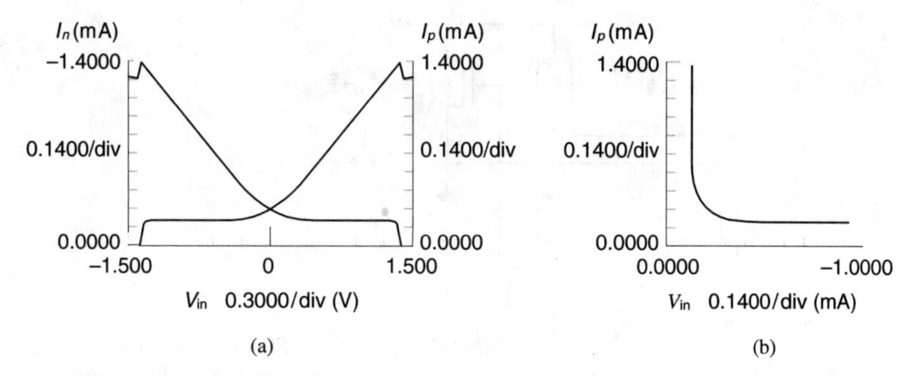

Figure 8.31 (a) Measured output currents I_n and I_p as function of the input voltage V_{in} for amplifier connected in unity-gain configuration. (b) Relation between I_n and I_p corresponds very well to theoretical relation of Figure 8.29(b).

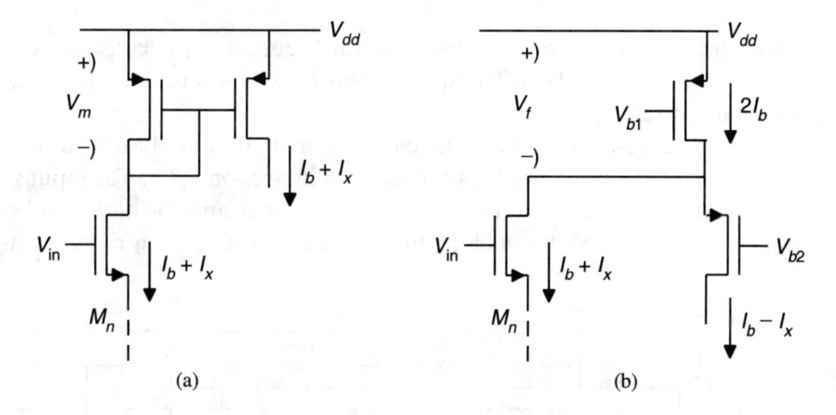

Figure 8.32 (a) At low supply voltages, a simple current mirror cannot be used to obtain output current of input stage because of large voltage drop V_m across current mirror's input transistor, which limits the maximum value of V_{in}. (b) Folded-cascode structure requires a much smaller voltage drop V_f. Now V_{in} can even exceed the supply rail V_{dd}.

folded-cascode structure as indicated in Figure 8.32(b) has to be used. The input voltage V_f of this structure can be much smaller than the input voltage of a current mirror and now V_{in} may even exceed the supply rails. A drawback of the folded-cascode topology is that the transistor providing the constant current $2I_b$ contributes significantly to the offset and noise of the complete amplifier.

A property of all the rail-to-rail input stages discussed in this chapter is that the tail currents of the input differential pairs are dependent on the common-mode input voltage. Figure 8.33 shows a simple circuit based on the folded-cascode topology which can be used to eliminate the tail currents and simultaneously obtain the differential output current of the differential pairs. This current can then be fed to

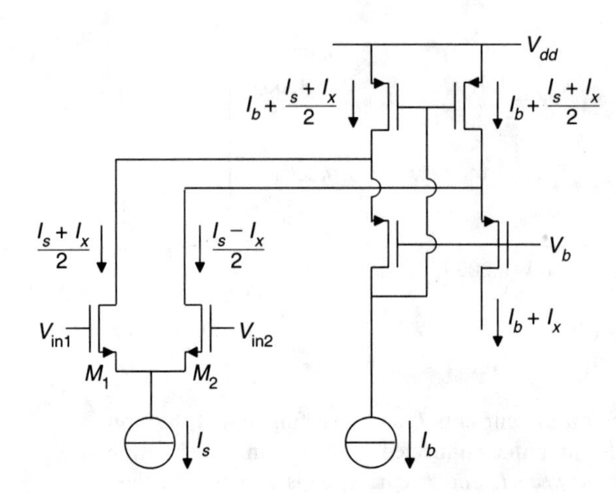

Figure 8.33 Circuit based on folded-cascode configuration of Figure 8.32(b), which can be used to eliminate (variable) tail current I_s of input differential pairs.

the output stage. The bias current I_b (in Figure 8.33) is constant and equal to the maximum value of the differential current I_x, which is usually the maximum value of the tail current I_s.

Figure 8.34 shows a complete operational amplifier based on the input stage of Figure 8.14 and the output stage described in Section 8.3.2. The input stage is slightly changed compared to Figure 8.14. The resistor R and the additional current source I_0 provide a level shift which further reduces the minimum required supply voltage.

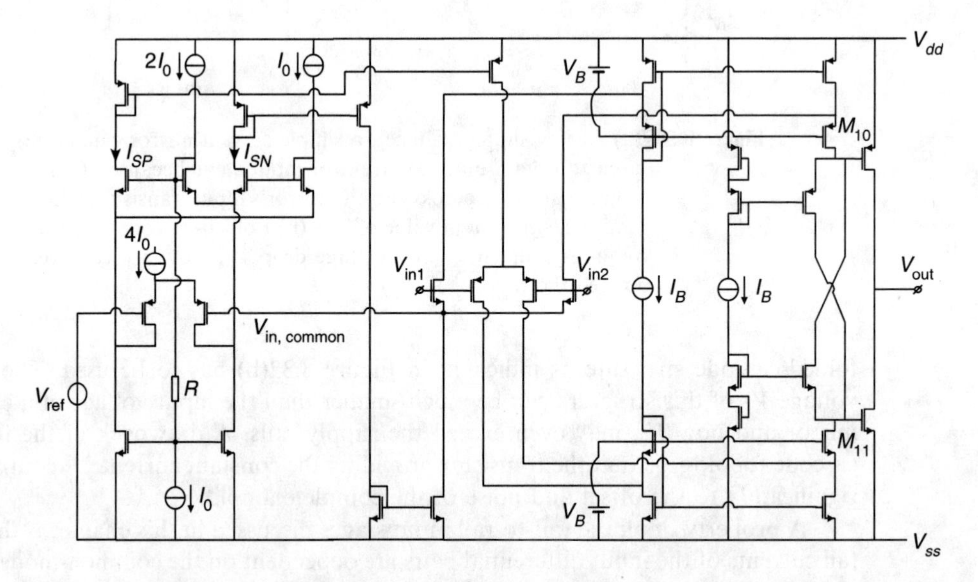

Figure 8.34 Complete operational amplifier based on constant-g_m input stage of Figure 8.14 and class AB output stage described in Section 8.3.2.

Two versions of the operational amplifier have been realized. The only difference between the two versions was the type of frequency compensation used. The first amplifier had a simple Miller compensation, consisting of two capacitors of 1 pF connected between the output and the gates of the output transistors. The second amplifier had 0.5-pF compensation capacitors connected from the output to the sources of transistors M_{10} and M_{11}. In practice, both amplifiers appeared to have comparable frequency characteristics. Figure 8.35 shows the measured open-loop gain and phase characteristics of the amplifier with simple Miller compensation and a load resistance of 1 kΩ. The DC gain is more than 70 dB. The unity-gain bandwidth was 6 MHz, with a phase margin of approximately 50°.

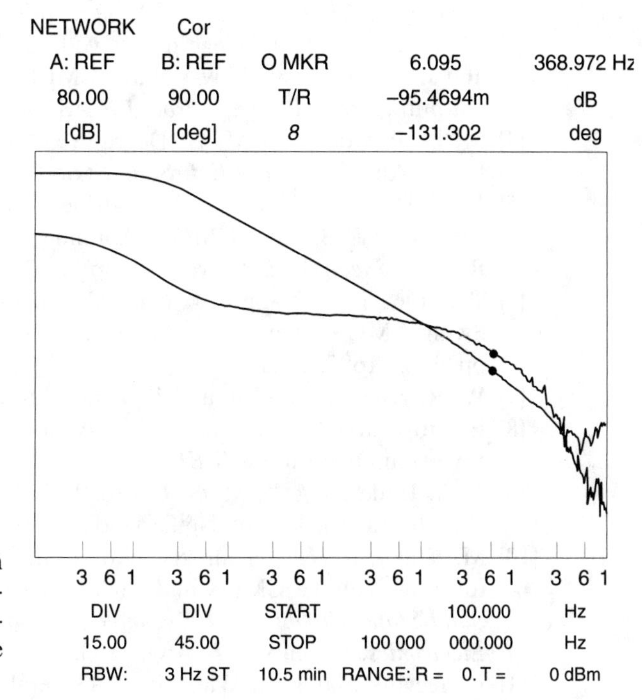

Figure 8.35 Measured open-loop gain and phase characteristics of operational amplifier of Figure 8.34 at supply voltage of 3 V and load resistance of 1 kΩ.

8.5. CONCLUSION

In this chapter several rail-to-rail input and output stages for CMOS operational amplifiers have been discussed, all capable of operating at supply voltages below 3 V (with transistor threshold voltages of approximately 0.7 V). The presented input stages have a transconductance g_m which is independent of the common-mode input voltage. The output stages are based on a common-source circuit topology with class AB controlled output transistors.

Acknowledgments

The authors thank E. A. M. Klumperink for his help with the design of the chips and P. Holzman for providing the measurement results.

References

[1] J. H. Huijsing and D. Linebarger, "Low-Voltage Operational Amplifier with Rail-to-Rail Input and Output Ranges," *IEEE J. Solid-State Circuits*, vol. 20, pp. 1144–1150, 1985.
[2] J. A. Fisher and R. Koch, "A Highly Linear CMOS Buffer Amplifier," *IEEE J. Solid-State Circuits*, vol. SC-22, pp. 330-334, 1987.
[3] R. Hogervorst, R. J. Wiegerink, P. A. L. de Jong, J. Fonderie, R. F. Wassenaar, and J. H. Huijsing, "Low-Voltage CMOS Opamp with Rail-to-Rail Input and Output Voltage Range," *Proc. ISCAS*, pp. 2876–2879, 1992.
[4] M. D. Pardoen and M. G. Degrauwe, "A Rail-to-Rail Input/Output CMOS Power Amplifier," *IEEE J. Solid-State Circuits*, vol. 25, pp. 501–504, 1990.
[5] J. F. Duque-Carrillo, R. Pérez-Aloe, and A. Morillo, "Push-Pull Current Circuit for Biasing CMOS Amplifier with Rail-to-Rail Common-Mode Range," *Electron. Lett.*, vol. 27, pp. 2122–2125, 1991.
[6] T. S. Fiez, H. C. Yang, J. J. Yang, C. Yu, and D. J. Allstot, "A Family of High-Swing CMOS Operational Amplifiers," *IEEE J. Solid-State Circuits*, vol. 24, pp. 1683–1687, 1989.
[7] W. R. Krenik, W. Hsu, and R. Nail, U.S. Patent 4,887048, Jan. 21, 1988.
[8] E. Vittoz and J. Fellrath, "CMOS Analog Integrated Circuits Based on Weak Inversion Operation," *IEEE J. Solid-State Circuits*, vol. 12, pp. 224–231, 1997.
[9] P. A. L. de Jong, "CMOS Rail to Rail Op Amp," M.Sc thesis, University of Twente, Enschede, The Netherlands, 1990.
[10] M. Ismail, R. Wassenaar, S. Sakurai, and R. Wiegerink, "Design Techniques for Low Voltage CMOS and BiCMOS Analog ICs," presented at the *IEEE Solid-State Circuits and Technology Committee Workshop on Low Power Electronics*, Phoenix, AZ, Aug. 1993.
[11] E. Seevink and R. J. Wiegerink, "Generalized Translinear Circuit Principle," *IEEE J. Solid-State Circuits*, vol. 26, pp. 1098–1102, 1991.
[12] R. J. Wiegerink, *Analysis and Synthesis of MOS Translinear Circuits*, Kluwer Academic, Dordrecht, The Netherlands, 1993.
[13] J. H. Botma, R. F. Wassenaar, and R. J. Wiegerink, "A Low-Voltage CMOS Op-amp with a Rail-to-Rail Constant-g_m Input Stage and a Class AB Rail-to-Rail Output Stage," *Proc. ISCAS*, pp. 1314–1317, 1993.
[14] E. Seevinck, W. de Jager, and P. Buitendijk, "A Low-Distortion Output Stage with Improved Stability for Monolithic Power Amplifiers," *IEEE J. Solid-State Circuits*, vol. 23, pp. 794–801, 1988.
[15] F. N. L. Op't Eynde, P. F. M. Ampe, L. Verdeyen, and W. M. C. Sansen, 'A CMOS Large-Swing Low-Distortion Three-Stage Class AB Power Amplifier," *IEEE J. Solid-State Circuits*, vol. 25, pp. 265–273, 1990.

[16] D. M. Monticelli, "A Quad CMOS Single-Supply Op Amp with Rail-to-Rail Output Swing," *IEEE J. Solid-State Circuits*, vol. 21, pp. 1026–1034, 1986.

[17] J. N. Babanezhad, "A Rail-to-Rail CMOS Op Amp," *IEEE J. Solid-State Circuits*, vol. 23, pp. 1414–1417, 1988.

Chapter 9

Low-Voltage/Low-Power Amplifiers with Optimized Dynamic Range and Bandwidth

Johan H. Huijsing,
Klaas-Jan de Langen,
Ron Hogervorst, and
Rudy G. H. Eschauzier
Delft Institute of Microelectronics and Submicron Technology (DIMES), Delft University of Technology, Mekelweg 4, 2628 CD Delft, The Netherlands

9.1. INTRODUCTION

The trend toward smaller element dimensions in digital very large scale integrated (VLSI) circuits, first, will lead to lower allowable supply voltages enforced by lower breakdown voltages across the isolation barriers. Supply voltages will go down from the present 5 V to around 3 V, further to 2 V, and ultimately to 1 V. Second, the accompanying trend toward higher element densities will lead to lower allowable power per functional circuit cell. Moreover, the increasing use of battery, or solar-powered electronics will also lead to lower supply voltages, such as 1.8 V or even 0.9 V, as well as lower power.

These trends touch the fundamental limits in the design of analog circuits. The dynamic range (DR) is squeezed down between lower signal voltages and higher noise voltages caused by lower supply voltages and lower supply currents. Also, the bandwidth (B) is restricted to minimum supply currents.

Less fundamental, but nevertheless real, are the problems which must be solved in designing complete new analog circuit architectures which allow supply voltages down to 1.8 or even 0.9 V. This implies the design of voltage-efficient rail-to-rail (R-R) input stages and power-efficient R-R class AB output stages as well as efficient overall topologies for bandwidth and gain.

In this chapter, first we consider in Section 9.2 how the dynamic range is squeezed down by the supply power. To obtain the maximum dynamic range at a certain supply power, we present in Section 9.3 voltage-efficient R-R input stages and in Section 9.4 voltage- and current-efficient R-R class AB output stages. Second, we show in Section 9.5 how the bandwidth is limited by the supply current. To obtain the maximum bandwidth at a certain current, we present in Section 9.6 parallel, Miller, and multipath nested Miller compensation structures. Finally, we summarize conclusions in Section 9.7.

9.2. DYNAMIC RANGE–SUPPLY POWER RATIO

The dynamic range of low-voltage low-power amplifiers is squeezed between low signal voltages and high noise voltages caused by low currents. The maximum top value of a single-phase signal voltage is equal to half the supply voltage $V_{sst} = \frac{1}{2} V_{\text{sup}}$, as is shown in Figure 9.1(a). Its root-mean-square (RMS) value is $V_{ss} = V_{\text{sup}}/2\sqrt{2}$. If this signal is present across a signal-processing resistor R_s, the supply power needed to drive this resistor in the class AB mode is $P_{\text{sup}} = V_{\text{sup}} I_{\text{av}} = V_{\text{sup}}^2 / 2\pi R_s$. The thermal noise voltage across this resistor is $V_N = (4kTB_e R_s)^{1/2}$, in which k is Boltzmann's constant, T the absolute temperature, and B_e the effective bandwidth. The maximum dynamic range as a function of the supply power can now be calculated as

$$\text{DR}_{\text{max}} = \frac{V_{ss}^2}{V_N^2} = \frac{\pi}{4} \frac{P_{\text{sup}}}{4kTB_e} \tag{9.1}$$

from which we can find the dynamic range–supply power ratio as

$$\frac{\text{DR}_{\text{max}}}{P_{\text{sup}}} = \frac{\pi}{4} \frac{1}{4kTB_e} \tag{9.2}$$

Exactly the same expressions are found for the balanced case, where the top value of a balanced signal voltage is equal to the full supply voltage $V_{sbt} = V_{\text{sup}}$, instead of half the supply voltage in the single case. To consume the same power, the value of the balanced resistor R_b must be taken four times that of the single one: $R_b = 4R_s$. A

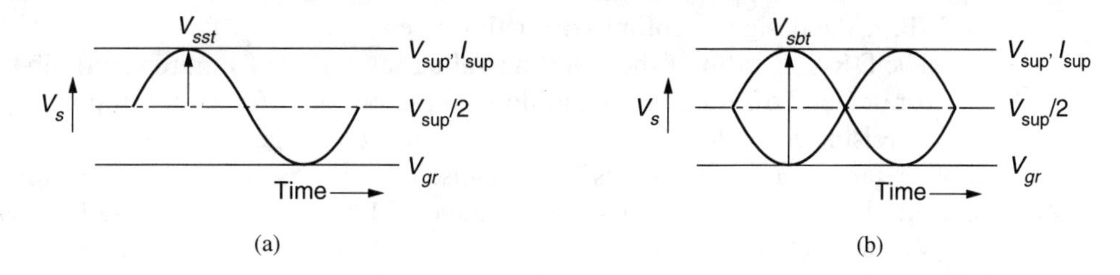

Figure 9.1 (a) Single R-R signal voltage $V_{sst} = \frac{1}{2} V_{\text{sup}}$ across class B-driven signal-processing resistor has maximum $DR = (\pi/4)P_{\text{sup}}/4tKB_e = 89\,\text{dB}$ at $P_{\text{sup}} = 16\,\mu\text{W}$, $B_e = 1\,\text{MHz}$. (b) Balanced R-R signal voltage $V_{sbt} = V_{\text{sup}}$ across class-B-driven signal-processing resistor has a maximum $DR = (\pi/4)P_{\text{sup}}/4kTB_e = 89\,\text{dB}$ at $P_{\text{sup}} = 16\,\mu\text{W}$, $B_e = 1\,\text{MHz}$.

simple example is shown in the single and balanced current-to-voltage converter shown in Figures 9.2(a) and (b) with a single resistor $R_s = 10\,\text{k}\Omega$ or a balanced resistor $R_b = 40\,\text{k}\Omega$, respectively, at a supply voltage of 1 V in a bandwidth of 1 MHz. In this case the supply power $P_{\text{sup}} = 16\,\mu\text{W}$ at a maximum sinusoidal signal. The result is a maximum dynamic range DR_{max} of 89 dB. This maximum can only be obtained when the signal-processing resistors can be driven in class AB and R-R and when the amplifier is noise free.

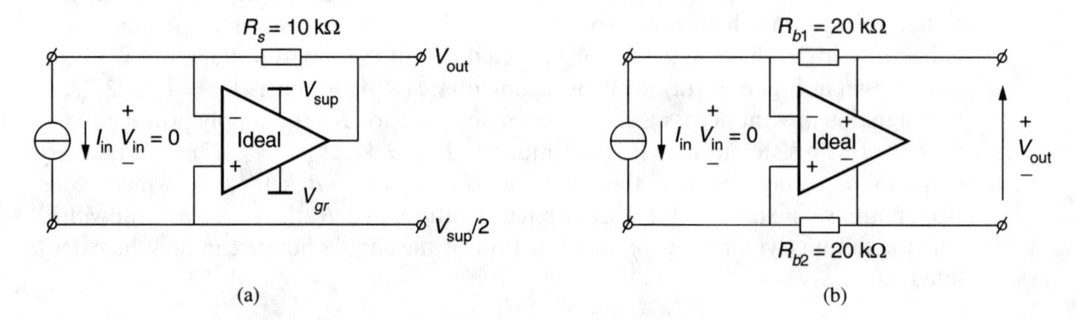

Figure 9.2 (a) Single current-to-voltage converter, $R = 10\,\text{k}\Omega$, with DR = 89 dB at $V_{\text{sup}} = 1$ V and $P_{\text{sup}} = 16\,\mu\text{W}$, R-R, class B, 1 MHz. (b) Current-to-voltage converter, $R = 40\,\text{k}\Omega$, with DR = 89 dB at $V_{\text{sup}} = 1$ V and $P_{\text{sup}} = 16\,\mu\text{W}$, R-R, class B, 1 MHz.

If the signal-processing resistor is driven in class A instead of class AB, the DR/P_{sup} ratio loses minimally a factor of π, or 5 dB from its maximum value. This loss for class A in regard to class AB may easily be a factor of 100, or 40 dB or more, in many cases where the signals are much lower than their maximum values. This is the case in audio, telecommunications, hearing aids, and the like. If the signals are restricted to one-third of the supply voltage, for instance, when a diode voltage V_{BE} is lost at a supply voltage of 1 V, the DR/P_{sup} ratio loses another factor of 3, or 5 dB, if the design is optimized for this case.

The DR/P_{sup} ratio of the inverting voltage amplifier of Figure 9.3(a), first, loses a factor of 2 or 3 dB because an additional input buffer is needed to supply the power in the resistor R_1; otherwise this power has to be supplied by the source. Second, another factor of 2, or 3 dB, is lost because of the noise of the two resistors. The resulting DR_{max} is 83 dB at a supply voltage of 1 V and a bandwidth of 1 MHz.

When we choose a gain of 10 in the inverting voltage amplifier shown in Figure 9.3(a) with $R_1 = 2\,\text{k}\Omega$ and $R_2 = 20\,\text{k}\Omega$, we first lose a factor of 2, or 3 dB, into the input buffer and second, another factor of 10, or 10 dB, because resistor R_1 only uses one-tenth of the supply voltage range. The resulting DR_{max} is 76 dB. The same result is obtained with the balanced version given in Figure 9.3(b).

In the noninverting voltage amplifiers shown in Figure 9.4(a) and (b) with a gain of 10, we only lose a factor of 10, or 10 dB, proportional to the gain. We do not lose

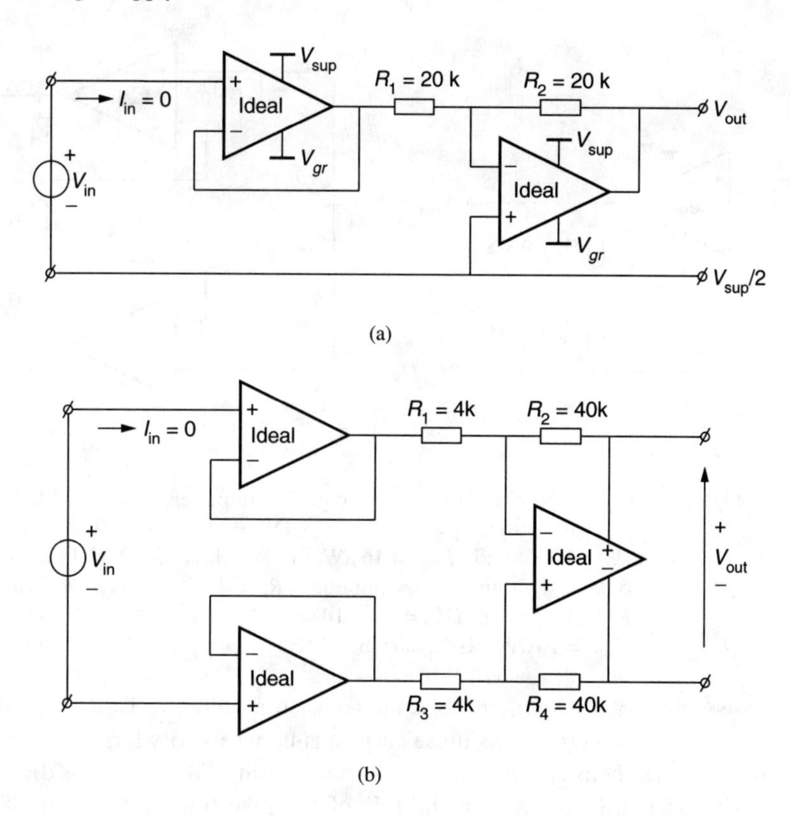

Figure 9.3 (a) Inverting voltage amplifier, $R_1 = R_2 = 20\,\text{k}\Omega$, $A = -R_2/R_1 = -1$, with $\text{DR} = 89 - 6 = 83\,\text{dB}$, at $V_{sup} = 1\,\text{V}$, $P_{sup} = 16\,\mu\text{W}$. (When $R_1 = 2\,\text{k}\Omega$, $R_2 = 20\,\text{k}\Omega$, we obtain $A = -10$ with $\text{DR} = 89 - 3 - 10 = 76\,\text{dB}$, R-R, class B, 1 MHz.) (b) Balanced inverting voltage amplifier, $R_1 = R_3 = 4\,\text{k}\Omega$, $R_2 = R_4 = 40\,\text{k}\Omega$, $A = -(R_2 + R_4)/(R_1 + R_3) = -10$ with $\text{DR} = 89 - 13 = 76\,\text{dB}$, at $V_{sup} = 1\,\text{V}$ and $P_{sup} = 16\,\mu\text{W}$, R-R, class B, 1 MHz.

the factor of 2, or 3 dB, because we do not need an additional input buffer. The DR_{max} is 79 dB in a frequency band of 1 MHz and a supply voltage of 1 V.

The balanced inverting voltage integrator shown in Figure 9.5(a) only loses the factor of 2, or 3 dB, because of the use of input buffers. The capacitors do not add to the noise. Within the effective bandwidth $B_e = 1/2\pi RC$, with $R = R_1 = R_2 = 40\,\text{k}\Omega$ and $C = C_1 = C_2$, at a supply voltage of 1 V, the DR_{max} is 86 dB. The resistor values have been chosen such that the supply power is again $16\,\mu\text{W}$ at a maximum sinusoidal signal.

A very severe loss of the DR is found in current mirrors. The current mirror of Figure 9.5(b) loses first, a factor of π, or 5 dB, in the DR_{max} because the circuit operates in class A and, second, a factor of 40, or 16 dB, because the signal is compressed in a voltage range of $V_T = kT/q \approx 25\,\text{mV}$ across the gain-setting

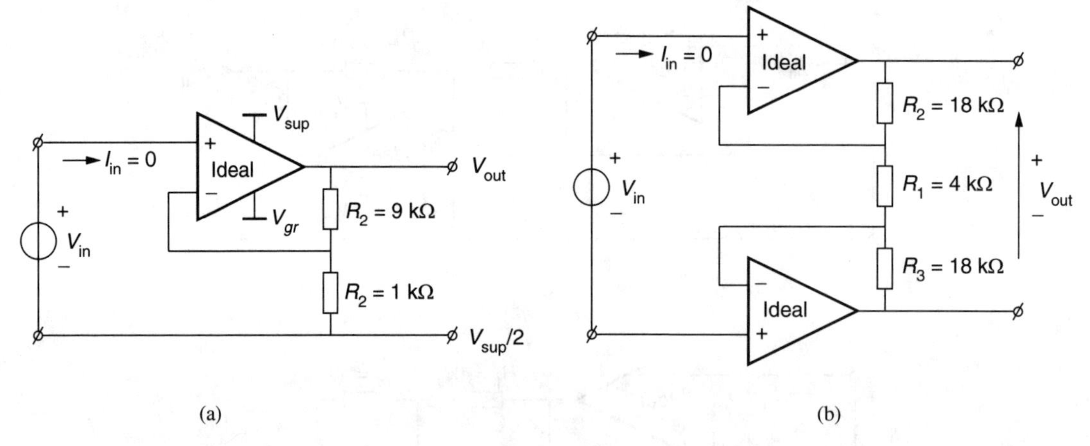

Figure 9.4 (a) Noninverting voltage amplifier, $R_1 = 1\,\text{k}\Omega$, $A = (R_1 + R_2)/R_1 = 10$, with $\text{DR} = 89 - 10 = 79\,\text{dB}$, at $V_{sup} = 1\,\text{V}$ and $P_{sup} = 16\,\mu\text{W}$, R-R, class B, 1 MHz. (b) Balanced noninverting amplifier, $R_1 = 4\,\text{k}\Omega$, $A = (R_2 + R_3)/R_1 = 10$, with $\text{DR} = 89 - 10 = 79\,\text{dB}$, at $V_{sup} = 1\,\text{V}$ and $P_{sup} = 16\,\mu\text{W}$, R-R, class B, 1 MHz.

base-emitter resistor of the bipolar transistors. This resistor is small in regard to V_{sup}/I_{sup}, and therefore its noise current is unnecessarily large. Emitter degeneration resistors will help decrease the loss. The resulting DR is only 68 dB. This is a factor of 30, or 15 dB, lower than the DR of the inverting voltage amplifier. A complementary metal-oxide-semiconductor (CMOS) mirror will do somewhat better, because better use of the supply voltage range is made because of the larger source resistances.

If we take into account the nonidealities of the amplifiers, then the DR/P_{sup} ratio is further reduced. Important nonidealities are caused by shot noise in bipolar transistors or thermal noise in field-effect transistors. The problem is that on the one hand we do not want to spill supply current into the input stages but on the other hand we need a significant amount of current to keep the input voltage noise low. Moreover, this current must continuously be drawn whether there is a large or small signal. The equivalent input voltage noise resistance is, for bipolar transistors, $R_{eq,v} = \frac{1}{2}r_e = kT/2qI_e = V_T/2I_e$ or, for field-effect transistors in strong inversion, $R_{eq,v} = \gamma/g_m = \gamma/\,[2\mu C_{ox}(W/L)I_D]^{1/2}$, which is of the order of $R_{eq,v} \approx 10\gamma/(I_D)^{1/2}$ for transistors with a W/L ratio of 100, while γ is of the order of 2. The value of W/L of 100 was selected that large in regard to digital circuits in order to lower the input voltage noise and offset voltage and increase the g_m of the input stage at a low value of the quiescent current. If we use a current of 0.1 of the maximum supply current of 16 µA in the examples of Figures 9.2–9.5, we still lose 3 dB in the DR in a single or balanced input stage with bipolar transistors or 5 dB with field-effect transistors.

If we need to extend the DR to direct currents (DC), the situation is even worse. The offset voltage in balanced input stages reduces $\text{DR}_{\text{max,DC}} = V_{sup}^2/V_{offs}^2$. At a supply voltage of 1 V for bipolar transistors with an offset of 0.3 mV $\text{DR}_{\text{max,DC}}$ is

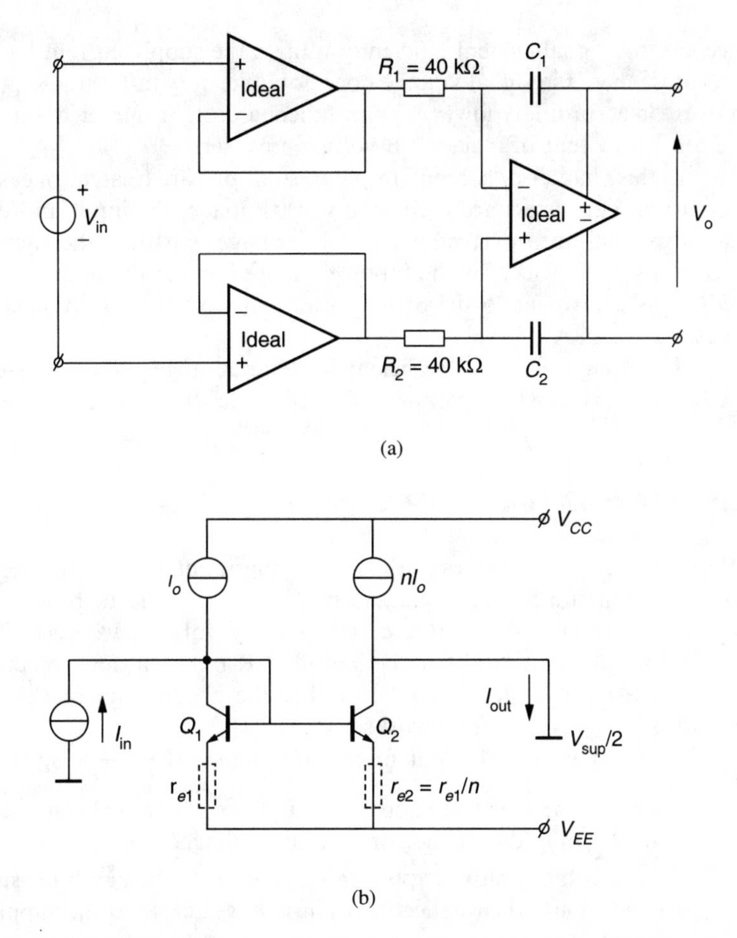

Figure 9.5 (a) Inverting voltage integrator, $R_1 = R_2 = 40\,\text{k}\Omega$, $A = -(1/j\omega C_1 + 1/j\omega C_2)\ (R_1 + R_2)$, with $\text{DR} = 89 - 3 = 86\,\text{dB}$, at $V_{\text{sup}} = 1\,\text{V}$, $P_{\text{sup}} = 16\,\mu\text{W}$, class B, 1 MHz. Generally $\text{DR} = (\pi/4)P_{\text{sup}}/4kTB_e = (\pi^2/2)P_{\text{sup}}RC/4kT = \pi V_{\text{sup}}^2 C/4kT$, with $B_e = 1/2\pi RC$, $R = R_1 = R_2$, $C = C_1 = C_2$. (b) Current mirror, $I_{\text{out}}/I_{\text{in}} = -n$, with $\text{DR} = [n/(n+1)^2](V_T/V_{\text{sup}})P_{\text{sup}}/4kTB_e$, $\text{DR} = (1/4)(1/40)\quad P_{\text{sup}}/4kTB_e = 89 - 5 - 16 = 68\,\text{dB}$, with $V_T = kT/q = 1/40\,\text{V}$, $n = 1$, $V_{\text{sup}} = 1\,\text{V}$, $P_{\text{sup}} = 16\,\mu\text{W}$, class A, 1 MHz.

limited to 70 dB, and for field-effect transistors with an offset of 3 mV, $\text{DR}_{\text{max,DC}}$ is limited to 50 dB regardless of how small we choose the bandwidth. Only chopping can elevate this limit.

Conclusion. The dynamic range of an amplifier operation is limited to a maximum value of $\text{DR}_{\text{max}} = (4/\pi)P_{\text{sup}}/4kTBe$ because of thermal noise in the gain-setting resistors. At a supply power of 16 μW the DR_{max} is 89 dB.

This maximum can only be obtained if, first, the signal voltage swing across the gain-setting elements efficiently utilizes the supply voltage from rail to rail and, second, the signal current efficiently utilizes the supply current by class AB biasing. In cases where the signal voltage does not efficiently utilize the supply voltage range, DR_{max} is accordingly lower. For instance, a current mirror has a ratio 30 times, or 15 dB, below that of a class AB voltage inverter.

In this chapter only regular opamp solutions are treated. In cases where voltage-to-current converters are used with passive loads, the input and output signal voltages can only use part of the supply voltage. On the other hand, the same bias current is used twice, for the input g_m as well as for the output load. The resulting DR_{max} is equivalent to the opamp approach, except that these circuits are usually biased in class A [1].

The design of voltage-efficient R-R input stages and power-efficient class AB R-R output stages for opamps is shown in the following two sections.

9.3. VOLTAGE-EFFICIENT INPUT STAGES

For sensing input voltages without drawing input current we need a noninverting opamp configuration or a voltage buffer. To be able to process signals with the maximum signal voltage at a certain supply voltage, we need R-R input stages. Even if we do not actually need the full R-R range, it may sometimes be preferred to process input voltages close to either the ground rail or the single-supply rail without using level shift networks.

The design of R-R input stages must satisfy the following requirements:

1. To reach the negative supply rail, PNP or P-channel transistors must be used while keeping their collector or drain voltages close to the ground voltage.
2. To reach the positive supply rail, NPN or N-channel transistors must be used while keeping their collector drain voltages close to the supply voltage.
3. To achieve the full R-R range, the signals of the P- and N-type input transistors must be summed and attenuated in such a way that the transconductance of the complete input stage is constant over the full R-R range. If the transconductance should change, the frequency behavior would be suboptimal. This would require more quiescent current in the output stage, as we find in Section 9.5.

This section presents designs that satisfy the above requirements.

The common-mode (CM) input voltage range of a P-channel differential CMOS input stage is restricted to a range from the negative-rail voltage up to a level of that of the positive-rail voltage minus the gate-source voltage V_{GS} and the saturation voltage V_{sat} of the tail current source, as shown in Figure 9.6. The CM input range of an N-channel input stage is restricted to a range from the positive-rail voltage down to V_{GS} and V_{sat} above the negative-rail voltage. If we want to obtain a full R-R input range, we must combine both complementary stages and allow at least one of the stages to function. This gives a lower limit to the supply voltage V_{supmin} of R-R input stages of

$$V_{supmin} = 2V_{GS} + 2V_{sat} \tag{9.3}$$

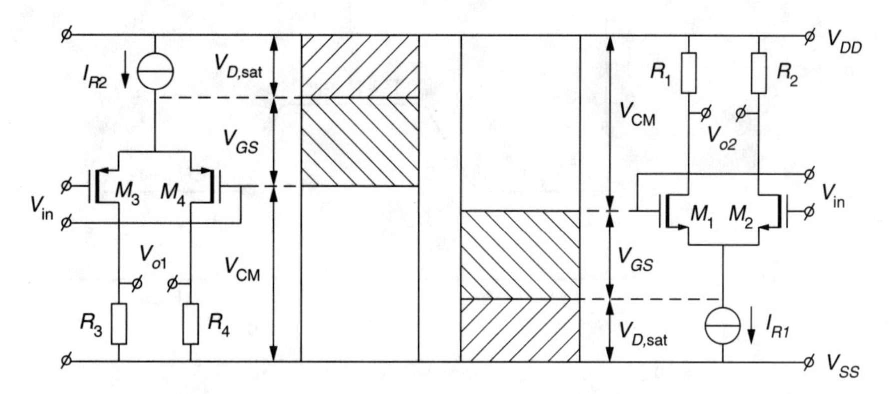

Figure 9.6 Common-mode input voltage range of P- and N-channel differential CMOS input stage.

The minimum supply voltage for R-R operation is 2.0 V for CMOS, depending on the technology, and 1.8 V for bipolar technology. At supply voltages down to 1 V, the CM input voltage range can still include one of the rail voltages.

The complementary input stages can be combined if their output currents are added by a summing circuit, as shown in Figure 9.7. The four transistors M_5–M_8 function as two folded current followers, while the pair M_6, M_8 simultaneously functions as a current mirror. The resistors R_5–R_8 provide bias current and degenerate the summing-circuit transistors in order to lower their contribution to the input noise voltage and the offset voltage.

However, one problem remains. The transconductance of the combination changes from that of the P-channel pair, up to that of the sum of both pairs, and down to that of the N-channel pair, when going from the negative-rail voltage V_{SS} toward the positive-rail voltage V_{DD}, as shown in Figure 9.8.

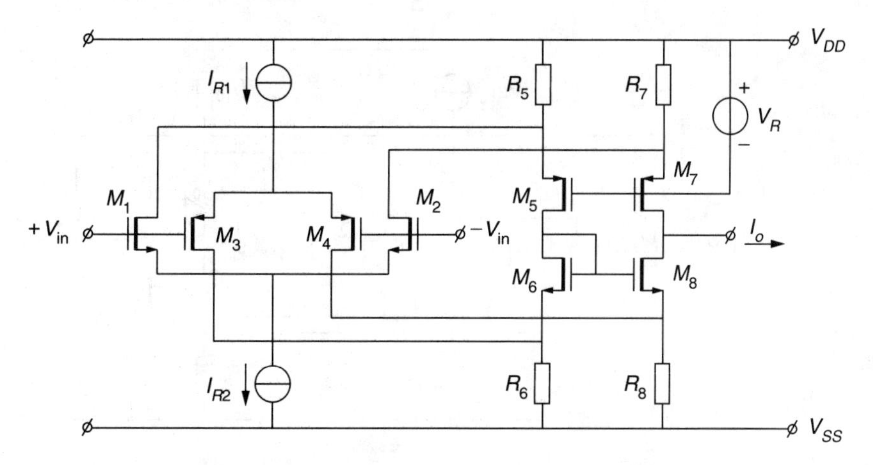

Figure 9.7 Rail-to-rail CMOS input stage consisting of complementary input stage and summing circuit.

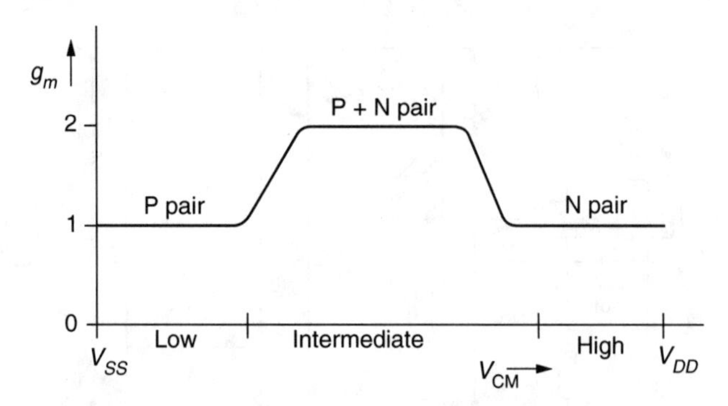

Figure 9.8 Transconductance g_m vs. common-mode input voltage of CMOS or bipolar R-R input stage.

In bipolar technology, the transconductance of the combination can be elegantly kept constant by keeping the sum of the tail currents of the complementary stages constant. The reason is that the transconductance g_m is proportional to the current, according to $g_m = qI_E/kT$. Figure 9.9 shows a reliable realization in which a current switch Q_5 guides one tail current I_{B1} either to the PNP pair Q_3, Q_4 or to the NPN pair Q_1, Q_2 through a current mirror Q_6, Q_7 [2, 3]. If Q_5 has the same emitter area as Q_3 plus Q_4, half of the tail current is steered to the NPN pair at a common-mode input voltage equal to V_{R1}. As all nodes of the current switch have a low impedance, the switch does not introduce glitches when the input voltage is swinging from rail to rail. Also, the current mirror can be designed fast enough not to cause any glitches.

If we were to apply CMOS technology to the circuit shown in Figure 9.9, the total transconductance would also be constant if the input transistors were biased in weak inversion. However, if they were biased in strong inversion a 40% higher

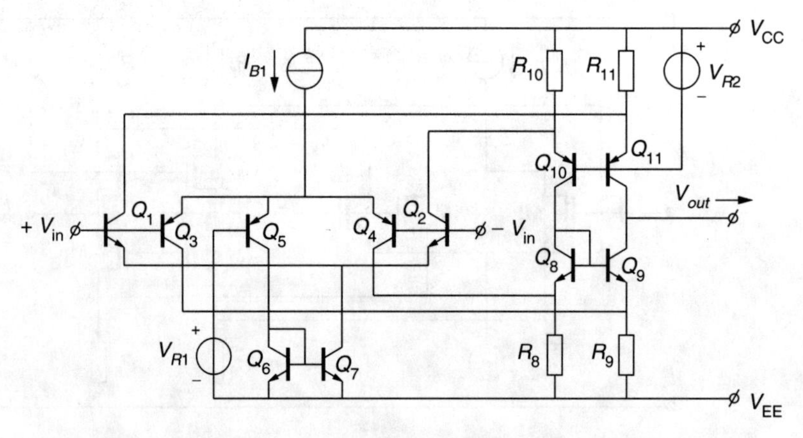

Figure 9.9 Rail-to-rail bipolar input stage with equalization of the transconductance by 1 : 1 tail current control.

transconductance would result in the situation where Q_5 conducts half of its current to one pair and the other to the second pair. This is shown in Figure 9.10. The reason is that the g_m of CMOS transistors is proportional to the square root of its current, according to $g_m = (2\mu C_{ox}(W/L)I_D)^{1/2}$.

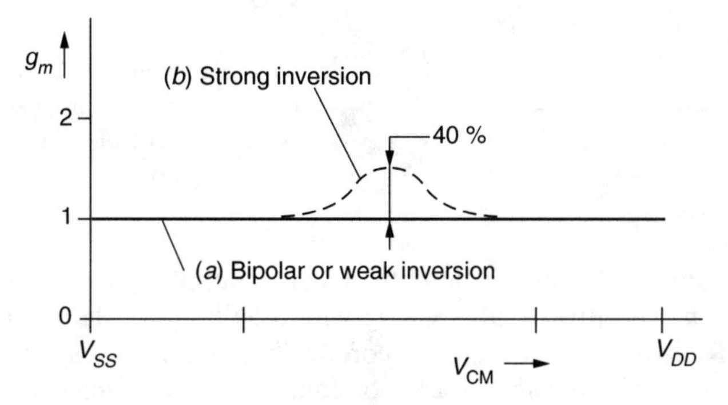

Figure 9.10 Relative transconductance of R-R input stage with 1 : 1 tail current control: (a) with bipolar or CMOS in weak inversion; (b) with CMOS in strong inversion.

To keep the total transconductance constant in CMOS technology, we would ideally have to keep the sum of the gate-source voltages of both pairs constant according to $g_m = \mu C_{ox}(W/L)(V_{GS} - V_{Th})$, which is more complicated to achieve [4].

A simpler solution is shown in Figure 9.11, where two current switches M_5 and M_8 and two 1 : 3 current mirrors M_6, M_7 and M_9, M_{10} supply each of the pairs with four times the normal tail current when the other pair is switched off [2, 4]. The resulting transconductance is shown in Figure 9.12. The variation in g_m is reduced to 14%, and with modified mirrors to lower values. Both mirrors must be prevented

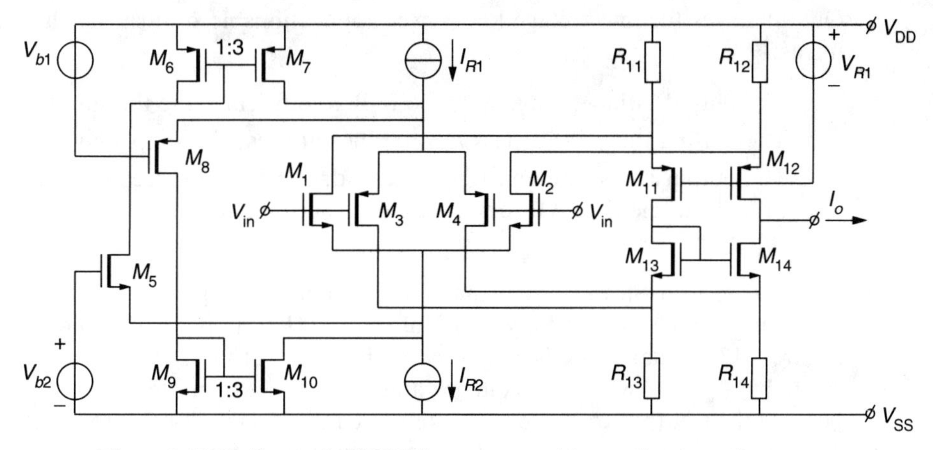

Figure 9.11 Rail-to-rail CMOS input stage with equalization of transconductance by 1 : 3 tail current control.

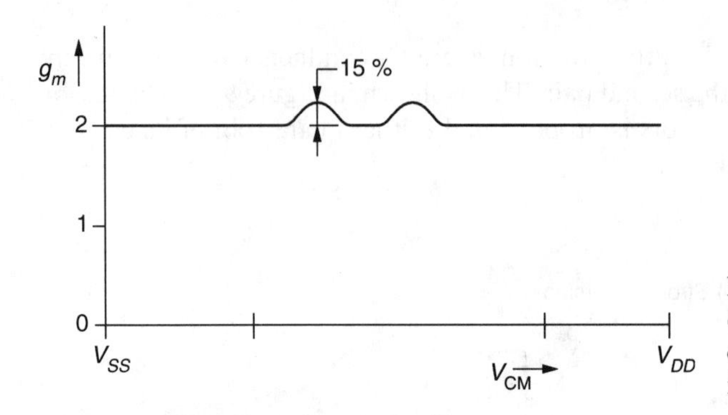

Figure 9.12 Equalized transconductance of R-R input stage with 1 : 3 tail current control with CMOS transistor in strong inversion.

from functioning at the same time at low supply voltage conditions, as this would result in a positive-feedback loop with a gain larger than unity.

Rail-to-rail operation is theoretically possible at a supply voltage down to 1 V. However, this is at the expense of complicated and noisy solutions [5].

It should be noted that the input offset voltage of R-R input stages varies from that of the P pair to that of the N pair when crossing the current-switching reference voltage. This change in offset deteriorates the CMRR to a certain extent in the CM range around the reference voltage.

Conclusion. High-performance input stages can be designed both in CMOS and in bipolar technology, both of which feature a rail-to-rail CM input voltage range at a constant transconductance. A minimum supply voltage of about 1.8 V is needed for R-R operation. At supply voltages down to 1 V, the CM range can still include one of the rail voltages.

9.4. VOLTAGE- AND CURRENT-EFFICIENT OUTPUT STAGES

Output stages for low-voltage low-power applications need to satisfy three requirements:

1. The output voltage range must be R-R to efficiently use the supply voltage.
2. The biasing must be in class AB to efficiently use the supply current.
3. The output transistors must be directly driven by the preceding stages without delay from the class AB control circuit to accommodate the highest bandwidth–supply power ratio.

This section shows designs that satisfy the above requirements.

Conventional output stages, such as in the 741 opamp family, which is shown in Figure 9.13, do have an efficient class AB biasing, but fall short of the R-R voltage range by at least two diode voltages and two collector-emitter saturation voltages because their output transistors are connected in a common-collector configuration. This allows the simple class-AB biasing circuit with a translinear loop consisting of diodes D_1, D_2 and transistors Q_P and Q_N.

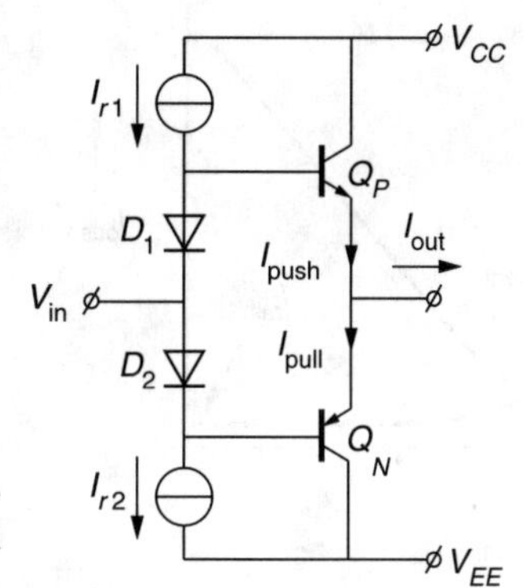

Figure 9.13 Conventional bipolar common-collector stage with feedforward class AB bias control.

The output transistors of an R-R output stage must be connected in a common-emitter configuration to the ground and to the supply rail, respectively. This complicates the class-AB biasing circuit. Figure 9.14 shows a common-emitter circuit derived by a transformation from Figure 9.13 in which the lower and upper halves are interchanged. Two floating voltage sources at half the supply voltage $\frac{1}{2}V_s$ are needed to reshape the translinear loop. But they are not so simple to realize.

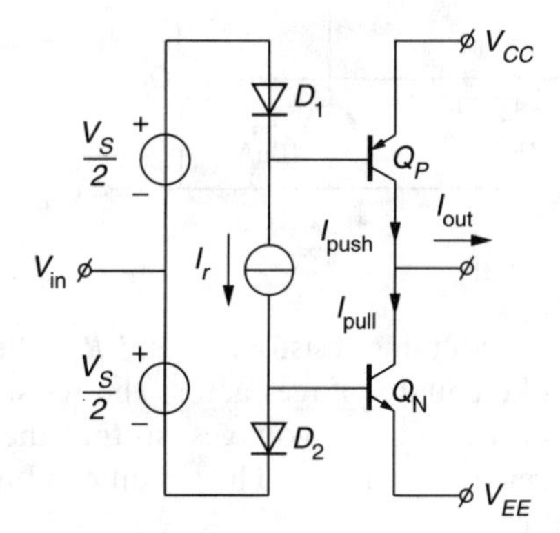

Figure 9.14 Rail-to-rail bipolar common-emitter stage with feedforwad class AB bias control.

The class AB biasing must satisfy: high I_{max}/I_{quiesc} ratio for high efficiency, high I_{min}/I_{quiesc} ratio to avoid HF distortion, smooth AB transition for avoiding distortion, and direct-driven output transistors for an optimum HF behavior. Figure 9.15 shows the desired characteristics we wish to obtain.

These aims can be achieved with the circuits principally shown in bipolar technology (BiCMOS) in Figures 9.13 and 9.14. We call the biasing of these circuits feedforward bias control. The translinear loop of the four diodes D_1, D_2, Q_P, and Q_N ensures that the sum of the base-emitter voltages of Q_P and Q_N remain constant so that the product of the push and pull currents I_{push} and I_{pull} remains constant in the circuits of Figures 9.13 and 9.14. The question remains of how to realize the floating voltages in Figure 9.14.

The first approach to the principle depicted in Figure 9.14 is given in Figure 9.16. The output transistors Q_P and Q_N are biased with diodes D_1 and D_2, respec-

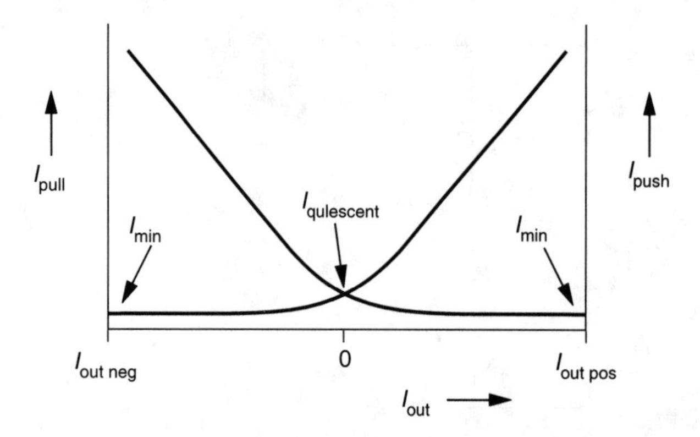

Figure 9.15 Desired characteristic of push and pull currents as function of output current of class AB stage.

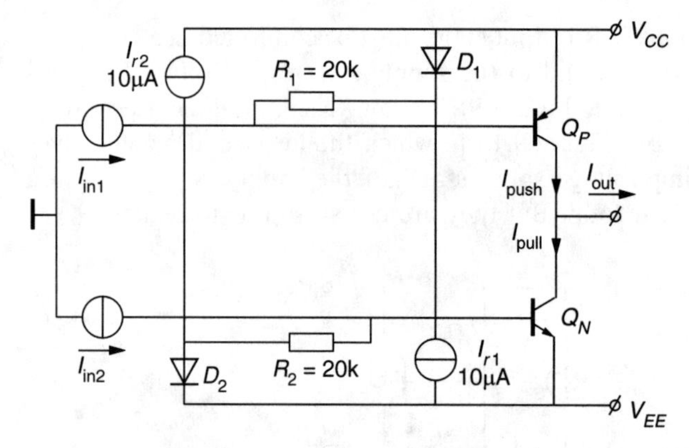

Figure 9.16 Rail-to-rail bipolar output stage with separate resistive class AB control, $V_{sup} = 0.9\,V$.

tively, through coupling resistors R_1 and R_2. The two equal-input current sources I_{in1} and I_{in2} invoke equal voltages across the resistors R_1 and R_2 in driving the output transistors with two equal voltages, so that the sum of the base-emitter voltages of Q_P and Q_N remains constant. The circuit can function with supply voltages as low as 0.9 V.

A disadvantage of the circuit shown in Figure 9.16 is the relatively large loss of driving current in the resistors R_1 and R_2. These resistors must be on the one hand large to minimize the signal current loss and on the other hand small to determine the base-emitter voltages. Otherwise, the output transistors may easily be fully cut off.

To conserve the current loss in the resistors R_1 and R_2, this current can be caught by the emitters of Q_1 and Q_2 and reused to drive the complementary output transistors, as shown in Figure 9.17. Now all the input signal current is used to drive either one or the other output transistor. The circuit has a higher gain than that in Figure 9.16, but the driving voltages are still too inaccurately determined by the resistors R_1 and R_2.

When we eliminate the resistors completely, we obtain the circuit given in Figure 9.18(a) [6]. The sum of the base-emitter voltages is thoroughly fixed by two translinear loops D_1, D_2, Q_2, Q_P and D_3, D_4, Q_1, Q_N. The two transistors Q_1 and Q_2

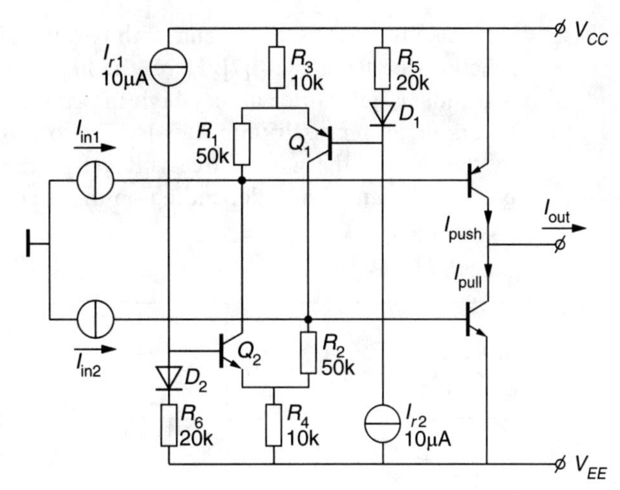

Figure 9.17 Rail-to-rail bipolar output stage with resistive-coupled feedforward class AB control, $V_{sup} = 1.1\,\text{V}$.

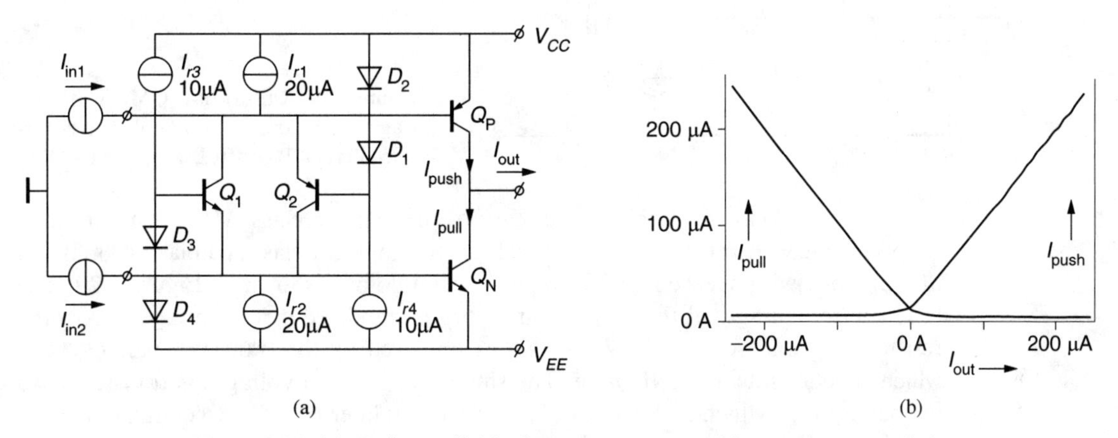

Figure 9.18 (a) Rail-to-rail bipolar output stage with transistor-coupled feedforward class AB control, $V_{sup} = 1.8\,\text{V}$. (b) Simulation result of push and pull currents as function of output current of bipolar output stage with transistor-coupled feedforward class AB control.

which are connected head to tail comprise a mesh between their emitters and collectors which does not allow driving current to escape or to be lost. Therefore, though it looks like driving current is lost in the emitter of either Q_1 or Q_2, there is no loss because the driving current is recycled in the loop of Q_1 and Q_2. All driving current which is not used for one output transistor is used for the other one. The circuit represented in Figure 9.18(a) is nearly ideal, as shown by the simulation result given in Figure 9.18(b). The class AB relation is so well fixed that the circuit can also be driven with one input current source instead of with two. The only drawback is that the applied series connection of two diodes does not allow the circuit to operate at supply voltages lower than 1.8 V.

Also in CMOS, the circuit with two translinear loops, as shown in Figure 9.19, functions perfectly well [7]. Here, the high gate impedances of the mesh transistors M_3 and M_4 guarantee a very high impedance level at the gates of the output transistors for in-phase driving signals. The lowest supply voltage V_{supmin} is determined by the sum of the gate-source voltages of the output transistors. This sum can be between 1.6 and 2.8 V, depending on the maximum allowable output current and on technology.

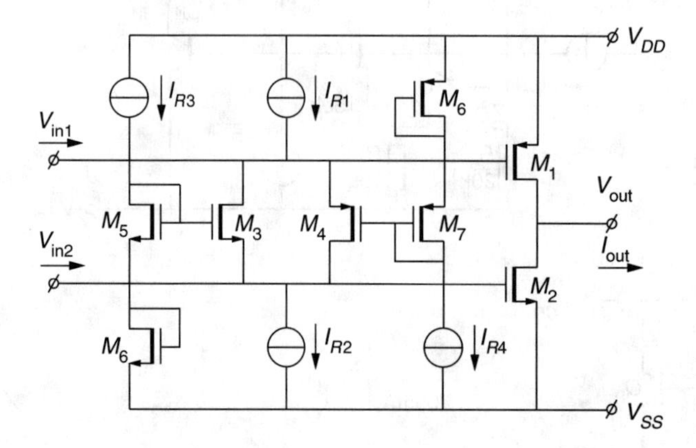

Figure 9.19 Rail-to-rail CMOS output stage with transistor-coupled feedforward class AB control, $V_{sup} = 1.6\text{–}2.8$ V.

When we want to combine the features of accurate class AB biasing and low supply voltage, we must apply a feedback control loop for class AB biasing [3, 8]. An output stage with such control in bipolar technology is shown in Figure 9.20. The base-emitter voltages of the output transistors are modeled by the voltages across the resistors R_{35} and R_{34}. These voltages are compared by the transistor pair Q_{35}, Q_{36} which functions as an AND gate. The smaller of the two voltages is transferred to the base of Q_{32}, which is the input of a control amplifier Q_{31}, Q_{32} to control the bias. In this way the smaller of the two push or pull output currents is regulated above a

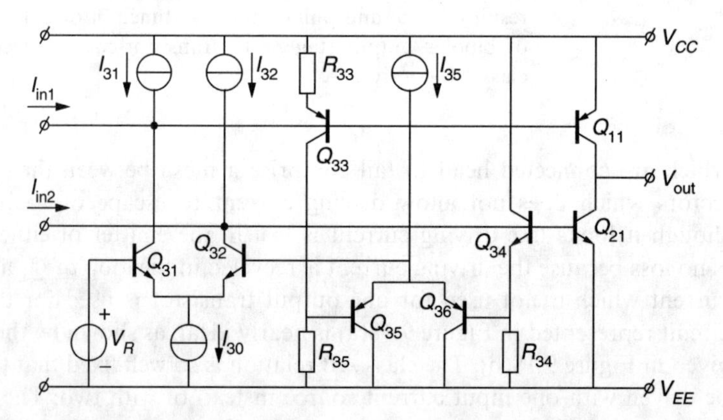

Figure 9.20 Rail-to-rail bipolar output stage with minimum selector and feedback class AB control, $V_{sup} = 1.0$ V.

minimum value [5, 9]. The class AB control is so firm that the circuit can also be driven with one input current source instead of with two. The driving current which is not needed at one side is automatically steered to the other side through the control amplifier. A ratio of more than 100 between the maximum output current and quiescent current can easily be obtained.

The same principle of feedback bias control can be applied in CMOS technology, as shown in Figure 9.21 [4]. This results in a combination of accurate class AB biasing at low supply voltages of 1.0–1.6 V, depending on the maximum output current needed.

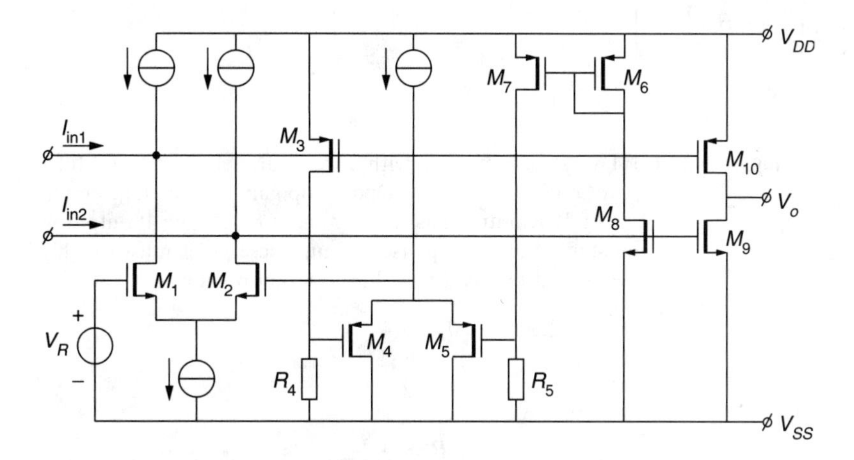

Figure 9.21 Rail-to-rail CMOS output stage with minimum selector M_4, M_5 and feedback class AB control $M_1, M_2, V_{sup} =$ 1.0–1.6 V.

Conclusion. We have seen that current-efficient R-R output stages can be realized in both bipolar and CMOS technology. The supply voltages can be as low as 1 V. Accurate class AB feedforward or feedback bias control results in a ratio of 100 or higher between the maximum output current and the quiescent current.

9.5. BANDWIDTH–SUPPLY POWER RATIO

The bandwidth of a one-stage amplifier is simply the ratio between the transconductance g_m and the load capacitance C_L:

$$B = \frac{g_m}{2\pi C_L} \tag{9.4}$$

For a bipolar one-stage amplifier like the output stage Q_1 of Figure 9.22(a), we can relate the bandwidth to the collector bias current with $g_m = qI_C/kT$ as

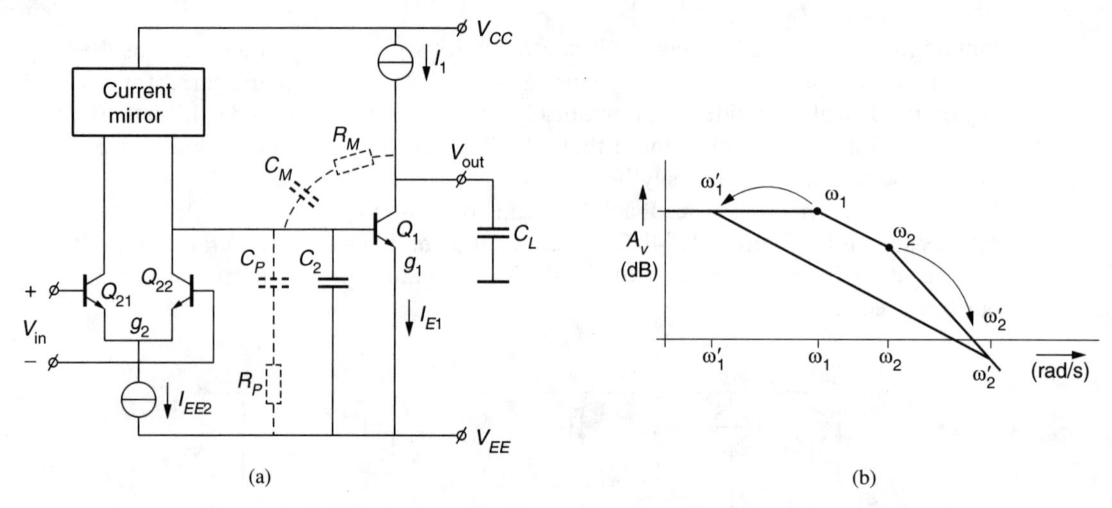

Figure 9.22 (a) Two-stage amplifier with either parallel C_p, R_p or Miller C_m, R_m compensation. (b) Open-loop gain of two-stage amplifier with R-R output has two dominant poles, one at output of each stage. For a 60° phase margin, these poles must roughly be separated by two times the closed-loop gain.

$$B = \frac{qI_C}{2\pi kTC_L} \approx 6\,\frac{I_C}{C_L} \tag{9.5}$$

Using Eq. (9.5) the bandwidth–supply power ratio is found as

$$\frac{B}{P_{\text{sup}}} = \frac{q}{2\pi kTV_{\text{sup}}C_L} \approx \frac{6}{V_{\text{sup}}C_L} \tag{9.6}$$

For a current of 10 μA and $C_L = 10\,\text{pF}$, the bipolar stage has a bandwidth of 6 MHz. It is clear that a low supply voltage is in favor of the B/P_{sup} ratio.

If we would regard the differential input stage of Figure 9.22(a) as a single-stage operational transconductance amplifier (OTA), we lose a factor of 2 because of the splitting of the tail current between Q_{21} and Q_{22}. Additionally, we lose a part of the output voltage swing by the voltage across the tail current source I_{EE2}. If we would like to have R-R swing at the output, we may use a folded-mirror stage using three current mirrors. But in that case we lose a second factor of 2 because of the additional bias current.

For CMOS transistors in strong inversion, we can relate the bandwidth to the bias current with $g_m = [2\mu C_{\text{ox}}(W/L)I_D]^{1/2}$, where μ is the mobility of the charge carriers, C_{ox} is the specific oxide capacitance, and W/L is the width-to-length ratio of the channel. For $W/L = 100$ (for a high g_m at a certain bias current), $\mu C_{\text{ox}} = 50\,\mu\text{A/V}^2$, and a current of 10 μA, we find

$$B = \frac{\sqrt{2\mu C_{ox}(W/L)I_D}}{2\pi C_L} \approx \frac{\sqrt{I_D}}{60C_L} \tag{9.7}$$

and

$$\frac{B}{P_{sup}} = \frac{\sqrt{2\mu C_{ox}(W/L)}}{2\pi\sqrt{I_D}\,V_{sup}C_L} \approx \frac{1}{60\sqrt{I_D}\,V_{sup}C_L} \tag{9.8}$$

At 10 μA and $C_L = 10\,\text{pF}$ the CMOS stage has a bandwidth of 5 MHz. At higher currents in strong inversion, the bandwidth is proportional to the root of the current. This means that CMOS output stages have a lower B/P_{sup} ratio than their bipolar counterparts, the lower for currents larger than 10 μA. Below this value, the bipolar stage is still favorable.

The bandwidth of a two-stage amplifier is equal to the geometric mean of the bandwidths B_1 and B_2 of the two stages [10]. However, for a phase margin of 60°, preferable in multistage feedback amplifiers, we have to reduce the overall bandwidth B_0 by a factor of 2. This gives

$$B_0 = \tfrac{1}{2}\sqrt{B_1 B_2} \tag{9.9}$$

For the bipolar two-stage amplifier in Figure 9.22(a) we find

$$B_0 = 2\sqrt{\frac{I_1 I_{EE2}}{C_L C_2}} \tag{9.10}$$

and

$$\left(\frac{B_0}{P_{sup}}\right)_{opt} \approx \frac{1}{V_{sup}\sqrt{C_L C_2}} \quad \text{at } I_1 = I_{EE2} \tag{9.11}$$

The $(B_0/P_{sup})_{opt}$ ratio has a broad optimum for ratios of I_1/I_{EE2} centered at $I_1/I_{EE2} = 1$.

For CMOS the bandwidth is given by

$$B_0 = \frac{1}{140}\frac{\sqrt[4]{I_1 I_{EE2}}}{\sqrt{C_L C_2}} \tag{9.12}$$

and the B_0/P_{sup} ratio by

$$\left(\frac{B_0}{P_{sup}}\right)_{opt} \approx \frac{1}{180}\frac{1}{V_{sup}\sqrt{I_{sup}}\sqrt{C_L C_2}} \quad \text{at } I_1 = \tfrac{1}{3}I_{EE2} \text{ or } I_1 = 3I_{EE2} \tag{9.13}$$

where $I_{sup} = I_1 + I_{EE2}$.

The phase margin of 60° is obtained when the poles ω_1 and ω_2 of the two stages are split apart by a factor twice the DC gain of the whole amplifier to new positions ω_1' and ω_2', as shown in Figure 9.22(b). The splitting of the poles can be accomplished by either a parallel pole-zero cancellation network C_p, R_p or a Miller splitting network C_m, R_m, which are depicted in Figure 9.22(a).

The parallel network has the advantage that the bandwidth of both stages can optimally be used, according to (9.9). The bandwidth of a two-stage amplifier with parallel compensation may even be larger than that of a single-stage amplifier, because the internal capacitor C_2 may be much smaller than the external load capacitor C_L. However, the pole-zero cancellation results in process- and signal-dependent pole-zero doublets, which can only be allowed in some cases.

The local Miller feedback network has the advantages that no pole-zero doublets occur, that the output impedance is lowered at high frequencies, and that the local Miller feedback in the frequency band between ω_1' and ω_2' is used to linearize the output stage in this frequency band. However, the bandwidth of the whole amplifier is restricted to half the bandwidth of the output stage with its load capacitor C_L:

$$B_0 = \tfrac{1}{2} B_1 = \frac{g_m}{4\pi C_L} \tag{9.14}$$

Conclusion. The bandwidth B_0 of an amplifier stage is determined by its load capacitance C_L and the current I_{sup} through the output device. The bandwidth of a bipolar stage is $B_0 = 3I_C/C_L$. The maximum B_0/P_{sup} ratio of a bipolar stage is $B_0/P_{sup} \approx 3/C_L V_{sup}$. The ratio for a CMOS stage is lower than that of a bipolar stage, particularly at higher currents in strong inversion. For that reason we use large W/L ratios for the output transistors. The maximum B_0/P_{sup} is inversely proportional to the supply voltage, which is in favor of low-voltage operation. The overall stability of two-stage amplifiers can be secured by either a parallel or a Miller compensation network. The parallel network offers the highest bandwidth but can only be used in some cases. The Miller network offers a robust solution.

9.6. GAIN

More gain can be obtained by either cascading more stages or improving each stage. In this section we will first show how the high-frequency (HF) stability of amplifiers with more than two cascaded stages can be secured while preserving bandwidth by nested Miller compensation and multipath nested Miller compensation structures. Second, we will show how we can improve single stages in bipolar, CMOS, and BiCMOS technology. Finally, as an example a two-stage CMOS amplifier with high gain is discussed.

A simple way to secure the HF stability of a three-stage amplifier is the nested Miller compensation [3, 12]. A simplified schematic of a nested Miller compensated amplifier is shown in Figure 9.23(a). Starting from the right-hand side, we find an output stage Q_1, an intermediate stage Q_{21}, Q_{22}, and an input stage Q_{31}, Q_{32}. Each

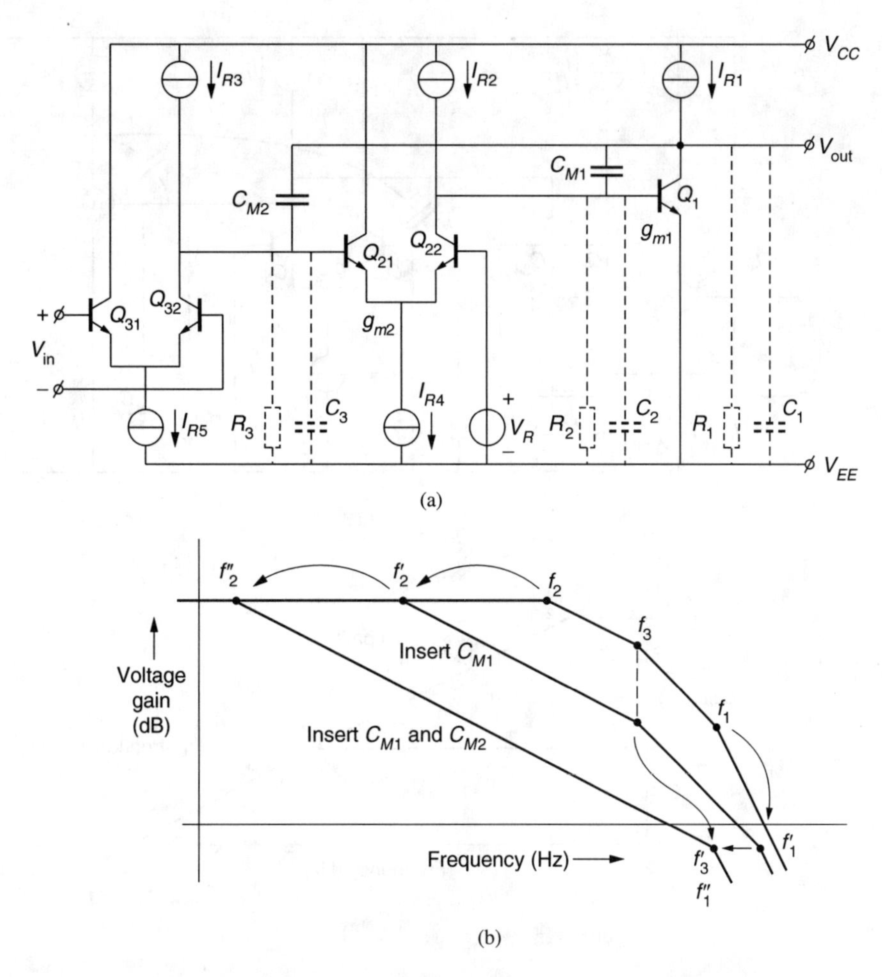

Figure 9.23 (a) Cascaded three-stage bipolar amplifier, 1 V, with nested Miller compensation structure, losing a factor 2 in bandwidth. (b) Open-loop frequency characteristic of three-stage amplifier with nested Miller compensation structure.

stage has a dominating pole at its output. The open-loop frequency characteristic is shown in Figure 9.23(b). The output and intermediate stage comprise a two-stage amplifier, which is unity-gain compensated by a Miller capacitor C_{m1}. This splits the poles at f_1 and f_2 so that f_1' disappears 3 dB below 0 dB, as shown in Figure 9.23(b). Only f_2' remains. The process of Miller splitting can now be repeated by C_{m2}, which splits apart the third pole at f_3 and f_2', so that f_3' disappears 3 dB below 0 dB. Only f_2'' remains as a dominating pole frequency.

The nesting procedure is quite simple and can be further repeated. The only drawback is that we lose a factor of 2 in bandwidth each time we nest, because we need to bury a pole 3 dB below 0 dB to obtain a phase margin of 60 dB.

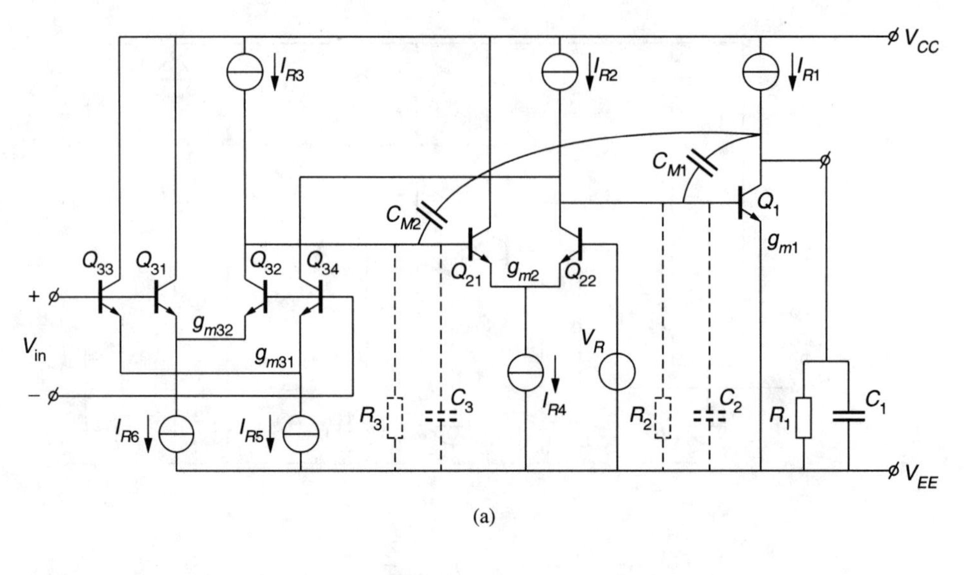

(a)

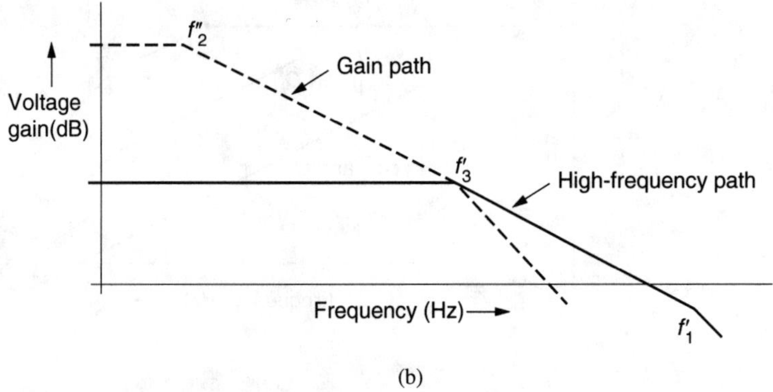

(b)

Figure 9.24 (a) Multipath nested Miller compensation three-stage bipolar amplifier. (b) Open-loop frequency characteristic of three-stage amplifier with multipath nested Miller compensation structure.

To prevent the loss of a factor of 2, we can apply a multipath nested Miller compensation structure [11, 13, 14], as shown in Figure 9.24(a). The addition to the circuit of Figure 9.23(a) is a second input stage the output of which surpasses the intermediate stage. The amplifier now consists of a two-stage HF path in parallel to a three-stage gain path. The open-loop frequency characteristic given in Figure 9.24(b) clearly shows the two paths. To obtain a good matching of both paths, the ratio of the two Miller capacitors C_{m1}/C_{m2} must be equal to the ratio of the transconductances g_{m31}/g_{m32}. This requirement can accurately be fulfilled, independently of process parameter changes.

Now we see how we can improve single stages in order to obtain more gain without giving in to the B_0/P_{sup} ratio.

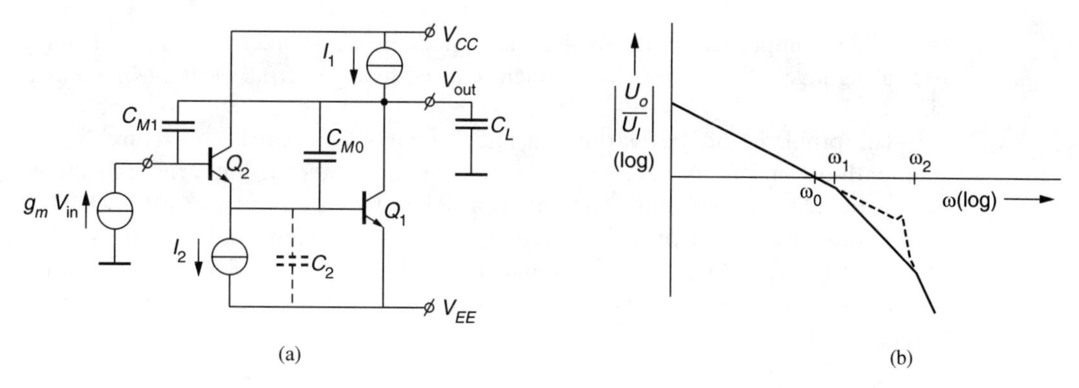

Figure 9.25 (a) Darlington output stage, 1.8 V, with internested Miller capacitor C_{m0}, and limiting frequency $\omega_L = 1/r_{e2}C_2$. (b) Frequency response of Darlington output stage. Dashed line shows "output bump" at large output currents as result of complex poles without C_{m0}.

The first example is to improve the current gain of bipolar stages by the well-known Darlington transistor Q_2, as shown in Figure 9.25(a). However, we find two disadvantages. First, the supply voltage needs to be larger than 1.8 V to accommodate at least two diodes in series with a saturation voltage. Second, the base-emitter capacitor C_2 of the output transistor loads the emitter of Q_2 resulting in a secondary pole frequency ω_2. Using Miller compensation around the whole stage with C_{m1} easily results in peaking of the frequency characteristic, as shown in Figure 9.25(b). This can be suppressed by an inner loop nesting of another Miller capacitor C_{m0} [5] at the expense of a two times lower bandwidth.

The next example is the Widlar bipolar output stage [15] shown in Figure 9.26 with high current gain. It consists of two meandered Darlington transistors Q_2 and Q_3 and a currrent booster Q_4, Q_5, Q_6. The stage has the advantage that it can function at supply voltages as low as 0.9 V. A severe disadvantage, however, is

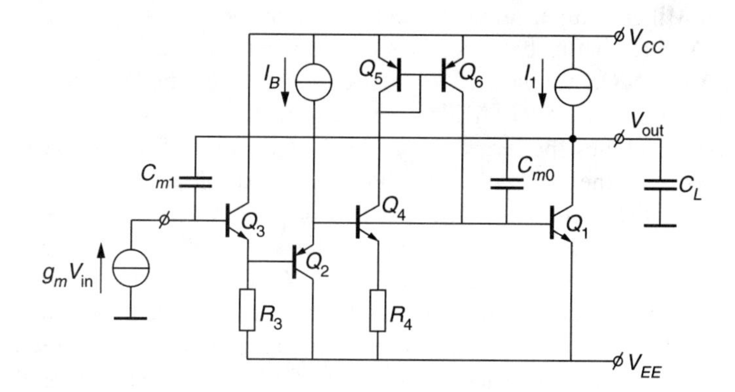

Figure 9.26 Widlar low-voltage output stage, 1 V, with boost circuit and with two meandering Darlington transistors having a low limiting frequency ω_L.

that Miller compensation across the whole stage by C_{m1} makes the circuit strongly oscillating unless the bandwidth is much reduced by an internal nested Miller capacitor C_{m0}.

The problems of the Darlington and Widlar stages can be overcome by the multipath-driven output stage with a nested Miller compensation structure, shown in Figure 9.27. The HF path consists of a direct-driven output transistor Q_1. The gain path consists of a preceding stage Q_2 coupled through a mirror Q_3, Q_4. The multipath nested output stage combines the high bandwidth of a single transistor with the current gain of a two-stage amplifier. The circuit functions well at supply voltages down to 0.9 V.

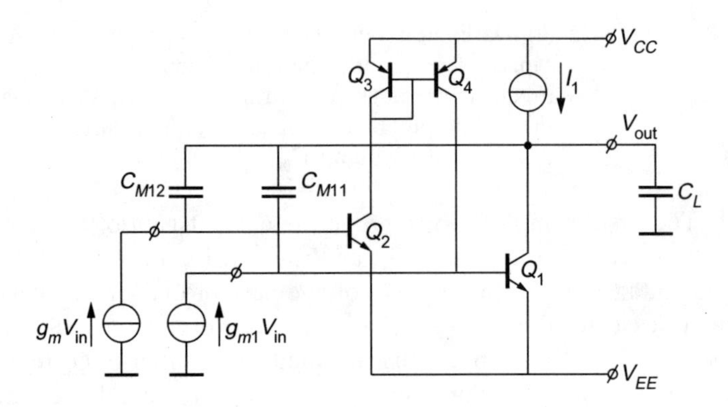

Figure 9.27 Multipath-driven output stage with nested Miller compensation structure, with highest possible bandwidth.

A very interesting combination is using a CMOS Darlington transistor M_1 to boost tremendously the current gain of a bipolar transistor Q_1, as shown in Figure 9.28. However, the disadvantages of the Darlington circuit mentioned with that of Figure 9.25, namely the higher minimum supply voltage and the oscillatory behavior when Miller compensated, are even stronger here.

A much better behavior is achieved when we combine the bipolar transistor Q_1 and the CMOS transistor M_1 in the two-stage configuration shown in Figure 9.29 with multipath nested Miller compensation, as discussed with Figure 9.27. The circuit combines the largest possible B_0/P_{sup} ratio of a direct-driven bipolar output transistor in the HF path with the largest possible current gain of the CMOS transistor in the gain path. The circuit does not waste supply voltage because both transistors are connected with the emitter and source to the negative rail. The circuit functions at supply voltages down to 0.9 V, depending on the desired output current.

To boost the voltage gain of bipolar or CMOS transistors, they can be cascoded. In the BiCMOS stage shown in Figure 9.30 an example is given where the gain of bipolar transistors is boosted by CMOS cascodes. Bipolar transistors Q_1, Q_2 are placed at the input where a low offset and noise and a high g_m are needed and in the current sources $Q_3, \ldots, Q_6$, where a high source impedance is needed. Complementary MOS transistors M_3–M_6 are cascode transistors to increase the

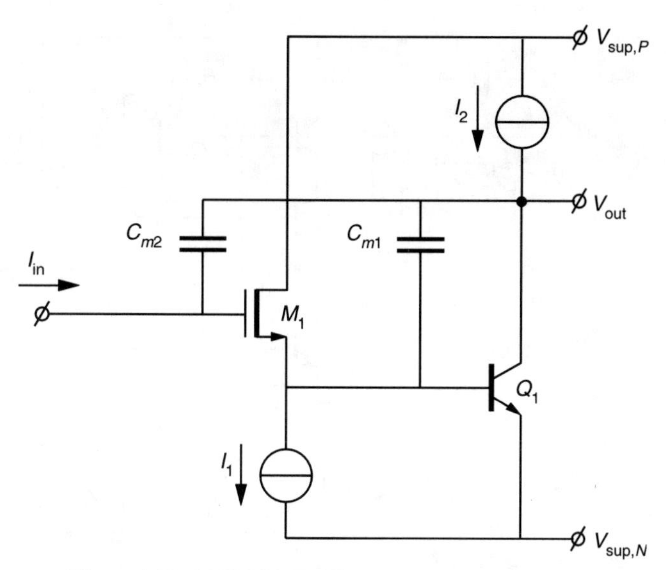

Figure 9.28 BiCMOS Darlington output stage, 2.8 V.

voltage gain of the whole stage. Transistor M_7 is used to mirror the current in M_3 and M_5 into M_6. The use of the mirror transistor M_7 increases the lowest supply voltage to a level of about 2.8 V.

Another example of increasing the voltage gain by cascoding is shown in an example of a two-stage CMOS opamp in Figure 9.31. The circuit consists of a R-R input stage M_1–M_4, a summing and cascoding stage M_{11}–M_{16}, and an output stage M_{25}, M_{26}, with a feedforward class AB control circuit M_{19}–M_{24} which was already discussed in Section 9.4. A g_m control circuit as discussed in Section 9.3 is needed to equalize the g_m of the R-R input stage over the CM input voltage range but is not shown in order to simplify the schematic. The folded cascodes M_{13}–M_{16} boost

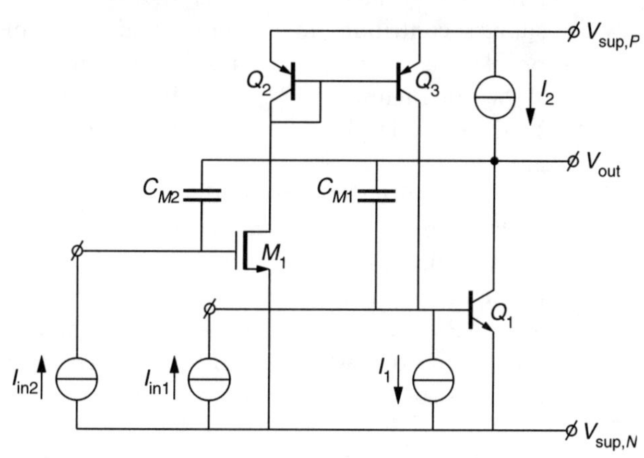

Figure 9.29 BiCMOS multipath-driven output stage 0.9–1.4 V, with CMOS gain stage and bipolar HF stage.

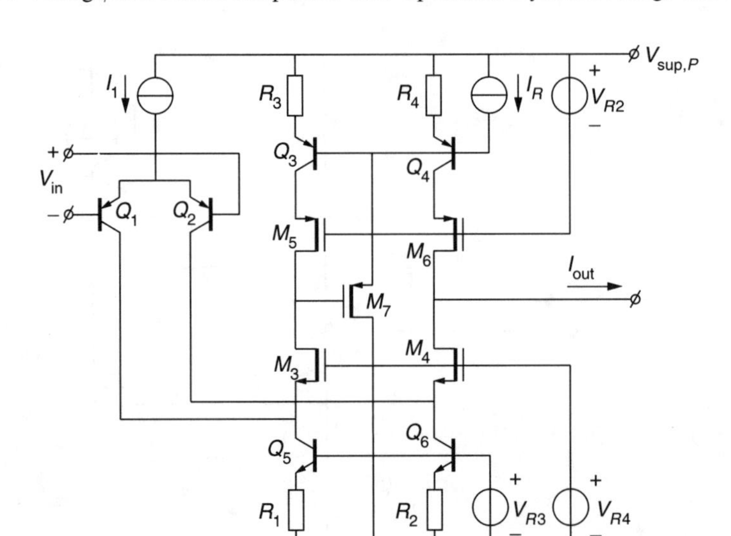

Figure 9.30 Cascoded BiCMOS input stage, 2.8 V, with bipolar input stage and current sources and CMOS cascodes.

the gain of the amplifier, while this gain is preserved by the high input impedance of the class AB output stage. The bandwidth is only slightly reduced by inserting the cascodes, since the bandwidth of the cascodes is usually much higher than the bandwidth of the amplifier with its load capacitor C_L. If we connect (not shown) the Miller capacitors C_{M1} and C_{M2} between the output and the sources of M_{14} and M_{16}, the bandwidth will even increase because of the higher voltage gain in the Miller loop [9, 18]. The lowest allowable supply voltage is increased by the saturation voltage of the cascodes, which is of the order of 200 mV.

Although the gain of the opamp shown in Figure 9.31 is reasonable, it is still reduced by the bias current sources of the class AB control, I_{b6} and I_{b7}, which are in parallel with the cascodes M_{14} and M_{16} of the summing circuit. Also, these current sources contribute to the noise and offset of the amplifier. A way to solve this problem is to shift the class AB control M_{19}–M_{20} into the summing circuit, as shown in Figure 9.32 [17, 18]. In this way, the floating class/AB control is biased by the cascodes M_{14}–M_{16} of the summing circuit and no additional current sources are needed.

A drawback of shifting the class AB control transistors into the summing circuit is that the quiescent current of the output transistors depends on the CM input voltage. Because the tail currents of the input pairs I_{b1} and I_{b2} vary as a function of the CM input voltage, the currents through the cascodes M_{14}–M_{16}, which bias the class AB control transistors M_{19}, M_{20}, change. This causes a varying quiescent current in the output stage. The bias current of the class AB control can be made constant by using a summing-circuit architecture with two current mirrors, M_{11}–M_{14} and $M_{15} - M_{18}$, which are biased by a floating current source I_{b3}, as shown in Figure 9.33. The NMOS current mirror M_{17}, M_{18} is now biased by the

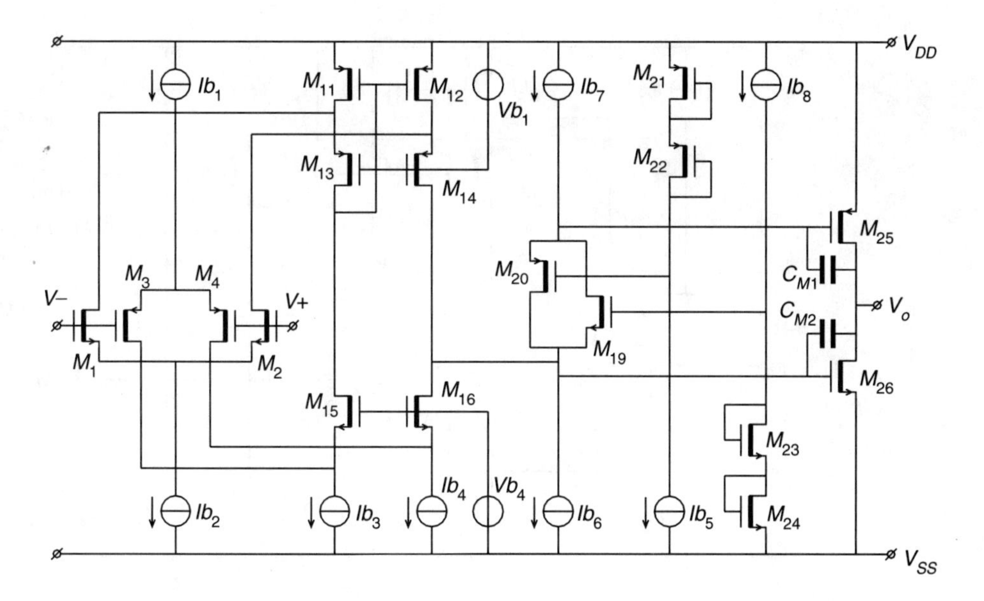

Figure 9.31 Two-stage cascoded opamp.

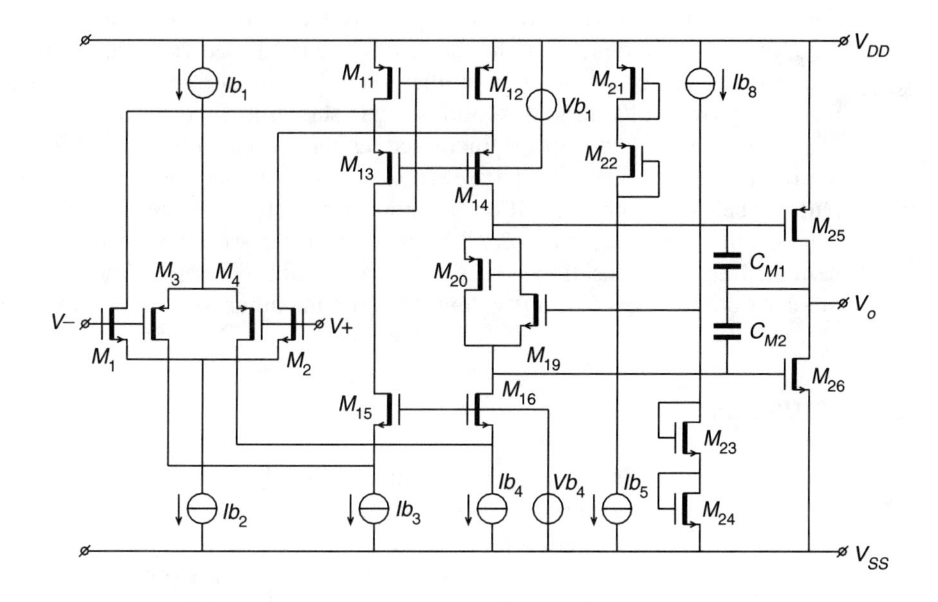

Figure 9.32 Compact two-stage opamp with floating class AB control which is biased by cascodes of summing circuit.

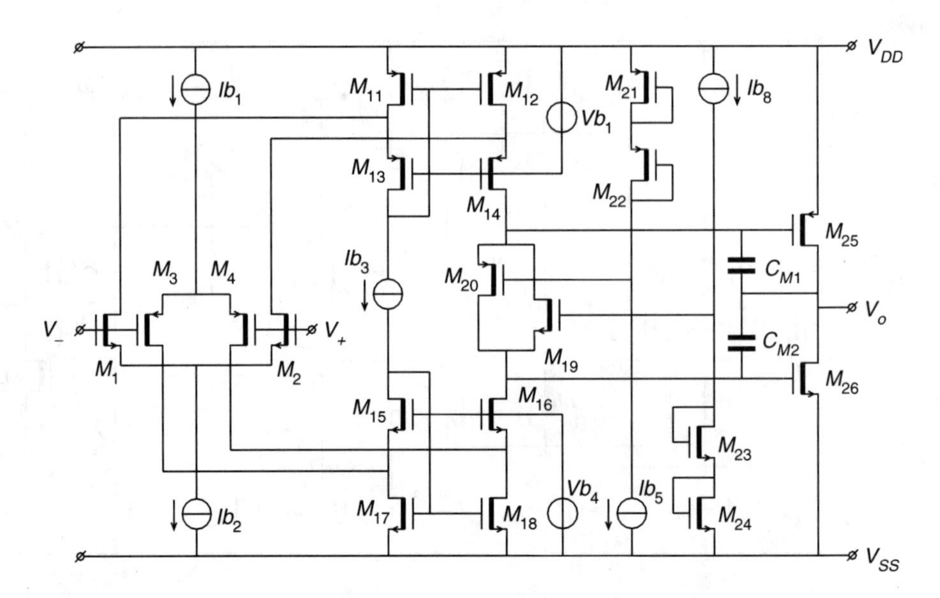

Figure 9.33 Compact two-stage opamp with two current mirrors in summing circuit which are biased by floating current source.

constant current source I_{b3} and the changing bias current of the input-stage transistor M_3. At the output of the current mirror, the drain of M_{18}, the input-stage bias current represented by the drain current of M_4 is subtracted again, and thus M_{16} and the class AB control transistors M_{19}, M_{20} are biased by a constant current equal to I_{b3}. Because of the floating architecture of the current source, it does not contribute to the noise and offset of the amplifier. The floating current source can be implemented by two transistors, M_{27} and M_{28}, as shown in Figure 9.34 [17].

Conclusion. Gain can be increased either by cascading more stages or by improving single stages. We have seen that nested Miller compensation and multipath nested Miller compensation are good structures to increase the gain without deteriorating the maximum B_0/P_{sup} ratio. Further we have seen that multipath nested Miller compensation is also a powerful tool to improve the current gain of bipolar output stages without giving in to the maximum B_0/P_{sup} ratio.

9.7. CONCLUSIONS

The R-R input stages have been presented in bipolar and CMOS technology with constant transconductance g_m. The R-R output stages have been shown with feedforward and feedback bias control in class AB.

Also it is shown that the maximum bandwidth–supply power ratio is restricted by physical properties to a value of $B_{0,\max}/P_{\text{sup}} \approx 3/C_L V_{\text{sup}}$ for bipolar transistors. The CMOS transistors biased in strong inversion have a lower ratio. The maximum ratio can be surpassed in two-stage amplifiers by parallel compensation at the cost of

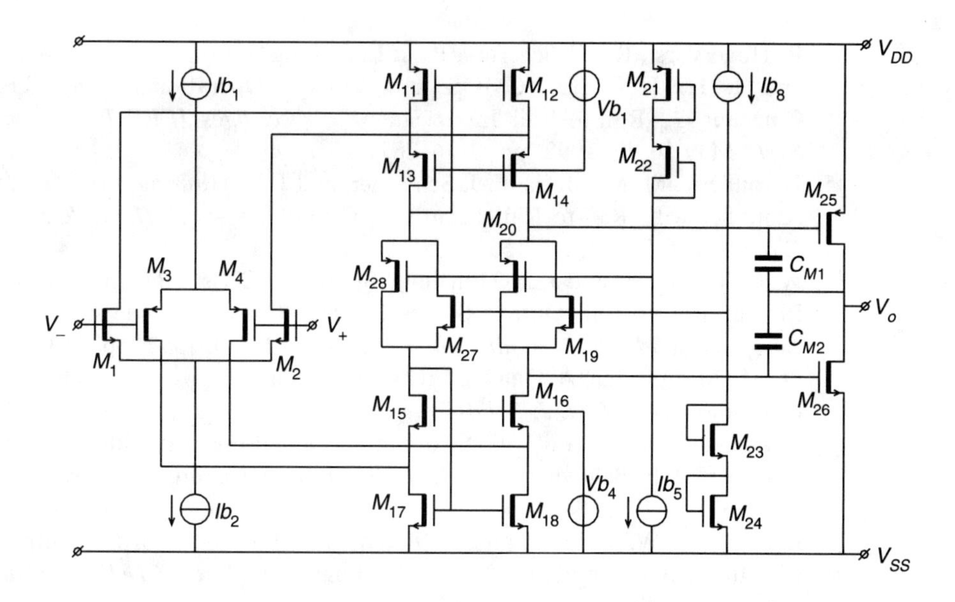

Figure 9.34 Compact two-stage opamp with floating current source implemented by transistors M_{27} and M_{28}.

inflexibility. The maximum ratio can be obtained in two-stage amplifiers with Miller compensation and in three-stage amplifiers with a multipath nested Miller compensation structure. This structure provides an abundance of gain.

A multipath nested Miller structure also proves to be a powerful tool to improve the current gain of a bipolar (output) stage. A BiCMOS version combines the ultimate high $B_{0,\max}/P_{\text{sup}}$ ratio of a bipolar transistor with the ultimate high current gain of a CMOS transistor.

Acknowledgments

This work was supported by the Technology Foundation (STW).

References

[1] G. Groenewold, "Optimal Dynamic Range Integrated Continuous-Time Filters," Ph.D Thesis, Delft University of Technology, Delft, The Netherlands, 1992.

[2] J. H. Huijsing and R. J. v. d. Plassche, "Differential Amplifier with Rail-to-Rail Input Capability and Controlled Transconductance," U.S. Patent No. 4, 555, 673, Nov. 26, 1985.

[3] J. H. Huijsing and D. Linebarger, "Low-Voltage Operational Amplifier with Rail-to-Rail Input and Output Ranges," *IEEE J. Solid-State Circuits*, vol. SC-20, no. 6, pp. 1144–1150, 1985.

[4] R. Hogervorst, R. J. Wiegerink, P. A. L. de Jong, J. Fonderie, R. F. Wassenaar, and J. H. Huijsing, "CMOS Low-Voltage Operational Amplifiers with Constant G_m Rail-to-Rail Input Stage," *Proceedings IEEE Int. Symp. Circ. Syst.*, May 10–13, 1992, pp. 2876–2879.

[5] J. Fonderie, M. M. Maris, E. J. Schnitger, and J. H. Huijsing, "1-V Operational Amplifier with Rail-to-Rail Input and Output Ranges," *IEEE J. Solid-State Circuits*, vol. SC-24, pp. 1551–1559, 1984.

[6] W. C. M. Renirie and J. H. Huijsing, "Simplified Class-AB Control Circuits for Bipolar Rail-to-Rail Output Stages of Operational Amplifiers," *Proceedings European Solid-State Circuit Conference*, Sept. 21–23, 1992, pp. 183–186.

[7] D. M. Monticelli, "A Quad CMOS Single-Supply Opamp with Rail-to-Rail Output Swing," *IEEE J. Solid-State Circuits*, vol. SC-21, pp. 1026–1034, 1986.

[8] J. H. Huijsing and F. Tol, "Monolithic Operational Amplifier Design with Improved HF Behavior," *IEEE J. Solid-State Circuits*, vol. SC-11, no. 2, pp. 323–328, 1976.

[9] E. Seevinck, W. de Jager, and P. Buitendijk, "A Low-Distortion Output Stage with Improved Stability for Monolithic Power Amplifiers," *IEEE J. Solid-State Circuits*, vol. SC-23, pp. 794–801, 1988.

[10] E. M. Cherry and D. E. Hooper, *Amplifying Devices and Low-Pass Amplifier Design*, Wiley, New York, 1988, pp. 690–701.

[11] J. Fonderie and J. H. Huijsing, "Operational Amplifier with 1-V Rail-to-Rail Multipath-Driven Output Stage," *IEEE J. Solid-State Circuits*, vol. 26, no. 12, pp. 1817–1824, 1991.

[12] J. H. Huijsing, "Multi-Stage Amplifier with Capacitive Nesting for Frequency Compensation," U.S. Patent No. 4, 559, 502, Dec. 17, 1985.

[13] J. H. Huijsing and M. J. Fonderie, "Multi-Stage Amplifier with Capacitive Nesting and Multipath Forward Feeding for Frequency Compensation," U.S. Patent No. 5, 155, 447, Oct. 4, 1992.

[14] R. G. H. Eschauzier, L. P. T. Kerklaan, and J. H. Huijsing, "A 100-MHz 100-dB Operational Amplifier with Multipath Nested Miller Compensation Structure," *IEEE J. Solid-State Circuits*, vol. 27, no. 12, pp. 1709–1717, 1992.

[15] R. J. Widlar, "Low Voltage Techniques," *IEEE J. Solid-State Circuits*, vol. SC-13, 1978, pp. 838–846.

[16] J. H. Huijsing, R. Hogervorst, M. J. Fonderie, K. J. de Langen, B. J. van den Dool, and G. Groenewold, "Low-Voltage Analog Signal Processing," in *Analog VLSI Signal and Information Processing*, M. Ismail and T. Fiez (eds.), McGraw-Hill, New York, 1993.

[17] R. Hogervorst, J. P. Tero, R. G. H. Eschauzier, and J. H. Huijsing, "A Compact Power-Efficient Rail-to-Rail Input/Output Operational Amplifier for VLSI Cell Libraries," *IEEE J. Solid-State Circuits*, vol. 29, no. 12, pp. 1505–1512, 1994.

[18] R. Hogervorst and J. H. Huijsing, *Design of Low-Voltage Low-Power Operational Amplifier Cells*, Kluwer Academic, Dordrecht, The Netherlands, 1996.

Chapter 10

Low-Voltage Analog CMOS Filter Design

Michel Steyaert, J. Crols, and S. Gogaert
Departement Elektrotechniek Afdeling ESAT, Katholieke Universiteit Leuven, Kardinaal Mercierlaan 94, B-3001 Heverlee, Belgium

10.1. INTRODUCTION

The demand for system mobility, greater packing densities, smaller size, and lower power drain has resulted in a demand for analog signal-processing circuits with low and very low operating power supply voltages (3 and 1.5 V). Therefore design techniques for low-voltage analog filters in complementary metal–oxide–semiconductor (CMOS) technologies are discussed in this chapter.

The implementation of analog filters in CMOS can be realized with several techniques, such as active-*RC*, MOS field-effect transistor (MOSFET), operational transconductance amplifier (OTA-C), and switched-capacitor (SC) techniques. However, these techniques are not all equally well suited for achieving low-power-supply voltages and at the same time low-distortion specifications. In the second part of this chapter the implication of low-power-supply voltages on filter performance is discussed for the different implementation techniques and a comparison between the techniques is made.

The OTA-C filters are preferred when a low-power-supply voltage (3 V) is used. Their main advantage lies in the low distortion, even at low voltages, and in the low power consumption. The main problem in these OTA-C filters is the design of a low-voltage low-distortion OTA. In the third part of the chapter possible structures for such an OTA will be reviewed and a full CMOS low-distortion structure based on source degeneration and input signal folding techniques is fully discussed.

Classic implementation techniques for SC filters can be easily adapted for use at low-power-supply voltages (3 V), and due to the high filter accuracy, their use is very attractive. Problems occur, however, for the switches when very low power supply voltages (1.5 V) are used. The fourth part of the chapter deals with the presentation and comparison of different techniques to overcome this problem. Design techniques using low-V_T transistors, single-device switches, voltage multipliers, and the so-called switched-opamp technique are analyzed and discussed.

10.2. IMPLICATIONS OF LOW-POWER-SUPPLY VOLTAGES ON TECHNIQUES FOR ANALOG CMOS FILTERS

The design of analog filters can be performed with several techniques, such as *RC*-active, MOSFET-C, OTA-C, and SC techniques. The *RC*-active technique is the most basic implementation technique of filters: The integrator structure consists of a passive resistor, usually polysilicon resistance or a high ohmic resistance, in combination with an opamp with the feedback capacitance. Because the only active component is the opamp, low-voltage filter designs based on this technique are mainly concentrated on the design of low-voltage opamps. The design of low-voltage CMOS opamps can be performed in standard CMOS technologies even down to 1.5 V [1]. The only extra requirement compared to OTA-C or SC techniques is that the opamp should be able to drive resistive loads. This last fact is, however, also the drawback of this type of filter, especially if not only low voltage is required but also low power drain. The power drain of the amplifiers and the filter is related to the resistor values in the integrator. However, large resistor values in standard CMOS technologies would result in unacceptable chip areas, and so higher power drains are the result. A second problem with active-*RC* filters is the inaccuracy of the integrator time constant. This is because the time constant is related to the *RC* product, which depends on the absolute value of the polysilicon sheet resistance and the absolute value of the capacitance. However, those absolute values are not accurate in CMOS technologies. The result is that the cutoff frequency of these filters has only an accuracy of less than 50%. Recent developments using discrete programmable capacitive banks have demonstrated accuracies down to 5% at the cost of chip area [2]. Because of this and because several other chapters of this book present detailed discussions of low-voltage opamps, this type of filter is not discussed further here. However, it has to be said that if very low distortion specifications are required (< -80 dB) the active-*RC* technique is the only remaining integration technique for analog integrated filters [2].

In MOSFET-C filters the resistor of the active-*RC* structure is replaced by a MOSFET in its linear region. The advantage is that by controlling the gate voltage, the equivalent *R* value can now be tuned. Hence the main drawback of active-*RC* filters can be overcome by integrating an automatic tuning technique on chip [3]. Hence high filter accuracies can be achieved even at high frequencies. The main drawback of using the MOSFET is that the transistor is a nonlinear component. The distortion can be improved by using fully differential structures. However, this requires not only opamps which can drive resistive loads but also fully differential opamps with a common-mode feedback system. The distortion specification is a function of the gate voltage with respect to the DC input voltage and bulk voltage [4]. To achieve −60 dB distortion with input signals of 0.6 $V_{p\text{-}p}$ (peak to peak), a minimum power supply of ±3 V is required. So in this technique large power supply voltages are very often common. (This technique will not be discussed further in this chapter.)

10.3. LOW-VOLTAGE OTA-C TECHNIQUES

10.3.1. Introduction

The design of full CMOS continuous-time filters can be realized with MOSFET-C or OTA-C techniques. However, as already discussed before, to achieve low-power-supply voltages and to achieve at the same time low-distortion specifications, the OTA-C technique is usually preferred. The main problem in OTA-C filters is the design of a low-voltage, low-distortion OTA. A full CMOS low-distortion structure, based on source degeneration and input signal folding techniques, will be discussed. Moreover, to fulfill high accuracy in the cutoff frequencies of the filter, an active tuning system is necessary. An automatic time-domain tuning circuit will be proposed.

10.3.2. Very Low Distortion Fully Differential OTA

Limitations of Easy Structures. The major limitation in ordinary OTA structures is the noise generated by the intrinsic nonlinearities of the electronic devices. Due to the nonlinear nature of the relationship between the input signal and the output signals, several unwanted signals are generated. In the case of a sinusoidal input, these signals are at multiple frequencies of the incoming signal frequency and are called the harmonic distortions of the input signal [5]. This kind of noise is accounted for by the total harmonic distortion (THD), which is the root-mean-square (RMS) value of the contribution of all the harmonics.

The THD of a differential pair can easily be calculated as

$$\mathrm{HD2} = 0 \qquad \mathrm{HD3} = \frac{1}{32}\frac{V_{\mathrm{in}}^2}{(V_{GS} - V_T)^2} \tag{10.1}$$

Because of the differential structure, the second harmonic distortion (HD2) is a function of mismatches between the transistors. By careful lay-out, this component can be made smaller than −65 dB in CMOS technologies.

However, to realize a third harmonic distortion (HD3) of −60 dB at a differential input voltage of 0.5 V, $V_{GS} - V_T$ should be 3 V, which requires high (±5 V) supply voltages.

Source-Degenerated Topologies. A technique that allows natural linearization due to its intrinsic feedback is the source degeneration technique. Other important advantages of this structure are that it exhibits excess lead phase in its phase response and that it is not very sensitive to transistor mismatches. An ordinary resistor can be used for this purpose. But this approach requires large resistors which increase both the silicon area and the parasitic capacitance. Those resistors have to be driven by large transconductances; then the power consumption increases as well. However, a major drawback of passive resistors is lack of tuneability. For this reason, several CMOS realizations of tuneable source-degenerated topologies have been proposed.

The topology by Krummenacher and Joehl [6], shown in Figure 10.1(a), which uses transistors biased in the triode region is basically a source-degenerated structure for small signals with an additional internal mechanism which increases its transconductance for large signals. This last fact increases the linear range of this topology on the order of a factor of 2.

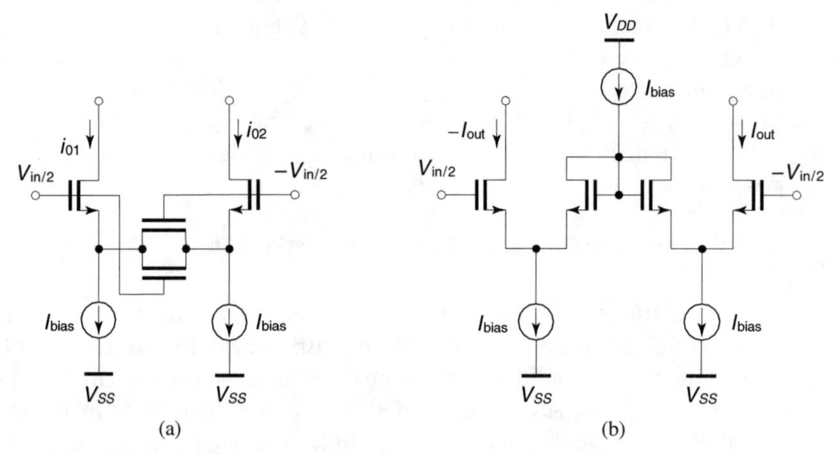

Figure 10.1 CMOS source degenerated topologies using (a) transistors in triode region and (b) saturated transistors.

Another approach using source degeneration was proposed by Torrance et al. [7]. In this technique, the resistors used for source degeneration are simulated by saturated transistors with their drain connected to their gate. In this case, the factor n corresponds to the number of additional pairs used. This topology is shown in Figure 10.1(b) for the use of two additional sections. Because the input voltage for each section is V_{in}/n, the transconductance is also reduced by the same factor n. The third-harmonic distortion, the most important one, is decreased by a factor of n^2.

Using both techniques, the design of OTAs with HD3 lower than -60 dB is feasible [8, 9]. However, not only is the harmonic distortion reduced but also the linear range for the input signals is almost twice that of the ordinary differential pair.

The major disadvantages of this structure are additional transistors and increased power consumption. The first drawback is not really important but it represents additional silicon area. The second shortcoming has been overcome using a cross-coupled technique as presented in Figure 10.2. The transconductance is increased as

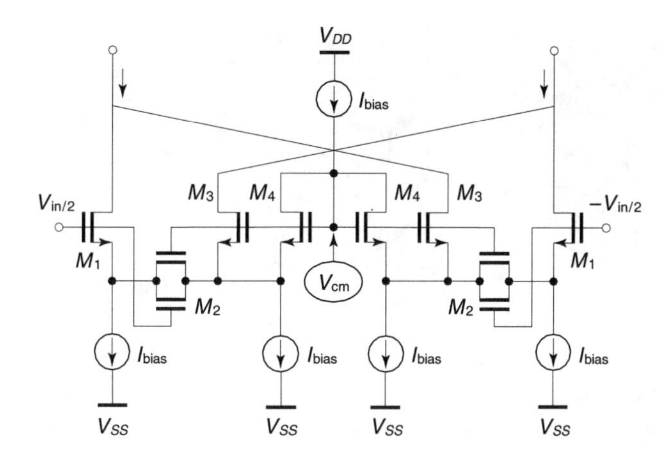

Figure 10.2 Cross-coupling to reduce power consumption.

$$K_C = 1 + \frac{1}{1 + B_4/B_3} \tag{10.2}$$

When we implement the same transconductance as before, the THD is reduced by a factor of K_C^2.

10.3.3. Low-Distortion Approach for Common-Mode Loop

As mentioned above, the linearity of the system can be further increased by implementing a fully differential signal path. In that case, because of increased symmetry, all common-mode signals and even-order distortions cancel out, except for the influence of component mismatches. A fully differential signal path leads to lower distortion, higher common-mode rejection ratio (CMRR), and higher power supply rejection ratio (PSRR). Differential structures have one major disadvantage. Because the signal is no longer referred to ground, the operating point of the amplifier cannot be stabilized with the differential feedback loop. A common-mode feedback loop (CMFB) has to be added as well. This loop can become quite complicated because it should reduce the common-mode signals over a wide frequency range without affecting the differential performance of the OTA. With the presented OTA a satisfactory CMFB can easily be realized. To understand the principle, it has to be mentioned that in high-order filters, several OTA-C integrator stages are placed in series to realize the filter structure, as shown in Figure 10.3. The problem is to control and to set the common output voltage of the OTAs (OTA1 in Fig. 10.3). This can easily be arranged because in the low-distortion OTA structure (see Fig. 10.2) an extra terminal is available: The common-mode voltage at the input of an OTA can be sensed at the V_{CM} terminal. The CMFB for the OTA1 can thus be generated from the V_{CM} of the next OTA (OTA2 in Fig. 10.3). The only extra component required is a simple opamp to regulate the measured common-mode voltage to the reference voltage (in this case ground). Low distortion of the

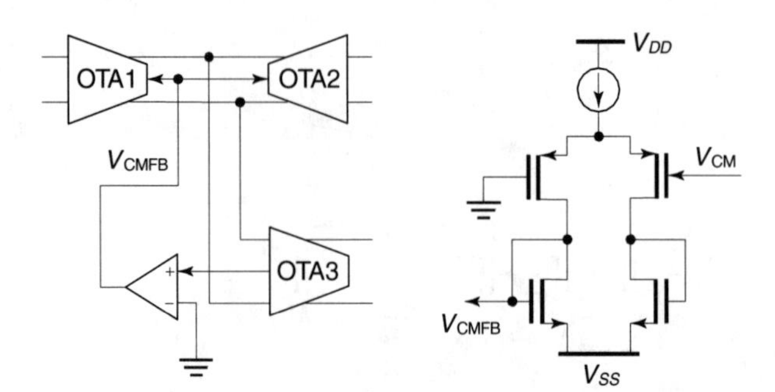

Figure 10.3 Block diagram of CMFB system.

common-mode detector is automatically guaranteed because the common-mode voltage is sensed by the low-distortion input stage of the OTA.

Performance of a Practical Realized Structure. Using both techniques described above and adding the biasing circuitry, we come to the low-distortion OTA displayed in Figure 10.4 [10].

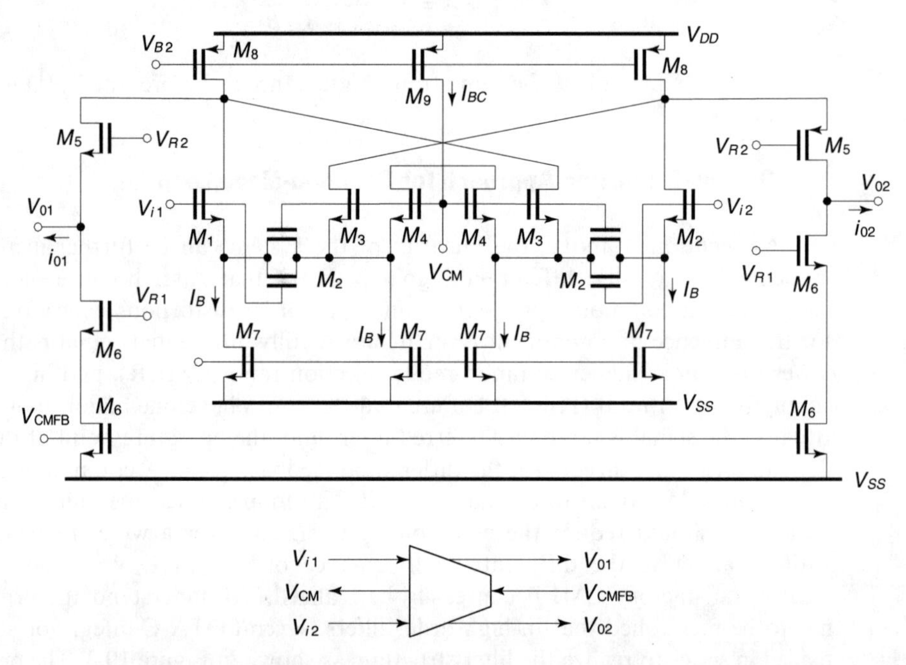

Figure 10.4 Realized folded-cascode operational transconductance amplifier.

With this OTA we realized a fourth-order Legendre filter at a supply voltage of 3 V with a total current consumption of 400 μA. The realized transfer function is given in Figure 10.5. Figure 10.6 shows measurement of the THD with an input voltage of $0.6\,V_{p\text{-}p}$ at a frequency of 10 kHz. It is clear that, using the explained

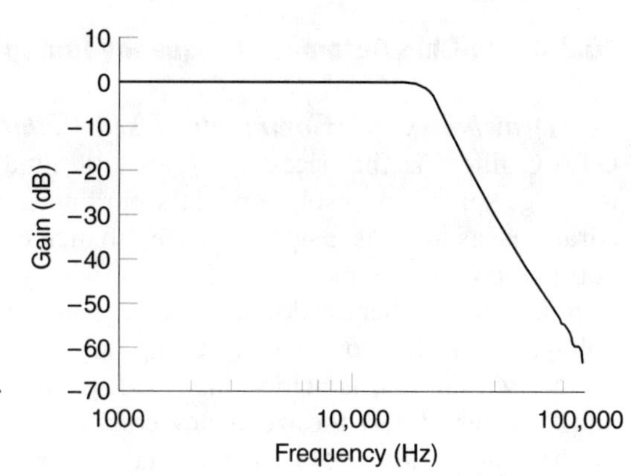

Figure 10.5 Transfer characteristic of realized fourth-order Legendre filter.

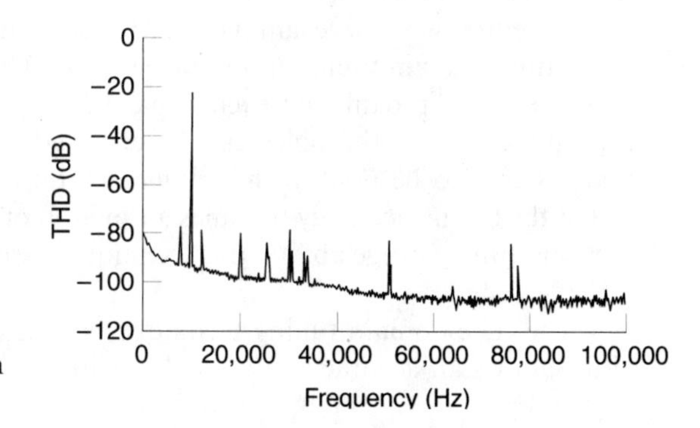

Figure 10.6 Total harmonic distortion $V_{in} = 0.6V_{p\text{-}p}$.

techniques for the design of the low-distortion OTA, a THD lower than −60 dB is feasible even at a supply voltage of 3 V. The measured THD as a function of the input swing is shown in Figure 10.7. The measured noise of the OTA used was 110 μV_{rms}.

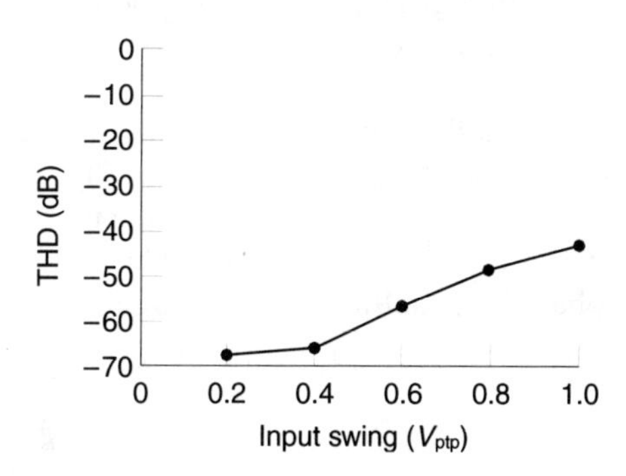

Figure 10.7 THD as function of input swing.

10.3.4. On-Chip Automatic Frequency Tuning

Drawbacks of Commonly Used Solutions. An important drawback of OTA-C filters is the necessity of an additional automatic tuning system. Due to the large spread in absolute process parameter values, the time constant C_L/g_m accuracy is as low as ±40%. Hence, to achieve and design high-accuracy filters, automatic tuning is required. The most widely used technique is based on information in the frequency domain. A second-order "slave" filter is arranged as a voltage-controlled oscillator. Using phased-locked loop techniques, accurate tuning, within 1%, is achievable [11–13]. For good matching between the "slave" and "master" filter, the frequency component of the slave is located at the poles of the "main" filter. This means that the frequency component is just located in the passband of the master filter. Due to parasitic coupling between the slave and master filter, the tuning system noise feedthrough is much higher than the noise level of the filter itself [11–14].

Recently two new techniques have been proposed: a discrete capacitor bank [15] and a time-domain tuning technique [16, 17]. The discrete capacitor bank technique uses a capacitor bank for each capacitor. By automatically selecting the proper capacitance value, the poles can be adjusted. The advantage of this technique is that it can also be used in the implementation of active-*RC* filters. The drawback is that the tuning accuracy becomes a function of the capacitor bank. Regarding chip area and tuning range ability, this technique leads to an accuracy on the order of 5% [15].

The time-domain tuning technique is based on charge comparison. This technique allows an accuracy down to 1% without large tuning noise feedthrough.

Automatic Frequency Tuning Technique Based on Charge Comparison. The poles of the filter transfer function are approximately determined by the C_L/g_m ratios of the OTA-C integrators. In the proposed tuning system the C_L/g_m ratio of a "replica" OTA-C stage is tuned by locking it to an external clock frequency and the ratio of two current sources. An accurate clock frequency can easily be made and is very often available in the overall system. The ratio of two current sources can also accurately be controlled in CMOS because it is only a function of the transistor geometries and relative process parameters. For a low-pass filter, tuning the frequency of the poles is equivalent to tuning the filter's cut-off frequency. Therefore tuning the C_L/g_m ratios is referred to as frequency tuning in this text.

The block diagram of the automatic frequency tuning system is given in Figure 10.8. Here, V_{B1} of the OTA (see also Fig. 10.4) is used to control its transconductance g_m. By connecting the OTA in the unity-gain mode and applying a current I_R, a voltage $V_{O1} = I_R/g_m$ is generated. This voltage is stored on the capacitor C_1 during clock phase ϕ_1, resulting in a stored charge (q) given by

$$q = \frac{C_1 I_R}{g_m} \tag{10.3}$$

During clock phase ϕ_2 this charge is transferred from C_1 to C_H, with the opposite sign. This gives rise to a negative jump in voltage V_{O2} (see Fig. 10.8). Meanwhile a constant current source (NI_R) discharges the C_H capacitor. Hence the negative jump in V_{O2} is followed by a linear slope. The averaged value (V_{freq}), which is a measure of the error between the negative jump and the charge extracted from C_H during one clock period T, is extracted with a low-pass filter and fed back to the OTA. In a steady state the charge injected on C_1 and the charge extracted from C_H should be equal due to the feedback mechanism. The charge extracted from C_H during one clock period T due to the constant current source is given by

$$q = NI_R T \tag{10.4}$$

Combining Eqs. (10.3) and (10.4) gives

$$\frac{g_m}{C_1} = \frac{f_{\text{clock}}}{N} \tag{10.5}$$

From this equation it is clear that in steady state the ratio g_m/C is locked to the clock frequency and the current ratio N. The control voltage (V_{freq}) is used to control the bias of the OTAs in the filter. Consequently the g_m/C ratios of the integrators in the filter are also locked to f_{clock}/N. From this it follows that the accuracy of the proposed tuning technique is limited by the relative matching between the capacitors and between the transconductances of the OTAs on the chip. The factor N, the ratio between the current sources, adds some versatility in the system. In a low-pass filter, for example, it can be used to determine the relative position of the corner frequency and the clock frequency. For instance, it is very useful to put the clock frequency in the stopband of the filter to reduce the clock feedthrough of the frequency tuning system. This will result in lower in-band noise of the filter, and hence a higher dynamic range can be obtained.

The proposed structure for the automatic tuning circuit based on charge comparison is extremely suited for use at low voltages. It only uses the low-distortion

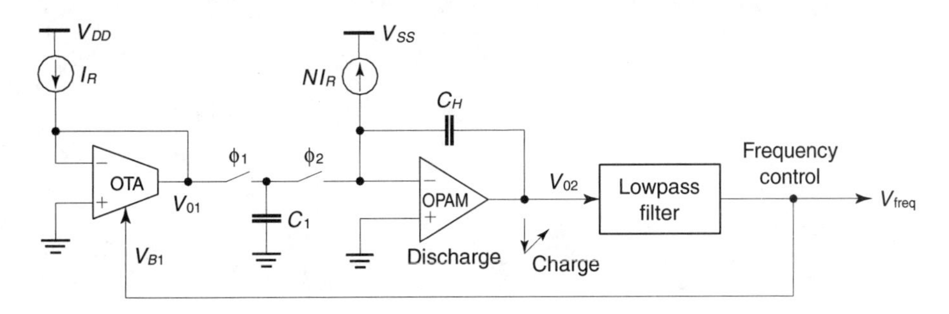

Figure 10.8 Block diagram of charge-based automatic frequency controller.

OTA and a SC-like structure. The OTA, which is the same as used in the filter, was designed to function at 3 V. Switched capacitors can be designed to work even below 3 V. This will be explained in the next section.

10.3.5. Conclusions

From this it can be concluded that OTA-C filters are the most suitable technique to combine low power consumption and high filter accuracy specifications. The main drawback is the higher distortion specification. To overcome the problem, a very linear transconductor and the applied linearization techniques are discussed. Measurements on such an OTA have shown that a dynamic range (signal to noise) of 78 dB with a THD less than −60 dB can be achieved. Because single-chip continuous-time filters suffer from absolute filter specification accuracies, a new on-chip active tuning system has been presented. This techique allows a large tuning range, a frequency accuracy as low as ±1%, and a low tuning signal feedthrough which ensures a large dynamic range. A photomicrograph of the fourth-order Legendre filter using the low-distortion OTA-C and the automatic frequency tuning is shown in Figure 10.9.

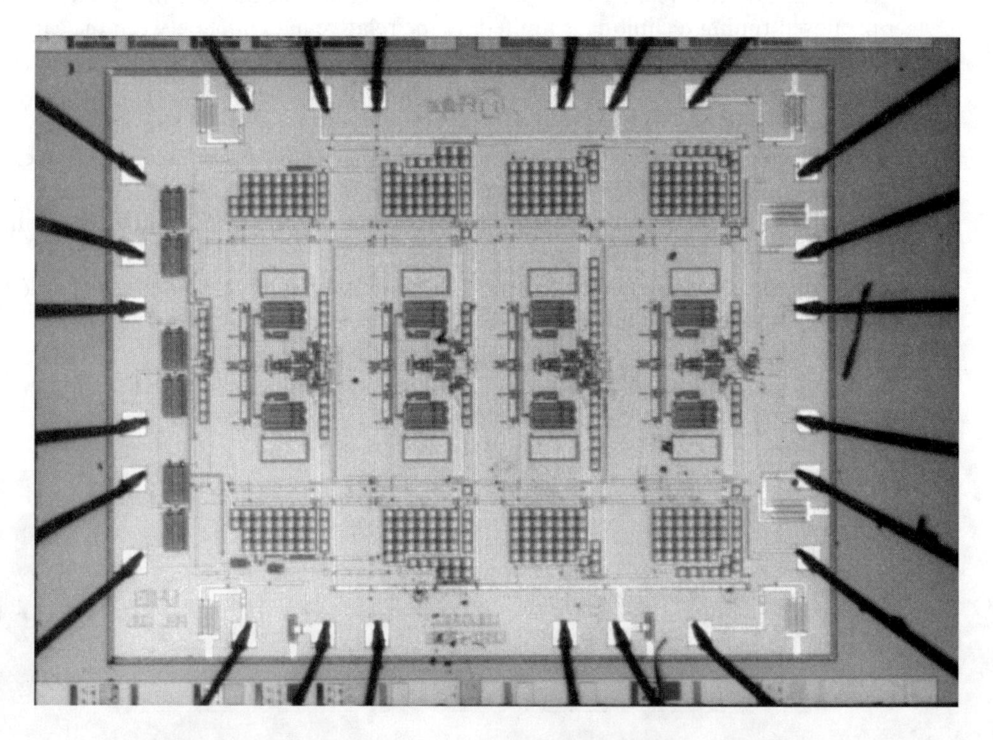

Figure 10.9 Photomicrograph of fourth-order Legendre OTA-C filter.

10.4. VERY-LOW-VOLTAGE SC TECHNIQUES

10.4.1. Introduction

Switched-capacitor circuits are well suited for fully integrated low-power applications. The reason for this is that the resistor of the integrator is replaced by a SC structure which makes the accuracy of the time constant only a function of capacitor ratios and a reference clock frequency. This allows the realization of high-performance filters and circuits fully integrated in standard CMOS technologies with a pole position accuracy of better than 2%.

An extra advantage of SCs is that a large number of different building blocks, such as filters, amplifiers, and multipliers, can be realized with the same basic implementation technique. The use of SC filters is very attractive because they achieve a high filter accuracy with a low distortion. Even at low-power-supply voltages the quality of both parameters is fairly independent of the power supply. This in contrast, to for example, OTA-C filters where the linearity of the OTAs is highly dependent on the available power supply voltage [18] and where the distortion increases rapidly when the power supply voltage is decreased to very low voltages. When the amplification of the OTA is high enough, the distortion in SC filters is mainly determined by the linearity of its capacitors. These capacitors are very linear and their linearity is not influenced by the power supply level. Consequently a THD of −70 dB is achievable down to a 2.5-V power supply for the full output swing of the OTAs, that is, the output range in which they deliver a high amplification [19]. However, at a very-low-power-supply voltage the performance reduces drastically due to the reduction of the signal swing caused by the switches. Special techniques, different from the standard SC design, have to be used to overcome this problem.

In the first part of this section of the text the limitations of SC circuits for low-voltage and very-low-voltage applications are discussed. In the second part three very low voltage SC techniques are presented and their possibilities are discussed. These three techniques are the use of a special process with extra low V_T transistors, the use of a voltage multiplier for the clock signals, and the use of the switched-opamp technique.

10.4.2. Limitations of SC at Low and Very Low Voltages

The following devices must be available for the realization of an SC circuit: a capacitor, an OTA or opamp, and a switch. The operation of a polysilicon-polysilicon capacitor is always independent of the power supply level. The OTAs for SCs can be realized in CMOS down to very low power supply voltages. Figure 10.10 shows the most simple input and output stage for an OTA. Normally the input stage is the limiting factor for an OTA [18]. A differential pair needs a voltage drop of at least V_T between the gate and source of the input transistors, and this can only be realized when it is used in weak inversion. However, due to the virtual ground principle, used in SCs, no input swing is necessary and, as a result, a V_T is sufficient for operation of the input transistors. The minimum value for the reference level V_{ref}

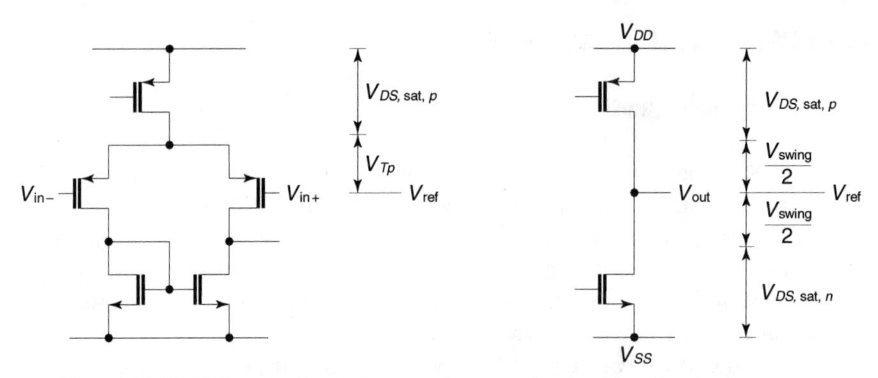

Figure 10.10 DC operating point requirements for input and output stages of OTA or opamp.

is therefore $V_T + V_{DS,\text{sat}}$. Another boundary is set by the output stage. The output stage must be able to deliver an output signal V_{swing} centered around the reference level. The minimum power supply voltage is therefore $V_{\text{ref}} + V_{\text{swing}}/2 + V_{DS,\text{sat}}$. In Eq. (10.6) an approximate formula is given for an OTA with respectively an N- or P-type (NMOS, PMOS) input pair. In a low-voltage design a $V_{DS,\text{sat}}$ will be between 0.1 and 0.15 V. The reference level is high enough to bias the input pair if $\frac{1}{2}V_{\text{swing}} > V_T$. The minimal power supply voltage is then completely determined by the output voltage swing and rail-to-rail operation is possible. The minimal power supply is given in Eq. (10.7). The main problem in a very low voltage SC is the switches. An NMOS transistor, for instance, needs at least a V_T on top of the highest possible signal that it has to switch. This is given in Eq. (10.8):

$$V_{DD,\text{OTA,min}} = V_{DS,\text{sat},n} + V_{Tn} + \tfrac{1}{2}V_{\text{swing}} + V_{DS,\text{sat},p}$$
$$\text{or} \quad V_{DS,\text{sat},p} + |V_{Tn}| + \tfrac{1}{2}V_{\text{swing}} + V_{Ds,\text{sat},n} \tag{10.6}$$
$$V_{DD,\text{OTA,r-to-t,min}} \approx V_{DS,\text{sat},n} + V_{\text{swing}} + V_{DS,\text{sat},p} \tag{10.7}$$
$$V_{DD,\text{n-switch,min}} > V_{Tn} + V_{\text{swing}} + V_{DS,\text{sat},p} \tag{10.8}$$

A typical SC switch consists of complementary driven NMOS and PMOS transistors of minimal size. A MOS transistor does not turn on unless the voltage between gate and source is higher than the V_T. This is no problem for the complementary structure and rail-to-rail operation is possible. The NMOS transistor conducts where the PMOS does not and vice versa. Equation (10.9) gives the conductivity of the transistors as a function of the power supply voltage V_{DD} and the source voltage V_S, which is equal to the signal that has to be switched. The conductivity of a minimum-size complementary switch is given in Figure 10.11 as a function of the voltage to switch. It is calculated with the parameters of a standard 2.4-μm process. The power supply voltage is 5 V, and the picture shows clearly that rail-to-rail operation is possible. In that case the maximal applicable swing is only limited by the output swing of the OTA given by Eq. (10.8):

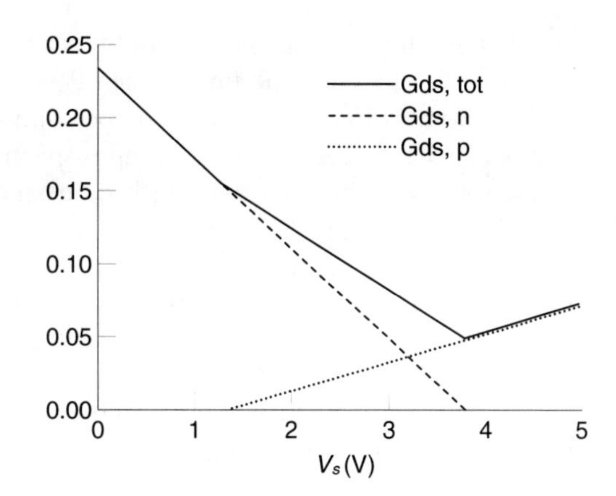

Figure 10.11 Conductivity of minimum-size complementary switch for $V_{DD} = 5\,\text{V}$.

$$G_{DSn} = KP_n \frac{W}{L}\left[V_{DD} - V_S - V_{Tn0} - g_n\left(\sqrt{2f_f + V_S} - \sqrt{2f_f}\right)\right]$$

$$G_{DSp} = KP_p \frac{W}{L}\left[V_S - |V_{Tp0}| - g_p\left(\sqrt{2f_f + V_{DD} - V_S} - \sqrt{2f_f}\right)\right] \tag{10.9}$$

In the curve of Figure 10.11 the conductivity is minimal at $V_{DD} - T_{Tn0}$ (3.8 V for $V_{DD} = 5\,\text{V}$). The conductivity must be high enough because it has considerable influence on the speed performance of the SC circuit. The switch resistance combined with the value of the SC is the time constant that determines the maximal clock speed. A maximum resistance of about 10 kΩ is accepted for most applications. Figure 10.12 represents the maximal resistivity of a complementary switch as a function of the power supply voltage. The resistance of the complementary switch increases when V_{DD} is reduced. The use of a switch which is not minimum size can compensate for this effect. First the PMOS must be enlarged until both transistors have equal conductivity and from then on both transistors must be scaled up. However, larger switches give rise to a large clock feedthrough an often to unwanted

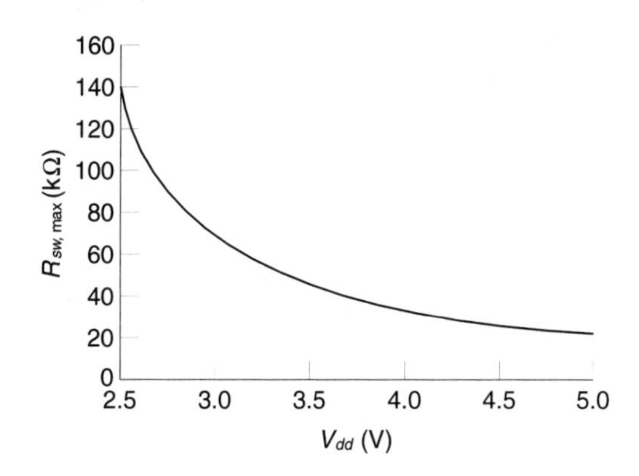

Figure 10.12 Maximum resistance of a complementary switch as function of power supply voltage V_{DD}.

effect. This effect is unavoidable, but it is limited because the clock signal voltage has been lowered at the same time. When designing a low-voltage SC circuit, one must make a compromise between high speed and high clock feedthrough or low speed and low clock feedthrough, resulting respectively in the use of large or small switch transistors. Making this trade-off becomes harder and harder for lower power supplies because the resistance increases faster than the power supply decreases. When the power supply reaches $V_{Tn} + |V_{Tp}|$, the resistance becomes infinite and rail-to-rail operation is not possible anymore. The conclusion is that the standard SC techniques can be used down to a power supply voltage which is equal to $V_{Tn} + |V_{Tp}| + 0.5\,\text{V}$. For a technology with $V_T = \pm 0.9\,\text{V}$ this is about 2.5 V. Hence, as a rule of thumb, SC circuits can be realized quite well in standard CMOS technologies down to a 3-V power supply. Lower voltages will require the use of a special design technique to overcome the problems of increased switch resistance.

The maximum resistance becomes infinite because there is no overlap region anymore where both transistors conduct. Consequently either the NMOS or the PMOS must be used as a switch and this can only be done in a limited signal range. Figure 10.13 depicts this situation. It is the same as Figure 10.11 but now with $V_{DD} = 1.5\,\text{V}$. The use of an NMOS device as a single-transistor switch is preferable because it has a better conductivity than a PMOS and in almost all processes V_{Tn} is lower than, or at least equal to, V_{Tp}. Equation 10.10 is the necessary clock signal voltage required by an NMOS switch. The minimum power supply level can be derived from the minimal conductivity that is needed for the NMOS switch. Again, as in Eq. (10.7), $V_{DS,\text{sat},n}$ is the boundary set by the output swing of the OTA. The $V_{DS,\text{sat},n} + V_{\text{swing}}$ is the maximal source voltage V_S that has to be switched with a conductivity larger than $g_{DSn,\text{min}}$. A W/L of 10 makes the term $g_{DSn,\text{min}}/(W/L \cdot KP_n)$ equal to $V_{DS,\text{sat}}$ for a conductivity of 50 μS. These are acceptable values for a single-device NMOS switch used in a very low voltage application. A W/L of 10 results in a clock feedthrough that is only 1.5 times higher than the clock feedthrough of a complementary switch driven at 5 V. A single-device switch has only half the parasitic capacitance of a complementary switch and the power supply is about three times lower:

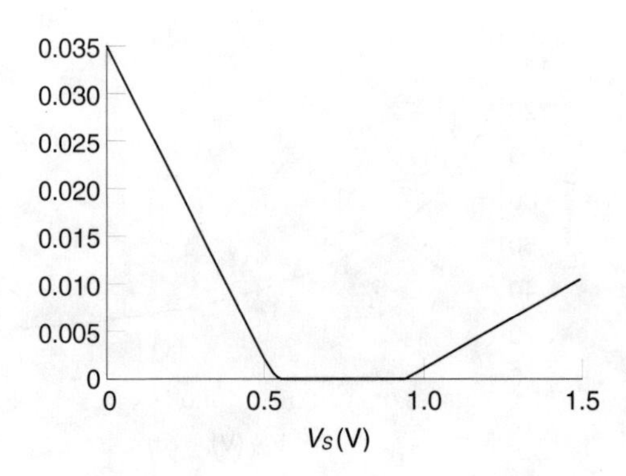

Figure 10.13 Conductivity of minimal size complementary switch for $V_{DD} = 1.5\,\text{V}$.

$$\begin{aligned} V_{DD,\text{n-switch,min}} &= \frac{g_{DSn,\min}}{(W/L)KP_n} + V_{Tn} + V_{S,\max} \\ &= \frac{g_{DSn,\min}}{(W/L)KP_n} + V_{Tn} + V_{\text{swing}} + V_{DS,\text{sat},n} \\ &\approx V_{DS,\text{sat}} + V_{Tn} + V_{\text{swing}} + V_{DS,\text{sat},n} \end{aligned} \tag{10.10}$$

Equation (10.10) is a more precise formulation of Eq. (10.8). A comparison between Eqs. (10.6) and (10.7) shows that it is possible to achieve rail-to-rail operation with a CMOS OTA if V_{Tn} or $|V_{Tp}| < \frac{1}{2} V_{\text{swing}}$. Equation (10.10) shows that a single-device NMOS switch does not allow rail-to-rail operation. An extra voltage drop equal to V_{Tn} is always needed. An NMOS transistor is the only switch that is suited for very low voltage applications. These circuits can therefore only be realized with a reduced swing. The swing that can be applied at very low voltages is, however, far too small for any useful application. Only the use of special SC techniques makes very low voltage operation possible.

10.4.3. Switched Capacitor at Very Low Voltages

Use of a Dedicated Low-V_T Process. From the discussion in the previous section it is clear that the use of low-V_T transistors makes the realizable voltage swing larger and thus, if the voltage swing is fixed, the minimal required power supply voltage smaller. To do this, a dedicated process with extra low V_T implants must be used [20].

An extra NMOS transistor with a V_T of 0.2 V is available in the process described in [20]. This NMOS transistor in combination with a normal PMOS transistor is used to make a complementary switch. The maximum resistance of the switch is relatively high (60 kΩ). The minimum required power supply voltage is 1.4 V, which enables a signal swing of 1 $V_{\text{p-p}}$.

A problem to which special attention must be given is the high leakage current of the low-V_T transistor. This causes a deformation of the frequency transfer characteristic. This effect can be decreased by applying a proper prescaling to the theoretical transfer function.

The main disadvantage of this technique is the high cost of a dedicated process which is only needed for one single type of building block.

Clock Signal Voltage Multiplication. The on-chip generation of a power supply voltage higher than the externally applied power supply voltage [21, 22] seems an obvious solution to any problem caused by the use of very low power supply voltages. This used to be the only known technique for the implementation of an SC at very low voltages. It can be seen from the discussion in the previous section that it is not necessary to let the complete SC circuit run on this higher voltage. Only the switches have to be driven on a higher voltage. This technique has been demonstrated in [23] and [24]. There are two main advantages. One is the fact that the voltage multiplier does not need to deliver the power for the opamps anymore. This greatly simplifies its design and makes its size acceptable.

The second advantage is the power consumption of the opamps which is much lower when they are running on a very low power supply voltage. The voltage multiplier required for these circuits is derived from the erased electrically programmable read-only memory (EEPROM) technology [25, 26].

Any SC circuit can be implemented in this way. Figure 10.14 shows the photomicrograph of a fourth-order SC low-pass filter. It operates on a power supply voltage of 1.8 V. It contains two fully differential correlated double-sampling low-pass biquads, shown in Figure 10.15. The switches are clocked with a 3.6-V signal. The clock frequency is 115 kHz. The bandwidth is 1.5 kHz.

The design of these SC filters is in the first place straightforward and does not differ from a standard SC design [19]. The required gain-bandwidth product (GBW) and open-loop amplification of the opamp are determined by the dynamic and static error specifications. However, special care has to be given to the design of the switches and the design of the common-mode feedback of the fully differential opamp. Figure 10.16 gives a circuit description of the opamp used. A bipolar CMOS (BiCMOS) opamp has been used, but this is imposed by specifications on a higher system level. A full CMOS version could be used as easily. For the input DC level any value between 0 V and the power supply voltage can be chosen. The switches will operate over this region. The value of this bias voltage is therefore not determined by the switches, but by the minimum required value for $V_{GS,M1} + V_{DS,\text{sat},M4}$. This highly depends on the value of V_{Tn} (or V_{Tp} when PMOS input transistors are used). The closer the bias level is to half the power supply voltage, the higher the voltage swing which can be applied.

In this technique single-device NMOS switches are used. The use of complementary switches would be pointless because the working region (from 0 to 1.8 V) is only a fraction of the clock signal (in this case 3.6 V). An NMOS switch is preferred at all times for its better conductivity for the same size of switch transistor. The large clock signal makes use of a minimum-size transistor possible. Consequently, the parasitic capacitances of the switch have no influence on the speed performance of the switch. The speed is mainly determined by the switch resistance R_{on} in combination with the capacitors which have to be switched. Still, the use of small switch transistors is necessary when a small clock feedthrough is required. Higher clock signal levels or larger switch transistors make the switch conductivity and the speed performance larger, but they also increase the clock feedthrough.

A problem in the design of opamps for very low voltage applications is the design of a buffer stage. A buffer stage needs to have an input signal capability which is as large as its output swing. This is very hard to achieve because of the minimum V_{GS} value which is required by the input transistor. Luckily the opamp is used not in a buffer configuration but in a virtual ground feedback configuration, meaning that no input swing is required. However, in fully differential systems there is a problem. The signals at the relatively high impedance output nodes have to be measured and their common-mode value has to be determined through a two-resistor bridge. In a continuous-time solution this would require the use of two buffers at the output nodes. This is not possible in a very low voltage design. In [23] a SC common-mode feedback has been proposed. Figure 10.17 shows this SC circuit for the determina-

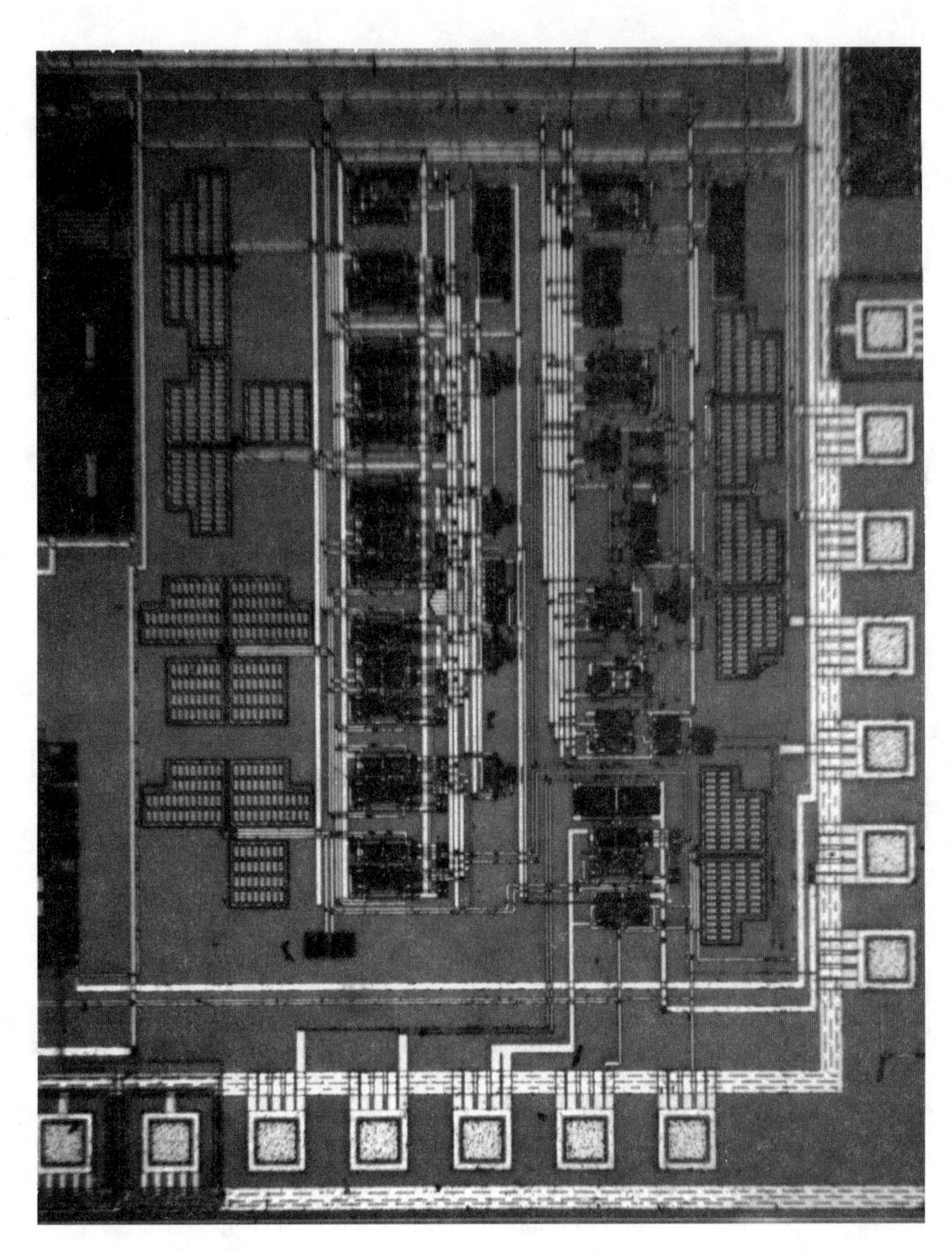

Figure 10.14 Chip photomicrograph of 1.8-V SC filter operated with 3.6-V clock signals.

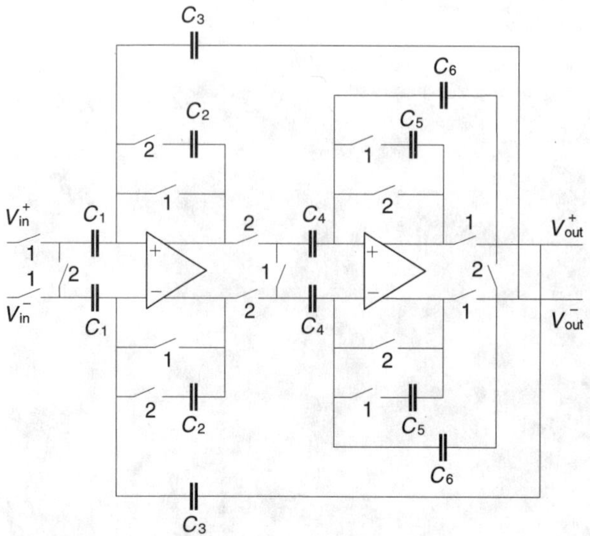

Figure 10.15 Correlated double-sampling low-Q biquad.

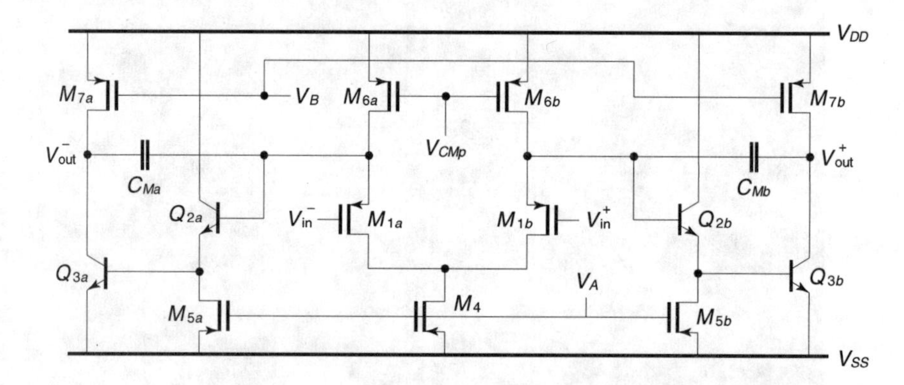

Figure 10.16 The 1.8-V BiCMOS opamp.

tion of the common-mode component of the output signal and its equivalent continuous-time model. This system uses the same switches as the overall SC circuit and it does not load the output stages resistively. The drawback of this common-mode suppression system is that it can only suppress low-frequency signals.

The realized very low voltage SC filter is designed for a power supply voltage of 1.8 V and operates well down to 1.4 V. The THD for a 1-V_{rms} signal is −71 dB. The measurement is shown in Figure 10.18. The equivalent input noise is 18 μV_{rms}.

Switched-Opamp Technique. The basic idea behind the switched-opamp technique [27] is that Eq. (10.10) does not hold for all switches in an SC circuit. Most switches do not require more power supply than an OTA. Those which do require more can be replaced by switchable OTAs. Figure 10.19 gives the topology of an SC inverting integrator, the basic building block for SC filters. Switches S_2, S_3, and S_4 are on one side connected to either the reference voltage V_{ref} or a

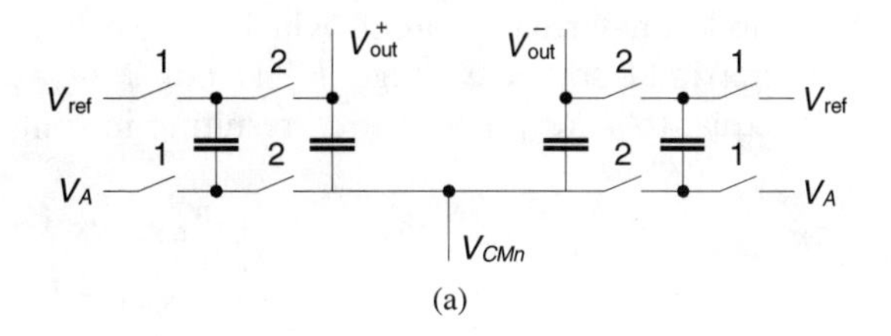

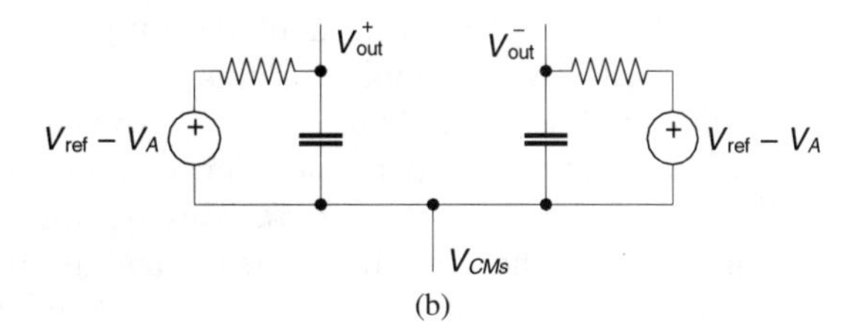

Figure 10.17 (a) The SC common-mode circuit and (b) its equivalent description.

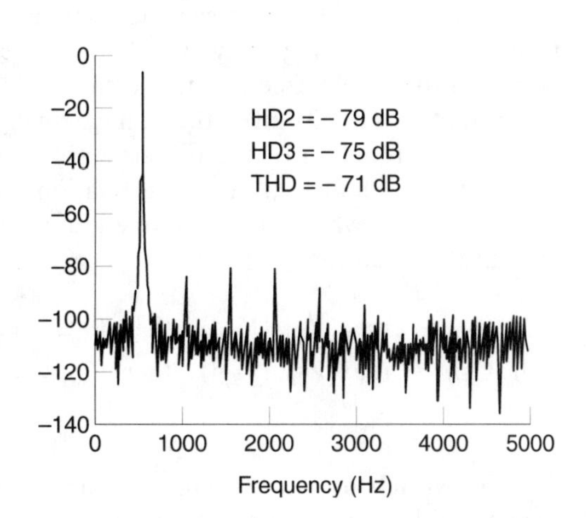

Figure 10.18 Distortion measurement of fourth-order CDS low-pass filter.

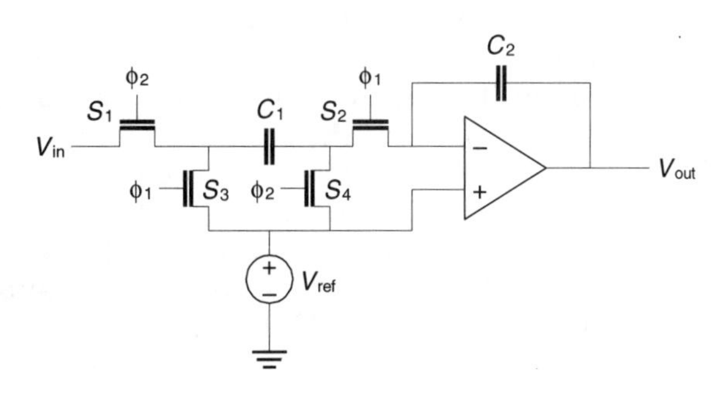

Figure 10.19 Switched-capacitor inverting integrator with NMOS switches.

node on virtual ground, which is also kept on V_{ref} by the feedback system. The maximal source voltage V_S is not $V_{DS,\text{sat},n} + V_{\text{swing}}$ for these switches but V_{ref}, equal to $V_{DS,\text{sat},n} + \frac{1}{2} V_{\text{swing}}$, resulting in (only for S_2, S_3, and S_4)

$$V_{DD,S_{2,3,4},\min} = V_{DS,\text{sat}} + V_{Tn} + \tfrac{1}{2} V_{\text{swing}} + V_{DS,\text{sat},n} \tag{10.11}$$

$$V_{DD,S_1,\min} = V_{DS,\text{sat}} + V_{Tn} + V_{\text{swing}} + V_{DS,\text{sat},n} \tag{10.12}$$

The only switch that has to be able to switch the total signal swing is the input switch S_1. It is connected to the output of the OTA of the previous stage. The minimum power supply for this switch is given by Eq. (10.12). Just eliminating this switch is not possible. Switch S_3 would then short the output of the preceding OTA during clock phase ϕ_2. This can, however, be solved by disabling the driving force of the OTA during ϕ_2. This is only possible if the OTA has to drive and integrate in the same phase. It is obvious that integrating it not possible when the OTA is turned off. As a result, a noninverting delay element has to be inserted everywhere this problem occurs, resulting in an overhead of about 1.5 times more opamps. However, the power consumption decreases to 0.75 of its original value because the opamps are switched off for 50% of the time. In Figure 10.20 the topologies for an SC and a switched-opamp low-Q low-pass biquad are presented. One extra noninverting delay of half a clock period, realized with amplifier A_2, has to be added to the switched-opamp version.

Equation (10.13) gives the voltage swing that is achievable with a switched-opamp as a function of the power supply and the V_T's. The bulk effect must be included in V_{Tn} and V_{Tp}. From this relation it is clear that a swing of $V_{DD} - 2V_{DS,\text{sat}}$ can be achieved when rail-to-rail operation is possible. An NMOS input pair will be used in this situation:

$$\begin{aligned} &\text{If } V_{DD} > 2\max(V_{Tn}, |V_{Tp}|) \quad \text{then } V_{\text{swing}} = V_{DD} - 2V_{DS,\text{sat}} \\ &\text{If } V_{DD} < 2\max(V_{Tn}, |V_{Tp}|) \quad \text{then } V_{\text{swing}} = V_{DD} - 2V_{DS,\text{sat}} \\ &\qquad\qquad\qquad\qquad\qquad\qquad\qquad - [2\max(V_{Tn}, |V_{Tp}|) - V_{DD}] \end{aligned} \tag{10.13}$$

The switched-opamp technique can be used for all SC circuits in which all switches either switch the output voltage of an OTA or else are connected to the reference voltage, directly or virtually. This means that almost all SC circuits can be converted to an equivalent switched-opamp circuit.

The realization of a switchable OTA or opamp can be performed by switching its bias currents. Figure 10.21 gives the circuit description of a switchable Miller opamp. The switch transistor M_{10} can short circuit the current mirror M_8, M_5, M_7. This PMOS switch operates with any power supply voltage larger than $|V_{Tp}| + V_{DS,\text{sat}}$ because its source is connected to the power supply. A high clock signal makes the conductivity of M_{10} equal to 0 and the current mirror will act as if M_{10} is not present. A low clock signal makes the conductivity high. Then $V_{DS,M10}$ is about 10–20 mV and no current flows through M_5, M_7, and M_8 anymore. The opamp is switched off. For a high switching speed M_8 must be relatively large because it has to charge the capacitance $C_{GS,\text{tot}}$ of the total current mirror when

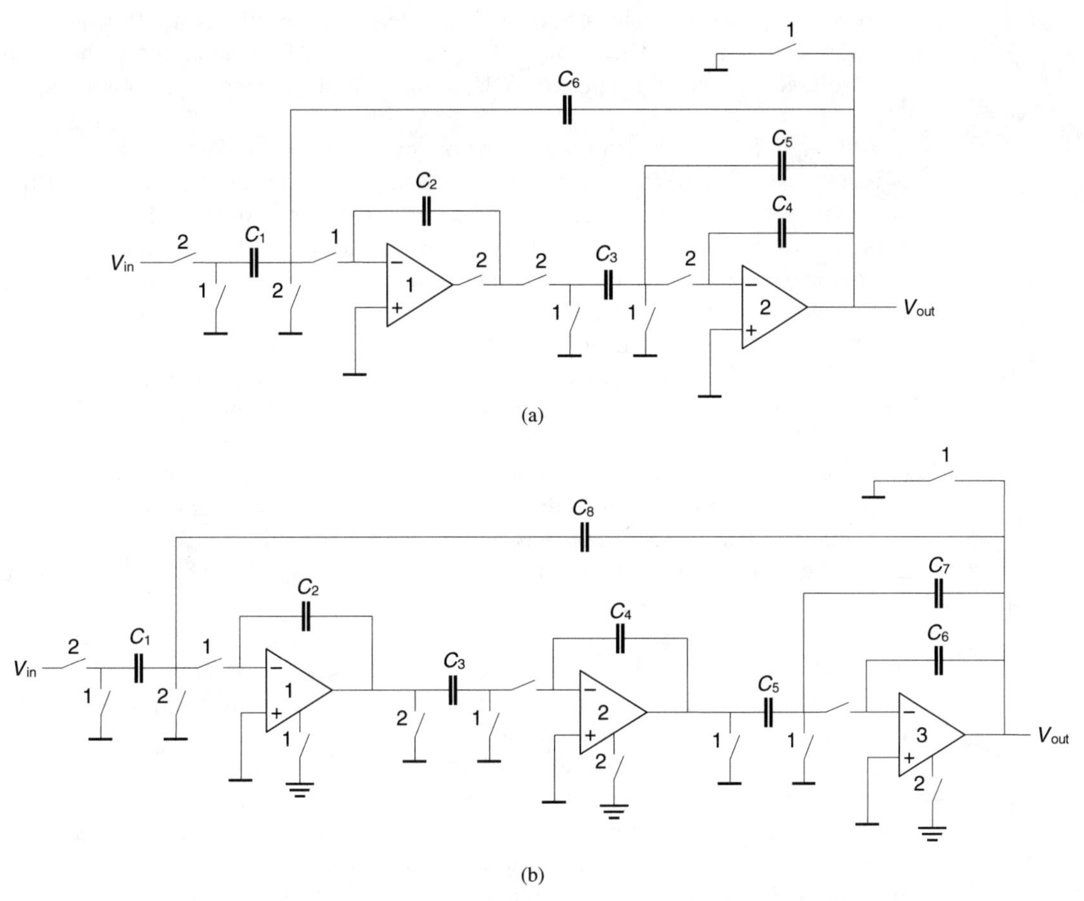

Figure 10.20 (a) Switched-capacitor and (b) switched-opamp low-Q biquad.

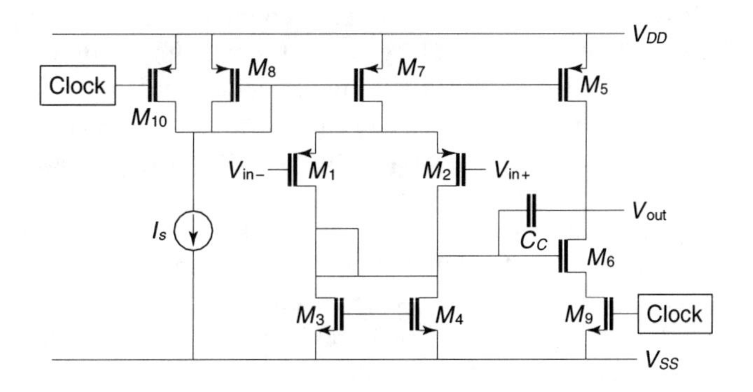

Figure 10.21 Topology of switchable Miller opamp.

the opamp is turned on. The size of M_{10} is determined by its resistivity when it is turned on. This must be a fraction of the resistivity of M_8. Due to the higher gate source voltage V_{GS} which is applied to M_{10}, this is fulfilled when M_{10} is about equal to M_8.

Although it is possible to turn off the opamp by turning off its supply currents, this is not enough. Special care has to be given to the output impedance and the compensation capacitors. The output impedance of the opamp has to be high in the off state. Compensation capacitors may not be charged or discharged during switching because their time constant is very high and consequently the maximal switching speed would then be very low. The symbol M_6 represents the amplifying transistor. The DC current flowing through this transistor is not changed when the current source M_5 is turned off and the capacitors C_C and C_L will be discharged during the off period. This unacceptable behavior can be avoided if the current path through M_6 is interrupted as well. This is the function of M_9, a small switchable resistor. The size of M_9 is defined by two specifications. First the voltage over M_9 must be small in the on state because the swing is directly decreased with twice this value. The maximal voltage that can be accepted is about 20 mV. Second, this resistor may not degrade the amplification of the output stage significantly. Therefore $g_{DS,M9}$ must be a lot higher than $g_{m,M6}$.

The Miller-compensated structure is preferred for this very low voltage application over the one-stage folded-cascode OTA structure which is often used in standard SC circuits [28]. This structure has only two transistors in the output stage, and this gives, apart from the switch transistor, the maximum output swing.

The second-order low-Q low-pass biquad from Figure 10.20 has been implemented, as an example, with a bandwidth of 1.5 Khz and a clock frequency of 115 kHz. It is implemented in a 2.4-μm CMOS process with V_T's equal to ±0.9 V. Figure 10.22 shows the chip photomicrograph. The chip is designed for a power supply voltage V_{DD} of only 1.5 V. According to Eq. (10.13), rail-to-rail operation is not possible and the reference level V_{ref} has to be chosen below $\frac{1}{2}V_{DD}$. The reference level is set by the maximum resistance of the single-device NMOS switch, which has been chosen to be 20 kΩ. The switches have as a result a W/L of 10. With Eq. (10.11) the reference level is found as 0.425 V. The switchable Miller opamp can only handle this value when its input transistors M_1 and M_2 are used in weak inversion. The voltage swing, set by the lower output boundary of the opamp, is 0.55 $V_{\text{p-p}}$.

Special attention has to be given to the switch resistor at the input of the circuit. It cannot be replaced with a switchable opamp because it is not preceded by one. Figure 10.23 shows a solution. The input switch and capacitor are replaced by a large external resistor, making the input current driven. In this design the input resistor is 1 MΩ. The circuit now starts with an active-RC part which is not tuned. However, this has no influence on the frequency response of filters. Only the overall amplification depends on the absolute value of the RC time constant. This replacement of a simulated resistor by a real resistor is not possible when there is a capacitor or the combination of a capacitor and a resistor required at the input. In these cases the input structure with the resistor can still be used if an extra noninverting amplifier is added at the input.

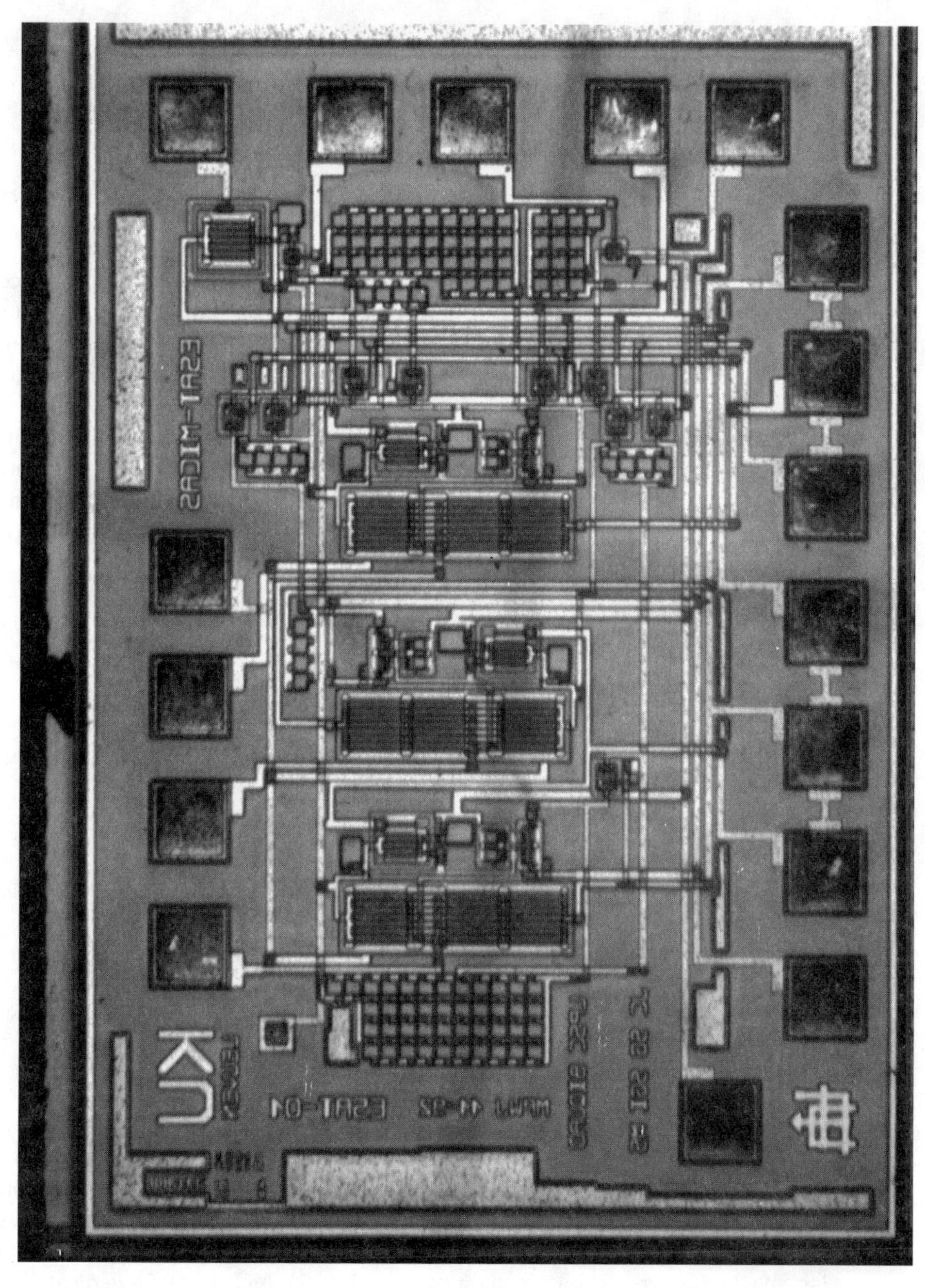

Figure 10.22 Chip photomicrograph of switched-opamp second-order low-pass filter.

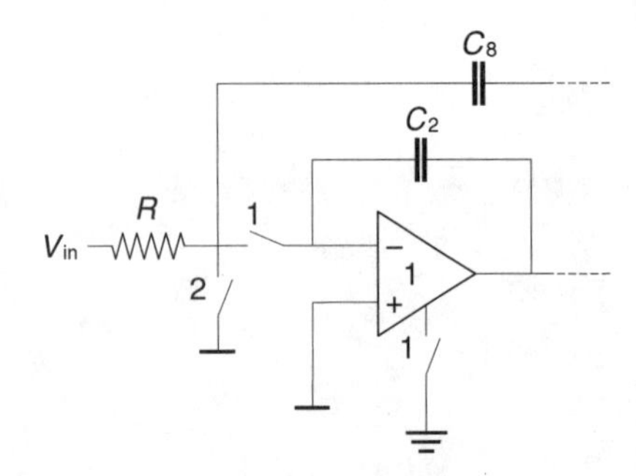

Figure 10.23 Switched-opamp input structure with resistor.

The total chip area of the implemented low-pass biquad is 2.1 mm^2. It contains three opamps, switches, capacitor banks, and an output buffer. The simulated filter bandwidth is 1534 Hz, and the measured bandwidth is 1541 Hz. This indicates that the transfer function has a very high accuracy, a property that can be expected from a SC filter.

According to Eq. (10.11), the swing is 0.55 $V_{p\text{-}p}$. Figure 10.24 is a plot of the THD as a function of the applied signal swing. The THD is −64 dB for the predicted signal swing. The measured equivalent input noise is 140 μV_{rms}. This gives a dynamic range of 69 dB. The power consumption is only 35 μA per pole. The filter operates down to a power supply voltage of only 1.3 V.

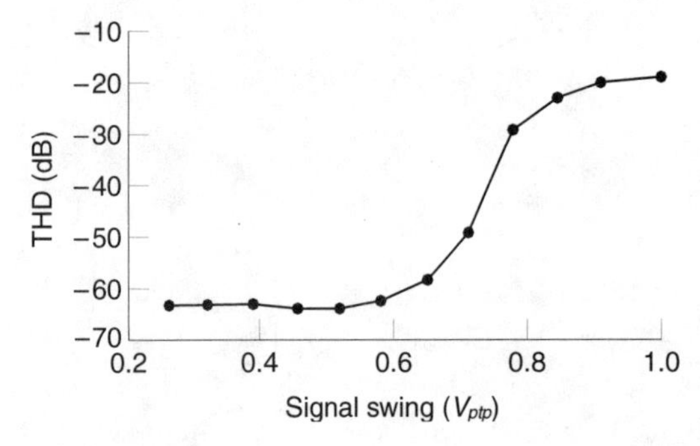

Figure 10.24 THD as a function of applied signal swing.

10.4.4. Conclusions

The design techniques for switched-capacitor circuits at very low power supply voltages have been presented in this section. The use of CMOS switches is not possible below 2.5 V because from that point on there is a region in which neither the NMOS or the PMOS conduct anymore. Three techniques to overcome this problem have been discussed. One is the use of a special dedicated process with extra low V_T NMOS transistors. It suffers from quality degradation caused by large leakage currents, but its main drawback is the high cost of a dedicated process. The second and up to now most used technique is the use of an on-chip voltage multiplier to generate clock signals which are larger than the power supply voltage level. The main disadvantage is the need for a voltage multiplier which consumes extra power and area. A third technique is the switched-opamp technique. It is derived from the classical SC capacitor technique in which critical switches, which cannot operate with very low clock signals, are replaced by opamps which can be switched on and off. This is a true very low voltage technique which can be implemented in any standard CMOS process. A switched opamp requires about 1.5 times more opamps than usual, but the power consumption is only 0.75 of what it normally would be.

References

[1] R. Eschauzier, R. Hogervorst, and J. Huijsing, "A Programmable 1.5 V CMOS Class-AB Operational Amplifier with Hybrid Nested Miller Compensation for 120 dB Gain and 6 MHz UGF," *Proc. IEEE ISSCC*, pp. 246–247, Feb. 1994.

[2] A. M. Durham, W. Redman-White, and J. B. Hughes, "High-Linearity Continuous Time Filter in 5 V VLSI CMOS," *IEEE J. Solid-State Circuits*, vol. SC-9, pp. 1270–1276, 1992.

[3] Y. Tsividis, M. Banu, and J. Houry, "Continuous-Time MOSFET-C Filters in VLSI," *IEEE J. Solid-State Circuits*, vol. SC-21, pp. 15–30, 1986.

[4] M. Banu and Y. Tsividis, "Detailed Analysis of Nonidealities in MOS Fully Integrated Active RC Filters Based on Balanced Networks," *Proc. IEEE*, vol. 131, pt. G, no. 5, pp. 190–196, 1984.

[5] P. R. Gray and R. G. Meyer, *Analysis and Design of Analog Integrated Circuits*, Wiley, Singapore, 1984.

[6] F. Krummenacher and N. Joehl, "A 4 MHz CMOS Continuous-Time Filter with On Chip Automatic Tuning," *IEEE J. Solid-State Circuits*, vol. SC-23, pp. 750–757, 1988.

[7] R. R. Torrance, T. R. Viswanathan, and J. V. Hanson, "CMOS Voltage to Current Transducers," *IEEE Trans. Circuits Systems*, vol. Cas-32, pp. 1097–1104, 1985.

[8] J. Silva-Martínez, M. Steyaert, and W. Sansen, "A High-Frequency Large-Signal Very Low-Distortion Transconductor," *Proc. IEEE ESSCIR-90*, Grenoble, France, pp. 169–172, Sept. 1990.

[9] J. Silva-Martínez, M. Steyaert, and W. Sansen, "A Large-Signal Very Low-Distortion Transconductor for High-Frequency Continuous-Time Filters," *IEEE J. Solid-State Circuits*, vol. SC-26, pp. 946–955, 1991.

[10] J. Silva-Martínez, "Design Techniques for High Performance CMOS Continuous-Time Filters," Ph.D. dissertation, Kath. Univ. of Leuven, April 1992.

[11] V. Gopinathan and Y. P. Tsividis, "A 5 V 7th-Order Elliptic Analog Filter for Digital Video Applications," paper presented at the IEEE International Solid-State Circuits Conference, San Francisco, CA, pp. 208–209, Feb. 1990.

[12] R. Schaumann and M. A. Tan, "The Problem of On-Chip Automatic Tuning in Continuous-Time Integrated Filters," *Proc. IEEE ISCAS-89*, Portland, OR, pp. 106–109, May 1989.

[13] P. M. Van Peteghem and S. Rujiang, "Tuning Strategies in High Frequency Integrated Continuous-Time Filters," *IEEE Trans. Circuits Systems*, vol. CAS-36, pp. 136–139, 1989.
[14] H. Khorramabadi and P. R. Gray, "High Frequency CMOS Continuous-Time Filters," *IEEE J. Solid-State Circuits*, vol. SC-19, pp. 939–948, 1984.
[15] J. B. Hughes, N. C. Bird, and R. S. Soin, "A Novel Digitally Self-Tuned Continuous-Time Filter Technique," *Proc. IEEE-ISCAS-86*, San Jose, pp. 1177–1180, May 1986.
[16] M. Steyaert and J. Silva-Martínez, "A 10.7 MHz CMOS OTA-R-C Bandpass Filter with 68 dB Dynamic Range and On-Chip Automatic Tuning," paper presented at the IEEE International Solid-State Circuits Conference, San Francisco, pp. 66–67, Feb. 1992.
[17] J. Silva-Martínez, M. Steyaert, and W. Sansen, "A Novel Approach for the Automatic Tuning of Continuous-Time Filters," *Proc. IEEE ISCAS-91*, Singapore, pp. 1452–1455, June 1991.
[18] M. Steyaert, J. Crols, S. Gogaert, and W. Sansen, "Low-Voltage Analog CMOS Filter Design," *Proc. ISCAS*, pp. 1447–1450, May 1993.
[19] R. Gregorian, K. W. Martin, and G. C. Temes, "Switched Capacitor Circuit Design," *Proc. IEEE*, vol. 71, pp. 941–966, 1983.
[20] T. Adachi, A. Ishinawa, A. Barlow, and K. Takasuka, "A 1.4 V Switched Capacitor Filter," *Proc. CICC*, Boston, pp. 8.2.1–8.2.4, May 1990.
[21] F. Calias, F. H. Salchi, and D. Girard, "A Set of Four IC's in CMOS Technology for a Programmable Hearing Aid," *IEEE J. Solid-State Circuits*, vol. SC-20, no. 2, pp. 301–312, 1989.
[22] R. Becker and J. Mulder, "SIGFRED: A Low-Power DTMF and Signalling Frequency Detector," *Proc. ESSCIRC*, Grenoble, pp. 5–8, Sept. 1990.
[23] R. Castello and L. Tomasini, "1.5-V High-Performance SC Filters in BiCMOS Technology," *IEEE J. Solid-State Circuits*, vol. Sc-26, no. 7, pp. 930–936, 1991.
[24] M. Steyaert, J. Crols, and G. Van der Plas, "A High Performance RDS-Detector for Low Voltage Applications," *Analog Integrated Circuits and Signal Processing*, special issue on low-voltage low-power analog integrated circuits, vol. 8, no. 1, 1995.
[25] J. Dickson, "On-Chip High-Voltage Generation in NMOS Integrated Circuits Using an Improved Voltage Multiplier Technique," *IEEE J. Solid-State Circuits*, vol. SC-11, pp. 374–378, 1978.
[26] B. Gerber, J. C. Martin, and J. Fellrath, "A 1.5 V Single-Supply One-Transistor CMOS EEPROM," *IEEE J. Solid-State Circuits*, vol. SC-16, no. 3, p. 195, 1981.
[27] M. Steyaert, J. Crols, and S. Gogaert, "Switched-Opamp, a Technique for Realising Full CMOS Switched-Capacitor Filters at Very Low Voltages," *Proc. EESCIRC*, pp. 178–181, Sept. 1993.
[28] T. Choi, R. T. Kaneshiro, R. N. Brodersen, and P. R. Gray, "High-Frequency CMOS Switched-Capacitor Filters for Communications Application," *IEEE J. Solid-State Circuits*, vol. SC-18, pp. 652–663, 1983.

Chapter 11

Continuous-Time Low-Voltage Current-Mode Filters

Edgar Sánchez-Sinencio
Department of Electrical Engineering, Texas A&M University, College Station, Texas 77843-3128
Sterling L. Smith
Texas Instruments, Dallas, Texas 75243

11.1. INTRODUCTION

The low-voltage current-mode filter was motivated by the need to have high-frequency filters with low-power-supply voltages suitable for portable equipment and multimedia applications. Some of the most influential research that preceded the present current-mode techniques were the transconductance-capacitor ($G_m - C$) [1–24, 26, 28, 29], the current conveyor building block [25, 59–61], and the switched-current techniques [33–35]. The natural core of continuous-time, current-mode (CTCM) circuits is the current mirror [44, 54] due to its simplicity, reduced silicon area, and versatility. The fundamental building block is a pseudodifferential current integrator [36–40] with its time constant set by the integrating capacitor and the inverse of the small-signal metal-oxide-semiconductor (MOS) transistor transconductance. The pseudodifferential building blocks cannot reject the common mode of the input signals. This problem can be tackled by using common-mode feedback circuits. We will see that several implementation alternatives exist, that is, cross-coupled configurations of feed forward, common-mode cancellation circuits. The current-mode techniques are often used to avoid the use of the conventional floating differential pair and its voltage drop across the bias current sources; they provide circuits that can operate at low-power-supply voltage levels. The silicon area of these CTCM filters is often significantly smaller than other approaches. The filters that use these integrators in a 2-μm complementary MOS (CMOS) technology consume about 0.7 mV per pole.

In this chapter we will discuss the key building blocks, their limitations and strengths, filter architectures, and examples of test-chip current-mode filters.

11.2. BASIC BUILDING BLOCKS

The fundamental building blocks for current-mode circuits are the current mirror and the transistor driver transconductor shown in Figure 11.1.

The basic current mirror shown in Figure 11.1(a) consists of transistors M_1 and M_2. The ideal current gain A_I is given by

$$A_I = \frac{i_{o_1}}{i_{\text{in}}} = -\frac{(W/L)_2}{(W/L)_1} = -\frac{(W/L)k_1}{(W/L)} = -k_1 \tag{11.1}$$

Also, observe that current copies can be simply expressed as

$$i_{o_i} = -k_i i_{\text{in}} \quad \text{for } i = 12, \ldots, n \tag{11.2}$$

The above expressions are valid if the input and output impedance are very low and high, respectively. Furthermore, it is assumed that $V_{DS_1} = V_{DS_2}$. In real implementations, the output current error is given by

$$\varepsilon_{I_{\text{out}}} = \lambda(V_{DS_2} - V_{DS_1}) \tag{11.3}$$

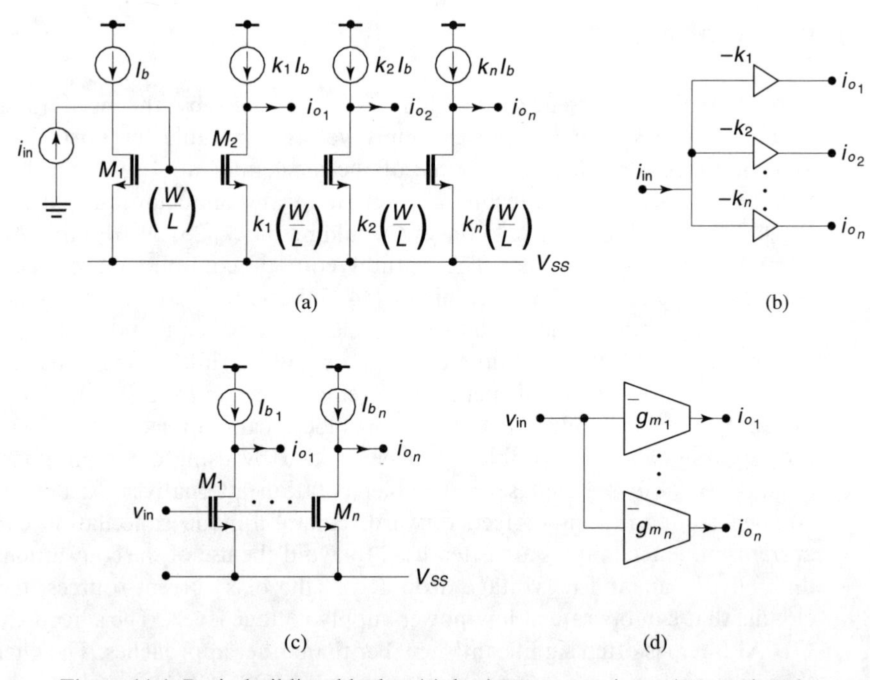

Figure 11.1 Basic building blocks: (a) basic current mirror (M_1 and M_2) and copies; (b) block diagram representation of (a); (c) single-transistor driver transconductor and copies; (d) block diagram representation of (c).

Here, $\varepsilon_{I_{out}}$ can be reduced by using a large transistor length L_2 and by keeping equal V_{DS} values of transistors M_1 and M_2. The channel length modulation parameter λ is a function of the length L of the transistor.

The output resistance $r_{o_2} = 1/g_{ds_2} = 1/g_{o_2} \cong 1/\lambda I_{DS_2}$. This resistance can be increased by increasing the transistor length L_2. However, large values of r_o are difficult to obtain, as therefore, are other, more complex configurations. The input conductance is approximately $g_{m_1} + g_{ds_1}$; thus, the nonideal input and output resistances (conductances) can cause performance degradation. For instance, consider two simple current mirrors in cascade; then the input current of the second mirror (referred to as II) can be expressed as

$$i_{in_2} \cong \frac{i_{in_1}}{1 + g_{o_I}/(g_{m_{II}} + g_{o_{II}})} \cong i_{in_1}\left(1 - \frac{g_{o_I}}{g_{m_{II}} + g_{o_{II}}}\right)$$

Since ideally, $\Delta i_{in} - i_{in_2} - i_{in_1} = 0$, the error involved becomes

$$\Delta i_{in} = i_{in}\left(\frac{g_{o_I}}{g_{m_{II}} + g_{o_{II}}}\right)$$

There are additional errors in a simple current mirror due to $K' = \mu_o C_{ox}$ and threshold voltages V_T assuming $k_1 = 1$; in this case (11.1) becomes

$$\Delta_I = \frac{i_{o_1}}{i_{in}} \cong -\left(1 + \frac{\Delta K'}{K'} - \frac{2\,\Delta V_T}{V_{GS} - V_t}\right)$$

This error can be significant in some applications. For instance, consider $\Delta K'/K' = \pm 4\%$ and $\Delta V_T/(V_{gs} - V_T) = \pm 6\%$; this yields an error of $\pm 20\%$. Furthermore, to reduce the aspect ratio errors, appropriate layout techniques are required [65], as we will show later. The output capacitance is imply $C_{DS_2} + C_{DG_2}$. The minimum output voltage at which M_2 remains in saturation is

$$\min V_{out} = V_{D,sat_2} = V_{GS_1} - V_T = \sqrt{\frac{I_{out}}{K_n'(W/L)_2}} \tag{11.4}$$

The dominant pole usually occurs at high frequencies due to the low input impedance (around $1/g_{m_1}$) and is located at

$$f_p = \frac{g_{m_1}}{2\pi C_p} \tag{11.5}$$

where the parasitic capacitance C_p is given by $C_p = C_{GS_1} + C_{GS_2} + C_{DS_1}$.

This simple current mirror is fast and easy to design; however, for high-precision circuits a more elaborate circuit [44, 52, 54, 63] is needed.

The transconductor shown in Figure 11.1(c) is the simplest, and thus, it can be suitable for high-frequency applications and moderated accuracy. The output resistance and capacitance of this transconductor are the same as for the current mirror discussed before. In a number of applications, the yielding accuracy is not sufficient and other higher output impedances [63] are required, as shown in Figure 11.2.

The output resistor in the cascode configuration [Fig. 11.2(a)] is typically of the order of $r_{o_1}(1 + g_{m_2}r_{o_2})$, which is an increase of roughly $g_{m_2}r_{o_2}$ (voltage gain of M_2). This approximation is valid assuming that M_1 and M_2 operate in strong inversion. One additional advantage of the cascode configuration is that the equivalent input noise of the cascode stage is the same as that of a single transistor of Figure 11.1(c). Besides, for the cascode configuration, the Miller capacitance effect is drastically reduced. Assuming an ideal current bias source, the approximate bandwidth for a capacitive load C_l becomes

$$f_{\text{BW}} \cong \frac{1}{2\pi r_{o_1}(1 + g_{m_2}r_{o_2})C_l} \tag{11.6}$$

An alternative to this cascode is the so-called folded-cascode configuration shown in Figure 11.2(c). Observe that in comparison with the regular cascode, the AC current circles around M_1, M_2, and ground. Furthermore, the AC current flow through M_2 is equal in magnitude but opposite in direction of the current through M_1. The drawbacks of the cascode configuration is the large power supply requirements to operate, in particular with I_b replaced by a cascode bias current, and the limited input (and output) signal swing. The folded-cascode transconductor needs less bias head room at the cost of additional current sources and swing. Until now, we have assumed the MOS transistors are to be operated in saturation. One possibility is to bias the driving transistor M_1 in the liner (ohmic) region [15, 16, 19, 64–67]. This will be translated into an improved input linear range. The corresponding bias of M_2 in Figure 11.2(a) must obey the inequality [40]

$$V_{\text{bias}} < V_{GS_1} + \sqrt{\frac{2I_b}{\mu_n C_{\text{ox}}(W/L)_2}} \tag{11.7}$$

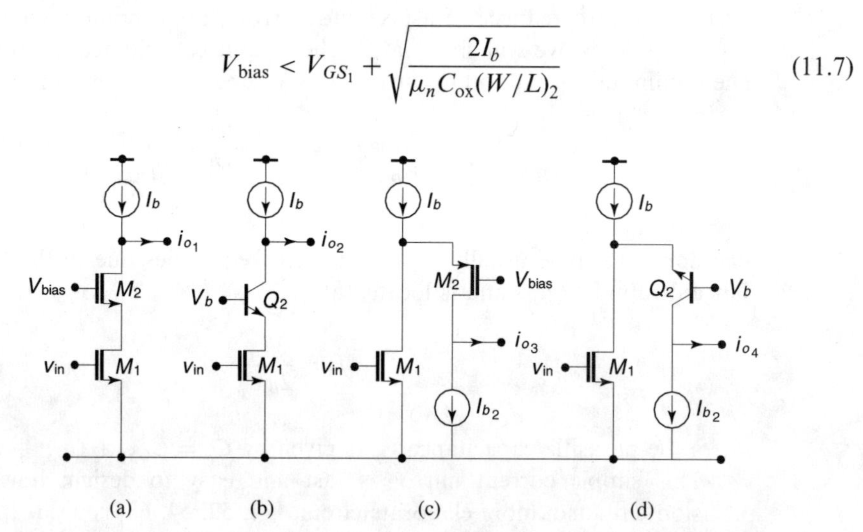

Figure 11.2 Cascode transconductors: (a) MOS cascode; (b) BiCMOS cascode; (c) MOS folded cascode; (d) BiCMOS folded cascode.

The small-signal output current can be roughly approximated to

$$i_{o_1} \cong -g_m v_{\text{in}} = -\mu_n C_{\text{ox}} \left(\frac{W}{L}\right)_1 V_{DS_1} v_{\text{in}} \tag{11.8a}$$

Observe that a key issue in this configuration is to keep $V_{DS_1}(= V_{\text{bias}} - V_{GS_2})$ constant. This transconductance can be tuned within a decade [29] by adjusting V_{DS_1} in the range of 0.1–1 V. One suitable variation is to substitute the MOS transistor M_2 [64] in Figure 11.2(a) by a bipolar transistor Q_2, as shown in Figure 11.2(b); in this case, V_{DS_1} will be easily fixed by V_b. The corresponding folded configuration is illustrated in Figure 11.2(d). All the configurations shown in Figure 11.2 are suitable for low-voltage power supplies, except the one in Figure 11.2(a) with both M_1 and M_2 operating in strong inversion. Furthermore, the complementary counterpart of the structures of Figure 11.2 can be obtained by changing the NMOS and NPN transistors by PMOS and PNP transistors, respectively.

For current-mode operation, we need to deal with both input and output current signals. The current mirror is already a useful inverting current amplifier for the current mode. Let us consider again Figure 11.1; this time a capacitor is connected from the gate of M_1 to ground. When the input signal is i_{in}, the single-transistor transconductor with the capacitor ideally becomes a current-mode integrator, that is,

$$I_{o_1}(s) = \frac{g_{m_1}}{sC} I_{\text{in}}(s) = -\frac{\omega_u}{s} I_{\text{in}}(s) \tag{11.8b}$$

where ω_u is the unity-gain frequency. In practice, (11.8b) is not a good approximation due to the finite $g_m R_o$ and the fact that g_m is frequency dependent. The improvement of the output resistance R_o can be obtained by using the structures of Figure 11.2. For the cases that a very high impedance is required, a gain-boosting scheme can be used [63], as illustrated in Figure 11.3. In this case, the output resistance becomes approximately $r_{o_1}[(1 + A_{V_o})g_{m_2} r_{o_2}]$; in addition, V_{DS_1} can be easily fixed by V_b. Typically, a simple voltage amplifier A_{V_o} is used, and a particular case is the regulated cascode configuration [32]. Observe that to keep a large output resistance, I_b should also have a large resistance.

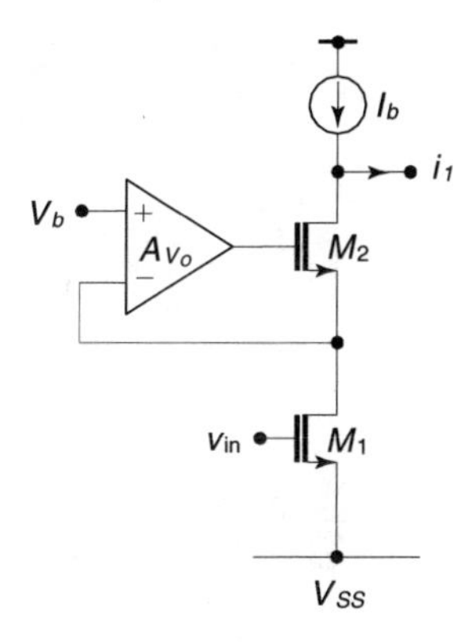

Figure 11.3 Cascode transconductor with improved output resistance.

The frequency compensation of $g_m(j\omega)$ will be discussed later. The improved linear input range can be obtained by using linear transconductors [i.e., Figs. 11.2(b) and (d)]. Fully differential configuration can cancel all the even-order harmonics and will be presented later.

A practical low-voltage bipolar CMOS (BiCMOS) pseudodifferential transconductor reported recently by Rezzi et al. [67] and shown in Figure 11.4 can be judiciously used. Observe that M_1 and M_2 operate in the triode region with a corresponding transconductance value of

$$G_m = \frac{\mu C_{\text{ox}}(W/L)V_{DS}}{1 + g_o/g_{mBJT}} = \frac{\mu C_{\text{ox}}(W/L)V_{DS}}{1 + V_t/V_{DS}} \tag{11.9}$$

where V_t is the thermal voltage. The transconductance G_m can be tuned by varying V_C (which changes V_{DS}). Here, g_{mC} is used for rejecting the common-mode input signal and $g_{md} = g_{mC}$. Thus, I_o^+ and I_o^- ideally (for perfect matching) will not have a common-mode component. Also, observe that this feedforward cancellation approach will not take care of the output offset.

Let us now consider the current mirror of Figure 11.1(a) with M_1, M_2, and a capacitor connected between the gate of M_1 and ground. The corresponding transfer function yields

$$\frac{I_{o_1}(s)}{I_{\text{in}}(s)} = -\frac{g_{m_2}/g_{m_1}}{1 + sC/g_{m_1}} \tag{11.10a}$$

$$f_{3_{\text{dB}}} = \frac{g_{m_1}}{2\pi C} \tag{11.10b}$$

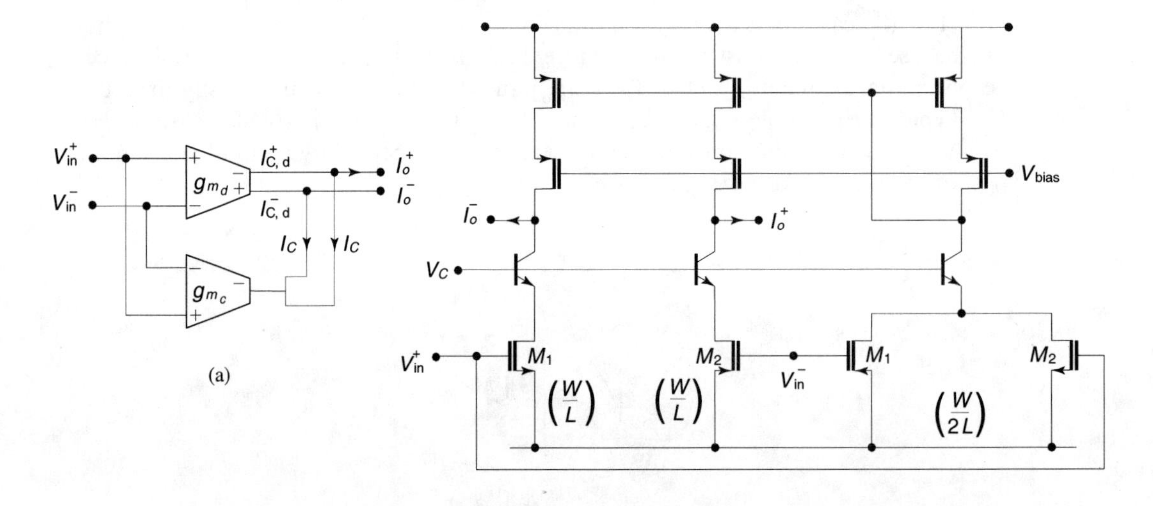

Figure 11.4 Pseudodifferential BiCMOS transconductor with feed-forward common-mode cancellation. (a) Conceptual idea. (b) BiCMOS implementation.

Transfer function (11.10) can be easily obtained from the block diagram shown in Figure 11.5, representing the current mirror from Figure 11.1(a) with an additional capacitor C at the input. Furthermore, notice the similarity of the integrator described by (11.8b) and the lossy integrator by (11.10), if the capacitor C in (11.8a) is substituted by a parallel RC combination with $R = 1/g_{m_1}$ Eq. (11.10) is obtained.

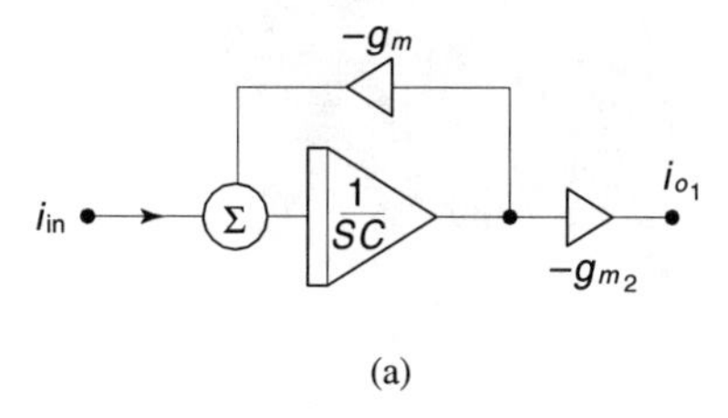

(a)

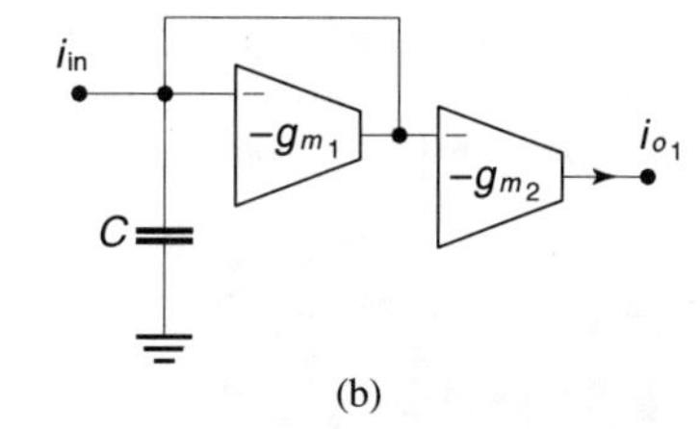

(b)

Figure 11.5 Lossy integrator: (a) block diagram representation; (b) transconductance representation.

A versatile integrator basic building block would be one to be electronically programmable to yield a lossless integrator or a lossy integrator with a programmable 3-dB cutoff frequency. This structure is shown in Figure 11.6. This structure is based on Figure 11.5, with a negative feedback implemented by g_{m_3} and g_{m_4}. Two identical representations are shown in this figure, one with a feedback emphasis representation and the other as a cross-coupled configuration.

The output current from Figure 11.6 yields

$$I_{\text{out}}(s) = \frac{-g_{m_o}(I_{\text{in}_2}(s) - (g_{m_4}/g_{m_3})I_{\text{in}_1}(s))}{[g_{m_1} - (g_{m_4}/g_{m_3})/g_{m_2}] + sC} \tag{11.11a}$$

$$f_{3_{\text{dB}}} = \frac{g_{m_1} - g_{m_2}(g_{m_4}/g_{m_3})}{2\pi C} \tag{11.11b}$$

Note that the trajectories from the inputs ($I_{\text{in}_1}, I_{\text{in}_2}$) to the output are different. These different trajectories can be significantly different when taking into consideration the transistor nonidealities and component matching.

Thus, we can see that for $(g_{m_4}/g_{m_3}) = 1$, a versatile integrator can be designed, yielding

$$f_{3_{\text{dB}}}\Big|_{g_{m_4}=g_{m_3}} = \frac{g_{m_1} - g_{m_2}}{2\pi C} \tag{11.12}$$

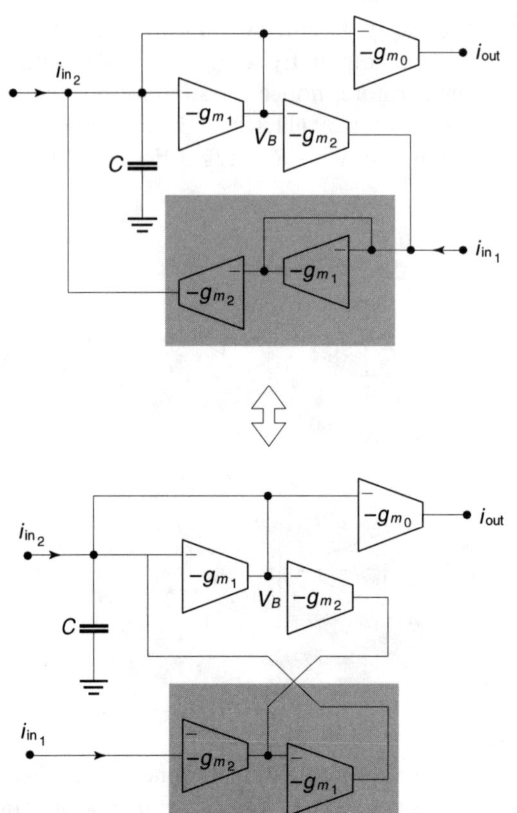

Figure 11.6 Programmable single-ended lossy integrator.

A lossy integrator is obtained for $g_{m_1} > g_{m_2}$, and an ideal lossless integrator for $g_{m_1} = g_{m_2}$. Thus, comparing (11.11a) with (11.8b) yields $\omega_u = g_{m_0}/C$ for $M_1 = M_2$, and $M_3 = M_4$. Next, a fully differential programmable integrator based on the single-ended integrator of Figure 11.6 is discussed. This fully differential structure is obtained by adding a capacitor and a transconductance g_{m_0} to the single-ended configurations. Both circuit [36, 39, 62] and block diagram representations are shown in Figure 11.7.

If the aspect ratios of transistors M_1, M_2, and M_0 are $A_N(W/L)$, $A_p(W/L)$, and (W/L), respectively, then the ideal differential- and common-mode current gains can be obtained by analyzing the first-order model of the integrator of Figure 11.8:

$$A_{dd}(s) = \frac{I_o^+(s) - I_o^-(s)}{I_{\text{in}}^+(s) - I_{\text{in}}^-(s)} = \frac{\omega_u}{s + (A_N - A_p)\omega_u} \tag{11.13a}$$

$$A_{CM}(s) = \frac{I_o^+(s) + I_o^-(s)}{I_{\text{in}}^+(s) + I_{\text{in}}^-(s)} = \frac{\omega_u}{s + (A_N + A_p)\omega_u} \tag{11.13b}$$

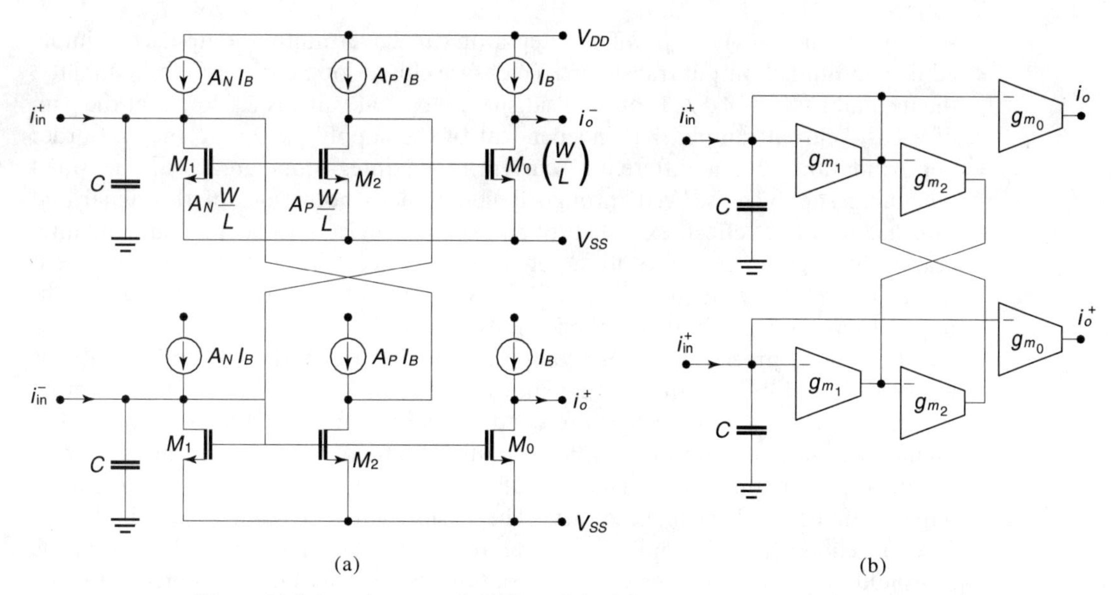

Figure 11.7 Lossless pseudodifferential of lossy current-mode integrator: (a) transistor level; (b) block diagram level.

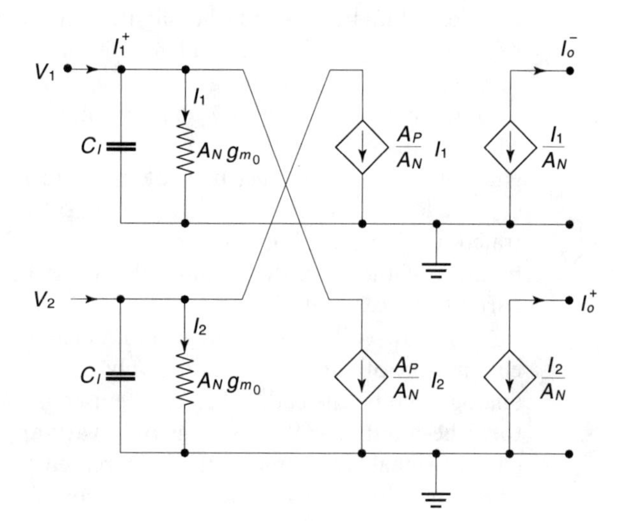

Figure 11.8 Pseudodifferential first-order current-mode model.

where

$$\omega_u = \frac{g_{m_0}}{C} = \frac{g_{m_N}}{A_N C} \tag{11.13c}$$

$$g_{m_0} = \mu C_{\text{ox}} \frac{W}{L} (V_{GS_o} - V_T) = \sqrt{2 I_B \mu C_{\text{ox}} (W/L)} \tag{11.13d}$$

Note that this structure inherently poses a common-mode current gain as well as a differential current gain. The ideal (lossless) integrator is obtained when $A_N = A_p$

and results in $A_{dd}(s) = \omega_u/s$. The integrator can have multiple outputs by simply adding additional output transistors. To ensure closed-loop common-mode stability, the inequality $A_p + A_N > 1$ must be satisfied; a typical value is 2. Note that the gain of a basic current amplifier is independent of the supply voltage to the first-order approximation. When differential current-mode integrators are used, the small remaining supply noise feed through is common to both sides of the signal and thus has no direct effect, except through random device mismatch. Thus, the integrators have good immunity to supply noise. Device mismatch can be minimized through careful layout [65, Chapter 16] (i.e., common centroid) and design techniques to around 1% in many applications.

For higher precision, a more complex fully differential (FD) lossy integrator is required. Based on the architecture shown in Figure 11.7(b) and by implementing g_{m_1}, g_{m_2}, and g_{m_0} by any of the transconductors of Figures 11.1(c), 11.2, or 11.3, a number of useful integrators can be generated. Some of these circuits are shown in Figure 11.9. To enhance the DC gain of the basic integrator circuit, cascode techniques can be used when necessary. The basic cascode circuit, shown in Figure 11.9(a), requires a 5-V supply to operate, due to the larger overhead bias from the threshold voltages. However, several variations of the cascode technique exist (each of which has advantages in particular applications) which can be used with low-voltage supplies. The high-swing cascode circuit [Fig. 11.9(b)] offers a similarly high accuracy, but because of the slightly different connection between transistors, it needs far less head room and has fewer internal parasitic pole-generating nodes between input and output, giving it a better frequency response. The simple transistor integrator of Figure 11.7 has zero internal nodes, resulting in no parasitic poles. The disadvantage of the improved cascode is that due to biasing constraints, the gate-source voltages must be kept small, resulting in larger devices for a given bias current level and smaller unity-gain frequencies. The folded-cascode circuit illustrated in Figure 11.9(c) offers a substantial improvement in biasing flexibility because of the increased drain voltage of the transistors at the cost of additional current sources and bias voltages.

The low-voltage advantage of the current-mode integrator in part comes from the elimination of the analog midsupply level normally associated with conventional analog circuits. Instead, the reference used is the bias current supplied to the integrator. The internal voltage swings only vary approximately as the square root of the current signal. Dynamic range is increased by raising the current (signal and bias) levels and capacitance, which can be accomplished independently of the bias voltages by scaling devices. In this section, we discuss the low-voltage design constraints of the proposed family of integrators. Both high-frequency performance and dynamic range improve with higher gate-source voltages. This applies to both the current sources/sinks and the driver devices. Table 11.1 decribes, in terms of voltage, the overhead required to bias the circuits and the voltage swing (due to the signal current) available for the integrators. It is assumed that all current sources and sinks use optimally biased cascoded circuits. The maximum signal swing must be limited to keep all devices in the saturation region. The maximum signal limit will be a function of the supply voltage, the overhead required to bias the transistors in the correct region of operation, and to an extent, the characteristics of the pseudodifferential circuit.

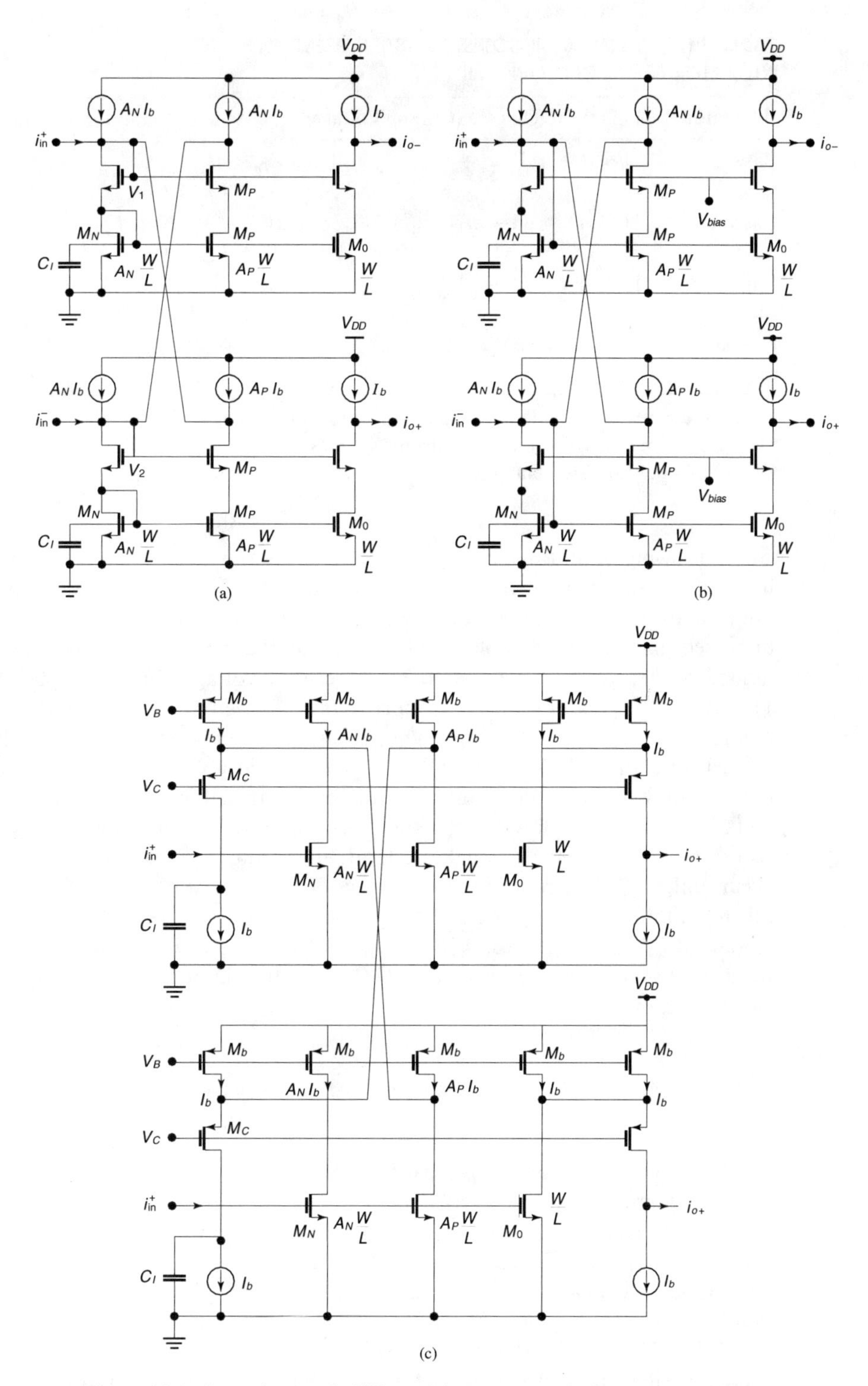

Figure 11.9 Cascode pseudodifferential current-mode integrators: (a) cascode; (b) high-swing cascode; (c) folded cascode.

TABLE 11.1 SIGNAL AND BIAS CONSTRAINTS OF CURRENT-MODE INTEGRATORS OF FIGURES 11.7 AND 11.9

Integrator	Figure	Quiescent Input Voltage	Voltage Swing Limitations
Basic	11.6(a)	$V_{d,\text{sat}} + V_{Tn}$	$V_{Tn} < V_1, V_2 < V_{DD} - V_{\text{sat},Ib}{}^{\dagger}$ $\lvert V_1 - V_2 \rvert < V_{Tn}$
Cascode	11.8(a)	$(1+m)V_{d,\text{sat}} + 2V_{\text{in}}{}^{*}$	$2V_{Tn} < V_1, V_2 < V_{DD} - V_{\text{sat},Ib}$ $\lvert V_1 - V_2 \rvert < V_{Tn}$
Improved cascode	11.8(b)	$(1+m)V_{d,\text{sat}} + V_{Tn}{}^{*}$	$\max(V_{Tn}, V_b - V_{Tn}) < V_1, V_2$ $< V_{DD} - V_{\text{sat},Ib}$
Folded cascode	11.8(c)	$V_{d,\text{sat}} + V_{Tn}$	$\max(V_{Tn}, V_{\text{sat},Ibn}) < V_1, V_2 < V_c{}^{\ddagger} + V_{T_p}$

*Bottom devices of cascode pairs have aspect ratio m^2 times aspect ratio of cascode device.
†$V_{\text{sat},Ib}$ is minimum voltage across current sources.
‡V_c is cascode transistor bias voltage.

The swing limitations given in Table 11.1 are those required to keep all devices in the saturation region, that is, $V_{GS} > V_T$ and $V_{DS} > V_{GS} - V_T$. These limitations are in addition to the maximum signal swing allowed for a given distortion, to be discussed later. The available voltage swing at the input of the basic integrator is limited to a peak value equal to the threshold voltage. For example, in Figure 11.9(a), if V_1 is greater than V_2 by more than V_{TN}, then M_{P1} will leave the saturation region. For this reason, the input voltage swing is increased when using MOS field-effect transistors (MOSFETs) with higher threshold voltages (as long as other limitations allow). Two other limitations affect the available voltage swing. The input voltages cannot go below the threshold voltage, which would put the NMOS transistors into the cutoff, or weak/moderate-inversion, regions. This is not likely to occur unless the current signal swing goes beyond twice the bias current. The other limitation is that the input voltages must not go above the saturation limit of the bias current sources, $V_{DD} - V_{\text{sat},IB}$. Assuming a worst-case full differential signal swing of twice the bias current, the resulting limit on the voltage bias levels yields

$$\sqrt{\frac{2I_b}{\beta}} + V_{Tn} + V_{\text{sat},Ib} < V_{DD} \tag{11.4}$$

Common values for $V_{\text{sat},Ib}$ are 0.5–1.0 V. For a minimum supply voltage of 3.0 V and a maximum threshold voltage of 1.2 V, the resulting quiescent bias value $V_{d,sat_{nom}} = \sqrt{I_b/\beta}$ can be designed to between 0.3 and 1.0 V. The limits for the integrators are summarized in Table 11.2, where it has been assumed for simplicity that in the folded cascode the maximum voltage at the drain of M_c is $V_{DD} - V_{\text{sat},Ib}$, as for the other three integrators.

Table 11.2 shows that the cascode integrator is clearly unsuitable for low-voltage design due to the second threshold voltage. The improved cascode can be

TABLE 11.2 SUPPLY VOLTAGE LIMITATIONS ON BIAS

Integrator	Limit	Typical $\sqrt{2I_b/\beta}$	Typical Gain
Basic	$\sqrt{2I_b/\beta} + V_{TN} + V_{\text{sat},Ib} < V_{DD}$	0.5–1.0	15–50
Cascode	$(1+m)\sqrt{2I_b/\beta} + 2V_{Tn} + V_{\text{sat},Ib} < V_{DD}$	0.1–0.2	100–1000*
Improved cascode	$(1+m)\sqrt{2I_b/\beta} + V_{Tn} + V_{\text{sat},Ib} < V_{DD}$	0.3–0.5	100–1000*
Folded cascode	$\sqrt{2I_b/\beta} + V_{Tn} + V_{\text{sat},Ib} < V_{DD}$	0.5–1.0	100–1000*

Note: Assume a 0.1% mismatch.
*Depends on device modeling.

designed using low-voltage supplies but with decreased override due to the need to keep the additional cascoded transistor in saturation. The folded cascode has the same basic bias and swing characteristics as the basic integrator. In conclusion, the analysis clearly shows that the basic, improved cascode and folded-cascode integrator circuits can be designed using 3-V supplies and that the cascode integrator circuit was clearly not suitable for low-voltage design. The design equations given can be used to minimize the parasitic effects (smallest devices possible at given current) and maximize signal to noise for a given capacitor area (highest current for a given transconductance g_m). Matching (threshold voltage) and extrinsic noise immunity are also improved with larger $V_{d,\text{sat}}$.

From previous discussion and Figure 11.1, we know that the basic building blocks are the integrator and the amplifiers (multiplication of the input signal by a constant). We have presented pseudodifferential integrator families and will identify their nonidealities next. Before doing that, we will present a *pseudodifferential current-mode amplifier*. One possible implementation is the circuit shown in Figure 11.7, the pseudodifferential integrator without the capacitor. In that case,

$$K_I = \frac{I_{od}}{I_{\text{in},d}} = \frac{1}{A_N - A_p} \qquad A_N > A_p \tag{11.15}$$

Thus, for a given K_I and A_p, the value of A_N can be determined as $A_N = A_p + 1/K_I$. However, notice that even though we do not connect an integrating capacitor, a parasitic input capacitor C_p, exists; this capacitor will determine the amplifier bandwidth given by (11.12) with C substituted by C_p, or equivalently see (11.4).

Noise. The wide-band input referred-noise power spectral density of the basic integrator is given as

$$S_{I_{\text{in}}} = 8KTg_m\left[A_N + A_p + \left|\frac{j\omega + \omega_u/A_{dd}}{\omega_u}\right|^2\right] \tag{11.16a}$$

where A_{dd} is the DC gain of the integrator and ω_u is the unity-gain frequency of the integrator. Equation (11.16a) makes the assumption that the transconductances of the bias current source transistors are equal to $\frac{1}{2}g_m, \frac{1}{2}A_{Ng_m}$, and $\frac{1}{2}A_p g_m$, respectively, for simplicity and that $1/f$ noise can be neglected due to the white-noise dominance in the frequency range of interest. The derivation of (11.16a) assumes device noise follows standard equations found in Berkeley SPICE models, which relates device noise to g_m. This is an industry standard, but an incorrect model [53] which for long-channel devices can underestimate the noise power by 20% or more and can lead to gross underestimation errors of a factor of 2–5 in relatively short channel length devices, such as those used in high-frequency designs. The noise of the folded-cascode integrator is 4 dB higher than the other integrators due to the extra current sources. Using (11.16a), the total output noise power of a two-integrator loop biquad is found to be approximately

$$\overline{i_{nlp}^2} = 12\pi kTg_m f_c(2Q+1) \tag{11.16b}$$

for the low-pass output. Assuming that the goal is to obtain a given maximum signal-to-noise ratio (dynamic range), the noise output equation can be manipulated to yield

$$\overline{i_{nlp}^2} = 12\sqrt{2}\pi kTf_c(2Q+1)\frac{I_b}{V_{d,\text{sat}}} = P_n I_b \tag{11.17}$$

where P_n is a constant. Due to probable supply voltage constraints, $V_{d,\text{sat}}$ is a fixed quantity. (As I_b increases, β must increase to maintain a fixed $V_{d,\text{sat}}$. The parameter C must increase by the same factor at β and I_b to maintain a constant filter cutoff frequency.) The maximum signal at the output will be limited to $I_{od,\max} = \eta 2I_b$, where $0 < \eta < 1$ is dependent on the maximum tolerable distortion, from THD and the third intermodulation distortion (to be derived later). The dynamic range can then be expressed as

$$\text{DR} \equiv \frac{I_{o,\max}}{\sqrt{\overline{i_{n,\text{out}}^2}}} = \frac{2\eta i_b}{\sqrt{P_n I_b}} = 2\eta\sqrt{\frac{I_b}{P_n}} \tag{11.18}$$

The dynamic range can thus be increased by increasing the bias current and corresponding device and capacitor areas, just as in other continuous-time techniques.

11.2.1. High-Frequency Parasitic Effects

When the parasitic capacitances and finite output resistance of the basic pseudodifferential current-mode integrator of Figure 11.7 are considered, the transfer function

$$A_{I_{dd}}(s) = \frac{I_o^+(s) - I_o^-(s)}{I_{\text{in}}^+(s) - I_1(s)} = -\frac{C_{gdo}}{C_T}\,\frac{s - g_m(s)/Cg_{do}}{s + [g_{oT} - (A_n - A_p)g_m(s)]/C_T} \tag{11.19}$$

is obtained, assuming nonideal input current sources, with conductance g_i and capacitance C_i, but the outputs can be assumed to go into an ideal short-circuit load without loss of accuracy, where

$$C_T = C_i + C + 4Cg_{dp} + Cg_{do} \qquad g_{oT} = g_i + g_{on} + g_{op}$$

g_{on} and g_{op} being the output transconductance of transistors M_N and M_p, respectively. Here, Cg_{dp} and Cg_{do} are the gate-drain capacitance of M_p and M_o, respectively.

Observe from (11.19) that the ideal pole has moved to a nonideal pole S_p,

$$S_p = \frac{g_{oT} + (A_N - A_p)g_m(s)}{C_T} \tag{11.20a}$$

Now, even for $A_N = A_p$ a residual pole is placed at

$$S_p\Big|_{A_n=A_p} = -\frac{g_{oT}}{C_T} \tag{11.20b}$$

instead of at $S_{p_{\text{ideal}}} = 0$. Furthermore, for the ideal integrator, the zero at infinity is located at the right half-plane (RHP) at

$$S_z = \frac{g_m(s)}{Cg_{do}} \tag{11.20c}$$

and the low-frequency current gain becomes finite, that is,

$$A_{I_{dd}}(0) \cong -\frac{g_m(0)}{g_{oT}} \tag{11.21}$$

One design strategy to enhance the integrator is to make $(A_N - A_p)g_m(s) \cong -g_{to}$. This can be done by either carefully changing the aspect ratios (A_N, A_p) or the biasing current for one of them and then verifying the stability via a Monte Carlo analysis. The transconductance of a MOS transistor [53] can be approximated by

$$g_m(s) = \frac{g_{m0}}{1 + s\tau} \cong g_{m0}(1 - s\tau) \tag{11.22}$$

where

$$\tau \approx \frac{4L^2}{15\mu(V_{gs} - V_T)} \approx \frac{4c_{g_o}}{15g_{m0}} = \frac{6c_{g_{so}}}{16g_{m0}} \tag{11.23}$$

To analyze the second-order frequency dependencies in the integrator, (11.22) may be substituted into (11.19). Setting $A_N = A_p$ for the lossless integrator, the transfer function becomes

$$A_{I_{dd}}(s) = \frac{g_{m0}}{sC_T + g_{oT} + g_m\delta A_{np}}\left(\frac{1 - s\tau_z - s^2\tau_p\tau_z}{1 + s\tau_p}\right) \approx \frac{g_{m0}}{sC_T + g_{oT} + g_m\delta A_{np}}\left(\frac{1 - s\tau_z}{1 + s\tau_p}\right) \tag{11.24}$$

where $\tau_z = C_{gdo}/g_{mo_1}$, $\tau_p = \tau$, and $\delta A_{np} = A_N - A_p$ is the transistor mismatch causing $A_N - A_p$ to have a small nonzero value. From (11.24), the Q of the integrator [5] can be obtained as

$$\frac{1}{Q_{\text{int}}} = \frac{g_{oT}}{g_{m0}} + \delta A_{np} - \frac{g_{m0}}{C_T}(\tau_p + \tau_z) = \frac{g_{oT}}{g_{m0}} + \delta A_{np} - \frac{3C_{gso}}{7C_T} - \frac{C_{gdo}}{C_T} \tag{11.25}$$

The same analysis can be applied to the improved and folded-cascode by substituting the frequency response of the cascode pairs into the equations for $g_m(s)$. Then, the quality factors of the integrators will be approximated by

$$\frac{1}{Q_{\text{int}}} \approx \left(\frac{g_{oT}}{g_{m0}}\right)^2 + \delta A_{np} - \frac{3c_{gs0}}{7C_T} - \frac{c_{gd0}}{C_T} - \frac{c_p g_{m0}}{C_T g_{moc}} \tag{11.26}$$

where c_p is the parasitic capacitance at the drain of the M_0 node between the cascoded transistors. From (11.25) and (11.26), we can see that the quality factors of the presented integrators available at a given frequency of operation approximate those available using high-frequency fully differential transconductance amplifier circuits [8]. The effect of mismatch in the common-mode feedback on the differential gain, while easily shown here, often applies to other fully differential integrator circuits, such as an operational transconductance amplifier (OTA-C).

Regarding the common-mode current gain of the basic pseudodifferential current-mode integrator assuming $A_N + A_p = 2$, ideally (11.13b) yields

$$A_{\text{CM}}(s) = -\frac{\omega_u}{s + 2\omega_u} \tag{11.27a}$$

$$A_{\text{CM}}(0) = -\tfrac{1}{2} \tag{11.27b}$$

If a better common-mode gain is required at the expense of additional circuitry, the circuit shown in Figure 11.10 will provide lower $A_{\text{CM}}(0)$. The triangular block "$-k$" represents a simple current mirror with a gain k as the one shown in Figure 11.1(a). This structure can be suitable for programmable filters and low frequency. The corresponding $A_{I_{\text{CM}}}$, assuming equal output conductance and a parasitic capacitance C_p for the amplifier k, yields

$$A_{I_{\text{CM}}}(s) = \frac{g_o}{g_m}\frac{1 + s(C_p/2g_o)}{1 + s(C/2kg_m)} \tag{11.28}$$

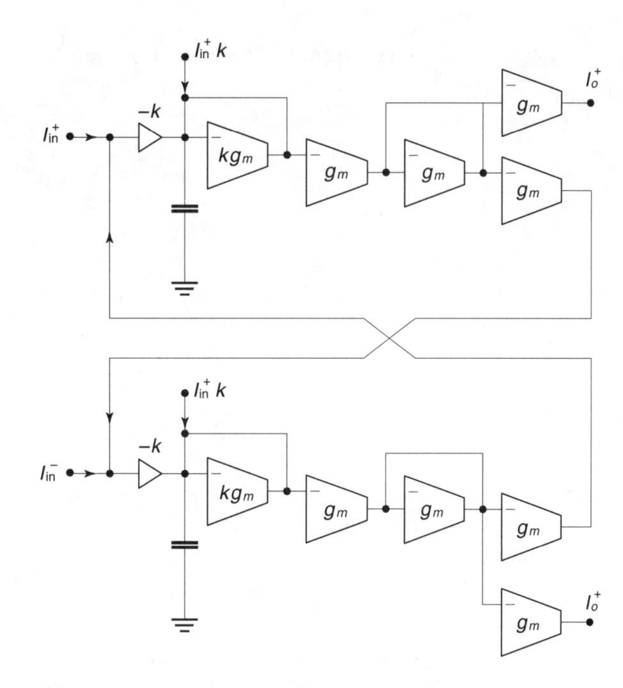

Figure 11.10 Pseudodifferential integrator with more degrees of freedom.

where

$$A_{I_{CM}}(0) = \frac{g_o}{g_m} \quad \text{and} \quad A_{I_{CM}}(\infty) = k\,\frac{C_p}{C}$$

At low frequency ($\omega < g_o/0.5C_p$), the gain is significantly smaller than the integrator of Figure 11.7. However, at higher frequencies both integrators behave similarly, even though the simplified expression (11.27a) does not show any high-order zero. One possible strength of the pseudodifferential integrator is the extended degrees of freedom to program (tune) the circuit. The ideal $A_{dd}(s)$ given by (11.13a) for this pseudodifferential integrator becomes

$$A_{dd}(s) = \frac{kg_m/C}{s + k[(A_N - A_p)g_m + 2g_o]/C} \tag{11.29}$$

Observe that the original ω_u becomes $k\omega_u$; also the 3-dB cutoff frequency has been changed by a k factor, while the $A_{dd}(0)$ gain remains constant.

11.2.2. Large-Signal Operation

The ideal differential output small-signal integrating current can be expressed as [62]

$$i_{oi} = \frac{K'W}{L}(V_{GS} - V_T)\int_{-\infty}^{t} \frac{i_{id}}{C}\,dt = \omega_u \int_{-\infty}^{t} i_{id}\,dt \tag{11.30}$$

The large-signal current can be derived [62] as

$$i_{od} = i_{oi}\sqrt{1 - \frac{1}{4}\left(\frac{i_{oi}}{2I_{\text{bias}}}\right)^2} \tag{11.31}$$

Note that the nonlinearity error is dependent only on the ratio of integrated signal current to bias current (I_{bias}). This expression resembles the differential output voltage of a CMOS differential pair [65]. The third intermodulation distortion and THD of the integrator can be approximated as

$$\text{IM}_3 \cong \frac{3}{32}\left[\frac{i_{oi}}{2I_{\text{bias}}}\right]^2 \quad \text{and} \quad \text{THD} \cong \frac{1}{32}\left[\frac{i_{oi}}{2I_{\text{bias}}}\right]^2 \tag{11.32a}$$

For small values of IM_3, the maximum current for a given third-order intermodulation becomes

$$i_{od,max} \cong 8I_{\text{bias}}(\tfrac{2}{3}\ \text{IM}_3)^{1/2} \tag{11.32b}$$

For instance, for $\text{IM}_3 = 1\%$ and $I_{\text{bias}} = 7.66\,\mu\text{A}$, $i_{od,\text{max}} \cong 5\,\mu\text{A}$. Furthermore, for an ideal $A_{DD}(0)(A_N = A_p)$, $\text{THD}_3 \leq 1\%$ limits $i_{oi}/2I_{\text{bias}}$ to be less than 0.54.

11.2.3. Effects of Transistor Mismatch on Distortion

The small-signal differential output current of the integrator for threshold voltage deviation $\Delta V_T = V_{T_n} - V_{T_p}$ yields a third-order harmonic distortion given by

$$\text{HD}_3\Big|_{\Delta V_T} = \frac{\Delta V_T}{32(V_{GS} - V_T)}\left(\frac{i_{oi}}{I}\right)^2 \tag{11.33}$$

This distortion can be made small by biasing the transistors at large values of $V_{GS} - V_T$. This could limit the minimum power supply voltage. This also implies that transistors must be used with small W/L ratios and operate deeply in strong inversion. The typical design of differential pairs implies opposite considerations.

11.2.4. Phase Compensation Techniques

The ideal integrator has the following properties:

a. The output current I_{od} always lags I_{id}, the input current, by exactly 90°. The term ϕ_E is the (excess) phase exceeding the 90°.
b. The integrator has an infinite DC gain due to its single pole at the origin.
c. The gain becomes 1 at $s = j\omega_u$.
d. The integrator is a linear building block.

In reality, the integrator has approximately these properties over a limited signal and frequency range. Because of the finite output resistance, the pole at the origin will move to a real pole in the left half-plane (LHP). Partial solutions to this problem consist of matching [see (11.19)] $g_{TO} = (A_p - A_N)g_m(0)$ as close as possible while keeping the integrator stable. Another way to improve the finite DC gain is to increase the output resistance by means of a folded cascode or a high-swing cascode [32]. To keep an ideal $-90°$ phase for the transfer function of the integrator, the integrator high-frequency parasitic pole ω_p [ideally $\omega_p = \infty \Rightarrow A_{dd}(j\omega_p) = 0$] must be several orders of magnitude higher than the unity-gain frequency of the integrator. Taking into account high-order terms ω_p can be expressed as the high-frequency (effective) parasitic pole involving a number of high-frequency real poles and zeros which can be approximated [52] as

$$\omega_p = \left(\sum_i \frac{1}{\omega_{pi}} - \sum_j \frac{1}{\omega_{zj}} \right)^{-1} \tag{11.34}$$

One possible solution to cancel the excess phase ϕ_E is shown in Figure 11.11. Here, C is the integrating capacitor and R_C is the compensation resistor.

This method consists of using a phase-lead network in which R_C is implemented by a transistor operating in the triode region with its drain-to-source resistance becoming R_C. The corresponding input admittances of Figure 11.1, denoted by Y_1 and Y_2, are

$$Y_1 = sC \, \frac{1 + sCR_C}{sCR_C} \tag{11.35a}$$

$$Y_2 = sC \, \frac{1 + sR_C C/4}{1 + sR_C C/2} \tag{11.35b}$$

The ideal integrator from (11.13a) for $A_N = A_p$ becomes

$$A_{dd}(s) = \frac{g_m(s)}{Y_i(s)} \qquad i = 1, 2 \tag{11.36a}$$

$$A_{dd}(s) = \frac{g_m(s)}{sC} \, \frac{sCR_C}{1 + sCR_C} \qquad i = 1 \tag{11.36b}$$

$$A_{dd}(s) = \frac{g_m(s)}{sC} \, \frac{1 + SR_C C/2}{1 + sR_C C/4} \qquad i = 2 \tag{11.36c}$$

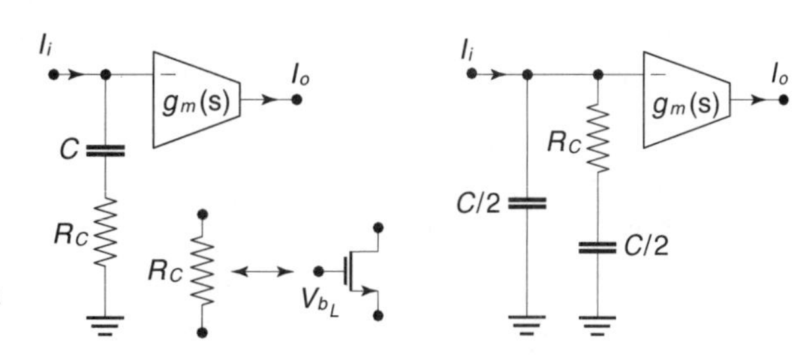

Figure 11.11 Passive phase compensation of current integrator.

From (11.36) one can see that by tuning R_C, a variable LHP zero can cancel the phase lag at the unity-gain frequency due to the parasitic poles of $g_m(s)$ and the output impedance. The product $R_C C$ is proportional to τ [see (11.22)], where $1/\tau$ is the high-frequency parasitic pole of $g_m(s)$.

One additional problem occurring with the basic current-mode integrator is the feedthrough due to the gate drain of the integrator output. This effect can be canceled via the cross-coupled capacitances C_x shown in Figure 11.12.

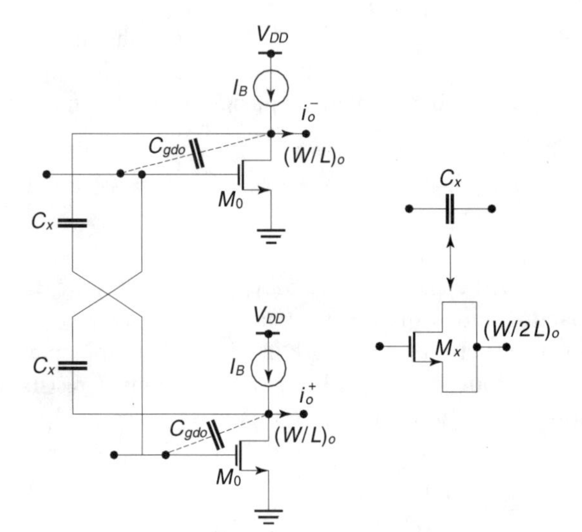

Figure 11.12 Gate-drain capacitance cancellation scheme and capacitor C_x implementation.

Observe that by feeding C_x, exactly equal to the gate-drain capacitance C_{gdo} to the output from the opposite node, the feedthrough through the additional capacitor C_x will have the same magnitude. But be 180° out of phase with the undesired feedthrough, thus canceling. Capacitance C_x can be implemented by a transistor with drain and source nodes connected together. Here, M_x must be in cutoff so the capacitance C_x is equal to the gate overlap capacitance.

11.3. BIQUADRATIC CURRENT-MODE FILTERS

One of the most convenient ways to design a high-order filter consists of cascading m second-order blocks. In the case of an odd-order filter, an additional first-order filter is cascaded with the biquads. Several desirable features for a biquadratic filter should be as follows:

i. Reduced silicon area.
ii. Independent tuning of f_o, Q, and gain.
iii. Capability to internal signal scaling without changing the original transfer function. This will allow maximization of the dynamic range.
iv. Flexibility to program any complex pole and complex zero in the s-plane.

In real filters, the features discussed above cannot be met simultaneously. There are some trade-offs that can be employed in the filter design. We will discuss both single-ended and pseudodifferential filter architectures. The block diagram of a general single-ended filter structure is shown in Figure 11.13. The corresponding output current is given by

$$I_o(s) = \frac{1}{s^2 + \omega_o/Q + \omega_o^2}\left[k_1\omega_o^2 I_{\text{in}_1} + k_2\omega_o s I_{\text{in}_2} + k_3\frac{\omega_o}{Q}sI_{\text{in}_3} - k_4 s^2 I_{\text{in}_4}\right] \tag{11.37}$$

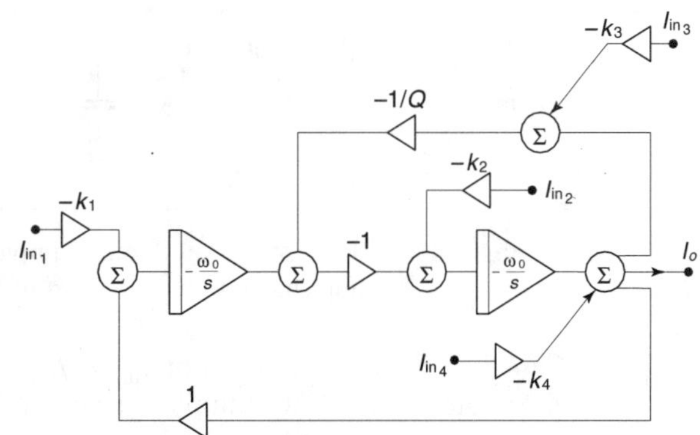

Figure 11.13 Block diagram of single-ended general biquad.

From this equation, one can see that a number of filter types can be implemented. A few possibilities are shown in Table 11.3.

TABLE 11.3 FILTER TYPES OF FILTER IN FIGURE 11.13

Input(s)	I_{in_1}	I_{in_2}	I_{in_3}	I_{in_4}
Filter type	LP	BP	BP	−HP

Observe that the $-k_1, -k_2, -k_3, -k_4$, and -1 coefficients can be implemented by a current mirror; the inverting integrator can be implemented by any of the transconductors of Figure 11.1(c), 11.2, or 11.5 (for $g_{m_3} = g_{m_4}$ and $g_{m_1} = g_{m_2}$) with an integrating capacitor. The coefficient 1 can be implemented by a direct connection (short circuit). Also, notice the fact that three output replica currents I_o are needed for the two feedback loops and the output to be used. The same consideration applies for the implementation of $-k_4$; three replicas are involved.

Figure 11.14 illustrates the simplest transistor implementation of the biquad filter. We have included only $-k_3$ for simplicity.

By combining different inputs and polarities of the coefficients, a host of filter types can be obtained. For example, consider the following cases where all inputs but the ones indicated are zero.

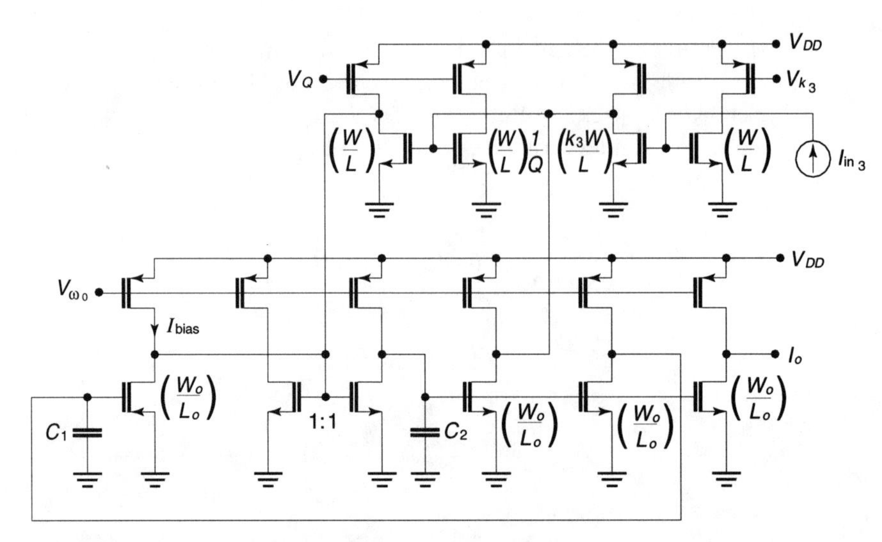

Figure 11.14 Transistor-level biquad implementation with only $-k_3 I_{\text{in}_3}$ implementation.

Case 1: $s^2 \pm k_1\omega_o^2$ *(zeros)*. For $I_{\text{in}_1} = I_{\text{in}_3} = I_{\text{in}}, k_4 \to -k_4$ and $k_1 > 1$ or $k_1 < 1$ yield notch filters. For example, a symmetric notch is obtained for $k_1 = 1$, a low-pass notch for $k_1 > 1$, and a high-pass notch for $k_1 < 1$. If k_4 is as shown in Figure 11.13, a symmetric real zero to boost amplitude can be obtained.

Case 2: $s^2 + k_3(\omega_o/Q)s + \omega_o^2$ *(zeros)*. For $I_{\text{in}_1} = I_{\text{in}_3} = I_{\text{in}_4} = I_{\text{in}}, |k_1| = 1$, and the polarity of k_4 changed and set to 1, an amplitude equalizer is obtained.

Let us now consider fully differential biquad filter structures. In single-ended structures, the coefficients and integrators are, in their simplest form, negative. To obtain a positive integrator (or coefficient), an additional current mirror is usually required. When using a fully differential version, both outputs, negative and positive, are available. It should also be noticed that the inputs are required to be positive and negative. A fully differential filter block diagram version is shown in Figure 11.15. Notice we have assumed, in this diagram, that all the negative k coefficients are equivalent to those of Figure 11.13. However, if any k coefficient polarity needs to be changed, this can be done by simply interchanging the output terminal connections. The k coefficients could be implemented by (11.15). If the transconductors are not based on differential pairs, these transconductors are denominated pseudodifferential since they do not inherently react to the common-mode component of the input signals.

One variation of the single-ended biquad of Figure 11.13 that is often reported in the literature is shown in Figure 11.16. Notice that we have included only one input signal for simplicity. Furthermore, the lossy integrator with a local feedback $(1/Q, \omega_o/s)$ loop has been replaced by a self-loop.

The coefficients $-k_1$ and -1 can be implemented by current mirrors [Fig. 11.1, Eq. (11.15)] and the transconductor integrators by the circuits illustrated in Figures 11.6 and 11.9. Observe that $\omega_o^2 = g_{m_o}^2/C_1C_2$ and $\omega_o/Q = g_Q/C_2$ and

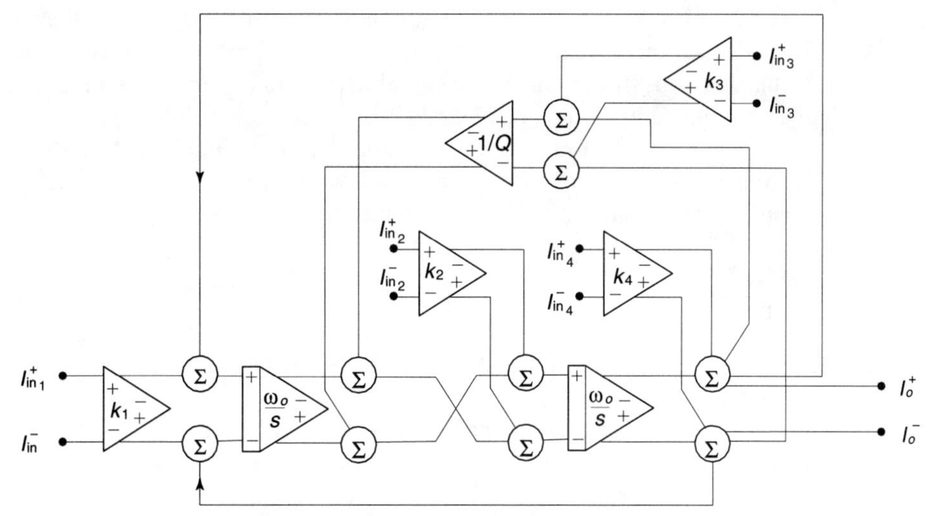

Figure 11.15 Fully differential general biquad.

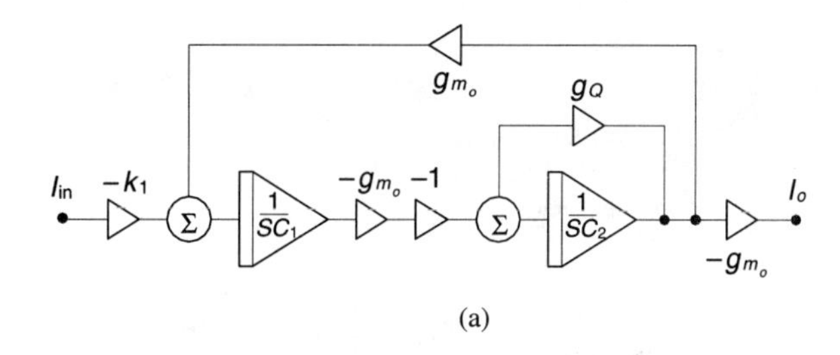

(a)

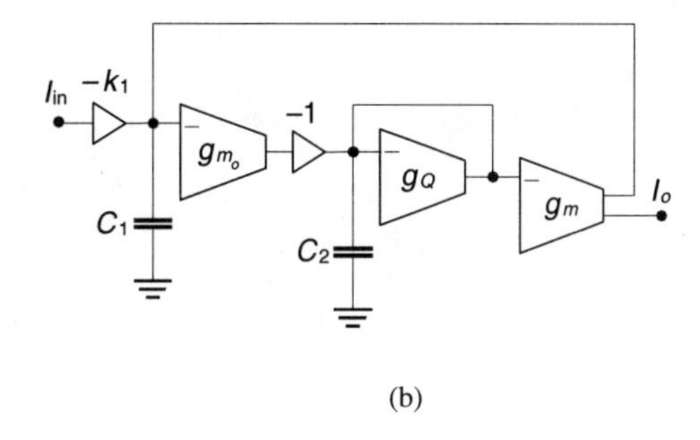

(b)

(c)

Figure 11.16 Biquad structure with self-loop integrator: (a) single-ended block diagram; (b) single-ended transconductance implementation; (c) fully differential transconductance implementation.

$Q = (g_{m_o}\sqrt{C_2/C_1})/g_Q$; thus, ω_o and Q can be independently tuned. The structures of Figures 11.12 and 11.15 are not dependent on any particular transconductor implementation. Thus, for instance, any suitable transconductor from Figure 11.2 can be used to implement Figure 11.13 or 11.16.

In some applications, the discussed architectures might not be the most suitable. For example, in some applications a low-Q sensitivity with respect to the component values is required. The architecture of Figure 11.12 yields

$$\mathrm{BW} = \frac{\omega_o}{Q} \quad \text{and} \quad S_Q^{\mathrm{BW}} = -1 \tag{11.38}$$

This property might be desirable for wide Q-tuning range but not for low sensitivity.

One possible biquad architecture with lower sensitivity is shown in Figure 11.17. The transfer function for I_{o_2} in Figure 11.17(a) yields

$$\frac{I_{o_2}(s)}{I_{\mathrm{in}}(s)} = \frac{-k}{s^2 + (\omega_o/Q) + \omega_o^2} \tag{11.39}$$

where

$$k = K_1 K_{o_1} K_{o_2} \tag{11.40a}$$

$$K_{o1} = \frac{g_{mo_1}}{C_1} \qquad K_{o_2} = \frac{g_{mo_2}}{C_2}$$

$$K_{Q_1} = \frac{g_{q_1}}{C_1} \qquad K_{Q_2} = \frac{g_{q_2}}{C_2}$$

$$\mathrm{BW} = \frac{\omega_o}{Q} = K_{Q_1}K_{o_1} + K_{Q_2}K_{o_2} \tag{11.40b}$$

$$\omega_o^2 = K_{o_1}K_{o_2}(+K_{Q_1}K_{Q_2}) \tag{11.40c}$$

$$S_{K_{o_1}}^{\mathrm{BW}} = S_{K_{Q_1}}^{\mathrm{BW}} = \frac{1}{1 + K_{Q_2}K_{o_2}/K_{o_1}K_{Q_1}}; \qquad S_{K_{o_2}}^{\mathrm{BW}} = S_{K_{Q_2}}^{\mathrm{BW}} = \frac{1}{1 + K_{Q_1}K_{o_1}/K_{Q_2}K_{o_2}} \tag{11.41}$$

$$S_{K_{o_1}}^{\omega_o^2} = S_{K_{o_2}}^{\omega_o^2} = \frac{1}{1 + K_{Q_1}K_{Q_2}}; \qquad S_{K_{Q_1}}^{\omega_o^2} = S_{K_{Q_2}}^{\omega_o^2} = \frac{1}{1 + 1/K_{Q_1}K_{Q_2}} \tag{11.42}$$

For the transconductance implementation of Figure 11.17(b) the corresponding values are

$$\mathrm{BW} = \frac{g_{q_1}}{C_1} + \frac{g_{q_2}}{C_2} \qquad k = \frac{k_1 g_{m_{o_1}} g_{m_{o_2}}}{C_1 C_2} \tag{11.43a}$$

$$\omega_o^2 = \frac{g_{m_{o_1}} g_{m_{o_2}} + g_{q_1} g_{q_2}}{C_1 C_2} \tag{11.43b}$$

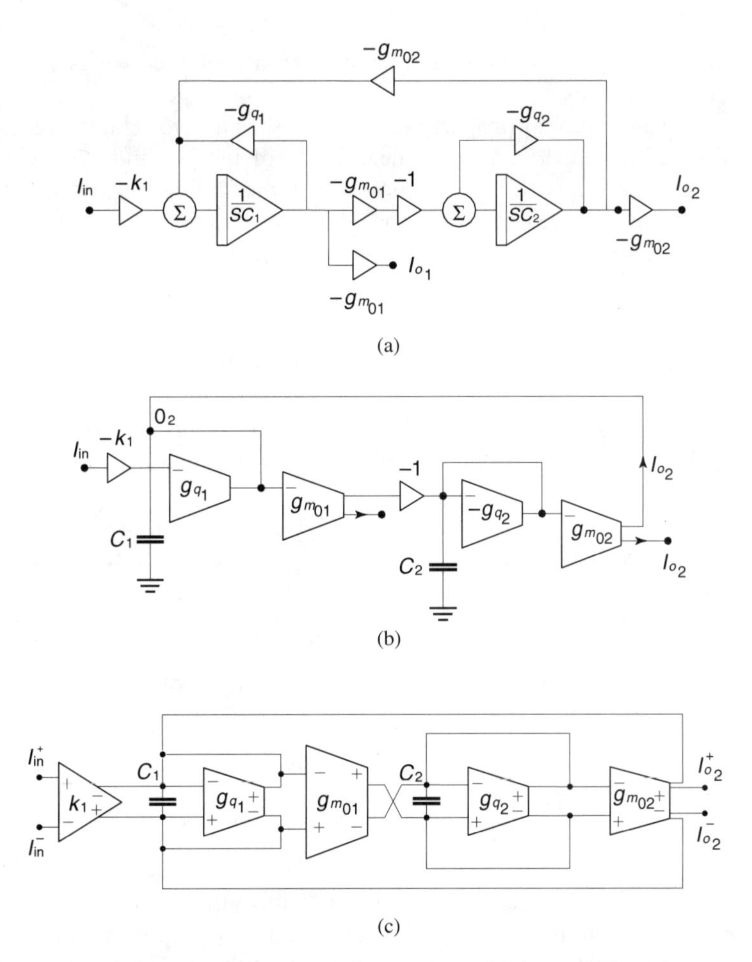

Figure 11.17 Low-sensitivity biquad structure with two self-loop integrators: (a) single-ended block diagram; (b) single-ended transconductance implementation; (c) fully differential transconductance implementation.

The corresponding sensitivities yield

$$-S_{C_1}^{BW} = S_{g_{q_1}}^{BW} = \frac{1}{1 + C_1 g_{q_2}/C_2 g_{q_1}}; \quad -S_{C_2}^{BW} = S_{g_{q_2}}^{BW} = \frac{1}{1 + C_2 g_{q_1}/C_1 g_{q_2}}; \quad S_{g_{m_{o_i}}}^{BW} = 0 \tag{11.44a}$$

$$S_{g_{m_{o_1}}}^{\omega_o^2} = S_{g_{m_{o_2}}}^{\omega_o^2} = \frac{1}{1 + g_{q_1} g_{q_2}/g_{m_{o_2}} g_{m_{o_1}}}; \quad -S_{C_1}^{\omega_o^2} = S_{g_{q_1}}^{\omega_o^2} = \frac{1}{1 + g_{m_{o_1}} g_{m_{o_2}}/g_{q_1} g_{q_2}} \tag{11.44b}$$

For a given BW and ω_o, optimal values of C_1/C_2, g_{q_1}/g_{q_2}, $g_{q_1}/g_{m_{o_1}}$, and $g_{q_2}/g_{m_{o_2}}$ can be determined to yield low sensitivities. Typical values [67] for C_1/C_2 and g_{q_1}/g_{q_2}

are 1; $g_{m_{o_2}}/g_{m_{o_1}}$ is limited by the internal voltage swings and the ratio between channel widths.

An additional property of this low-sensitivity architecture is that the transconductor connected in a closed negative feedback (g_{q_1} and g_{q_2}) yields a low impedance to ground for both the common-mode and differential-mode signals [67, 68]. This structure inherently controls the common-mode output voltage. This architecture is suitable for low-pass applications and resonator (a zero not at the origin) applications.

11.3.1. Biquadratic Architecture Nonidealities

The BW and Q of a real biquad, taking into account the finite output impedance and frequency dependence of the transconductors, can be severely distorted. Fortunately, the ω_o deviations are small, even for relatively low transconductor parasitic poles. From (11.22), $\tau \approx 1/\omega_p$, and assuming $\omega_p > 20\omega_o$, $\Delta\omega_o$ is lower than 0.15%. Commonly, ω_o is known as the center frequency and is often used when discussing bandpass filters or as the cutoff frequency of low-pass filters. The actual bandwidth (BW_a) and actual quality factor (Q_a) can be approximated as

$$\text{BW}_a \cong \text{BW}(1 - Q\phi_T) \tag{11.45a}$$

and

$$Q_a \cong \frac{Q}{1 - Q\phi_T} \qquad \max Q < 1/\phi_T \tag{11.45b}$$

where BW and Q are the ideal bandwidth and quality factor for the filter under consideration, respectively. Stability conditions imply that the maximum Q is lower than $1/\phi_T$. The total phase factor ϕ_T is given as

$$\phi_T \cong \frac{\omega_{u_1}}{\omega_{p_1}} + \frac{\omega_{u_2}}{\omega_{p_2}} - \frac{1}{A_{V_{o_1}}} - \frac{1}{A_{V_{o_2}}} \tag{11.46a}$$

$$\phi_T \cong \phi_{E_1} + \phi_{E_2} - \frac{1}{A_{V_{o_1}}} - \frac{1}{A_{V_{o_2}}} \tag{11.46b}$$

where ω_{u_i} is the ideal unity-gain frequency of integrator i; ϕ_{E_i} is the excess phase (phase lag) and represents the integrator phase in excess of $-90°$ at the center frequency (ω_{u_i}). Here, $A_{V_{o_i}}$ is the voltage gain ($g_m R_o$) of the transconductor i. It can be seen that the Q deviation can be large for low ϕ_T, that is,

$$\frac{\Delta Q}{Q} = \frac{Q_a - Q}{Q} \cong \frac{Q\phi_T}{1 - Q\phi_T} \tag{11.47}$$

These deviations can be reduced by using either phase compensation, as shown in Figure 11.11, or a prewarped quality factor [67] given by

$$Q_{p_\omega} \cong \frac{Q - \phi_T}{1 + Q\phi_T} \tag{11.48}$$

In low-Q applications yielding low-ΔQ deviations, the use of a tuning circuit could be avoided. However, in general, tuning circuits are required [9, 29, 41, 43, 56–59] to yield practical filters.

11.4. CMOS EXPERIMENTAL RESULTS: EXAMPLE

The example of a monolithic filter using the integrator of Figure 11.5 was fabricated in a standard 2-μm digital CMOS process using MOSIS [62]. the capacitors were implemented using MOSFET gates with the source/drain tied to ground. The IC microphotograph is shown in Figure 11.18. The filter bias was manually adjusted to correct the cutoff frequency over process and temperature variations. Only one bias control was used for the correction such that automatic tuning methods could easily be applied.

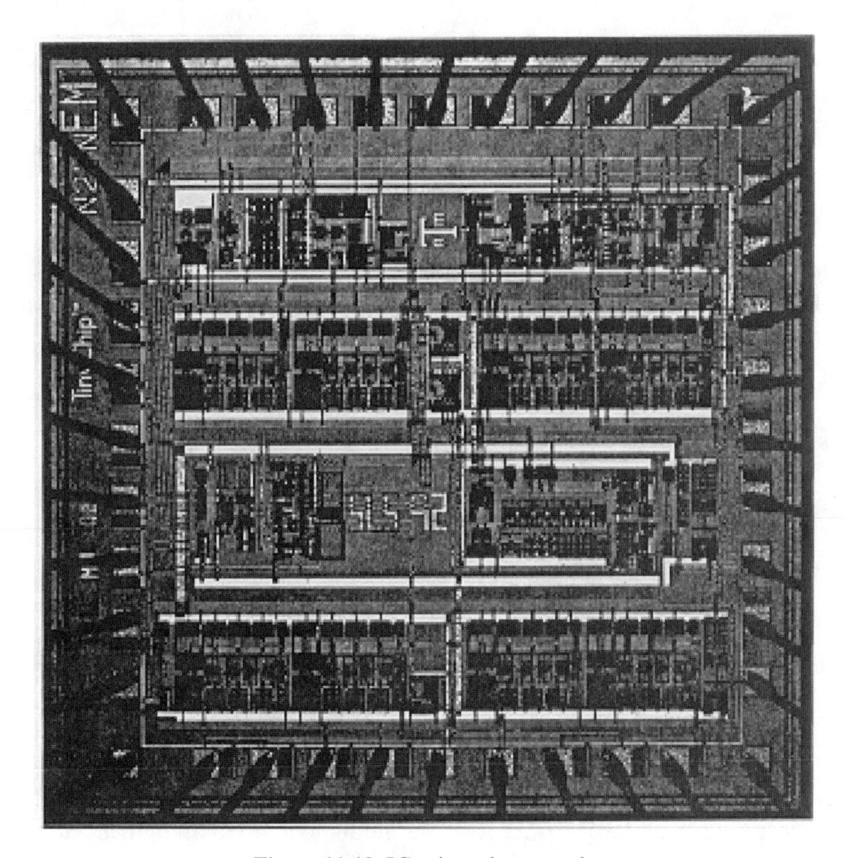

Figure 11.18 IC microphotograph.

Figure 11.19 shows the resulting nominal magnitude response of the filter, which follows closely the ideal and simulated responses. In Figure 11.20, group delay characteristics with some variation in the passband are shown. This variation has been attributed to a combination of layout techniques and the small size of the devices. The layout had matching devices in close proximity but not cross coupled. This, together with the fact that the devices were quite small, reuslted in large mismatches between the devices due to random width and length variations and implant shadowing. This also degraded the power supply rejection ratio (PSRR) results from the expected 40–50 dB. From measurements of "matched" devices which were accessible from the IC pins, the observed mismatch was over 5%. The offset was measured at the IC analog input and averaged about 40 mV for the four chips received. This 40-mV offset represented over 7% of the $v_{gs} - v_t$ of the devices. The measured current at the output of the calibration path of the IC also correlated with this mismatch value. The group delay of the experimental filter dropped off to lower values in the upper region of the passband, which is quite characteristic of mismatch-related problems. High-frequency parasitic problems generally result in a *peaking* in the group delay response at or around the cutoff frequency.

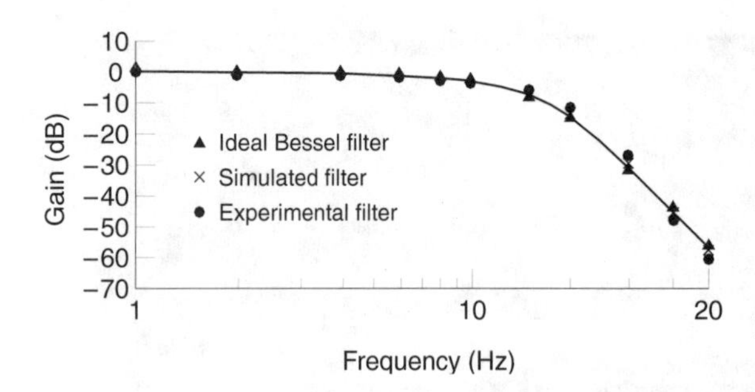

Figure 11.19 Experimental magnitude frequency response of sixth-order filter.

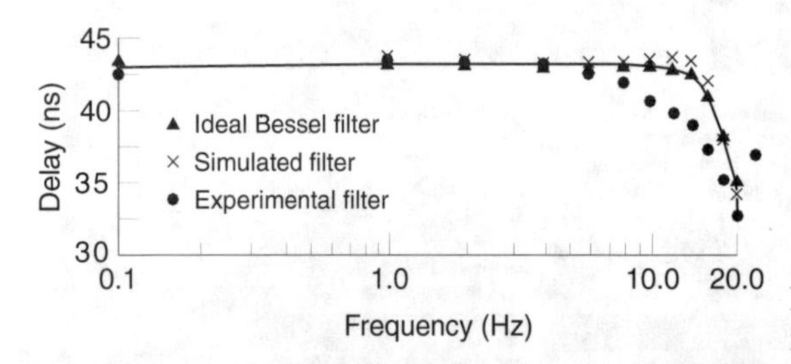

Figure 11.20 Experimental group delay frequency response of sixth-order filter.

To accurately measure the distortion characteristics of the filter, a two-tone intermodulation distortion test was performed with the two tones close to the outer edge of the filter passband, one at 9 MHz and the other at 11 MHz. The filter cutoff frequency was adjusted to 10 MHz for the intermodulation test. Figure 11.21 shows the output spectrum of the intermodulation test when the output signal level reached nearly 440 μA peak (corresponding to a differential filter output at 55 μA, or $I_{\text{out},i}/2I_b = 0.86$, before amplification). The intermodulation distortion was 1% at

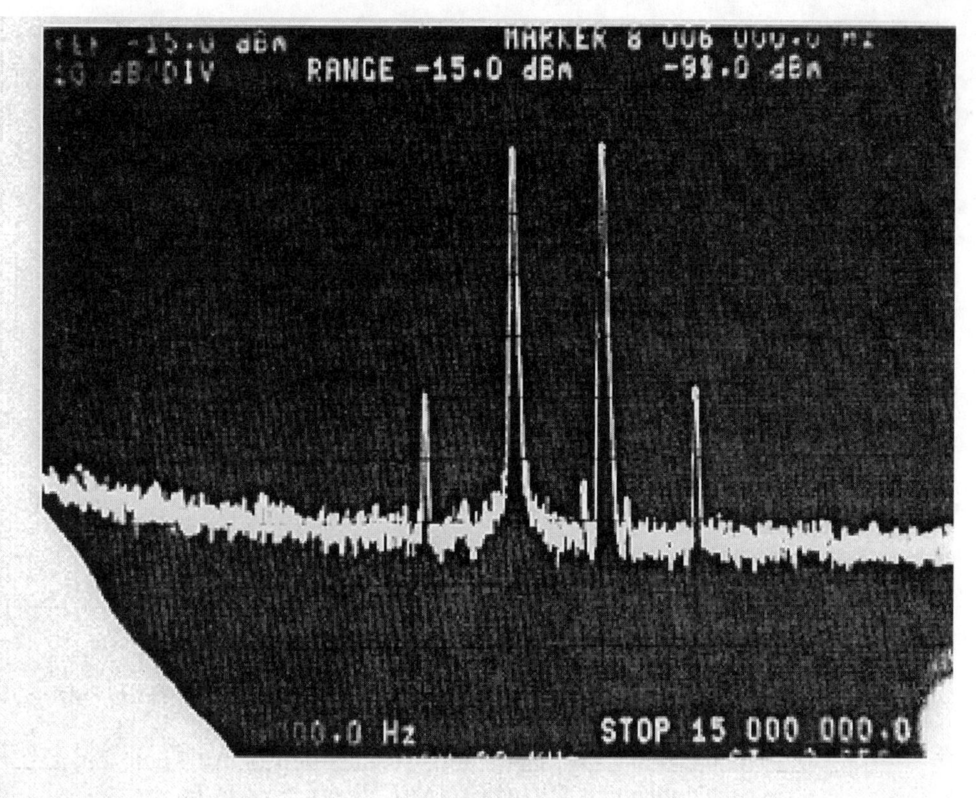

Figure 11.21 Experimental spectral components of intermodulation distortion test.

this signal level. This result is better than the open-loop distortion characteristics of the integrators, where the intermodulation distortion would reach 1% before the normalized signal level reached half that demonstrated here. The reason for this "improvement" is that the actual integrator unity-gain frequencies in the Bessel filter are outside the filter passband. Thus, feedback still provides some reduction of the distortion throughout the passband and somewhat beyond. The integrated output noise, integrated from DC to 12 MHz, was 714 NA_{rms}. This gives a dynamic range for the filter and input/output circuit of close to 53 dB.

An increase in matching would have been obtained by using longer devices and standard linear layout techniques such as common centroiding and cross-coupling. The filter had a cutoff frequency range of 7.5–13.5 MHz. The remaining performance results are presented in Table 11.4. The dynamic range [maximum root-mean-square (RMS) signal where distortion is less than 1% divided by the RMS noise output] is very close to the predicted value. All measurements were performed with the filter using a supply voltage of 3.3 V.

TABLE 11.4 BESSEL FILTER PERFORMANCE SUMMARY

Parameter	Simulated[a]	Experimental	Comment
Frequency range	6–13 MHz	7.5–13.5 MHz	
Delay ripple	1.5 ns	5 ns	To 1.5 F_C
Power consumption	—	4 mW	Filter only
Supply voltage	3.3 V	3.3 V	
IC area	—	1200 mil[b]	
PSRR			
V_{DD}	55 dB[b]	24 dB[c]	
V_{SS}	40 dB	24 dB	Passband
Dynamic range	55 dB	52 dB	1%

[a]Using HSPICE simulator and MOSIS supplied level 2 and BSIM MOS models.
[b]Monte Carlo analysis assuming 1% 3-sigma mismatch in transistors.
[c]Observed mismatches in transistors were close to 5%.

References

[1] R. L. Geiger and E. Sánchez-Sinencio, "Active Filter Design Using Operational Transconductance Amplifiers: A Tutorial," *IEEE Circuits Devices Mag.*, vol. 1, pp. 20–32, 1985.

[2] M. Mialko and R. W. Newcomb, "Generation of All Finite Linear Circuits Using the Integrated DVCCS," *IEEE Trans. Circuit Theory*, vol. CT-18, pp. 733–736, 1971.

[3] A. Nedungadi and T. R. Viswanathan, "Design of Linear CMOS Transconductance Elements," *IEEE Trans. Circuits Sys.*, vol. CAS-31, pp. 891–894, 1984.

[4] C. Plett, M. A. Copeland, and R. A. Hadaway, "Continuous-Time Filters Using Open-Loop Tunable Transconductance Amplifiers," *Proc. IEEE/ISCAS*, vol. 3, pp. 1172–1176, 1986.

[5] H. S. Malvar, "Electronically Controlled Active-C Filter and Equalizers with Operational Transconductance Amplifiers," *IEEE Trans. Circuits Sys.*, vol. CAS-31, pp. 645–649, 1984.

[6] A. Nedungadi and R. L. Geiger, "High Frequency Voltage Controlled Continuous Time Low-Pass Filter Using Linearized CMOS Integrators," *Electron. Lett.*, vol. 22, pp. 729–731, 1986.

[7] E. Sánchez-Sinencio, R. L. Geiger, and H. Nevarez-Lozano, "Generation of Continuous-Time Two Integrator Loop OTA Filter Structures," *IEEE Trans. Circuits Sys.*, vol. 35, pp. 936–946, 1988.

[8] H. Khorramabadi and P. R. Gray, "High Frequency CMOS Continuous-time Filters," *IEEE J. Solid-State Circuits*, vol. SC-19, pp. 939–948, 1984.

[9] F. Krummenacher and N. Joehl, "A 4MHZ CMOS Continuous-Time Filter with On-Chip Automatic Tuning," *IEEE J. Solid-State Circuits*, vol. SC-23, pp. 750–757, 1988.

[10] R. R. Torrance, T. R. Viswanathan, and J. V. Hanson, "CMOS Voltage-to-Current Transductors," *IEEE Trans. Circuits Sys.*, vol. CAS-32, pp. 1097–1104, 1985.
[11] E. Seevink and R. W. Wassenaar, "A Versatile CMOS Linear Transconductor/Square-Law Function Circuit," *IEEE J. Solid-State Circuits*, vol. SC-22, pp. 366–377, 1987.
[12] P. M. Van Peteghem, B. J. Haby, H. M. Fossati, and G. L. Rice, "A Very Linear CMOS Transconductance Stage for OTA-C Filters," paper presented at the IEEE Custom Integrated Circuits Conference, pp. 25.3, 1–25.3.4, San Diego, CA, May 1989.
[13] Z. Wang and W. Guggenbuhl, "A Voltage-Controllable Linear MOS Transconductor Using Bias Offset Technique," *IEEE J. Solid-State Circuits*, vol. SC-25, pp. 315–317, 1990.
[14] J. Silva-Martínez, M. Steyaert, and W. Sansen, "A Large-Signal Very Low-Distortion Transconductor for High-Frequency Continuous-Time Filters," *IEEE J. Solid-State Circuits*, vol. SC-16, pp. 946–955, 1991.
[15] J. L. Pennock, "CMOS Triode Transconductor for Continuous-Time Active Integrated Filters," *Electron. Lett.*, vol. 21, pp. 817–818, 1985.
[16] R. Allini, A. Baschirotto, and R. Castello, "Tunable BiCMOS Continuous-Time Filter for High-Frequency Applications," *IEEE J. Solid-State Circuits*, pp. 1905–1915, 1992.
[17] J. Ramírez-Angulo and I. Grau, "Wide g_m Adjustment Range, High Linear OTA with Linear Programmable Current Mirrors," *Proc. IEEE/ISCAS-92*, pp. 1372–1375, May 1992.
[18] W. B. Adams and J. Ramírez-Angulo, "OTA Linearization via Electronically Programmable Current Mirrors," *Proc. IEEE/ISCAS-91*, pp. 2553–2556, June 1991.
[19] M. I. Ali, M. Howe, E. Sánchez-Sinencio, and J. Ramírez-Angulo, "A BiCMOS Low Distortion Tunable OTA for Continuous-Time Filters," *IEEE Trans. Circuits Sys. I*, vol. 40, pp. 43–49, 1993.
[20] E. J. van der Zwan, E. A. M. Kumperink, and E. Seevinck, "A CMOS OTA for HF Filters with Programmable Transfer Function," *IEEE J. Solid-State Circuits*, vol. 26, pp. 1720–1723, 1991.
[21] M. Qu and M. A. Styblinski, "Phase Compensation of an OTA-C Integrator for High-Frequency Continuous-Time Filters," *Electron. Lett.*, vol. 29, pp. 1814–1816, 1993.
[22] G. A. de Veriman and R. G. Yamasaki, "Designs of a Bipolar 10 MHz Programmable Continuous-time 0.005 Degree Equiripple Linear Phase Filter," *IEEE J. Solid-State Circuits*, pp. 324–331, Mar. 1992.
[23] Y. Tsividis, Z. Czarnul, and S. C. Fang, "MOS Transconductors and Integrators with High Linearity," *Electron. Lett.*, vol. 22, pp. 245–246, 619, 1986.
[24] D. L. Hiser and R. L. Geiger, "Impact on OTA Nonlinearities on the Performance of Continuous-Time OTA-C Bandpass Filters," *Proc. IEEE/ISCAS-90*, New Orleans, pp. 1167–1170, June 1990.

[25] C. Toumazou, F. J. Lidgey, and D. G. Haigh (eds.), "Analogue IC Design: The Current-Mode Approach," *IEEE Circuits and Systems Series*, vol. 2, Peter Peregrinus, London, 1990.
[26] H. Nevárez-Lozano and E. Sánchez-Sinencio, "Minimum Parasitic Effects Biquadratic OTA-C Filter Architectures," *Analog Integration Circuits Signal Process.*, vol. 1, pp. 297–319, 1991.
[27] J. Ramírez-Angulo and E. Sánchez-Sinencio, "High-Frequency Compensated Current-Mode Ladder Filters Using Multiple Output OTA's," *IEEE Trans. Circuits Sys. II: Analog Digital Signal Process.*, vol. 41, pp. 581–586, 1994.
[28] A. C. M. de Queiroz, L. P. Caloba, and E. Sánchez-Sinencio, "Signal Flow Graph OTA-C Integrated Filters," *Proc. IEEE/ISCAS-88*, pp. 2165–2168, 1988.
[29] J. Silva-Martínez, M. Steyaert, and W. Sansen, *High-Performance CMOS Continuous-Time Filters*, Kluwer Academic, Boston, 1993.
[30] J. Ramírez-Angulo and E. Sánchez-Sinencio, "Active Compensation of Operation Transconductance Amplifier Filters Using Partial Positive Feedback," *IEEE J. Solid-State Circuits*, vol. SC-25, no. 4, pp. 1024–1028, 1990.
[31] G. W. Roberts and A. S. Sedra, "Adjoint Network Revisited," *IEEE/ISCAS-90*, pp. 540–544, May 1990.
[32] E. Säckinger and W. Guggenbüh, "A High-Swing High-Impedance MOS Cascode Circuit," *IEEE J. Solid-State Circuits*, vol. SC-25, pp. 289–298, 1990.
[33] J. B. Hughes, N. C. Bird, and I. C. Macbeth, "Switched-Currents—A New Technique for Analog Sampled-Data Signal Processing," *Proc. IEEE/ISCAS-87*, pp. 1584–1587, 1989.
[34] T. S. Fiez, G. Liang, and D. J. Allstot, "Switched-Current Circuit Design Issues," *IEEE J. Solid-State Circuits*, vol. 26, pp. 192–202, 1991.
[35] J. Richardson, J. B. Hughes, K. Moulding, M. Bracey, W. Redman-White, J. Bennet, and R. S. Soin, "An Integrated Design and Synthesis Systems for High Performance Switched-Current Analogue Filters," *Proc. E. Solid-State Circuits Conf. ESSCIR '95*, Lille, France, Sept. 1995.
[36] S. S. Lee, R. H. Zele, D. J. Allstot, and G. Liang, "A Continuous-Time Current-Mode Integrator," *IEEE Trans. Circuit Sys.*, vol. 38, pp. 1236–1238, 1991.
[37] S. S. Lee, R. H. Zele, D. J. Allstot, and G. Liang, "A 40 MHz CMOS continuous-Time Current-Mode Filter," *Proc. IEEE Custom Integrated Circuits Conference*, pp. 24.5, 1–24.5.4, 1992.
[38] J. Ramírez-Angulo, M. Robinson, and E. Sánchez-Sinencio, "Current Mode Continuous Time Filters: Two Design Approaches," *IEEE Trans. Circuits Sys.*, vol. 39, no. 6, pp. 337–341, 1992.
[39] S. Smith and E. Sánchez-Sinencio, "3V High-Frequency Current Mode Filters," *IEEE/ISCAS-93*, vol. 2, pp. 1251–1254, 1993.
[40] J. T. Nabicht, E. Sánchez-Sinencio, and J. Ramírez-Angulo, "A Programmable 1.8–18 MHz High-Q Fully-Differential Continuous-Time Filter with 1.5–2 V Power Supply," *IEEE/ISCAS-94*, vol. 5, pp. 653–656, 1994.

[41] B. Nauta, "A CMOS Transconductance-C Filter Technique for Very High Frequencies," *IEEE J. Solid-State Circuits*, vol. 27, pp. 142–153, 1992.
[42] A. T. Behr, M. C. Schneider, S. N. Filho, and C. G. Montoro, "Harmonic Distortion Caused by Capacitors Implemented with MOSFET Gates," *IEEE J. Solid-State Circuits*, vol. 27, pp. 1470–1475, 1992.
[43] B. Nauta, "Analog CMOS Filters for Very-High Frequencies," Ph.D. Dissertation, University of Twente, Enschede, The Netherlands, 1991.
[44] Z. Wang, "Current-Mode CMOS Integrated Circuits for Analog Computation and Signal Processing: A Tutorial," *Analog Integrated Circuits and Signal Processing*, vol. 1, no. 4, pp. 287–295, 1991.
[45] F. Goodenough, "Lone 3-V Rail Power IC Disk-Drive Read Channel," *Electron. Design*, pp. 43–51, June 25, 1992.
[46] Silicon Systems (SS), *SSI 32F8030 Programmable Electronic Filter*, Preliminary Data Sheet, SS, Tustin, CA, May 1991.
[47] International Microelectronics Products (IMP), *IMP42C555-30 Programmable Tracking Filter*, Preliminary Data Sheet, IMP, San Jose, CA, June 1991.
[48] T. G. Kim and R. L. Geiger, "Monolithic Programmable RF Filter," *Electron. Lett.*, vol. 24, pp. 1569–1571, 1988.
[49] S. Noceti Filho, M. C. Schneider, and R. N. G. Robert, "New CMOS OTA for Fully Integrated Continuous-Time Circuit Applications," *Electron. Lett.*, vol. 25, pp. 1674–1675, 1989.
[50] E. Seevinck and R. F. Wassenaar, "A Versatile CMOS Linear Transconductor/Square-Law Function Circuit," *IEEE J. Solid-State Circuits*, vol. 22, pp. 366–377, 1987.
[51] G. A. de Veirman and R. G. Yamasaki, "Design of a Bipolar 10 MHz Programmable Continuous-Time 0.05 Degree Equiripple Linear Phase Filter," *IEEE J. Solid-State Circuits*, vol. 27, pp. 324–331, 1992.
[52] A. S. Sedra and K. C. Smith, *Microelectronic Circuits*, Holt, Rinehart and Winston, New York, 1987.
[53] Y. P. Tsividis, *Operation and Modeling of the MOS Transistor*, McGraw-Hill, New York, 1987.
[54] K. D. Peterson and R. L. Geiger, "Area/Bandwidth Tradeoffs for CMOS Current Mirrors," *IEEE J. Solid-State Circuits Sys.*, vol. 33, pp. 667–669, 1986.
[55] Silicon Systems (SS), *SS32F8011 Programmable Electronic Filter*, Preliminary Data Sheet, SS, Tustin, CA, Feb. 1990.
[56] C.-G. Yu, W. G. Bliss and R. L. Geiger, 'A Tuning Algorithm for Digitally Programmable Continuous-Time Filters," *IEEE/ISCAS-91*, pp. 1436–1439, May 1991.
[57] D. Senderowicz, D. A. Hodges, and P. R. Gray, "An NMOS Integrated Vector-Locked Loop," *IEEE/ISCAS-82*, pp. 1164–1167, May 1982.
[58] R. Schaumann and M. A. Tan, "The Problem of On-Chip Automatic Tuning in Continuous-Time Integrated Filters," *IEEE/ISCAS-89*, pp. 106–109, May 1989.
[59] P. M. Van Peteghem and S. Rujiang, "An Investigation of the Stability of Tuning Loops in High-Frequency, High-Q Continuous-Time Filters," *Proc. of 31st Midwest Symposium on Circuits and Systems*, Aug. 1988.

[60] K. C. Smith and A. S. Sedra, "The Current Conveyor—A New Circuit Building Block," *Proc. IEEE*, vol. 56, pp. 1368–1369, 1968.

[61] A. S. Sedra and K. C. Smith, "A Second-Generation Current Conveyor and Its Applications," *IEEE Trans. Circuit Theory*, vol. CT-17, pp. 132–134, 1970.

[62] S. L. Smith and E. Sánchez-Sinencio, "Low Voltage Integrators for High-Frequency CMOS Filters Using Current Mode Techniques," *IEEE Trans. Circuits Sys. II*, vol. 43, pp. 39–48, 1996.

[63] T. Serrano and B. Linares-Barranco, "The Active-Input Regulated-Cascode Current Mirror," *IEEE Trans. Circuits Sys. I*, vol. 41, pp. 464–467, 1994.

[64] J. Ramírez-Angulo and E. Sánchez-Sinencio, "Programmable BiCMOS Transconductor for C-TA Filters," *Electron. Lett.*, vol. 28, pp. 1185–1187, 1992.

[65] M. Ismail and T. Fiez, *Analog VLSI: Signal and Information Processing*, McGraw-Hill, New York, 1994.

[66] U. Gatti, F. Maloberti, and G. Torelli, "A Novel CMOS Linear Transconductance Cell for Continuous-Time Filters," *Proc. IEEE/ISCAS-90*, pp. 1173–1176, 1990.

[67] F. Rezzi, A. Bashirotto, and R. Castello, "A 3V 12-55 MHz BiCMOS Pseudo-Differential Continuous-Time Filter," *IEEE Trans. Circuits Sys. I*, vol. 42, pp. 896–903, 1995.

[68] A. Wysynski and R. Schaumann, "Avoiding Common-Mode Feedback in Continuous-Time G_m–C Filters by Use of Lossy Integrators," *IEEE/ISCAS*, pp. 281–284, June 1994.

Chapter 12

High-Efficiency Low-Voltage DC-DC Conversion for Portable Applications

Anthony J. Stratakos, Charles R. Sullivan, Seth R. Sanders, and Robert W. Brodersen
Department of Electrical Engineering and Computer Sciences, University of California, Berkeley, California 94720-1772

12.1. INTRODUCTION

Portable electronic equipment demands ultra-low-power hardware to maximize system run time. Perhaps the most effective way to reduce power dissipation and maintain computational throughput is to operate each subsystem at its optimum supply voltage and compensate for the resulting decrease in performance by exploiting parallelism and pipelining [1]. This low-power design strategy assumes that the supply voltage is a free variable and can be set to any arbitrarily low level with little penalty. In portable systems, high-efficiency low-voltage DC-DC conversion is required to efficiently generate each low-voltage supply from a single battery source.

While high-efficiency DC-DC conversion can substantially improve system run time in virtually any battery-operated application, this same enhancement of run time may also be achieved by simply increasing the capacity of the battery source. However, especially if voltage conversion is performed by highly integrated converters custom-designed for their individual loads, their volume will typically be much smaller than the additional battery volume required to achieve the equivalent extension of run time [2]. This chapter describes the application-specific design techniques necessary to meet the rigorous size and efficiency requirements of a portable device, focusing on the complementary metal-oxide semiconductor (CMOS) implementation of a low-output-voltage buck converter. The CMOS is usually the technology of choice for low-power, low-voltage applications. It allows control circuitry with extremely low power consumption and low-drop power devices to be integrated in a standard process. However, many of the techniques described here are applicable to other technologies as well.

12.2. PULSE-WIDTH-MODULATED DC-DC CONVERTER

The switching regulator shown in Figure 12.1 converts the unregulated battery source voltage V_{in} to the desired regulated DC output voltage V_o. A single-throw, double-pole switch chops V_{in}, producing a rectangular wave having an average voltage equal to the desired output voltage. A low-pass filter passes this DC voltage to the output while attenuating the AC ripple to an acceptable value. The output is regulated by comparing V_o to a supply- and temperature-independent reference voltage, V_{ref}, and adjusting the fraction of the cycle for which the switch is shorted to V_{in}. This pulse-width modulation (PWM) controls the average value of the chopped waveform and thus desensitizes the output voltage against input, load, and temperature variations. Unlike a switched-capacitor converter (see Section 12.7), a switching regulator has an efficiency which approaches 100% as the components are made more ideal. In practice, efficiencies above 75% are routine, and efficiencies above 90% are attainable.

There are several simple alternative arrangements of the switching and filter components that can be used to produce an output voltage larger or smaller than the input voltage with the same or opposite polarity. Some of these are discussed in [2]. However, many of the design issues are similar, so one topology, the step-down (buck) converter, will be discussed here in detail. The reader is referred to [3] and the references contained therein for more discussion of other topologies.

12.2.1. Buck Converter

The power train of a CMOS low-output-voltage buck circuit, which can produce any arbitrary output voltage $0 \leq V_o \leq V_{\text{in}}$, is shown in Figure 12.2. The basic operation is as follows: The power transistors (pass device M_p and rectifier M_n) chop the battery input voltage V_{in} to reduce the average voltage. This produces a square wave of duty cycle D and period $T_s = f_s^{-1}$ at the inverter output node v_x. A typical periodic steady-state $v_x(t)$ waveform is shown in Figure 12.3. This chopped signal is filtered by the second-order low-pass output filter, L_f and C_f. In the ideal case, the DC output voltage is given by the product of the input voltage and the duty cycle:

$$V_o = V_{\text{in}} D \tag{12.1}$$

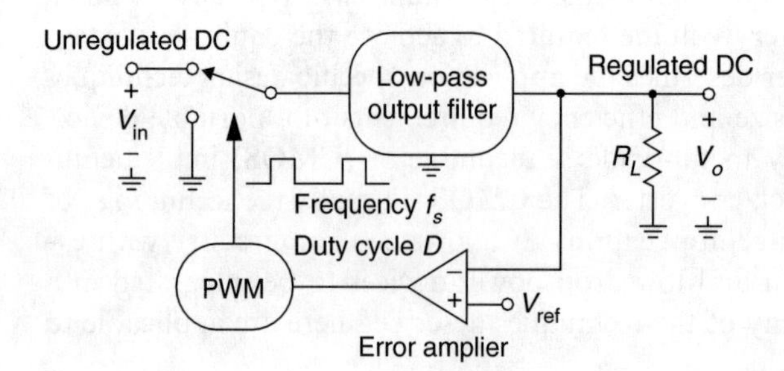

Figure 12.1 Block diagram of pulse-width-modulated switching regulator.

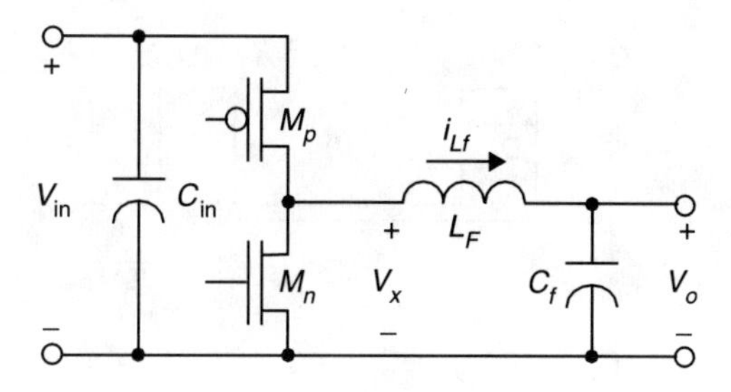

Figure 12.2 CMOS low-output-voltage buck circuit.

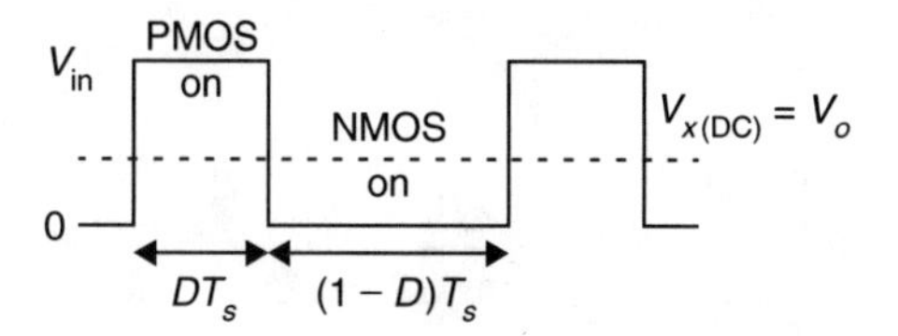

Figure 12.3 Nominal periodic steady-state $v_x(t)$ buck circuit waveform.

The switching pattern of M_n and M_p is pulse-width modulated, adjusting the duty cycle of the rectangular waveform $v_x(t)$ and, ultimately, the DC output voltage to compensate for input and load variations. The PWM is controlled by a negative-feedback loop, shown in the block diagram of Figure 12.1, but omitted from Figure 12.2 for simplicity.

Output Filter Design. In Figure 12.4, the rectangular wave of the inverter output node is applied to the second-order low-pass output filter of the buck circuit (L_f and C_f), which passes the desired DC component of $v_x(t)$ while attenuating the AC component to an acceptable ripple value. Load R_L draws a DC current I_o from the output of the filter. Figure 12.5 shows the nominal steady-state $i_{L_f}(t)$ and $v_o(t)$ waveforms for a rectangular input $v_x(t)$.

In order to achieve the large attenuation needed in a practical power circuit, $L_f C_f \gg \omega_s^{-2}$, where $\omega_s = 2\pi f_s$ and f_s is the switching frequency of the converter. In this case, the filter components may be sized independently using time-domain analysis, rather than frequency-domain analysis. Neglecting the effects of output voltage ripple ($v_{o,\mathrm{AC}} \ll v_{x,\mathrm{AC}}$), for a rectangular input with period T_s, the AC inductor current waveform is triangular with period T_s and peak-to-peak ripple ΔI symmetric about the average load current I_o. The peak-to-peak current ripple may

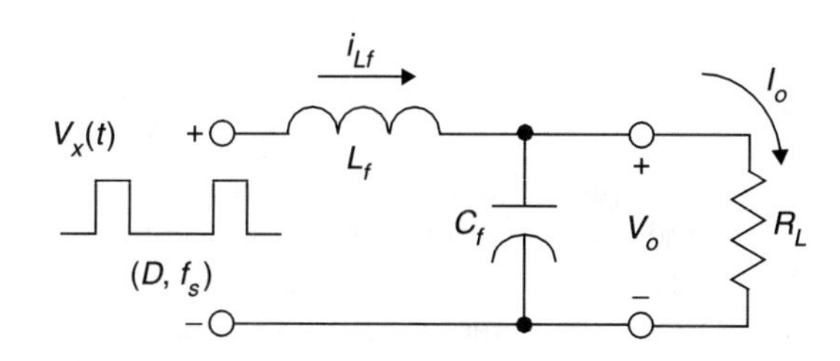

Figure 12.4 Output filter of buck circuit (L_f and C_f) with load R_L.

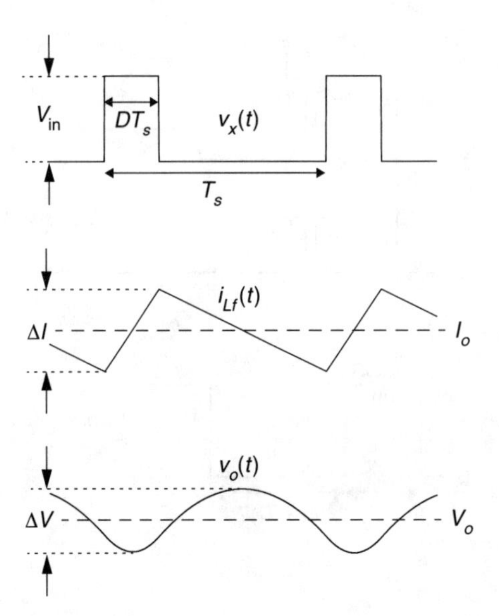

Figure 12.5 Nominal steady-state waveforms of buck circuit output filter.

be found by integrating the AC component of the $v_x(t)$ waveform over a fraction D of one cycle, yielding

$$\Delta I = \frac{V_{\text{in}} D(1 - D)}{L_f f_s} = \frac{V_o(1 - D)}{L_f f_s} \tag{12.2}$$

The output filter capacitor is selected to ensure that its impedance at the switching frequency, including its equivalent series resistance (ESR), is small relative to the load impedance. Thus, the AC component of the inductor current flows into the filter capacitor rather than into the load. For many capacitor technologies at frequencies above several hundred kilohertz, the resistive impedance dominates over the capacitive impedance. In high-current-ripple designs, a primary design goal is to minimize ESR to reduce both output voltage ripple and conduction loss (see below). For this reason, a high-Q capacitor technology, such as multilayer ceramic, is typically used, and ESR may often be neglected in calculating output voltage ripple. Considering only capacitive impedance, the peak-to-peak output voltage ripple may be found through charge conservation. Assuming the AC inductor current flows only into the filter capacitor,

$$\Delta V = \frac{\Delta I}{8 C_f f_s} = \frac{V_o(1 - D)}{8 L_f C_f f_s^2} \tag{12.3}$$

This output voltage ripple is symmetric about the desired DC output voltage V_o and, for the $v_x(t)$ waveform shown in Figure 12.5, is piecewise-quadratic with period T_s.

Equations (12.2) and (12.3) illustrate the two principal means of miniaturizing a DC-DC converter. First, the values of filter inductance and capacitance decrease with f_s^{-1}. Thus, a higher operating frequency typically results in a smaller converter. Second, because the requirement of interest is output voltage ripple, it is the $L_f C_f$ product, rather than the values of the individual components, that is important. Through choice of a higher current ripple ΔI, a lower filter inductance solution may be obtained, often resulting in smaller overall volume and cost.

Buck Converter Efficiency. The power train of a CMOS low-output-voltage buck circuit, including parasitic capacitance C_x, stray inductance L_s, and drain-body diodes of the power transistors, is shown in Figure 12.6. Listed below are the chief sources of dissipation that cause the conversion efficiency of this circuit to be less than unity. In Section 12.4, methods which reduce these losses are described.

CONDUCTION LOSS. Current flow through nonideal power transistors, filter elements, and interconnections results in dissipation in each component:

$$P_q = i_{\text{rms}}^2 R, \tag{12.4}$$

where i_{rms} is the root-mean-squared current through the component and R is the resistance of the component.

FILTER-ELEMENT LOSS. Resistance of conductors in inductors and capacitors contributes to $i^2 R$ losses. Additionally, capacitors may have significant dielectric losses, inductors may have substantial core losses, and conduction losses in inductors may be greatly increased by skin effect and proximity effect losses at high frequency.

GATE-DRIVE LOSS. Raising and lowering the gate of a power transistor each cycle dissipates an average power:

$$P_g = E_g f_s \tag{12.5}$$

where E_g is directly proportional to the gate energy transferred per cycle (which can include some energy due to the Miller effect) and includes dissipation in the drive circuitry (see the discussion of CMOS gate-drive design in Section 12.4.5).

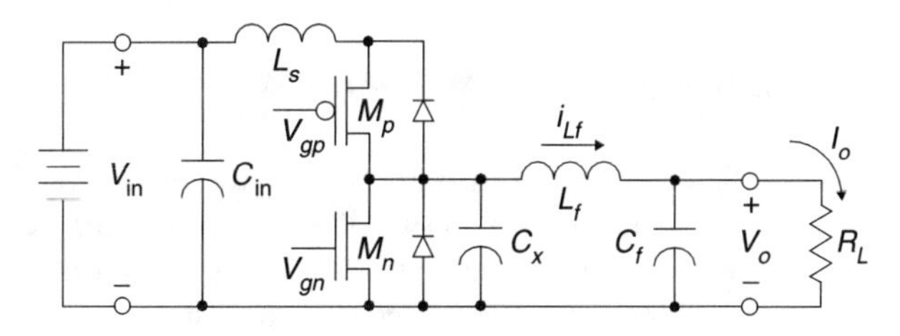

Figure 12.6 CMOS low-output-voltage buck circuit, including parasitics.

CAPACITIVE SWITCHING LOSS. In a hard-switched converter, the MOS field-effect transistor (MOSFET) M_p charges parasitic capacitance C_x to V_{in} each cycle, dissipating an average power:

$$P_{C_x} = \tfrac{1}{2} C_x V_{\text{in}}^2 f_s \tag{12.6}$$

where C_x includes reverse-biased drain-body junction capacitance C_{db} and some or all of the gate-drain overlap (Miller) capacitance C_{gd} of the power transistors, wiring capacitance from their interconnection, and stray capacitance associated with L_f. In ultra-low-power monolithic converters, C_x may be dominated by parasitics associated with the connection of an off-chip filter inductor, which include a bonding pad, bond wire, pin, and board interconnect capacitance. In circuit topologies which do not recover the energy stored on C_x once per cycle through the inductor, the factor of $\frac{1}{2}$ is removed from (12.6).

SHORT CIRCUIT LOSS. A short-circuit path may exist temporarily between the input rails during switch transitions. To avoid potentially large short-circuit losses, it is necessary to provide dead times in the conduction of the MOSFETs to ensure that the two devices never conduct simultaneously.

BODY DIODE REVERSE RECOVERY. If the durations of the dead times are too long, the body diode of the N-type MOS (NMOS) power transistor may be forced to pick up the inductor current for a fraction of each cycle. If the forward-bias body diode voltage is comparable to the converter output voltage, its conduction loss, given by (12.4), is likely to be significant. Furthermore, when the P-type MOS (PMOS) device is turned on, it removes the excess minority-carrier charge from the body diode, dissipating an energy bounded by

$$E_{rr} = Q_{rr} V_{\text{in}} \tag{12.7}$$

where Q_{rr} is the stored charge in the body diode.

STRAY INDUCTIVE SWITCHING LOSS. Energy stored by the stray inductance L_s in the loop formed by the input decoupling capacitor C_{in} and the power transistors causes dissipation.

QUIESCENT OPERATING POWER. The PWM and other control circuitry consume power. In low-power applications, this control power may contribute substantially to the total losses, even at full load.

12.3. CONVERTER MINIATURIZATION

Since the portability requirement places severe constraints on physical size and mass, the volume and mass of a converter can be a critical design consideration. This section introduces several design techniques that may be used to reduce the size and cost of a PWM DC-DC converter.

12.3.1. High-Frequency Operation

As indicated by (12.2) and (12.3), there are inherent size and cost advantages associated with higher frequency operation. The reactive filter components are likely to be the major contributors to the volume of a highly integrated converter. For the same impedance, $j\omega_s L$ or $1/j\omega_s C$, a higher switching frequency $f_s = \omega_s/2\pi$ enables the use of reactive components with smaller value and smaller physical size. Ideally, the size of these components will decrease with f_s^{-1}. However, as will be described in Section 12.4.4, if the operating frequency of the circuit is increased, the sum of the losses in the power transistors and drive, if optimized, will increase roughly with $\sqrt{f_s}$. Thus, the general theoretical relationship between the size of a DC-DC converter and its losses is as illustrated in Figure 12.7. Here, operating frequency is used as a parameter, and the sum of the losses in the power transistors and drive is plotted against the volume of the converter.

If the cost and volume of the converter are decreased, additional space and resources are left for a larger or better battery, compensating for lower conversion efficiency. The system requirements and battery characteristics will help to determine which point on this curve is optimal for a specific application. For example, in systems designed for shorter run times, the volume of the converter can become comparable to the volume of the battery, particularly if a battery with a relatively high volumetric energy density is used [2]. Then, it will be worthwhile to operate the converter at a higher frequency, sacrificing efficiency while leaving space for additional battery capacity.

In Section 12.4, circuit-level optimizations are described which significantly reduce the frequency-dependent losses in the power train, yielding a class of miniature yet highly efficient converters that are well suited for portable applications. In practice, higher frequency operation is limited not only by frequency-dependent losses in the power train and controller but also by diminishing returns in the miniaturization of the filter components. Frequency limitations in inductive filter components will be addressed in Section 12.6.3.

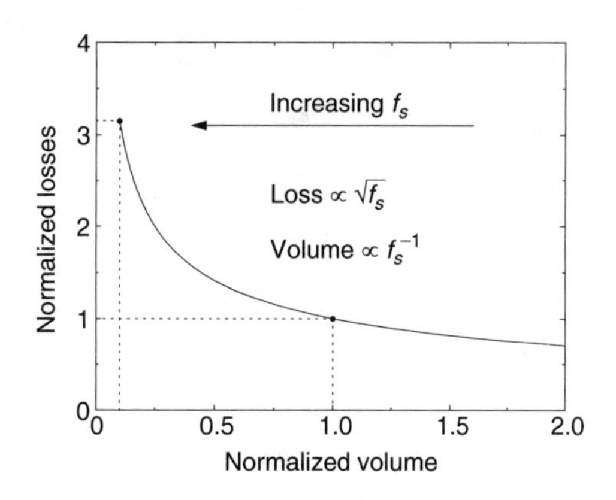

Figure 12.7 Theoretical relationship between power transistor losses and volume of DC-DC converter.

12.3.2. High Current Ripple

Since the $L_f C_f$ product determines the output voltage ripple (12.3), the relative size and cost of inductance versus capacitance should be considered in the selection of these components. As the size, cost, and commercial availability of low-voltage multilayer ceramic chip capacitors are often superior to those of inductors, using large-value capacitors and small-value and small-size inductors is preferred. This decision is restricted primarily by the increasing rms current in the inductor, which circulates throughout the power train, increasing conduction loss in proportion to $i^2_{L_f,\mathrm{rms}}$.

The inductor current is approximated as a triangular AC waveform with peak-to-peak ripple ΔI superimposed on the DC output current I_o (see Fig. 12.5). In Figure 12.8, ΔI is varied, and its effects on three key circuit parameters are shown. As illustrated by (12.2), the value of filter inductance decreases with ΔI^{-1}. However, the physical size of L_f is roughly proportional to its peak energy storage, which in turn is given by

$$E_{L_f} = \tfrac{1}{2} L_f (I_o + \tfrac{1}{2}/\Delta I)^2 \tag{12.8}$$

and is minimized for $\Delta I = 2I_o$. The rms current is

$$i_{L_f,\mathrm{rms}} = \sqrt{I_o^2 + \tfrac{1}{3}(\tfrac{1}{2}/\Delta I)^2} \tag{12.9}$$

and for $\Delta I = 2I_o$, the AC component of the current accounts for 25% of the overall conduction loss in the power train.

Although the preferred value of ΔI will depend slightly on the trade-off between size and loss in a particular application, it can be concluded that a peak-to-peak current ripple in the range $I_o < \Delta I < 2I_o$ is optimal for many applications. As ΔI is decreased, the ripple current contribution to total rms current (and so to conduction loss) decreases. However, below $\Delta I = I_o$, further decreases in ΔI make little differ-

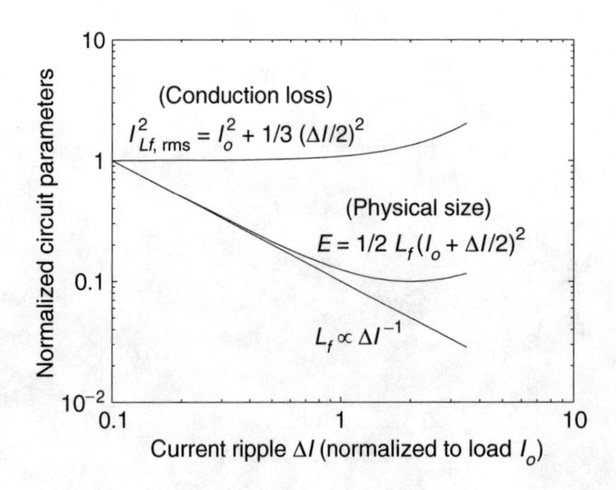

Figure 12.8 Effect of increased current ripple on value of L_f, physical size of L_f, and $i^2_{L_f,\mathrm{rms}}$.

ence in conduction loss at full load and do not justify the larger inductor that would be required. There is no obvious benefit for $\Delta I > 2I_o$, but this will be seen to be advantageous for one mode of operation in Section 12.4.2.

12.3.3. High Integration

A completely monolithic supply (active and passive elements) would meet the severe size and weight restrictions of a hand-held device. Because most portable applications call for low-voltage power transistors, their integration in a standard logic process is tractable. However, existing monolithic magnetics technology cannot provide inductors of suitable value and quality for efficient power conversion [4]. Emerging magnetics technology may allow completely monolithic supplies (see Section 12.6), but currently, magnetics, capacitors, and silicon circuitry are fabricated separately and assembled at the board level or in a multichip module (MCM). The extent of integration is the use of a monolithic silicon circuit, including all power transistors with their drive and all control circuitry.

Such a highly integrated solution not only results in a more compact and cost-effective design, but also gives the designer more latitude in physical design and device sizing, allowing application-specific optimizations which are likely to yield a more efficient converter. Parasitics from both the active devices and interconnect may be orders of magnitude lower on an integrated circuit (IC) than on a printed circuit board. Many of the frequency-dependent losses in a power circuit increase in direct proportion to the energy storage of these parasitics; thus, integration enables higher efficiency at high operating frequencies than that obtained by a board-level solution.

12.4. CIRCUIT TECHNIQUES FOR HIGH EFFICIENCY

The chief mechanisms of dissipation in a CMOS low-ouput-voltage buck converter have been summarized in Section 12.2.1 under Buck Converter Efficiency. In this section, design techniques to eliminate, minimize, or reduce the dissipation due to these mechanisms are described. While the following discussion is sometimes specific to the buck circuit, all of the techniques presented here can be applied to maximize the efficiency of boost- and buck-boost-type converters, each of which may be required in the power distribution scheme of a battery-operated system.

12.4.1. Synchronous Rectification

The focus of this chapter is the CMOS low-voltage buck converter, in which the switching elements, modeled by the single-throw, double-pole switch in the block diagram of Figure 12.1, are implemented by complementary MOSFETs. The more conventional implementation consists of one controlled switch and one uncontrolled switch (a diode). The pure CMOS implementation allows an important advantage.

Consider the conventional buck circuit of Figure 12.9. Even if all other losses in the circuit are made negligible, the maximume efficiency is limited by the forward-

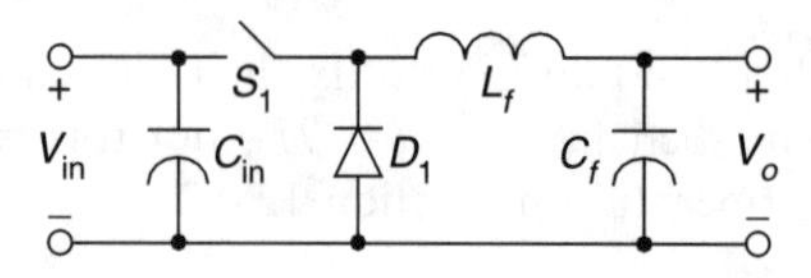

Figure 12.9 Conventional buck circuit with pass device S_1 and diode.

bias diode voltage V_{diode}. Since the diode conducts for a fraction $(1 - D)$ of the switching period, the maximum efficiency this circuit can obtain is given by

$$\eta_{\max} = \frac{V_o}{V_o + (1 - D)V_{\text{diode}}} \tag{12.10}$$

For example, consider a conventional buck circuit used to generate an output voltage of 1.5 V from a single lithium ion cell. Even using a low-voltage Schottky diode with a forward drop of 0.3 V, at the nominal cell voltage of $V_{\text{in}} = 3.6$ V, $\eta_{\max}$ is lower than 90%. With a silicon bipolar diode, $V_{\text{diode}} = 0.7$ V and $\eta_{\max} = 0.79$.

If the diode in Figure 12.9 is replaced by an NMOS device which is gated when the diode would have conducted (M_n in Fig. 12.6), the forward drop can be made arbitrarily small by sizing the device sufficiently large. In this way, the NMOS device, used as a synchronous rectifier, can more efficiently perform the same function as the diode. Assuming all other losses, including the gate-drive for the synchronous rectifier, are still negligible, the maximum efficiency of the low-voltage buck converter approaches unity.

Synchronous Rectifier Control. Although the synchronous rectifier can reduce conduction loss at low output voltage levels, it comes at the expense of an additional gate-drive signal and its associated loss. In addition, as mentioned earlier in the discussion on buck converter efficiency, without proper control of the rectifier, a short-circuit path may exist temporarily between the input rails during transients. In the rectifier control scheme described in Section 12.4.3, the dead times, which ensure that M_p and M_n never conduct simultaneously, are adjusted in a negative-feedback loop to achieve nearly ideal zero-voltage switched turn-on transitions of both power MOSFETs (see below).

12.4.2. Zero-Voltage Switching

When the low-voltage buck circuit of Figure 12.6 is hard switched, it dissipates power in proportion to $C_x V_{\text{in}}^2 f_s$ as a result of the step charging of parasitic capacitance C_x through a resistive path (M_p). In addition, it is likely to exhibit either substantial short-circuit loss (if no dead time is provided) or reverse recovery loss (if a dead time is provided). In a soft-switched circuit, the filter inductor is used as a current source to charge and discharge this capacitance in an ideally lossless manner, allowing additional capacitance to be shunted across C_x, slowing the inverter output node transitions. In this way, appropriate dead times may be set such that the power transistors are switched with $v_{ds} = 0$, essentially eliminating the associated switching loss.

Figures 12.6 and 12.10 show the low-voltage buck circuit and associated periodic steady-state waveforms for ideal zero-voltage switching (ZVS) operation. The soft-switching behavior is similar to that described in [5] and by other authors. Assume that at a given time (the origin in Fig. 12.10), the rectifier (M_n) is on, shorting the inverter output node to ground. Since, by design, the output is DC and greater than zero, a constant negative potential is applied across L_f and i_{L_f} is linearly decreasing. If the value of filter inductance is small enough, the zero-to-peak current ripple exceeds the average load ($\Delta I > 2I_o$) and i_{L_f} ripples below zero. As illustrated in Section 12.3.2, for ΔI slightly larger than $2I_o$, the physical size of the inductor is close to minimum.

If the rectifier is turned off after the current reverses (and the PMOS device, M_p, remains off), L_f acts approximately as a current source, charging the inverter output node. To achieve a lossless low-to-high transition at the inverter output node, the PMOS device is turned on when $v_x = V_{\text{in}}$. In this scheme, a pass device gate transition occurs exactly when $v_{ds_p} = 0$.

With the PMOS device on, the inverter output node is shorted to V_{in}. thus, a constant positive voltage is applied across L_f, and i_{L_f} linearly increases, until the high-to-low transition at v_x is initiated by turning M_p off. As indicated by Figure 12.10, at this time, the sign of current i_{L_f} is positive. Again, L_f acts as a current source, this time discharging C_x. If the NMOS device is turned on with $v_x = 0$, a lossless high-to-low transition of the inverter output node is achieved and M_n is switched at $v_{ds_n} = 0$.

In this scheme, a form of soft switching, the filter inductor is used to charge and discharge all capacitance at the inverter output node (and supply all Miller charge) in a lossless manner, allowing the addition of a shunt capacitor at v_x to slow these

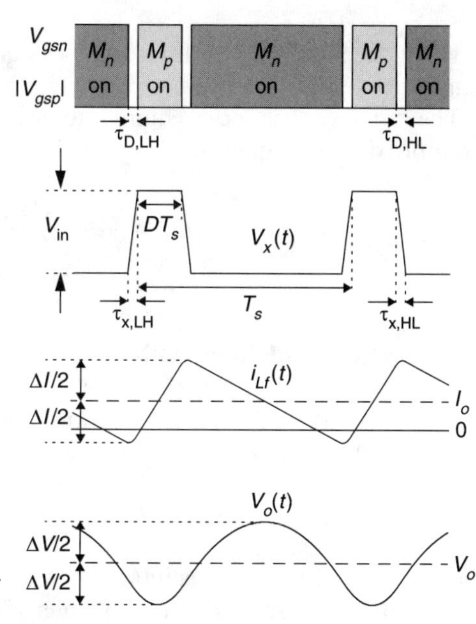

Figure 12.10 Nominal periodic steady-state ZVS waveforms.

transitions. Since the power transistors are switched at zero drain-source potential, this techique is known as zero-voltage switching and essentially eliminates capacitive switching loss. Furthermore, because the inductor current in a ZVS circuit reverses, if the body diode conducts for a portion of the cycle, it is turned off through a short circuit (rather than through a potential change of V_{in}), nearly eliminating the dissipation associated with reverse recovery, a factor which might otherwise dominate switching loss, particularly in low-voltage converters.

Design of a ZVS Buck Circuit. As discussed earlier in the section on output filter design, the inductor current waveform in a buck circuit is assumed triangular with maximum and minimum values $I_o + \frac{1}{2}\Delta I$ and $I_o - \frac{1}{2}\Delta I$ which are relatively constant over the entire dead time. Under ZVS operation, $\Delta I > 2I_o$, allowing L_f to charge and discharge the inverter output node between $v_x = 0$ and $v_x = V_{\text{in}}$. The ratio of these 0–100% soft-switched transition times is given by the ratio of the magnitude of the currents available for each commutation,

$$\frac{\tau_{x,\text{LH}}}{T_{x,\text{HL}}} = \frac{(\Delta I)/2 + I_o}{(\Delta I)/2 - I_o} \tag{12.11}$$

and approaches unity for large inductor current ripple. Here, τ_x indicates a soft-switched inverter output node transition interval, with subscripts LH and HL denoting low-to-high and high-to-low transitions, respectively. For a given ripple ΔI, the maximum asymmetry in transition times occurs at full load.

Using (12.11), the inductor current ripple ΔI is chosen to limit the maximum asymmetry in the durations of the soft-switched transitions to a reasonable value. From this value of ΔI, L_f is selected according to (12.2). Given specifications on the maximum tolerable output voltage ripple ΔV, the value of C_f is then chosen using (12.3).

To slow the soft-switched transitions to durations for which dead times may be programmed or adjusted, extra capacitance may be added at the inverter output node. The total capacitance required to achieve a given low-to-high transition time is approximately equal to

$$C_x \approx \frac{\tau_{x,\text{LH}}(\Delta I/2 - I_o)}{V_{\text{in}}} \tag{12.12}$$

12.4.3. Adaptive Dead-Time Control

To ensure ideal ZVS of the power transistors, the periods when neither conducts (the dead times), τ_D, must exactly equal the inverter output node transition times:

$$\tau_{D,\text{LH}} = \tau_{x,\text{LH}} \qquad \tau_{D,\text{HL}} = \tau_{x,\text{HL}} \tag{12.13}$$

In practice, it is difficult to maintain these relationships. As indicated by Figure 12.10, the inductor current ripple is symmetric about the average load current. As the average load varies, the DC component of the i_{L_f} waveform is shifted and the current available for commutating the inverter output node is modified. Thus, the inverter output node transition times are load dependent.

In one approach to soft switching, a value of average load may be assumed, yielding estimates of the inverter output node transition times. Fixed dead times are based on these estimates. In this way, losses are reduced, yet perhaps not to negligible levels.

In portable applications where battery capacity is at a premium, this approach to soft switching may not be adequate. To illustrate the potential hazards of fixed dead-time operation, Figure 12.11 shows the impact of nonideal ZVS on conversion efficiency through reference to a high-to-low transition at the inverter output node. In Figure 12.11(a), the dead time is too short, causing the NMOS device to turn on with $v_{ds_n} > 0$, partially discharging C_x through a resistive path and introducing losses. Since, as indicated by (12.12), shunt capacitance with a value much larger than the intrinsic parasitics may be added to slow the soft-switched transitions in a ZVS circuit, this loss may be substantial. In Figure 12.11(b), the dead time is too long, and the inverter output node continues to fall below zero until the drain-body junction of M_n becomes forward biased. In low-voltage applications, the forward-bias body diode voltage is a significant fraction of the output voltage; thus, body diode conduction must be avoided for efficient operation. When the rectifier (M_n) turns on, it removes the excess minority-carrier charge from the body diode and charges the inverter output node back to ground, dissipating additional energy.

To provide effective ZVS over a wide range of loads, an adaptive dead-time control scheme for a 1-MHz ZVS buck circuit has been outlined in [6]. Figure 12.12 shows a block diagram of the approach. A phase detector updates an error signal based on the relative timing of v_x and the gate-drive signals of the power transistors. A delay generator adjusts the dead times relative to the clock based on these error signals. Using this technique, effective ZVS is ensured over a wide range of operating conditions and process variations.

Figure 12.13 shows a circuit implementation of a $\tau_{D,\mathrm{HL}}$ adaptation scheme which is similar in principle to a delay-locked loop. The phase detector consists of two *SR* flip-flops and controls the complementary switches of a charge pump. An error voltage proportional to the difference between the high-to-low soft-switched

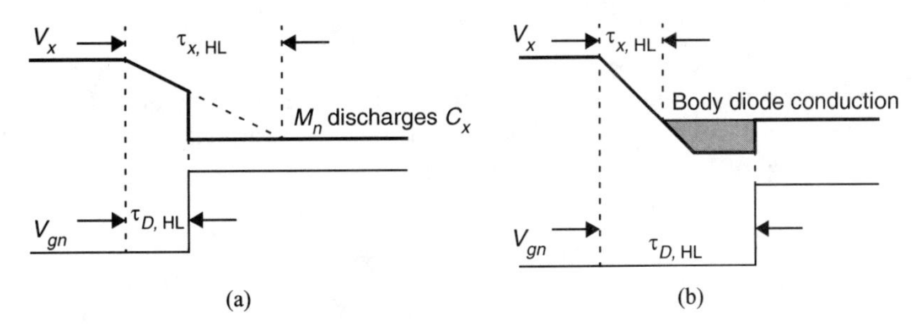

Figure 12.11 Nonideal ZVS and its impact on conversion efficiency: (a) dead time is too short; (b) dead time is too long.

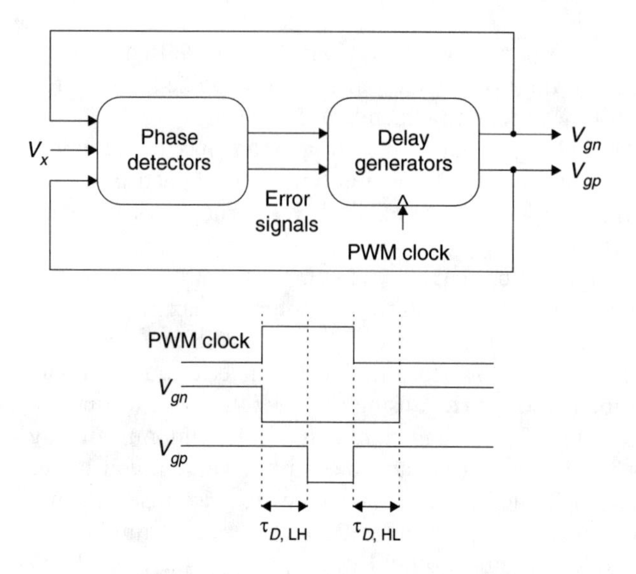

Figure 12.12 Conceptual representation of dead-time adaptation scheme.

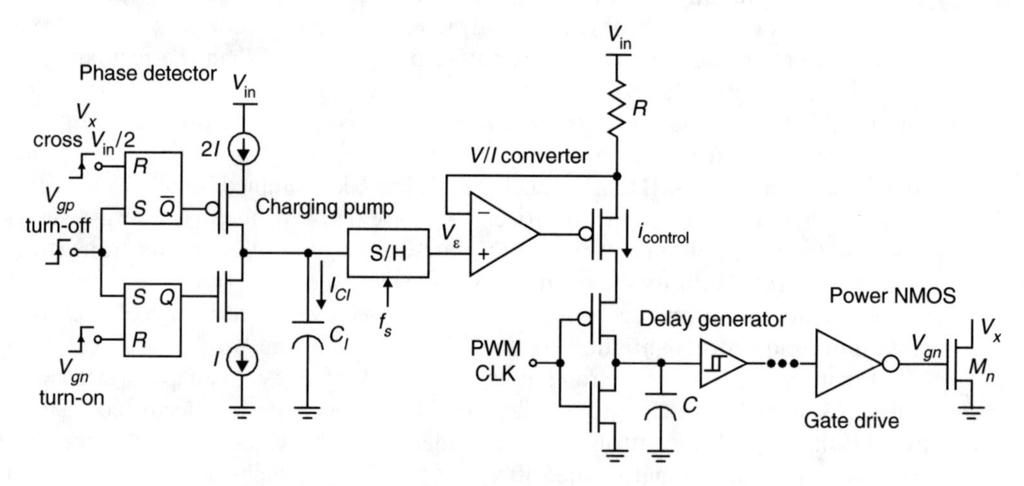

Figure 12.13 Rectifier turn-on delay adjustment loop.

inverter output node transition time and its corresponding dead time is generated on integrating capacitor C_I. This error voltage is sampled and held at the switching frequency of the converter, such that

$$v_\varepsilon[(n+1)T_s] \approx v_\varepsilon(nT_s) + \frac{I}{C_I[\tau_{x,\mathrm{HL}}(nT_s) - \tau_{D,\mathrm{HL}}(nT_s)]} \tag{12.14}$$

The delay generator, which is implemented by a V/I converter and a monostable multivibrator, produces a dead time with a length that is controlled by $v_\varepsilon(nT_s)$.

In periodic steady state, the error voltage and thus the gate timing errors are forced to zero, nulling propagation delays in the control and drive circuitry. Figure 12.14 shows the periodic steady-state waveforms associated with an ideal ZVS rectifier turn-on.

A similar loop is used to adjust the dead time between the turn-off of M_n and the turn-on of M_p, $\tau_{D,\mathrm{HL}}$.

12.4.4. Power Transistor Sizing

Through use of ZVS with adaptive dead-time control, switching loss is essentially eliminated. If the filter components in the buck circuit of Figure 12.6 are ideal and series resistance and stray inductance in the power train are made negligible, the fundamental mechanisms of power dissipation will include on-state conduction loss and gate-drive loss in the power transistors. When sizing a MOSFET for a particular power application, the principal objective is to minimize the sum of the dissipation due to these mechanisms. This minimization is performed at the operating point where high efficiency is most critical. This is typically at full load, at high temperature, and in portable applications at the nominal battery source voltage, but a similar optimization may be performed for other operating conditions, if necessary to maximize overall battery life.

During their conduction intervals, the power transistors operate exclusively in the triode region, where $r_{ds} = R_0 W_0 / W$ (the channel resistance is inversely proportional to gate width with constant of proportionality $R_0 W_0$ and R_0 is the channel resistance of a power MOSFET with minimum gate width W_0). Thus, at a given operating point, the on-state conduction loss in a FET is given by

$$P_q = \frac{i_{ds,\mathrm{rms}}^2 R_0}{W/W_0} \tag{12.15}$$

Since the device parasitics generally increase linearly with increasing gate width, the gate-drive loss can be expressed as a linear function of gate width W:

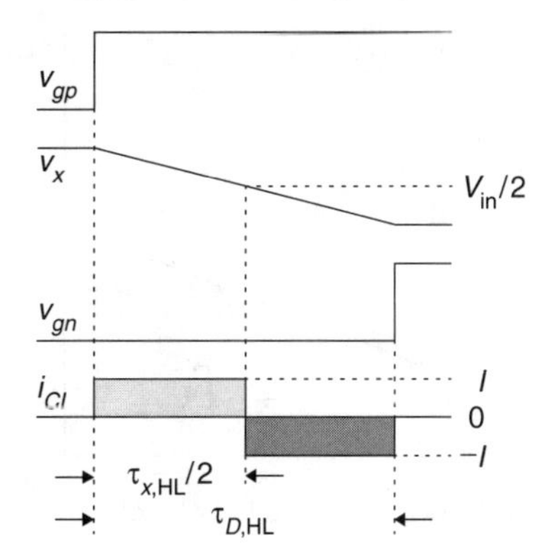

Figure 12.14 Ideal steady-state waveforms for the $\tau_{D,\mathrm{HL}}$ adjustment loop.

$$P_g = E_{g0} f_s \frac{W}{W_0} \tag{12.16}$$

where E_{g0} is the total gate-drive energy consumed in a single off-to-on-to-off gate transition cycle of a power MOSFET with minimum gate width W_0 (see the section on CMOS gate-driven design that follows later for more detail) and f_s is the switching frequency of the converter. In a ZVS circuit, the filter inductor supplies all of the Miller charge, so E_{g0} contains no dissipation due to Miller effect.

Using an algebraic minimization at the most critical operating point, the optimal gate width of the power transistor

$$W_{\text{opt}} = W_0 \sqrt{\frac{i^2_{ds,\text{rms}} R_0}{E_{g0} f_s}} \tag{12.17}$$

is found to balance on-state conduction and gate-drive losses, where

$$P_{q,\text{opt}} = P_{g,\text{opt}} = \sqrt{i^2_{ds,\text{rms}} R_0 E_{g0} f_s} \tag{12.18}$$

and $P_t = P_q + P_g$ is at its minimum value, $P_{t,\text{min}}$. Figure 12.15 illustrates normalized power transistor losses as a function of gate width.

12.4.5. Reduced-Swing Gate-Drive

To ensure that the duration of the low-to-high soft-switched transition is kept reasonably short in a ZVS buck circuit, the inductor current ripple must be made substantial. This gives rise to large circulating currents in the power train and, therefore, when the power transistors are sized according to (12.17), to increased gate-drive losses. Since gate-drive losses increase in direct proportion with f_s, this proves to be the limiting factor to higher frequency operation of soft-switched converters. To reduce gate-drive losses, a number of resonant gate-drives have been proposed

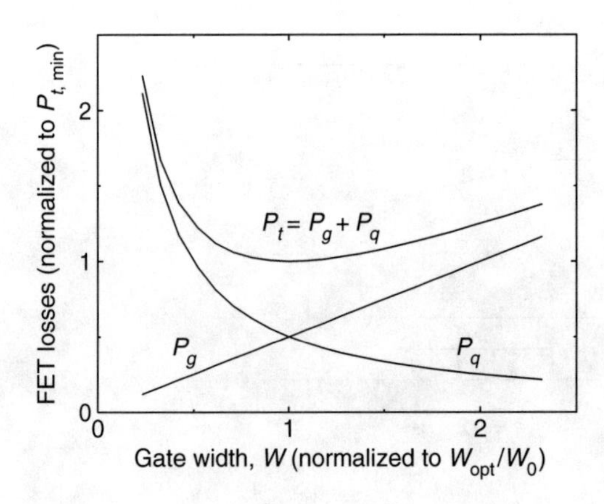

Figure 12.15 Power transistor losses vs. gate width.

[7–9]. Several such techniques have demonstrated the ability to recover a significant fraction of the gate energy at lower frequencies. However, due to the resistance of the polysilicon gate of a power transistor, none are likely to be as practical in the 1-MHz frequency range.

Rather than attempting to recover gate energy in a resonant circuit, another approach to reducing gate-drive dissipation is to reduce the gate energy consumed per cycle. By decreasing the gate-source voltage swing between off-state ($V_{GS} = 0$) and on-state ($V_{GS} = V_g$) conduction, for $V_g \gg V_t$, where V_t is the device threshold voltage, gate energy may be quadratically reduced. This is an attractive alternative in portable systems where a number of low-voltage supplies are typically available for the gate-drive. However, because the channel resistance of the device increases with $(V_g - V_t)^{-1}$, gate swing cannot be arbitrarily reduced, implying the existence of an optimum V_g (see the Section on V_g selection below).

To further reduce the total dissipation of a power MOSFET with a given gate voltage swing V_g, the off-state voltage $V_{GS(\text{off})}$ can be made greater than zero [Fig. 12.16(a)] to increase the gate overdrive, reducing the device channel resistance without increasing gate-drive loss. This scheme is equivalent to that shown in Figure 12.16(b), where $V'_{GS(\text{off})} = 0$ and the device threshold voltage is scaled such that $V'_t = V_t - V_{GS(\text{off})}$ [2]. Threshold voltage scaling is limited primarily by subthreshold current conduction in the power MOSFETs, which increases exponentially with decreasing V_t and with increasing temperature. For sufficiently low V_t and/or sufficiently high temperature, subthreshold leakage can result in significant static power dissipation in the power train of the converter.

CMOS GATE-DRIVE DESIGN. In CMOS circuits, a power transistor is conventially driven by a chain of N inverters which are scaled with a constant tapering factor u such that

$$u^N = \frac{C_g}{C_i} \tag{12.19}$$

Here, C_g is the gate capacitance of the power transistor and C_i is the input capacitance of the first buffering stage. This scheme, depicted in Figure 12.17, is designed

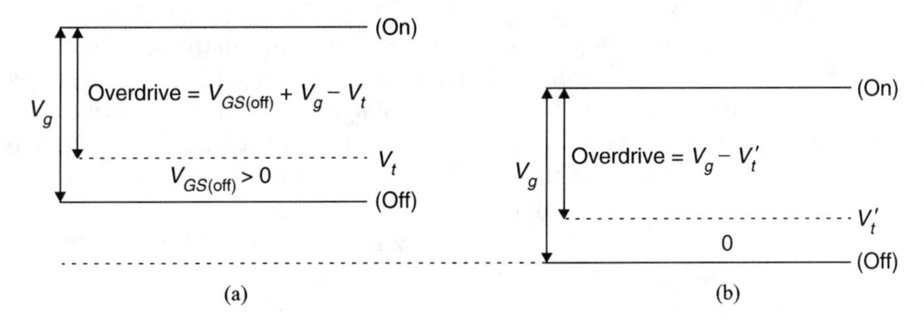

Figure 12.16 Two equivalent schemes to further reduce total power transistor losses: (a) gate-source voltage is not brought to zero; (b) lower V_t.

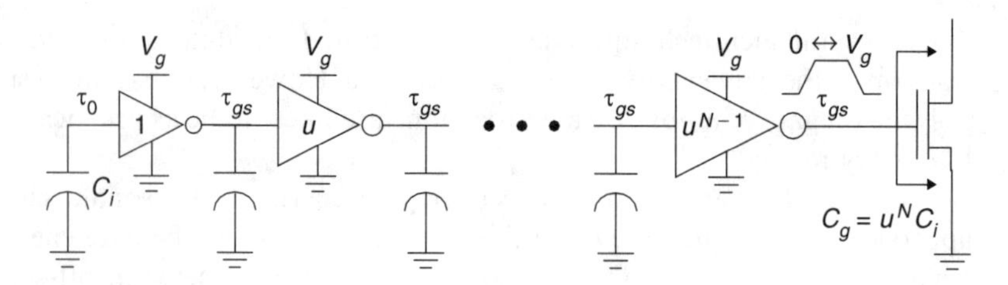

Figure 12.17 CMOS gate-drive scheme.

such that the ratio of average dynamic current to load capacitance is equal for each inverter in the chain. Thus, the delay of each stage and the rise/fall time at each node are identical. It is a well-known result that under some simplifying assumptions, the tapering factor u that produces the minimum propagation delay is the constant e [10]. However, in power circuits, the chief concern lies not in the propagation delay of the gate-drive buffers but in the energy dissipated during a gate transition.

In a ZVS power circuit, the following timing constraint is desired:

$$\tau_{gs} \ll \tau_x \tag{12.20}$$

where τ_x is the soft-switched inverter output node transition time and τ_{gs} is the maximum gate transition time which ensures effective ZVS of the power transistor. For a given output transition time (rise/fall time) τ_0 of a minimal inverter driving an identical gate and a tapering factor u between successive inverters in the chain. τ_{gs} can be approximated by

$$\tau_{gs} \approx u\tau_0 \tag{12.21}$$

In general, it is desirable to make τ_{gs} as large as possible (yet still a factor of 5–10 less than τ_x), minimizing gate-drive dissipation. Given τ_{gs} and τ_0, if there exists some $u > e$ such that the criteria given by (12.20) and (12.21) are met, the buffering scheme of Figure 12.17 will provide a more energy-efficient CMOS gate-drive than that obtained through minimization of delay.

DETERMINATION OF THE INVERTER CHAIN. In this analysis, a minimal CMOS inverter has an NMOS with minimum dimensions (W_0/L) and threshold voltage V_{tn} and a PMOS device with threshold voltage V_{tp} whose gate width is $\mu_n/\mu_p \approx 3$ times that of the NMOS device. It has lumped capacitances C_i at its input and C_o at its output. Given that the pull-down device operates exclusively in the triode region during the interval of interest, and assuming it is a long-channel device, it can be shown [11] that the output fall time of a minimal inverter driving an identical gate from $V_{\text{out}} = V_g - |V_{tp}|$ to $V_{\text{out}} = V_{tn}$ is

$$\tau_0 = \frac{C_o + C_i}{W_0}\kappa \tag{12.22}$$

which is linearly proportional to the capacitive load, inversely proportional to the gate width of the N-channel device, and directly related to the application- and technology-dependent constant:

$$\kappa \equiv \frac{2L}{\mu_n C_{ox}(V_g - V_{tn})} \ln\left[\frac{2V_g - 3V_{tn}}{V_{tn}} \frac{V_g - |V_{tp}|}{V_g - 2V_{tn} + |V_{tp}|}\right] \tag{12.23}$$

In [12], a similar expression is found for the output fall time assuming a heavily velocity-saturated pull-down device.

The factor u which results in an output signal transition time τ_{gs} is found by solving

$$\tau_{gs} = \frac{\kappa(C_o + uC_i)}{W_0} \approx u\tau_0 \tag{12.24}$$

yielding a corresponding tapering factor of

$$u = \frac{\tau_{gs} W_0 - \kappa C_o}{\kappa C_i} \tag{12.25}$$

between successive buffers. Given u, the number of inverters in the chain is

$$N = \frac{\ln(C_g/C_i)}{\ln(u)} \tag{12.26}$$

The inverter chain guarantees a gate transition time of τ_{gs} with minimum dissipation and a propagation delay of

$$t_p \approx Nut_{p0} \tag{12.27}$$

where t_{p0} is the propagation delay of a minimal inverter loaded by an identical gate.

Loss Analysis. The taper factor u is constant throughout the inverter chain, providing equal transition times τ_{gs} at each node. Thus, short-circuit dissipation is a small fraction of the overall dissipation [13] and may be assumed negligible compared to the dynamic dissipation at low supply voltages [2, 14]. The total gate energy, including the drive, is dominated by the dynamic dissipation:

$$E_g = C_T V_g^2 \tag{12.28}$$

where the total switching capacitance in the inverter chain, including the loading gate capacitance of the power MOSFET, is

$$\begin{aligned} C_T &= (1 + u + u^2 + \cdots + u^{N-1})(C_o + C_i) + C_g \\ &= \left[\frac{u^N - 1}{u - 1}\right](C_o + C_i) + C_g \end{aligned} \tag{12.29}$$

Since u^N is the constant given by (12.19), C_T, and thus the dynamic dissipation, is minimized for large u.

To make a first-order estimate of the total energy consumed in a single off-to-on-to-off gate transition cycle of a minimal power MOSFET (with gate width W_0) (12.28) and (12.29) are used in conjunction with the values of u and N derived in (12.25) and (12.26), giving

$$E_{g0} \approx C_{go} V_g^2 \left[\frac{\kappa(C_o + C_i)}{\tau_{gs} W_o - \kappa C_o - \kappa C_i} + 1 \right] \tag{12.30}$$

where C_{go} is the gate capacitance of a power transistor with minimum gate width W_0 linearized over $0 \leq V_{GS} \leq V_g$. To obtain (12.30), it is assumed that the short-circuit dissipation in the inverter chain is negligible compared to the dynamic dissipation, that all capacitances scale linearly with gate width, and that $u^N \gg 1$. Under these simplifications, gate-drive losses are expressed as a linear function of gate width, identical in form to (12.16).

SCALING V_g. The practical limit to gate-drive supply voltage scaling is set by increasing delays in the CMOS drive circuitry. As the supply voltage for the CMOS drive, V_g, is reduced, the tapering factor between successive inverters must be decreased to compensate for increasing delays in the drive circuitry. This not only diminishes returns in the reduction of gate energy consumption, but actually reverses the trend and increases overall gate energy consumption as V_g approaches the sum of the threshold voltages of the complementary devices used in the gate-drive buffers, $V_{tn} + |V_{tp}|$. Using a linearized first-order model to a CMOS inverter delay [15], it can be shown that, for $V_g \gg V_t$, τ_0 increases with V_g^{-1} for long-channel devices and is roughly independent of V_g for heavily velocity-saturated short-channel devices. However, as $V_g \rightarrow V_{tn} + |V_{tp}|$, the delays increase rapidly [1].

This phenomenon is illustrated in Figure 12.18, where the output signal rise and fall times of a CMOS inverter with $W_p/W_n \approx \mu_n/\mu_p$ in a 1.2-μm technology are plotted versus the supply voltage V_g. For $V_g > 3\,\text{V}$, delays are indeed relatively independent of supply voltage, and the rise and fall times are nearly equal.

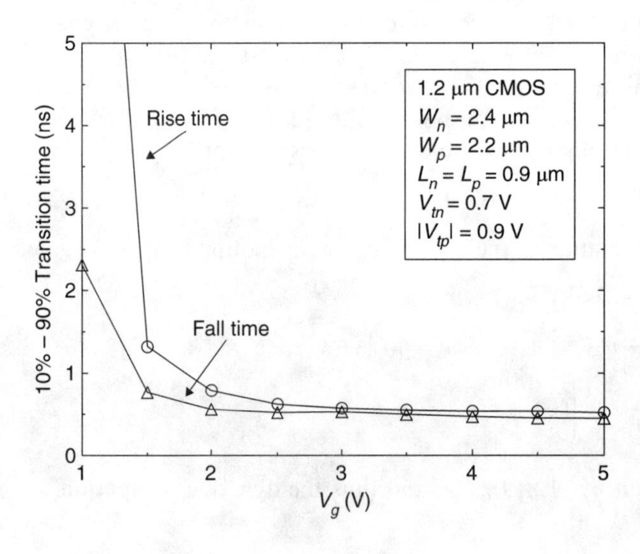

Figure 12.18 Simulated output rise and fall times for a minimal CMOS inverter driving an identical gate as a function of supply voltage, V_g.

However, as the supply is dropped below 2 V, it becomes comparable to $V_{tn} + |V_{tp}|$, and inverter output signal transition times increase rapidly. Furthermore, because $|V_{tp}| > V_{tn}$ in this technology, the output rise time increases more quickly than the output fall time. To achieve balanced rise and fall times at the output of a CMOS inverter with a supply voltage comparable to V_t, the difference in threshold voltages of N- and P-channel MOSFETs must be considered in the ratioing of the devices.

Figure 12.19 plots the total minimal power MOSFET gate energy consumed per cycle, E_{g0}, as a function of the gate-drive supply voltage. Here, the gate of an N-channel power transistor with minimum gate width W_0 is driven from 0 to V_g in a time $\tau_{gs} = 5\,\text{ns}$ (a suitable gate transition time for a 1-MHz ZVS power circuit), which is held constant through independent sizing of P- and N-channel devices in the CMOS inverter chain. For $V_g \gg V_t$, there is an approximately quadratic reduction in E_{g0} with decreasing supply voltage. However, at lower voltages, particularly as $V_g \to V_{tn} + |V_{tp}|$, the dynamic energy consumed by the gate-drive buffering increases dramatically and begins to dominate over that required by the gate capacitance of the power transistor. This is due to the increase in inverter delays, the increase in buffer input and output capacitances associated with larger P-channel device ratioing, and the resulting need to decrease the tapering factor u between successive inverters in the chain to achieve the desired output signal transition times. Thus, when the dissipation in the inverter chain is considered in gate-drive supply voltage scaling, at ultralow voltages, E_{g0} *increases* as V_g decreases.

V_g Selection. In Section 12.4.4, it was shown that in a ZVS power circuit, if a power transistor is sized according to (12.17), its total dissipation is minimized under a specific set of operating conditions and that this minimum dissipation is related to the gate energy and channel resistance of a minimal power MOSFET as follows:

$$P_{t,\min} \propto \sqrt{R_0 E_{g0}} \tag{12.31}$$

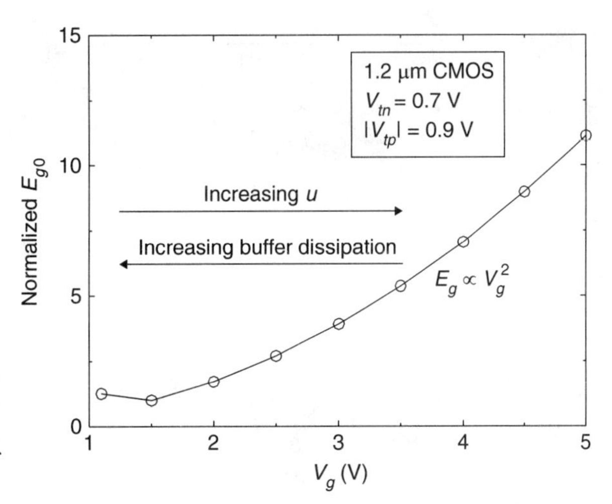

Figure 12.19 Extrapolated gate energy (including the CMOS drive) per cycle versus gate-drive supply voltage for a minimal power MOSFET and equal inverter output rise and fall times of $\tau_{gs} = 5\,\text{ns}$.

A further optimization might be to minimize this expression with respect to V_g, yielding an optimum gate voltage swing [2]. Yet, because it is common to have at least one low-voltage supply already available that could be used for the gate-drive, it is typically more important to compare the minimum achievable FET losses and the corresponding gate-width requirement for $V_g = V_{in}$ and for V_g equal to a given available low-voltage supply.

Figure 12.20 provides data to enable such a comparison for a 1.2-μm CMOS technology. Here, W_{opt} and $P_{t,min}$ are plotted versus V_g. A large-area N-channel MOSFET has been simulated to find device parameters R_0 and E_{g0} at each data point. Dissipation in the drive circuitry is included in E_{g0}. From this plot, it can be seen that the greatest power savings with reduced V_g are achieved in the range where $V_g \gg V_t$: Since E_{g0} decreases quadratically while R_0 increases linearly, if the gate width of the power device is appropriately scaled ($W_{opt} \propto V_g^{-3/2}$), $P_{t,min}$ decreases as $\sqrt{V_g}$. However, since both R_0 and E_{g0} increase as V_g is brought below the sum of the threshold voltages in the gate-drive buffers, $P_{t,min}$ increases with further decrease in V_g. It may be concluded that the choice for V_g for minimum energy dissipation is

$$V_{g,opt} \approx V_{tn} + |V_{tp}| \tag{12.32}$$

Note that V_{tp} enters into the expression for the drive voltage even for an N-channel power device, because of its role in the buffer circuitry. Note also that the loss penalty for a slightly higher V_g is insignificant and the area requirement drops substantially. For a particular application, the power loss and area constraints or costs should be evaluated at the available voltages that might be used for V_g.

For example, consider a converter in a portable system operating from a lithium ion battery souce. From Figure 12.20, the total loss in each power FET at $V_g = 1.5\,V$ is 20% lower than at $V_g = 3.6\,V$ (the nominal battery source voltage). However, the gate width of each device must be increased by a factor greater than 4.7 to achieve this reduced dissipation.

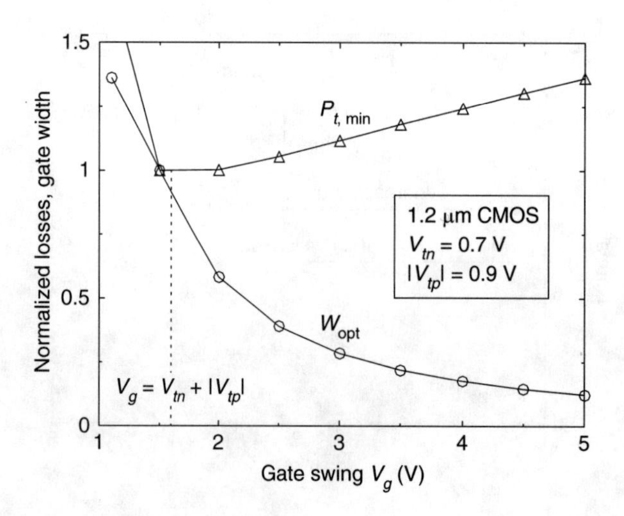

Figure 12.20 Optimal gate width and minimum total dissipation for power N-channel MOSFET vs. gate swing in 1.2-μm CMOS technology.

Reduced Gate-Swing Circuit Implementation. Figures 12.21 and 12.22 show a circuit implementation of a reduced-swing gate-drive and its associated waveforms in a low-output-voltage ZVS CMOS buck circuit. The gate of M_n is actively driven from 0 to V_g by its CMOS gate-drive. The gate of the P-channel power MOSFET is driven from V_{in} to approximately $V_{in} - V_g$ with an AC-coupled gate-drive. The PMOS device M_{off}, whose gate swings from rail to rail, provides a low-impedance path from the gate of M_p to V_{in}, ensuring that M_p remains fully off during its off state. In ultra-low-power applications, the AC-coupling capacitor $C_c \gg C_{gp}$ might be implemented on-chip.

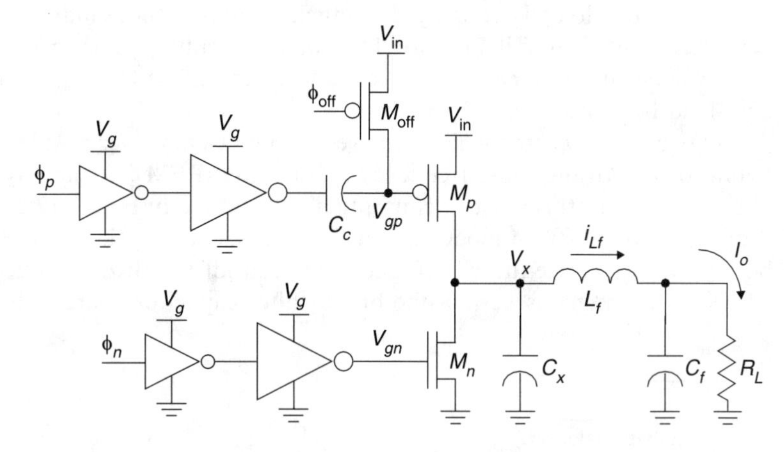

Figure 12.21 Reduced gate-swing CMOS buck circuit implementation with gate supply voltage V_g.

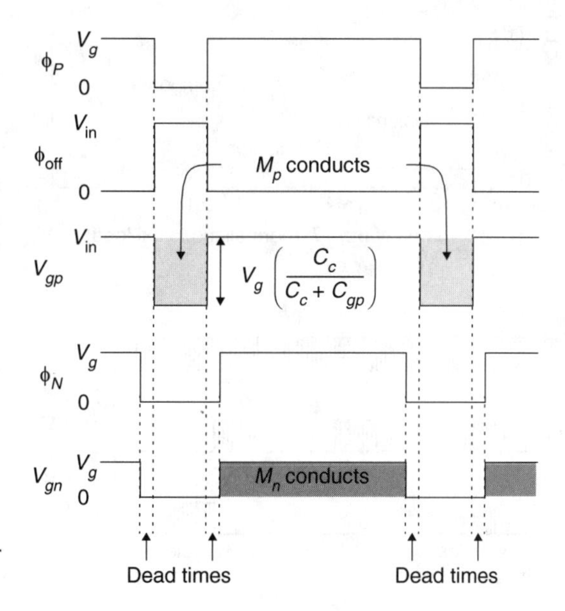

Figure 12.22 Waveforms for reduced-swing gate-drive.

12.4.6. PWM-PFM Control for Improved Light-Load Efficiency

While a PWM DC-DC converter can be made to be highly efficient at full load, many of its losses are independent of load current, and it may therefore dissipate a significant amount of power relative to the output power at light loads. Figure 12.23 plots power transistor losses (conduction and gate-drive) versus a 1000 : 1 load variation for a ZVS PWM buck converter with a peak-to-peak inductor current of two times the full-load current. The power transistors are optimized for full-load operation. From this plot, it may be concluded that a PWM converter which is 95% efficient at full load is roughly 3% efficient at one thousandth full load. If the converter is used at full load for little of its operating time, energy loss at light load will be the dominant limitation on battery life and improving efficiency at light load becomes essential.

One control scheme which achieves high efficiency over a wide load range is a hybrid of PWM and pulse-frequency modulation (PFM), commonly referred to as "burst mode." In this scheme, conceptually illustrated in Figure 12.24, the converter is operated in the PWM mode only in short bursts of N cycles each. Between bursts, both power FETs are turned off, and the circuit idles with zero inductor current for M cycles. During this period, the output filter capacitor sources the load current.

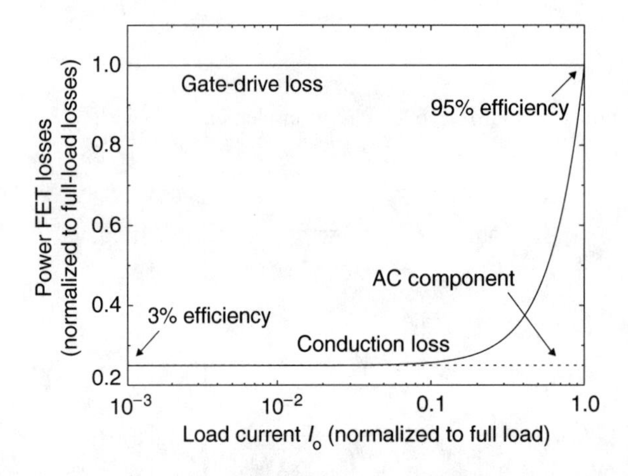

Figure 12.23 Power transistor losses vs. 1000 : 1 load variation for ZVS PWM buck converter.

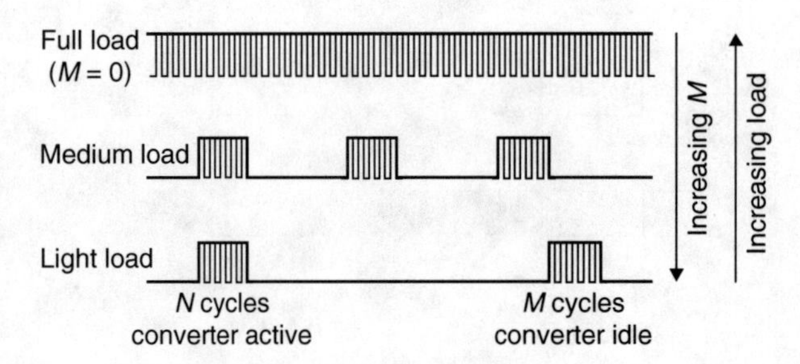

Figure 12.24 Conceptual illustration of PWM-PFM control.

When the output is discharged to a certain threshold below V_o, the converter is operated as a PWM converter for another N cycles, returning charge to C_f. Thus, the load-independent losses in the circuit are reduced by the ratio $N/(N+M)$. As the load current decreases, the number of off cycles, M, increases.

During the N cycles that the converter is active, it may be zero-voltage switched. However, the transitions between idle and active modes require an additional energy overhead. When the idle mode is initiated by turning both FETs off, the body diode of M_n picks up the inductor current until i_{L_f} decays to zero. In low-voltage applications, body diode conduction is highly dissipative. When the converter is reactivated, M_p charges C_x to V_{in}, introducing additional losses. The minimum number of off cycles, M_{min}, can be chosen to ensure that the energy saved by idling is substantially larger than this loss overhead.

Circuit implementations of PWM-PFM control may be found in [16] and [17].

12.5. EXAMPLE DESIGN

In this section, the design techniques of Sections 12.3 and 12.4 are applied to the 6-V to 1.5-V, 500-mA buck converter presented in [6]. All of the active devices are integrated on a single die and fabricated in a standard 1.2-μm CMOS process. The circuit exhibits nearly ideal ZVS using an adaptive dead-time control scheme similar to that described in Section 12.4.3.

Figures 12.25 and 12.26 show the power train of the low-output-voltage CMOS buck converter with device sizes and component values and its associated periodic steady-state waveforms. The inverter output node voltage is quasi-square with a nominal duty cycle $D = V_o/V_{\text{in}}$ of 25% and an operating frequency $f_z = 1\,\text{MHz}$ which allows a compact, yet highly efficient converter. The inductor current reverses to allow ZVS transitions of the power transistors. The maximum asymmetry in the soft-switched transition intervals, $\tau_{x,\text{LH}}/\tau_{x,\text{HL}} = 4$ at full load, is chosen to make the timing constraint of (12.20) feasible in a 1.2-μm technology. From (12.11), this requires a minimum peak-to-peak inductor current ripple of $\Delta I = \frac{10}{3} I_o = 1.66\,\text{A}$ and, using (12.2), results in a filter inductor of value $L_f = 675\,\text{nH}$. Alowing for a 2% peak-to-peak AC ripple in the output voltage, $C_f = 13.9\,\mu\text{F}$ is selected using (12.3). To slow the soft-switched transitions at v_x, snubber capacitance is shunted across the inverter output node. The total capacitance required to achieve

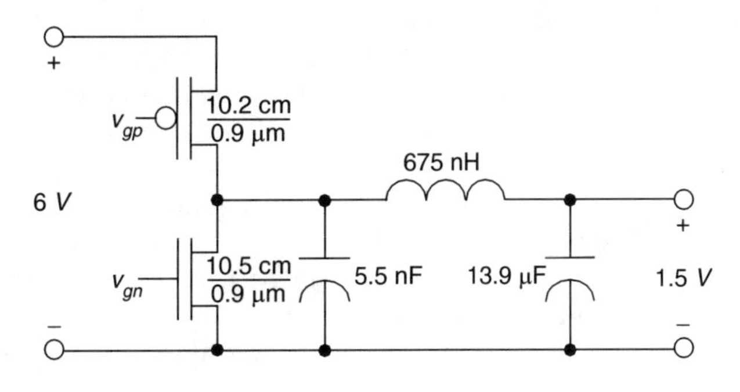

Figure 12.25 Circuit schematic for example buck circuit.

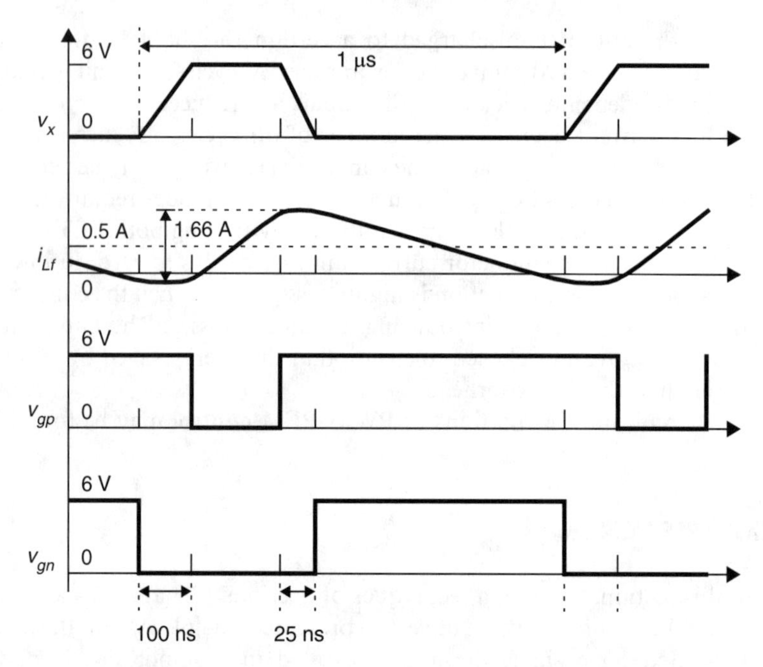

Figure 12.26 Ideal periodic steady-state waveforms for example buck circuit.

$\tau_{x,\text{LH}} = T_s/10 = 100\,\text{ns}$ is $C_x = 5.56\,\text{nF}$, where C_x includes the snubber and all parasitic capacitance at v_x.

The power transistors are sized according to (12.17) to minimize their total losses in periodic steady state at full load. The minimum effective channel length $L_{\text{eff}} = 0.9\,\mu\text{m}$ is used. Device parameters R_0 and Q_{g0}, which represent the effective channel resistance and gate charge of a minimum gate-width device, are found at $V_g = V_{\text{in}} = 6\,\text{V}$ by interpolating results obtained from circuit simulations performed on extracted layout of large-geometry FETs. Here, $W_0 = 0.6\,\mu\text{m}$, the minimum feature size in the 1.2-μm process, is chosen to represent the gate width of a minimal power MOSFET. Substituting $C_{g0} = Q_{g0}/V_{\text{in}}$ and all necessary application- and technology-specific parameters into (12.30), a first-order estimate to E_{g0} is made. Approximate power transistor gate widths are found by substituting this estimate and the interpolated value of R_0 into (12.17). A prediction of the gate-drive design is effected through selection of u and N with (12.25) and (12.26). Iteration using circuit simulations on an extracted layout is beneficial to refine the design. From (12.18), total FET losses at full load can be estimated. The design is summarized in Table 12.1.

The circuit as presented in [6] uses the full input voltage to drive the gates of the power transistors. To improve efficiency, the reduced-swing gate-drive implementation of Figure 12.21 may be used to bootstrap the gate-drive from the regulated output of the converter. This requires additional circuitry to deliver charge to C_f during start-up, avoiding the stable state in which $V_o = 0$. With $V_g = V_o = 1.5\,\text{V}$,

TABLE 12.1 EXAMPLE DESIGN SUMMARY

Parameter	M_p	M_n
R_0	23.7 kΩ	6.2 kΩ
Q_{g0}	8.6 fC	9.7 fC
E_{g0}	58.7 fJ	68.8 fJ
Gate width, W	10.2 cm	10.6 cm
Buffering, u	5.6	5.2
Buffering, N	4	4
Estimated loss	2.7%	3.2%

total FET losses may be reduced from 5.9% to roughly 4% at full load, but at the expense of considerable silicon area—the total gate width would be increased by a factor greater than 10.

12.6. PHYSICAL DESIGN CONSIDERATIONS

12.6.1. Power Transistor Layout

The power transistors are arrayed as a number of parallel fingers whose length is determined by the maximum tolerable distributed *RC* delay of the gate structure (Fig. 12.27). In order to reduce series source and drain resistances, the diffusion is heavily contacted, with source-body and drain running horizontally within each finger in metal 1 and vertically throughout the array in metal 2. The gate runs horizontally in polysilicon and is strapped vertically in metal 1 such that each finger is driven from both sides.

Because the ZVS operation of the power devices effectively eliminates capacitive switching losses associated with the drain-body junction, increased drain diffusion area may be accepted in return for more reliable and efficient body diode conduction. Rather than sharing diffusion between adjacent fingers in the array, well or substrate contacts may be placed at a minimum distance from drain contacts to minimize both the series resistance and transit time of the body diode. This is the approach taken in [6]. In [18], a more area-efficient layout is used. Here, body and drain diffusion of adjacent fingers is shared, eliminating the small spaces between each finger. Although this tiling strategy results in an inferior body diode structure, the length of source-body, drain, and gate straps are minimized for a given gate width, reducing the physical resistances in series with the terminals of the device as well as utilizing die area more efficiently.

Due to the potentially large magnitude of substrate current injection, substantial well or substrate contacts should be placed throughout the structure and a guard ring should surround each power device, eliminating latch-up and decoupling supply bounce from the control circuitry.

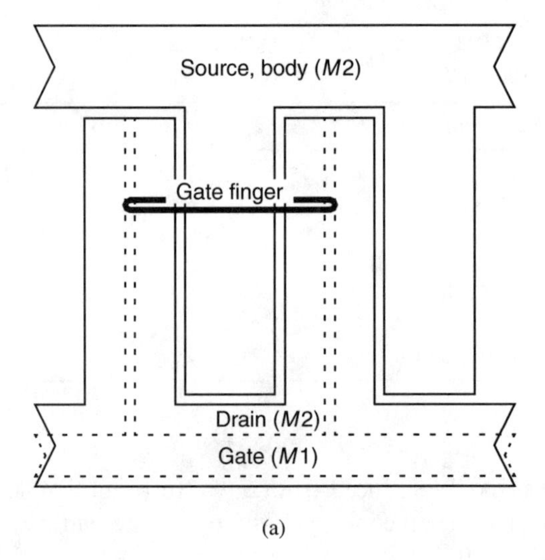

(a)

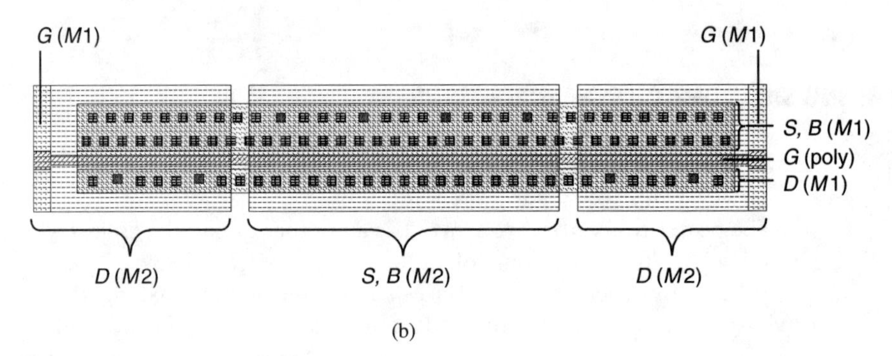

(b)

Figure 12.27 Layout of power MOSFET: (a) metallization pattern; (b) one gate finger.

12.6.2. Board-Level Assembly

It may soon be possible to build a completely monolithic converter including control circuits, power semiconductors, magnetics, and capacitors. However, it appears that for some time it will remain more technically and economically feasible to combine separate semiconductor, magnetic, and capacitive components. A PWM converter may have only four separate components—an IC with power devices and control functions, two multilayer ceramic capacitors for input and output filters, and a filter inductor.

Parasitic inductance, capacitance, and resistance in interconnects may cause substantial losses. The packaging and interconnect technology may dominate the physical size, especially in a low-power converter. There are also likely to be significant effects on cost and reliability. Many of these considerations point toward the use of a MCM or similar technology. The elimination of a separate package for the silicon circuitry may significantly decrease physical size, and parasitics associated with the package are eliminated.

The effect of parasitics on converter efficiency is determined by the ratio of the parasitic impedance to the effective load impedance. Series impedances such as inductances and resistances are most important at low voltages, although at very low current levels, they become less important again. Shunt impedances, such as stray capacitance, become important at low current levels, but ZVS can eliminate some of the losses associated with them.

Decreasing the interconnect resistance is effected by careful layout using wide, short conductor paths and thick, low-resistivity conductors. Stray inductance is reduced by minimizing the area of loops in critical paths. In a multilayer interconnection technology, the lowest stray inductance is achieved by using a conductor that overlaps a return path in a different layer, with thin dielectric separating the layers [19].

Numerous MCM technologies suitable for low-power converters exist. For higher power converters, a substrate with good thermal conductivity is a necessity, but for lower power converters, the requirements are much less stringent. Perhaps the most important parameter is low sheet resistance, to reduce series interconnect resistance. Thin dielectric between conductive layers is also desirable for minimizing stray inductance, but most technologies use thin enough dielectric that other stray inductances, such as bond wires and package leads, will dominate when careful layout is used. MCM technologies are surveyed in [20, 21].

12.6.3. Magnetic Components

Magnetic components (inductors and transformers) are essential for power electronics. With a few exceptions (such as switched-capacitor converters, described in Section 12.7.2), switching power converters ubiquitously require inductors and/or transformers. Inductors are required as components of low-loss filters and resonant circuits. Transformers are used for isolation, converting voltage and current levels, and energy storage and transfer. Magnetic devices serve these and other functions in both traditional and newly introduced converter topologies.

Magnetic devices often are the physically largest components in a converter and can be the most expensive. Typically, they must be custom designed for a particular circuit, either because suitable standard commercial parts do not exist or because of the size, cost, and performance advantages possible in a custom design. For these reasons, it is important for a power converter designer to have some knowledge of magnetics design. A complete review of magnetics design is beyond the scope of this text. However, some of the special issues in magnetics for low-power portable applications will be addressed. A basic tutorial introduction to magnetics for power electronics can be found in [3]. A more complete work on magnetics design is [22].

Magnetic Cores. In some cases, a coil (or pair of coils for a transformer) may have sufficient inductance for a power circuit application, even with no magnetic core material. Such an air-core coil is attractive because it can be fabricated on a PC board or an MCM substrate simply by patterning a spiral, ideally in multiple layers. However, there are many advantages to adding a core, and there are potential pitfalls in using an air-core coil.

An air-core coil is typically characterized at a given frequency by its inductance and its quality factor, $Q = \omega L/R$, where R and L are the effective values at a given frequency ω. A high value of Q, in the range of 20–200, is required for efficient operation. The addition of a magnetic material can, ideally, increase the inductance of a coil by a factor equal to the relative permeability of the material, μ_r, without affecting the resistance of the coil. Since practical magnetic materials for power applications often have permeabilities in the thousands, the value of Q can be greatly increased by the addition of a magnetic core. However, the increase is not as large as might be expected from this naive analysis. Power is dissipated in the core, contributing to the effective value of R. In order to combat the resulting lowering of Q and/or to prevent magnetic saturation of the core, an air gap is often introduced in the magnetic path through the core. This decreases the gain in L. Nonetheless, the increase in Q is substantial, and it is rarely possible to make practical magnetic components for power circuits operating below 1 MHz without the use of a magnetic core.

An additional advantage of a magnetic core is that it may serve to contain the flux. The external high-frequency flux from an air-core coil may cause radiofrequency interference (RFI) problems in nearby circuits and will induce eddy currents in any nearby metal objects. The resulting losses can severely lower the Q of the coil and even decrease its inductance. It may be possible to place the coil in a volume empty of metal components to avoid eddy current problems, but this additional volume becomes another reason to prefer using a component with a core, which is typically smaller than the equivalent air-core coil with the same inductance, even without this extra volume.

The most commonly used magnetic material for power applications is MnZn ferrite. Although it has a lower saturation flux density than typical magnetic metal alloys, it has much higher resistivity ($\approx 10\ \Omega$-cm, vs. $20 \times 10^{-8}\ \Omega$-cm for NiFe alloy). This allows operation at high frequencies (between 20 kHz and 1 MHz) without the severe eddy current losses that would result from using a lower resistivity material.

Despite the high resistivity of ferrites, significant losses do occur in high-frequency operation, largely due to hysteresis. Since the hysteresis losses increase rapidly with frequency and flux level, it is often necessary to reduce flux level at high frequency in order to keep losses under control. This typically limits the flux density to well below the saturation flux density, and thus loss at high frequency is the most important parameter in determining the power-handling capability, as well as the efficiency, of a magnetic component based on a ferrite core.

Although power-handling density of magnetic components should ideally increase proportionally to frequency, the losses in ferrites increase faster than this, and thus power handling is a weaker function of frequency, typically reaching a maximum around 500 kHz to 1 MHz for MnZn ferrites. To further improve power density, different materials are needed. The NiZn ferrites may give a factor of 2 or 3 improvement in power density at frequencies in the range of 10–30 MHz [23]. More substantial improvements require a superior technology.

One emerging technology is microfabrication using thin-film magnetic metal alloy core materials. A process similar to that used for IC manufacture (more closely related to the process used for thin-film magnetic recording heads) is used to deposit and pattern thin films of magnetic alloys and copper coils. These magnetic materials can have very low hysteresis loss and can be operated near their saturation flux density (≈ 1 T) without disproportionate losses. The use of thin films and fine patterning can control eddy current losses in both the magnetic material and the coil. Experimental devices have been demonstrated in principle [24–29], and calculations show that much higher power density will be possible [30]. Although fabricating the magnetics on a separate substrate from the silicon circuitry and combining the two in a MCM is more likely to be economically viable, it is also possible to fabricate the magnetics and silicon circuitry on a single substrate, as has been experimentally demonstrated in [31].

Either because of the use of planar microfabrication techniques or for packaging and thermal dissipation considerations, magnetics for low-power converters are often designed in a planar configuration. While this presents no particular difficulties for transformer design, it can be a problem for inductors that use a gap in the magnetic path. The resulting magnetic field distribution can introduce severe eddy current losses in the conductor. For a careful description of this problem and some remedies, see [30, 32].

12.7. ALTERNATIVES TO SWITCHING REGULATORS

For ultra-low-power applications, the complexity of a switching regulator may prove prohibitive. In particular, the necessity of including a magnetic component may preclude the use of a PWM DC-DC converter in many applications. Two alternatives that do not require magnetic components are linear regulators and switched-capacitor converters. Both types of circuits can be advantageous in ultra-low-power applications and in a limited range of other specialized applications.

12.7.1. Linear Regulators

Linear regulators, illustrated conceptually in Figure 12.28, are limited by two principal constraints: The output voltage V_o must be less than the input voltage V_{in} and the efficiency η can never be greater than V_o/V_{in}. However, linear regulators have the advantage of requiring few or no reactive components, and they can be very small and simple. This makes them attractive for portable applications.

A linear regulator can be efficient only in applications that require an output voltage just slightly below the input voltage. This requirement may be incompatible with other system design constraints, but in some systems it is practical, and, in this case, a linear regulator may be highly efficient. The achievable efficiency then

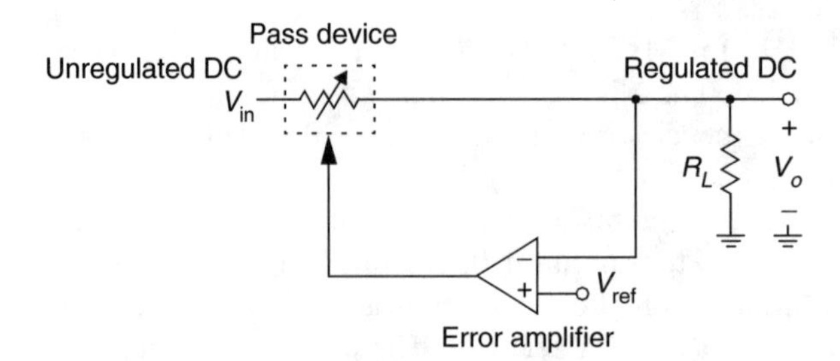

Figure 12.28 Block diagram of linear (series-pass) regulator.

depends on two parameters of the regulator: quiescent current and dropout voltage. The quiescent current determines the regulator's dissipation when the load is not drawing current, and in ultra-low-power applications, it may also contribute significantly to dissipation at full load.

If the input voltage of a linear regulator drops below a certain threshold, regulation is lost, and the output voltage will sag below the nominal regulation point. Dropout voltage is this minimum voltage difference between input and output required to maintain regulation. If it is not very low, it can conflict directly with the design requirement of having the output voltage only slightly less than the input voltage and will therefore preclude high efficiency. This becomes especially important in low-voltage systems. With 10 V output, 2 V dropout voltage represents only a 20% increase in the minimum input power over what would be required with zero dropout voltage. However, with a 2 V output, 2 V dropout voltage doubles the minimum input power.

Linear regulator circuits with low quiescent power and PNP or MOSFET pass devices to allow low dropout voltage are now commercially available. In the limited class of circuits that require a regulated voltage just below the input voltage of the regulator, these can provide a high-efficiency solution.

12.7.2. Switched-Capacitor Converters

Switched-capacitor converters (also known as charge pumps) are widely used in ICs where a voltage higher than or of opposite polarity to the input voltage is needed. Unlike a PWM converter, a switched-capacitor converter requires no magnetic components. In addition, it is often possible to integrate the necessary capacitors, though applications are typically those in which poor efficiency and very low output power are adequate.

Figure 12.29 illustrates the basic principle of operation of a switched-capacitor voltage doubler. The switches are closed in pairs, alternately. First the switches labeled ϕ_1 are closed, charging capacitor C_s to the input voltage V_{in}. Then the ϕ_1 switches are opened and the ϕ_2 switches are closed. This places C_s, which is now charged to V_{in}, in series with the input voltage, producing a voltage of $2V_{\text{in}}$ across the output. The cycle then repeats. The output capacitor maintains the output voltage near $2V_{\text{in}}$ during ϕ_1. The same converter topology can be used as a step-down converter, producing an output voltage of half the input voltage, by exchanging the input and output terminals. By using more complex configurations, it is possible to

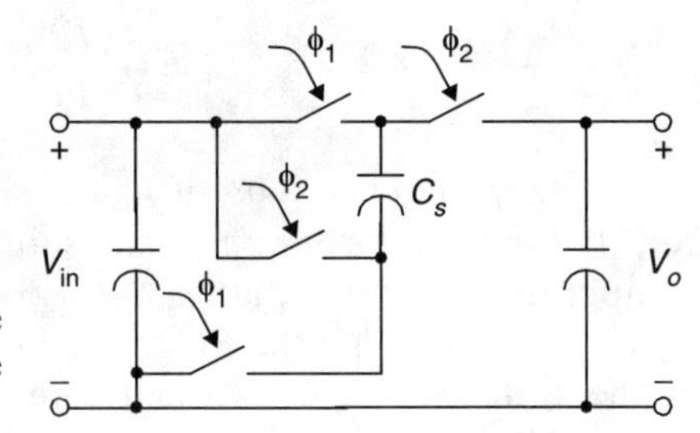

Figure 12.29 Switched-capacitor voltage doubler. Switches labeled ϕ_1 and ϕ_2 are closed alternately.

produce any rational conversion ratio, for example by first stepping the voltage up by one integer ratio and then stepping down by another integer ratio. Some of the many possible topologies are discussed in [33–35].

Like a PWM DC-DC converter, a switched-capacitor converter may be built entirely of theoretically lossless elements—in this case, only switches and capacitors. However, a switched-capacitor converter is not ideally lossless. As the parasitic resistances in the capacitors and switches approach zero, the loss in the converter approaches a nonzero limit. This is in contrast to a PWM converter, in which the losses approach zero as parasitic effects are reduced.

The inherent losses in a switched-capacitor converter are due to unavoidable dissipation which occurs when a pair of capacitors, charged to different voltages, are shorted together through a switch. If two capacitors with values C_1 and C_2, initially charged to voltages V_{1o} and V_{2o}, respectively, are shorted together through a parasitic resistor R, the energy dissipated in the resistor will be

$$E_{\text{diss}} = \frac{1}{2}\frac{C_1 C_2}{C_1 + C_2}(V_{1o} - V_{2o})^2 \tag{12.33}$$

Note that is is independent of the value of R.

To better understand these losses, consider the efficiency of the voltage doubler shown in Figure 12.29. During ϕ_2, the equivalent circuit is as shown in Figure 12.30. The charge flowing to the output is supplied by both the input and C_s. During ϕ_1, this same quantity of charge must be supplied from the input and stored on C_s for the next cycle. Since all the charge that flows out of the output must be supplied twice by the input, the average input current must equal twice the average output current, that is, $I_{\text{in}} = 2I_o$. Thus, the efficiency is

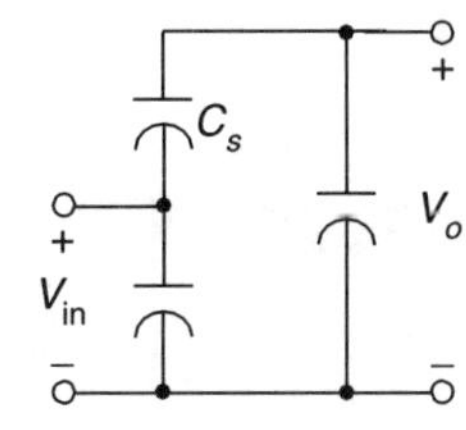

Figure 12.30 Equivalent voltage doubler circuit during ϕ_2.

$$\eta = \frac{V_o I_o}{V_{\text{in}} I_{\text{in}}} = \frac{V_o}{2V_{\text{in}}} \tag{12.34}$$

The efficiency would be 100% if V_o were in fact twice V_{in}. However, in order for a charge Q to flow into C_s during ϕ_1 and subsequently flow out of C_s during ϕ_2, the voltages applied across C_s during the two phases must differ by an amount $\Delta V = Q/C_s$. Assuming that the RC time constant determined by the parasitic resistance of the switches and C_s is small compared to the switching period so that the charge on C_s reaches its steady-state value before the end of each phase and that the input and output capacitors are large enough to maintain constant V_{in} and V_o, the voltage drop $\Delta V = 2V_{\text{in}} - V_o$. With a switching period of T_s, $Q = I_o T_s$, and so

$$2V_{\text{in}} - V_o = \frac{I_o T_s}{C_s} \tag{12.35}$$

The circuit may be modeled as shown in Figure 12.31, with an ideal doubler (shown as an ideal transformer) followed by an effective resistance

$$R_{\text{eff}} = \frac{T_s}{C_s} \tag{12.36}$$

that accounts for the voltage drop ΔV. The effective resistance also accounts for the loss; calculating the dissipation in this resistor gives a result identical to that found from (12.33). Constraints on R_{eff} for a wide class of converters are analyzed in [33].

In general, the model of a switched-capacitor converter includes an ideal transformer with a fixed rational turns ratio N and an effective resistance. The conversion ratio N can be chosen to bring V_o near the desired output voltage; to precisely regulate V_o, R_{eff} is varied through changes in the switching frequency. Using R_{eff} for regulation is undesirable, since increasing it to lower the output voltage produces additional power dissipation. However, N is fixed by the topology and cannot be used to regulate the output.

This is the main limitation of switched-capacitor converters: they can efficiently *convert* voltages, but they cannot *regulate* these converted voltages any more efficiently than a linear regulator. Thus, their efficient application is limited to situations in which a voltage must be converted to another rationally related voltage but regulation is not necessary or to situations in which the required regulation range is limited and so the efficiency $\eta = V_o/(NV_{\text{in}})$ is adequate.

In practice, there are several other considerations that limit efficiency in a CMOS implementation of a switched-capacitor converter. In order for (12.36) to

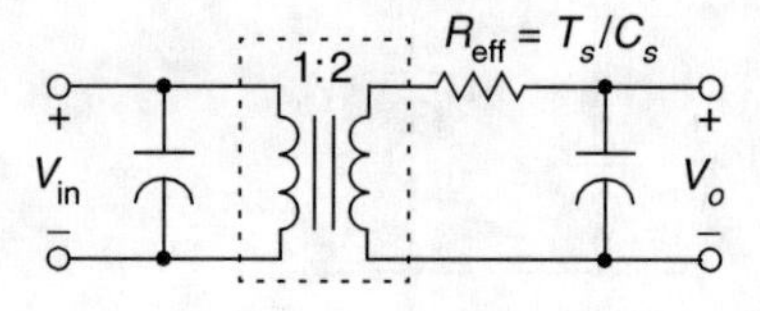

Figure 12.31 Equivalent circuit for switched-capacitor voltage doubler.

hold, it is necessary for the time constant of the switched capacitor and the on resistance of the switch to be much less than the switching period, that is, $C_s R_{\text{on}} \ll T_s$. This requires the use of a large MOSFET to implement the switch, but the gate-drive for that device then requires substantial power, especially if a high switching frequency is used to minimize the required size of C_s. Thus, gate-drive loss must be considered in the design.

If an on-chip capacitor is used for C_s, the stray capacitance from one of its plates to ground will be a substantial fraction of its terminal capacitance. This introduces $C_{\text{stray}} V^2 f_s$ loss, further hampering efficiency. Technologies for fabricating capacitors with low stray capacitance to ground or off-chip capacitors are necessary to achieve high efficiency.

12.8. CONCLUSION

The DC-DC converters in portable electronic equipment allow a variety of power supply voltages to be provided from a single battery source. They can also provide load circuitry with substantial energy efficiency advantages, permitting extended battery run time.

Important design objectives for DC-DC converters in portable applications include small size, low cost, and high efficiency. Small size and low cost can be obtained through a high level of integration and through design to minimize requirements for reactive components, particularly inductors. Considerations for high efficiency include synchronous rectification with careful timing control, low-loss magnetic components, zero-voltage switching, minimized power consumption in control circuitry, careful sizing and design of power-switching devices, and selection of gate-drive voltage. Many applications also require careful attention to light-load efficiency. Through an examination of these issues and an example design, an approach to realizing converters that meet the stringent requirements of portable applications has been demonstrated.

References

[1] A. Chandrakasan, S. Sheng, and R. Brodersen, "Low-Power CMOS Digital Design," *IEEE J. Solid-State Circuits*, vol. 27, no. 4, 1992.

[2] A. Stratakos, C. Sullivan, S. Sanders, and R. Brodersen, "DC Power Supply Design in Portable Systems," Technical Report ERL Memorandum Number M95/4, University of California, Berkeley, 1995.

[3] J. G. Kassakian, M. F. Schlecht, and G. C. Verghese, *Principles of Power Electronics*, Addison-Wesley, Reading, MA, 1991.

[4] W. Baringer and R. Brodersen, "MCMs for Portable Applications," in *Proc. IEEE Multi-Chip Module Conf.*, 1993.

[5] D. Maksimovic, "Design of the Zero-Voltage Switching Quasi-Square-Wave Resonant Switch," in *Proc. IEEE Power Electronics Specialists Conf.*, 1993.

[6] A. Stratakos, S. Sanders, and R. Brodersen, "A Low-Voltage CMOS DC-DC Converter for a Portable Battery-Operated System," in *Proc. IEEE Power Electronics Specialists Conf.*, pp. 619–626, 1994.

[7] D. Maksimovic, "A MOS Gate Drive with Resonant Transitions," in *Proc. IEEE Power Electronics Specialists Conf.*, pp. 96–105, 1990.

[8] P. Theron, P. Swanepoel, J. Schoeman, J. Ferreira, and J. van Wyk, "Soft Switching Self-Oscillating FET-Based DC-DC Converters," *Proc. IEEE Power Electronics Specialists Conf.*, vol. 1, pp. 641–648, 1992.

[9] S. Weinberg, "A Novel Lossless Resonant MOSFET Driver," *Proc. IEEE Power Electronics Specialists Conf.*, vol. 2, pp. 1002–1010, 1992.

[10] C. Mead and L. Conway, *Introduction to VLSI Systems*, Addison-Wesley, Reading, MA, 1980.

[11] M. Elmasry, "Digital MOS Integrated Circuits: A Tutorial," in *Digital MOS Integrated Circuits*, M. Elmasry (ed.), IEEE Press, New York, 1991, pp. 3–33.

[12] A. Chandrakasan, "Low-Power Digital CMOS Design," Ph.D. thesis, University of California, Berkeley, 1994.

[13] H. Veendrick, "Short-Circuit Dissipation of Static CMOS Circuitry and Its Impact on the Design of Buffer Circuits," *IEEE J. Solid-State Circuits.*, vol. SC-19, no. 4, pp. 468–473, 1984.

[14] T. Burd, "Low-Power CMOS Library Design Methodology," Master's thesis, University of California, Berkeley, 1994.

[15] J. Rabaey, *Digital Integrated Circuits: A Design Perspective*, Prentice-Hall, Englewood Cliffs, NJ, 1995.

[16] J. Williams, "Achieving Microamp Quiescent Current in Switching Regulators," *Linear Technol. Design Notes*, vol. DN-11, no. 11, 1988.

[17] J. Locascio and W. Cho, "New Controllers for Battery Systems Increase Systems Efficiency," *Power Quality USA*, 1993.

[18] B. Acker, C. Sullivan, and S. Sanders, "Synchronous Rectification with Adaptive Timing Control," in *Proc. IEEE Power Electronics Specialists Conf.*, 1995.

[19] B. A. Miwa, L. F. Casey, and M. F. Schlecht, "Copper-Based Hybrid Fabrication of a 50 W, 5 MHz 40 V-5 DC/DC Converter," *IEEE Trans. Power Electron.*, vol. 6, no. 1, pp. 2–10, 1991.

[20] L. Ginsberg and D. P. Schnorr, *Multichip Modules and Related Technologies: MCM, TAB, and COB Design*, McGraw-Hill, New York, 1994.

[21] F. W. Kear, *Hybrid Assemblies and Multichip Modules*, 2nd ed., Dekker, 1993.

[22] E. C. Snelling, *Soft Ferrites, Properties and Applications*, 2nd ed., Butterworths, 1988.

[23] Th. G. W. Stijntjes, "Power Ferrites; Performance and Microstructure," *Crystal Properties Preparation*, vol. XX, no. 1, pp. 587–594, 1989.

[24] K. Yamasawa, K. Maruyama, I. Hirohama, and P. Biringer, "High-Frequency Operation of a Planar-Type Microtransformer and Its Application to Mutilayered Switching Regulators," *IEEE Trans. Magnetics*, vol. 26, no. 3, pp. 1204–1209, 1990.

[25] K. Yamaguchi, E. Sugawara, O. Nakajima, and H. Matsuki, "Load Characteristics of a Spiral Coil Type Thin Film Microtransformer," *IEEE Trans. Magnetics*, vol. 29, no. 6, pp. 3207–3209, 1993.

[26] K. Yamaguchi, S. Ohnuma, T. Imagawa, J. Toriu, H. Matsuki, and K. Murakami, "Characteristics of a Thin Film Microtransformer with Spiral Coils," *IEEE Trans. Magnetics*, vol. 29, no. 5, pp. 2232–2237, 1993.

[27] M. Yamaguchi, S. Arakawa, H. Ohzeki, Y. Hayashi, and K. I. Arai, "Characteristics and Analysis of a Thin Film Inductor with Closed Magnetic Circuit Structure," *IEEE Trans. Magnetics*, vol. 28, no. 5, 1992.

[28] T. Yachi, M. Mino, A. Tago, and K. Yanagisawa, "A New Planar Microtransformer for Use in Micro-Switching-converters," in *22nd Annual Power Electronics Specialists Conf.*, pp. 1003–1010, June 1991.

[29] T. Yachi, M. Mino, A. Tago, and K. Yanagisawa, "A New Planar Microtransformer for Use in Micro-Switching-Converters," *IEEE Trans. Magnetics*, vol. 28, no. 4, pp. 1969–1973, 1992.

[30] C. R. Sullivan and S. R. Sanders, "Microfabrication of Transformers and Inductors for High Frequency Power Conversion," *IEEE Trans. Power Electron.*, in press.

[31] M. Mino, T. Yachi, A. Tago, K. Yanagisawa, and K. Sakakibara, "Microtransformer with Monolithically Integrated Rectifier Diodes for Micro-Switching Converters," in *24th Annual Power Electronics Specialists Conf.*, pp. 503–508, June 1993.

[32] W. M. Chew and P. D. Evans, "High Frequency Inductor Design Concepts," in *22nd Annual Power Electronics Specialists Conf.*, pp. 673–678, June 1991.

[33] M. S. Makowski and D. Maksimovic, "Performance Limits of Switched-Capacitor DC-DC Converters," in *Proc. IEEE Power Electronics Specialists Conf.*, pp. 1215–1221, 1995.

[34] I. Oota, F. Ueno, and T. Inoue, "Analysis of a Switched-Capacitor Transformer with a Large Voltage-Transformer-Ratio and Its Applications," *Electron. Commun. Japan, Pt 2 (Electron.)*, vol. 73, no. 1, pp. 85–96, 1990.

[35] I. Harada, F. Ueno, T. Inoue, and I. Oota, "Characteristics Analysis of Fibonacci Type SC Transformer," *IEICE Trans. Fund. Electron. Commun. Computer Sci.*, vol. E75-A, no. 6, pp. 655–662, 1992.

Chapter 13

Two New Directions in Low-Power Digital CMOS VLSI Design

Vitit Kantabutra
Department of Mathematics,
Idaho State University,
Pocatello, Idaho 83209

13.1. INTRODUCTION

This chapter introduces two methods of low-power complementary metal-oxide-semiconductor (CMOS) very large scale integrated (VLSI) design. The first involves eliminating races in asynchronous sequential circuits while keeping the circuit size reasonable. Using this method, we will show how dynamic power dissipation may be reduced due to reduced switching activity. The second method of low-power design involves using complex gates and is applicable to general combinational functions (and therefore to sequential circuits as well). Specifically, we will give strong statistical evidence that CMOS complex gates are more suitable than standard cells for computing multivariate Boolean functions. We will then reason that automatically generated complex gates should therefore be integrated with standard cells in the next generation of CAD tools.

The first method is applicable to asynchronous circuits, which are used extensively in digital integrated circuits. For example, in the watch industry it is well recognized that asynchronous circuits are preferable to synchronous ones because asynchronous circuits tend to have a minimal number of nodes in transit per unit time [29]. This limits the dynamic power dissipation and reduces the switching noise. Heuristic procedures for designing asynchronous CMOS divider circuits were presented by Vittoz et al. [30, 31]. Those circuits were designed based on earlier work by Huffman [14, 15] and consume only a small amount of power because only one state variable switches at a time.

With the increasing interest in mixed analog/digital VLSI circuits and the trend toward lower voltage power supplies, and thus lower noise margins, issues related to switching noise and power dissipation are becoming even more important. For example, asynchronous, self-timed circuits have been considered for interchip communication in analog VLSI processing interfaces [25].

Thus, in Section 13.2 we will briefly present a procedure for state assignment in asynchronous circuits so that for each circuit state transition only one state variable switches. (This procedure was presented in greater detail in [17].) Just as it was in the case of Vittoz et al.'s method [30, 31], we can achieve a reduction in power because of the reduced switching activities. In terms of circuit size, we tend to get smaller circuits than the circuits obtained by existing procedures. In fact, in the example picked, our circuits were almost as small as the ones obtained by using Tracey's state assignment procedure and yet used far less power. Tracey's [27] circuits used more power because they were not race free and involved more switching activities.

Our second method of low-power design is based upon a small but convincing study that shows that complex gates should be used widely in low-power digital static CMOS design. We shall present the results of this study in Section 13.3. Some researchers seem to believe that complex gates consume less power and are smaller than standard cells, yet others suspect the opposite, since before now there have been no scientific studies that indicate which is correct. As a consequence, there are no widely available computer-aided design (CAD) tools that include complex gates. The standard-cell VLSI design methodology remains the dominant one for low-power applications.

The goal of this comparison study of complex gates and standard cells is to attempt to spur interest among VLSI designers and CAD tool developers to consider complex gates. It appears that the main reason industry has continued to use standard cells in such an exclusive manner in low-power design is not because of real technical difficulties but because of pure inertia [26].

Ideally, we envision a CAD system that can automatically generate complex gates from behavioral or logical description and can integrate these complex gates into a design that includes standard cells [8]. This entails a system that can generate complex gates that are standard-cell compatible, that is, that have, for example, the same power-rail widths as the standard cells in the same circuit. The routing of the wires within each complex gate should not be much more difficult than the routing of wires between cells, since it appears that the two types of routing involve similar theoretical underpinnings.

13.2. STATE ASSIGNMENT APPROACH TO LOW-POWER ASYNCHRONOUS CMOS CIRCUIT DESIGN

13.2.1. Introduction

Our design procedure is based on the following strategy:

- Starting from a flow table, our state assignment algorithm produces optimum-speed race-free circuits that are usually smaller (and never larger) than the race-free circuits of the same speed produced by existing means.

- In doing so, we may also be able to produce circuits that are low power. Since our circuits are race free, there would be only one memory element or state variable switching for each transition between two flow-table rows. This leads to reduced noise due to the switching events and may also lead to lower power dissipation.

The main component of our design procedure is a state assignment algorithm that produces a race-free assignment. Our race-free circuits have the following key properties.

- Our circuits are optimum speed in the sense that for each transition between rows of the flow table the circuit goes through only one state change. This is different from many of the other state assignment algorithms such as the ones in [14]. No intermediate (unstable) states are used. The true speed of our circuits, of course, depend on the capacitances that they have to drive. In the randomly picked example in this chapter, however, these capacitances are reasonable, leading to fast circuits.
- If N is the number of rows in the flow table and $n = [\log_2 N]$, then a circuit produced by our state assignment algorithm can be of size anywhere from n to $2^n - 1$, where "size" here refers to the number of state variables (or memory elements). The size will be at the maximum when the undirected graph underlying the flow graph that represents the flow table is a complete graph. Although no theory has been developed for characterizing the sizes of our circuits any further than just giving the upper and lower bounds, and to say intuitively that our circuits are probably only very large if the flow graph is more or less complete, from a few examples it seems that our circuits are usually much smaller than the worst possible size.
- We use custom-designed complex gates. However, the procedure for designing these gates is quite standard (see Section 13.3) and has been automated in several different ways. As the results in Section 13.3 indicate, complex gates are the preferred choice in terms of power dissipation and size.

Our state assignment algorithm is an improvement over Huffman's well-known state assignment algorithm [15] that gives optimum-speed race-free circuits (without intermediate states between flow-table rows) because Huffman's circuits *always* have $2^n - 1$ (secondary) state variables, a very high number attained by our circuits only in the very worst cases.

13.2.2. Background and Motivation

State assignment procedures date back to the 1950s. In classic papers on sequential circuit designs [14, 15], Huffman presented several state assignment algorithms. In the 1950s, asynchronous circuits were the standard type of sequential circuit, and thus Huffman's state assignment algorithms were all race free. (Freedom from races is a sufficient, but not necessary, condition for guaranteeing freedom from *critical races*.) We may classify Huffman's state assignment algorithms into two major types as follows:

1. *Algorithms that yield relatively small but slow circuits*. These algorithms produce circuits that have approximately $2n$ (secondary) state variables, where n, the log of N defined earlier, is the minimum number of state variables needed just to encode all the flow table rows, that is, all the *internal* states (to use Huffman's terminology). These circuits are thus small but slow, as for each transition between flow-table rows they need to traverse through approximately $2n$ intermediate (unstable) states.
2. *Algorithms that yield fast circuits, going through zero or one intermediate state for every transition between two rows of the flow table*. Unfortunately, these algorithms always produce very large circuits: Their sizes (number of state variables) are about 2^n. The algorithm that yields optimum speed circuits, that is, the one using no intermediate states, yields circuits with $2^n - 1$ state variables. This algorithm is the one mentioned earlier. The algorithm that yields circuits using one intermediate state per internal state transition is the trivial one that uses one-hot state encoding.

Since Huffman's first type of algorithm yields slow circuits, such algorithms are not comparable with ours. Hence we will focus on the second type of algorithm, particularly the optimum-speed one.

Basic Idea behind Huffman's Algorithm: Each flow table row or adjacency graph node A will correspond to a large *set* of (secondary) states A' in the circuit. Such state sets in the circuit will be assigned numbers so that if A' and B' are any two state sets, then each member of A' will have a state number that differs by only 1 bit from the number of *some* member state of B'. This property is called the *adjacency property*. With the adjacency property in place, any transition in the flow table can be mimicked in the circuit by a secondary state transition in which at most one bit changes.

The reason Huffman needed an exponential number of state variables is that he *always* made sure that the circuit had enough states so that a transition between any pairs of nodes in the adjacency graph would have its counterpart in the circuit. Thus, even when the numbered adjacency graph had only very few races, Huffman's algorithm would still output an exponentially large number of secondary state variables.

In the 1960s Liu [20] and Tracey [27] came up with new state assignment algorithms for asynchronous circuits. Unlike Huffman, these researchers made use of the fact that circuits with *noncritical* races would still operate correctly: Not all races would have to be eliminated. Their circuits were as fast as Huffman's and ours in the sense that no intermediate states were used. However, as we will show in the random example that we picked, Tracey's circuits can use significantly more power than our circuits, precisely because Tracey's circuits were not race free (and thus had more simultaneous switching activities than ours).

13.2.3. General, Race-Free State Assignment Algorithm

Our state assignment algorithm differs fundamentally from Huffman's: Huffman starts out by defining a large set of A' of states to correspond to each flow-table row. On the contrary, we start out small and attempt to use *the given adjacency graph* as the state transition diagram, if possible. If not, we add as few copies of the original diagram to the original set of states as possible.

To be more explicit, note that it may be possible to number the nodes of the original adjacency graph so that every pair of adjacent nodes has numbers that differ only by 1 bit. In this ideal case, we are done, and the resulting circuit would be of size n. In ordinary, less ideal cases, the best numbering will still leave us with some transitions joining states that differ by more than 1 bit. Such transitions are called *bad transitions*. A bad transition causes a race condition, which is probably accompanied by high energy consumption.[1]

Now we define the *type* of bad transition to be the *set* of bit positions at which the numberings of the two states at the ends of the transition differ. For example, suppose we count the bit positions from left to right. Then a transition between state 1010 and 0000 would be a bad transition of type $\{1, 3\}$. Finally, we can say that our circuit will be of size $n + b$ (size refers to the number of state variables), where b is the total number of *types* of bad transition. (Note that b is not the number of bad transitions!) thus, if the original, numbered adjacency graph does not have very many types of bad transitions, then the resulting circuit will not be very large. We suspect that many circuits fall into this fortunate category or can be modified to fit this category.[2]

As stated earlier, if the original adjacency graph had no bad transitions, then we will just use this graph as the final state transition diagram, and the circuit will just have n (secondary) states.

Now let us consider the second-best case, in which there is *exactly one* type of bad transition. In this case, we would just add one state variable to the existing n variables. Adding just one state variable gives us twice as many states as we had previously and will enable us to use two binary numbers to encode each flow-table row. After having encoded all the secondary states, we will *replace* all the bad transitions by good ones, that is, by transitions such that only 1 bit in the secondary state encoding changes. Such replacements are possible because there are now two secondary states equivalent to each internal state (flow-table row), and so for each transition there are two choices of destination states. It turns out (by design) that one of these choices always results in a good transition. Next we will study an example.

Consider the adjacency graph in Figure 13.1. The nodes (internal states) of this diagram have not been numbered, and there is no way to number the nodes in order to get a race-free circuit. (This is due to the "triangle" in the graph.)

[1]We do not offer an algorithm for numbering the state of the original adjacency graph, but such an algorithm would be quite useful and is an open area of research.

[2]Further research is needed to determine how best to perform such modification.

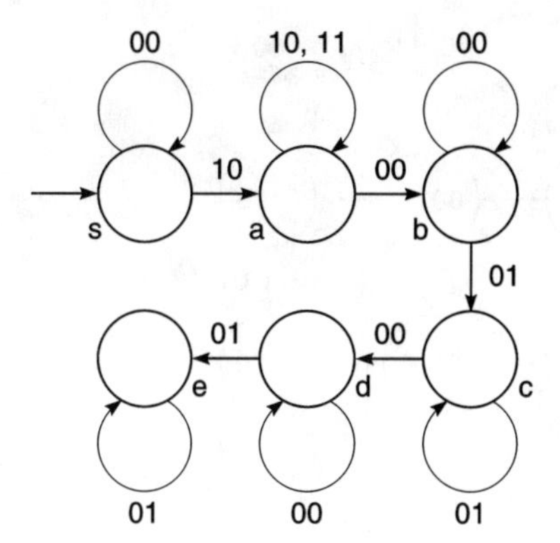

Figure 13.1 Adjacency graph.

We "convert" the adjacency graph in Figure 13.1 into a race-free state transition diagram in Figure 13.2 as follows:

1. We number the states in the adjacency graph in such a way that the number of bad transitions is minimized or kept small. This minimization is not necessary for our algorithm to work correctly; the purpose of the minimization is to help us minimize the circuit size. (We recognize that this minimal initial number is a complex and important problem by itself. However, this chapter does not cover that problem.) For example, the "gray-code-like" numbering shown in Figure 13.2 can be used.
2. Note that the transition from c to a is bad. Thus we create a copy of the adjacency graph below the original graph and for now label the states of the new copy in the same way as in the old graph. Next, for each state of the new copy, we *flip the second and third bits from the left. The positions of the flipped bits are defined by the type of the bad transition; that is, we flip the second and the third bits because the bad transition in the original graph is of type* {2, 3} *(i.e., the transition involves flipping bits 2 and 3)*. Now we *augment* each state number in the old copy by writing a "0" to the right of the old number and augment each state number in the new copy with a "1" in the same fashion. This new state

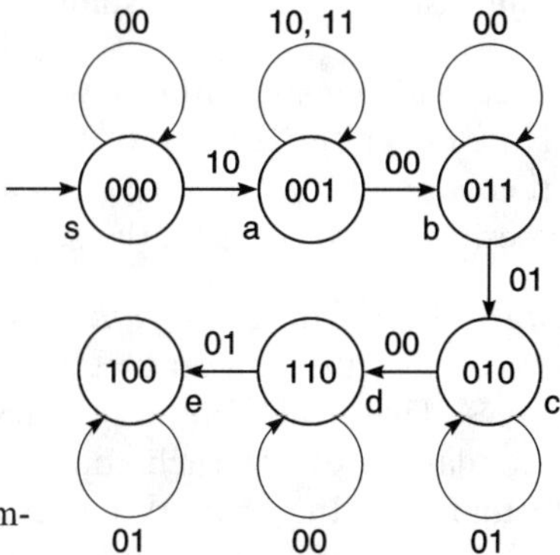

Figure 13.2 Gray-code-like state numbering for graph in Figure 13.1.

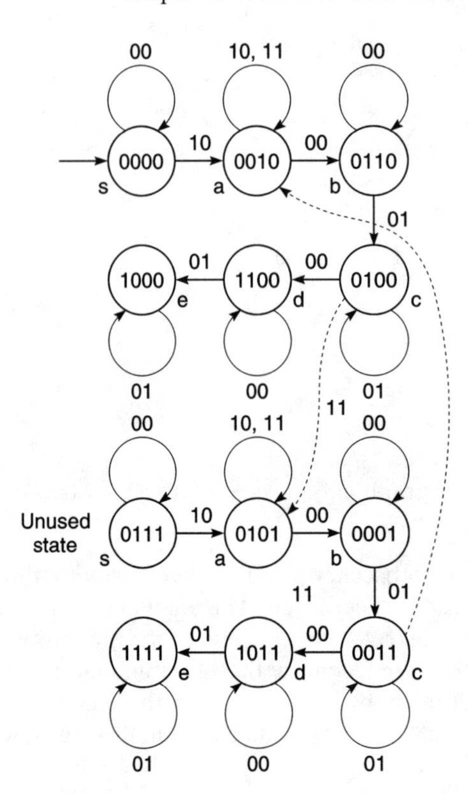

Figure 13.3 Race-free state transition diagram.

numbering is shown in Figure 13.3. Now we will specify the state transitions in the resulting state transition diagram. The "good" transitions in the original graph are just carried over to both copies. However, instead of carrying the bad transitions over to the two copies, we connect state *c* of each copy to state *a* of the *other* copy. Both these new transitions will be good transitions, as we can see from Figure 13.3.

3. Finally, note that if we had used the state assignment algorithm by Huffman that yields circuits as fast as ours, then we would have obtained a seven-state-variable circuit instead of the four-state-variable circuits that we obtained.

It is easy to see that our algorithm always works when there is only one bad transition type. Now we will generalize it so that it works no matter how many bad transition types there are. In [17] we have given an illustrative example with multiple bad transition types. Here we will give the state assignment algorithm, as follows:

States of State Transition Diagram. The state transition diagram will contain the original nodes of the adjacency graph plus two types of copies: *primary* and *secondary* copies. There will be one primary copy for each type of bad transitions and one secondary copy for each combination of two or more types of bad transitions. Note that only the primary copies, and not the secondary copies, involve an increase in the number of secondary state variables.

Naming the Copies. The copies must be named for later reference. Let C (also called C_ϕ, where ϕ is the empty set) be the set of original states, that is, the states represented by the nodes of the original adjacency graph. The primary copies will be called $C_{\{1\}}, C_{\{2\}}, \ldots, C_{\{b\}}$, where b is the number of *types* of bad transition. The copy $C_{\{i\}}$ will be thought of as the copy associated with the bad type i. The secondary copies will be called C_s, where s is the set of bad transition types associated with the particular copy. Note that the naming scheme of the copy associated with the original states as well as the primary copies is consistent with the naming scheme of the second states. In fact, mathematically speaking, there is no need to define three different types of copies. We only define these different types in order to improve the presentation. Note also that all these copies (including the original) naturally form the elements of a Boolean algebra, although we have no need at present to use that fact.

Numbering the States. Originally each state in every copy is numbered exactly like its counterpart in the original copy C_ϕ. Obviously these numbers will have to be modified. In the modification, we will not only flip some of the bits in the original n-bit positions but will also have to add b bits to the state numbers. (Recall that b is the number of bad transition types.)

Modifying Numbering of States in C_ϕ. Simply augment the original numbers by concatenating to the right end of these numbers the string 0^b.

Modifying Numbering of States in Each Primary Copy $C_{\{j\}}$. Perform the following steps:

1. If the badness type associated with the copy is $\{j_1, j_2, \ldots, j_k\}$, then flip bits $j_1, j_2, \ldots, j_k$ (counting from the left end) of every state number in that copy.
2. Now augment the n-bit number of each state in the copy by adding to the right end the string $0^{j-1}10^{b-j}$. This string is just a string of length b that consists of zeros everywhere except at position j from the left.

Modifying Numbering of States in Each Secondary Copy $C_{\{i_1, i_2, \cdots, i_k\}}$. Perform the following steps:

1. If bit position i (in the original n bits of the state numbering) occurs as a member of an *odd* number of the bad transition types among the types $i_1, i_2, \ldots, i_k$, then flip bit position i in the original n bits. Else do not flip bit position i.
2. Now augment the state numbering by adding to the right a suffix which will be a string of b bits that are all 0's *except* for the 1's at the position $i_1, i_2, \ldots, i_k$, counting from the left of the suffix string.

Replacing Bad Transitions by Good Ones. Let transition t from state A to state B be bad. We will replace t with another transition arrow that goes from A to B's counterpart in a different copy of the original adjacency graph C. This new destination of the transition will be picked so that the prefix (leftmost n bits, i.e., the bits at the original bit positions) will be exactly like A's prefix, and its suffix will differ from A's suffix by exactly 1 bit. The destination state will, of course, also be equivalent to state B. It is an important property of our algorithm that all these requirements on the destination state can always be met.

Choosing Destination State. Note that t, being a bad transition, must have a bad transition type, say i_t. Now suppose that state A is in the copy called $C_{\{i_1, i_2, \ldots, i_k\}}$. (*Note*: A may be in the original diagram C_{ϕ}.) Pick the new destination state B^* to be B's counterpart in the copy call C_{σ}, where σ is defined to be the set of integers that differs from $\{i_1, i_2, \ldots, i_k\}$ exactly by whether i_t is present or absent.

Correctness of Algorithm. Our original paper [17] contains a proof that this algorithm is correct; that is, it always produces a race-free circuit with all the properties claimed. We will skip the proof here and go directly to a randomly picked example in which we show how our algorithm can produce a low-power circuit that is not much larger than the less economical equivalent circuit produced by Tracey's algorithm.

13.2.4. Comparison between Race-Free Circuits and Circuits with Races: Experiment using Circuit-Level Simulation

We picked an arbitrary flow table from a digital design textbook ([9], Fig. 4.26) and used both our state assignment algorithm and Tracey's [27] on it in order to compare the energy consumption by the two circuits as well as the circuits' sizes. Our circuit does not take much more than half the energy per (internal) state transition. We find that this situation agrees surprisingly well with the fact that our circuit experiences *half* the amount of switching activities as Tracey's. Note that in both our circuit and Tracey's, we used complex gates as the drivers for the secondary state variables. If standard cells had been used, the results may have been more unpredictable, although intuitively, it would appear that our circuits on average might consume less energy because of fewer secondary variables switching. We note, however, that according to the second half of this chapter, complex-gate technology should be the technology of choice for low-power design. Thus, it is irrelevant how our circuits would compare to Tracey's if complex gates were not used.

We are convinced from our tests that, in general, race-free circuits may take less energy to run than circuits with races, and thus the issue of race-free versus raced circuits deserves further scrutiny. The reason for this belief is as follows: When a state variable switches, it takes a certain amount of energy (in some reasonably narrow range) provided that complex gates that do not vary too much in size are used to drive the state variables and the transistors are of the same size. Thus, a race-free circuit would take less energy to run because fewer state variables switch at once.[3]

The flow table tested is shown in Figure 13.4. Our state assignment algorithm would give the assignments shown in terms of a transition diagram in Figure 13.5. Tracey's algorithm on the same flow table is shown in Figure 13.6. Note that both our circuit and Tracey's circuit have the same number of state variables (three). Note also that a transition between any two rows of the flow table entails switching two state variables in Tracey's circuit but *only one* state variable in our circuit. Liu's state assignment algorithm would yield four (secondary) state variables ([9], Fig. 4.26), which is more than our algorithm or Tracey's algorithms would yield.

Internal state \ $x_1 x_0$	00	01	11	10
1	(1)	2	(1)	3
2	1	(2)	3	(2)
3	1	4	(3)	(3)
4	(4)	(4)	1	2

Figure 13.4 Flow table to be input into state assignment algorithms.

The complex gates for driving each of the three state variables in each circuit were designed manually. In particular, the pull-up and pull-down circuits were designed separately to minimize the lengths of transistor chains. (The design of this type of complex gate can be automated.) The complex-gate design procedure used here is essentially the same as the one presented in standard VLSI texts such as [13].

The driver circuit for state variable y_1 of our algorithm's circuit is shown in Figure 13.7, and the driver circuit for the state variable y_1' in Tracey's circuit is shown in Figure 13.8. Figure 13.9 shows the drivers for the other two state variables in each of the two circuits under comparison. Observe that Tracey's drivers tended to be smaller than ours, but not always. The reason Tracey's circuits tended to be smaller is because his state assignments were of the *unicode* type (one numbering per internal state), leading to many don't-care entries in the Karnaugh maps used in the gate designs. It is important to remember, however, that Tracey used 60 transistors, whereas we used 74, only 1.23 times more. This is quite good when we take into account that we only needed one complex gate switching every time Tracey needed two.

[3]In the particular case of the complex gates tested in this section, it takes a certain amount of energy in a fairly narrow range when a gate switches. The reason for this is as follows: It turns out that the complex gates have comparable input capacitances and switching speeds, because (1) the *average* number of transistors in a path from V_{dd} to the output or from the output to ground does not vary much and (2) the total number of transistors in all the complex gates in the race-free circuit is not much larger than in the circuit with races (74 versus 60, a ratio of 1.23).

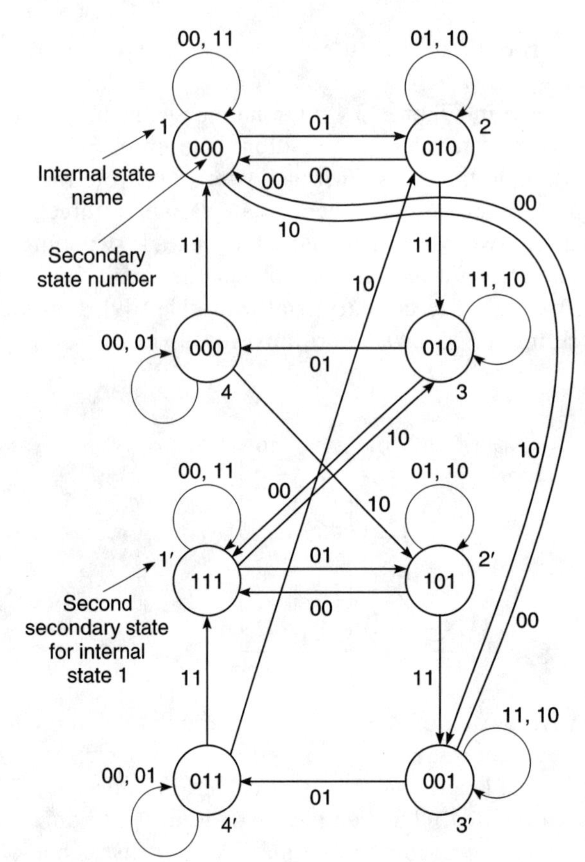

Figure 13.5 Our algorithim's state assignment for flow table of Figure 13.4.

Secondary state variables

Internal state	y'_2	y'_1	y'_0
1	0	0	0
2	0	1	1
3	1	1	0
4	1	0	1

Figure 13.6 Tracey's state assignment for flow table of Figure 13.4.

For all circuits under test, we used the transistor modes *nss* and *pss* from [13], with channel width and length of each transistor of 15 and 2 μm, respectively. Each complex gate was simulated separately with a parallel load of 10 MΩ and 0.2 pF between V_{dd} and the output, with the same load between the output and ground. Note that these load values are not too large to serve as test loads for CMOS gates of this size. (For example, see Fig. 1.42, p. 34, in [13].) In comparing energy consumption figures, we have compensated for the fact that our circuits are larger by a factor

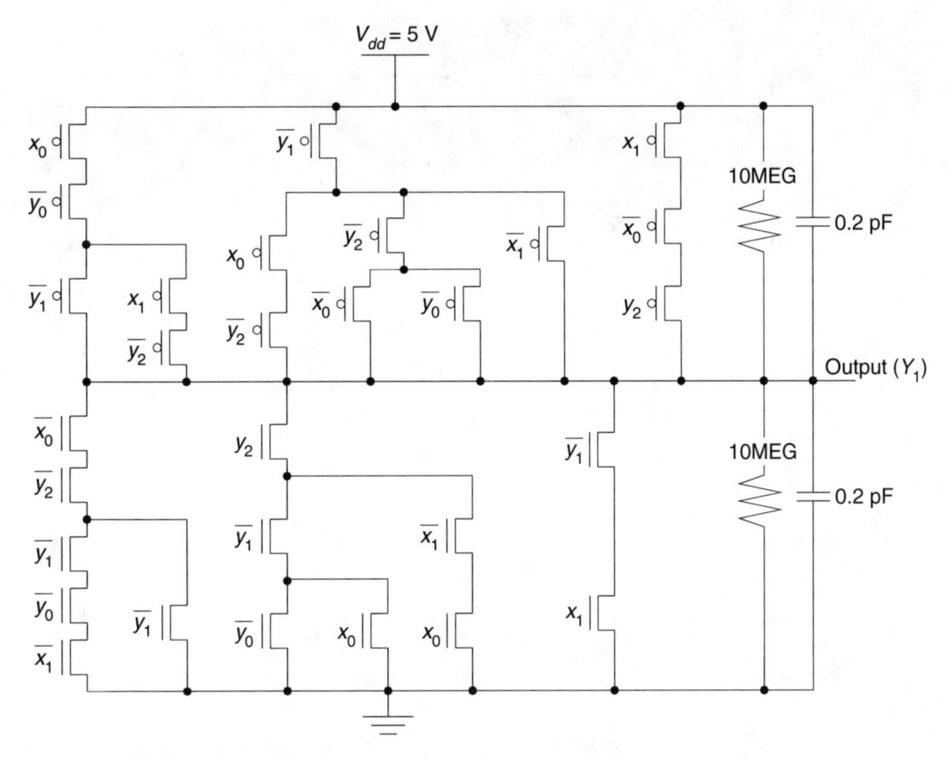

Figure 13.7 Complex gate for computing Y_1; note that average number of stacked NMOSs or PMOSs is not so high, even though it may seem otherwise at first glance.

of 1.23 by *multiplying* the energy figures of our circuit by 1.23 before comparing them to those of Tracey's circuit. Of course, such normalization does not give precise comparison but is sufficient for our purposes.

13.2.5. Simulation Results and Discussion

We will now observe these circuits in action. Refer to the table in Figure 13.10. Consider the transition in each of the two circuits from internal state 3 to internal state 4. In our circuit, only y_1 switches, whereas in Tracey's circuit, y_0' and y_1' both switch. In our circuit, the driving gate for y_1 consumes -5.424 pJ (before normalization by the size factor), whereas the nonswitching state variables y_0 and y_2 take -0.6158 and 0.01302 pJ, respectively. Summing all three variables yields -6.026 pJ as the total energy consumed by our circuit in this transition.[4]

[4]Following the convention in SPICE, a negative energy means the circuit is taking energy from the power supply, whereas a positive energy means the power supply is being charged by the circuit. We note that small amounts of power are transferred between the power supply and the complex gate driven by the supply even when there is no switching because the conduction paths between V_{dd} and ground may change even though the output does not. In any case, these small amounts of energy are insignificant.

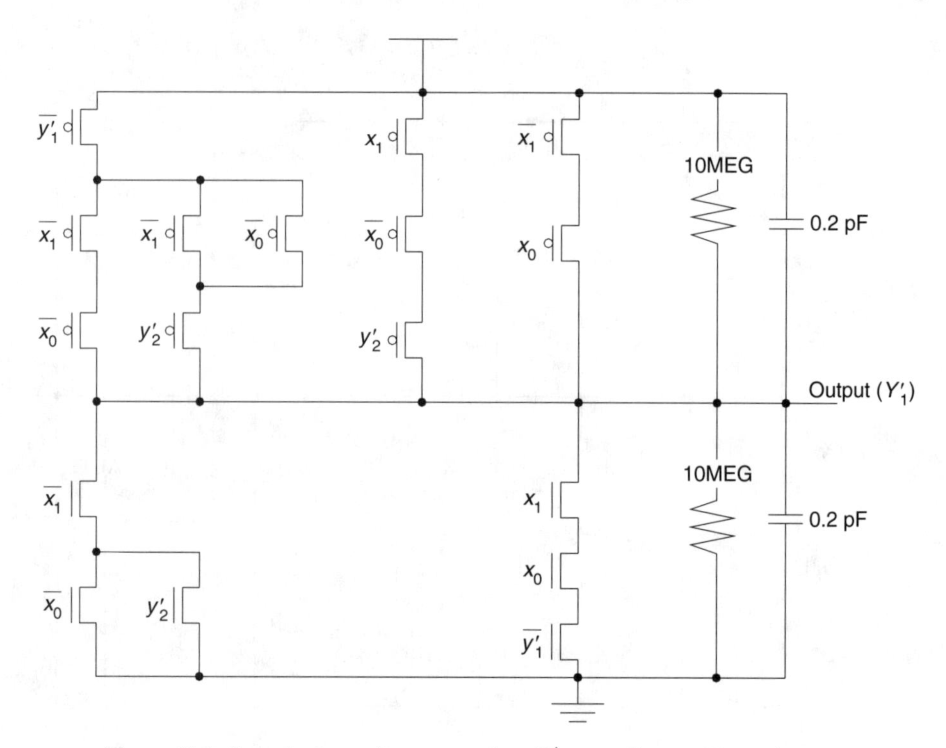

Figure 13.8 Complex gate for computing Y_1' according to Tracey's state assignment algorithm. This circuit is much smaller than circuit in Figure 13.7, but other two circuits needed by Tracey's scheme are not so small.

In Tracey's circuit, the two switching state variables y_0' and y_1' take -9.11 and -3.793 pJ, respectively. The nonswitching variable y_2' takes -0.855 pJ. Thus the total energy consumed by Tracey's circuit comes to -12.0 pJ, about twice the amount consumed by our circuit for the same *internal* state transition. Even after normalization by the circuit size factor, our circuit still consumes only -6.67 pJ, which is not much more than half the amount consumed by Tracey's circuit undergoing the same internal state transition. After we tested all transitions between two transition table rows (randomly picking a secondary source state for each such transition), we found that our circuit dissipates only about 58% (a little more than half) of the power dissipated by Tracey's circuit, even after normalization by the circuit size factor.

We observe that, in general, when a secondary state variable switches in either circuit, the amount of energy dissipated by the complex gate driving that variable is roughly between -3.7 and -11 pJ (unnormalized). In contrast, a complex gate driving a nonswitching state variable usually consumes only a fraction of a picojoule (except in rare cases) and sometimes even charges the power supply somewhat. *Thus the total amount of energy used by a circuit is roughly proportional to the number of state variables, or complex gates, that switch.* Of course, since the range of energy

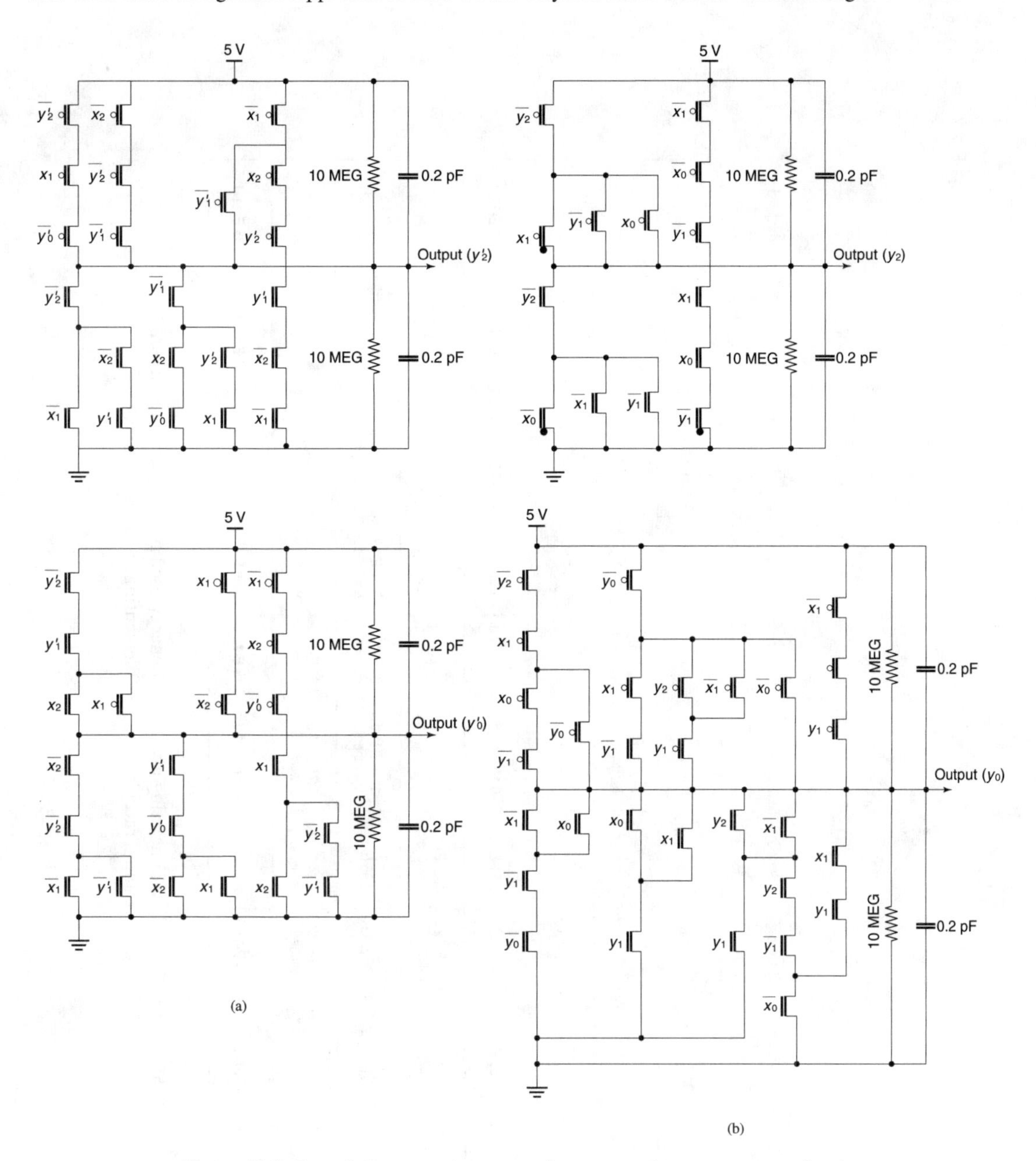

Figure 13.9 Remaining complex gates for computing next state: circuits designed according to (a) Tracey's algorithm and (b) our algorithm.

used by a switching complex gate is greater than 2, it is conceivable that a circuit with two switching gates may take less energy than a circuit with a single switching gate. However, this never happened in our tests. Note that, in general, a circuit with races

	State variable \ Transition	State 3 to state 4	State 1 to state 3 (Use 1' and 3 in our circuit)	State 4 to state 1	State 2 to state 3	State 1 to state 2 (Use 1' and 2' in our circuit)	State 2 to state 1	State 3 to state 1 (Use 3' and 1' in our circuit)	State 4 to state 2 (Use 4' and 2 in our circuit)	Average energy	Normalized average energy
Our circuit	y_2	+1.302 E-14 J	+2.853 E-14 J	–5.504 E-12 J (switched)	–9.97 E-12 J (switched)	–1.93 E-14 J	–4.70 E-13 J	–1.2 E-13 J	–3.96 E-13 J		
	y_1	–5.424 E-12 J switched	+6.58 E-14 J	+8.189 E-13J	+7.53 E-13 J	–4.99 E-12 J (switched)	–5.62 E-12 J (switched)	–2.87 E-13 J	+6.9 E-13 J		
	y_0	–6.158 E-13 J	–4.646 E-12 J (switched)	+5.38 E-13 J	+4.11 E-13 J	–1.09 E-12 J	+5.80 E-13 J	–1.02 E-11 J (switched)	–5.27 E-12 J (switched)		
	Total	**–6.026 E-12 J**	**–4.55 E-12 J**	**–4.147 E-12 J**	**–8.81 E-12 J**	**–6.10 E-12 J**	**–5.51 E-12 J**	**–1.06 E-11 J**	**–4.98 E-12 J**	**–6.34 E-12 J**	**–7.82 E-12 J**
Tracey's	y'_2	–8.55 E-13 J	–8.24 E-12 J (switched)	–6.31 E-12 J (switched)	–1.12 E-11 J (switched)	–6.22 E-13 J	+5.72 E-13 J	–4.50 E-12 J (switched)	–6.31 E -12 J (switched)		
	y'_1	–3.793 E-12 J (switched)	–7.687 E-12 J (switched)	–2.206 E-13 J	–2.55 E-12 J	–1.04 E-11 J (switched)	–3.84 E-12 J (switched)	–3.67 E-12 J (switched)	–7.24 E-12 J (switched)		
	y'_0	–9.11 E-12 J (switched)	–2.16 E-13 J	–4.726 E-12 J (switched)	–7.24 E-12 J (switched)	–6.75 E-12 J (switched)	–4.92 E-12 J (switched)	–8.22 E-13 J	+8.72 E-13 J		
	Total	**–1.20 E-11 J**	**–1.61 E-11 J**	**–1.13 E-11 J**	**–2.10 E-11 J**	**–1.78 E-11 J**	**–8.19 E-12 J**	**–8.99 E-12 J**	**–1.27 E-11 J**	**–1.35 E-11 J**	**–1.35 E-11 J**

Figure 13.10 Table summarizing amounts of energy used by complex gates in our circuit and Tracey's circuit implementing same flow table.

may make state transitions that involve the switching of three or more state variables, in which case it seems even less likely that a race-free circuit would require more energy than the circuit with races.

Our state assignment algorithm is not limited to asynchronous circuits and can be applied to synchronous circuits as well. In the case of synchronous circuits, however, the clock also dissipates power. In fact, as is now well known, the clock may dissipate as much or more power than the other logic circuits [21]. Since our circuits may have more memory elements than a circuit in which many memory elements switch simultaneously, it is unclear whether the synchronous version of our circuit would dissipate less power than the conventional type.

13.3. USING COMPLEX GATES IN LOW-POWER CMOS DESIGN

13.3.1. Introduction

Most static CMOS low-power devices are currently implemented with standard-cell components[5] for two main reasons: (1) companies have very large standard-cell libraries of ready-to-use circuits and (2) well-integrated CAD software exists for standard-cell design. For instance, software for automatic placement and wire routing is quite common and accessible even to very small design shops. Some researchers believe that when computing multivariate Boolean functions, complex gates are superior in terms of power consumption and perhaps also in terms of layout area, yet others suspect just the opposite as far as power consumption is concerned. However, the apparent belief in the superiority in terms of power and area efficiency of complex gates over standard cells has never been studied or expressed scientifically. As a result, there is a great deal of inertia in the CAD tool industry that prevents the existence of well-known commercial tools that allow complex gates to be automatically generated and integrated into a chip design.

We now present a small but convincing piece of scientific evidence that complex gates are superior to standard cells in terms of power consumption and area and should therefore be used liberally in low-power circuits. Specifically, we will show statistically that if we were to compute multivariate (more accurately, four-variable) Boolean functions, then on average (assuming all four-variable functions are equally likely to be computed) we would save significant amounts of switching energy and chip area if we performed the computation using complex gates. The speed sacrifice, if any, would be insignificant.

More specifically, we will generate several four-variable Boolean functions at random. Then for each function, we will lay out a pair of circuits to compute them: one complex-gate circuit and one circuit made of standard-cell gates.[6]

[5]As an example of such a device, see [23].

[6]More accurately speaking, we will generate four circuits per function. But we will discuss this matter in detail later.

For each circuit, we will measure its size as well as its average delay and average energy per one-variable input transition. For each circuit, we will define the following measure of efficiency:

$$e = \text{layout size} \times \text{average delay} \times \text{average energy}$$

where the average delay is measured from the 50% V_{dd} point of the input to the 50% V_{dd} point of the output. (Delay is zero if the output does not switch.) We will then compute the efficiency ratio of the two circuits, that is, $\rho = e_{sd}/e_{cg}$, where e_{sd} and e_{cg} are the e values of the standard-cell and the complex-gate circuits, respectively. Over all pairs of circuits, the estimated average value of ρ is as high as 4.7. More precisely, we will show that if ρ is normally distributed, then with a 95% level of confidence, we can say statistically that the average ρ lies in between 2.06 and 7.42. Therefore it is highly unlikely that the average standard-cell circuit would be superior to the average complex-gate circuit.

This study is the first study to compare static CMOS complex gates to standard-cell circuits for computing general Boolean functions. In fact, few existing papers on low-power circuits quantitatively compare any two design styles for computing general functions of any sort. However, several papers compare design styles for computing specific functions such as addition (e.g., see [3, 4, 19]) and contain discussions on the computation of more general functions. Chandrakasan et al. [7] compared the energy consumption of (a) dynamic versus static logic and (b) conventional static versus pass-gate logic. Their study was qualitative rather than quantitative.[7] In other words, they discussed physical properties of a circuit that might make one type of circuit consume less power than another type but did not come up with enough numerical data to conclude to what extent one type of circuit is better than the other. For example, when they were comparing two types of circuits in terms of energy consumption, one type may have had properties that made it superior to the other type as well as properties that made it inferior. Thus it is often not clear which circuits consume less energy on average.

13.3.2. Method of Study

In [17] we recognized that it is impossible to say which Boolean functions occur more frequently than others in practice. But this does not mean there is no meaningful way to perform a convincing statistical study of circuits that compute Boolean functions in general.

One of the best ways to perform such a study is to generate combinational functions at random. Then, ideally, to compute each such function, we would make a complex-gate circuit and a circuit from standard cells and compare the two. However, the standard-cell library to which we have access is of the regular-power and not the low-power type, which means that the library consists of transistors that are wider than the low-power type. So we must compare the

[7]Except for their reference to a paper by Yano et al. [36] which only studied the power consumption of full adders and not circuits that compute Boolean functions in general.

circuits built from these standard cells with complex-gate circuits with the same transistor widths. We also must simulate circuits made of low-power circuits by editing the netlists of both the standard-cell circuits and the complex-gate circuits to narrow the transistors down to typical widths used for low-power design. In editing the netlists, it is clearly unnecessary to edit the lengths of the interconnection wires because such lengths are so small that their capacitances are trivial compared to the capacitances of the transistors and the load. The diffusion capacitances have been ignored in all cases.

We will limit our study to random functions of four variables and assume that all functions of up to four variables are equally likely to occur. Each random function is generated by randomly filling a truth table. That is, we will simply toss a coin to determine how to fill each truth table entry. Surprisingly, it turns out that we only have to study six random functions to obtain sufficient data to make statistically significant conclusions.

After generating each function f by randomly filling a truth table with 0's and 1's, we simplify the algebraic expression for f by hand. One obvious problem that arises from Boolean minimization is that the resulting circuit's energy consumption, delay, and layout area depend on the particular algebraic expression obtained. To overcome this problem, we will perform the minimization in a nonprejudiced manner, which simply means that we will perform Karnaugh map minimization without thinking about whether the resulting algebraic expression for each function is going to be better for the complex-gate circuit or the standard-cell circuit in any particular way.

The Circuits. The next step is to convert each simplified algebraic expression into circuits. Since the standard-cell library to which we have access is Tanner Research's 2-μm library, which is not a low-power library, we will work first with non-low-power complex gates as well in order to compare only circuits with the same transistor widths. Following is a list of assumptions to make our comparisons fair: Unless stated otherwise, the assumptions are made in terms of non-low-power circuits. However, also unless stated otherwise, the assumptions will remain the same when we later work with low-power circuits.)

- In any pair of circuits to be compared, the P-type MOS (PMOS) transistors must all be of the same size, and likewise for the N-type MOS (NMOS) transistors. Tanner's standard-cell library is basically a 2-μm library in which all PMOS and NMOS transistors are 28 μm wide. [The transistor aspect ratio (width over length) of 14 is a typical aspect ratio for non-low-power standard cells; such wide transistors make the timing fairly immune to the capacitive delay effects of moderately long wires.] When we consider low-power circuits, the PMOS and NMOS transitor width will be 8 and 4 μm, respectively. (These aspect ratios are typical aspect ratios for low-power standard cells.)
- For any particular Boolean function studied, each of the two circuits computing that function must have the same number of inputs. This rule ensures that both circuits present the same load to the outside world. An important

consequence of following this rule is that it is then fair to load both circuits with the same output load. There are two possible ways to follow this rule:

1. Make the standard-cell circuit compute the function under study, f, and make a complex gate to compute f'. To complete the complex-gate circuit, invert f' using an inverter.
2. Alternatively, we can compute f itself with a complex gate. In this case we would compute f' with a circuit made of standard-cell gates and then invert the output to get f. (Or just use the inverted version of the gate at the output instead of the noninverted version, or vice versa, whichever is appropriate.)

For the sake of fairness, we will use the former alternative for $f_1, \ldots, f_3$ and the latter alternative for $f_4, \ldots, f_6$.

- The complex gates must be laid out in an orderly fashion, as if done automatically. We do not have access to software for laying out complex gates, but we manually follow a well-known algorithm [28] that gives a single row of each type of transistor. This rule helps to ensure that, in principle, one could switch from the standard-cell methodology to the complex-gate methodology (or include complex gates in a system that already uses standard cells) without too much difficulty because the complex gates can be laid out automatically. Note that many other references on complex-gate layout exist, including one on performing complex-gate layout from a high-level behavioral description [1, 2, 10, 11, 18, 24, 22, 35]. Many of these describe experimental or research CAD systems.

The functions that we randomly generate are

$$f_1 = x[w(z + y') + yz] + x'z'(w' + y)$$

$$f_2 = y(z + x + w'z') + w(xz + x'y') + w'xz'$$

$$f_3 = (y + w'x')z$$

$$f_4 = (w + x + y'z)[z + (x + y')(w' + x' + y)]$$

that is

$$f_4' = w'x'(y + z') + z'(x'y + wxy')$$

$$f_5 = [x' + y(w' + z')](wx + y' + z)$$

that is,

$$f_5' = x(y' + wz) + (w' + x')yz'$$

$$f_6 = (w' + x' + y)(w + y + z')[y' + (x' + wz)(w' + x + z')]$$

that is,

$$f_6' = wxy' + w'y'z + y[x(w' + z') + wx'z]$$

We will now discuss the layouts of some of the circuits. The most interesting layouts are the ones for computing f_6 (normal power). Figures 13.11(a) and (b) show the complex-gate and the standard-cell versions, respectively, of the layout. First we notice that the complex-gate layout is much smaller than the standard-cell counterpart, which is the case for all six randomly generated functions under study. Next we notice that the complex-gate circuit is in standard-cell format in the sense that the power rails are standard-cell compatible. However, some wires are run *outside* the

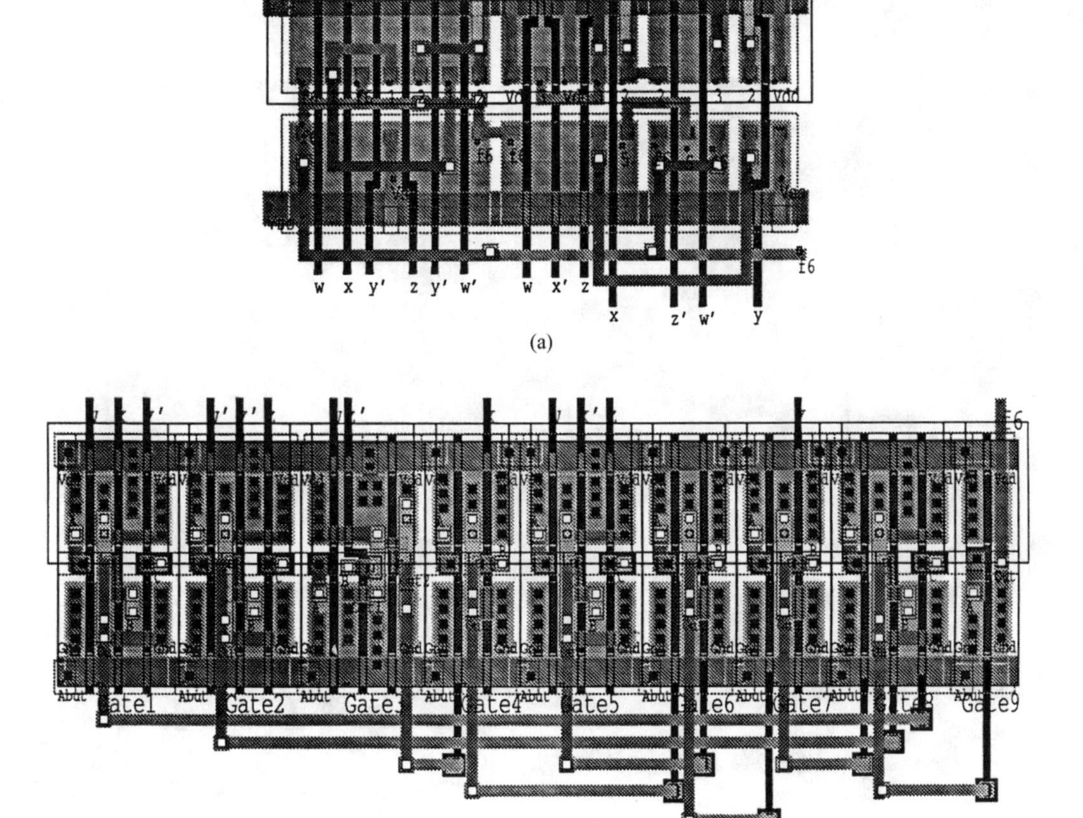

Figure 13.11 (a) Complex-gate circuit for computing f_6. (b) Standard-cell circuit for computing f_6.

power rails, and so any routing algorithm to be used with such complex gates must be able to accommodate such a situation. Both circuits have transistors of the same width, except for a few transistors in the complex-gate circuit that have a slight jog to avoid the substrate contacts. (These transistors have an effective length of 30λ instead of 28λ, which makes practically no difference.) Finally, we note that both circuits have the same inputs and thus the same number of inputs. Therefore the circuits present the same load to the outside world. The layouts of the complex-gate and standard-cell circuits for computing f_1, f_2, and f_3 are shown in Figures 13.12, 13.13, and 13.14, respectively. (All the layouts shown are of the standard-power type, since the circuits of the low-power type are only simulated and not laid out.) Figure

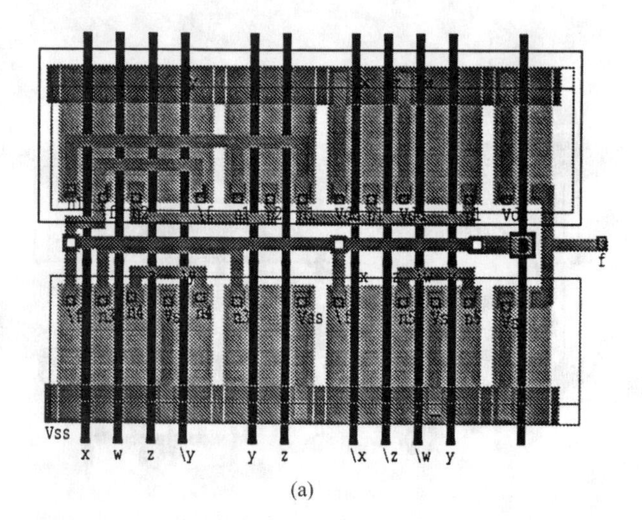

(a)

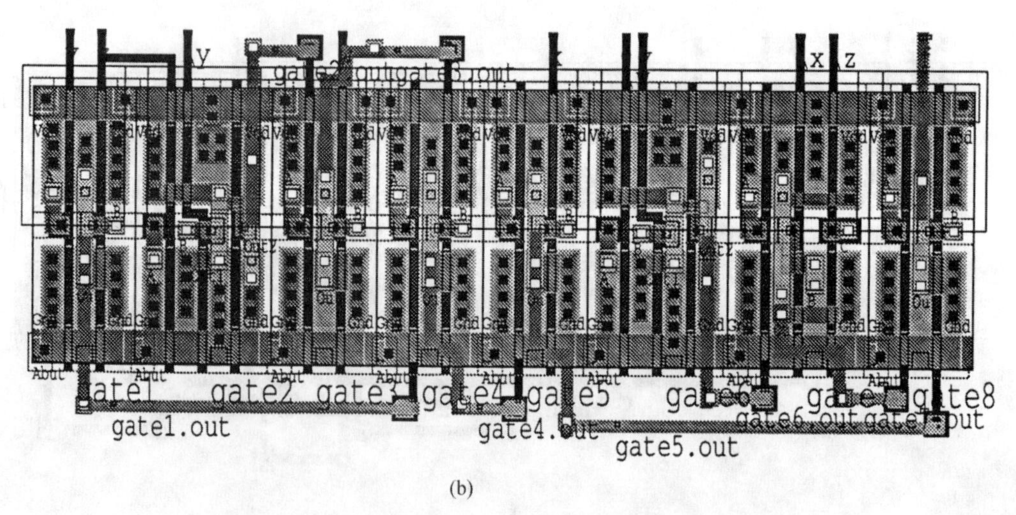

(b)

Figure 13.12 (a) Complex gate for computing f_1. (b) Standard-cell circuit for computing f_1.

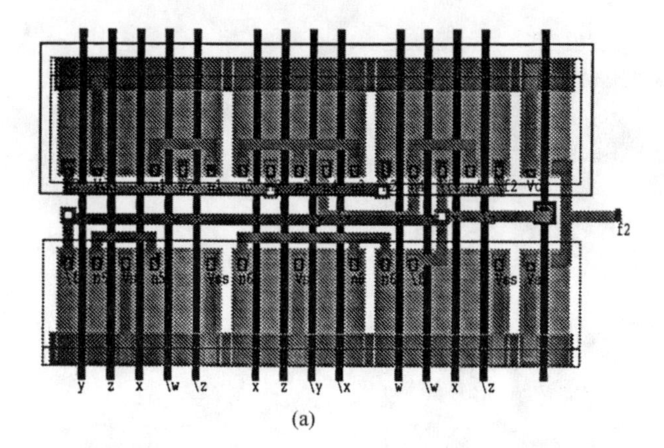

(a)

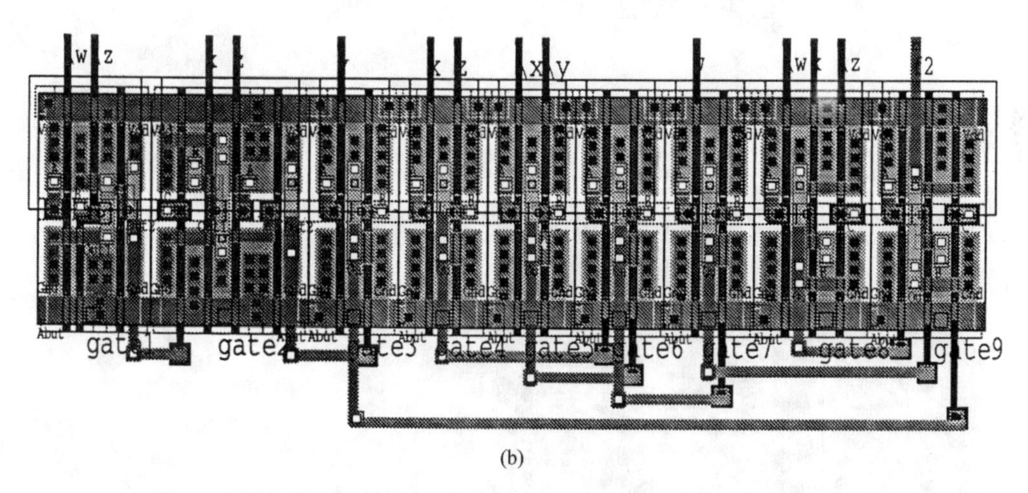

(b)

Figure 13.13 (a) Complex gate for computing f_2. (b) Standard-cell circuit for computing f_2.

13.15 shows the circuit diagrams of the circuits that compute f_1. The circuits for computing f_4 and f_5 are similar to the rest and thus not shown.

Now we will discuss the low-power circuits. these circuits do not really exist except in simulation, because we do not have access to a low-power standard-cell library. However, according to a design engineering (Korey Brown, 1994) at AMI (a semiconductor company in Pocatello, Idaho, specializing in standard cells, among other technologies) and text by Weste and Eshraghian ([34], p. 414), a typical low-power standard cell can have PMOSs and NMOSs with aspect ratios of 4 : 1 and 2 : 1, respectively. (Of course, these aspect ratios may differ from library to library, but both sources happen to quote more or less the same aspect ratios.) Thus, we obtained our low-power netlists by globally changing the transistor widths to obtain the aforementioned transistor aspect ratios, making provisions for the wire jogs in the complex gates. As stated before, there is no need to change the wire capacitances in the new netlists because these capacitances are trivial. Now note that we also need

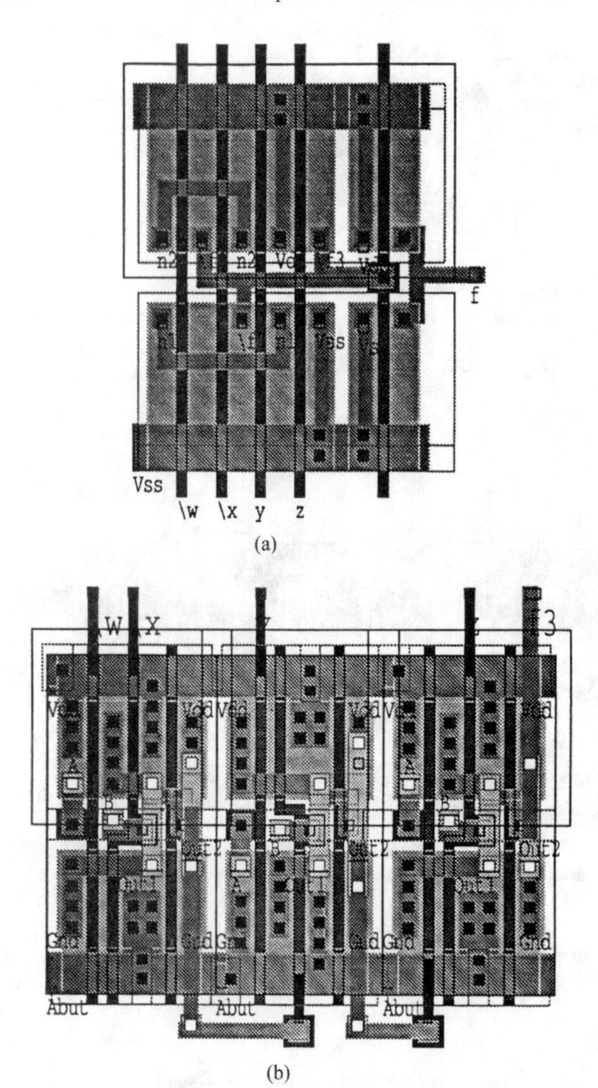

Figure 13.14 (a) Complex gate for computing f_3. (b) Standard-cell circuit for computing f_3.

a good estimate for the sizes of the low-power circuits. This is not hard to do. We begin by first looking at the layout of each regular-power standard cell and then shrinking the PMOS and NMOS width from 28 μm to 8 and 4 μm, respectively. Now if we examine the layout of a standard cell closely, we will notice that the cell height can be shrunk by 24λ without violating any design rules. (We cannot shrink the cell height any more because the contacts are in the way.) Thus we conclude that the layout area of a standard-cell circuit is obtained simply by multiplying the width of old (regular-power) circuit with the old height less 24λ. In order to estimate the size of a low-power complex-gate circuit, we repeat what we did for the standard cell, except that the amount of height shrinkage for each complex-gate circuit is often

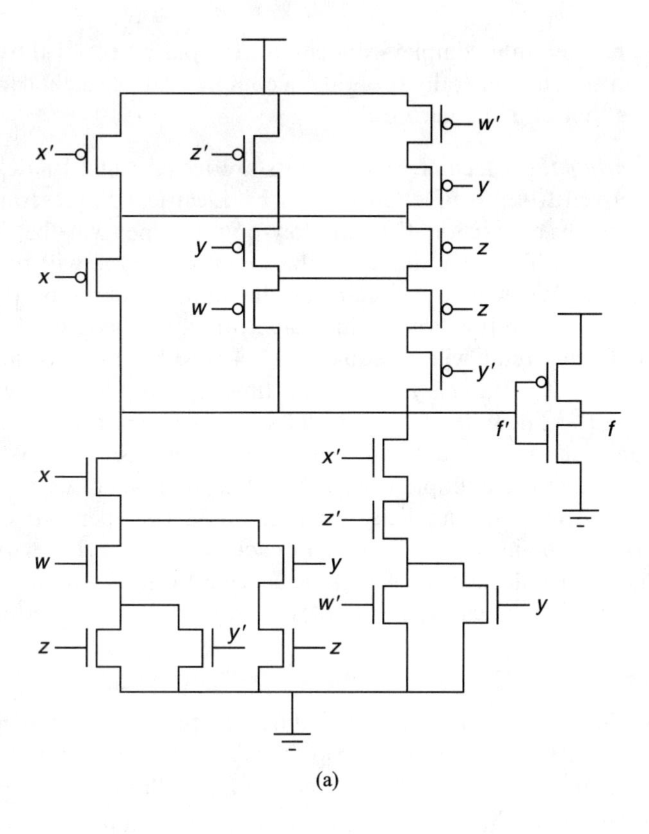

(a)

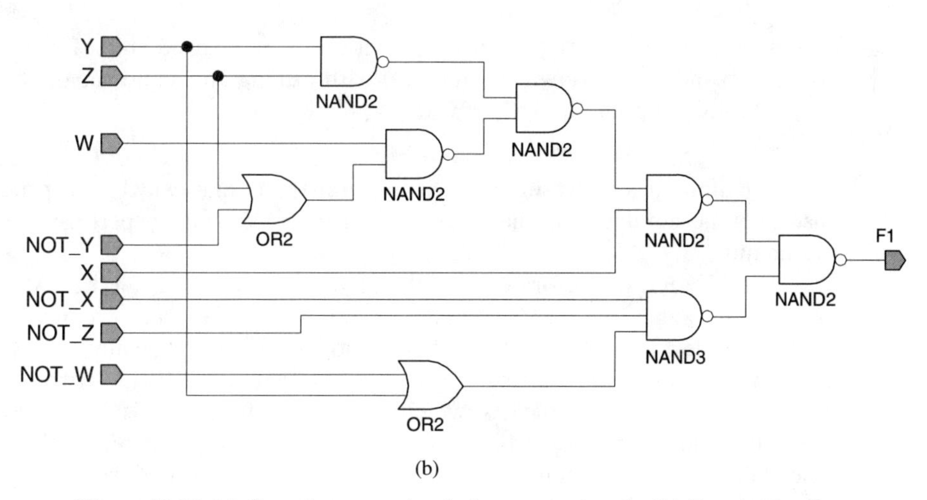

(b)

Figure 13.15 (a) Complex-gate circuit for computing f_1. (b) Standard-cell circuit for computing f_1, shown as logic diagram.

unique because in a complex-gate circuit, the placement of the contacts varies from circuit to circuit. Generally speaking, a complex gate's height cannot be shrunk by as much as that of a standard cell.

Testing the Circuits. The circuits are tested using PRECISE, a circuit simulator derived from SPICE2G6 offered by Electrical Engineering Software of California. Each circuit of the non-low-power type will be loaded with 0.5 pF capacitance, while each circuit of the low-power type will be loaded with 0.1 pF capacitance. We will test circuits of the same power type (low or regular) and compute the same function against each other, so there will be 12 pairs of circuits to test. Each circuit will be subject to 64 possible up and down transitions with only one input variable switching at a time. The input signal will be a linear ramp with a rise or fall time of 5 ns (0–100% or vice versa), which is of the same order of magnitude as the rise or fall time of a typical response. (We have no reason to believe that letting multiple variables switch simultaneously would lead to more interesting results. For simplicity we will run 32 tests per pair of circuits instead of 64 tests, which simply means that in each test one variable will be made to both rise and then fall, instead of making 2 separate tests for the input variable rising and falling. Let us call the variables w, x, y, and z. We perform the tests as follows:

- First fix x, y, and z at 0, 0, and 0 and let w form a ramp up from 0 V at 0 ns to 5 V at 5 ns. Then let w drop from 5 V at 20 ns to 0 V at 25 ns. Measure the total power supplied to the circuit for both switching events.
- Repeat the above, but with x, y, and z fixed at 0, 0, and 1, then 0, 1, and 0, all the way to 1, 1, and 1.
- Repeat the two steps above, but with w, y, and z fixed and x forming a ramp up and a ramp down. Then repeat with y going up and down, and finally with z going up and down.

The input files to the simulator are generated automatically by a C program that also calls the simulator, so that the 32 tests for each circuit are performed using one command.

The 32×6 tests performed are tabulated and analyzed using the SAS statistical software package. In Tables 13.1 and 13.2, we only tabulate a few representative test results. Table 13.1 contains results from testing low-power circuits, and Table 13.2 contains results from testing regular-power circuits. In the two tables the FUNCTION column indicates which of the six functions is being tested. The VARIES column indicates which of the four variables (w, x, y, or z) is being varied. The OTHER VARS column entry, when converted to a three-digit binary number, indicates the values of the three fixed variables in alphabetical order. (For example, in the fourth row of Table 13.1 x varies, and so OTHER VARS $= 4$ means $w, y, z = 1, 0, 0$.) The CPX DELAY and STD DELAY columns tabulate the delays of the complex gate and standard-cell circuits, respectively, in nanoseconds, while the CPX ENRGY and STD ENRGY columns tabulate the energy used by the two types

TABLE 13.1 SELECTED TEST RESULTS FOR LOW-POWER CIRCUITS

Function	Varies	Other Vars	CPX Delay	STD Delay	CPX Energy	STD Energy
1	w	0	3.31	2.95	8.81	13.94
1	w	6	0.00	0.00	0.04	0.05
1	y	5	0.00	0.00	−0.01	7.24
5	x	4	2.14	2.33	5.56	12.31
5	y	1	0.00	0.00	−0.13	4.47
6	z	7	3.66	3.73	6.57	13.54

Note: Channel length = 2 μm, PMOS aspect ratio = 4, NMOS aspect ratio = 2, load = 0.1 pF.

TABLE 13.2 SELECTED TEST RESULTS FOR REGULAR-POWER CIRCUITS

Function	Varies	Other Vars	CPX Delay	STD Delay	CPX Energy	STD Energy
1	w	0	2.72	2.71	35.38	60.13
2	y	5	0.00	0.00	8.35	39.78
3	z	2	0.00	0.00	−0.30	−0.31
4	x	4	2.36	2.96	19.17	47.99
6	y	5	3.31	2.08	40.60	34.17

Note: Channel length = 2 μm, PMOS aspect ratio = NMOS aspect ratio = 14, load = 0.5 pF.

of circuits in picojoules. Each energy entry is the average between the energy used when the varying input variable rises and when it falls. Note that when there is *no switching* in the *output* of a circuit, then we use zero as the delay figure. Note also that when there is no switching activity at all, then the energy figure could be negative. However, this amount of energy is insignificant.

Now we will examine the entries of these two tables in detail. The first row of Table 13.1 is a very typical row for the case in which the main output switches. As we can see, the complex-gate circuit has a slightly longer delay than the standard-cell circuit but dissipates significantly less power. The second, third, and fifth rows are typical when the main output does not switch. Row 2 is different from rows 3 and 5 in a very important way, however: In row 2, neither the complex-gate circuit nor the standard-cell circuit consumes much energy from the supply. This means that neither circuit has a gate that switches. But in rows 3 and 5, the standard-cell circuits, have some switching gates despite the fact that the main output does not switch. These are situations that further deteriorate the power consumption of standard-cell circuits. Figures 13.16 and 13.17 show the simulation results pertaining to row 3. The bottom two graphs of Figure 13.17 are the outputs of the two gates that switch, namely, gates 1 and 6 [see also the circuit diagram, Fig. 13.15(b)]. Finally, rows 4 and 6 seem to be slightly atypical because the complex-gate circuits are *faster* than their

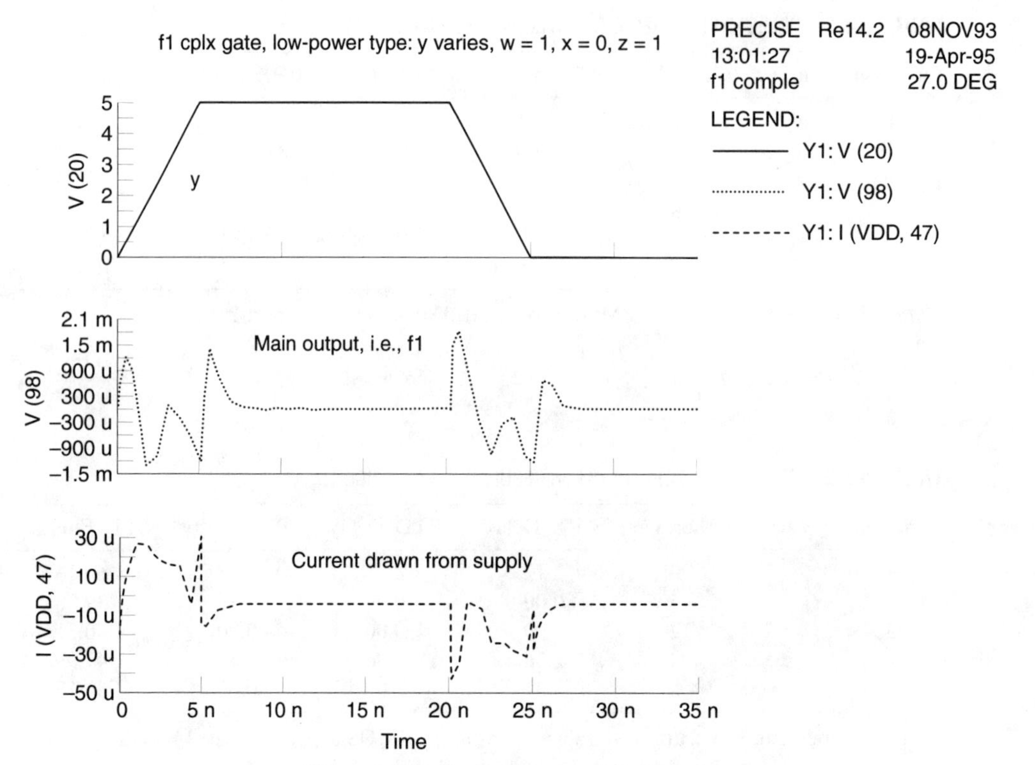

Figure 13.16 Particular simulation result.

standard-cell counterparts, even though the complex gates still win in terms of energy consumption.

In Table 13.2 rows 1–4 offer no new cases, but row 4 is an unusual case in which the standard-cell circuit clearly wins with respect to both delay and power. Such a situation is quite unusual.

Statistical Analysis. The simulation results as well as the circuit sizes are subject to statistical study. In particular, we are interested in comparing the complex-gate circuits against functionally equivalent standard-cell circuits in terms of energy consumption, delay, and size. Tables 13.3 and 13.4 contain the data for the low-power and the standard-power circuits, respectively. In both tables, the ratio ρ (the ratio of size $\times$ delay $\times$ energy of a standard-cell circuit to that of a complex-gate circuit, as defined earlier) appears on the last column. It is quite interesting to note at this time that the value of ρ in each row of Table 13.3 barely differs from the value of ρ in the same row of Table 13.4, even though the transistor sizes and loads used in the low-power and standard-power simulation runs are quite different from each other. Somehow, we find this observation encouraging because it seems to indicate that ρ may be reasonably independent of transistor size and load.

We now perform our statistical analysis on the low-power circuits using the data in Table 13.3. We find the average ρ (i.e., the sample mean $\bar{x}$) to be 4.743 and the

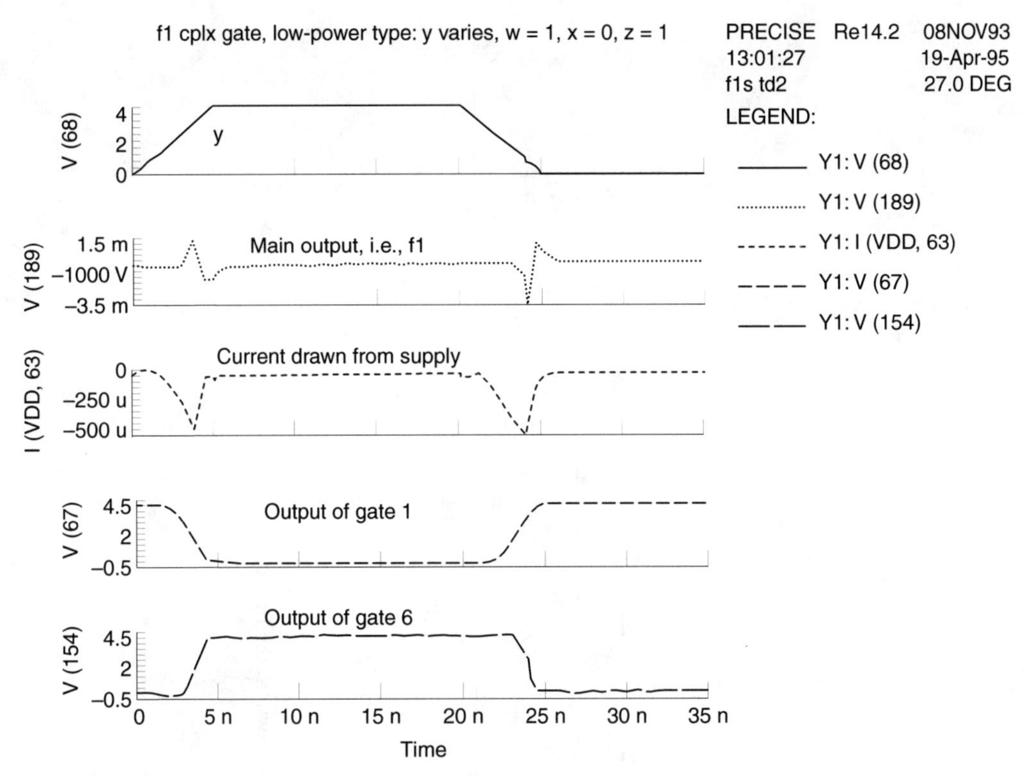

Figure 13.17 Another simulation result.

sample standard deviation s to be 2.515. If the population of ρ is normal, then the sampling distribution of the statistic $(\bar{x} - \mu)/(s/\sqrt{n})$, where μ is the population mean and n is the sample size, would be a t distribution with the degree of freedom $\mathrm{df} = n - 1$ ([32], p. 311). Thus the maximum error of estimate E when the sample mean $\bar{x}$ is used to estimate the population mean μ is given by the equation

$$E = t_{\alpha/2}(df)\,\frac{s}{\sqrt{n}}$$

To be 95% confident that μ lies within $\pm E$ of $\bar{x}$, we let $\alpha = 0.05$. From a t distribution table, we have $t_{.025}(5) = 2.571$, which then implies that $E = 2.640$.

The final conclusion is that if the population of ρ is normal, then with 95% confidence, we can say that the population mean of ρ is in the range [2.103, 7.383].

If we do the same analysis for standard-power circuits, we will find that the sample mean $\bar{x} = 4.7395$. Furthermore, we can say with 95% confidence that the population mean of ρ is in the range [2.0648, 7.4142].

Hence, in both the low-power and standard-power tests, we can say that the average ρ (sample mean) is approximately 4.7, and we can say further that with 95% of confidence, the population mean of ρ is within the range [2.06, 7.42].

TABLE 13.3 TEST RESULTS FOR LOW-POWER CIRCUITS

	Complex Gate Circuits				Standard-Cell Circuits				
Function	Size (λ^2)	Average Delay (ns)	Average Energy (pJ)	Size×Delay×Energy (×1000)	Size (λ^2)	Average Delay (ns)	Average Energy (pJ)	Size×Delay×Energy (×1000)	ρ
f_1	9,514	1.52	4.01	57.99	17,374	1.43	7.53	187.1	3.226
f_2	10,048	1.55	4.01	62.45	21,660	1.34	7.71	223.8	3.584
f_3	4,488	0.59	1.85	4.899	6,322	0.70	3.87	17.13	3.497
f_4	8,268	1.07	2.27	20.08	18,480	1.30	6.66	160.0	7.968
f_5	8,056	1.12	2.57	23.19	17.679	1.41	7.35	183.2	7.900
f_6	14,652	1.68	3.82	94.03	22.428	1.41	6.79	214.7	2.283

Note: For those circuits with small transistors. If the output does not switch, then delay = 0.

TABLE 13.4 TEST RESULTS FOR CIRCUITS WITH STANDARD-SIZED TRANSISTOR

	Complex Gate Circuits				Standard-Cell Circuits				
Function	Size (λ^2)	Average Delay (ns)	Average Energy (pJ)	Size×Delay×Energy (×1000)	Size (λ^2)	Average Delay (ns)	Average Energy (pJ)	Size×Delay×Energy (×1000)	ρ
f_1	12,194	1.34	17.04	287.43	23,086	1.26	31.71	922.39	3.209
f_2	14,287	1.39	16.82	334.03	28,500	1.15	32.58	1067.8	3.197
f_3	5,508	0.57	8.63	27.094	8,938	0.65	16.80	97.603	3.602
f_4	9,116	1.03	9.86	92.580	24,024	1.14	28.06	768.49	8.301
f_5	9,116	1.11	11.11	112.42	22,791	1.23	30.75	862.01	7.668
f_6	15,836	1.73	20.95	573.95	28,836	1.46	33.54	1412.1	2.460

Note: If the output does not switch, then delay = 0.

13.4. CONCLUSIONS AND FURTHER DISCUSSIONS

We have presented two new approaches to low-power CMOS VLSI design. In the first half of the chapter, we showed how our race-free state assignment algorithm can be used to obtain circuits that consume less energy due to reduced switching activity. The advantage that our algorithm has over existing race-free state assignment algorithms is that our circuits need not be so large.

After some additional work in the same area, it begins to appear that the same strategy could be used to eliminate selected races in a non-race-free circuit rather than to create a completely race-free circuit. With this new approach, it may be possible to start out by using Tracey's state assignment algorithm, thus ensuring a fairly small circuit and eliminating only a few often-traversed races, incurring a cost of only one extra secondary state variable per race condition eliminated. Work is under way to determine whether this new approach would lead to smaller or lower power circuits than the existing approaches.

Then we presented strong statistical evidence that static CMOS complex gates of the type that compute arbitrary functions should be widely used in low-power digital systems. Digital systems that use complex gates are likely to be superior to systems using only standard cells when we take into account energy consumption, delay, and layout area. We further argued that the best strategy might be to combine complex gates with standard cells in order to benefit from both types of technology. However, since it will take a very large investment to incorporate complex gates into CAD systems, it would be worthwhile to perform a more extensive study of complex gates versus standard cells as well to determine the types of complex gates to generate. The following are some ideas for further study:

1. A study of more than six functions would most probably sharpen our statistical results. In other words, with a larger study, we would know with greater certainty whether complex gates are better than standard cells and by how much. Also, all complex gates tested should be laid out in a standard-cell compatible fashion.
2. It is most probably the case that the comparison results are fairly independent of the transistor channel length and width. However, this should be confirmed more precisely. In particular, the inclusion of diffusion capacitance would yield more accurate results, especially when comparing regular-power to low-power styles.
3. Since complex gates would be compiled on the fly, it seems to be unnecessarily restrictive to require all device sizes to be fixed. A more realistic alternative may be to fix the complex-gate delay at the same value as that of the standard-cell solution and then select the device sizes for smallest power or smallest area. This means that the compiler would have to compute the best device sizes based on the user's specifications.
4. Layout area depends on layout style. (For instance, there are algorithms for laying out complex gates that contain more than one row of transistors of each kind.) Since we used a figure of merit that depends on area in our study, the

result may vary according to the layout style used. It is still true, though, that the energy × delay × area figure of merit incorporates three of the most important quantities of interest in VLSI design. While it is not as meaningful a measure as the power-delay product, it is certainly much more meaningful than such measures as the popular measures of performance of computers (MIPS, MFLOPS, and SPEC marks), for example, or many other cost measures in other fields.

5. A way to perform minimization that is less biased than hand minimization would be to generate the complete set of solutions and then pick the best solution for each design style. An alternative would be to pick a random solution.
6. The entire study should be extended to include more than just four-variable functions.

In closing, we note that the material presented in this chapter on two new directions in low-power digital CMOS VLSI represents only the beginning of a wide open avenue for future exploration.

Acknowledgments

I thank Lars "J." Svensson of the Information Sciences Institute of the University of Southern California for his extensive comments on the work on complex gates. Many of his ideas have been incorporated into the second half of this chapter. I also thank Andreas G. Andreou of Johns Hopkins University and CalTech (one of the editors of this book) and Silvio P. Eberhardt of Swarthmore College for helpful ideas.

References

[1] K. Asada and J. Mayor, "An MOS Leaf-Cell Generation System from Boolean Expressions," *IEEE Proc. Custom IC Conf.*, pp. 21–24, May 1987.
[2] D. Adolphson and T. C. Hu, "Optimal Linear Ordering," *SIAM J. Appl. Math.*, vol. 25, pp. 403–423, 1973.
[3] W. C. Athas and N. Tzartzanis, "Energy Recovery for Low-Power CMOS," *1995 VLSI Conf.*, Chapel Hill, NC, pp. 415–429.
[4] M. Borah, R. M. Owens, and M. J. Irwin, "High-Throughput and Low-Power DSP Using Clocked-CMOS Circuitry," *1995 ISLPD*, pp. 139–144.
[5] R. K. Brayton, G. D. Hachtel, and A. L. Sangiovanni-Vincentelli, *Logic Minimization Algorithms for VLSI Synthesis*, Kluwer, Boston, MA, 1984.
[6] J. Buchanan, "Guidelines for Reducing Inductance and Transient-Current Effects," *BiCMOS/CMOS Systems Design*, McGraw-Hill, New York, 1991, section 4.4, pp. 75–77.
[7] A Chandrakasan, S. Sheng, and R. W. Brodersen, "Low-Power CMOS Digital Design," *IEEE JSSC*, vol. 27, no. 4, pp. 473–484, 1992.
[8] S. P. Eberhardt, personal communication, Nov. 1994.

[9] A. Friedman, R. Graham, and J. Ullman, "Universal Single Transition Time Asynchronous State Assignments," *IEEE Trans. Comput.*, vol. C-18, pp. 541–547, 1969.
[10] A. Friedman and P. Menon, *Theory and Design of Switching Circuits*, Computer Science, Woodland Hills, CA, 1975.
[11] L. Glasser and D. Dobberpuhl, *The Design and Analysis of VLSI Circuits*, Addison-Wesley, Reading, MA, 1985.
[12] B. Guan and C. Sechen, "Large Standard Cell Libraries and Their Impact on Layout Area and Circuit Performance," *Proceedings of the 1996 IEEE Int. Conf. on Computer Design (ICCD)*, Austin, TX.
[13] M. Hanan, P. K. Wolff, and B. J. Agule, "A Study of Placement Techniques," *J. Design Automation Fault Tolerant Computing*, vol. 2, pp. 28–61, 1978.
[14] D. A. Huffman, "The Synthesis of Sequential Switching Circuits," *J. Franklin Inst.*, vol. 257, nos. 3 and 4, pp. 161–190 and 275–303, 1954.
[15] D. A. Huffman, "Study of the Memory Requirements of Sequential Switching Circuits," Technical Report 293, MIT Research Laboratory for Electronics, Cambridge, MA, Mar. 14, 1955.
[16] T. Indemaur and N. Horowitz, "Evaluation of Charge Recovery Circuits and Adiabatic Switching for Low Power CMOS Design," *Symp. on Low Power Electronics*, Oct. 1994.
[17] V. Kantabutra and A. G. Andreou, "A State Assignment Approach to Asynchronous CMOS Circuit Design," *IEEE Trans. Comput.*, vol. 43, no. 4, pp. 460–469, 1994.
[18] T. Lengauer, *Combinational Algorithms for IC Layout*, Wiley-Teubner, New York, 1990.
[19] H. Lindkvist and P. Andersson, "Dynamic CMOS Circuit Techniques for Delay and Power Reduction in Parallel Adders," *1995 VLSI Conf.*, Chapel Hill, NC, pp. 121–130.
[20] C. N. Liu, "A State Variable Assignment Method for Asynchronous Sequential Switching Circuits," *J. ACM*, vol. 10, pp. 209–216, 1963.
[21] T. G. Noll and E. De Man, "Pushing the Performance Limits due to Power Dissipation of Future ULSI Chips," *Proc. Int. Symp. Circuits Syst.*, San Diego, CA, May 1992.
[22] C. Piguet, "Logic Synthesis of Asynchronous Circuits," *Proc. 1990 Custom Integrated Circuits Conf.*, 1990, pp. 29.6.1–29.6.4.
[23] D. Regenold, "A Single-Chip Multiprocessor DSP Solution for Communications Applications," *Proc. 7th Annual IEEE Int. ASIC Conf.*, Rochester, NY, 1994, pp. 437–440.
[24] Y. Shiraishi, J. Sakemi, M. Kutsuwada, A. Tsukisoe, and T. Satoh, "A High Packing Density Module Generator for CMOS Logic Cells," *Proc. IEEE 25th Design Automation Conf.*, pp. 439–444, 1988.
[25] M. Sivilotti, "Wiring Considerations in Analog VLSI Systems with Applications to Field-Programmable Networks," Ph.D. dissertation, California Institute of Technology, Pasadena, CA, 1991.
[26] L. "J." Svensson, personal communication, 1995.

[27] J. Tracey, "Internal State Assignments for Asynchronous Sequential Machines," *IEEE Trans. Electron. Comput.*, vol. EC-15, no. 4, pp. 551–560, 1966.

[28] T. Uehara and W. M. van Cleemput, "Optimal Layout of CMOS Functional Arrays," *IEEE Trans. Comput.*, vol. C-30, no. 5, pp. 305–311, 1981.

[29] E. Vittoz, "Micropower Techniques," in *Design of MOS VLSI Circuits for Telecommunications*, Y. Tsividis and P. Antognetti (eds.), Prentice-Hall, Englewood Cliffs, NJ, 1985.

[30] E. Vittoz, B. Gerber, and F. Leuenberger, "Silicon-Gate CMOS Frequency Divider for the Electronic Watch," *IEEE J. Solid-State Circuits*, vol. SC-7, no. 2, pp. 100–104, 1972.

[31] E. Vittoz and H. Oguey, "Complementary Dynamic MOS Logic Circuits," *IEE Electron. Lett.*, vol. 9, no. 4, pp. 77–78, 1973.

[32] R. C. Weimer, *Applied Elementary Statistics*, Brooks/Cole, Monterey, CA, 1987.

[33] N. H. E. Weste and K. Eshraghian, *Principles of CMOS VLSI Design*, 2nd ed., Addison-Wesley, Reading, MA, 1992.

[34] S. Wimer, R. Y. Pinter, and J. A. Feldman, "Optimal Chaining of CMOS Transistors in a Functional Cell," *IEEE Trans. CAD Integrated Circuits*, vol. CAD-1, no. 1, pp. 795–801, 1987.

[35] Q. Wu and S. R. Ramirez-Chavez, "Layout Synthesis of Combinational Blocks from Behavioral Hardware Descriptions," *Proc. IEEE Int. ASIC Conf.*, 1992, pp. 38–41.

[36] K. Yano, "A 3.8-ns CMOS 16 × 16-b Multiplier Using Complementary Pass-Transistor Logic," *IEEE JSSC*, vol. 25, no. 2, pp. 388–395, 1990.

Chapter 14

Low-Power CMOS Data Conversion

Marcel J. M. Pelgrom
Philips Research, Nederlandse Philips Bedrijven B.V., Eindhoven, The Netherlands

14.1. INTRODUCTION

Low power starts as a customer requirement. The end users of electronic equipment require features that can be translated into low-power issues: long-life battery operation, limited heat production, no fan cooling, low-cost packaging of very large scale integration (VLSI) electronics or low standby power consumption required by environmental regulations, and so on. Therefore the need for low-power electronic circuits is embedded in a low-power design of the complete system, including the mechanical parts involved, the efficiency of transducers, and energy conversion.

In this chapter the impact of low power will be restricted to the electronic choices that are taken on the level of an equipment and integrated-circuit (IC) designer and that relate specifically to analog-to-digital (A/D) conversion. Low-power aspects of complementary metal-oxide-semiconductor (CMOS) data conversion will be discussed using four levels of abstraction (see Fig. 14.1).

The *system* level (for ease of definition) is interpreted as everything related to the general choices for an IC. Decisions in the system area have most influence on the power optimization of electronic equipment. This is the level where boundary conditions for all underlying levels are set. The first task on a system level is to fit the specifications as closely as possible to the desired functionality. It is on a system level, too, that the choices are made that determine the way in which the signal is processed: accuracy, bandwidth, sample clock, chip interfacing, power supply, and the like. Aspects of functional partitioning will also be discussed: functions can be shifted from the analog side of the converter to the digital side, or vice versa.

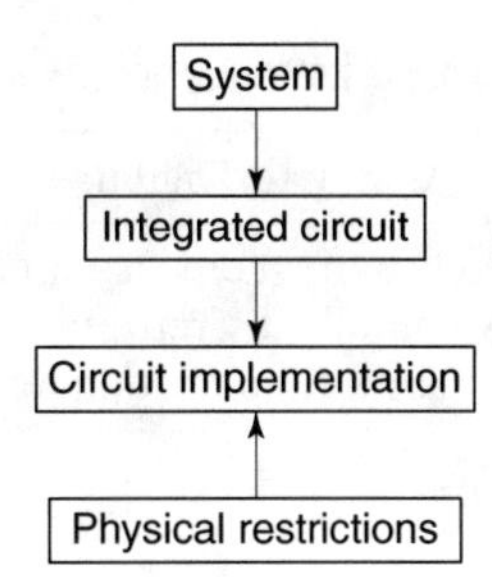

Figure 14.1 Electronic equipment: four levels of abstraction.

The *integrated circuit* level is concerned with the subdivision into basic functions within a chip. On this level, the influence of technological choices on power will be evaluated: the use of technological options and CMOS scaling. The effect of parameter tolerances on power consumption will be analyzed. A summary of A/D and D/A converters will be given for the purpose of examining trade-offs.

Although most power-saving operations surrounding A/D conversion depend on choices concerning the system and IC, attention is focused mostly on the *circuit implementation* of the converter building block (for an overview of A/D conversion see [1]). In the implementation section a comparison will be made of several realization options for A/D and D/A converters.

Several implementation design decisions are based on *physical restrictions*. Physical restrictions determine the limits that can be achieved due to physical and technological phenomena. For many A/D and D/A converters the effect of component (mis)match is critical: Limits to data conversion related to matching problems will be discussed.

Low power consumption in data converters will be discussed within the scope of large digital system ICs, which implies that the emphasis will be on CMOS implementation.

14.1.1. Analog-to-Digital Conversion: Basic Terminology

Analog-to-digital converters and D/A converters form the link between the world of physical quantities and the abstract realm of bits and numbers. In the process of conversion, a number of key functions can be distinguished (see Table 14.1).

Both A/D and D/A conversion require referencing to a conversion unit. The digital value is the ratio of the physical signal value and the conversion units (Fig. 14.2).

In many practical systems a reference voltage is used as a conversion unit, which is subdivided by means of resistors, capacitors, or transistors. Current, charge, and time units can also be used. In A/D terms: the reference value, and consequently the maximum input signal, of an N-bit converter is subdivided into 2^N least significant bits (LSBs), where V_{LSB} is the physical value corresponding to one LSB. During operation of an A/D or D/A conversion the unit is subdivided, copied, or multiplied, which causes various deviations (see Section 14.5).

TABLE 14.1 KEY FUNCTIONS IN A/D AND D/A CONVERSION

Analog to Digital	Digital to Analog
Reference to a conversion unit	Reference from a conversion unit
Amplitude discretization	Amplitude restoration
Time discretization	Holding

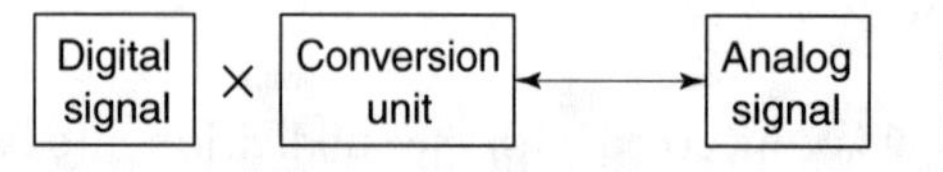

Figure 14.2 A conversion unit is needed for A/D conversion.

During the conversion from analog to digital, the quantization in both amplitude and time is the dominant signal-disturbing mechanism. The quantization error in an A/D conversion with a small number of quantization levels results in a signal with odd harmonics. When the signal is quantized with a larger ($N > 6$) number of quantization levels, the resulting error is approximated as a uniformly distributed error, with an energy content of

$$E_{\text{err}} = \tfrac{1}{12} V_{\text{LSB}}^2 \tag{14.1}$$

This error energy results in a white-noise spectrum and leads to a maximum obtainable signal-to-noise ratio for a full-scale signal:

$$\text{SNR} = 1.76 + 6.02N \qquad \text{dB} \tag{14.2}$$

All converters suffer from more errors than this quantization error alone. In order to characterize the converter, the effective number of bits is calculated by reversing the above formula:

$$\text{ENOB} = \frac{\text{SINAD} - 1.76}{6.02} \tag{14.3}$$

where signal-to-noise-and-distortion (SINAD) stands for the ratio of the signal power to all the unwanted components: quantization noise, thermal noise, distortion, and so on.

A well-known component is the total harmonic distortion (THD), the ratio between the signal and its harmonics. Usually the first 5 or 10 harmonics are counted as THD, while higher order components and folded products are counted as SINAD contributions. The spurious free dynamic range is the distance between the signal and the largest single unwanted component.

The dynamic range (DR) is not equivalent to SNR or SINAD as it represents the ratio of the full-scale input signal and the noise floor at a small signal input. The difference between DR and SNR is clearly present in, for example, range-switching configurations.

Two other important specification points are integral and differential nonlinearity (INL, DNL). Two succeeding digital codes should be spaced in the physical domain at a distance of 1 V_{LSB}; a deviation from this measure is the differential nonlinearity curve:

$$\text{DNL} = \left| \frac{V_{j+1} - V_j}{V_{\text{LSB}}} - 1 \right| \quad \text{for all } 0 < j < 2^N - 2 \tag{14.4}$$

where V_j indicates the physical value that corresponds to the digital code j.

Integral nonlinearity is defined as the real conversion value's deviation (measured in LSBs) from the ideal conversion value:

$$\text{INL} = \left| \frac{V_j - V_0}{V_{\text{LSB}}} - j \right| \quad \text{for all } 0 < j < 2^N - 1 \tag{14.5}$$

where V_0 is the zero code value. For A/D and D/A converters, the DNL and INL are specified as a graph for all codes or as a number which represents the maximum value occurring in the entire range. The INL and DNL are parameters that are mostly measured at near-DC conditions: They highlight the single strongest deviation of an ideal transfer curve. The ENOB curve as a function of input or sampling frequency represents an average measure but reflects more accurately the behavior of a converter in dynamic circumstances.

In the frequency domain, the bandwidth (BW) that can be ideally converted is limited to half of the sampling rate f_s (Nyquist criterion); note that the conversion bandwidth usually, but not necessarily starts at zero hertz (baseband). During reproduction of the signal in the D/A converter, the signal is held at the value of its last sample moment. This operation results in a characteristic signal attenuation and phase delay, which are given by

$$\frac{\sin(\pi f_{\text{signal}}/f_s)}{\pi f_{\text{signal}}/f_s} e^{-j\pi f_{\text{signal}}/f_s} \tag{14.6}$$

Often this effect is named $\sin(x)/x$ and the delay term is left out. Note that this attenuation is an average value over all possible phases of the signal to the sampling frequency. In the case of a signal that is locked in some way to the sample frequency, any attenuation between 0 and 1 may occur.

14.2. SYSTEM

Figure 14.3 shows the typical set-up for consumer electronic equipment. Dedicated analog interface circuits (sensors, tuners, etc.) provide signals to the processing core. The results of the processing are fed to analog output interfaces (power amplifiers, display drivers, etc.). The CMOS signal processing formed by, for example, A/D converters, digital signal processors (DSPs), and application-specific integrated circuits (ASICs) is controlled by central processing units (CPUs). Many system chips

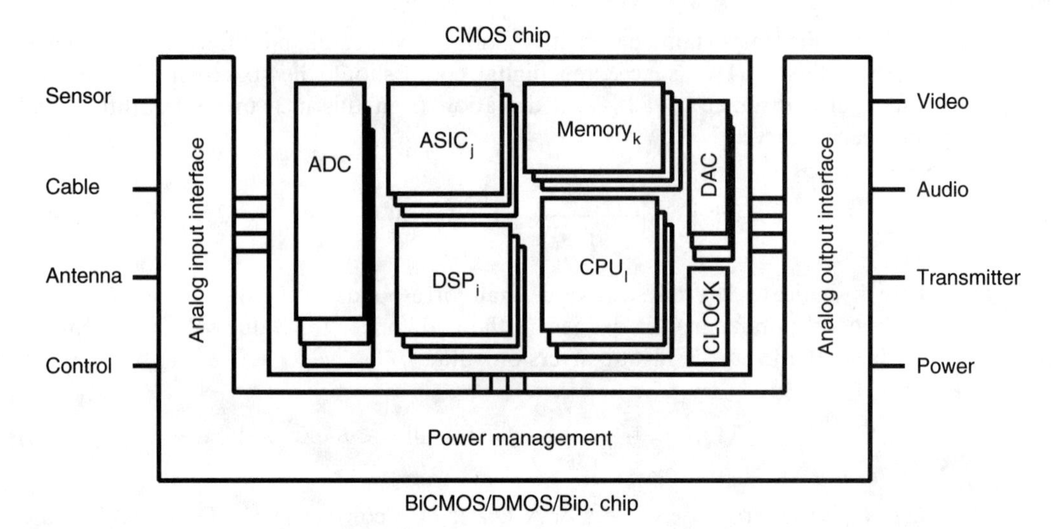

Figure 14.3 General set-up of electronic system.

are designed in standard digital CMOS. The addition of analog circuitry results in a strong focus on mixed-signal implementation. Digital signal interfacing between chips can be avoided, which helps to minimize the I/O pin count and reduce electromagnetic compatibility problems. The requirement for low-power A/D conversion is dominantly present, as the power for the analog part is typically in the order of 10–50% of the required power for the total system chip.

Technology allows full integration of the system functionality, including most of the analog interfacing, for many fields of application (video, audio, telecom, etc.). As a result of this trend, more and more functionality has to fit within the maximum power budget of an IC package. Low-cost packages for large-volume markets (consumer, telecom) allow around 1–5 W dissipation. This requirement has led to a general trend to lower the power needed for analog and digital circuits. The use of low power for digital systems and the reduction of technological dimensions have led to power supply reductions from 5 V via 3.3 V to 2.5 V and will culminate in around 1.5 V in the year 2000 [2].

A second force toward low power comes from portability requirements. The entire system power budget must be lowered for longer battery lifetimes. This form of low power requires additional considerations; for example, stand-by power, caused by leakage currents, must be lowered to a few microwatts. Specific details can be found in [3–5].

14.2.1. Specification of Functionality

Many trade-offs have to be made during the specification process of a complex system (e.g., a television set, hand-held telephone, medical imaging equipment). Today's equipment is so complicated that design decisions have to be taken by large teams of experts. Specification points for every individual function have to be fixed. Overspecification particularly leads to unnecessary power consumption and must be avoided. Analog-to-digital conversion is particularly vulnerable to over-specification due to its natural position between the analog and digital disciplines. A frequently encountered approach is to specify a converter by asking the team working on the analog side of the system to determine signal bandwidth and amplitude, SNR, and THD specifications and combine the data obtained with the specifications of the team working on the digital side (clock speed and duty cycle, word width, etc.). This "collision" of the analog and digital worlds must be avoided in the designing of the system

Figure 14.4 shows three types of A/D converters used in video signal processing [high-definition TV (HDTV), picture in picture (PIP), quadrature phase shift key (QPSK), Camcorder]. The plot shows the relation between the required ENOB and the signal frequency. The charge-coupled device (CCD) interface A/D converters require a relatively high degree of accuracy at low frequencies in order to meet the CCD dynamic range requirement. Because of the use in portable cameras, the focus is on low power and low voltage [6]. Data transmission applications (Teletext and QPSK) need some 6–7 effective bits but require proper conversion at high signal frequencies.

Between these two extremes is the baseband video converter, used in the signal paths of high-end television sets. Its requirements are good DC linearity (DNL = 0.5 LSB) in order to prevent contouring, some 7.5-bit resolution at the color carrier frequency (3.57–4.43 MHz) for color decoding, and a minimum 7-bit performance up to 10–12 MHz signal frequencies for [quadrature amplitude modulation (QAM)] HDTV-like applications. Although all three converters are used for video signal processing, the specifications are hardly interchangeable.

Several suboptimizations exist for each of these three converters which have resulted in a wide range of power-performance combinations. Some simple benchmarking can, however, be derived from the combination of bandwidth and accuracy requirements[1] (see also Fig. 14.37 and [4]):

$$\frac{\text{Power}}{10^{\text{SNR}/20}\text{BW}} = \frac{\text{Power}}{2^{\text{ENOB}}\text{BW}} \tag{14.7}$$

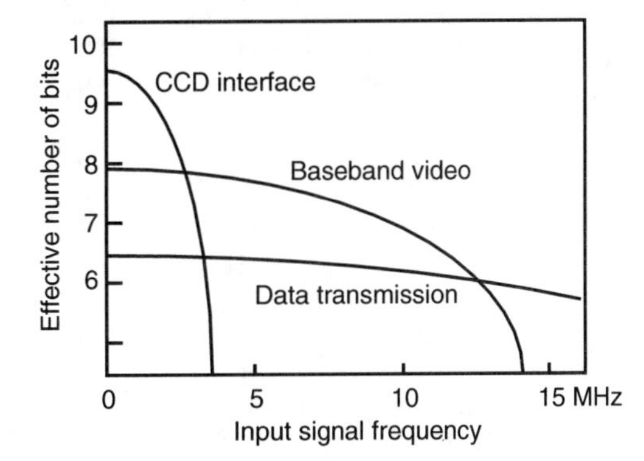

Figure 14.4 Analog-to-digital converters for CCD interfaces, baseband video, and data transmission applications.

[1]Other benchmarks use DC accuracy and sampling rate, ignoring the signal performance.

where the ENOB is valid throughout the entire bandwidth. For converters which keep their performance up to the Nyquist criterion, BW can be substituted by half of the sampling rate f_s. The proportionality constant still depends on technology, architecture, and other specifications. Table 14.2 compares some recent A/D converters. The differences are attributable to architecture (number of comparators), technology, noise or limited matching, and the like. See e.g. Section 14.5.

The trade-off on system level demands the optimization of the entire signal chain, not only of the lower power A/D converter. An important aspect of this optimization is flexibility. Unlike specification tolerances, flexibility demands the adaptation of the system to predetermined (large) system parameter shifts. Some of these shifts are necessitated by the wish to serve several product lines of a manufacturer (e.g., compatibility with several power supply voltages, availability of various output formats). In other cases the flexibility is implemented in the system itself, as the system has a multistandard nature, comprising of several transmission standards, interfaces with different sources, and so on. Flexibility mostly implies more hardware and more power than an optimized solution. Sometimes the increase in power can be reduced by designing a part in such a way that losses are avoided (e.g., by bonding or externally setting like control of current). A proper balance has to be found in incorporating flexibility in a system with minimum power or component overheads.

Power Shifts. One of the pitfalls in optimizing power on the system level is the effect of power shifts. Power can be reduced in certain circuits, but this may very well lead to greater power consumption in other parts of the system. Major power shifts occur via the interfaces in the specification (input and output terminals, clock requirements, references, output representation, etc.).

TABLE 14.2 POWER EFFICIENCY OF RECENTLY PUBLISHED A/D CONVERTERS

Bits	ENOB	BW at ENOB	Power (mW)	Power/ 2^{ENOB}BW (pJ)	Remark*
7	6.5	10 MHz	100	110	Embedded, 0.8 μm CMOS (128)
8	6.5	10 MHz	90	100	ISSCC89, 1 μm CMOS (57)
	7.0	10 MHz	65	50	TDS 8792, 0.8 μm CMOS (56)
9	8.2	5 MHz	80	40	Embedded, 0.8 μm CMOS (64)
	8.7	12 MHz	240	50	0.5 μm CMOS, ISSCC97 (128)
	9.0	1 MHz	3	6	Embedded, 0.8 μm CMOS (1)
	9.0	5 MHz	100	40	SPT7840, CMOS
	9.2	20 MHz	850	80	TDA8760, ISSCC92, BiCMOS (30)
12	10.0	15 MHz	300	20	TDA8767, ISSCC97, BiCMOS (50)
	10.5	10 MHz	1650	110	AD9027, BiCMOS
13	12.5	5 kHz	0.2	7	ISSCC96, 0.5 μm CMOS (1)
16	15	20 kHz	2.3	4	ISSC97, 0.5 μm, CMOS (1)

Note: The three embedded converters are discussed in Section 14.4.3.
*Numbers in parentheses are number of comparators.

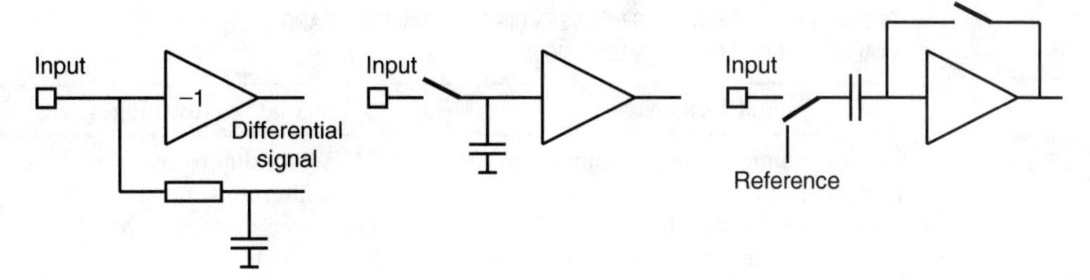

Figure 14.5 Three input circuits of A/D converters.

As an example Figure 14.5 shows three input circuits that are found in commercially available A/D converters. The first circuit is used to create a differential input signal. The signal is inverted, and the in-phase signal is applied to an *RC* circuit with (hopefully) an identical pole. At high frequencies this input solution works only at a very well-defined source impedance. The second circuit is used to generate a sample-and-hold (S/H) function at the input. Again the input source is forced to deliver large signal currents and to cope with strong clock kickback. The last circuit uses input offset cancellation. The switching order leads to a pulsed current which is to be delivered from the input source.

All three circuits require a driver with a 20–50-Ω impedance over a wide voltage range: Application circuit diagrams show bipolar drivers biased at several milliamperes that often require the same amount of power as the A/D converter or even more! From a system point of view, it may be more efficient to use a buffered input converter than a low-power A/D converter with high input-circuit demands.

14.2.2. Signal-Processing Strategy

Sampling Rate and Bandwidth. The main parameters of analog signal processing on a system level are dynamic range, SNR, and bandwidth. These quantities translate into resolution and sampling rate requirements in the digital domain.

On the system level the sampling rate of an A/D converter is locked to or derived from the system clock. The choice of clock (and sampling) frequency used on the system level is important with respect to specification and power. The main criterion for the sampling frequency is given by the Nyquist theorem ($f_s > 2\text{BW}$). For high-bandwidth circuits the sampling frequency is in practice 20–50% higher than the minimum rate required by the Nyquist theorem. The main reasons for this are the trade-off between analog signal properties and the consequences for digital signal processing, where a higher clock rate has disadvantages (see Table 14.3). Power and area in the digital domain are balanced by signal quality and filter complexity in the analog domain.

In the case of low signal bandwidths the ratio of the sample rate and the bandwidth is chosen so that it optimizes the entire conversion chain. Figure 14.6 shows how a trade-off can be found between the filter order and the sample rate–

TABLE 14.3 BALANCING BETWEEN DIGITAL DRAWBACKS AND ANALOG BENEFITS OF HIGH SAMPLING RATE

Digital Drawbacks	Analog Advantages
More dynamic power consumption	SNR gain due to oversampling
Longer digital filter structures	Alias filtering simpler
Less ripple through time for logic	Less steep voltage steps (slewing)
Larger storage units for fixed-time delays	Less $\sin(x)/x$ loss

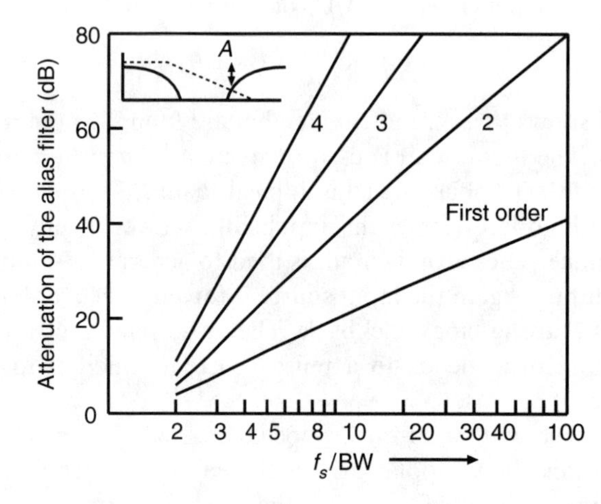

Figure 14.6 Alias suppression as function of filter order.

bandwidth ratio. A higher ratio of the sample rate and signal bandwidth allows a lower order alias filter.

Another example of a very specific sampling frequency will be given below in the section on sampling of modulated signals.

Sample Rate and Accuracy. In the processing strategy it is very important to choose the appropriate conversion position in order to minimize the power required. There will usually be some freedom in choosing the position of the data converter within the signal chain. As A/D[2] conversion is quantization in the amplitude and in time domains, trade-offs can be made between these two domains.

In the amplitude domain the optimum signal SNR and the dynamic range have to be determined. In those cases in which a lower SNR can be used with respect to the dynamic range, the preferred system solution is a gain-controlled amplifier followed by a minimum SNR A/D converter; see Figure 14.7. Gain control does require some form of signal analysis, which is mostly a cheap function in the digital domain. Another option in this case is a companding A/D converter.

A certain degree of freedom will also exist if the (potential) sample rate of the A/D converter is much larger than the required bandwidth. An exchange can then

[2]The arguments here presented apply to A/D conversion but mostly hold also for D/A conversion.

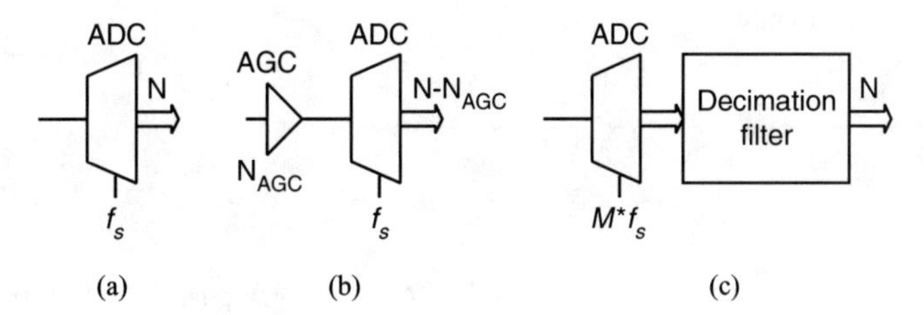

Figure 14.7 (a) Standard A/D converter, (b) A/D converter with gain control, and (c) A/D converter with oversampling and decimation filter.

be made between SNR and sample rate. If, for a given bandwidth, the sample rate of an N-bit converter is increased by a factor of 4, the noise energy will spread over a frequency band that is four times as large and the amplitude of the noise in that bandwidth will be reduced by a factor of 2. An appropriate decimation filter can increase the effective resolution by 1 bit. The general relation between an increase in resolution and an increase in sample rate is

$$\Delta N = \tfrac{1}{2}\log_2 \frac{f_s}{2\text{BW}} \tag{14.8}$$

This solution allows us to use converters with less resolution, resulting in less power, but accuracy (e.g., comparator trip voltages) will still be required at the final resolution depth. Another disadvantage lies in the white-noise assumption for the quantization. In the case of low signal amplitudes quantization becomes harmonic distortion and oversampling will only work in the presence of dither signals or noise fed back from the converter (delta modulation).

In specific system architectures the advantages and disadvantages of these solutions have to be investigated; generally a reduction in A/D resolution outweighs the costs of gain control or of a digital filter.

Sampling of Modulated Signals. Input signals containing carrier-modulated components imply additional difficulties: Aliases of the harmonic distortion of the signals will fold back during the sampling process in the converter. If the sampling rate is close to an integer multiple of that carrier frequency, the harmonics will interfere with the original signal (see Fig. 14.8). This occurs in many transmission systems [e.g., in phase alternating lines (PAL) video, in which the color modulation frequency is 4.433 MHz while the preferred sampling rate is 13.5 Ms/s (megasamples per second)]. A sample rate close to an integer of the modulation frequency may necessitate additional suppression of the amplitude of the distortion component. This will be achieved at the expense of more power in the converter.

14.2.3. Implementing System Functions

Proper signal analysis allows us to use the conversion properties in a functional way. Examples are:

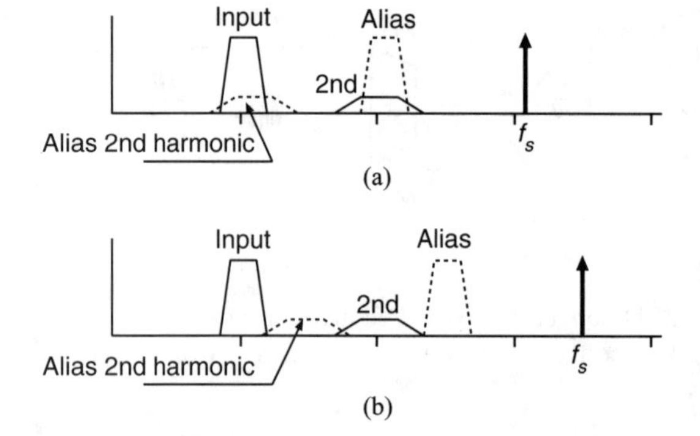

Figure 14.8 Spectrum of sampled modulation signal with distortion: (a) sample rate equals 3.1 times modulation frequency; (b) sample rate equals 3.5 times modulation frequency.

- The inherent sampling performs demodulation (see the section that follows), which can be implemented in several A/D architectures. Particularly suitable are structures with high-performance S/H circuits.
- The inverse process is equally applicable; use upper bands generated by the D/A function for, for example, direct digital synthesis (DDS) of radio frequency signals.
- Multiple input signals can be multiplexed on the S/H input of an A/D converter.
- Alias filtering can be combined with system-required filtering.
- A local sample rate increase can be used to relax the requirements in other parts of the system (see the section below on oversampled D/A converters).

Conversion of Modulated Signals. Figure 14.9 shows an example of the use of the inherent S/H function for realizing a system need: down mixing of modulated signals. The information content of the input signal is rather band limited, but it is modulated on a relatively high carrier. In this case it is power efficient to implement the conversion starting with an S/H. The sampling function is used to modulate the signal band to a much lower frequency. In this example down modulation is performed around twice the clock frequency. Now the conversion task for the A/D converter core is much simpler. Note that the remark on power shifts applies to the circuit of Figure 14.9.

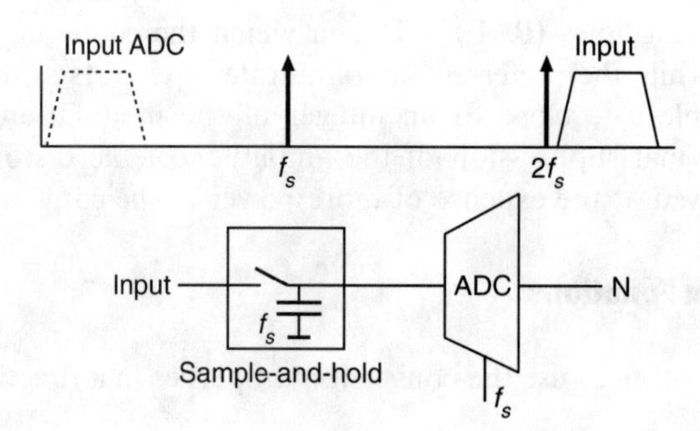

Figure 14.9 Converter arrangement for IF conversion.

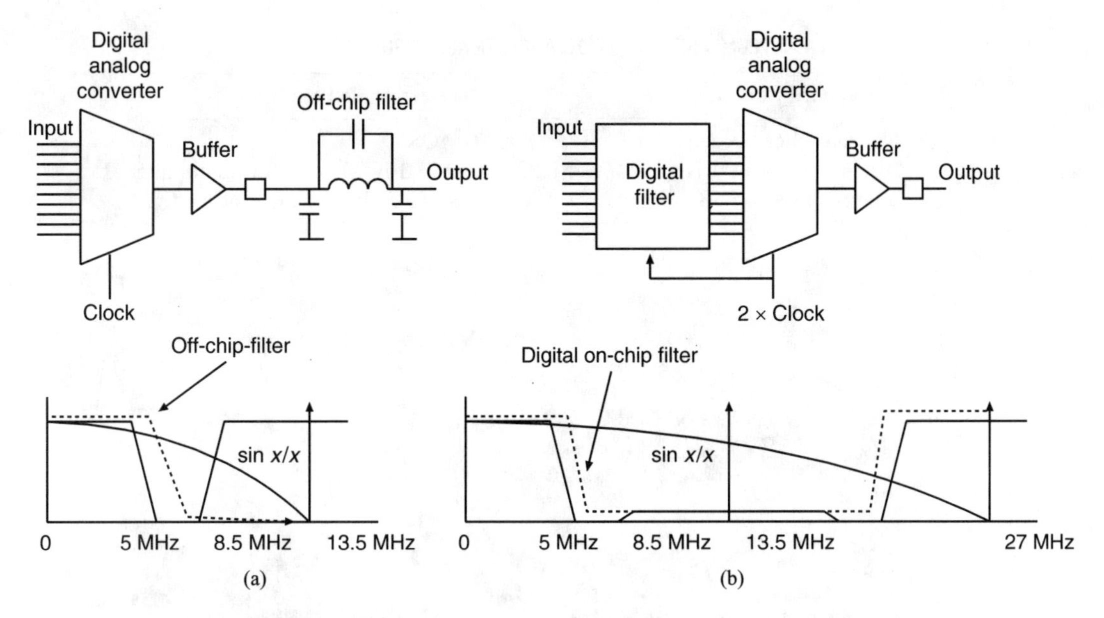

Figure 14.10 (a) D/A converter with output filtered by external filter and (b) D/A driven by digital prefilter.

Oversampled D/A Converter. Figure 14.10 shows a practical example of local oversampling. In Figure 14.10(a) a normal D/A conversion is followed by an off-chip filter. Due to the rather low ratio of sample rate and high signal frequency, large transient steps will occur at the on-chip buffer and the output pin. This will usually cause slewing and distortion. The driver has to be designed with additional bandwidth and the input stages will need large bias currents. This setup moreover has a relatively poor high-frequency performance due to $\sin(x)/x$ signal loss. The passive filter requires some three to seven poles and is expensive to produce, especially if $\sin(x)/x$ compensation is also needed.

Figure 14.10(b) shows an IC solution: The sample rate is locally doubled and the odd alias terms in the frequency spectrum are removed by digital filtering. Now the large transients in the output are more than halved in amplitude, and relatively simple noncritical postfiltering (first order) is sufficient to restore the analog signal. The inherent $\sin(x)/x$ is reduced by an order of magnitude, so no compensation is required. Figure 14.11 shows a chip photograph of an oversampled filter with a D/A converter. From a power point of view a trade-off must be made between the additional power and area for the filter and the quality loss and additional power in the buffer (see Table 14.4).

In this section several aspects have been shown of system choices that influence the power budget of A/D conversion. Low power requires thorough signal and system analysis.

TABLE 14.4 COMPARISON OF TWO D/A SOLUTIONS IN FIGURE 14.10

Standard Solution	Oversampled solution
External filter needed	$\approx 1\,\text{mm}^2$ CMOS
10 mA current in driver	5 mA current in driver power for digital filter
$\sin(x)/x$ loss $= 4\,\text{dB}$	$\sin(x)/x$ loss $= 0.5\,\text{dB}$, $2 \times$ clock needed

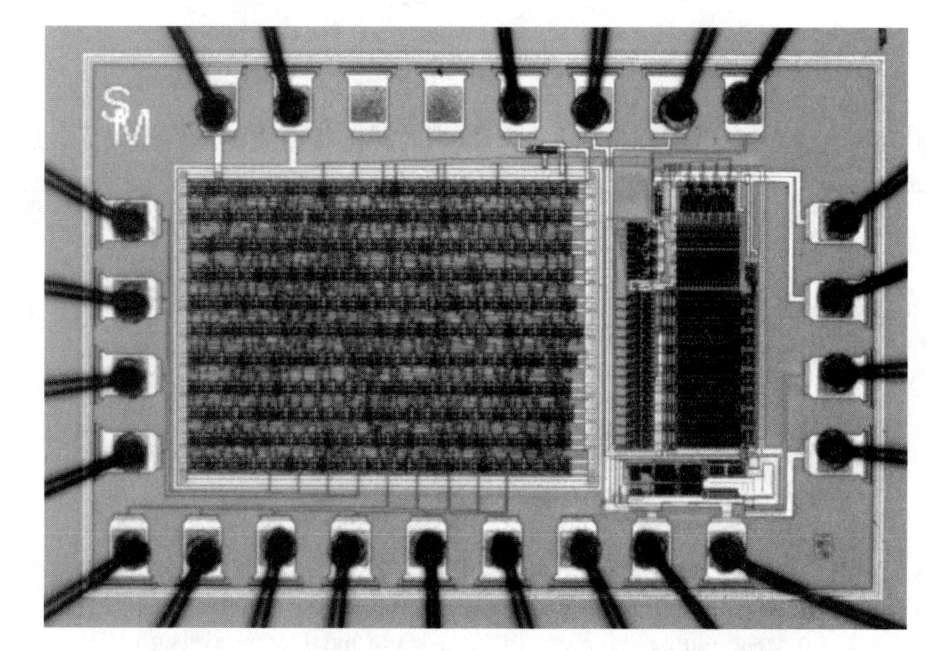

Figure 14.11 Chip photograph of an 8-bit D/A and oversample filter. (Designed by S.Menten/M. Pelgrom.)

14.3. INTEGRATED CIRCUIT

Once the system aspects of an IC have become clear and the system designer has fixed several boundary conditions, some important options are still open that will influence the power dissipation in and around the A/D converter. In this section the choices in technology and the basic A/D and D/A architecture will be discussed. The focus will be on CMOS technology, but many arguments will also hold equally for other technologies.

14.3.1. Technology

The technological items that are important for a designer are signal swing, feature size (speed), process options, and tolerances (see also [7]).

Signal Swing. Many analog circuits are limited by signal swing. Figure 14.12 gives the expected power supply voltage according to the Semiconductor Industry

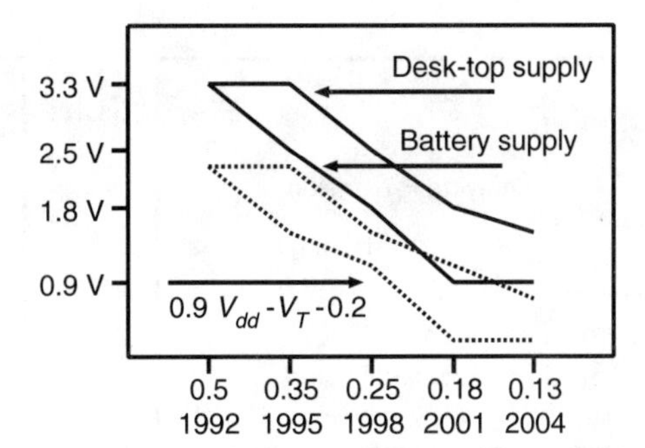

Figure 14.12 Power supply evolution as function of process generation. Signal swing is based on minimum power supply minus threshold and bias voltages.

Association (SIA) (bold line) for desktop and battery applications (see [2] for a popular version of the SIA roadmap). The available signal swing (dotted line) is derived by taking 90% from the nominal power supply (minimum power supply) and subtracting a maximum threshold voltage (0.5 V) and 0.2 V gate bias. The graph clearly shows that this form of signal handling has reached its limits. Besides a low-V_T option or a local V_{DD} boost, the most suitable form of representation for signals on mixed-signal chips is the differential mode. This form of signal representation is less sensitive to substrate noise and has inherent cancellation of even harmonic distortion but requires differential inputs and outputs: two-pin analog input/outputs (I/Os). Despite these measures, signal swing is expected to decrease from 2 to some 0.5–0.75 V peak to peak in minimum power supply applications. Further limits with respect to the reduced signal swing will be discussed in Section 14.5.

Feature Size. In digital complementary metal-oxide-semiconductor (CMOS) technology the drive for smaller feature sizes is of course a dominant factor. For low-power operation immediately the development in capacitances is of particular importance, because low capacitance values lower the required power. For digital power consumption the general law in CMOS for a single sample operation is

$$\text{Power} = N_i(C_{\text{load}} + C_{\text{gate}})V_{DD}^2 + P_{\text{sc}} + P_{\text{leak}} \tag{14.9}$$

The power is composed of the capacitor-charging component and (usually of minor importance) a short-circuit and leakage component. The number of transitions (activity) N_i, in a worst case, may exceed 1 due to, for example, carry mechanisms. The close relation between the power and the supply voltage V_{DD} explains the drive for lower voltages (see Fig. 14.12). Here, C_{load} represents the node capacitance, which is strongly affected by the technology and the feature size.

In analog circuits feature size reduction promises a higher cutoff frequency for the same current. Figure 14.13(a) shows the evolution of gate and diffusion capacitors. These capacitors dominate the internal performance of analog building blocks, whereas in digital cells the wiring is of prime importance. The gate capacitance, 1 μm wide and of minimum gate length, shows that the reduction in gate length is largely compensated by the thinner gate oxide. The diffusion capacitance is determined at zero junction voltage by a 1-μm-wide slice of diffusion which allows one contact hole to be placed using minimum-dimension design rules. Incorporated are 1 μm of gate

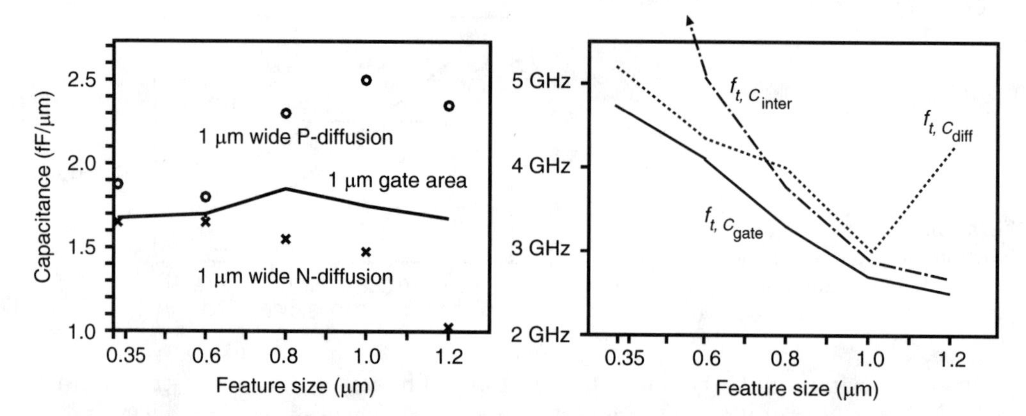

Figure 14.13 Evolution of gate and diffusion capacitance and cutoff frequency f_t for minimum-gate-length NMOS transistor biased at 5 μA current per micrometer gate width.

edge and local oxidation of silicon (LOCOS) edge capacitors. The improvement of the P-diffusion in the 0.6-μm process due to the lower doped twin well is clearly visible. The N-type MOS transistor (NMOST) diffusions, however, increase in capacitance.

The ratio of transconductance and load capacitance results in a cutoff frequency ($f_t = g_m/2\pi C$). In Figure 14.13(b) the cutoff frequencies for diffusion, gate, and 10-μm densely spaced interconnects are compared. The speed performance determined by diffusion capacitance shows only little improvement. Note that these cutoff frequencies were determined at the low designers' V_{GT} ($\approx$ 0.2–0.4 V), whereas technologists usually use much higher V_{GT} (>2 V) to show the maximum cutoff frequency of their technology. Such high V_{GT} values are used only in specific circuits like CMOS low-noise amplifiers (LNAs).

The improvement of the speed for wiring is notable—some 30% of speed for every generation. The so-called interconnect crisis is therefore more of a design automation (placement) issue than a technological one. To a lesser extent the speed improvement also holds for minimum-length gate capacitance. If however, DC gain must be kept constant to maintain a circuit's accuracy specification, the gate length cannot be scaled with technology and the gate capacitance cutoff frequency will not improve. These observations must be taken into account when comparing digital and analog speed performance.

Process Options. A design team is rarely able to choose the main line of technology to be used to design a function: Issues like specific demands, availability, experience, and cost dominate the choice of the main process line. One level below that choice there is much more freedom: Most manufacturers offer two or more generations of a process—a mature process with a large feature size, an advanced process recently released for production, and experimental processes. Baseline processes also allow a variety of process options—in wiring layers, in additional elements (capacitors, resistors, fuses), and in device parameters. Process options are often used to circumvent the inherent drawbacks of baseline digital CMOS processing. Options moreover allow designers to map the function better on the process, which allows reducing the amount of power consumed. Options are defined as everything offered that is above the baseline process and can be subdivided into four categories:

- *No extra masks, no process adaptations.* Analog characterization and monitoring of specific process parameters is most important. Examples are MOST noise, matching, temperature and voltage dependencies of passive and active components. Proper characterization (of, e.g., matching) immediately allows design for minimum power consumption. The reduction in process spreads also falls in this category (see the next section).
- *No extra masks, minor process adaptations.* Redefinition of MOST thresholds is crucial for low-voltage operation. Analog designers also want to use structures that are normally only parasitics in the baseline process. This option does not introduce new mask steps but may require another combination of masks. Examples are vertical PNP transistors and large MOSTs as capacitors, resistors, and three-layer interconnect capacitors. Stacked-layer capacitors show a minimum parasitic component, which saves power.
- *Extra masks, minor process adaptations.* Modification of parameters of elements, especially multithresholds, in order to optimize transistors for analog switching. Lightly doped drain (LDD) suppression on the source side and definition of resistors in a silicide process fit into this category.
- *Extra masks, major process adaptations.* The addition of new elements may necessitate new mask levels. This is certainly the most critical category of process options as it affects the costs. Examples are second gate oxide for high analog voltages and second polysilicon layer. A double-polysilicon capacitor has a decade less parasitics than a stacked-layer capacitor, so the same function is realized at a lower power consumption.

The lists of examples in the above categories can of course be expanded. The first category is needed in all industrial analog design; in that sense analog design always requries options. The second category is also very often needed for A/D designs. The third and particularly the fourth categories (adding masks) are disputable; here the analog designer's wishes will inflict a cost penalty upon the digital part of the design. This will rarely be acceptable for a commercially viable application.

Tolerances. The absolute tolerances of the components, the power supply variation, and the temperature span in which a circuit must operate strongly influence the power level. Standard specifications for the power supply tolerances are ±5, ±10, and ±20%. Similar variations are observed in the spreads of process capacitances and MOST currents. The temperature range may vary from 0–70°C to −40–140°C. The circuit nevertheless has to function and meet the specifications. Although these three factors influence all aspects of the IC design, here the attention will be on the effect on speed performance versus power.

One of the critical points in a design is speed optimization. Suppose a MOS has to (dis)charge a load capacitor C_{load} within a fraction $\alpha \ll 1$ of the clock period:

$$R_{\text{on}} C_{\text{load}} \approx \frac{C_{\text{load}}}{(W/L) C_{\text{ox}} \mu (V_{DD} - V_T)} = \alpha T_{\text{clock}} \tag{14.10}$$

which leads to the following relation for the minimum gate area needed in the nominal case:

$$C_{\text{gate,nom}} = WLC_{\text{ox}} = \frac{C_{\text{load}} L^2}{\alpha T_{\text{clock}} \mu (V_{DD} - V_T)} \tag{14.11}$$

If not restricted by other demands, the designer will minimize the load capacitance and transistor gate length and maximize the available time and drive voltage. The speed optimization of this component becomes critical when worst-case figures replace the ideal-case figures:

$$C_{\text{gate}} > \frac{C_{\text{load,max}} L^2_{\text{max}}}{\alpha T_{\text{clock,min}} \mu_{\text{min}} (V_{DD,\text{min}} - V_{T,\text{max}})} \tag{14.12}$$

where L_{max} is the maximum value of the minimum gate length. A worst-case inspection of the required overdesign factor adds up the tolerances on the load capacitor, the gate length, the clock skew or the clock duty cycle variation, the mobility as a function of temperature, and the power supply. In Table 14.5 a comparison of large, medium, and small tolerances is given with respect to nominal conditions for power, temperature, and parameters.

In view of the fact that the power per transition is the well-known relation [power $= (C_{\text{load}} + C_{\text{gate}}) V_{DD}^2$], it is obvious that the increase in C_{gate} due to tolerances is a major power problem in both analog and digital designs. This argument plays a role in the specification of both the system and the IC.

TABLE 14.5 INCREASE IN GATE CAPACITANCE OF DRIVING TRANSISTORS DUE TO PARAMETER TOLERANCES

Parameter Variation	Nominal	Small	Medium	Large
Wiring/Diffusion C_{load}	0%	+10%	+20%	+30%
Minimum length L	0%	+10%	+20%	+30%
Clock skew T_{clock}	0%	−10%	−20%	−30%
Temperature T	25°C	85°C	120°C	140°C
Mobility, $\mu \propto T^{-2}$	0%	−30%	−42%	−48%
Gate drive $V_{DD} - V_T$	0%	−5 %	−10%	−15%
Effect on C_{gate}	$1 \times C_{\text{gate,nom}}$	$2.2 \times C_{\text{gate,nom}}$	$4.1 \times C_{\text{gate,nom}}$	$7.1 \times C_{\text{gate,nom}}$

14.3.2. Conversion Architectures

D/A Converters. In a D/A converter the digital number is transformed into a physical quantity by means of a conversion unit. A conversion voltage or current is subdivided by passive or active elements, such as resistors, CMOS transistors, or capacitors, or it is subdivided in time. A few comments on each of these elements are given below:

Resistor	The relative accuracy is on the order of 10^{-3}–10^{-4}; the absolute value will suffer from (large) process variations and temperature. The resistor value is determined by the successive loading. Capacitive coupling to substrate may be a noise pick-up. Constant current is always needed.
Capacitor	The relative accuracy is on the order of 10^{-3}–10^{-4}; the absolute value is usually well defined in a double-polysilicon process. An opamp configuration is needed to manipulate charges. Minimum size is determined by parasitics or the kT/C noise floor. Often seen as low-power solution, but requires large peak currents during charge transfers. Sensitive to different parasitic couplings [8].
Transistor	Relative accuracy is on the order of 10^{-3}; the absolute value is sensitive to temperature and process spread. Mostly used as current source or currrent divider. Back-gate modulation and $1/f$ noise must be considered.
Time	With more or less fixed variation (30–100 ps_{rms}), the best accuracy is at low signal bandwidths; converters are always several orders of magnitude slower than other systems. Used mostly in $\Sigma\Delta$ approaches. Interference enters via clock buffers.

Two representations are used for the conversion itself: unary and binary. In a unary format 2^N copies of the conversion unit of an LSB are present and the conversion is performed by selecting the proper number of units in a particular manner. Examples are resistor strings (thermometer coding) and parallel current sources (segmentation) (see the upper part of Figure 14.14). There are similar techniques with capacitor arrays and timing (counting D/A). The alternative is the direct use of the binary information. Instead of 2^N copies, there are physical values that correspond to the binary powers of the digital signal. Forward straight use of those binary numbers (in case of a direct binary representation) results in the required conversion.

All the schemes involve the problem of accurately reproducing the unit value. Several improvements have been proposed (see the later discussion on resistor string D/A converters in Section 14.4.2 and e.g., [9–14]). Often the binary representation offers the lowest area use. Most of these schemes do not directly affect the converter's overall power consumption.

In signal quality the difference occurs at transitions at which many bits flip (01111 $\rightarrow$ 10000). While the unary organized D/A conversion adds another unit, the binary converter changes from one group of elements to a completely different one. Imperfections in the different groups of elements directly result in errors at those code transitions in the form of differential nonlinearity. Most D/A (sub)-

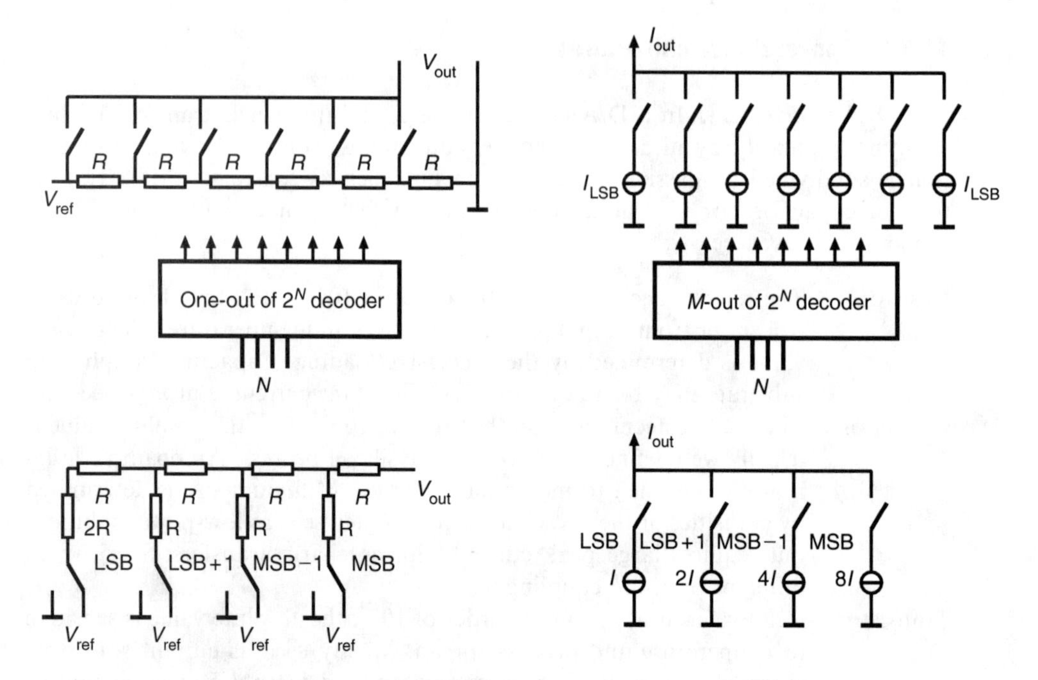

Figure 14.14 Unary and binary forms of D/A conversion.

schemes can be classified along the above lines, though there are a few deviating forms such as ternary coding (+1, 0, −1), which is sometimes used in combination with sign/magnitude representation.

In the case of converters with a high resolution the problem with the above schemes is the large number of units involved or the wide range of binary values. Subranging is generally applied in order to circumvent these problems; a converter of N-bit resolution is subdivided into a cascade of two subconverters of M and $N - M$ bits. Subranging can be extended to an N-bit converter that is split into N subconverters of 1 bit. Figure 14.15 shows some subranging schemes. The reduction in area is compensated with additional control or buffering circuits and for $N > 8, \ldots, 10$ the need exists to provide some form of overlap on the edges of the subranges.

In many A/D and D/A converters there is a direct match between the required accuracy and the number of levels in the converter. In an examination of the fundamental properties of A/D and D/A conversion, Section 14.1.1 showed that quantization errors are energy components. Feedback techniques allow to shift the quantization energy out of desired signal frequency ranges. Figure 14.16 shows the basic principle of $\Sigma\Delta$ modulation [15, 16]: The feedback loop shapes the noise energy (symbolized by noise entering into the dashed summing point) from a flat spectrum into a shape which is the complement of the loop filter $H(\omega)$. When the quantizer's effective gain is modeled as A_q, the I/O relation becomes

$$\text{Out}(\omega) = \frac{A_q H(\omega)}{1 + A_q H(\omega)} \text{In}(\omega) + \frac{1}{1 + A_q H(\omega)} \text{Noise} \tag{14.13}$$

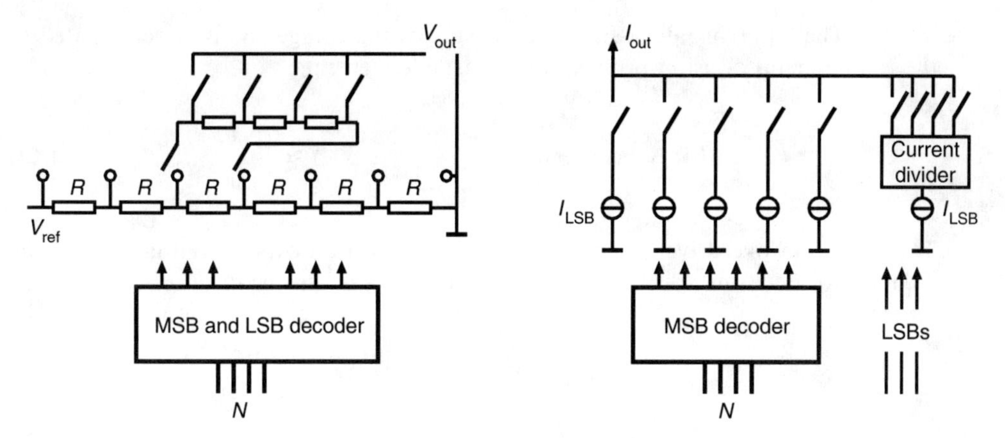

Figure 14.15 Voltage and current subranging D/A converters.

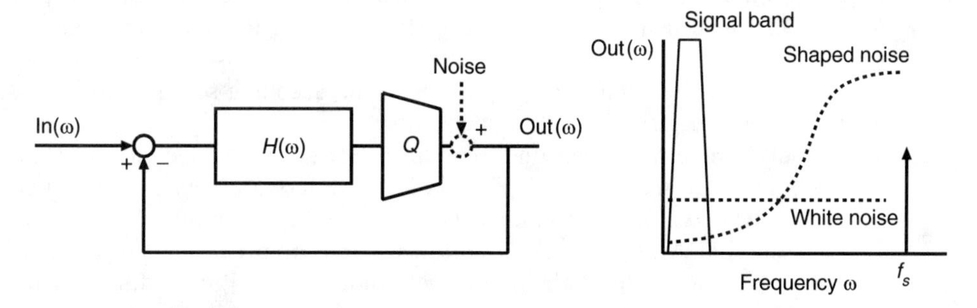

Figure 14.16 Sigma-delta modulation.

At high values of $A_q H(\omega)$ the quantization noise is suppressed and the output signal equals the input signal. This type of noise shaping can be used in A/D, D/A, and D/D conversions of signals. A prerequisite is a sufficiently high oversample factor of f_s over the signal bandwidth.

This kind of conversion allows us to use very small quantizers (e.g., a few switches for D/A conversion) and is therefore very power efficient (see, e.g., [5]).

A/D Converters. Every A/D converter consists of an inherent D/A function in combination with a comparison. The number of comparators in an architecture is a dominant parameter for the power budget. Control and references are usually of minor importance. The mutual equality of comparators is very important: Their input-referred errors add up to the errors related to the attached D/A converter. The next section will discuss the problem of comparator mismatch.

An A/D comparator needs a minimum overdrive voltage $V_{\text{overdrive}}$ to achieve a stable output level within the allowed decision time T_d. It is, however, possible to define an input difference voltage range ΔV in which the comparator cannot decide in time. The bit error rate (BER) is a measure of that range and is closely related to the fundamental decision problem in metastable elements:

$$\mathrm{BER} = \text{probability}(|V_{\text{overdrive}}| < \Delta V) \approx \frac{2\,\Delta V}{V_{\text{LSB}}} \tag{14.14}$$

The minimum overdrive ΔV can easily be derived if an exponential growth of the comparator decision toward the power supply V_{DD} is assumed:

$$\mathrm{BER} = \frac{V_{DD}\mathrm{e}^{\frac{-T_d}{\tau}}}{V_{\text{LSB}}} \approx 2^N e^{\frac{-T_d}{\tau}} \tag{14.15}$$

In a CMOS, τ is the time constant formed by the parasitic and gate capacitances and the achievable transconductance. A typical example with 8 bits is 5 fF total capacitance for 1 μm gate width, with 5 μA/V transconductance for the same gate width, and $T_d = 20$ ns, which results in a BER of 10^{-7}. This BER can be improved to better than 10^{-10} by means of more current in the latch transistors. Next to the mere improvement of the latch speed, measures in the decoding scheme can be taken to avoid serious code errors due to a metastable state.

A full-flash converter has one comparator for every possible reference value, so $2^N - 1$ comparators are needed. Full-flash converters are very power hungry but have the advantage of single clock delay conversion. A variant of a full-flash converter is the fold A/D converter [1, 17] in which a preprocessing stage "folds" the input signal, which reduces the number of comparators. Proper design of the preprocessing, combining high speed and high yield, is the critical issue. More recent techniques use interpolation for further optimization [17, 18].

An important question with respect to these converters is whether to use an S/H circuit in the converter's input stage. Experience has shown that a high-speed S/H of full-signal and bandwidth performance requires 10–30% of the total A/D power budget. The advantage of using an S/H is that signal propagation errors in the analog preprocessing are reduced and architectures can be used that make multiple use of the input signal.

As in the case of D/A converters, much use is made of subranging for A/D conversion, too (see Fig. 14.17). In 2, ..., N stages, smaller flash converters convert part of the signal and the remainder is amplified and passed to a next section. The amplification is crucial as it reduces the effect of errors in the succeeding stages. Subranging requires more time for the signal to propagate. Additional hardware is required for proper high-speed operation; with overranging circuits so that signals that have been misinterpreted in the earlier sections can also be converted. Subranging converters require multiple access to the original or intermediate signals; these converters consequently are equipped with S/H circuits between each section. Nevertheless, the smaller number of comparators lead to low-power realizations.

In an extreme case, N sections convert one bit in each section; this architecture is referred to as "pipelined." Despite the fact that a single comparator is used per

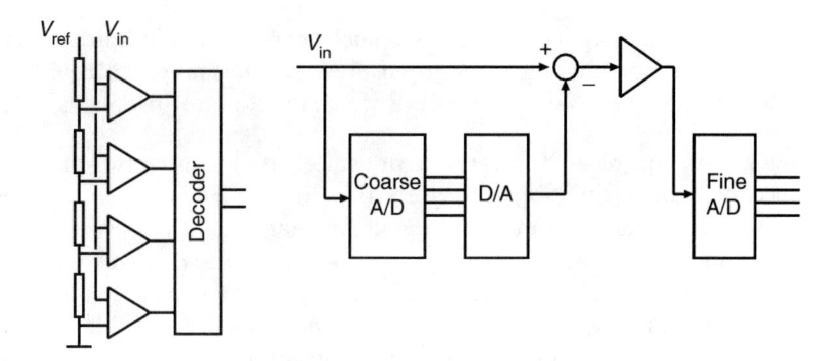

Figure 14.17 Full-flash converter and subranging converter.

section, equality problems between sections and between the comparator and the subtract circuit have to be solved.[3]

If more than 7-bit accuracy A/D converters in CMOS are needed, all these techniques require a form of random-offset cancellation. A number of basic techniques can be used:

- Trimming is used in converters for professional applications. In a special layer on top of the IC, resistors are formed which can be trimmed by means of laser cutting. Different offsets can be removed through proper application. Once it has been trimmed, the A/D converter can run without timing restrictions. A disadvantage is that testing is more expensive; the converter is characterized in a first run, then trimming is applied and the converter is remeasured. There are other forms besides resistor trimming, for example by means of fuses or a programmable read-only memory (PROM).
- The signal can be amplified with respect to the random offset. This can only be done via some form of amplitude adaptation. One form is the aforementioned folding technique. Another form is subranging. In both techniques a well-known part of the reference voltage is subtracted from the input signal. this subtraction has to be done accurately.
- The offset can be measured and stored. A popular diagram is shown in Figure 14.5(b). Another example will be given in Section 14.4.3. The disadvantage of these forms of offset cancellation is the time required to measure and store the offset. A more fundamental problem is that offset is never fully canceled: First there is a gain-bandwidth limitation of the cancellation loop and, second, the circuit in which the offset is canceled differs in some way from the circuit that is used to convert the signal (e.g., MOSTs are switched on or off). In Figure 14.5(b) the remaining error is equal to the ratio of the

[3]Of course many variants exist: with different D/A structures, combination of techniques, using differential or residue signals, etc. The reader is referred to [1].

mismatching part of the channel charge and the storage capacitor. This capacitor must be large to ensure a low mismatch voltage but must at the same time be small to reduce settling and power problems.

Single-comparator architectures do not suffer from comparator matching problems: Successive-approximation converters (see Fig. 14.30 later) go in a binary search through the reference range and require as many clock pulses as output bits.

Another form of single-comparator A/D conversion involves the use of a single comparator in a $\Sigma\Delta$ approach (see Fig. 14.16). The oversample ratio and the filter characteristics determine the bandwidth and resolution. Table 14.2 clearly shows how single-comparator designs (the three slowest in the table) exploit this advantage.

Yield. One of the major differences in the design of CMOS and bipolar A/D converters is the lack of accuracy in CMOS comparators. A bipolar differential pair has a random offset on V_{be} on the order of $\sigma_{Vbe} = 0.3\,\text{mV}$. A pair of CMOS transistors with small gate lengths show random offsets on the order of $\sigma_{Vt} = 2$–$6\,\text{mV}$ [19]. Because of the low CMOS gain, the offsets of the entire comparator accumulate, which may lead to $\sigma = 5$–$20\,\text{mV}$ as a total input-referred random offset.

The effect of random comparator offset on an N-bit full-flash A/D converter (with $2^N - 1$ comparators) is DNL or nonmonotonicity (DNL < -1LSB). The DNL for a converter with 2^N conversion level is specified as

$$\text{DNL} = \max_{j=0,2^N-2} \left| \frac{V_{j+1} - V_j}{V_{\text{LSB}}} - 1 \right| \tag{14.16}$$

Here, V_j is the value of the input signal which causes comparator j to flip and consequently reflects the comparator random offset. The voltage V_{LSB} is the physical value corresponding to an LSB. If the input random offset of every comparator is given by a Gaussian distribution with zero mean and standard deviation σ, then the probability of all the comparators being within the monotonicity limit can be calculated. This probability is an estimation of the yield of the A/D conversion. Ideally, $V_{j+1} - V_j - V_{\text{LSB}} = 0$. Nonmonotonicity occurs when comparator $j+1$ (fed with a rising input signal) switches before the adjacent comparator j with a lower reference voltage does. In mathematical formulation this becomes $p = \text{Probability}(V_{j+1} < V_j)$. then $1-p$ is the probability of comparators j and $j+1$ switching in the correct order; this condition must hold for all $2^N - 2$ pairs of comparators, so

$$\text{Yield} = (1-p)^{2^N-2}$$

with

$$p = \text{Probability}\left(\frac{V_{j+1} - V_j - V_{\text{LSB}}}{\sigma\sqrt{2}} < \frac{-V_{\text{LSB}}}{\sigma\sqrt{2}} \right) \tag{14.17}$$

The mean value of the argument in the probability function for p is 0, while the standard deviation of the argument corresponds to $\sqrt{2}\times$ the standard deviation of a single comparator σ in millivolts. The probability function is consequently normalized to a standard normal curve $N(0, 1)$.

Figure 14.18(a) shows the curves that relate the yield to the standard deviation σ of the comparator random offset for an input voltage range of 2 V. A 10-bit converter requires $\sigma < 0.5\,\text{mV}$. This is still achievable in bipolar technology [20]; for higher accuracy trimming, higher input voltages or other forms of offset correction are needed.

An 8-bit full-flash converter has to reach an input standard deviation of less than 1.6 mV to achieve an acceptable yield. In the case of a CMOS, straightforward A/D converters are replaced by offset-compensated A/D converters at the 7-bit level, where comparators with random offset better than 3.5 mV are sufficient. Table 14.6 shows that this seemingly small difference in random offsets between 7- and 8-bit A/D converters results in a substantial difference at the level of gate capacitance. In this table it is assumed that the comparator's design comprises three transistor pairs that contribute to the random offset: the input pair, a current source pair, and a latch pair.

In Figure 14.18(b) the requirements are more severe; now the probability of the converter achieving a DNL of less than 0.5 LSB has been calculated. The probability p of two adjacent comparators exceeding the DNL limits is

$$\text{Yield} = (1 - p)^{2^N - 2}$$

with

$$p = \text{Probability}\left(\left|\frac{V_j - V_{j-1} - V_{\text{LSB}}}{\sigma\sqrt{2}}\right| > \frac{\text{DNL} \times V_{\text{LSB}}}{\sigma\sqrt{2}}\right) \tag{14.18}$$

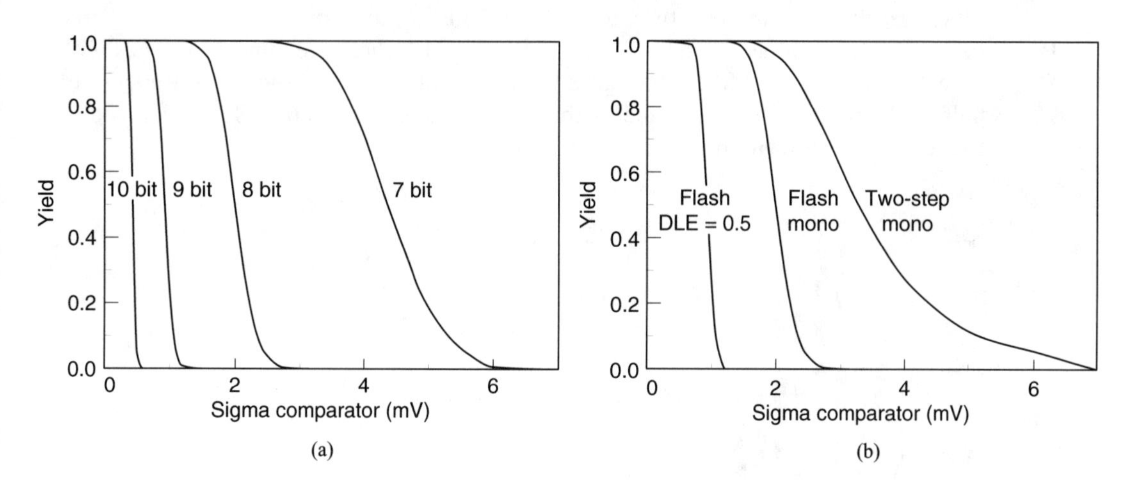

Figure 14.18 Yield on monotonicity vs. comparator random offset. Random offset is indicated in millivolts (a) with a signal swing of 2 V. (b) Yield on DNL = 0.5 LSB, on monotonicity for 8-bit full-flash and on monotonicity for 8-bit two-step subrange architecture.

TABLE 14.6 COMPARISON OF 7- AND 8-BIT FULL-FLASH A/D CONVERTERS

	7 bits	8 bits
Input range, V	2	2
LSB size, mV	16	8
Input random error 95% yield, mV	3.5	1.6
Relevant MOS pairs in comparator	3	3
Random error per pair, mV	2.0	0.9
Area per MOS (20 nm gate oxide), μm^2	49	242
Capacitance of 2^N gates, pF	11	107

As expected, the σ required for a DNL limit value of 0.5 LSB has been more than halved.

The demands for a two-stage subranging architecture are also shown for (a theoretical minimum of) 15 coarse and 15 fine comparators. Owing to the steep nature of the Gaussian distribution, the advantage of having only 15 critical comparators results in about 25% more tolerance on the input random offset voltage. This example indicates that these yield considerations for full-flash architectures give a good first-order approximation for more complex architectures.

So far, the DNL has been used as a boundary value for the yield probability. The actual value of the DNL, being the maximum of $2^N - 1$ differences of stochastic variables, is a random function itself. Figure 14.19 shows the distribution of the actual DNL of 1000 8-bit full-flash converters generated by means of Monte Carlo simulation. The shape of the distribution is characteristic of production measurements of various types of A/D and D/A converters. The comparator random offset was in this example chosen to be $\sigma = 0.15\ V_{\mathrm{LSB}}$, corresponding to $\sigma = 1.17$ mV for a 2-V signal range. As predicted by Figure 14.18, there are almost no trials with an actual DNL < 0.5 LSB, although all trials result in monotonicity. In the case of half of the trials the actual value of the DNL is lower than 0.64 LSB, which is also found in the following first-order calculation:

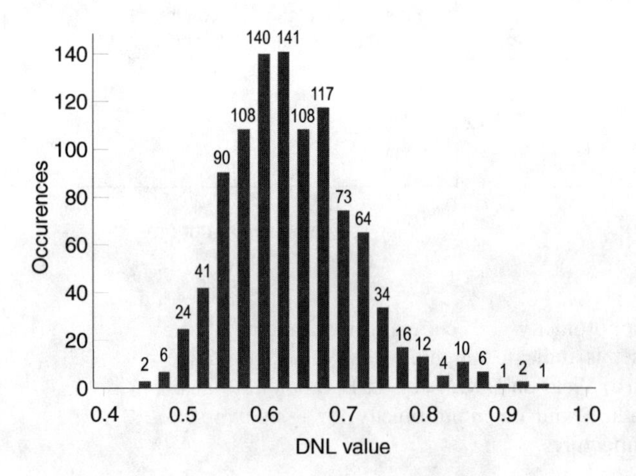

Figure 14.19 Typical distribution of DNL values for 8-bit full-flash architecture, with $\sigma = 0.15\ V_{\mathrm{LSB}}$. Mean DNL used in this simulation is 0.64 LSB.

$$\text{Yield} = (1 - p)^{2^N - 2} = 0.5 \rightarrow p = 0.9973$$

$$\text{From } N(0, 1) \text{ table: } p = 0.9973 \leftrightarrow \alpha = 3.0 \rightarrow \frac{\text{DNL} \times V_{\text{LSB}}}{\sigma\sqrt{2}} = 3.0 \tag{14.19}$$

$$\text{With } \sigma = 0.15 V_{\text{LSB}} \rightarrow \text{DNL} = 0.6$$

Noise. The above analysis also allows us to estimate the effect of comparator random offset on the noise performance. On the one hand, the process of comparator random offset results in a DNL which is the maximum allowed value of the difference in random offset of $2^N - 2$ comparator pairs. In a normal distribution α is the threshold value for the stochastic variable for which the probability of the maximum random offset of $2^N - 2$ comparator pairs remaining within the limits (−DNL, +DNL) is acceptable. For $N = 8, \ldots, 10$, α is in the range 3.0–3.4.

On the other hand, the random offset causes a nonuniform quantization. This effect can be modeled as an additive noise term at the input of the A/D converter influencing the SNR. If a full-swing sine-wave input is assumed, the SNR due to quantization and the DNL for N-bit resolution is

$$\begin{aligned} \text{SNR}_{Q+\text{DNL}} &= 10 \log 2^{(2N-3)} - 10 \log\left(\frac{1}{12} + \frac{\sigma^2}{V_{\text{LSB}}}\right) \\ &= 6.02N - 9.03 - 10 \log\left(\frac{1}{12} + \frac{\text{DNL}^2}{2\alpha^2}\right) \end{aligned} \tag{14.20}$$

For DNL = 0 the formula results in the well-known SNR = $6.02N + 1.76$ dB. In Figure 14.20 the SNR has been plotted in decibels versus the DNL for $N = 8$ and $\alpha = 3.0$. The squares indicate the results of Monte Carlo computer simulations for both SNR and DNL. At higher DNL the assumption of independence of "DNL noise" and quantization noise is less valid. Moreover, the simple analysis ignores the fact that sine-wave signals more often involve the quantization errors at the top and bottom than errors around midrange. The triangles indicate measurement points of the A/D converter described in the following sections.

Figure 14.20 shows that a DNL = 0.5 LSB yields a −49.2 dB noise level and a poor DNL = 1 LSB results in SNR = 47.7 dB, corresponding to a loss of 0.37 ENOB. The DNL and ENOB are therefore both important specification points for an A/D converter.

So far no frequency dependencies have been considered. Figure 14.21 gives an example of comparators during offset reduction. The offset reduction is accomplished by means of a gain stage before the latch stage. These elements can be identified in most offset reduction schemes. The latch stage is considered ideal except for a load capacitor C and a random offset source V_o. Both are related to the latch transistor dimensions by $C = WLC_{\text{ox}}$ and $\sigma_{Vo} = N_T A_{Vt}/\sqrt{WL}$. Here, A_{Vt} is the process constant for threshold matching (Section 14.5) and N_T is the number of

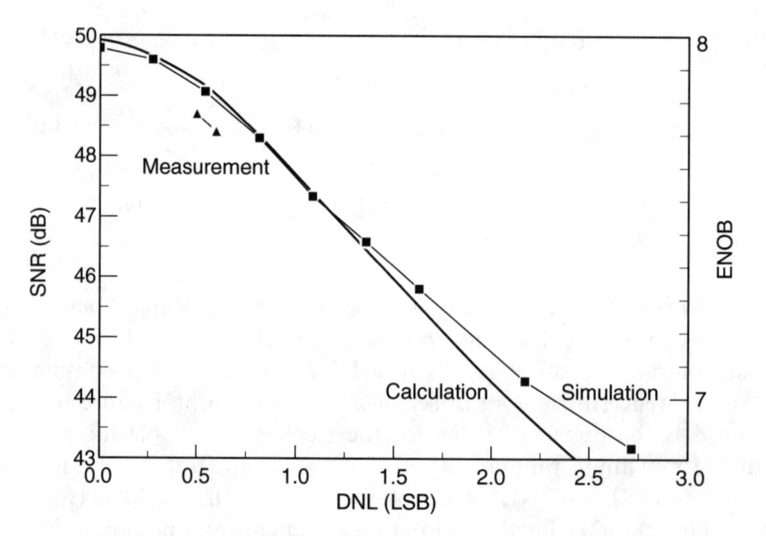

Figure 14.20 The SNR for quantization with Gaussian-distributed random offset errors vs. expected value of DNL. Squares indicate Monte-Carlo simulations, triangles refer to measurements obtained using A/D converter described in next section.

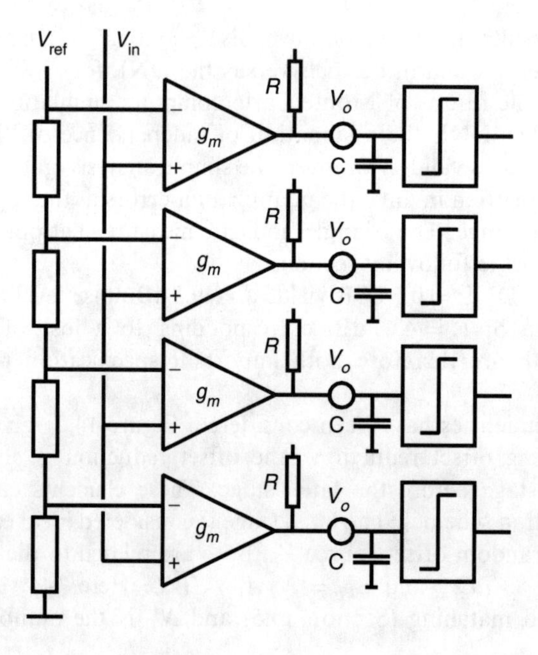

Figure 14.21 Random offset reduction by means of gain stage.

latch transistor pairs that contribute to the random offset. Now the input-referred comparator random offset σ is given by

$$\sigma = \frac{\sigma_{Vo}(1 + j\omega RC)}{g_m R}$$
$$= \frac{N_T A_{Vt} \omega C_{ox} \sqrt{W/L}}{g_m} \qquad \omega RC > 1 \qquad (14.21)$$

This is only a rough formula, which must be adapted for S/H operation, timing, and different architectures. It indicates that random offset reduction is frequency limited and depends on design (N_T, g_m), technology (A_{Vt}, W, L), and power (g_m, number of comparators). Example: In a 1-μm CMOS technology $\sigma_{Vo} \approx 20\,\text{mV}$, $g_m/2\pi C \approx 250\,\text{MHz}$, so the desired input random offset of 0.8 mV can be achieved up to 10 MHz.

Jitter. The systematic and random variations of time moments are referred to as "skew" and "jitter"—shifts of timing moments from the desired grid. Time deviations affect A/D and D/A conversion and are often characterized in terms of dependency of the conversion error on the signal speed. Examples of systematic offsets in timing are skews due to unequal propagation paths of clocks, interference from subdivided clocks, loading of clock lines, and clock doubling by means of edge detection. Random jitter variations occur during the generation of clock signals in noise-sensitive oscillators, phase-locked-loops (PLLs), long chains of clock buffers fed by noisy digital power supplies, and the like. A practical value for jitter on a clock edge in a CMOS environment is 30–100 ps_{rms}.[4] As time effects in A/D and D/A conversion are characterized in a dependency of the conversion error on the signal speed, the *additional* effect of a jittered timing pulse on the SNR of a sine-wave signal can be estimated:

$$\text{SNR}_\tau \approx -20 \log(2\pi f_{\text{signal}} \tau_{\text{rms}}) \qquad (14.22)$$

where f_{signal} is the signal frequency and τ_{rms} the effective value of the jitter.

14.4. IMPLEMENTATION

14.4.1. Status of Data Converters

Table 14.7 gives a status review of data converters for two major low-power A/D areas. High-resolution conversion is usually used in audio equipment such as CD players and telecom equipment. High-speed conversion is used in high-end television sets, desktop video systems and several digital transmission systems. The

[4]A peak-to-peak value is often used for jitter, but peak-to-peak values for stochastic processes have no significance if the process and the corresponding number of observations are not identified.

TABLE 14.7 HIGH-END CATALOG A/D AND D/A PRODUCTS

	Analog Power	Digital Power
High resolution:		
Stereo hifi 16-b A/D (SAA7360)	10 mW	40–400 mW
Stereo hifi 16-b D/A (TDA1545)	6 + 10 mW	< 1 mW
High speed:		
Video 8-b 30 Ms/s A/D (TDA8792)	45 mW	15 mW
Video 8-b 30 Ms/s D/A (TDA8772)	8 + 23 mW	3 mW

Note: The power supply voltage is 3.3 V nominal. From [21].

examples are stand-alone versions of converters that are used for large system chips, intended for low-cost markets.

In high-resolution audio A/D systems sigma-delta modulators dominate the conversion market. Signal-to-noise ratios at the converter's analog input determine the amount of noise allowed in the V/I transconductance [4]. This affects the analog power via the currents required in input differential pairs. As excessive noise may be folded back into the signal band due to the sampling process, proper design of the filter sections is essential for optimum performance. Moreover, efficient digital filtering is essential for low-power conversion. Table 14.7 shows a range of digital power values which are caused by the different requirements that are set for high-quality conversion. In some consumer equipment full 16-bit performance is required in the entire system, whereas in other (portable) equipment, audio specifications are traded off against the potential power reduction. In the case of these high-resolution converters the power problem is more a question of efficient digital filter structures than of analog circuitry. Many high-resolution D/A converters are also based on sigma-delta modulation; similar arguments hold as mentioned for the A/D converter. In the 16-bit D/A conversion of Table 14.7 (which is described in [14]), a segmented current approach is used and calibration is implemented to get rid of current source errors. The analog power is subdivided into a part consumed in the D/A converter itself and a part consumed in the output section. Digital requirements are limited to input circuitry.

Today's architectures for high-speed A/D conversion have multiple signal path architectures: Different hardware is active for different levels of the signal. In 8-bit CMOS converters this immediately leads to DNL and/or SNR problems due to mismatching in those signal paths. Circuit-designed offset compensation is necessary but becomes less effective at high signal frequencies. Offset-compensated comparators and A/D realization will be discussed in Section 14.4.3.

In video D/A converters (Section 14.4.2) the greater part of power is consumed in the output interface. The digital power is limited to input latching and some optimized forms of decoding. Both types of D/A converters in Table 14.7 are designed to deliver an off-chip signal, which in audio converters interfaces with an operational amplifier. In video the requirement of driving a 75-Ω load implies choosing between voltage-domain and current-domain conversion; this trade-off will be discussed in detail below.

14.4.2. CMOS Digital-to-Analog Converters

The requirements imposed on D/A converters for video applications become more severe with the introduction of new display standards in television and graphics. Some of the target specifications for video D/A conversion are a DC resolution of 10 bits, clock rates up to 50 MHz, good differential linearity, little distortion, signal bandwidths (−1 dB) of 20 MHz, minimum power consumption, and no external interfacing electronics toward CRT drivers. For reasons of costs these converters have to be integrated on digital signal-processing chips in CMOS technology. This requirement imposes additional demands on interference immunity of the A/D and D/A converters.

This section describes the design of the two most popular types of trimless D/A converters based on current sources and D/A converters based on resistor strings.

Current-Domain D/A Conversion. High-speed CMOS D/A converters designed with a current cell matrix allow fast and accurate settling. With a resolution of 10 bits, parameter gradients over the wafer and the mutual matching of current sources and switches result in yield losses. In the case of binary decoding, the gradient compensation and matching have to be very tight to avoid differential nonlinearities. If segmentation [10, 22] is used, the glitch matching of wiring and switches has to be done carefully.

Current-based circuits directly switch the required signal current in the load and dump the complementary part of the signal current to ground; the power supply current is equal to the maximum current, which is twice the average signal current. If a two-sided terminated transmission line has to be fed by the high-impedance output of the current cell D/A converter, the current must be doubled to obtain the required output swing. In this case, the power supply current is theoretically four times the average signal current. A triple-video D/A converter intended for supplying 1 V_{pp} to 75 Ω will consequently require an 80-mA power supply current.

The pole formed by the load resistor and the capacitance is the dominant pole in the current D/A; if the current is fed in an output buffer, this pole is the second pole in the transfer function.

A 100-Ms/s 10-bit D/A converter was designed with a segmented array of 64 current sources for the 6 most significant bits (MSBs). This structure was completed with a 4-bit binary-coded LSB section. Figure 14.22 shows the block diagram of a 10-bit design. The major design issue in this circuit is the switching of the 64 segmented current sources. Decoding delays in the 64 current sources are reduced by an additional latching stage just before the current switch. The current switch itself is designed with a low-swing (1-V) differential pair. This measure results in a low clock feedthrough on the output line, while in this case the switch transistors also act as a cascode stage, thus reducing the modulation of the output current by the output voltage. Figure 14.23 shows the chip photograph of this design.

Resistor String D/A Converter. In a system chip with A/D and D/A converters, it is advantageous to have similar references for the A/D and D/A converters: The tracking of input and output ranges for processing, for example,

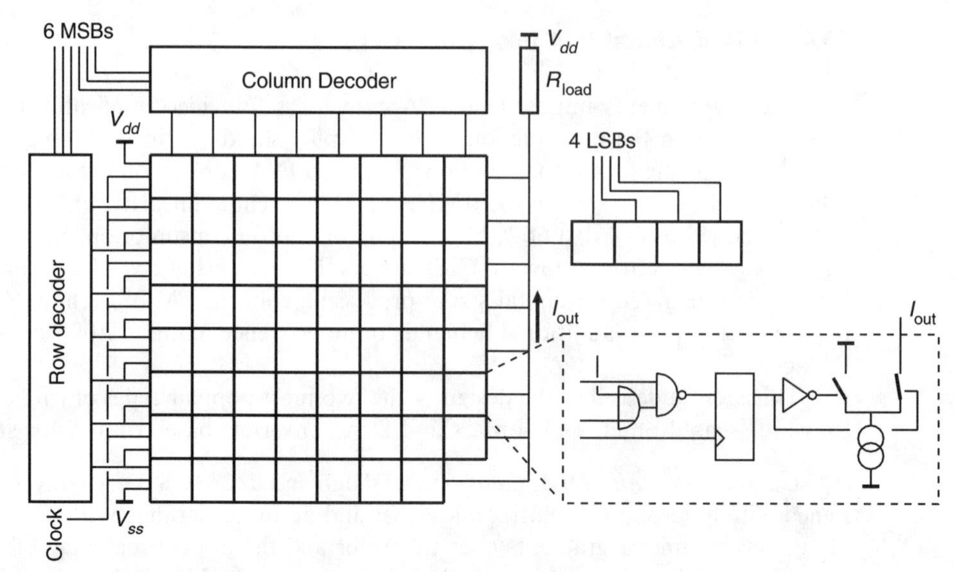

Figure 14.22 Block diagram of 10-bit current D/A converter. (Design by A. Waterman/M. Pelgrom.)

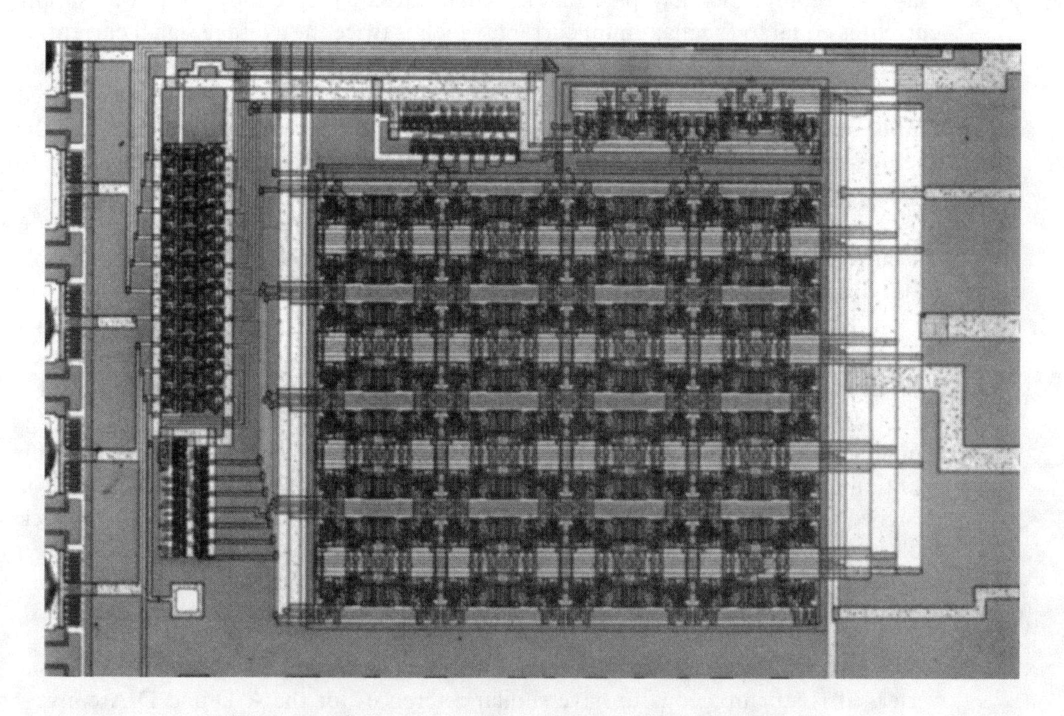

Figure 14.23 Chip photograph of 10-bit current D/A converter.

variations and temperature is then guaranteed and the overall gain of A/D and D/A conversion is better controlled. The voltage dependence and the mutual matching of large-area polysilicon resistors[5] allows the design of a converter with high integral and differential linearity. Basically, the variation in the polysilicon resistance value is determined by its geometry variations: The variations in length and width result in local mismatches and the variation in thickness results in gradients.

The design of a D/A converter with a single 1024-tap resistor ladder and sufficiently fast output settling requires tap resistors on the order of 6–10 Ω. The size of such resistors in conventional polysilicon technology is such that accurate resistor matching, and hence linearity, becomes a problem. The solution to this problem is to use a dual ladder [23] with a matrix organization [24, 25].

Figure 14.24 shows the ladder structure: The coarse ladder consists of two ladders, each with 16 large-area resistors of 250 Ω which are connected antiparallel to eliminate the first-order resisitivity gradient. The coarse ladder determines 16 accurate tap voltages and is responsible for the integral linearity. A 1024-resistor fine ladder is arranged in a 32 × 31 matrix, in which every 64th tap is connected to the coarse ladder taps. This arrangement makes it possible to increase the fine ladder tap resistance to 75 Ω without loss of speed. The effect of wiring resistances has to be related to the 75-Ω tap resistors and may therefore be neglected. There are only currents in the connections between the ladders in the case of ladder inequalities; this

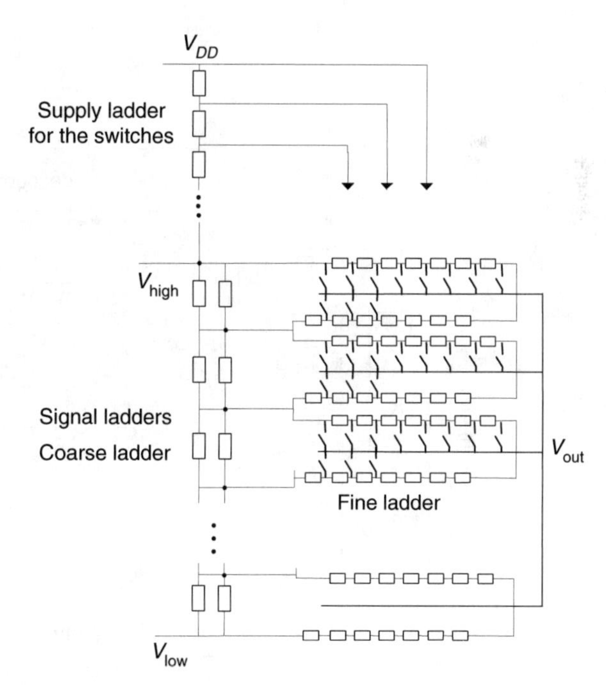

Figure 14.24 Resistor network for video D/A converter.

[5]Diffused resistors are a viable alternative in several processes.

reduces the effect of contact resistance variance. The current density in the polysilicon is kept constant to avoid field-dependent nonlinearities. The coarse ladder is designed with polysilicon resistors in order to avoid voltage dependence of diffused resistors. The fine ladder is designed in either polysilicon or diffusion, depending on secondary effects in the process implementation. The double-ladder structure is also used in all of the three A/D converters discussed in Section 14.4.3.

In a basic ladder design consisting of one string of 1024 resistors, the output impedance of the structure varies with the selected position on the ladder and therefore with the applied code. The varying output impedance in combination with the load capacitance results in unequal output charging time and consequently signal distortion of high-frequency output signals. This source of varying impedance has been eliminated by means of a resistive output rail. The insert in Figure 14.25 shows part of two rows of the matrix.

The second source of the varying output impedance is the switch transistor; usually its gate voltage equals the positive power supply, but the voltage on its source terminal is position dependent. The turn-on voltage doubles from one end of the ladder to the other. In this design an additional supply ladder is placed on top of the signal ladders to keep the turn-on voltage of the switches more constant. The turn-on voltage of each switch transistor is effectively made to correspond to the lowest turn-on voltage of a basic ladder D/A structure. Therefore there are no additional power supply constraints.

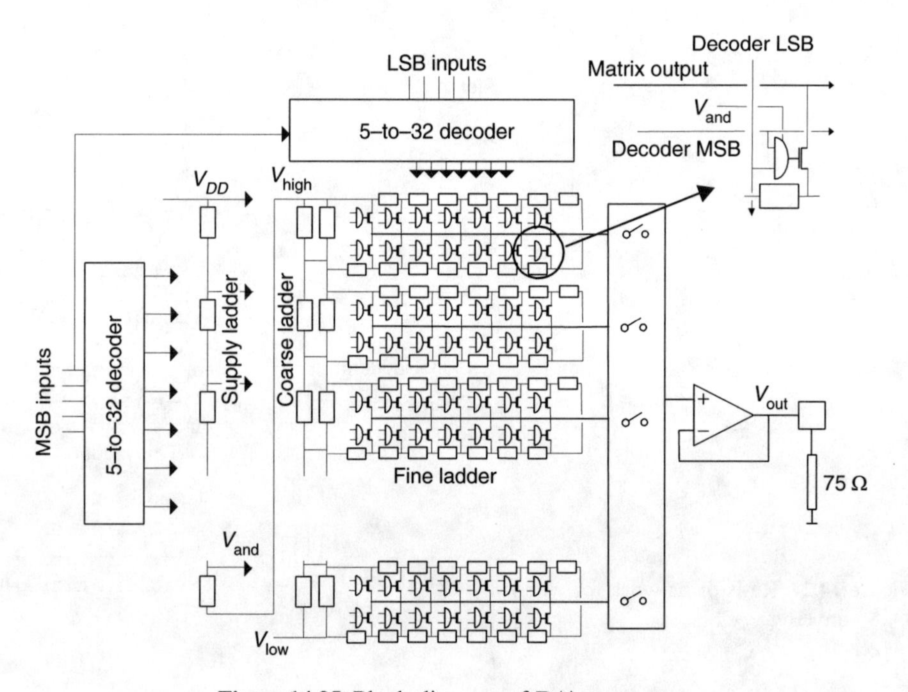

Figure 14.25 Block diagram of D/A converter.

The total ladder configuration can now be fed from the analog power supply; the signal ladders are in the range between ground level and 40% of the power supply, the supply ladder goes from 60% to V_{DD}.

The core of the D/A converter is formed by the 32×32 fine-resistor matrix.

The two decoders are placed on two sides of the matrix. The two sets of 32 decoded lines are latched by the main clock before running horizontally and vertically over the matrix. In the matrix, the 1024 AND gates perform the final decoding from the 32 horizontal MSB lines and the 32 vertical LSB lines.

A voltage-domain D/A converter generates only the necessary momentary current and is hence more efficient in power consumption. However, a voltage-domain D/A converter requires an on-chip buffer, which introduces two drawbacks: The output always needs some offset from the power supply rail and the opamp is inherently slower. The output buffer is a folded-cascode opamp (Fig. 14.26). The P-channel input stage operates on input voltages ranging from 0 to 2.2 V. The main current path through the output stage goes from the PMOS driver ($W/L = 1400$) down into the output load. A resistive load is consequently needed for optimum performance.

The on-chip resistor is on the order of 25–75 Ω; it keeps up a feedback path even at frequencies at which the bondpad capacitance shorts the circuit output. It also serves as a line termination. The swing of the output load resistor is consequently half of the buffer input voltage. The actual value of the stop resistor can be controlled to within 10%; the resulting gain error is no serious drawback in video equipment, as there is always a total gain adjustment.

The power distribution over the 10-bit D/A converter is dominated by the output stage: With a full-swing sine wave (0.1–1.1 V on 75 Ω), the average current through the driver transistor is 7.3 mA. The remaining part of the driver requires 1 mA. The ladder current is 1 mA while the digital part running at 50 MHz is limited to 0.7 mA, resulting in a total power supply current of 10 mA.

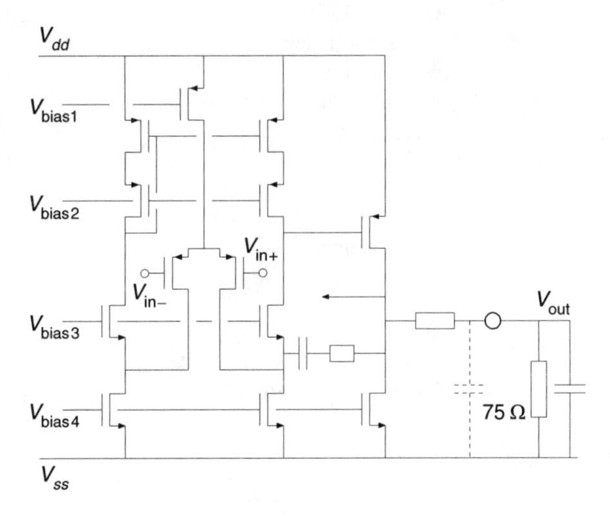

Figure 14.26 Folded-cascode opamp circuit used for buffer.

In the next section Table 14.8 summarizes performance that is more extensively reported for an older technology in [26]. Figure 14.27 shows a photograph of a triple 10-bit converter.

A Comparison. Table 14.8 compares the specifications of the two 10-bit video D/A converters.

Remarkable differences are the differential linearity error and the distortion. The DNL error in the current D/A is a direct effect of the current source mismatch, especially between the coarse and fine sections. In the voltage D/A converter this problem is circumvented by a fully segmented approach: 1024 resistors in series are used. The consequences are of course seen in a larger area. The harmonic distortion has different origins in the two converters: In current D/A converters the nonlinear behavior of the large output diffusion node and the output transconductance of the current sources is important. The distortion in the voltage D/A converter is caused by the limited performance of the driver stage in the output buffer. The modulation of the switch resistance in the resistive D/A is very effectively canceled by the ladder organization, while the reduced swing scheme of the current source switching limits the switch distortion in the current D/A.

The dynamic behavior of the D/A converter is determined by the output pole; in the current D/A converter this is the dominant pole with a 25-pF/75-Ω load. The buffer of the resistive D/A sees this pole as its second pole because the internal Miller compensation is the dominant pole. The buffered output is consequently slower than the current output, which is seen in differences in the rise/fall time. In most systems the values reported here will be sufficient. The minimum value for the output code is

TABLE 14.8 COMPARISON OF MEASURED SPECIFICATIONS OF LADDER D/A CONVERTER AND CURRENT D/A CONVERTER

Type	Ladder D/A	Current D/A
Process	1.0 μm CMOS	1.0 μm CMOS
DC resolution	10 bits	10 bits
Sample frequency	> 100 MHz	> 100 MHz
Area 1 μm CMOS	1.05 mm^2	0.7 mm^2
Differential linearity error	< 0.1 LSB	< 0.6 LSB
Integral linearity error	< 0.35 LSB	< 1 LSB
Glitch energy	100 ps-V	100 ps-V
Rise/fall time (10–90%)	4 ns	1 ns
Settling time (1 LSB)	20 ns	5 ns
Signal bandwidth (−1 dB)	20 MHz	> 20 MHz
Minimum power supply (THD > 40 dB)	3 V	3 V
Output in 75 Ω	1 V	1 V
Output in minimum code	100 mV	< 1 mV
Average current (50 MHz, 75 Ω)	10 mA	15 mA
Average current (50 MHz, 2 × 75 Ω)	10 mA	28 mA
THD $f_{signal} = 1$ MHz, $f_{clock} = 27$ Ms/s	58 dB	60 dB
THD $f_{signal} = 5$ MHz, $f_{clock} = 100$ Ms/s	50 dB	44 dB

Note: Both converters are loaded with 75 Ω and 25 pF.

Figure 14.27 Photograph of a triple 10-bit D/A converter intended for YUV (see Chapter 15) conversion in digital television.

limited for a buffer because a minimum saturation voltage is needed for the output stage.

The difference in power dissipation is less pronounced because there is more overhead for ladders and biasing in a voltage-domain D/A. Even so, a factor of 1.5–2.8 remains. Voltage-domain D/A converters route about 75% of their average current into the 75-Ω output load; with those (system-determined) output loads the potential for further power reduction seems to be low on an implementation level.

A high-speed D/A converter can be chosen if system requirements are mapped onto these specifications.

14.4.3. High-Speed Analog-to-Digital Converters

The Comparator. Random-offset reduction is necessary to achieve 8-bit DNL performance. In most offset reduction schemes the offset + signal and the offset are determined at different points in time, so at least one must be stored in a capacitor. References [6] and [25] describe a well-accepted method to perform offset reduction with the aid of a capacitor at the input of the comparators, which switches between input and reference [see Fig. 14.5(b)]. The disadvantage of this method is that the high capacitive input load requires a low-ohmic ladder (300 Ω) and poses high demands on the circuitry for driving the A/D converter. The single-sided input capacitor has a large parasitic capacitance to a bouncing substrate in a mixed-signal chip. The inherent capacitive input voltage division attenuates the signal. The basic inverter-type comparator has a poor power supply rejection [27], which affects the performance if power supply bouncing occurs between coarse and fine cycles. These comparator design points have to be solved for embedded operation.

Figure 14.28 shows the comparator that is used in the full-flash A/D converter (discussed in the next section). Then two other A/D converters are discussed, and use is made of modified comparators that allow larger signals to be stored.

With the exception of input and ladder terminals, the design is fully balanced and the PMOST current sources allow a good power supply rejection ratio (PSRR). The design consists of an input stage N_1, from which the signal is fed into an S/H stage N_2, S_1, C. Comparison involves three cycles: sampling, amplification, and latching. During the sampling phase switch S_1 is conducting and $V_{in} - V_{ref} + V_{off}$ is stored on both capacitors. These capacitors are grounded on one side and do not suffer from parasitic coupling to substrate. Here, V_{off} represents the sum of all the offsets in the comparator. During the sampling phase the negative conductance of latch stage N_3 is balanced by the positive conductance of the load stage N_4. Their combination acts as an almost infinite impedance, which is necessary for good signal + offset storage. The latch stage has twice the W/L of the load stage, but its current is only half due to the current mirror ratio N_5, N_6. For large differential signals the effect of the factor 2 in W/L of N_3 and N_4 becomes important: The conductance of N_3 reduces at a much higher rate that of N_4, so the effective impedance of N_3 and N_4 collapses. The large-signal response is consequently improved by the reduced time constant. The feedback of N_2 via switches S_1 allows a 150-MHz bandwidth, resulting in a high-quality S/H action.

After the sampling phase, the switch S_0 at the input connects to the reference voltage, effectively disconnecting the common input terminal from the comparator.

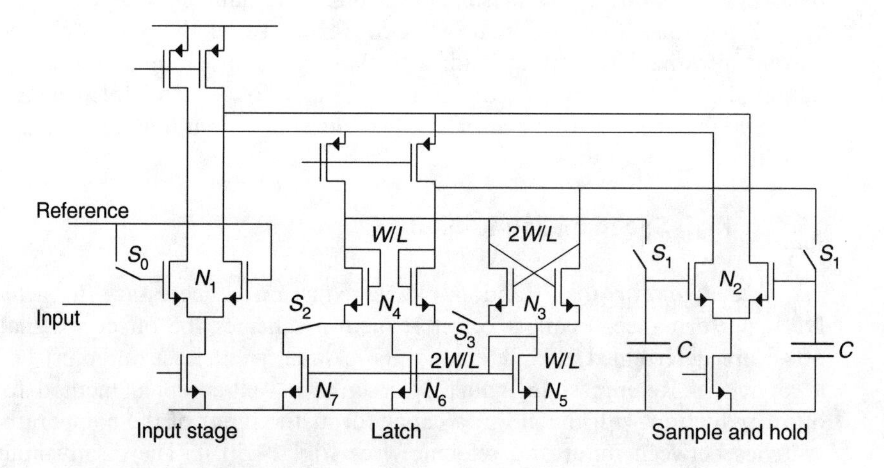

Figure 14.28 Schematic of basic comparator; switches shown for amplification phase.

As switches S_1 are disconnected, the S/H stage will generate a current proportional to $V_{in} - V_{ref} + V_{off}$, while the input stage and the rest of the comparator generate only the part proportional to V_{off}. The (differential) excess current is almost free of offsets and will be forced into the load-latch stage N_3 and N_4. Switch S_2 is made conductive, which increases the conductance of N_4 and decreases the conductance of N_3; the gain of N_2 on N_3 and N_4 is now about 8 for small differential signals.

Finally, S_3 is made conductive; the current now flows in a 2 : 1 ratio into N_3 and N_4, thereby activating the latch operation. The latch decision is passed on to the decoding stage and a new sample can be acquired. Remaining random offsets (0.4 mV) are caused by limited gain during the sampling phase, charge dump of S_1, and the difference in matching contributions of N_3 and N_4 in the sampling and amplification phases.

The comparator described in this section forms the basis of three types of converters: an 8-bit full-flash A/D converter, a 10-bit successive-approximation A/D converter, and the so-called multistep A/D converter, which combines a 3-bit full flash with a 5-bit successive approximation in a multiplexed approach.

Full-Flash A/D Converter. The full-flash A/D converter has been subdivided into 8 sections of 32 comparators; each section uses local 5-bit Gray-coding for the LSBs. This scheme allows fast and efficient (2 MOS/comparator) decoding. The 32nd, 64th, . . . comparators in a 3-bit MSB A/D converter decide which section output will be passed to the data rails.

The ladder structure is identical to that described earlier under Resistor String D/A Converter: a two-ladder structure provides a good-quality reference voltage. Due to the fact that the comparator uses an internal S/H, the input load and reference ladder load are limited to parasitic charges of the input switch. The reference supply current and the decoupling required in this design are both small. Care has been taken to avoid delay skew between the on-chip-generated clock scheme and the input signal; the generation of bias voltages for 255 comparators and the power distribution also requires careful design and layout. The main characteristics of the A/D converter are summarized in Table 14.9.

Figure 14.29 shows the measured SINAD in decibels; the ENOB has been plotted on the right axis. The converter achieves over 7.5 bits at low signal frequencies and 7.4 bits at 4.43 MHz and 25 Ms/s.

In order to test the substrate immunity, a 17.7-MHz, 200-mV_{pp} pulse wave was applied to the substrate of a 13.5 Ms/s running A/D converter–D/A converter combination. The resulting 4.2-MHz (17.7–13.5) disturbance was 45 dB below the 4.4-MHz converted signal at the D/A converter output. The A/D converter was experimentally used on several chips [28–31].

Successive-Approximation A/D Converter. Basically, the successive-approximation technique consists of comparing the unknown input voltage with a number of precise voltages generated by a D/A converter by means of a single comparator (see Fig. 14.30). As the input voltage may not change during the comparisons, there has to be a S/H circuit in front of the actual A/D. The offset-compensated comparator described in the previous section has a built-in S/H stage

TABLE 14.9 SPECIFICATIONS OF THREE A/D CONVERTERS

A/D Converter	Full Flash	Successive Approximation	Multistep
Resolution	8 bits	10 bit	8 bit
Sample rate	25 Ms/s	2 Ms/s	30 Ms/s
Differential linearity	0.6 LSB	0.5 LSB	0.5 LSB
Integral linearity	0.6 LSB	1 LSB	0.6 LSB
ENOB at input 4.43 MHz	7.4	8.5 (1 MHz)	7.4
SINAD (4.43 MHz)	46 dB		46 dB
Signal to distortion (2–5 harmonics, 4.43 MHz)	>52 dB		>52 dB
Input bandwidth (1 dB)	>70 MHz	20 MHz	70 MHz
Input signal swing	2 V	1.5 V	1.6 V
Ladder resistance	1200 Ω	4800 Ω	1200 Ω
Active area	2.8 mm^2	1.2 mm^2, 0.4 mm^2 (8 bits)	1.1 mm^2
Technology		0.8–1 μm (1 PS, 2 Al)	
Current A/D	55 mA	3 mA	13 mA
Current/comparator	200 μA	500 μA	200 μA
Number of comparators	256	1	56

Note: All converters are based on comparator shown in Figure 14.28. PS = polysilicon; Al refers to aluminum interconnect layer.

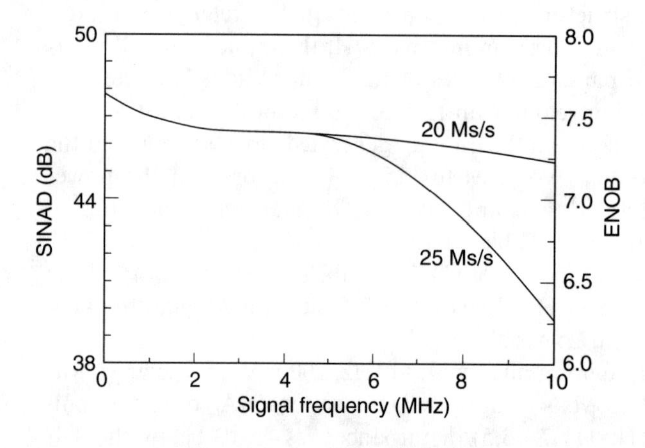

Figure 14.29 The SNAD at 20 and 25 Ms/s as a function of the input frequency.

and can therefore be used for comparison as well as for storage of the input signal. The successive approximation (SA) A/D converter realized consists of one offset-compensated comparator, a 10-bit D/A converter, and some control logic for controlling the D/A converter and storing the intermediate results. The converter requires 11 clock cycles for complete conversion. Therefore, the actual sampling frequency is limited to about 1–2 Ms/s. Inherent to the architecture, the

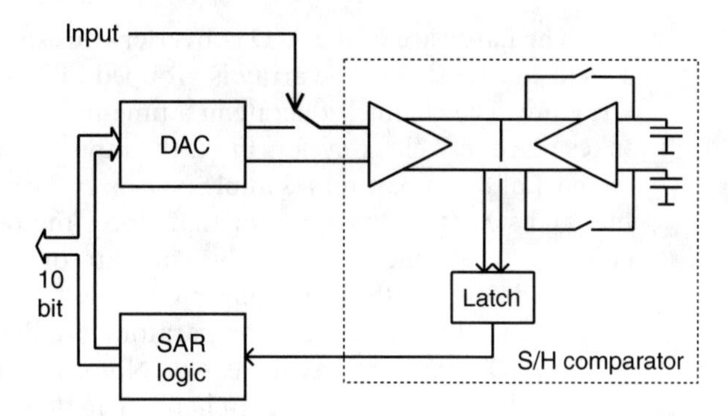

Figure 14.30 Successive-approximation A/D converter.

DNL of the A/D converter is good. A disadvantage of the converter is the relatively small signal input range (or the risk of INL errors in the case of large signal swings), which is a consequence of using long tail pairs at the input of the comparator.

The successive-approximation A/D converter design is used in servo applications and microcontroller inputs. Specifications are given in Table 14.9.

Multistep A/D Converter.[6] The multistep A/D (Fig. 14.31) is an 8-bit converter based on a technique involving a combination of successive approximation, flash, and multiplexing. The 3 most significant bits of the conversion are determined by means of a flash conversion; the remaining 5 bits are realized through successive approximation. Multiple time-interleaved signal paths have been used to increase the maximum sampling frequency.

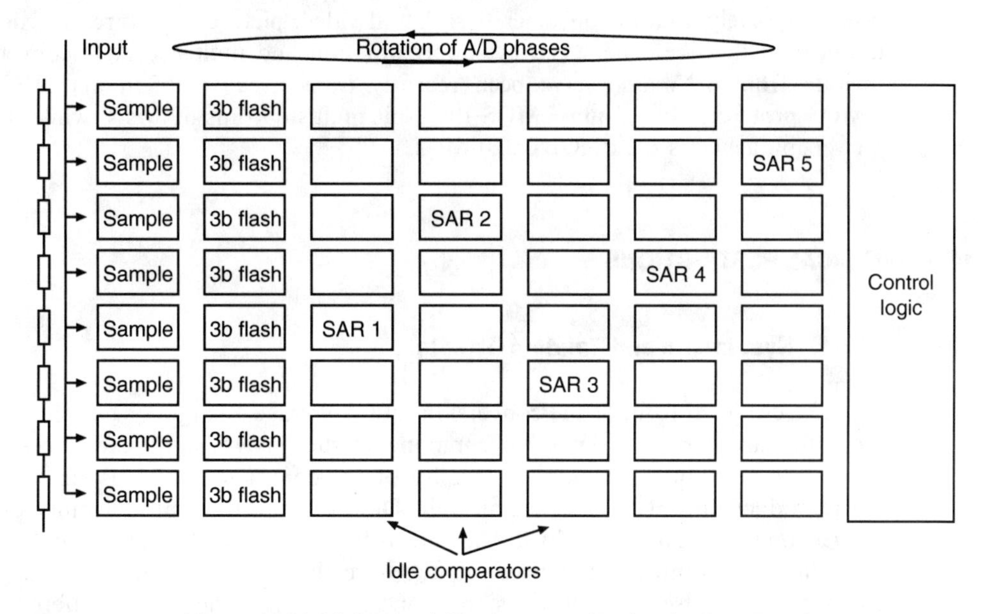

Figure 14.31 Multistep A/D converter. (Design by J. v. Rens.)

[6]This section has been contributed by Jeannet v. Rens.

The hardware of the A/D converter consists of an array of 56 comparators with a built-in S/H stage. The array is grouped in 7 channels of 8 comparators in a flash structure. The channels operate in a time-interleaved manner. The actual conversion takes place in 7 clock cycles. First, the input signal is sampled by and stored in one channel of 8 comparators (sample phase). A flash decision generates the three coarse bits and selects the comparator that stores the replica of the unknown input signal closest to a reference voltage. This comparator is used in a successive-approximation loop to determine the remaining bits, while the other comparators are idle.

Due to the use of eight comparators in a flash structure, the signal input range of the A/D converter has increased. Note that use of parallel signal paths can be successful only if the different channels match very well. Offset, gain, and timing mismatches between multiple channels give rise to fixed patterns which manifest themselves as spurious harmonic distortion in the frequency domain. The effect of offset is minimized by the use of the previously described offset-compensated comparators. Gain mismatch is minimized by the use of a common resistor ladder D/A converter and timing mismatch by the use of a master clock which determines the sampling moments of all the channels (see Fig. 14.32).

A Comparison. Table 14.9 shows the main specifications of the three A/D converters. Basic to all converters is the comparator of Figure 14.28, in which signal speed and accuracy meet with current. The decisive factor for the power comparison is the comparator current. In the successive-approximation design this current is higher because this comparator has to handle a larger signal span. The lower kick-back of the single comparator in the successive-approximation A/D also makes it possible to reduce the ladder impedance. All the converters have been extensively used in consumer ICs: digital video, picture-in-picture, instrumentation, and the like. The successive-approximation and multistep A/D converters operate from 3.3 V and have been recently transferred to 0.5-m and 0.35-μm CMOS processes. In 0.5-μm CMOS the 8-bit multistep runs 50 Ms/s, while the 9-bit version achieves 8.2 ENOB at 20 Ms/s.

14.5. PHYSICAL RESTRICTIONS

14.5.1. Systematic and Random Offsets

The electrical form of the conversion unit is voltage, charge, or current. These quantities are derived from basic components such as resistors, capacitors, and transistors and units of time. The quality of the conversion depends strongly on the reproducibility of the conversion unit. Deviations in the reproducibility of the conversion unit are described by "systematic offsets" and "random offset" or "(mis)-matching": unwanted differences in the effective value of equally designed components. Offset is defined as the non-zero-mean value of the difference in value between many pairs of components, while random matching refers to the stochastic spread σ (see Fig. 14.33).

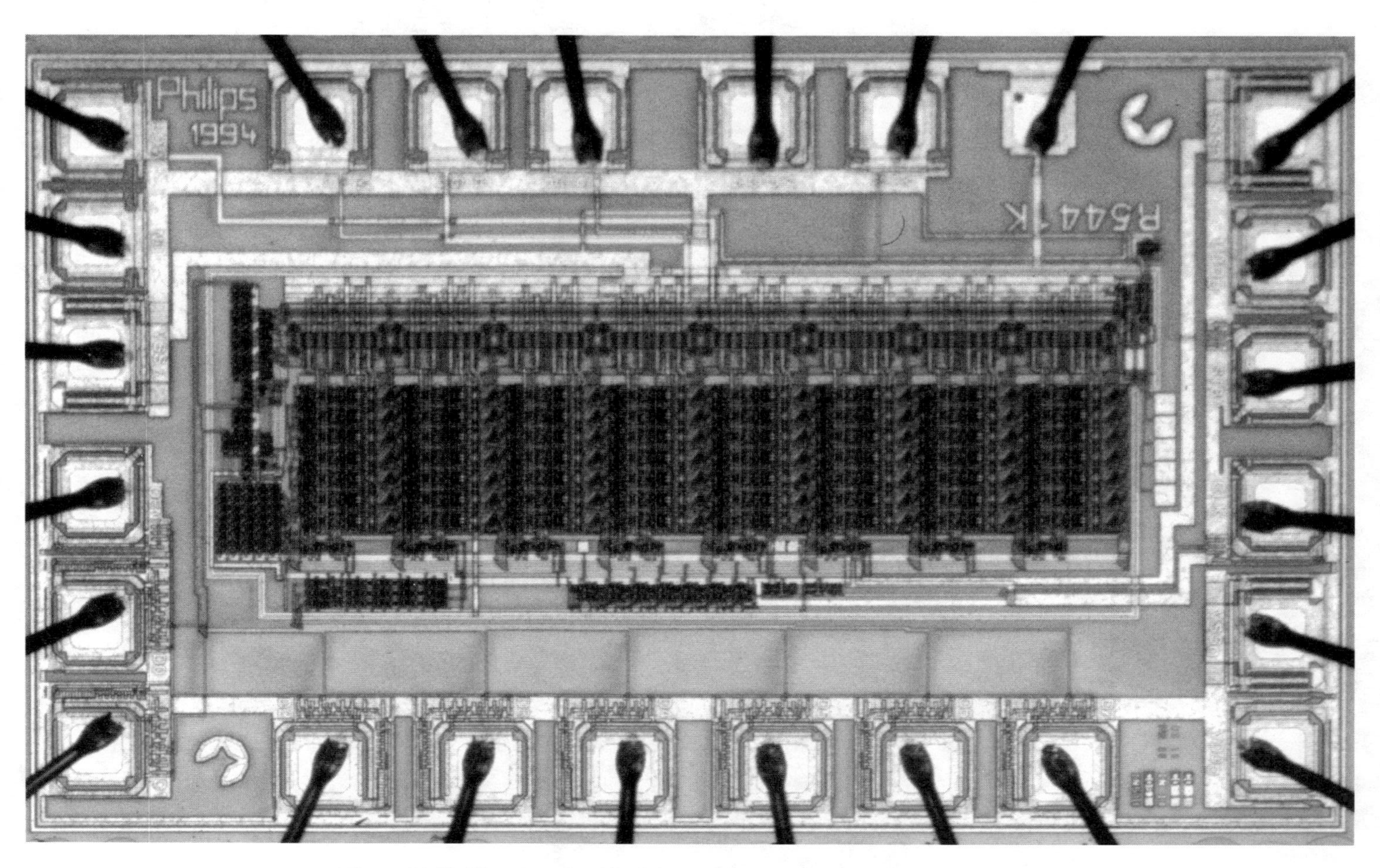

Figure 14.32 Photograph of multistep A/D converter. (Design by J. v. Rens.)

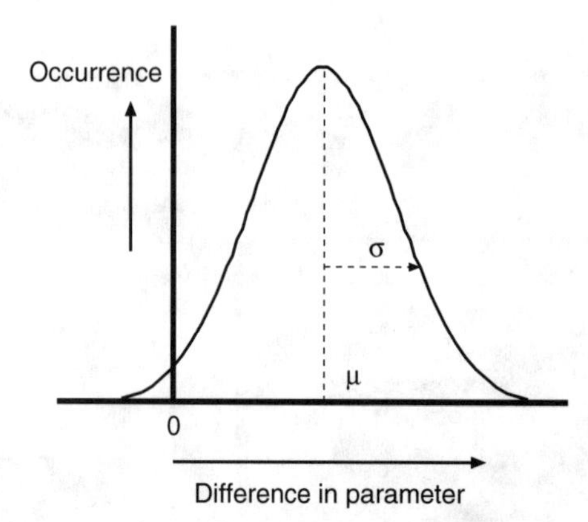

Figure 14.33 Offset μ and random matching σ of differences in parameters of two identically designed components.

Systematic and random offsets in the conversion process are caused by deviations in the conversion unit due to the fabrication process, deviations in the electrical conditions during use, or unequal timing moments.

For electrical matching of the conversion unit the voltages on all the elements must all be the same (affected by voltage drops in power lines, leakage currents in diodes), the substrate coupling must be the same, parasitical components must be matched, and so on. Electrically derived effects must also be considered (e.g., heat gradients due to dissipation).

Especially in high-frequency A/D conversion the conversion units are based on the reproduction of physical structures, while low-bandwidth converters rely strongly on timing accuracy. Systematic offsets are here due to lithographic and chemical effects in the fabrication process. Examples are:

- The proximity effect. The line width is enlarged by diffused light from neighboring structures. The standard procedure is to arrange dummy structures at distances of up to 20–40 μm.
- Gradients in doping, resistivity, and layer thickness: Although structures tend to decrease in dimensions, situations may occur in which equality is required in a distance on the order of 1 mm. In CMOS thresholds may deviate up to 5mV over this distance. Resistivity gradients may moreover reach the percentage level in this distance. Common centroid structures are used to reduce these effects.
- Patterns in one layer may affect the patterning in other layers. During the spinning of the resist fluid, resist may accumulate against altitude differences (from previous patterns) on a partly processed wafer. This effect results in circular gradients and is therefore often not recognized as a systematic offset.
- During the ion implantation steps the implantation beam is tilted about 5°–8° in order to prevent too deep penetration of the ions into the lattice; this effect is called "channeling." As a result of this nonperpendicular implantation, source and drain diffusions may be asymmetrical: One may extend further underneath the gate than the other. In order to prevent

inequalities in currents or overlap capacitors, the directions in which the MOS currents flow must be chosen run parallel, not rotated or antiparallel.

Systematic deviations will have to be identified and measures for overcoming them will have to be found in extensive study of the fabrication process involving process engineers.

14.5.2. Component Matching

Many forms of systematic offsets can be circumvented at the expense of design effort. Random component offsets are more difficult to cope with via design. The basic problem in random offset concerns the local variations that occur in the fabrication process. Granularity occurs on linear structures (resistors, capacitors) in the form of edge roughness of polysilicon lines. In MOS circuit design the effect of two-dimensional local variations is the dominant cause of random offset in transistors.

The basic principle behind random offset in the threshold definition of MOS transistors is the number of fixed charged atoms in the active and depletion regions. These charged atoms (dopants, dislocations, oxide charge, etc.) are implanted, diffused, or generated during the manufacturing process, but not in an atom-by-atom controlled manner. The average value is controlled by implantation dope levels or average substrate dopes. The *actual* number of carriers in a particular depletion region may well differ from this average. In this analysis it will be assumed that the presence of dopants is governed by a Poisson process: the presence or absence of an individual atom is independent of the presence of other charges. During the operation of a MOS transistor with well-controlled voltages the balance of all charges and potentials will determine the channel charge in that transistor. If that channel charge varies from one transistor to another due to a varying number of depletion or implanted charges, the threshold voltage will vary accordingly. The threshold voltage is given by

$$V_T - V_{FB} = \frac{Q_B}{C_{ox}} = \frac{qN_x x_d}{C_{ox}} = \frac{\sqrt{2q\varepsilon N_x \phi_b}}{C_{ox}} \tag{14.23}$$

where ε is the permittivity, N_x the dope concentration, and ϕ_b the built-in potential. Here, x_d is the depletion width $\sqrt{2\varepsilon\phi_b/qN_x}$. If the depletion area of a transistor (see Fig. 14.34) is defined by a width W, length L, and depletion region depth x_d, then the volume of the depletion region is (in first order) $W \times L \times x_d$. Different impurities are active in this region—$N_x \approx 10^{16}\,\text{cm}^{-3}$—$N_x$ containing acceptor and donor atoms from the intrinsic substrate dope, the well, threshold adjust, and punch-through implantations.[7] In the variance analysis it is important to note that the total number

[7]For ease of understanding only a uniformly distributed dopant is used here; more complicated distributions must be numerically evaluated.

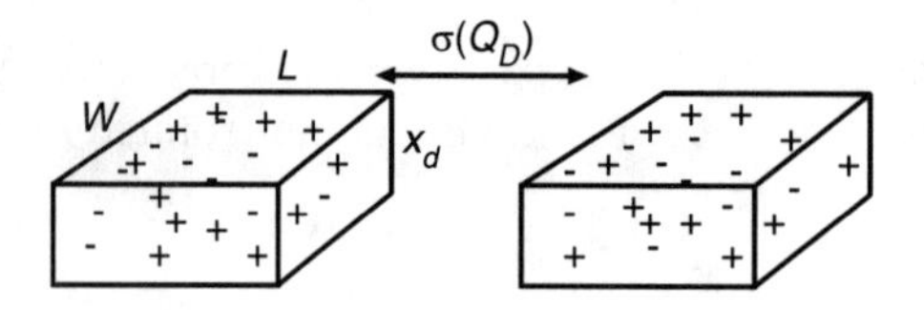

Figure 14.34 Depletion region of two MOS transistors.

of charged atoms must be considered, not the net resulting charge. The standard deviation of this amount of charge in a Poisson process is now approximated by

$$\sigma(WLx_dN_x) = \sqrt{WLx_dN_x} \tag{14.24}$$

The threshold variance can now be derived from Eq. (14.23) by considering that

$$\sigma_{\text{single}}(V_T) = \sigma(WLx_dN_x)\,\frac{\delta(V_T)}{\delta(WLx_dN_x)} \tag{14.25}$$

As matching usually occurs between pairs of transistors, the variance of the *difference between two transistors* is

$$\sigma(V_T) = \sqrt{2}\sigma_{\text{single}}(V_T) = \frac{qt_{\text{ox}}\sqrt{2N_xx_d}}{\varepsilon_{\text{ox}}\sqrt{WL}} = \frac{A_{VT}}{\sqrt{WL}} \propto \frac{t_{\text{ox}}\sqrt[4]{N_x}}{\sqrt{W/L}} \tag{14.26}$$

The linear relation between $\sigma(V_T)$ and $1/\sqrt{\text{area}}$ is well known [19]. Figure 14.35 shows the measured dependence for $\sigma(V_T)$ versus $1/\sqrt{\text{area}}$.

In order to test the hypothesis that depletion charge is the dominant factor in threshold matching, Table 14.10 compares the A_{VT} coefficients as measured and as calculated using the above formula. The quantity N_xx_d was derived in process simulation which was tuned with accurate C/V measurements [7, 32]. In the case of three out of the four coefficients, the fit is good: The deviation of the 0.6-μm NMOST is not yet understood. The large PMOST coefficient present in 0.8-μm PMOSTs is caused by the compensating implants: the NMOST and PMOST thresholds adjust and *n*-well implants are used. The quantity $N_x = (N_a + N_d)$ is relevant for matching, while control of the net value $(N_a - N_d)$ determines the threshold. In the 0.6-μm PMOST a twin-well construction with a single-well implant was used.

In Figure 14.35 the matching of the current factor is shown. The proportionality factor for the current factor $\beta = \mu C_{\text{ox}} W/L$ is defined as

$$\frac{\sigma(\beta)}{\beta} = \frac{A_\beta}{\sqrt{W/L}} \tag{14.27}$$

The relative matching of the current factor is also proportional to the inverse square root of the area. In [19] it was assumed that the matching of the current factor is determined by local variations of the mobility.

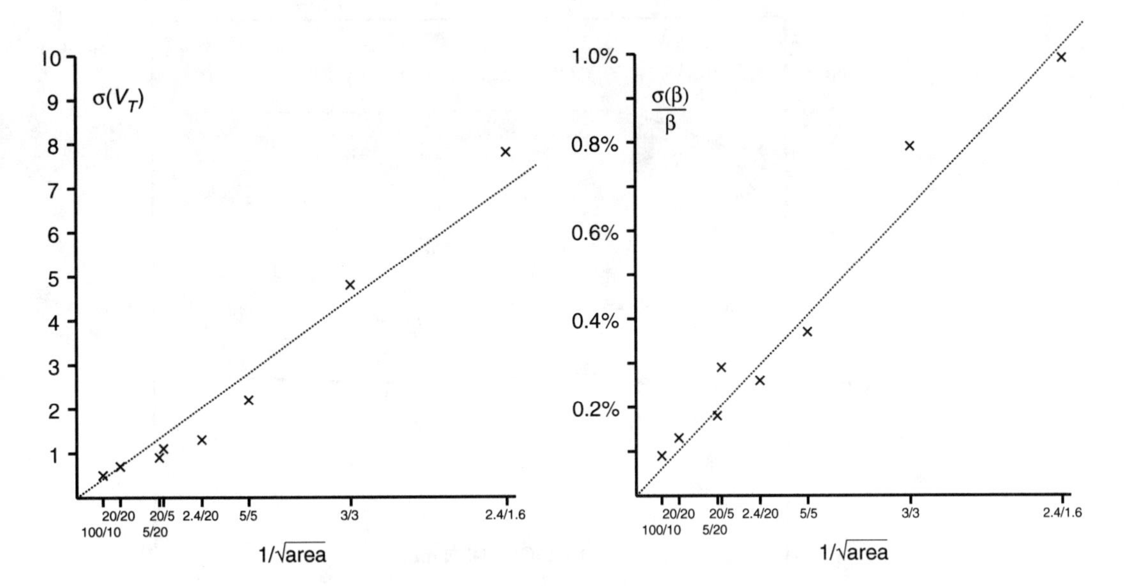

Figure 14.35 Standard deviation of NMOST threshold and relative current factor vs. inverse square root of area for 17.5-nm gate oxide process.

TABLE 14.10 COMPARISON OF MEASURED AND CALCULATED THRESHOLD MISMATCH COEFFICIENTS

	A_{VT} Measured	A_{VT} Calculated
0.8-μm data		
NMOST	10.7 mV μm	10.6 mV μm
PMOST	18.0 mV μm	18.6 mV μm
0.6-μm data		
NMOST	11.0 mV μm	7.5 mV μm
PMOST	8.5 mV μm	8.6 mV μm

Low Voltage and Matching. The analysis in the previous section on the origins of MOST matching allows us to analyze the development of matching in various processes. Figure 14.36 shows the development of power supply voltage and matching over several process generations. A 1/1 transistor was chosen, which in fact corresponds to the numerical value of the factor A_{VT} in the above analysis. During the process development from 2.5-μm processes to 0.35-μm processes, a slight matching improvement was observed. Equation (14.26) predicts a reduction in the matching factor proportional with the gate oxide thickness, which is confirmed by the measurements in Figure 14.36 and, for example [33, 7].

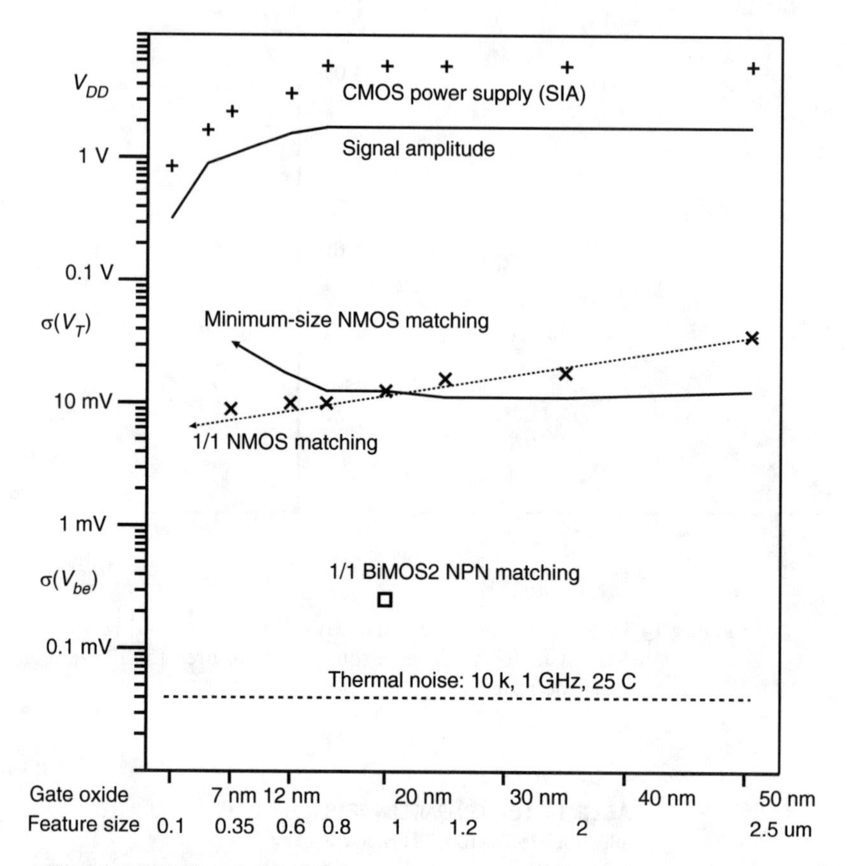

Figure 14.36 Development of power supply voltage and measured NMOS threshold matching factor A_{VT} through various process generations.

The power supply remained fixed at 5 V for many process generations, which led to signal swings of between 1 and 2.5 V. The part of analog CMOS performance that relied on the ratio between signal and component matching (high-speed converters) could improve in future process generations.

At the level of 0.6-μm CMOS processes the maximum electrical fields in intrinsic transistors were reached, both the vertical gate oxide field and the lateral field governing the charge transport. For this reason, and in order to reduce power consumption, efforts started to concentrate on lowering the power supply voltage. On the other hand, the need to tailor the internal fields in the transistor has led to more and higher implantation doses. As can be expected from the theoretical background, the threshold matching factor A_{VT} tends to increase in deep submicrometer processes, becoming especially pronounced when minimum dimensions are used. Shrinking of analog blocks in submicrometer processes is clearly an important issue.

The reduction in the signal-to-matching coefficient ratio in submicrometer CMOS will necessitate changes in the system or technology. In order to perform high-quality data conversion, data converters may shift from the digital CMOS domain to bipolar CMOS processes. Another possibility would be to use analog options in the technology (see Section 14.3.1).

Limits of Power and Accuracy. One of the main questions in low-power design is the ultimate conversion limit. There is no mathematical evidence showing that zero-power mapping of analog values on a discrete amplitude scale would not be possible.

In physics, however, a lower limit can be derived from quantum-mechanical effects, concerning the minimum number of electrons required to represent a bit. Another limit is given by Dijkstra [4, 34] for ΣΔ A/D converters; his analysis assumes that the thermal noise on the input impedance or transconductance is dominant. This approach results in an energy limit based on the product of SNR and thermal kT noise. These limits are, however, four to five decades removed from practical realizations. This is partly due to the fact that much "overhead" power has to be incorporated in real designs: Transconductances in MOS need large amounts of current, parasitics have to be overcome, and safety margins for all kinds of unexpected variations have to be accounted for.

In this section an approximation for power consumption in parallel-circuit structures as in high-speed A/D converters will be derived for conversion in general and A/D conversion in particular.

The starting point is the observation that component-random variations determine relevant specifications in circuits in which the signal passes over different pieces of hardware. This may happen in multiplexed circuits or in circuits in which the signal path is level dependent, as in full-flash A/D converters. The component variations between the various paths will result in unwanted signals (fractions of sample rates, fixed pattern noise, etc.). These component variations will decrease when the area of a MOST is increased. On the other hand, the loading of the signal will also increase when gate areas increase:

$$\begin{aligned} C_{\text{gate}} &= WLC_{\text{ox}} \quad \text{(capacitive load)} \\ \sigma(V_T) &= \frac{A_{VT}}{\sqrt{W/L}} \quad \text{(threshold variance)} \end{aligned} \tag{14.28}$$

The voltage uncertainty on the gate capacitance can be described as an energy term [5, 36],

$$E_{\sigma VT} = C_{\text{gate}}\sigma^2(V_T) = C_{\text{ox}}A_{VT}^2 = 4.5 \times 10^{-19}\ \text{J} \tag{14.29}$$

which is independent of the transistor size and corresponds to about $100kT$ at room temperature. This energy can be seen as the energy required to toggle a latch pair of transistors in the metastable condition into a desired position with a 1σ certainty. In circuits with parallel data paths unwanted signals resulting from component mismatch may hence dominate over more general noise mechanisms such as kT noise to a great extent in the voltage domain.

In A/D converters, mismatch manifests itself as noise but is also observable in other specification points. The most critical is differential nonlinearity, because this effect relates to the maximum error in the parallel structure, while noise is only a power average value over all the error terms.

The charge standard deviation [Eq. (14.24)] is linked to the A/D converter's DNL performance by deriving from it a minimum LSB size for the input signal. In a parallel structure a multitude of MOS devices will be involved in the decision process. A safety margin must be employed that an LSB change in the input signal is always detected. This safety margin is defined via an $N(0, 1)$ normal distribution as α. A sufficiently low probability is obtained at $\alpha = 7, \ldots, 10$. So the resulting minimum LSB size is

$$Q_{\mathrm{LSB}} = \alpha \times \sigma(Q_d) = \alpha q \sqrt{WLx_d N_a} \tag{14.30}$$

The minimum LSB size combined with the signal speed results in a current. It is assumed that no slewing of the comparator stage is allowed and the current is delivered in class A operation. In converter terminology, the DNL performance must be reached at a signal frequency f_{signal}. The minimum amount of time required for an input signal to change V_{LSB} is t_{LSB}. A distinction can be made between the time-continuous mode and the S/H mode:

$$\begin{aligned} t_{\mathrm{LSB}} &= \frac{V_{\mathrm{LSB}}}{\delta V_{\mathrm{signal}}/\delta t} \quad \text{(time-continuous mode)} \\ t_{\mathrm{LSB}} &= \frac{V_{\mathrm{LSB}}}{\Delta V_{\mathrm{SH}}/\tau_{\mathrm{SH}}} \quad \text{(S/H mode)} \end{aligned} \tag{14.31}$$

$$\begin{aligned} t_{\mathrm{LSB}} &= \frac{1}{2^N \pi f_{\mathrm{signal}}} \quad \text{(time-continuous mode)} \\ t_{\mathrm{LSB}} &= \frac{1}{2^N N \ln(2) f_s} \quad \text{(S/H mode)} \end{aligned} \tag{14.32}$$

In the context of this analysis the results of both modes are comparable, so the analysis will proceed with the time-continuous result. The current needed for supplying Q_{LSB} at the steepest point of an input signal at a frequency f_{signal} is

$$i_{\max} = \frac{Q_{\mathrm{LSB}}}{t_{\mathrm{LSB}}} = 2^N \pi f_{\mathrm{signal}} \alpha q \sqrt{WLx_d N_a} \tag{14.33}$$

When this current is calculated for an 8-bit resolution converter and a 10-MHz signal frequency ($N = 8, f_{\mathrm{signal}} = 10^7$ Hz, $\alpha = 10$, $N_a = 10^{16}\ \mathrm{cm}^{-3}$, $W = L = x_d = 1\ \mu\mathrm{m}$), this results in $i_{\max} = 1.3\ \mu\mathrm{A}$ per transistor.

Via multiplication with V_{DD} and rearrangement of terms, this formula can be linked to the power per resolution and bandwidth of Table 14.2. The proportionality constant is then

$$\frac{\text{Power}}{2^N \text{BW}} = \pi \alpha q \sqrt{W L x_d N_a} \tag{14.34}$$

which amounts to about 10^{-3} pJ per relevant transistor. On the assumption of some 4 to 6 matching critical transistors per comparator and 60–100 comparators per high-speed A/D converter, this (very rough) approximation results in 0.5 pJ, which is about two orders removed from the realized converters.

Figure 14.37 shows a graphical representation of the current consumptions of some comparators in recent A/D converters. The drawn line corresponds to a comparator with a sampling rate of 30 Ms/s, with 4 critical transistors.

Closer examination of this discrepancy shows that the following factors contributed to this gap:

- It was assumed that only the intrinsic transistor was charged. In the design the parasitic capacitors require roughly the same amounts of charge.
- More effects than the depletion charge alone contribute to the uncertainty of an LSB: W, L dependencies, mobility variations, and so on.
- The A/D converters are designed to operate under worst-case circumstances: Temperature, power supply, and process variations require an overdesign of about 2–3 times.
- In the analysis minimum power consumption was realized by using minimum transistors. The consequence, not accounted for, is that the LSB size in volts, $Q_{\text{LSB}}/WLC_{\text{ox}}$, and the corresponding input signal are high in amplitude. Because of this, additional current paths are required in a comparator.

Basically, offset compensation enables reductions in component variations (see [6, 23, 25, 29]). In practice, this technique does not cancel the offsets in the final decision element (the latch), but it reduces the random offset in the preamplifiers. The above analysis can also be used for this purpose because the basic dilemma

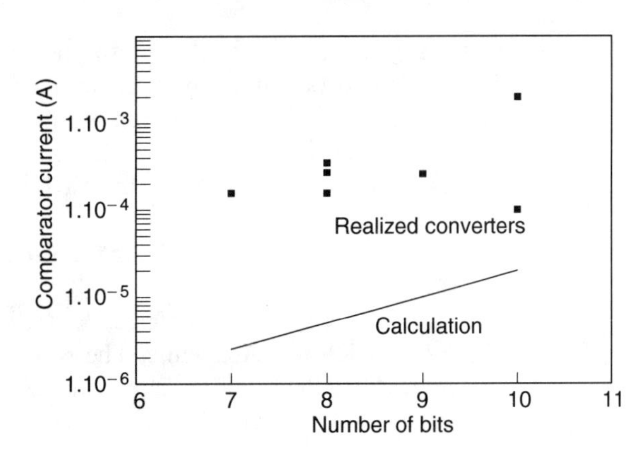

Figure 14.37 Comparison of power consumptions with derived approximation.

between accuracy and speed is still present. In other words: the above energy must be supplied during offset cancellation. Another point is the additional random offset that is introduced into the signal capacitors by the switches. The ratio of the switch gate capacitance and the signal storage capacitors is limited due to the fast settling requirement, which limits the offset reduction capabilities.

Real improvement may come from better intrinsic components, for example by means of trimming of sensitive stages or new silicon active elements (silicon on insulator?).

In the above analysis the basic problem of reaching accuracy at high speeds was located in the comparators. This implies that N comparators for an N-bit digital word is the minimum configuration for low-power A/D converters. Most designs employ 10 times more comparators. In the architecture the number of contributing transistors must be minimized. This can potentially be realized by means of (multiplexed) successive approximation. Another approach is to avoid the problems that parallel circuits introduce by going to single-comparator oversampling; then the trade-off with digital filtering becomes important again.

14.6. CONCLUSION

Several aspects of low power in CMOS data converters have been discussed. Essential for low-power consumption in general is that power is spent only in performing the required tasks. In low-power data conversion this goal is reached by tailoring the circuit to the needs of the signal, within the possibilities allowed by the technology and system. It is important to include all levels of design in the analysis: the system, the IC, the implementation, and the physics.

Special attention has been paid to some aspects that have so far been ignored in the literature: specification, clock strategy, the impact of options and tolerances in technology, and the position of the conversion. A new approximation to power in parallel structures is based on mismatch of components. In the implementation several trade-offs have been discussed with reference to practical circuits.

Acknowledgments

Many colleagues have contributed to the various topics presented. The author thanks A/D mentors Carel Dijkmans and Rudy vd. Plassche; co-workers Jeannet v. Rens, Pieter Vorenkamp, Kaarten Vertregt, and Johan Verdaasdonk; and students Adrie Waterman and Stefan Menten as well as many others in the Philips Research Laboratories and Philips Industries.

References

[1] R. van de Plassche, *Integrated Analog-to-Digital and Digital-to-Analog Converters*, Kluwer Academic, The Netherlands, 1994.

[2] P. Singer, "1995: Looking Down the Road to Quarter-Micron Production," *Semiconductor Int.*, vol. 18, pp. 46–52, Jan. 1995.

[3] E. Vittoz, "Low Power Low-Voltage Limitations and Prospects in Analog Design," in *Advances in Analog Circuit Design*, R. J. vd Plassche (ed.), Kluwer Academic, Eindhoven, The Netherlands, 1995, p. 3.

[4] E. Dijkstra, O. Nys, and E. Blumenkrantz, "Low Power Oversampled A/D Converters," in *Advances in Analog Circuit Design*, R. J. vd Plassche (ed.), Kluwer Academic, Eindhoven, The Netherlands, 1995, p. 89.

[5] E. J. van der Zwan and E. C. Dijkmans, "A 0.2 mW CMOS ΣΔ Modulator for Speech Coding with 80 dB Dynamic Range," *IEEE J. Solid-State Circuits*, vol. SC-31, pp. 1873–1880, 1996.

[6] K. Kusumoto, A. Matsuzawa, and K. Murata, "A 10 b 20 MHz 30 mV Pipelined Interpolating CMOS ADC," in *Int. Solid-State Circuits Conf. Dig. Tech. Papers*, vol. 36, pp. 62–63, 1993.

[7] M. J. M. Pelgrom and M. Vertregt, "CMOS Technology for Mixed Signal ICs," *Solid-State Electron.*, vol. 41, pp. 967–974, 1997.

[8] M. J. M. Pelgrom and M. Roorda, "An Algorithmic 15 bit CMOS Digital-to-Analog Converter," *IEEE J. Solid-State Circuits*, vol. SC-23, pp. 1402–1405, 1988.

[9] R. J. van de Plassche, "Dynamic Element Matching for High-Accuracy Monolithic D/A Converters," *IEEE J. Solid-State Circuits*, vol. SC-11, pp. 795–800, 1976.

[10] J. A. Schoeff, "An Inherently Monotonic 12 bit DAC," *IEEE J. Solid-State Circuits*, vol. SC-14, pp. 904–911, 1979.

[11] J. R. Naylor, "A Complete High-Speed Voltage Output 16-bit Monolithic DAC," *IEEE J. Solid-State Circuits*, vol. SC-18, pp. 729–735, 1983.

[12] D. J. Schouwenaars, E. C. Dijkmans, B. M. J. Kup, and E. J. M. van Tuijl, "A Monolithic Dual 16-bit D/A Converter," *IEEE J. Solid-State Circuits*, vol. SC-21, pp. 424–429, 1986.

[13] H. J. Schouwenaars, D. W. J. Groeneveld, and H. A. H. Termeer, "A Low-Power Stereo 16-bit CMOS D/A Converter for Digital Audio," *IEEE J. Solid-State Circuits*, vol. SC-23, pp. 1290–1297, 1988.

[14] D. W. J. Groeneveld, D. J. Schouwenaars, H. A. H. Termeer, and C. A. Bastiaansen, "A Self-Calibration Technique for Monolithic High-Resolution D/A Converters," *IEEE J. Solid-State Circuits*, vol. SC-24, pp. 1517–1522, 1989.

[15] J. C. Candy and G. C. Temes (eds.), *Oversampling Delta-Sigma Data Converters: Theory, Design and Simulation*, IEEE, New York, 1992.

[16] P. J. A. Naus and E. C. Dijkmans, "Multi-bit Oversampled ΣΔ A/D Converters as Front-End for CD Players," *IEEE J. Solid-State Circuits*, vol. SC-26, pp. 905–909, 1991.

[17] B. Nauta and A. G. W. Venes, "A 70 Ms/s 110 mW 8-b CMOS Folding and Interpolating A/D Converter," *IEEE J. Solid-State Circuits*, vol. SC-30, pp. 1302–1308, 1995.

[18] K. Bult et al., "A 170 mW 10 b 50 Ms/s CMOS ADC in 1 mm," in *Int. Solid-State Circuits Conf. Dig. Tech. Papers*, vol. 40, pp. 136–137, 1997.

[19] M. J. M. Pelgrom, A. C. J. Duinmaijer, and A. P. G. Welbers, "Matching Properties of MOS Transistors," *IEEE J. Solid-State Circuits*, vol. SC-24, pp. 1433–1440, 1989.

[20] P. Vorenkamp and J. P. M. Verdaasdonk, "A 10 b 50 MHz Pipelined ADC," in *Int. Solid-State Circuits Conf. Dig. Tech. Papers*, vol. 35, pp. 32–33, 1992.

[21] *Philips Data Handbook*, Philips Semiconductors, www.semiconductors.philips.com, 1997.

[22] T. Miki, Y. Nakamura, M. Nakaya, S. Asai, Y. Akasaka, and Y. Horiba, "An 80-MHz 8-bit CMOS D/A Converter," *IEEE J. Solid-State Circuits*, vol. SC-21, pp. 983–988, 1986.

[23] A. G. F. Dingwall and V. Zazzu, "An 8-MHz CMOS Subranging 8-bit A/D Converter," *IEEE J. Solid-State Circuits*, vol. SC-20, pp. 1138–1143, 1985.

[24] A. Abrial, J. Bouvier, J. Fournier, P. Senn, and M. Viellard, "A 27-MHz Digital-to-Analog Video Processor," *IEEE J. Solid-State Circuits*, vol. SC-23, pp. 1358–1369, 1988.

[25] N. Fukushima, T. Yamada, N. Kumuzawa, Y. Hasegawa, and M. Soneda, "A CMOS 40 MHz 8 b 105 mW Two-Step ADC," in *Int. Solid-State Circuits Conf. Dig. Tech. Papers*, vol. 32, pp. 14–15, 1989.

[26] M. J. M. Pelgrom, "A 10 b 50 MHz CMOS D/A Converter with 75 Ω Buffer," *IEEE J. Solid-State Circuits*, vol. SC-25, pp. 1347–1352, 1990.

[27] M. Haas, D. Draxelmayr, F. Kuttner, and B. Zojer, "A Monolithic Triple 8-bit CMOS Video Coder," *IEEE Trans. Consumer Electron.*, vol. 36, pp. 722–729, 1990.

[28] M. Vertregt, M. B. Dijkstra, A. C. v. Rens, and M. J. M. Pelgrom, "A Versatile Digital CMOS Video Delay Line with Embedded ADC, DAC and RAM," *ESSCIRC93*, Sevilla, Spain, 1993, pp. 226–229.

[29] M. J. Pelgrom, A. C. v. Rens, M. Vertregt, and M. B. Dijkstra, "A 25-Ms/s 8-bit CMOS A/D Converter for Embedded Application," *IEEE J. Solid-State Circuits*, vol. SC-29, pp. 879–886, 1994.

[30] G. Muller, "New Generation of an IDTV Chip Set," *IEEE Int. Conf. Consumer Electron. Dig. Tech. Papers* WPM14, vol. ICCE92, 1993.

[31] B. Murray and H. Menting, "A Highly Integrated D2MAC Decoder," *IEEE Int. Conf. Consumer Electron. Dig. Tech. Papers*, vol. TUAM4.3, pp. 56–57, 1992.

[32] M. J. v. Dort and D. B. M. Klaasen, "Circuit Sensitivity Analysis in Terms of Process Parameters," in *IEEE 1995 International Electron Devices Meeting, Technical Digest*, vol. 95, 37.3.1, 1995.

[33] T. Mizuno, J. Okamura, and A. Toriumi, "Experimental Study of Threshold Voltage Fluctuation due to Statistical Variation of Channel Dopant Number in MOSFETs," *IEEE Trans. Electron Devices*, vol. ED-41, pp. 2216–2221, 1994.

[34] E. J. Swanson, "Analog VLSI Data Converters—The First 10 Years," in *Proceedings ESSCIRC 95*, pp. 25–29, 1995.

[35] M. J. M. Pelgrom, "Low-Power High-Speed A/D Conversion," *ESSCIRC 94*, Low-Power Workshop, Ulm, Sept. 23, 1994.

[36] P. Kinget and M. Steyaert, "Impact of Transistor Mismatch on the Speed Accuracy Power Trade-Off,' *CICC96*, San Diego, May 1996.

Chapter 15

Low-Power Multiplierless YUV-to-RGB Converter Based on Human Vision Perception

Teresa H. Meng, Benjamin M. Gordon, and Navin Chaddha
Computer Systems Laboratory, Stanford University, Stanford, California 94305-4070

15.1. INTRODUCTION

Digital color images are represented as a set of three components representing the intensities of each of the three primaries: red, green, and blue (RGB). The color space obtained through combining the three colors can be determined by drawing a triangle on a special color chart with each of the base colors as an endpoint. Using the Commission Internationale d'Eclairage (CIE) [1] chart as a guideline, the National Television Standard Committee (NTSC) defines the transmission of color signals in a luminance and chrominance format called YIQ.

The YUV format, a variant of the YIQ format, concentrates most of the image information into the luminance (Y) and less in the chrominance (UV), allowing the chrominance to be specified less frequently than the luminance. Hence in most color image and video compression algorithms, only every other U and V elements in the horizontal direction are sampled (4 : 2 : 2 format). The missing elements are reconstructed by either interpolation or duplication.

The reduction in storage or data rate by using YUV over RGB is by a factor of 1.5. If the RB format is specified by 8 bits for each color, then each RGB pixel is represented by 24 bits, whereas after conversion and decimation, each YUV pixel is described by an average of 16 bits, 8 bits for the luminance and 8 bits for the chrominance per pixel.

Since most displays require RGB format while most image processing uses the more compact YUV format, a color conversion is required. Given the growing number of portable computer systems incorporating multimedia capabilities for displaying video data, a low-power implementation of the color converter would be very useful. Reductions in power consumption increase battery lifetime or decrease battery weight, which are key concerns in designing portable systems.

The actual conversion between the standard RGB format to the YUV format and vice versa is slightly different for digital signals than for analog signals. The digital conversion requires that if the RGB values are between 0 and 255, the YUV values should also be between 0 and 255. Written in a matrix form where all conversion coefficients are represented as unsigned characters, the inverse conversion from YUV to RGB is

$$\begin{bmatrix} R \\ G \\ B \end{bmatrix} = \begin{bmatrix} 1 & 0 & 1.402 \\ 1 & -0.34414 & -0.71414 \\ 1 & 1.772 & 0 \end{bmatrix} \begin{bmatrix} Y \\ U - 128 \\ V - 128 \end{bmatrix}$$

This YUV to RGB conversion requires four multiplications and six additions for each pixel in the image. In this chapter we describe a converter design that does not require any multiplication operations while retaining the color quality. We have approximated multiplications with shift and adds based on a human perception model [2] so that there is little perceived difference between the colors generated with multiplications and the colors generated using our approximation.

15.2. DISTORTION MEASURE

The efficiency of any color conversion algorithm depends on two criteria: implementation complexity and quality of the color-converted image. The first criterion is easy to evaluate, but the second criterion is difficult to quantify. Thus, we use a recently proposed objective distortion measure to determine image quality [2] for our study. This measure corresponds well to subjective assessment, and wherever necessary, the results have been verified with subjective tests.

The distortion model follows the human vision perception in that the distortion as perceived by a human viewer is dominated more by the compression error unrelated to the local features of the original image than by the compression error correlated with the local features [2]. As a result, we first separate the compression error signal, defined to be the difference between the pixels of the original and the compressed images, into two parts: the part of the error signal that is correlated with the local features of the original image and the part of the error signal that is uncorrelated with the local features of the original image. The splitting operation is performed by a two-dimensional adaptive filter using a modified version of the two-dimensional least mean square (LMS) algorithm [2]. Next, a logarithmic operation is performed on the absolute value of the errors. The presence of a logarithmic sensitivity in vision has been known for a long time, supported by the fact that light-sensitive neurons fire at a rate proportional to the logarithm of the incident energy on them. Next, a power operation squares the two parts of the error signal after the logarithmic stage. Finally, the two results are summed over the entire image. Since the human viewers are more sensitive to uncorrelated errors than correlated errors, the model gives more weight to the uncorrelated errors. This error model is used for each of the color components of the RGB color space. Finally the total error is obtained by weighting each component in the RGB space by its relative importance.

We would like to point out that this distortion measure is valid only for making binary decisions. In comparing two color-converted images, it accurately gives the ranking based on visual quality, but it does not quantify the distortion present in each image. However, this does not disqualify the distortion measure from being a design guideline, as even humans are unable to quantify something as abstract as visible distortion present in an image. Humans can only assign rankings, that is, binary decisions. In comparing images with this method, Chaddha and Meng [2] found that this distortion measure has a correlation coefficient of 0.92 with subjective results whereas the mean-square-error [and signal-to-noise ratio (SNR)] has a correlation coefficient of only 0.53.

15.3. COEFFICIENT SELECTION

Precise color conversion requires a significant amount of computation, as illustrated in the Brooktree Bt294 chip [3], which performs fixed-point multiplication supporting two different conversion matrices, color sample interpolation, dynamic rounding, and clipping. However, many of these features do not provide visually significant improvement in image quality, leaving room for a much simpler implementation for smaller silicon area and lower power.

First, color sample interpolation and dynamic rounding are discarded because of the little effect upon the final image quality. Instead of interpolation, the same UV values are used along with two horizontal Y values to produce two sets of RGB values. Clipping may also be eliminated, though for some images, the input values should be prescaled to prevent any under- or overflow at the output. Most importantly, the fixed-point multiplication can be replaced with a hard-wired shift-and-add implementation of the conversion coefficients.

Additionally, the precision of these coefficients can be greatly reduced without noticeable impact upon the final image. Determining the precision of the coefficients balances the trade-off between conversion accuracy, the hardware complexity, and the perceptual error between the color-converted image and the original image.

Based on the distortion model described in Section 15.2 we approximate the coefficients in such a way that the errors introduced are correlated to the local features of the image and hence correspond to little perceptual distortion in some cases. We analyze three different methods for the approximation of the coefficients and compare their performance in terms of the numerical error introduced, perceptual distortion, and implementation complexity of the corresponding design.

By multiplying the approximate YUV-to-RGB conversion matrix K by the exact inverse matrix K_{inv} the amount of numerical error can be determined. Note that for the actual color conversion matrix this product is an identity matrix.

We first chose a conversion matrix in which every coefficient contains at most three 1's in their binary representations. The approximate color conversion matrix becomes

$$K = \begin{bmatrix} 1 & 0 & 1.375 \\ 1 & -0.34375 & -0.6875 \\ 1 & 1.75 & 0 \end{bmatrix}$$

and the product of the approximate matrix and the exact inverse matrix is

$$K \times K_{\text{inv}} = \begin{bmatrix} 0.99 & 0.01 & 0.002 \\ 0.01 & 0.99 & -0.002 \\ 0.003 & 0.007 & 0.99 \end{bmatrix}$$

This indicates that even without psychovisual considerations, the precision would be sufficient. The hardware would require a 3-2 adder followed by a carry-propagate adder to compute the intermediate values which are then combined with two horizontal Y values to produce the final RGB values.

The simplest implementation would just use shifted versions of the color components. The conversion matrix introduces greater error, especially for the red component, whereby a visible color shift occurs. Thus we have

$$K = \begin{bmatrix} 1 & 0 & 1 \\ 1 & -0.50 & -1 \\ 1 & 2 & 0 \end{bmatrix} \qquad K \times K_{\text{inv}} = \begin{bmatrix} 1 & -0.07 & -0.29 \\ 0 & 1.17 & 0.16 \\ 0 & 0.05 & 0.92 \end{bmatrix}$$

The chosen conversion matrix for our design uses coefficients with two 1's which can be implemented with a carry-propagate adder and still provide sufficient accuracy. Another advantage of this precision is that all the coefficients have the same shift and add representation ($x + x \ll 1$). This allows one to use the same hardware for all the coefficient calculations and compute only two intermediate values instead of four. This approximation gives good visual quality of color images with peak SNR (PSNR) of above 40 dB.

The coefficients are approximated as

$$\begin{aligned} 1.402 &\approx \tfrac{11}{8} \approx \tfrac{6}{4} = \tfrac{3}{2} \\ 0.34414 &\approx \tfrac{11}{32} \approx \tfrac{6}{16} = \tfrac{3}{8} \\ 0.71414 &\approx \tfrac{11}{16} \approx \tfrac{6}{8} = \tfrac{3}{4} \\ 1.772 &\approx \tfrac{7}{4} \approx \tfrac{3}{2} \end{aligned}$$

Thus we have

$$K = \begin{bmatrix} 1 & 0 & 1.5 \\ 1 & -0.375 & -0.75 \\ 1 & 1.5 & 0 \end{bmatrix} \qquad K \times K_{\text{inv}} = \begin{bmatrix} 1.05 & -0.04 & -0.008 \\ -0.01 & 1.02 & -0.01 \\ 0.05 & 0.09 & 0.86 \end{bmatrix}$$

For comparing the distortion introduced by different conversion matrices we have used the following coefficients. For the 2-bit precision case, the coefficients are [1.5 – 0.375 – 0.75 1.5]. For the 1-bit precision case we tried four sets of coefficients: 1-bit.1, [1 – 0.5 – 0.5 2], 1-bit.2, [1 – 0.5 – 1.2 2], 1-bit.3, [2 – 0.5 – 0.5 2], and 1-bit.4, [2 – 0.5 – 1 2]. For the 3-bit precision case, the coefficients are [1.375 – 0.34375 – 0.6875 1.75].

Table 15.1 gives the perceptual distortion measure results of the above color conversion matrices in terms of image quality. The results were corroborated by perceptual ranking from human viewers. Table 15.2 shows PSNR results which, despite the visual similarity between 2- and 3-bit precision cases, give substantially different numbers. Table 15.3 gives the implementation complexity for the different methods of color conversion discussed.

From our distortion measure, the conversion matrix with 2-bit precision performs just as well as the matrix with 3-bit precision, indicating that for a low-power and low-complexity implementation, there is no need to use coefficients with more than 2 bits in YVU-to-RGB color conversion.

TABLE 15.1 PERCEPTUAL DISTORTION MEASURE RESULTS

Image	2-bit	1-bit.1	1-bit.2	1-bit.3	1-bit.4	3-bit
Air	1.313	3.648	1.872	4.137	1.361	1.212
Bike	0.917	1.876	1.294	2.197	1.615	0.886
Tennis	1.405	4.078	2.488	4.165	3.025	1.356
Mandrill	1.031	3.575	3.161	3.68	3.266	0.931
Flowg	1.28	2.734	1.924	2.974	2.164	1.149

TABLE 15.2 PSNR (DB) RESULTS

Image	2-bit	1-bit.1	1-bit.2	1-bit.3	1-bit.4	3-bit
Air	41.3	33.07	36.96	34.66	35.49	51.22
Bike	45.37	38.68	40.67	37.16	39.15	51.33
Tennis	40.86	31.62	33.38	30.3	32.06	50.16
Mandrill	39.04	29.35	30.53	29.34	30.53	47.92
Flowg	41.92	33.40	36.08	32.30	35.01	50.13

TABLE 15.3 IMPLEMENTATION COMPLEXITY

Details	2-bit	1-bit.1	1-bit.2	1-bit.3	1-bit.4	3-bit
Adder	5	4	4	4	4	6
3-2 Add	0	0	0	0	0	0

15.4. ARCHITECTURE

The data flow of the converter circuit is shown in Figure 15.1 and the data timing in Figure 15.2.

The YUV data arrive on an 8-bit input bus in UYVY order and is latched into the corresponding register. Subtracting 128 from U and V can be handled simply by inverting the top bit of the value. Next, the color data (U or V) is added to itself shifted by 1, to produce an intermediate value which is stored in one of two 10-bit registers (K_1 for V and K_2 for U). Here, K_1 and K_2 are computed using the equations

$$K_1 = (V - 128) + (V - 128) \ll 1 \qquad K_2 = (U - 128) + (U - 128) \ll 1$$

The K_1 and K_2 values are combined with the two corresponding Y values to produce two sets of RGB output in every four cycles. We compute the RGB values based on the equations

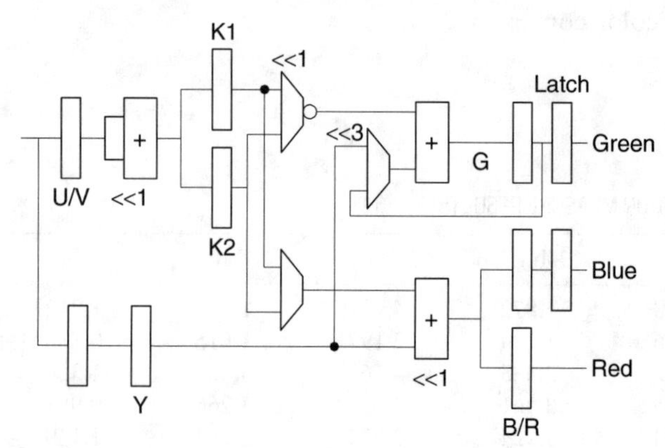

Figure 15.1 Block diagram of architecture.

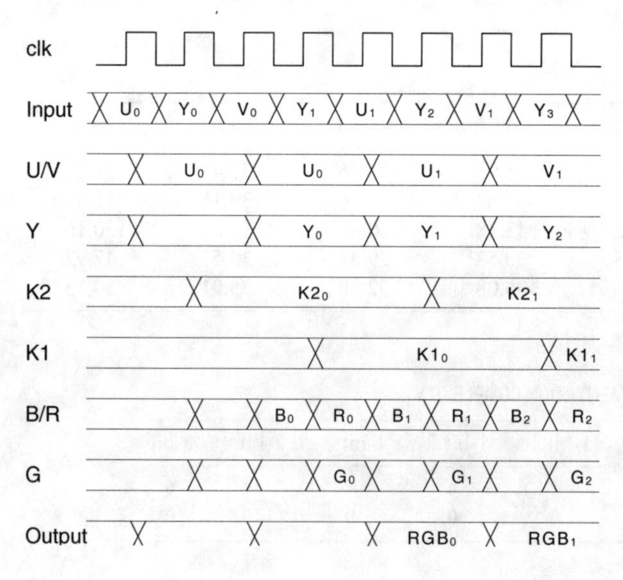

Figure 15.2 Timing waveforms for design.

$$R = ((Y \ll 1) + K_1 + 1) \gg 1$$
$$G = ((Y \ll 3) - (K_1 \ll 1) - K_2) \gg 3$$
$$B = ((Y \ll 1) + K_2 + 1) \gg 1$$

First, K_2 is selected and combined with Y to produce the blue value. In the same cycle, the Y value is selected and K_2 is subtracted to produce the first half of the green calculation. The next cycle, K_1 is added to Y to produce red, and a shifted K_1 is subtracted from the intermediate green value to produce the final green value. The output registers align the data so that all the RGB values change at the output at the same time. Additional hardware could be added after the adders to detect under/overflow and clip the output value accordingly. However, for many images this is not necessary.

The control of the data path is very simple, requiring only a 2-bit state counter to form the mux-select and register-enable signals. Thus the total implementation uses minimal area and power, requiring only three carry-propagate adders performing five adds per RGB output.

15.5. LOW-POWER ADDER DESIGN

Compared with the Brooktree Bt294 color converter chip [3], dissipating 900 mW at 5 V and 13.5 MHz RGB output rate, our converter is approximately 1000 times more power efficient. Within this factor of 1000, a factor of 10 can be easily obtained by voltage scaling of the power supply. Reduced supply voltage, however, increases circuit delay. This increase in delay needs to be compensated for by duplicating hardware, or chip area, to maintain the same real-time throughput.

15.5.1. Low-Power Design Guidelines

The dynamic power consumption of a complementary metal-oxide-semiconductor (CMOS) gate with a capacitive load C_{load} is given by

$$P_{\text{dynamic}} = p_f C_{\text{load}} V^2 f \tag{15.1}$$

where f is the frequency of operation, p_f the *activity factor*, and V the voltage swing [4]. Reducing power consumption amounts to the reduction of one or more of these factors. Analysis in [5] indicated that for energy-efficient design, we should seek to minimize the *power-delay product* of the circuit, or the *energy* consumed per operation.

A more complete discussion on low-power CMOS circuit design is given in [4] and Chapter 6. This section only briefly describes the guidelines used in designing our low-power circuit library. The CMOS circuits operated at extremely low supply voltages allow a very small noise margin. A safe design is therefore of ultimate importance. The static CMOS logic style was chosen for its reliability and relatively good noise immunity. The transistors were sized in the ratio of 2 : 1 for PMOS and NMOS, multiplied by the number of transistors in series. A library of standard cells has been designed with these specifications in mind.

Scaling Factor in a Buffer Chain. Well-known analysis of a multistage buffer chain indicates that, to minimize the total delay, the optimal scaling factor of the inverters in a buffer chain should be *e* (the exponential constant) [4]. This number is usually too small for practical use, however, as many stages of inverter buffers would be needed to drive a large capacitive load such as clock lines, consuming more power than necessary. Through simulations we have found that a scaling factor of 5–6 is optimal for low-power buffer chains operated at 1–1.5 V, where the total delay would be comparable to that of a buffer chain with a scaling factor of *e* while consuming only two-thirds of its power. Table 15.4 shows the energy consumed by a buffer chain at a supply voltage of 1.5 V with different scaling factors. The energy consumption of a buffer chain with a scaling factor of 6 remains a constant fraction of the energy consumed by buffer chains with a scaling factor of 4 or less over a wide range of loads, and its delay relative to the buffer chains of smaller scaling factors only increases slightly for large loads (over 5 pf). This is attributed to the fact that if we can save one or more stages in a buffer chain by increasing the scaling factor by a reasonable margin, the net result is a low-power design with a comparable delay.

With a larger scaling factor, the slope of the voltage transfer function of each buffer is smaller. This may potentially increase power consumption due to short-circuit currents when both the PMOS and NMOS branches may be simultaneously on for a longer period of time. However, hSPICE simulations indicated that the factor of short-circuit currents is negligibile, consuming less than 5% of the total power.

Minimizing the Occurrence of Spurious Switching. Spurious switching, which consumes power but does not generate correct output signals, was quoted to account for from 20 to 70% of total circuit switching power [6]. Spurious switching is caused by unmatched delays such as glitches in combinational circuits and input signals not arriving at the same time. Power-optimized logic synthesis and careful circuit design can limit spurious switching to a minimum.

In designing our library cells, extra care was taken to ensure that unwanted switching is never allowed to drive a buffer chain, as the power consumed is further amplified by the large capacitive load in successive buffer stages. To prevent spurious switching from driving a buffer chain, the input signal to the first buffer is either latched or gated. The former limits the switching to occur only when the latch is on, holding the signal constant when it is off. The latter is used if a completion signal can be identified from the logic module that drives a buffer chain to gate its outputs.

Equalizing the Delays Between Logic Paths. In our design, we tried to equalize the delays of all logic paths so that there is no single critical path. The reason is that if the delays of different logic paths are not matched, some logic paths will operate faster than others. As a result, energy will be wasted in those

TABLE 15.4 COMPARISON OF BUFFER CHAINS WITH DIFFERENT SCALING FACTORS

C_{load} (pF)	Scaling Factor = e			Scaling Factor = 4			Scaling Factor = 6						
	Number of Buffer Stages	Energy/ Switch (pJ)	Average Delay (ns)	Number of Buffer Stages	Energy/ Switch (pJ)	Average Delay (ns)	Number of Buffer Stages	Energy/ Switch (pJ)	Average Delay (ns)	Normalized to Factor e		Normalized to Factor 4	
										Energy	Delay	Energy	Delay
0.5	4	1.15	4.44	3	0.90	4.07	2	0.75	3.83	0.66×	0.86×	0.84×	0.94×
5	6	11.69	7.27	4	8.99	6.24	3	7.66	6.24	0.66×	0.86×	0.85×	1.00×
20	7	47.09	8.62	5	36.22	7.96	4	30.67	8.63	0.66×	1.00×	0.85×	1.08×
50	8	116.87	9.88	6	90.57	9.69	5	76.64	10.11	0.66×	1.02×	0.85×	1.04×

Note: In 0.8-μm CMOS technology.

paths with shorter delays by delivering a current larger than necessary. Since switching speed is a function of transistor sizing, which determines switching currents and capacitive loads, we can effectively "slow down" the faster paths by using smaller transistors and fewer buffer stages to reduce both driving currents and transistor capacitive loads.

15.5.2. Design Example: The Adder

We use the design of an adder to illustrate the performance trade-off between low-power and high-speed designs. The choice of an adder design is determined by the area, speed, and power budget available. Carry-propagate adders are compact and useful for adding numbers of short word lengths (fewer than 4 bits) but too slow for adding numbers of wider word lengths. On the other hand, tree adders perform both carry and summation computations in parallel, each carry taking only $\lceil \log_2 n \rceil$ gate delays, where n is the word length. The area needed for a full tree adder, however, is relatively large, requiring long interconnects and large capacitive loads. Hence, tree adders only yield significant advantage for adders with more than 32 bits. As our system requires adders of a word length between 8 and 10 bits, the adders to be considered are carry-select and carry-lookahead adders. By paying close attention to layout and the loading on input signals that an adder introduces, we can identify the most energy-efficient adder class meeting a given speed requirement.

The adder used in our decoder design, shown in Figure 15.3, is a variant of the standard carry-select adder. Each adder block has two speed-optimized ripple-carry chains, one with a zero carry-in, the other with a one carry-in. The output carry bit is chosen by the actual carry-in to the adder block and then used in the final XOR to produce the sum bit. The transistor sizing in each adder block was designed to hide the delay of ripple-carry chains, guaranteeing that all output carry bits will be ready by the time the carry-in reaches the adder block, eliminating the probability of spurious switching mentioned earlier. By simulation, the different adder block sizes are chosen to be 2, 4, 5, 6 and 8 bits respectively.

We use a 17-bit adder design to illustrate our design strategies. The 17-bit example shown in Figure 15.3(d) consists of four adder blocks. The layout of each bit slice of this carry-select adder using the MAGIC layout tool occupies an area of $332\lambda \times 84\lambda$. HSPICE simulations of a 17-bit adder indicated a worst-case delay of 35 ns at 1.5 V, with an energy consumption of only 19pJ per add operation.

For applications that require a clock frequency less than 25 MHz, as in the case of our color conversion circuit, the supply voltage can be further reduced to minimize the energy per operation. If the supply voltage is reduced to 1 V, the worst-case delay is increased by approximately 5 times, with the energy per add operation reduced to 8 pJ.

A comparative study of various adder classes is given in Table 15.5, illustrating the types of adders compared and the sizes of their layouts in MAGIC. The performance comparisons were obtained by simulations of the extracted layouts using

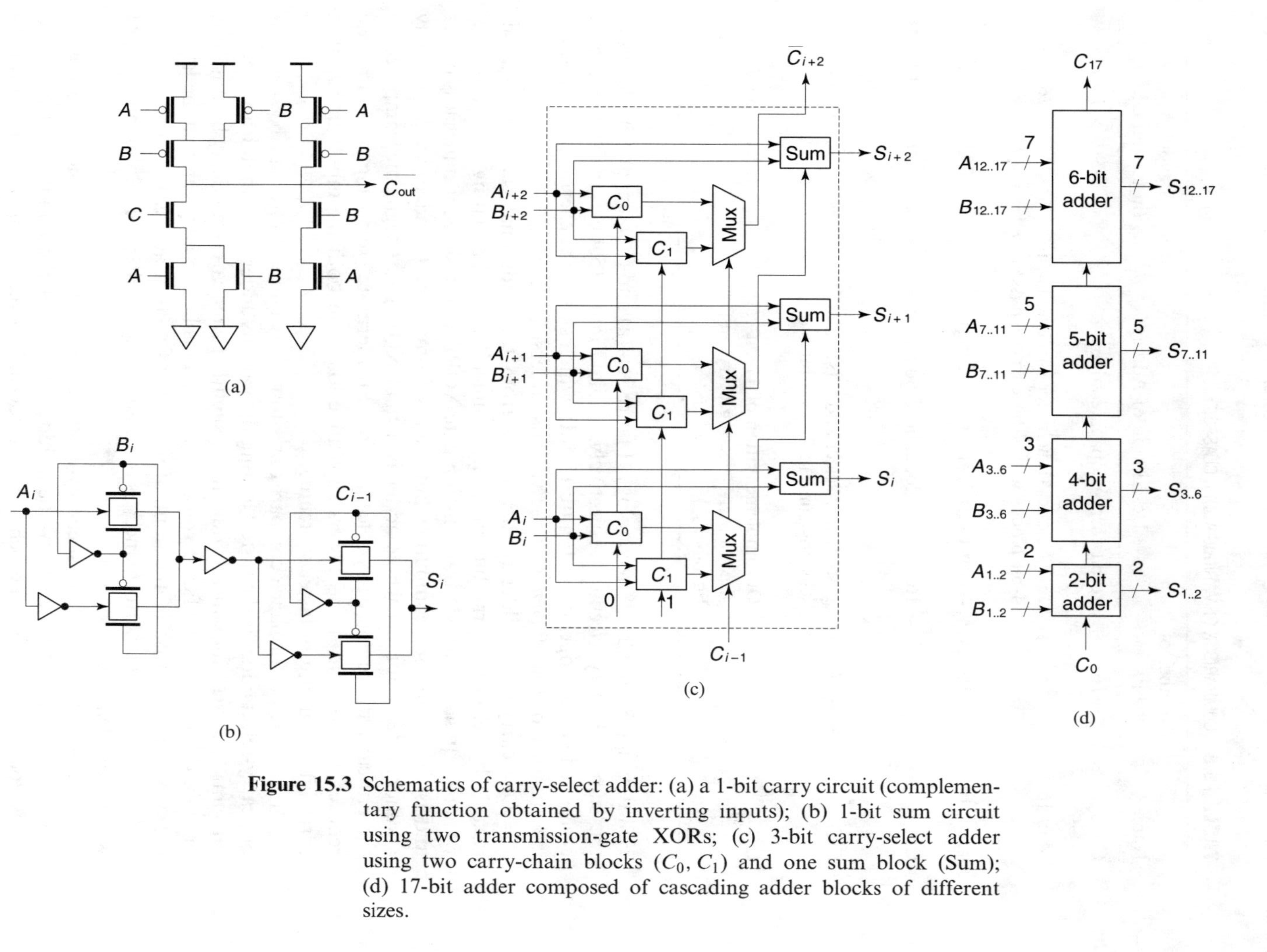

Figure 15.3 Schematics of carry-select adder: (a) a 1-bit carry circuit (complementary function obtained by inverting inputs); (b) 1-bit sum circuit using two transmission-gate XORs; (c) 3-bit carry-select adder using two carry-chain blocks (C_0, C_1) and one sum block (Sum); (d) 17-bit adder composed of cascading adder blocks of different sizes.

TABLE 15.5 COMPARISON OF VARIOUS ADDER CLASSES

Name	Number of bits	Description of Adders	Layout Area (in 0.8-μm CMOS)
Cp16	16	16-bit carry-propagate adder	890λ × 190λ
Man 16	16	16-bit static Manchester carry adder, in 4-bit blocks with 4-bit carry-bypass	1284λ × 265λ
Tree 16	16	16-bit tree adder (Brent-Kung)	1292λ × 545λ
Cla16	16	16-bit carry-lookahead adder, in 4-bit blocks	1260λ × 484λ
Csf17	17	17-bit carry-select adder (standard), selecting sum, in 2-4-5-6 bit blocks	1236λ × 440λ
Cs17	17	Our implementation of 17-bit carry-select adder, selecting carry, in 2-4-5-6 bit blocks	1236λ × 332λ

HSPICE in 0.8-μm CMOS technology. Figure 15.4(a) and (b) compare the delays of the sum and carry circuits, respectively, while Figure 15.4(c) graphs the energy consumed per add operation. Figure 15.4(d), which shows the energy-delay product, is a measure of the *efficiency* of the various adders.

Our carry-select adder (Cs17) differs from the traditional carry-select adder (Csf17) by selecting carry bits instead of sum bits. By eliminating one sum gate, which is equivalent to two transmission gate XORs, our design consumes less power and is faster. By not computing both sums from carry 0 and carry 1, the only additional delay for the whole adder is a final XOR gate, negligible compared to the critical path of the carry-select circuit. Furthermore, removing one sum gate reduces area overhead and lowers capacitive loads on both the inputs and output carry bits, resulting in faster switching.

From the performance comparisons shown in Figure 15.4, it can be seen that our carry-select adder is the fastest among the various adders to compute a 16- or 17-bit sum. The carry-lookahead adder has a slightly faster carry-out delay because it is optimized for generating fast carries. Its sum delay is larger, however, and the layout is asymmetric because of the nonuniform carry function across the bits. The tree adder delivers slower performance, as its design is more suited for adding numbers of wider word lengths.

Our design is not the most energy-efficient one. The reason for not choosing the more efficient Manchester carry-chain adder is to meet the timing constraint of a 40-ns clock cycle at 1.5 V. This choice is a trade-off between maintaining the throughput required by the system and the energy needed to achieve that throughput. A slower adder implies not only that more adders need to be implemented to meet a throughput requirement, but also that other faster logic blocks will be idle waiting for some addition to complete, resulting in unmatched delays in logic paths.

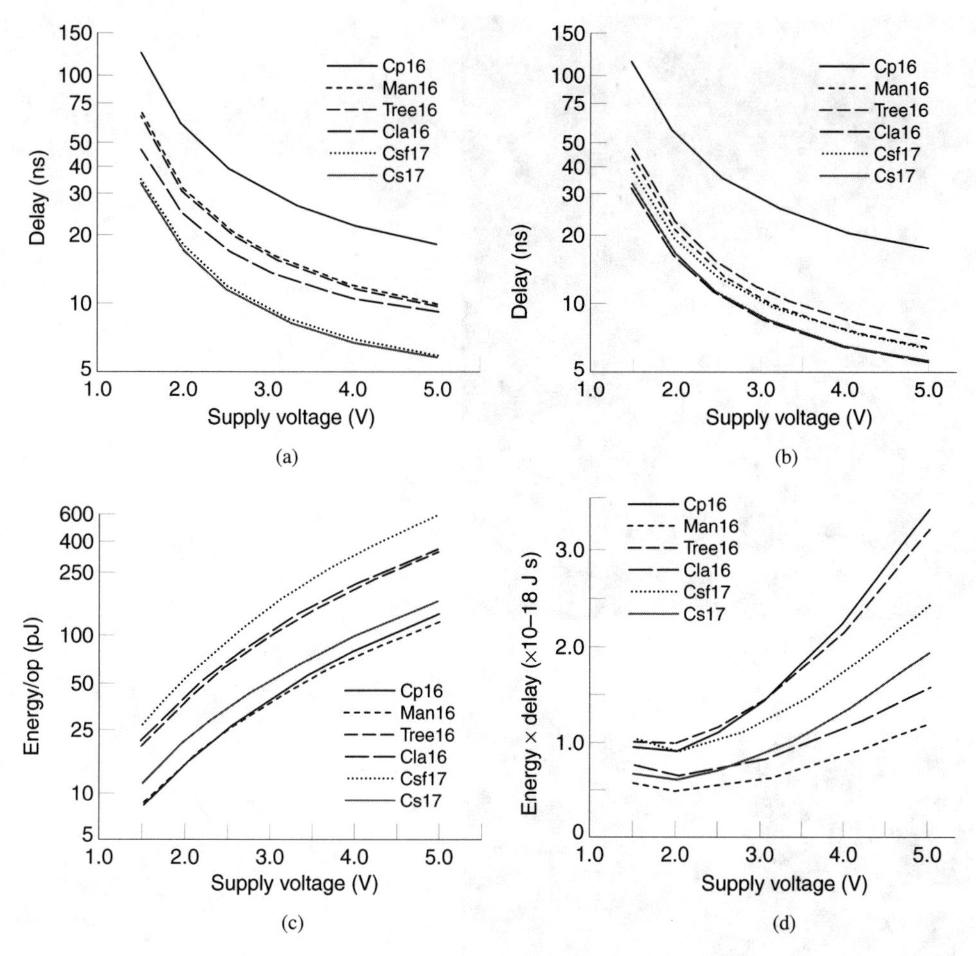

Figure 15.4 Performance comparisons among different adder implementations: (a) sum delay; (b) carry-out delay; (c) energy per operation; (d) energy-delay product.

15.6. PERFORMANCE

The measured power dissipation at 1.5 V for this converter design is 2.1 mW at 25 MHz at a 1.5-V supply, with the energy per conversion per pixel under 100 pJ. This low-power color converter has been incorporated into a low-power subband decoder chip for real-time video decompression [7, 8]. Compared to a direct implementation of the exact color conversion matrix using multipliers, we obtained a power saving of three orders of magnitude. The total area for the design in a 0.8-μm CMOS process is 1.5 × 1 mm. The image quality resulted from this color conversion design appears comparable in visual quality to the original image, as shown in Figure 15.5.

(a)

(b)

Figure 15.5 Comparison of (a) original color image and (b) color image regenerated by approximate color conversion matrix.

For smaller liquid crystal displays (LCDs) where the resolution is 180× 240 pixels/frame at 30 frames/s, the required color conversion rate is below 1.5 Mpixels/s. Therefore our converter can be operated at a supply voltage lower than 1.5 V and still deliver real-time color conversion. The power consumption of the color conversion circuit is measured to be less than 110 μW at a supply voltage of 1 V, with a conversion rate in excess of 1.5 Mpixels/s.

15.7. CONCLUSIONS

Color conversion can be performed with minimum complexity, and therefore at minimum power, if psychovision effects are used to judge color quality. For many portable applications a low-power design is far more important than exact color reproduction, especially when the images are regenerated after lossy compression. We have demonstrated a converter design of standard YUV to RGB conversion at the NTSC video rate at a power level of 2 mW while maintaining high visual quality.

References

[1] A. Netravali and B. Haskell, *Digital Pictures*, Plenum, New York, 1989.
[2] N. Chaddha and T. H. Meng, "Psycho-Visual Based Distortion Measure for Monochrome Image Compression," *SPIE Proc., Visual Commun. Image Process.*, Nov. 8–11, 1993.
[3] "A 27 MHz VideoNet YCrCb to RGB Converter for 4 : 2 : 2 Video Applications," Brooktree Corporation Product Description.
[4] N. Weste and K. Eshragian, *Principles of CMOS VLSI Design: A Systems Perspective*, 2nd ed., Addison-Wesley, Reading, MA, 1992.
[5] A. Chandrakasan, S. Sheng, and R. W. Brodersen, "Low Power CMOS Digital Design," *IEEE. J. Solid-State Circuits*, vol. 27, no. 4, pp. 685–691, 1992.
[6] A. Shen, A. Ghosh, S. Devadas, and K. Keutzer, "On Average Power Dissipation and Random Pattern Testability of CMOS Combination Logic Networks," *IEEE ICCAD Dig. Tech. Papers*, pp. 402–407, 1992.
[7] B. M. Gordon and T. H. Meng, "A Low Power Subband Video Decoder Architecture," *Proc. ICASSP 94*, Apr. 19–22, 1994.
[8] B. M. Gordon and T. H. Meng, "A 1.2 mW Video-Rate 2D Color Subband Decoder," *IEEE Solid-State Circuits*, vol. 30, no. 12, pp. 1510–1516, 1995.

Chapter 16

Micropower Systems for Implantable Defibrillators and Pacemakers

Marwan Jabri and
Richard Coggins
University of Sydney, System Engineering and Design Automation Laboratory, Department of Electrical Engineering, J03 NSW 2006, Australia

16.1. INTRODUCTION

Implantable defibrillators and pacemakers are devices implanted in patients with heart disease. These devices monitor the heart and apply a therapy when an arrhythmia[1] is detected. Although defibrillators and pacemakers are intended for different therapeutic functions, we will call both devices implantable cardiac therapy systems (ICTS) for the sake of simplicity.

The ICTSs are battery powered. The average lifetime of a battery is about 5 years. Battery replacement requires surgery, which is presently rather expensive. As ICTSs are implanted devices, there is currently an increasing drive in the industry to reduce their size (volume). Due to the limitation in space and battery energy, the computational algorithms used in pacemakers tend to be very simple. There are also constraints on the use of magnetic devices such as inductors. Put simply, the therapy provided by ICTSs consists of either high-voltage/high-energy (HVHE) shocks or high-voltage/low-energy (HVLE) shocks. The HVHE shock is applied when a critically dangerous arrhythmia such as ventricular fibrillator (VF) is detected. The HVLE therapy is applied as a train of pulses to the heart tissue with the aim of inducing a regular rhythm.

Electric energy is consumed by ICTSs in the two processes of diagnosis and therapy. The first process is constantly performed while the second is considered (and hoped to be) a much less frequent event. Note that a proper diagnosis will reduce the number of false-positive events and hence the energy consumed in

[1]An arrhythmia is an abnormal heart rhythm.

therapy. With the increasing complexity of the diagnosis performed by pacemakers, greater emphasis is being placed on reducing the energy usage of the implementation circuits.

Compared to other low-power applications, ICTSs can be distinguished by other features:

1. The temperature of the operation environment (human body) can be considered stable.
2. Although an ICTS has to operate in real time, the events it responds to are very low frequency (about 3 Hz).

These two features provide opportunities not available in most other low-power systems. For example, the stable temperature is an opportunity for temperature-sensitive implementation technologies, such as subthreshold complementary metal-oxide-semiconductors (CMOS). The low event frequency provides an opportunity for the exploitation of adiabatic computing paradigms.

In this chapter, we describe very low power pattern recognition systems based on CMOSs that are aimed at enhancing the arrhythmia detection capabilities of present ICTSs. We demonstrate subthreshold microelectronic neural networks which perform morphology analysis tasks that are too energy expensive to implement by present ICTS controllers.

In Section 16.2, we review MOS energy dissipation fundamentals and highlight the differences with the energy reduction strategies used in other applications. In Section 16.3 we briefly review CMOS subthreshold operation fundamentals. Section 16.4 describes pattern recognition requirements in ICTSs. In Section 16.5 we describe two implementation approaches for analog subthreshold multilayer perceptrons and assess their performance.

16.2. ENERGY MINIMIZATION STRATEGIES FOR DIGITAL CIRCUITS

Digital CMOS is the dominant implementation technology in ICTSs. In a MOS circuit, and depending on the logic approach, one can distinguish the following types of power used in the operation:

Static power. This is the power used to maintain the present state of the system in the absence of logic changes. It is usually due to gate leakage.

Short-circuit power. This is the power used when there is a resistive path from the supply to ground and is over a very short period of time. Typically in a CMOS circuit, it is the integral over the time that both the NMOS and the PMOS devices (in series) are conducting. It can be in the order of hundreds of nanowatts (peak).

Dynamic power. This is the power used to propagate logic levels from one gate to the next. It is therefore expressed in terms of the charge or discharge of the gate capacitance. The charging or discharging resistance and the gate capacitance account for the time needed to propagate a logic level.

Dissipation power. This power is dissipated in terms of heat. The heat is the result of the collisions of the charge carriers with the lattice structure of the semiconductor. These collisions have a double-sided effect: Energy is wasted and is dissipated in the form of heat and the heat needs to be extracted or could lead to damage to the circuit (this is especially the case for high-speed circuits).

In CMOS, the logic family known for its low power usage, the dissipation and dynamic energies amount to about CV_{DD}^2, where C is the capacitance of the nodes and V_{DD} is the supply voltage. Thus reducing node capacitance and/or the supply voltage is an approach to reduce energy usage. However, there are limits to the reduction of capacitance and supply voltage. Hence, one has to look at other methods for energy reduction. Over the last few decades, extensive work on low-power design techniques has been performed for calculators, watches, and implantable devices such as pacemakers [1–5] and trade-offs between chip area, power, and speed have been identified [6, 7]. A review of low-power digital techniques up to 1985 can be found in [8]. Reference [9] provides an analysis of the different consequences of minimizing energy alone versus energy-time products. Work has been done, as in [9], to produce single gates minimizing the energy criteria; however, there are limits to techniques based on lowering the MOS threshold and supply voltage.

Another approach to reducing power in digital CMOS has been the investigation of different logic families and design styles. This has included designs using complementary pass gate logic and self-timed logic [5, 10].

In recent years, the portable computer and notebook industries have focused on "power management" aspects. Although notebooks and portable computers share many similarities with implantable devices in terms of the reduction of energy usage, there are a number of important differences:

1. Notebooks and portable computers are not "real-time" systems in that their power management strategies rely on minimizing any energy usage when they are inactive. On the other hand, an implantable device such as a pacemaker has to perform real-time operations.
2. A notebook microprocessor is expected to run at a very high clock speed (tens of megahertz) whereas the controller of a pacemaker runs at clock speeds of tens of kilohertz given they deal with signals of the order of 3 Hz.
3. Area is not as critical for notebooks as it is for implantable devices. Notebooks can rely on facilities like long-term storage on hard disks to provide capabilities such as virtual memory. It is known that most of the battery energy used by a notebook is for the mechanical drives of the disks.

The differences above highlight the reason why ICTSs cannot exploit much of the low-power techniques used in notebook processors.

16.3. BRIEF REVIEW OF CMOS SUBTHRESHOLD OPERATION

In this section the main principles, features, and design considerations of analog subthreshold circuits are presented. For more detailed treatments see [8, 11].

16.3.1. Characteristics

Sub-threshold operation of a MOS transistor is defined by sufficiently low gate voltages such that the major conduction mode is diffusion. This leads to an exponential relationship between gate voltage and drain current rather than the usual square law and the transistor behaves more like a bipolar device. The exponential relationship between gate voltage and drain current also results in the fact that the transistor reaches saturation in just a few multiples of the thermal voltage. In the strong-inversion mode the temperature dependence of a MOS transistor is usually dominated by the mobility dependence on temperature such that the drain current decreases with increasing temperature. In subthreshold operation the drain current increases exponentially with temperature. At high temperatures or very low currents, junction leakage also adds to the drain current.

The conduction mode also affects the bandwidth of the transistor. This may be seen by considering the small signal current gain of the device in both weak inversion and strong inversion. If we define the cutoff frequency of the transistor by that frequency at which the small-signal drain current is equal to the small-signal gate current, then we see that this is the frequency limit of useful operation of the transistor. In strong inversion the cutoff frequency is proportional to the square root of the drain current. In weak inversion it is proportional to the drain current.

Weak-inversion operation also influences the noise characteristics of the transistor. This can be seen by considering the equivalent noise resistance at the gate of the transistor. At low frequencies the noise resistance is dominated by $1/f$ flicker noise and is essentially independent of drain current. At high frequencies in strong inversion it is dominated by thermal noise, whereas in weak inversion a shot noise model applies. This means that for a given drain current the high-frequency noise characteristic can be minimized by operating the transistor in the weak-inversion mode, which may be achieved by increasing the width-to-length ratio of the device sufficiently to achieve subthreshold operation.

Another important consideration in analog circuits is transistor matching. Consider the drain current matching of two transistors at the same gate voltage and the gate voltage matching of another two transistors at the same drain current. In the first case drain current matching is improved by driving the transistors further into strong inversion. In the second case, gate voltage matching is maximized in weak inversion.

16.3.2. Advantages

Analog circuits inherently have a number of advantages over their digital counterparts. These include reduced device count due to the direct mapping of functions and algorithms onto the physical characteristics of the device and lower energy usage. The resultant gains in chip area and power, however, come with the penalties of increased design time and limits of precision of arithmetic operations, to be discussed in the following section.

To see how these advantages may be realized in low-power implementations of neural networks, we consider the aforementioned characteristics of the subthreshold operation of the MOS transistor. First, functions of exponentials (e.g., tanh, log) are easily mapped to silicon, so the usual nonlinearities used in neural networks are easily implemented in subthreshold analogs. For low-bandwidth applications such as intracardiac electrogram (ICEG) classification, energy usage overheads can be reduced by lowering the bias currents of circuits. Further, when the transistors are operated in the subthreshold regime at reduced bandwidth, energy usage is more efficient since the bandwidth of the transistor is increasing linearly with the drain current, whereas bandwidth increases with the root of drain current above threshold. Also, since in subthreshold the transistors saturate within a few multiples of the thermal voltage, current sources may be realized which operate over almost the full supply range. Finally, reducing bias currents so that transistors are in subthreshold ensures that the minimum high-frequency noise is generated, potentially increasing the precision of the system.

16.3.3. Design Considerations

Subthreshold analog circuit implementations require certain specific design considerations. As with any analog circuit design, precision and stability over temperature must be carefully considered. In the case of implantable systems, the fact that subthreshold analog designs can be quite sensitive to temperature variations is largely alleviated by the temperature stable environment of the human body. With regard to precision, the highly parallel neural network architecture has been quite robust to loss of precision through limited dynamic range and device offsets, and 6-bit precision has been shown to be adequate for a number of applications. Another important consideration is the reduced drain current matching between transistors when operated in subthreshold. This, for example, can lead to nonmonotonic operation of weighted current digital-to-analog converters (DACs) as the bias currents are reduced.

16.4. PATTERN RECOGNITION IN ICTS

The ICTSs monitor the heart state through leads attached to the heart tissue. The trace of the electrical activity of the heart appears as waveforms called intracardiac electrograms (ICEGs). These "near-field" waveforms are not to be confused with (external or surface) electrocardiograms (ECGs) which are far-field signals. The ICTSs generally make use of one or two leads. In a one-lead system (also called single chamber) the activity of the ventricle is monitored. In a two-lead system, the activity of the atrium is also monitored. The shape of the heart beat on the ventricle lead is shown in Figure 16.1 and is commonly referred to as the QRS complex with R

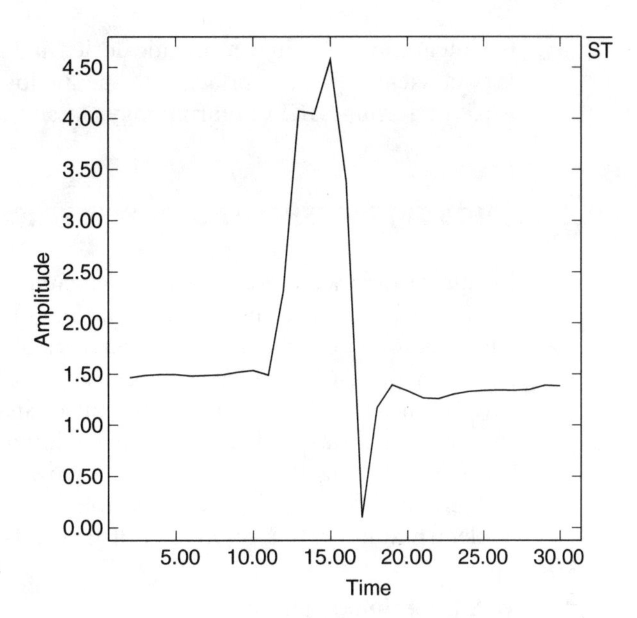

Figure 16.1 Heart beat signal. A beat is called a QRS with R being the peak.

being the peak of the beat. The ICTSs employ QRS detectors to detect heart beats and deduce the heart beat rate by measuring the time between consecutive R peaks.

There are over 17 arrhythmias of interest to cardiologists. The arrhythmias are normally grouped under a few general classes to which specific therapies are applied. With present ICTS technologies, not all arrhythmias can be detected because some require morphological (signal shape) analysis that is beyond power and space budgets. Arrhythmias that can be reliably detected are those that can be recognized according to the heart rate.

Hence, the availability of morphology analysis at very low power and space would make the reliable detection of some arrhythmias possible. An example is ventricular tachycardia with 1 : 1 retrograde conduction (VTR), which could be considered a dangerous arrhythmia. A patient who is developing a VTR could have a heart rate in the same range as when a sinus tachycardia (ST) is developing. However, ST is not considered dangerous, and hence on the basis of the heart rate alone, VTR could be confused with an ST, which could endanger the life of the patient. Fortunately, the shapes of the VTR and ST are very often different, and morphology analysis can easily distinguish these two classes.

However, such a morphology analysis would be too expensive (in terms of power) to perform using the ICTS controller or microprocessor. The analog subthreshold neural network classifiers we are presenting here have been demonstrated to perform such morphology analysis in ultra-low power. The advantages of an analog approach is borne out when one considers that an energy-efficient analog-to-digital converter such as [12] uses 1.5 nJ per conversion, implying 375 nW power consumption for analog-to-digital conversion of the ICEG alone (assuming a 250-Hz sampling rate). The Snake chip (to be described in the following section) consists of the integration of a bucket brigade device and analog neural network and provides a very efficient way of interfacing to the analog domain and consumes only 200 nW while performing ICEG morphology classification.

16.5. ANALOG SUBTHRESHOLD MULTILAYER PERCEPTRONS

In this section we present two implementations of analog multilayer perceptrons (AMLPs). The two chips are based on the same synapse design and have the same basic feedforward architecture. An analysis of this synapse design apears in [13]. The first chip presented, called Wattle, has switched-capacitor neurons and parallel-network inputs. The second chip, called Snake, has diode-based common-mode cancelling neurons and a serial analog input which feeds the first synaptic layer via a bucket brigade device. In the following, we first discuss the problems and methodology for training AMLPs, followed by a description of the two network implementations and a discussion of their relative merits.

16.5.1. Training Issues

The first issue to be addressed in the design of an analog multilayer perceptron is how to train it, that is, to adapt the synaptic weights to optimally perform the task at hand. Training algorithms that have been mathematically formulated for high-precision, noise-free, digital computing may not necessarily be practical for analog implementations [14]. This is in particular the case for analog subthreshold implementation where transistor mismatch could lead to severe problems. The easiest way to accommodate transistor mismatch is through the training process. This requires that the analog chip be used in the training loop so that its specific characteristics are taken into account to determine the weights. The implication of such in-loop training is that one cannot simply load weights calculated on a host computer (or for another chip) and expect the desired behavior. However, for morphology analysis applications, weights have to be determined for a particular patient anyway, since ICEG signal morphology varies significantly across patients.

The training algorithms developed for our AMLPs have been reported elsewhere [4, 15, 16]. They are based on approximating the error gradient by finite differences produced by the means of weight perturbations.

16.5.2. Wattle

This section describes a low-power analog very large scale integrated (VLSI) neural network called Wattle [17]. Wattle is a 10 : 6 : 4 three-layer perceptron with multiplying digital-to-analog converting (MDAC) synapses and on-chip switched-capacitor neurons fabricated in 1.2 μm CMOS. The on-chip neurons facilitate variable gain per neuron and lower energy/connection than for previous designs. The following sections briefly describe the architecture, implementation, and performance of the Wattle chip in the ICEG classification task.

Architecture. Figure 16.2 shows a floor plan of the Wattle chip. The 10 : 6 : 4 perceptron is formed by the 10×6 and 6×4 synapse arrays and the associated switched-capacitor neurons abutted to the right-hand side of each array. The current source in the left corner of the chip provides five current references to each synapse of the chip to form the weighted currents used in the synapse implementation. The weight storage is digital and is local to each synapse cell. The address and data for the weights access are serial and are implemented by the row address, column address, and data shift registers on the boundary of the chip. The hidden-layer multiplexer allows access to the hidden-layer neuron outputs in order to reduce pin count. Each neuron must be clocked for its output to be evaluated. Again, in order to reduce pin count, the neuron clock demultiplexer switches the neuron clocks between the hidden and output layers. The neuron outputs are buffered off-chip so as to avoid the effects of pad leakage, capacitive loading, and noise pickup at the expense of some power overhead.

Implementation. Signaling on Wattle is fully differential to reduce the effect of common-mode noise. The synapse is a multiplying DAC with six bit weights. The synapse is shown in Figure 16.3. The MDAC synapses use a weighted current source to generate the current references for the weights. The voltage inputs from the previous layer are then multiplied by the weighted current by the differential pair. Four-quadrant multiplication is achieved by the sign bit controlling the switches to swap the signals at the output of the synapse. The multiplication achieved by this circuit is nonlinear. Indeed, since the current source is biased in the subthreshold regime, the differential pair is also in subthreshold and hence the differential output currents show a tanh relationship to the differential input voltage. This feature of the circuit is used as the required nonlinearity in the network

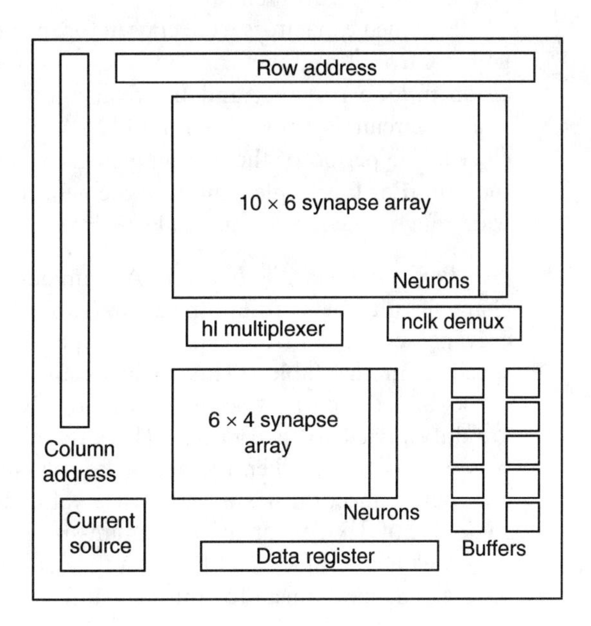

Figure 16.2 Floor plan of Wattle chip.

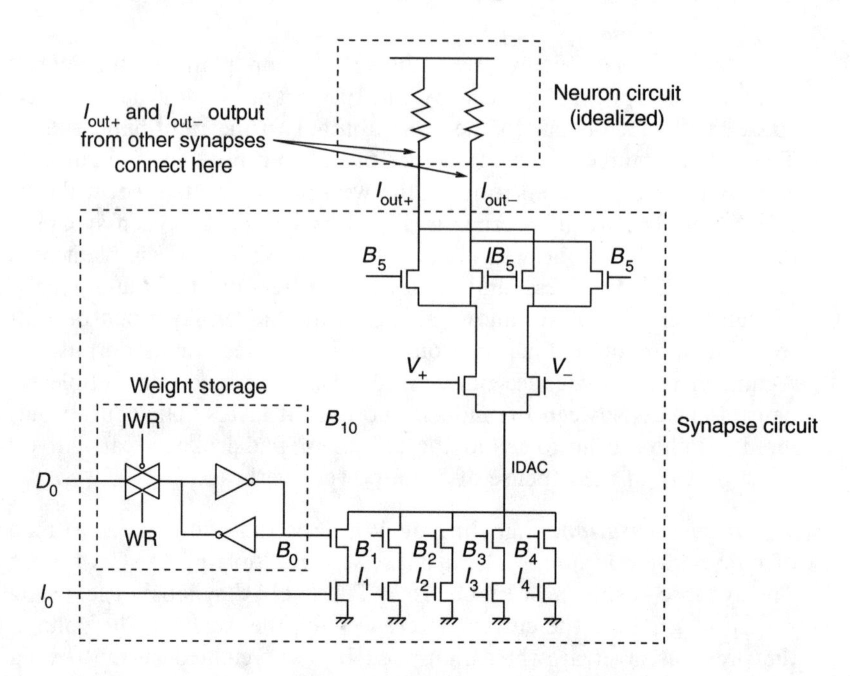

Figure 16.3 Synapse circuit diagram for Wattle and Snake. V_+ and V_- are the neuron (voltage) input to the synapse and I_{out+} and I_{out-} are the (current) outputs.

for nonlinear mappings. In a conventional architecture the nonlinearity is at the output of the neuron; here it is at the input of the synapse. Hence, with this design simpler linear neuron circuits can be used.

Switched capacitors were chosen for the neurons on Wattle as they allow flexible gain control of each neuron, investigation of gain optimization during limited precision in-loop training, and the realization of very high effective resistances. The neuron circuit is shown in Figure 16.4. The neuron requires reset and charging clocks. The period of the charging clock while the reset clock is inactive determines the gain. For this implementation the neuron clocks are generated off-chip so that a wide range of gain settings could be investigated.

Performance Evaluation. A summary of the electrical performance and characteristics of Wattle appears in Table 16.1. Wattle achieves a worst-case energy per connection of 43 pJ when operating continuously under the conditions specified in the table. (This includes the overheads of the neuron buffers and clocks and the current sources.) For the ICEG classification task the chip biases are duty cycled to save power. However, this introduces energy overheads due to bias settling times. when the network is classifying at a typical rate of 3 Hz, the energy per connection has a worst-case value of 120 pJ, which implies a power consumption of 3.3 nW for the maximum-size network.

A gain crosstalk effect between the neurons was discovered during the electrical testing. The mechanism for this crosstalk was found to be transients induced on the

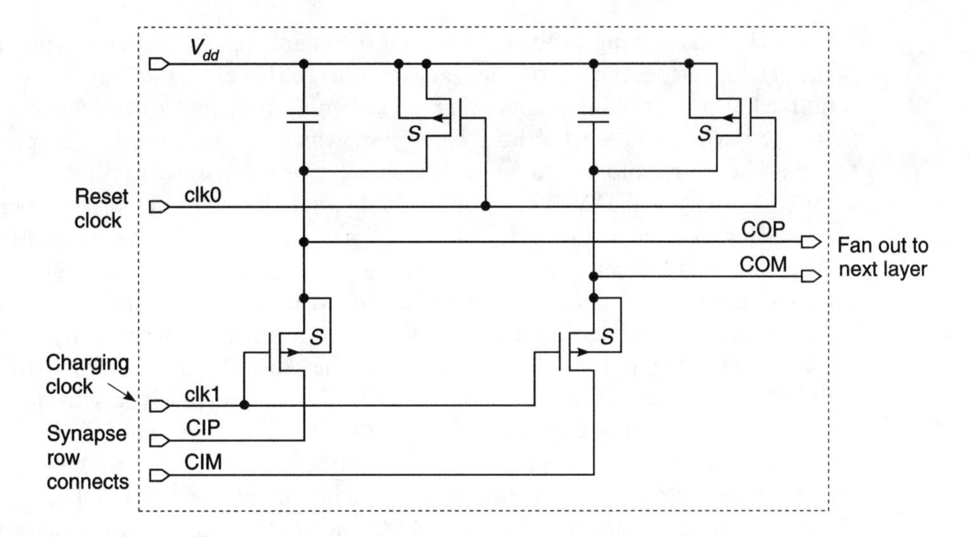

Figure 16.4 Wattle neuron circuit diagram.

TABLE 16.1 ELECTRICAL PERFORMANCE AND CHARACTERISTICS OF WATTLE

Parameter	Value	Comment
Area	$2.2 \times 2.2\,\text{mm}^2$	
Technology	1.2-μm N-well CMOS 2M2P	Standard process
Resolution	Weights 6 bits, gain 7 bits	Weights on chip, gains off
Energy per connection	43 pJ	All weights maximum
Least significant bit (LSB) DAC current	200 pA	Typical
Feedforward delay	1.6 ms	At 200 pA, 3 V supply
Synapse offset	5 mV	Typical maximum
Gain crosstalk delta	20%	Maximum

current source reference lines going to all the synapses as individual neuron gains timed out. The worst-case crosstalk coupled to a hidden-layer neuron was found to be a 20% deviation from the singularly activated value. However, the training results of the chip do not appear to suffer significantly from this effect.

Wattle has been trained in-loop using a variation on the Combined Search Algorithm (CSA) for limited precision training [18] (combination of weight perturbation and axial random search). The variation consists of training the gains in exactly the same way that the weights are trained rather thans scaling the gains based on the average value of the weights connected to a neuron. In this case, the gains are 7 bits and are implemented off-chip. The in-loop training results of the chip are compared with that of a numerical model. The numerical model used does not model noise, crosstalk, offsets, or common-mode saturation, and in this sense is an idealized model of the chip.

The ICEG training problem consisted of separating the ICEG morphology of patients with retrograde ventricular tachycardia. In this case, for Wattle, since there is no facility for sampling, segmenting, and aligning the input to the network, these tasks are carried out on the host computer, which is executing the training procedure. The morphologies to be separated were normal sinus rhythm (NSR) and ventricular tachycardia (VT) on an individual patient basis. In this case, 10 morphology samples are used as inputs to a 10 : H : 2 network where H was varied from 1 to 6. The chip was trained for each patient using trainable gains. The eight training patterns were selected and the network trained twice on each architecture. Table 16.2 summarizes the classification performance on data not previously seen by the network for simulated and in-loop chip training. The table shows that for most patients over 90% correct classification can be obtained for the previously unseen data. Note that this result has to be considered in light of the fact that current ICTS using only rate-based classification techniques cannot distinguish these arrhythmias at all and are therefore not used for patients known to have such a condition. Of course there are significant consequences for misclassifications. For false-positive detection of dangerous rhythms there are the problems of patient discomfort and battery depletion for administering unnecessary therapy and the possibility that therapy can induce an arrhythmia condition. For false-negative classification errors, the consequences can be heart muscle damage and possibly death. However, these risks have to be weighed against the availability, cost, effectiveness, and risks of alternative treatments such as open heart surgery and transplantation. Thus, there are two implications for this work, in that morphology classifiers provide a means to increase the specificity of classifications for patients that currently receive ICTSs and enable the availability of ICTSs to some patients who could not previously benefit from them.

TABLE 16.2 ICEG GENERALIZATION PERFORMANCE

Number of hidden units	Testing Patterns Correct, %			
	Simulated		Chip Training	
	NSR	VT	NSR	VT
1	24	100	91	99
	24	100	90	98
2	94	99	97	93
	94	99	96	93
3	97	87	100	84
	97	87	100	82
4	96	97	90	97
	96	97	91	99
5	100	99	100	97
	100	99	97	99
6	94	93	94	95
	94	93	96	97

16.5.3. Snake

Snake is an analog VLSI neural network designed to perform cardiac morphology classification tasks [19]. The network is a 10 : 6 : 3 multilayer perceptron with on-chip digital weight storage, a bucket brigade input to feed the ICEG to the network and a winner-take-all circuit at the output. The following sections briefly describe the architecture, implementation, and performance of the Snake chip.

Architecture. A floor plan of the chip appears in Figure 16.5. Similar to the Wattle chip, the network consists of two synapse arrays and associated neurons. However, instead of direct inputs from the pads, the analog input is serial and sampled in time by the bucket brigade device. The access to the weights is via the shift registers on the boundary of the chip. A winner-take-all module is provided for multiple-output applications, and as in the case of Wattle, a weighted current source is used to provide references to the synapse arrays. A functional overview of the Snake chip appears in Figure 16.6.

Implementation. The synapses used on the Snake chip are identical to those used on Wattle (see Fig. 16.3). The inputs of the synapses are connected to the bucket brigade device (BBD), one cell of which is shown in Figure 16.7. The BBD operates by transferring charge deficits from S to D in each of the cells. PHI1 and PHI2 are two-phase nonoverlapping clocks. The cell is buffered from the synapse array to maintain high charge transfer efficiency. A sample-and-hold facility is provided to store the input on the gates of the synapses. The BBD clocks are generated off-chip and are controlled by the QRS complex detector in the implantable cardioverter defibrillator (ICD).

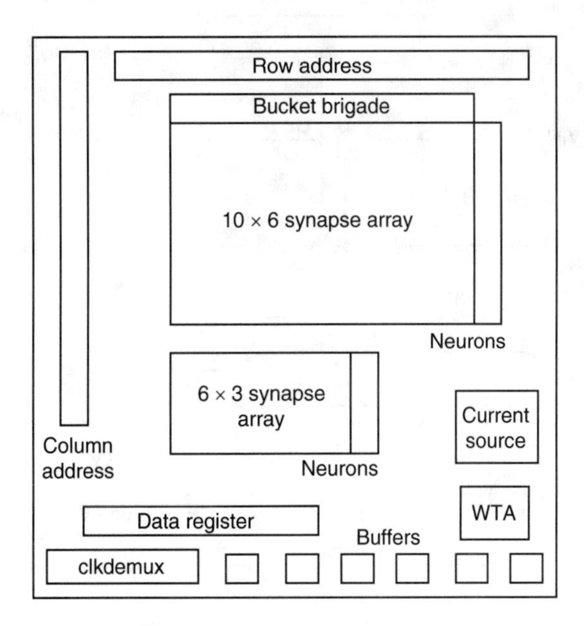

Figure 16.5 Floor plan of Snake chip.

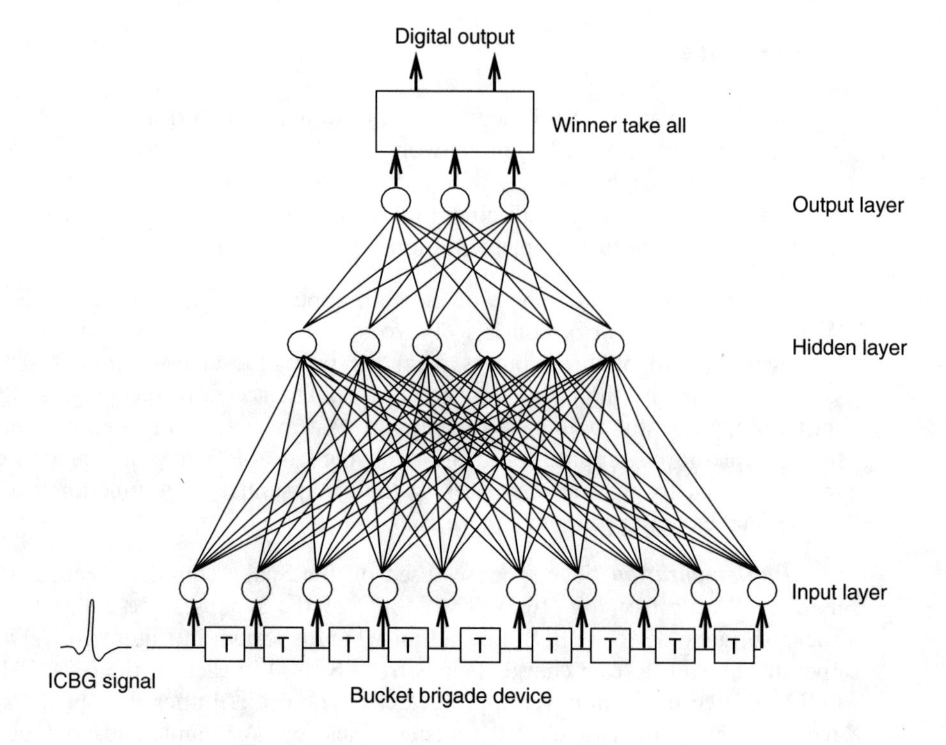

Figure 16.6 Functional diagram of Snake chip.

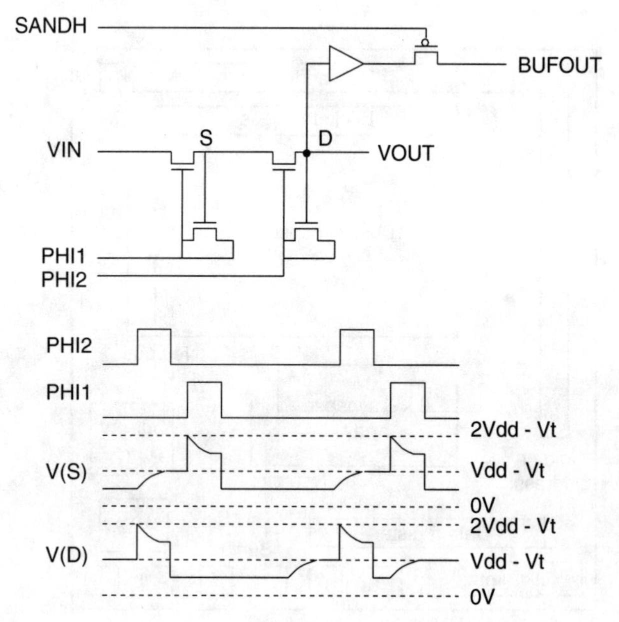

Figure 16.7 Circuit diagram of BBD cell.

Due to the low power requirements, the bias currents of the synapse arrays are of the order of hundreds of nanoamperes; hence the neurons must provide an effective resistance of many megaohms to feed the next synapse layer while also providing gain control. Without special high-resistance polysilicon, simple resistive neurons use prohibitive area. However, for larger networks with fan-in much greater than 10, an additional problem of common-mode cancellation is encountered. that is, as the fan-in increases, a larger common-mode range is required or a cancellation scheme using common-mode feedback is needed.

The neuron of Figure 16.8 implements such a cancellation scheme. The mirrors M_0/M_2 and M_1/M_3 divide the input current and facilitate the sum at the drain of M_7. Here, M_7/M_8 mirrors the sum so that it may be split into two equal currents by the mirrors formed by M_4, M_5, and M_6, which are then subtracted from the input currents. Thus, the differential voltage $V_p - V_m$ is a function of the transistor transconductances, the common-mode input current, and the feedback factor. The gain of the neuron can be controlled by varying the width-to-length ratio of the mirror transistors M_0 and M_1. The implementation in this case allows seven gain combinations using a 3-bit random-access memory (RAM) cell to store the gain.

The importance of a common-mode cancellation scheme for large networks can be seen when compared to the straightforward approach of resistive or switched-capacitor neurons. This may be illustrated by considering the energy usage of the two approaches. First, we need to define the required gain of the neuron as a function of its fan-in. If we assume that useful inputs to the network are mostly sparse, that is, with a small fraction of non-zero values, then the gain is largely independent of the fan-in, yet the common-mode signal increases linearly with fan-in. For the case of a neuron which does not cancel the common mode, the power supply voltage must be increased to accommodate the common-mode signal, thus leading to a quadratic increase in energy use with fan-in. A common-mode canceling neuron, on the other hand, suffers only a linear increase in energy use with fan-in since extra voltage range is not required and the increased energy use arises only due to the linear increase in common-mode current. A detailed analysis of this neuron appears in [20].

Performance Evaluation. The power consumption performance of the Snake chip is illustrated in Table 16.3. Comparing the power consumption of the Wattle chip to that of the Snake chip, we see that Snake uses considerably more

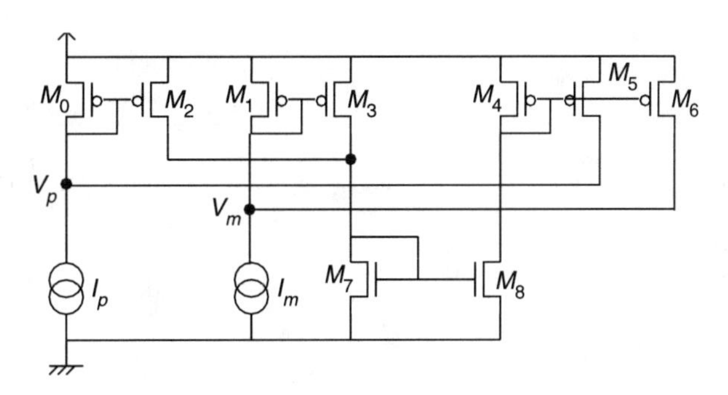

Figure 16.8 Circuit diagram of Snake neuron.

TABLE 16.3 POWER CONSUMPTION OF CHIP

Module	Power (nW)	Conditions
BBD	40	250 Hz rate
Network	128	75 nA LSB
Buffers	18	3 μA bias
Total	186	120 bpm

power while still being suitably low for an implantable device. There are a number of reasons for the increased power consumption. First, Wattle is not a complete system for ICEG classification in the sense that no facility was included for applying the ICEG to the inputs of the chip. Second, the neurons in the Snake chip were designed so that they would be useful in considerably larger networks than Snake and to overcome fan-in limitations of simpler designs. Hence, a power penalty has been paid for the increased durability of the neuron. Thus, Wattle plus the BBD would be optimal for small networks whereas the Snake implementation will scale to larger architectures.

The Snake chip was trained on seven different patients separately all of whom had VT with one-to-one retrograde conduction. Table 16.4 summarizes the training statistics by best training run obtained for the given patient. Notice that 100% training performance is obtained for all but two of the patients. (Each of the patients was trained with eight of each class of arrhythmia.) Once the morphologies to be distinguished have been learned for a given patient, the remainder of the patient data base is played back in a continuous stream and the outputs of the classifier at each QRS complex are logged and may be compared to the classifications of a cardiologist.

Table 16.5 summarizes the performance of the system on the seven patients. Most of the patients show a correct classification rate well over 90%, whereas a timing-based classifier cannot separate these arrhythmias. The network architecture used to obtain these results was 10 : 6 : 1, the unused neurons being disabled by setting their input weights to zero. The input signal was sampled at a rate of 250 Hz.

TABLE 16.4 TRAINING PERFORMANCE OF NETWORK ON SEVEN PATIENTS WITH ICD IN LOOP

		% Correct	
Patient	Training Iterations	ST	VT
1	56	100	100
2	200+	100	87.5
3	200+	87.5	100
4	46	100	100
5	200+	100	100
6	140	100	100
7	14	100	100

TABLE 16.5 CLASSIFICATION PERFORMANCE OF NETWORK ON SEVEN PATIENTS WITH ICD IN LOOP

	Number of Complexes		% Correct	
Patient	ST	VT	ST	VT
1	440	61	100	98
2	94	57	100	95
3	67	146	78	99
4	166	65	91	99
5	61	96	97	93
6	61	99	97	100
7	28	80	96	99

Referring to Table 16.4, we see that patients 2 and 3 were not trained to perfectly distinguish the classes. In the case of patient 2 this did not appear to affect the generalization performance significantly. Hence, this may be due to a rare VT morphology or, more likely, the training algorithm being trapped in a local minima. However, for the case of patient 3, a high false-positive error rate for VT is obtained; that is, many of the STs were classified as VTs. Inspection of this patient data showed that in fact, typically, the morphologies of the two rhythms were very close.

16.6. DISCUSSION

In this chapter we have presented techniques for implementing ultra-low-power neural networks for arrhythmia morphology analysis. The chips were successfully trained using perturbation-based algorithms. The energy usage of these chips is far less than what present state-of-the-art digital implementations can achieve.

An important aspect of the design methodology employed in Wattle and Snake chips is their tolerance for transistor mismatch and other fabrication imperfections. These imperfections are worked around by the training (in-loop) of these chips.

The inclusion of techniques presented in this chapter in present defibrillators is being studied and a number of engineering issues are currently being addressed.

Our present work aims at overcoming these engineering problems in order to make the implementation more flexible. In particular, we are addressing the problem of biasing and the improvement of the dynamic range of synapses and neurons. A longer term goal of this work is to make the system adapt on-line while implanted in the patient [21]. This is an important step since it is known that a patient's morphology can vary over time due to tissue growth and variations in drug treatments. Such a self-supervised classifier would obviate the need for long-term tuning of the classifier under clinical conditions.

16.6.1. Other Implementation Methods

We are also considering the competitiveness of two emerging digital implementation methodologies: low threshold and adiabatic computing. Both methodologies are still at the experimental stage, but preliminary results indicate they could have some potential for ICTS applications.

Adiabatic Logic. Adiabatic logic (AL) is a completely different low-energy logic design approach (there are a substantial number of references on adiabatic implementations; for a recent work the reader is referred to [22]. The principle behind AL is that of extracting each charge carrier at the lowest feasible voltage and returning it at the highest feasible voltage. To achieve that, power and clocking signals with soft edges (ramplike) are required. Due to the low-bandwidth requirements, ICTS could readily benefit from adiabatic computer techniques.

Adiabatic logic has received much attention recently because of its potential in significantly reducing power usage without requiring a specialized MOS fabrication process (in contrast with other approaches such as that of ultralow threshold and supply [9, 23]. This is of importance in implantable medical devices as the introduction of a new fabrication process in any part of the system requires additional verification by the US Food and Drug Administration (FDA). Another important aspect of AL is that it does not conflict with other energy minimization techniques such as the lowering of circuit supply voltage. Furthermore, it can be combined with energy restoration schemes to enable the recovery of transferred energy.

Low-Threshold Systems. By lowering the threshold of MOS transistors, still lower supply voltages may be used to design digital logic circuits [9, 23].[2] This could lead to significant energy savings being achieved. However, lowering the threshold of the transistor comes at the cost of increasing the leakage current and hence static power dissipation. Hence, there is a trade-off as to how low to set the logic threshold according to the logic activity. (The logic activity is a measure of the fraction of nodes being toggled in a circuit.) One approach has been to fabricate zero-threshold devices and then to create a threshold by biasing the substrate. This type of approach can be seen as an extension of techniques used to lower the supply voltage but as yet has not received as much attention due to the cost of modifying fabrication process parameters. As indicated in the previous section, a drawback of low-threshold techniques when applied to ICTS is the costly validation requirements for the FDA. However, if the energy reduction is substantial, FDA validation will not represent a serious obstacle in the long term.

16.7. CONCLUSIONS

In this chapter we have presented methods, architectures, and implementations of subthreshold CMOS neural networks for arrhythmia morphology analysis. One of the chips has been interfaced to a research version of a commercial defibrillator and

[2]As low as 200 mV has been suggested.

been demonstrated to train and generalize successfully. The work shows that analog subthreshold CMOS is an attractive and well-suited medium for ICTS. The outcome of this work is applicable to many other battery-operated implantable devices, for example behind the ear cochlear implants.

References

[1] L. J. Stotts, "Introduction to Implantable Biomedical IC Design," *IEEE Circuits Devices Mag.*, vol. 5, no. 1, pp. 12–18, Jan. 1989.

[2] C. Svensson and D. Lui, "Trading Speed for Low Power by Choice of Supply and Threshold Voltages, *IEEE J. Solid-State Circuits*, vol. 28, pp. 10–17, 1993.

[3] V. Von Kaenel, P. Macken, and M. G. R. Degrauwe, "A Voltage Reduction Technique for Battery Operated Systems," *IEEE J. Solid-State Circuits*, vol. 25, pp. 1136–1140, 1990.

[4] R. Brodersen, A. Chandrakasan, and S. Sheng, "Low Power Signal Processing Systems," in *VLSI Signal Processing*, vol. 5, K. Yao, R. Jain, and W. Przytula (eds.), IEEE, New York, 1992, pp. 3–13.

[5] K. Yano, T. Yamanka, T. Nishida, M. Saitor, K. Shimohigashir, and A. Shimizu, "A 3.8 ns CMOS 16 × 16-b Multiplier Using Complementary Pass Transistor Logic," *IEEE J. Solid-State Circuits*, vol. 25, pp. 388–395, 1990.

[6] R. F. Lyon, "Cost, Power, and Parallelism in Speech and Signal Processing," in *Proceedings of IEEE Custom Integrated Circuits Conference*, pp. 15.1.1–15.1.9, 1993.

[7] M. R. C. M. Berkelaar and J. F. M. Theeuwen, "Real Area-Power-Delay Trade Off in the Euclid Synthesis System," In *Proc. IEEE Custom Integrated Circuits Conf.*, pp. 14.3.1–14.3.4, 1990.

[8] E. A. Vittoz, "Micropower Techniques," in *Design of MOS VLSI Circuits for Telecommunications*, Y. Tsividis and P. Antognetti (eds.), Prentice-Hall, Englewood Cliffs, NJ, 1985.

[9] J. Burr and A. Peterson, "Ultra Low Power CMOS Technology," in *3rd NASA Symposium on VLSI Design*, 1991.

[10] G. M. Jacobs and R. W. Brodersen, "A Fully Asynchronous Digital Signal Processor Using Self-timed Circuits," *IEEE J. Solid-State Circuits*, vol. 25, no. 6, pp. 1526–1537, 1990.

[11] Y. P. Tsividis, *Operation and Modeling of the MOS Transistor*, McGraw-Hill, 1988.

[12] K. Kusumoto, K. Murata, A. Matsuzawa, S. Tada, M. Maruyama, K. Oka, and H. Konishi, "A 10 b 20 MHz 30mW Pipelined Interpolating CMOS ADC," in *Proceedings of the IEEE 1993 International Solid-State Circuits Conference*, pp. 62–63, 1993.

[13] M. A. Jabri, R. J. Coggins, and B. G. Flower, *Adaptive Analog VLSI Neural Systems*, Chapman & Hall, London, 1996.

[14] M. A. Jabri, "Practical Performance and Credit Assignment Efficiency of Analog Multi-Layer Perceptron Perturbation Based Training Algorithms," SEDAL Technical Report, July 1994.

[15] M. Jabri and B. Flower, "Weight Perturbation: An Optimal Architecture and Learning Technique for Analog VLSI Feedforward and Recurrent Multilayer Networks," *IEEE Trans. Neural Networks*, vol. 3, no. 1, pp. 154–157, 1992.

[16] B. G. Flower and M. A. Jabri, "Summed Weight Neuron Perturbation: A o(n) Improvement Over Weight Perturbation," in *Advances in Neural Information Processing Systems*, vol. 5, S. Jos Hanson, J. D. Cowan, and L. Giles (eds.), Morgan Kauffmann, San Mateo, CA, 1993, pp. 212–219.

[17] R. J. Coggins and M. A. Jabri, "Wattle: A Trainable Gain Analogue VLSI Neural Network," in *Neural Information Processing Systems*, Morgan Kauffman, 1994, pp. 874–881.

[18] Y. Xie and M. Jabri, "Training Limited Precision Feedforward Neural Networks," in *Proceedings of 3rd ACNN 92*, pp. 68–71, 1992.

[19] R. J. Coggins and M. A. Jabri, "Low Power Intracardiac Electrogram Classification Using Analogue VLSI," in *Microneuro '94*, pp. 376–382, 1994.

[20] R. J. Coggins, M. A. Jabri, B. G. Flower, and S. J. Pickard, "A Hybrid Analog and Digital VLSI Neural Network for Intracardiac Morphology Classification," *IEEE J. Solid-State Circuits*, vol. 30, no. 5, pp. 542–550, May 1995.

[21] M. A. Jabri, *Neural Networks for ECG Signal Interpolation, Handbook of Neural Computation*, Institute of Physics Publishing and Oxford University Press, G2.5, 1996.

[22] A. Kramer et al., "Adiabatic Computing with the 2n-2n2d Logic Family," in *Snowbird Neural Computing Workshop*, Utah, 1994.

[23] J. Burr and A. Peterson, "Energy Considerations in Multichip Module Based Multiprocessors," in *IEEE International Conference on Computer Design*, pp. 593–600, Oct. 1991.

Chapter 17

An Information Theoretic Framework for Comparing the Bit Energy of Signal Representations at the Circuit Level

Andreas G. Andreou
Department of Electrical and Computer Engineering, Johns Hopkins University, Baltimore, Maryland 21218
Paul M. Furth
Department of Electrical and Computer Engineering, New Mexico State University, Las Cruces, New Mexico 88003

17.1 INTRODUCTION

Over the last few years, new emerging opportunities in information technologies point toward markets for portable systems where battery operation, light weight, and small size will be in demand. From an economic perspective, miniaturization and high levels of system integration, with an implicit potential for large markets, are predicted to be the technology drivers in the decades ahead [1]. From a technology and engineering perspective, the development of these systems will be done with *energy efficiency* as the prime engineering constraint, taking a lead over other considerations. The cost and reliability of these portable systems are also important factors.

With such emphasis on portable operation and real-time processing, there has also been an intense discussion as to how much processing should be done in analog and how much in digital to achieve a high throughput while maintaining low power dissipation commensurate with battery operation. For example, it has been argued and demonstrated experimentally [2] that low precision analog may be the technology of choice in applications where the goal is not the precise restitution of information but rather some perceptual or sensory communication task in audition or vision.

In this chapter we outline a framework to formalize and quantify the notion of "low-power information processing," "low-power signal processing," and "low-power computation." We treat microelectronic systems as communication channels (see Fig. 17.1) transmitting messages in the presence of noise [3]. We aim at some fundamental treatment of the problem and employ the formalism of information

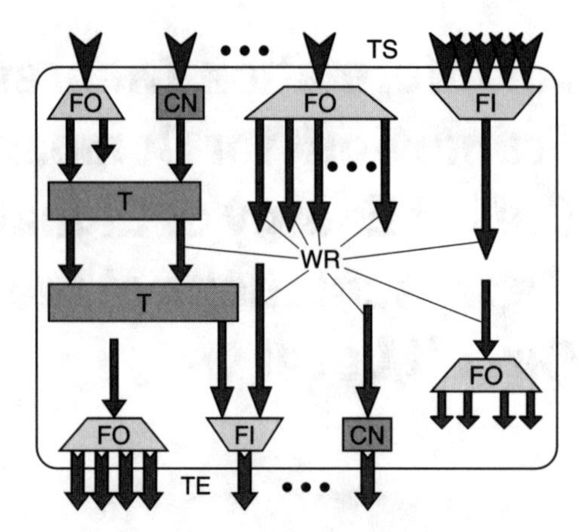

Figure 17.1 Communication network model for VLSI microsystems. There are sensors (TS) and actuators (TE), data converters with one input and one output (CN) or multiple inputs and outputs (FO, FI), wires, shift registers, latches (WR), gates, and ALU blocks (T).

theory and Shannon's entropy H as a measure of information [4]. "Bit energy," the ratio of Shannon's channel capacity to the power dissipated (bits per joule), is used as the goodness criterion to compare information flow in physical channels where encoding, transmission, and decoding of messages must be carried out under strict power supply/dissipation constraints.

Landauer [5] discussed the ultimate limitations of computing and employs the entropy reduction in a logical calculation, which is an irreversible logical mapping, as a measure of useful work done and for which there should be associated energy costs comparable to kT. He also suggests that all one-to-one mappings could be reversed and therefore, in principle, should not consume energy, i.e., could be physically reversible. Mead and Conway [6] further elaborate on the view of computation as entropy reduction and introduce the notion of *logical entropy* and *spatial entropy*. Logical entropy S_L is related to the logical operations performed on a given assemblage of data while *spatial entropy* S_S is a measure of the data being in the wrong place. It requires data communication to get the data in the right place to do the logical operations. In essence, our transfer channels relate to spatial entropy and transform channels relate to logical entropy introduced in [6]. Hosticka [7] employs information theoretic arguments to compare different signal representations for signal processing and our work in [8] improves and extends on the work of Hosticka. Information theoretic measures of complexity for computation—logical reduction of entropy—have been discussed in [9] and [10], and different groups have used information theoretic ideas to analyze analog [12] and digital synchronous [12] and asynchronous [13] very large scale integrated (VLSI) circuits.

This chapter is organized as follows. In Sections 17.2 and 17.3 we begin with a communication and information theory primer, introducing the reader to the key concepts and definitions. This is followed by Section 17.4 where four different methods of encoding data in VLSI systems are examined and their figures of merit discussed. In Section 17.5 we discuss the results and conclude the chapter.

17.2. A COMMUNICATION THEORY MODEL FOR VLSI

The connection between information theory and microelectronic hardware VLSI systems can be made by viewing the latter as information-processing, transformation/calculation, and communication channels in the presence of noise. We consider two kinds of channels—*information transfer* and *information transform*—in subsystems organized at different levels in a hierarchical network architecture. For example, shift registers, latches, digital inverters, analog delay lines, sampled data delay lines, bucket brigades, and charge-coupled device (CCD) shift registers are information transfer structures while logical gates, arithmetic logic units (ALUs), and analog and digital filters are information transform structures.

Formally, given a subsystem, the distinction is made by considering the entropy at the input $H(X)$ and output $H(Y)$. In an information transfer channel $H(Y) = H(X)$ while in an information transform channel $H(Y) < H(X)$, where entropy is reduced through an information-processing step that reduces the entropy of the data calculation or filtering.

Furthermore, we make the distinction between *logical* ξ_i and *physical* ϕ_i information transform/transfer steps (see Fig. 17.2). The ensembles $\Xi = \{\xi_1, \xi_2, \ldots, \xi_n\}$ and $\Phi = \{\phi_1, \phi_2, \ldots, \phi_n\}$ represent the architecture. Logical processing steps ξ_i are determined by the algorithms for data processing and noise is introduced because of quantization effects (both in data converters and in arithmetic transformations due to finite precision arithmetic).

Physical transformations ϕ_i correspond to transduction and actuation of physical stimuli and data conversion such as A/D and D/A, and noise represents thermal noise, shot noise, $1/f$ noise, structural variability, and processing-induced noise such as switching noise. Note that the classical linear amplifier circuits are simply "wires" performing a physical transfer of data (from voltage to voltage or from voltage to current) which does not change the information and therefore is an identity information transform.

In terms of classical communication theory concepts in macroscopic systems, the logical transformations correspond to data coding while the physical transformations correspond to modulation and physical transmission.

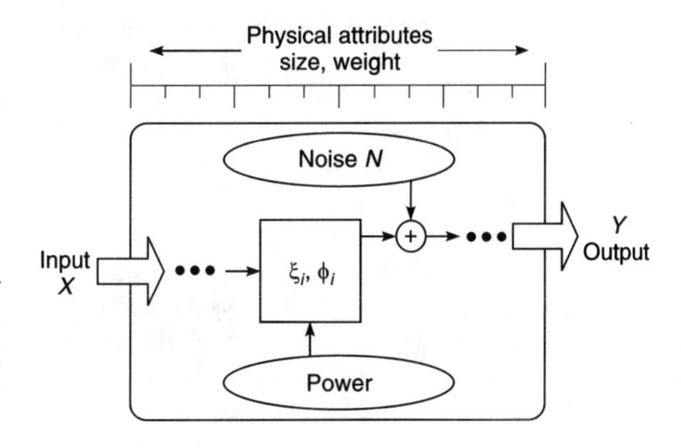

Figure 17.2 Canonical model of information processing in the presence of noise. We model channel noise as a collection of noiseless logical information transform/transfer steps followed by additive noise.

At each level in the system architecture and at the level of the subsystem, the goal is to attain a maximum or a prescribed rate of information, while, power, size, and weight as well as the complexity of the information-processing machine represent the cost. In the information transfer subsystems the information per symbol rate R_{xfer} in bits is determined by the source-input statistics. In the case of a transform channel, the information rate R_{xform} is a fundamental property of data processing *as well as* the source statistics. At this point it is obvious that a distinction between transform and transfer channels is rather superficial as the transfer channels are systems in a subspace of the transform channels where the mapping between the input and the output is the identity, i.e., the entropy at the output equals to the entropy of the input. However, we find the distinction between the two kinds of channels is intuitively appealing and therefore maintain it throughout our discussion.

Here we are interested in the fundamental limits and choose as input to the channel a signal whose distribution has the maximum entropy, and therefore we derive results that *are not* dependent on the statistics of the source. As such we revisit and extend earlier work by Hosticka [7] and investigate the performance, as measured in bit energy, of continuous/discrete-value and continuous/discrete-time signal representations as realized in electrical circuits (communication channels). Without loss of generality, we assume that signals are voltages which are limited by the power supply rails in the circuits and use simple but realistic models of noise and power dissipation. We consider information transfer channels, where the goal is to maintain a given information rate per unit of time R_t in bits per second, and we derive the signal-to-noise ratio, average power dissipation, and Shannon channel capacity while carefully stating the assumptions underlying our derivations. In previous work with analog filter banks [11] the derivation assumed a signal-dependent statistic (speech) for the input and obtained the optimal architecture for a system to perform a one-to-N encoding for information under the assumption that each of the output channels carries equal amounts of information.

17.3. INFORMATION AND COMMUNICATION THEORY PRIMER

The concept of information was originally conceived by Hartley [14], who was interested in the transmission of messages over telephony wires and radio links. Information theory as we know it today was fully developed 20 years later by Shannon [4]. The Hartley and Shannon definitions of information are measures of *uncertainty* based on classical sets and probability theory, respectively.

Information theory has had a profound impact on a wide range of fields from telemetry to speech recognition to image encoding. An essential aspect of information theory is that it requires an a priori definition of what is meant by "information." Once this definition is established, different means of encoding and transmitting messages in physical communications channels can be objectively evaluated. Ultimately, information theory provides the link between the logical and physical aspects of the computation and the particulars of the message that is transferred/transformed and hence the complexity of the problem that is solved.

17.3.1. Entropy and Mutual Information

Shannon's entropy $H(X)$, in bits, for a discrete source X generating symbols $\{X_1, X_2, X_3, \ldots, X_N\}$ with probability $P(X_i) = p_i$ is given by

$$H(X) = -\sum_{i=1}^{N} p_i \log_2 p_i \tag{17.1}$$

The above definition of information is essentially an average measure of uncertainty in the symbols generated by the source, where the most improbable symbols have the highest information content. The source which produces output symbols with equal probability achieves the maximum entropy [4].

The entropic measure of information employed for the discrete source can be extended to the case of continuous distributions [4], although there are technical difficulties involved in a formal treatment that are beyond the scope of this chapter. In the case of continuous variables, the definition of entropy is a relative one and the *differential* entropy $h(X)$ of a random variable X with probability density function $f(x)$ is defined as

$$h(X) = -\int_{-\infty}^{+\infty} f(x) \log_2 f(x)\, dx \tag{17.2}$$

We define the rate of information generated by the source per unit of time as $R_{ts} = f_{ts}H(X)$, where f_{ts} is the rate at which symbols are produced and has the unit of frequency (hertz).

We now consider channels where a transformation or transfer through it can be defined as a mapping $\Xi[X] \rightarrow Y$ or $\Phi[X] \rightarrow Y$, where X and Y are the input and output of the channel, respectively. Then $H(X, Y)$ is the joined entropy defined as

$$H(X, Y) = H(X) + H(Y|X) = H(Y) + H(X|Y) \tag{17.3}$$

where $H(X)$ and $H(Y)$ are the entropies at the input and at the output of the channel and the conditional entropies $H(Y|X)$ and $H(X|Y)$ represent the entropy of the output when the input is known and conversely. Conditional entropies are defined similarly to the entropy definition in Eqs. (17.1) and (17.2).

In the case of the continuous channel, the conditional differential entropy $h(x|y)$ of the input x conditioned on the output y is defined as

$$h(x|y) = -\iint f(x, y) \log_2 \frac{f(x, y)}{f(y)}\, dx\, dy \tag{17.4}$$

The actual information transmitted through the channel, known as the *mutual information* $I(X; Y)$, can be obtained by subtracting the conditional entropy $H(X|Y)$ from the information production $H(X)$. The conditional entropy $H(X|Y)$ is often referred to as the equivocation and is the average measure of ambiguity in the received signal:

$$I(X;\, Y) \equiv H(X) - H(X|Y) = H(Y) - H(Y|X) \tag{17.5}$$

In general, the mutual information $I(X;\, Y)$ is a function of both the input statistics and the transformation in the channel. In the case where the channel performs the identity transform, which will be the case of an information transfer channel, $\Xi = 1$ and $Y = X$, then $H(X|Y) = 0$, i.e, the uncertainty in X given Y is zero and $I(X;\, Y) = H(X)$.

Often, a system can be modeled using the identity transform followed by an additive noise source; this is the approach that we will take here. When the signal and noise are independent,

$$I(X;\, Y) = H(Y) - H(N) \tag{17.6}$$

where $H(N)$ is the entropy of the noise. This will be a good model for a scalar quantizer (data converter) or vector quantizer (classifier).

The rate R_t of information flow per unit time at the output of the channel is given by

$$R_t = f_{ts} I(X;\, Y) \tag{17.7}$$

where f_{ts} was defined as the temporal rate of symbol generation at the source.

17.3.2. Shannon Channel Capacity

The maximum information rate R_t of error-free transmission that can be processed is limited by the system Shannon capacity $\mathcal{C}$ [4], and furthermore it was shown that an information rate R_t is achievable with probability of error approaching zero through coding so long as $\mathcal{C} > R_t$. Although this is a remarkable result, the proof does not give an algorithm of how to find the best coding strategy.

Shannon capacity is defined through the mutual information $I(X;\, Y)$ between the input X and output Y defined in terms of probability density functions and in the continuous case takes the form [4]

$$\mathcal{C} \equiv I(X;\, Y) = \lim_{T\to\infty} \max_{f(x)} \frac{1}{T} \iint f(x, y) \log_2 \frac{f(x, y)}{f(x)f(y)}\, dx\, dy \tag{17.8}$$

which is given by the signal-to-noise ratio S/N and the message bandwidth f_p, as defined by [4]

$$\mathcal{C} = h(Y) - h(N) \tag{17.9}$$

where the entropy per second of the received ensemble is $h(Y) = f_p \log_2 2\pi e(S+N)$ and the noise entropy per second is $h(N) = f_p \log_2(2\pi e N)$, resulting in a channel capacity C given by

$$C = f_p \log_2\left(1 + \frac{S}{N}\right) \tag{17.10}$$

where S and N are the average signal and noise powers, respectively. The dimension of C is bits per second. This law applies to systems having an average signal power constraint S subject to additive white Gaussian noise of power N. According to [4], in order to approach this limiting rate of transmission, the transmitted signals must approximate, in statistical properties, white noise.

More generally, Shannon capacity is given in terms of the signal $S(f)$ and noise power $N(f)$ as

$$C = \int_0^\infty \log_2\left(1 + \frac{S(f)}{N(f)}\right) df \tag{17.11}$$

In the case of a peak signal power constraint, the channel capacity C in a frequency band f_p perturbed by white Gaussian noise of power N is limited by [4]

$$C \leq f_p \log_2\left(1 + \frac{2S_{\text{peak}}}{\pi e N}\right) \tag{17.12}$$

for sufficiently large S_{peak}/N, where the instantaneous signal power is limited to S_{peak} at every sample point of a *discrete-time* signal. Sample points are assumed to be taken at the Nyquist rate, i.e., every $1/2f_p$ seconds. The discrete-time signal which achieves this capacity is one for which the samples are independent and uniformly distributed in the range $-\sqrt{S_{\text{peak}}}$ to $+\sqrt{S_{\text{peak}}}$.

Let us assume now without loss of generality that the physical encoding of information takes the form of voltages. These voltages are constrained to operate in the range from 0 to V volts. Constraints on the maximum power supply voltages can be included in this formulation as they affect the maximum value of the voltage V. Referenced to $V/2$, then, the peak power is given by

$$S_{\text{peak}} = \tfrac{1}{4} V^2 \tag{17.13}$$

An independent uniformly distributed discrete-time signal achieves the channel capacity. The average power, or variance, of such a signal is given by [16]

$$S = \tfrac{1}{12} V^2 \tag{17.14}$$

Because each sample is independent, the signal power is evenly distributed across all frequencies in the range 0–f_p hertz, so that the power spectrum is given by

$$\frac{\overline{V_{\text{in},s}^2}}{\Delta f} = \frac{V^2}{12 f_p} \tag{17.15}$$

Noting that $S_{\text{peak}} = 3S$, we substitute for S_{peak} in (17.12) to obtain

$$\mathcal{C} \leq f_p \log_2\left(1 + \frac{6}{\pi e}\frac{S}{N}\right) \tag{17.16}$$

This is the equation for the capacity of a discrete-time system subject to a peak power constraint S_{peak} and additive white Gaussian noise of power N.

17.4. FOUR SIGNAL REPRESENTATIONS: PHYSICAL ENCODINGS

Four types of signal representations are distinguished in this work, ranging from what is commonly known as analog to what is commonly known as digital (see Fig. 17.3) and they are simply four different types of physical encoding or "modulation": continuous-value, continuous-time (CVCT), continuous-value, discrete-time (CVDT), discrete-value, discrete-time (DVDT), and discrete-value, continuous-time (DVCT).

For the first three signal representations, we compute the noise power of a circuit which, as closely as possible, implements a communication process over a unit distance of space. This is nothing but a delay function, a useful function in all digital and analog signal-processing algorithms and their implementations. We then compute the signal power of an input signal which achieves the maximum informa-

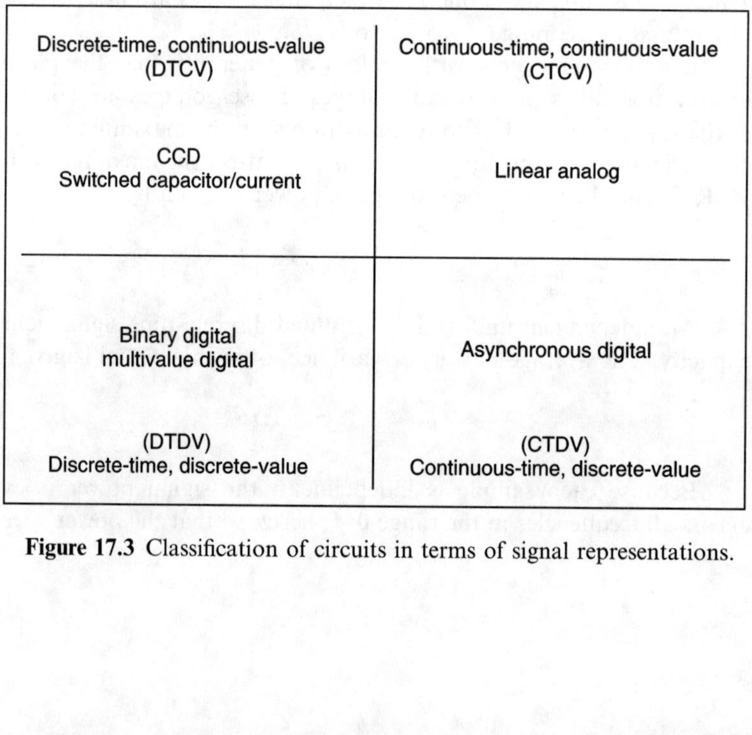

Discrete-time, continuous-value (DTCV) CCD Switched capacitor/current	Continuous-time, continuous-value (CTCV) Linear analog
Binary digital multivalue digital (DTDV) Discrete-time, discrete-value	Asynchronous digital (CTDV) Continuous-time, discrete-value

Figure 17.3 Classification of circuits in terms of signal representations.

tion rate, or capacity, of the circuit. We also derive the power dissipated in the channel using such an input signal. Further, the signal-to-noise ratio is written as a function of the power consumption. We compute the capacity of the circuit and relate this quantity to the power consumption so as to finally arrive at a measure of the bit energy.

The fourth signal representation, DVCT, is currently under investigation [16]. In the present work we set up the problem and summarize some of our preliminary results.

17.4.1. Continuous-Value, Continuous-Time Circuit

Figure 17.4 gives the CVCT analog of a delay function—a first-order RC low-pass filter. At both the input and output of the RC filter are ideal noiseless low-pass filters. These ideal filters define the message bandwidth f_p. The noise power spectral density of a nominal resistor of value R (one-sided) is

$$\frac{\overline{V_n^2}}{\Delta f} = 4kTR \tag{17.17}$$

where k is Boltzmann's constant and T is absolute temperature. Every physical embodiment of a resistor demonstrates a second source of noise at low enough frequencies, called $1/f$ or flicker noise, which we ignore here. The noise-equivalent bandwidth of this filter (ENBW) is $1/4RC$. The noise power across the capacitor C is the product of these two terms, i.e.,

$$\overline{V_{c,n}^2} = \text{ENBW} \cdot 4kTR = \frac{kT}{C} \tag{17.18}$$

Hence the mean-square noise on the capacitor is not a function of the resistance. This is a well-known result that can be inferred directly from the equipartition theorem. The noise in a system/circuit is equally split among the N degrees of freedom in the system. In this case the capacitor is the only energy storage element/degree of freedom, and thus the noise is simply given by the expression above.

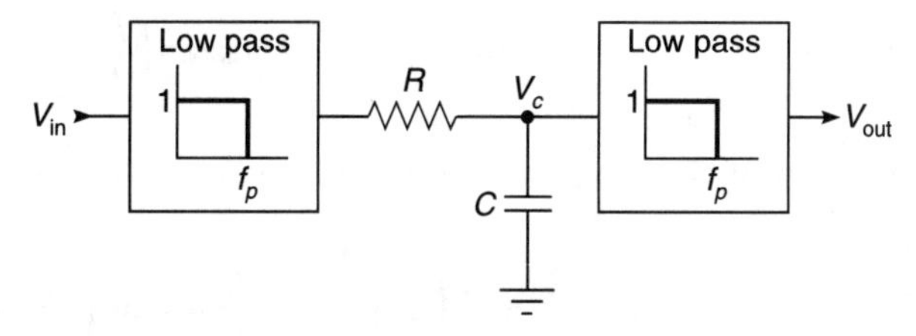

Figure 17.4 Continuous-value, continuous-time RC low-pass circuit approximating a delay function.

For the *RC* circuit, the presence of the capacitor has no effect on the theoretical channel capacity. The reason is that a filtering operation amounts to no more than a coordinate transformation [4]. Note that the bandwidth is limited by the brickwall filter at the input. As such, the analysis can be simplified by removing the capacitor from Figure 17.4. In this case, the output noise power is the resistive thermal noise times the message bandwidth f_p:

$$\overline{V_{\text{out},n}^2} = 4kTRf_p \tag{17.19}$$

Moreover, the output signal power is equal to the input signal power if the capacitor is removed. Assuming the signal to be uniformly distributed in the range 0–V and strictly band limited to f_p, the signal power is approximately $V^2/12$, from (17.14). Therefore, the signal-to-noise ratio is given by

$$\frac{S}{N} = \frac{V^2}{48kTRf_p} \tag{17.20}$$

The above equation for signal-to-noise ratio is valid even after the capacitor is reintroduced to Figure 17.4 because its effect on the signal and noise power is identical.

To compute the mean power dissipation, we derive the mean-square voltage *across* the resistor and divide by R. The power spectrum of the voltage across the resistor is found by multiplying the input power spectrum in (17.15) by the square magnitude of the transfer function from the input voltage to the voltage across the resistor. We obtain

$$\frac{\overline{V_R^2}}{\Delta f} = \frac{V_{\text{in}}^2}{\Delta f} \frac{(f/f_o)^2}{1+(f/f_o)^2} \tag{17.21}$$

where $f_o = 1/2\pi RC$ is the cutoff frequency of the low-pass filter. The mean power dissipated in the resistor is found by integrating over the message bandwidth and dividing by R:

$$\begin{aligned} P_m &= \frac{1}{R}\int_0^{f_p} \frac{V^2}{12f_p} \frac{(f/f_o)^2}{1+(f/f_o)^2}\,df \\ &= \frac{V^2}{12R}\left[1 - \frac{f_o}{f_p}\arctan\left(\frac{f_p}{f_o}\right)\right] \end{aligned} \tag{17.22}$$

As a function of the signal-to-noise ratio, the mean power dissipation is equal to

$$P_m = 4kT\left(\frac{S}{N}\right)f_p\left[1 - \frac{f_o}{f_p}\arctan\left(\frac{f_p}{f_o}\right)\right] \tag{17.23}$$

Similarly, one can write the signal-to-noise ratio as a function of P_m, as in

$$\frac{S}{N} = \frac{P_m}{4kTf_p} \frac{1}{1-(f_o/f_p)\arctan(f_p/f_o)} \tag{17.24}$$

Now we are also in a position to write the capacity of the channel as a function of the power dissipation. Substituting (17.24) into (17.16), the capacity is

$$\mathcal{C} = f_p \log_2\left(1 + \frac{6}{\pi e}\frac{P_M}{4kTf_p}\frac{1}{[1-(f_o/f_p)\arctan(f_p/f_o)}\right) \tag{17.25}$$

The ratio $P_m/\mathcal{C}$ is the bit energy.

17.4.2. Continuous-Value, Discrete-Time Circuit

The CVDT circuit to be analyzed is the sample-and-hold circuit of Figure 17.5. The input signal is band limited by an anti-aliasing filter to frequency $f_s/2$ where f_s is the sampling rate. It is assumed that the switch is closed for a period much longer than the correlation time $R_{sw}C/2$, so that a complete charge transfer occurs. In practice, this is always the case.

The thermal noise of the resistance associated with the MOSFET switch R_{sw} is aliased by the sampling process into the baseband $(0 \to f_s/2)$. In order to show this, one must convolve the noise spectrum at node V_c with a pulse train at frequencies nf_s, where $n = 0, \pm1, \pm2, \ldots$. The result is that all of the noise power which is outside the baseband is aliased *into* the baseband and that, in fact, none of the noise escapes this aliasing. Since the total noise on a capacitor is given by (17.18), it follows that the total noise in the baseband is also given by this equation and that it is very nearly white. The noise at node V_d becomes more flat as R_{sw} decreases relative to f_s. Thus, the noise spectrum at node V_d is approximately

$$\frac{\overline{V_d^2}}{\Delta f} = \frac{kT}{C}\frac{1}{f_s/2} \tag{17.26}$$

Since the output node V_{out} is an ideally filtered version of V_d at frequency f_p, the noise power at the output is

$$\overline{V^2}_{\text{out},n} = \frac{2kTf_p}{Cf_s} \tag{17.27}$$

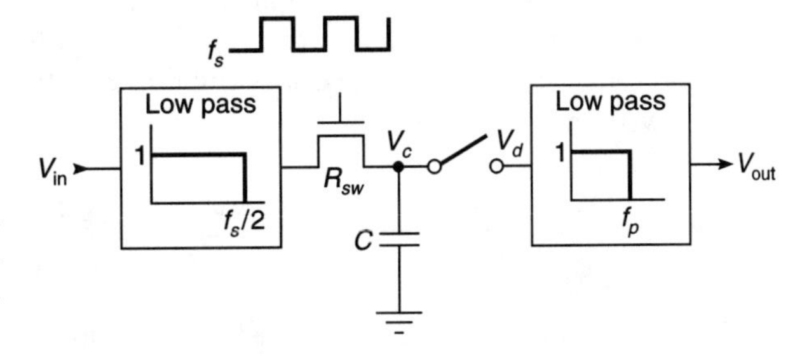

Figure 17.5 Continuous-value, discrete-time switched-capacitor sample-and-hold circuit approximating a delay function.

As before, we shall assume that the input signal is approximately white in the message bandwidth f_p and that the peak power S_{peak} is never exceeded at every sample point. Note that if the signal is sampled at a rate f_s which exceeds the Nyquist rate $2f_p$ or if the sampling clock is not phase locked to the input signal, one would find that some of the samples exceed the peak power limitation, as described earlier.

The average output signal power is equal to the average input signal power $V^2/12$. Dividing S by N in (17.27), we get

$$\frac{S}{N} = \frac{CV^2 f_s}{24kTf_p} \tag{17.28}$$

In order to compute the mean power dissipated in the switch, we need to find the mean-square voltage difference between samples. We perform this computation assuming that the input frequency is not phase locked to the clocked frequency. Let $V_c(n)$ be the voltage at time n/f_s and $V_c(n+1)$ be the next. We want to compute the expected power of the voltage difference $\overline{(\Delta V_c)^2}$, where

$$\begin{aligned}\overline{(\Delta V_c)^2} &\equiv \overline{(V_c(n) - V_c(n+1))^2} \\ &= \overline{V_c(n)^2} + \overline{V_c(n+1)^2} - 2\overline{V_c(n)V_c(n+1)}\end{aligned} \tag{17.29}$$

The first two terms are equal to the signal power. The last term is the input signal autocorrelation function sampled at $1/f_s$.

Now, the input signal is approximately white over the message bandwidth, with average power $V^2/12$. Therefore, the autocorrelation function $R(\tau)$ is a *sinc* function, $\sin(x)/x$, given by

$$R(\tau) = \tfrac{1}{12} V^2 \operatorname{sinc}(2\pi f_p \tau) \tag{17.30}$$

Substituting $\tau = 1/f_s$, the average voltage difference power is

$$\overline{(\Delta V_c)^2} = \tfrac{1}{6} V^2 \left(1 - \operatorname{sinc} \frac{2\pi f_p}{f_s}\right) \tag{17.31}$$

When the sampled voltage on a capacitor changes abruptly by an amount ΔV_c volts, the energy stored on the capacitor changes by $C(\Delta V_c)^2/2$ joules. That same amount of energy is dissipated in the switch, no matter how small the switch resistance. The average power dissipated in switching events occurring at a rate f_s can therefore be written as

$$P_m = \frac{C\overline{(\Delta V_c)^2}}{2} f_s = \frac{CV^2 f_s}{12}\left(1 - \operatorname{sinc}\frac{2\pi f_p}{f_s}\right) \tag{17.32}$$

As a function of S/N the mean power dissipation in the channel is

$$P_m = 2kT\left(\frac{S}{N}\right) f_p \left(1 - \operatorname{sinc}\frac{2\pi f_p}{f_s}\right) \tag{17.33}$$

Likewise, the signal-to-noise ratio can be written as

$$\frac{S}{N} = \frac{P_m}{2kTf_p} \frac{1}{1 - \operatorname{sinc}(2\pi f_p/f_s)} \tag{17.34}$$

The capacity of this circuit can be expressed as

$$C = f_p \log_2\left(1 + \frac{6}{\pi e} \frac{P_m}{2kTf_p} \frac{1}{1 - \operatorname{sinc}(2\pi f_p/f_s)}\right) \tag{17.35}$$

Again, the bit energy is the ratio of power dissipation to channel capacity.

17.4.3. Discrete-Value, Discrete-Time Circuit

For the case of the parallel M-bit register of Figure 17.6, the signal-to-noise ratio is a function of the number of bits used. The input samples must be quantized to fit the finite register length M; herein lies the major source of noise.

Assume that the quantization noise Q_n is uniformly distributed between the two nearest quantization steps. Let the distance between quantization steps be 1 bit. Then the average quantization noise power is $\frac{1}{12}$ bits^2 [15].

The signal power S is computed assuming independent uniformly distributed samples. With 2^M levels, ranging from 0 to $2^M - 1$, one can show that the signal power is

$$S = \frac{2^{2M} - 1}{12} \tag{17.36}$$

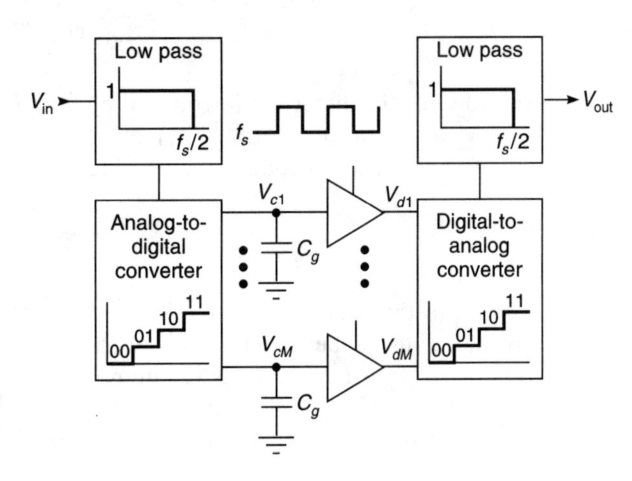

Figure 17.6 Discrete-value, discrete-time digital M-bit parallel register implementing a delay function.

Combining the last two results, the S/N ratio is therefore

$$\frac{S}{N} = 2^{2M} - 1 \tag{17.37}$$

The above computations assume fixed-point, rather than floating-point, representations.

If f_s is defined as the signaling rate at which the entire M-bit message is transmitted, the digital system capacity is

$$C = Mf_s \tag{17.38}$$

The above equation is correct if we only consider the error introduced by quantization noise. Should there be any additive noise in the channel, the binary transmission will exhibit a certain bit error rate P_e. If we then tried to reconstruct an analog waveform from the received digital signal, we would find the resulting S/N ratio to be degraded.

Suppose that the probability of a single-bit error is P_e while the probability of more than one error occurring in a single M-bit transmission is negligible. The expected square distance between the sent and received digital signal $\overline{D^2}$ is

$$\begin{aligned}\overline{D^2} &= P_e 1^2 + P_e 2^2 + \cdots + P_e(2^{M-2})^2 + P_e(2^{M-1})^2 \\ &= \sum_{k=0}^{M-1} P_e(2^k)^2 = P_e \frac{2^{2M} - 1}{3}\end{aligned} \tag{17.39}$$

If we now sum the noise contributions of the quantization step and the distortion introduced by the M-bit register, the total "noise" is $\frac{1}{12} + \frac{1}{3}P_e(2^{2M} - 1)$. Assuming the signal power to be approximately unchanged, we have

$$\frac{S}{N} = \frac{2^{2M} - 1}{1 + 4P_e(2^{2M} - 1)} \tag{17.40}$$

In order for the distortion introduced by the M-bit register to be negligible, we must satisfy the condition $4P_e(2^{2M} - 1) \ll 1$.

Presently, we introduce the weighted probability of error ε, where

$$\varepsilon = 4P_e(2^{2M} - 1) \tag{17.41}$$

For ε constant, the degradation in the S/N ratio due to the parallel register is held constant, independent of M. In order to achieve this goal, the probability of a single-bit error P_e must decrease as the number of bits M increases.

Now we wish to relate the probability of error P_e to the power consumption in the M-bit register. Let us consider a stream of binary symbols with two permissible states, 0 or V volts. At the receiver we are interested in knowing whether a pulse of fixed amplitude V is present or not within a certain time interval. Assuming Gaussian noise $\overline{V_n^2}$ which affects both states equally and a detection threshold set to $V/2$, the bit error rate is

$$\begin{aligned} P_e &= \int_{V/2}^{\infty} \frac{1}{\sqrt{2\pi \overline{V_n^2}}} \exp\left(\frac{-x^2}{2\overline{V_n^2}}\right) dx \\ &= \operatorname{erfc}\left(\frac{V}{2\sqrt{\overline{V_n^2}}}\right) \end{aligned} \tag{17.42}$$

where $\operatorname{erfc}(x) \equiv 1/\sqrt{2\pi} \int_x^{\infty} \exp(-y^2/2)\, dy$.[1]

Consider a logic gate consuming no quiescent current. The energy in an elementary switching event W_g is

$$W_g = \tfrac{1}{2} C_g V^2 \tag{17.43}$$

The noise power in a single gate is $\overline{V_c^2} = kT/C_g$. Therefore, we have

$$P_e = \operatorname{erfc}\left(\left[\frac{W_g}{2kT}\right]^{1/2}\right) \tag{17.44}$$

Substituting for P_e from the definition of ε in (17.41) and inverting the equation, one can show that

$$W_g = 2kT\left[\operatorname{erfc}^{-1}\left(\frac{\varepsilon}{4(2^{2M}-1)}\right)\right]^2 \tag{17.45}$$

If the digital samples are independent and uniformly distributed, then on average half of them will be switching and half will remain unchanged. For the case of an M-bit register and clock rate f_s, we have

$$\begin{aligned} P_m &= \frac{MW_g}{2} f_s \\ &= kTMf_s\left[\operatorname{erfc}^{-1}\left(\frac{\varepsilon}{4(2^{2M}-1)}\right)\right]^2 \end{aligned} \tag{17.46}$$

where f_s is the sampling rate.

Now the number of bits M relates to the S/N ratio of an analog signal by inverting equation (17.37) to obtain

[1] Unfortunately, the complementary error function has multiple definitions.

$$M = \tfrac{1}{2}\log_2\left(1 + \frac{S}{N}\right) \tag{17.47}$$

Substituting (17.47) into (17.46), we obtain

$$P_m = \frac{kTf_s}{2}\log_2\left(1 + \frac{S}{N}\right)\left[\operatorname{erfc}^{-1}\left(\frac{\varepsilon}{4S/N}\right)\right]^2 \tag{17.48}$$

Numerical techniques must be used to solve for S/N as a function of P_m.

Similarly, one can write an equation for P_m as a function of the capacity. The inverse looks very difficult. From (17.38), we substitute $M = C/f_s$ into (17.46) to obtain

$$P_m = kTC\left[\operatorname{erfc}^{-1}\left(\frac{\varepsilon}{4(2^{2C/f_s} - 1)}\right)\right]^2 \tag{17.49}$$

The equation for the energy dissipated per M-bit digital transmission is just $P_m f_s$, which can be computed easily from the above equations. The lower bound on the amount of energy dissipated in a digital gate was derived by Landauer. He estimated that dissipation of the order kT per logic step is required owing to thermodynamic limits [15]. How close one operates to the thermodynamic limit will directly influence the probability of error.

17.4.4. Discrete-Value, Continuous-Time Circuit

Here, the input signal originates in a binary source, which, at certain instants in time, changes state from 1 to 0 or from 0 to 1. The information to be conveyed is *the time between transitions*. This type of signal representation is at times referred to as "event-driven" or "zero-crossing" representation. Logan's theorem [17] states that if a signal is strictly band limited to within one octave, then a signal can be completely reconstructed from its zero crossings within a multiplicative constant. Therefore, a bank of noiseless single-octave bandpass filters comprise the first stage of a DVCT function, as seen in Figure 17.7.

The time delay in the discrete-value signal is implemented with a simple RC low-pass filter, as for the case of the CVCT circuit. The noise in the circuit will come from the thermal noise in the resistor. The energy dissipated per switching event will be the same as that for the M-bit register. An upper bound on the power dissipated in the channel effectively results in an upper bound on the frequency of events, or transitions.

The three major questions that are currently under investigation are [16]:

1. How does the thermal noise in the resistor relate to noise, or jitter, on the arrival time of the transition at node V_c?
2. What is the capacity of this circuit?

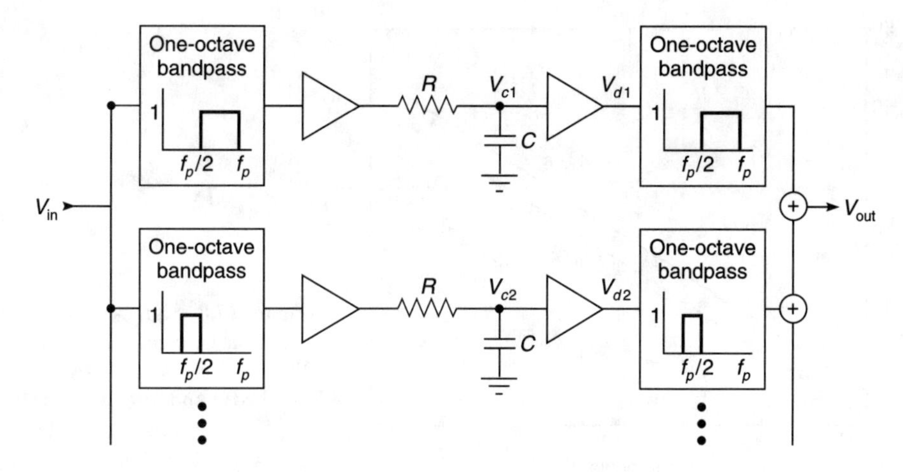

Figure 17.7 Discrete-value, continuous-time zero-crossing and *RC* low-pass filter approximation of a delay function.

3. What is the distribution of the input signal which achieves this capacity?

Partial answers to these questions are outlined in [16]. In brief, it appears that the jitter in the DVCT circuit is approximately Gaussian provided $V \gg 4kTR$. Also, using the entropy-power inequality [18], a lower bound can be found on the capacity of the DVCT channel. This bound becomes tight as R is reduced. Finally, the input signal which achieves the lower bound is one for which the time between transitions follows an exponential distribution, i.e., it is a Poisson process.

17.5. RESULTS

In Figure 17.8, the signal-to-noise ratio is plotted as a function of the mean power dissipation for the CVCT, CVDT, and DVDT circuits. In this plot, the message bandwidth f_p is 100 MHz, the low-pass filter cutoff frequency f_o is also 100 MHz, and the sampling rate f_s is equal to the Nyquist rate, or 200 MHz. For the digital circuit, the weighted bit error probability ε is taken as 10^{-6}.

For the continuous-value circuits, the signal-to-noise ratio grows linearly with the power dissipation. On the other hand, for the digital implementation, the signal-to-noise ratio grows *exponentially* with power dissipation. Herein lies the main advantage of digital signaling schemes. Note that the crossover point between the analog and digital circuits nominally occurs at 4 bits, which is low indeed. For another set of parameter values, this crossover point will change.

In Figure 17.9, the capacitance in bits/second is plotted as a function of the mean power dissipation in joules per second for the CVCT, CVDT, and DVDT systems. Parameter values are the same as for Figure 17.8.

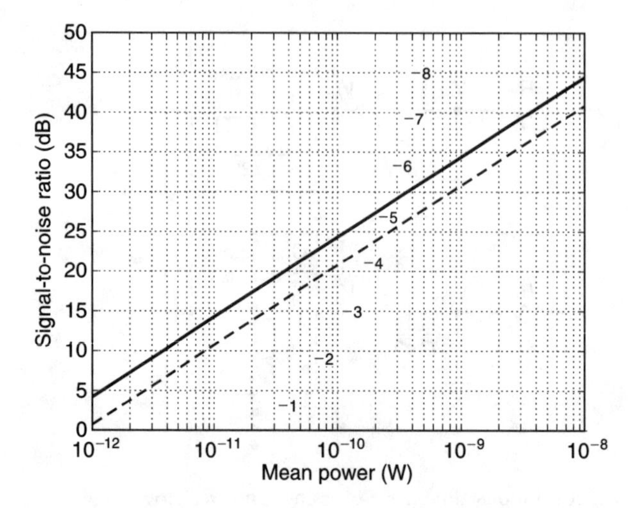

Figure 17.8 Signal-to-noise ratio as a function of mean power dissipation for the analog (solid), switched-capacitor (dashed), and digital (number of bits) implementations of a delay function ($f_p = f_o = 100$ MHz, $f_s = 200$ MHz, $\varepsilon = 10^{-6}$.

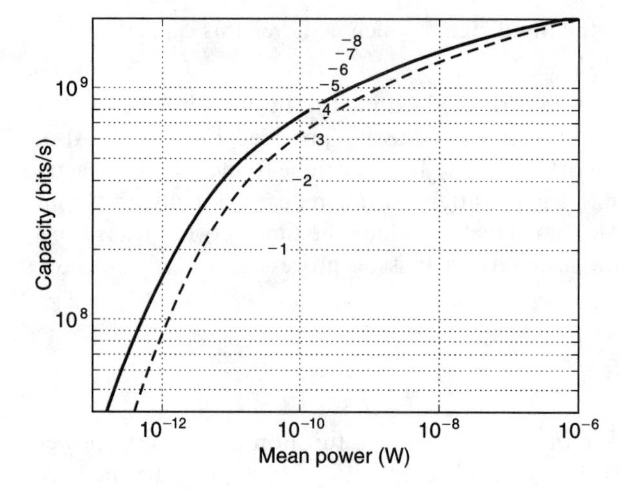

Figure 17.9 Channel capacity as a function of mean power dissipation for the analog (solid), switched-capacitor (dashed), and digital (number of bits) implementations of delay functions ($f_p = f_o = 100$ MHz, $f_s = 200$ MHz, $\varepsilon = 10^{-6}$.

For very low values of power dissipation, the two continuous-value circuits have superior performance in terms of channel capacity. However, as the number of parallel lines increases, the channel capacity of the digital system surpasses the continous-value circuits. The crossover is again in the neighborhood of 4 bits.

The trade-off between analog, switched-capacitor and digital implementations of a delay function is perhaps best demonstrated in Figure 17.10. For the same parameter values as before, we see that the continuous-value systems outperform the digital system at very low signal-to-noise ratios. However, as the signal-to-noise level increases, the bit energy of the continuous-value systems increases dramatically, whereas it remains virtually unchanged for the digital system.

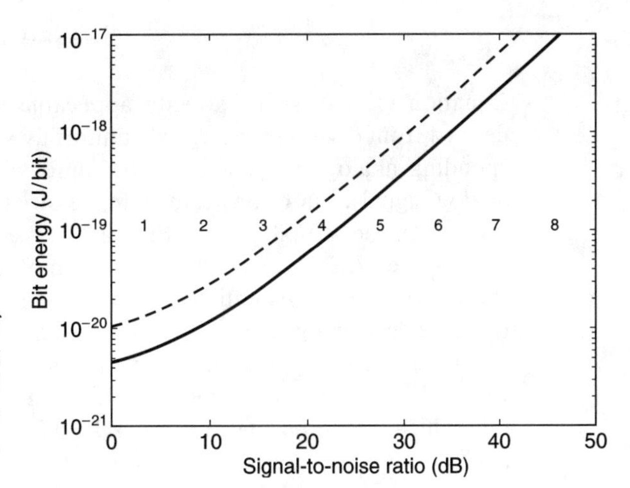

Figure 17.10 Bit energy as a function of signal-to-noise ratio for the analog (solid), switched-capacitor (dashed), and digital (number of bits) implementations of delay functions ($f_p = f_o = 100\,\text{MHz}$, $f_s = 200\,\text{MHz}$, $\varepsilon = 10^{-6}$.

17.6. DISCUSSION

The results presented in this chapter are qualitatively similar to those found in [7] when comparing analog and digital signal representations. However, quantitatively they are quite different. For example, the signal-to-noise ratio of the CVCT system as a function of power dissipation is 6.7 dB higher in this work than that computed in [7]. The discrepancy is due to the Hosticka's assumption that *all* of the signal power is dissipated in the resistor.

From the graphical results, it seems that digital systems outperform analog systems whenever "precise" information must be communicated. Why is this so? The answer lies in the exponential relationship among parallel lines of the digital system. Each bit gives information over a different range of input signal amplitude. One could build an analog system that encodes information in parallel lines so that each line conveys different amplitude ranges. In other words, the key here is the *encoding* and *decoding* strategies.

We have systematically ignored many issues in this work. We have not considered the power dissipation and noise introduced by the ideal low-pass filters in all of the circuits. In addition, we have examined neither algorithms nor implementations for encoding and decoding the input and output signals, respectively. Because these encoder and decoder blocks must operate within the energetic and physical constraints of the main system, the encoder and decoder functions cannot be arbitrarily complex.

Finally, the information theoretic framework presented in this chapter is valuable not only for establishing fundamental limits in signal representations but also for establishing practical limits for future VLSI microsystems. For example, instead of employing maximum entropy encoding for our signals to attain the Shannon capacity, we could use realistic statistics for our signals. Based on that and the necessary rate R_t at which the computation must be carried out and the bit error rates, we can choose the power supplies to attain the required information rate at the given degree of reliability.

Appendix A

Equation (17.16) is not directly applicable to continuous-time signals which are derived from discrete-time signals uniformly sampled in the range $\pm V/2$. The corresponding analog signal *will not* be uniformly distributed. More importantly, the analog signal is not constrained to lie strictly within the region of $\pm V$ at times other than the sample points. Figure 17.11 shows this phenomenon.

Figure 17.12 gives a histogram of samples from a uniform distribution taken on the interval ± 0.5. Essentially, this distribution is a boxcar function. A 16-times oversampled version of the signal is created by placing 15 zeros between each sample point and then low-pass filtering with a Kaiser window. The resulting histogram in Figure 17.13 shows that indeed the distribution shape is no longer uniform but has tails which extend beyond the interval ± 0.5. Just how far they extend determines the peak power S_{peak}. For the time being, we will ignore these tails. As such, the application of (17.16) to continuous-time signals yields results which are somewhat optimistic.

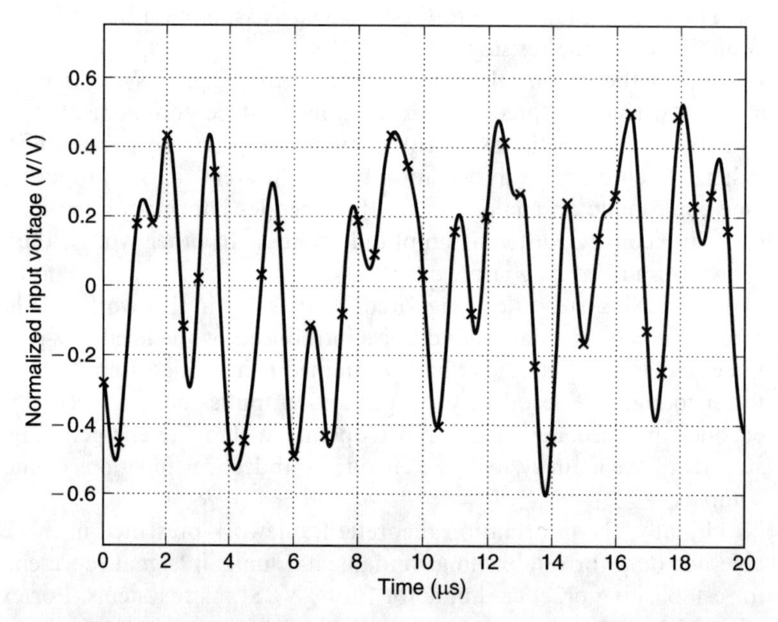

Figure 17.11 A 16-times oversampled version (solid line) of a discrete-time signal with uniformly-distributed samples (x's) on the interval ± 0.5. Note that the amplitude of the oversampled signal extends beyond the range of ± 0.5 at times other than the sample points.

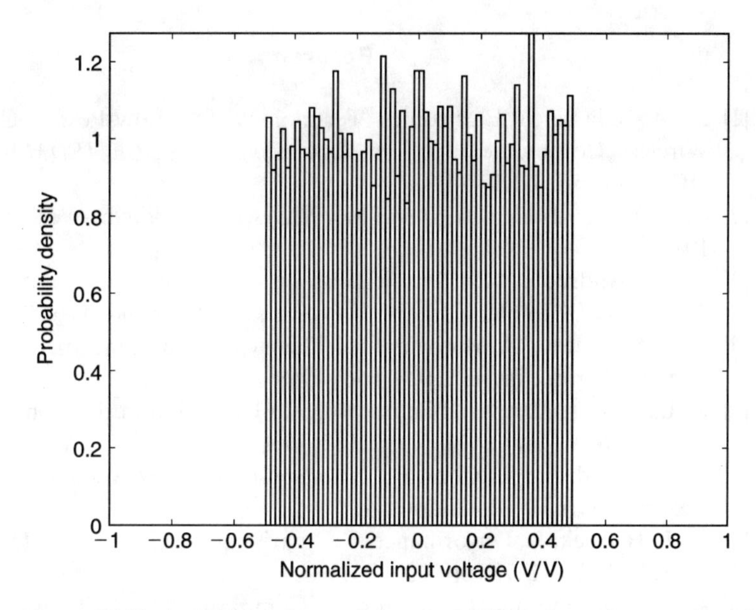

Figure 17.12 A 64-bin histogram of 8192 independent uniformly-distributed samples on the interval ±0.5.

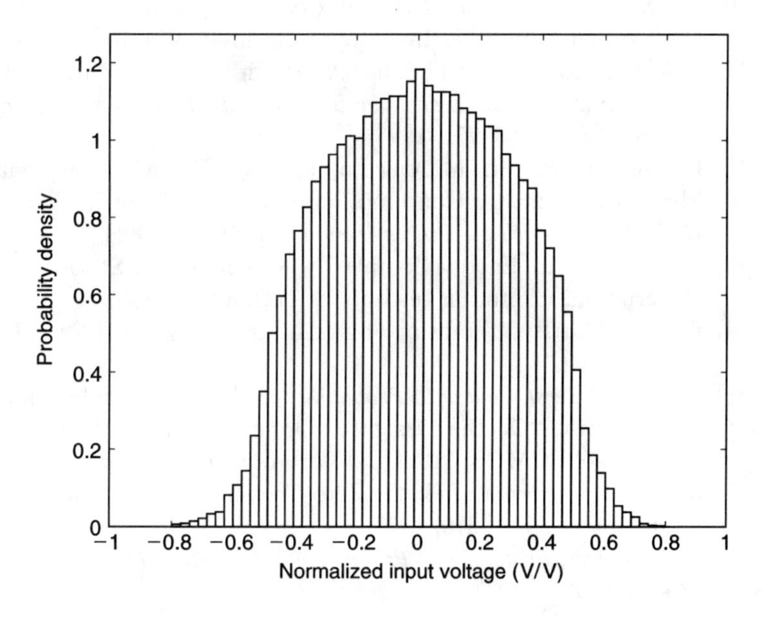

Figure 17.13 A 64-bin histogram of a 16-times oversampled version of the 8192-point uniformly-distributed discrete-time signal on the interval ±0.5.

References

[1] L. A. Glasser, "Electronics Technology for Low-Power Computing and Wireless Communication," in *Proceedings of IEEE IEDM-93*, Washington DC, 1993, pp. 1.1.1–1.1.3.

[2] C. A. Mead, *Analog VLSI and Neural Systems*, Addison-Wesley, Reading, MA, 1989.

[3] A. G. Andreou, "On Physical Models of Neural Computation and Their Analog VLSI Implementation," *Proceedings PhysComp94*, 1994, pp. 255–263.

[4] C. E. Shannon, "A Mathematical Theory of Communication," *Bell Sys. Tech. J.*, vol. 27, pp. 379–423, 623–656, 1948.

[5] R. Landauer, "Irreversibility and Heat Generation in the Computing Process," *IBM J. Res. Devel.*, vol. 5, pp. 183–191, 1961.

[6] C. A. Mead and L. Conway, *Introduction to VLSI Systems*, Addison-Wesley, Reading, MA, 1980, Chapter 9.

[7] B. J. Hosticka, "Performance Comparison of Analog and Digital Circuits," *Proc. IEEE*, vol. 73, pp. 25–29, 1985.

[8] A. G. Andreou and P. M. Furth, "Comparing the Bit-Energy of Continuous and Discrete Signal Representations," in *Proceedings of the Fourth Workshop on Physics and Computation (PhysComp96)*, New England Complex Systems Institute, Boston, MA, pp. 127–133, 1994.

[9] G. Chaitin, "Information-Theoretic Computational Complexity," *IEEE Trans.* Inform. Theory, vol. IT-20, pp. 10–15, 1974.

[10] Y. S. Abu-Mostafa, "Complexity of Information Extraction," Ph.D. Dissertation, California Institute of Technology, 1983.

[11] P. Furth and A. G. Andreou, "A Design Framework for Low Power Analog Filter Banks," *IEEE Trans. Circuits Systems, Pt I: Fund. Theory Applicat.*, vol. 42, no. 11, pp. 966–971, 1995.

[12] D. Marculescu, R. Marculescu, and M. Pedram, "Information Theoretic Measures of Energy Consumption at Register Transfer Level," *Proc. ACM/IEEE Int. Symp. Low Power Design*, pp. 81–86, 1995.

[13] J. Tierno, "An Energy-Complexity Model for VLSI Computations," Ph.D. Dissertation, California Institute of Technology, 1995.

[14] R. V. L. Hartley, "Transmission of Information," *Bell Syst. Tech. J.*, pp. 535–563, 1928.

[15] S. Ross, *A First Course in Probability*, 3rd ed., Macmillan, 1988.

[16] P. M. Furth, "On the Design of Optimal Continuous-Time Filter Banks in Subthreshold CMOS," Ph.D. Dissertation, Johns Hopkins University, 1996.

[17] B. Logan, "Information in the Zero-Crossings of Band Pass Signals," *Bell Syst. Tech. J.*, vol. 56, p. 510, 1977.

[18] R. E. Blahut, *Principles of Practice of Information Theory*, Addison-Wesley, pp. 280–282, 1987.

Chapter 18

Paul Vanoostende and
Geert Van Wauwe
Alcatel-Bell, Advanced CAD for VLSI, F. Wellesplein 1, B-2018 Antwerp, Belgium

A Synchronous Gated-Clock Strategy for Low-Power Design of Telecom ASICs

18.1. INTRODUCTION

This chapter shows how a synchronous gated-clock strategy can result in a reduction by 25% of the complete chip power [including logic, random-access memory (RAM) clock and input/output (I/O)] while retaining the ease and safety of fully synchronous design. The key point is that, after a limited extension of commercial logic synthesis tools, the clock-gating problem can be completely hidden for a designer who uses a methodology based on logic synthesis. The objective of this chapter is to show that including this extension in logic synthesis is a major added value.

Recently, the cooling problems of state-of-the-art integrated circuits (ICs) and the rapid growth of battery-operated electronics have caused a large interest in power reduction techniques. In the literature (e.g., [1–4]), the main focus has been the reduction of the fC product of the outputs of the logic gates, where f is the product of the clock period and the probability of having a rising transition at the node and C its load capacitance.

In contrast, our investigation started with an analysis of the importance of the different contributions (logic gates, clock, RAM, I/O) in a set of typical telecom ICs that were designed at Alcatel-Bell. Surprisingly, it appeared that the most important contribution is due to the clock circuitry inside the flip-flops (cf. Section 18.2). As a result, we developed a synchronous gate-clock strategy (cf. Section 18.3) that enables a significant power reduction (cf. Section 18.4) while retaining the ease and safety of synchronous design.

18.2. POWER CONTRIBUTIONS: IMPORTANCE

To determine the area where most power gain is possible, we started with an analysis of the importance of the different contributions to the power.

Our design methodology is based on logic synthesis and fully synchronous design, where only interfaces contain a small amount of asynchronous logic.

Figure 18.1 shows the contributions for a 125k-gate telecommunication application-specific IC (ASIC) designed in a 0.7-μm complementary metal-oxide-semiconductor (CMOS) technology whose power dissipation is 3.4 W. Similar results were obtained for a set of 12 CMOS designs.

The power data are obtained from a simulation with a Verilog-XL version we modified to report cell and net toggle counts.

The total capacitance due to the nonclock nodes is much larger than the capacitance of the clock circuitry. However, the clock circuitry causes the largest power contribution because of the large difference in switching activity between a clock node and an output of a logic gate. The probability of a rising transition during a clock period is 100% for the clock, whereas in the analyzed designs, it is only 6% on the average for the logic gate outputs (including glitches). For random data without time correlations and with an equal probability of observing a low and a high logic value at the considered node, this probability is 25% (neglecting glitches). This figure is obtained as the product of the probability that the node is low in the preceding clock cycle and the probability that it is high in the current clock cycle. The low probability observed is due to the fact that, in a digital telecommunication non-DSP ASIC (in which digital words are manipulated purely as data and not as signals), only a fraction of the logic is active every cycle of the clock period. The remainder of the logic is only active for a few cycles per data frame of x-bytes (e.g., 2400 bytes for the example of Fig. 18.1) or when an on-board controller command must be executed.

In DSP ASICs containing a pipeline of blocks that process signals, higher activities occur. However, even in such ASICs, blocks with a low activity are often present. An example is error correction logic in a channel decoder ASIC. Such logic typically has less activity when processing correct data than when processing incorrect data and thus consumes less power in the former case.

The capacitance of the clock circuitry is composed of two contributions:

- Clock-routing capacitance and input capacitance of the clock pins of the flip-flops
- Clock circuitry inside each flip-flop, as shown in Figure 18.2

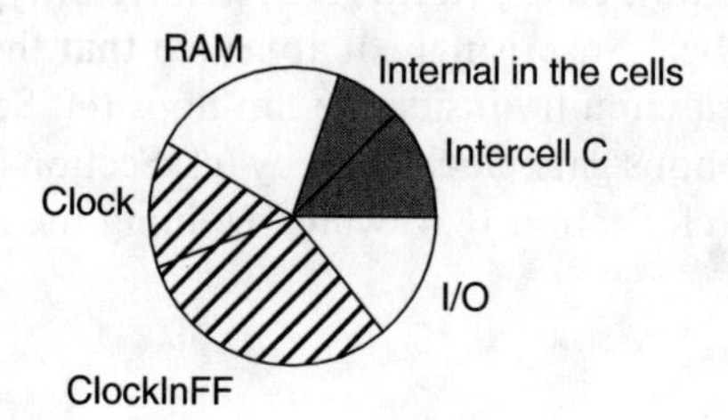

Figure 18.1 Power contributions for 125k-gate ASIC.

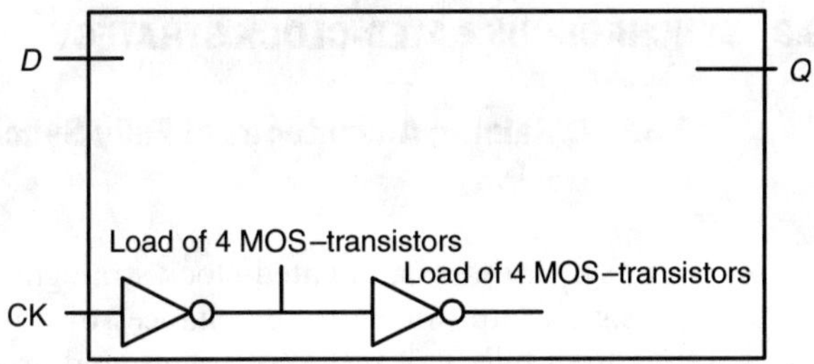

Figure 18.2 Clock circuitry inside flip-flop.

The importance of the second contribution is illustrated by the data for the resettable flip-flop in our 0.7-μm library (Table 18.1). In the table, Cin_Ck is the input capacitance of the clock pin of the flip-flop. Cequiv_internalClocknet is the total capacitance inside the flip-flop that switches when its clock pin switches but its output Q remains stable. Cequiv_Qrise is the difference of the total capacitance inside the flip-flop switching when the output Q switches after an active edge of the clock pin, and Cequiv_InternalClocknet.

The power dissipated by a flip-flop when its output does not switch is thus

$$220\,\text{fF} \times f_c V_{dd}^2$$

whereas the power dissipated when its output is at its most active (Qrise − prob = 25%, cf. above) is

$$(220\,\text{fF} \times f_c + 684\,\text{fF} \times \tfrac{1}{4} f_c) V_{dd}^2$$

the "nonactive" power is thus more than 50% of the "full active" power. Although the above figures are taken from one specific library, similar tendencies have been observed in other libraries.

The above observation provides the key toward power reduction: *If the output of a flip-flop does not have to switch, the flip-flop is not active, and does not have to be clocked.* This can be achieved by using a gated-clock strategy, as discussed in the next section.

TABLE 18.1 EXAMPLE EQUIVALENT CAPACITANCES RELATED TO SWITCHING OF SPECIFIC NODES

Cin_Ck	26 fF
Cequiv_internalClocknet	220 fF
Cequiv_Qrise	684 fF

18.3. SYNCHRONOUS GATED-CLOCK STRATEGY

18.3.1. Obtaining Advantages of Fully Synchronous Design

Asynchronous and gated-clock strategies are currently not heavily used because they significantly increase the chances of errors in the circuit, require more design effort than fully synchronous design, and are hard to automate.

As a consequence, the fully synchronous design methodology, in which each flip-flop is clocked continuously, has become our preferred design approach. However, as discussed in Section 18.2, it wastes power. The challenge thus lies in avoiding the power wasting while still obtaining the advantages of fully synchronous design.

This can be achieved by the following strategy:

- Replace *each* flip-flop by a flip-flop with a load pin L.
- Inside the flip-flop, combine the L and Ck signals in such a way that:
  On an active edge of Ck, Q becomes equal to D if L is high and does not switch when L is low.
  On the nonactive edge of Ck and when L is stable, Q does not switch.
  When Ck is equal to the value it has between the active and the nonactive edge (e.g., low for a neg-edge flip-flop), Q does not switch, regardless of the value of L.
  When Ck is equal to the value it has between the nonactive and the active edge (e.g., high for a neg-edge flip-flop), and L is stable, Q does not switch.
- Generate for each flip-flop a load signal that is high in at least the cases where the flip-flop output must toggle. This signal must be stable before the non-active edge of the clock.
- Match for each gated-clock flip-flop (e.g., the settable and resettable ones) the delays between the active edge of the Ck-pin and the sampling of the D-pin.

This is illustrated in Figure 18.3.

Traditional gated-clock strategies implement the gating of the clock together with the logic that implements the clock-gating signal. This circuitry is, for example, implemented in a standard cell that is part of the module that also contains the considered flip-flop. In that case, skew is introduced between the active edges of the clock pins of the flip-flops of the module, depending on the variations in the length of the wires connecting the clock pin of the considered flip-flip to the output of the standard cell that implements the gating logic. Note that the correct behavior is very sensitive to even small skews.

In contrast, the new strategy performs the gating of the clock internally in each flip-flop. Each clock-gating NAND gate (cf. Fig. 18.3) has thus exactly the same load, and the delay from Ck to Ckg is identical for all flip-flops. As a result, the above strategy does not introduce additional skew, compared to the case without the gated-clock flip-flops.

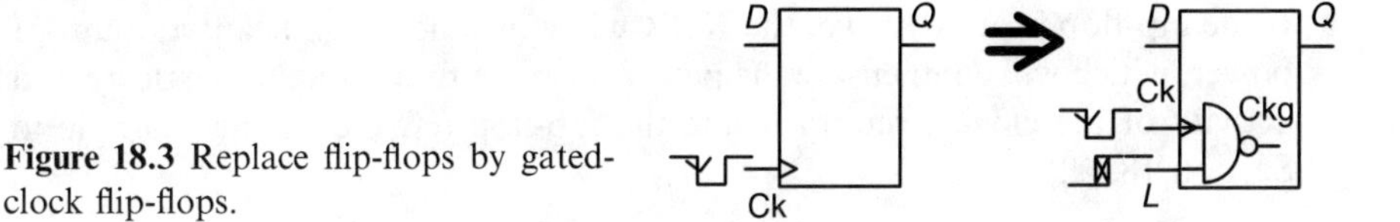

Figure 18.3 Replace flip-flops by gated-clock flip-flops.

As a consequence, the proposed methodology is as safe and easy to apply as the fully synchronous design strategy and can be driven from logic synthesis. The only remaining problem is the derivation of the load signal.

The proposed technique is also easily combined with the full-scan methodology, which is currently the most widespread technique for testing digital logic. This can be achieved by using the scan-enable pin to select inside the flip-flop between the clock and the gated clock.

18.3.2. Derivation of Load Signal

XOR-Based Approach. For each cycle in which the new data (D) differs from the stored data (Q), the flip-flop must be clocked. In the other cycles, this is not really needed. Therefore, the exor of D and Q is a suitable load signal. This is shown in Figure 18.4. This is a general solution which is easy to implement. However, it has the following disadvantages:

- The delay of L is equal to the delay of D plus an XOR delay, and L must be stable before the nonactive clock edge. For a symmetric clock, this means that L must be stable before half the clock cycle. As a consequence, the maximum operating frequency is more than halved.

  This problem can be alleviated by using an asymmetric clock (duty cycle $\gg 50\%$) and by not using the XOR-L for flip-flops for which the paths to D are long. For these flip-flops, V_{dd} could be used as the load signal. Note that it is not possible to replace such flip-flops, (i.e., flip-flops having V_{dd} as load signal) by flip-flops without gating logic, as the non-gated-clock flip-flops clock their internal latches on Ck as opposed to the gated-clock flip-flops where Ckg clocks the latches (cf. Fig. 18.3), which would result in skew betwen the gated-clock and non-gated-clock flip-flops.
- For each flip-flop, there is an overhead of 1 XOR to generate the load signal. This overhead can be reduced by incorporating the XOR in the flip-flop cell but still remains significant.

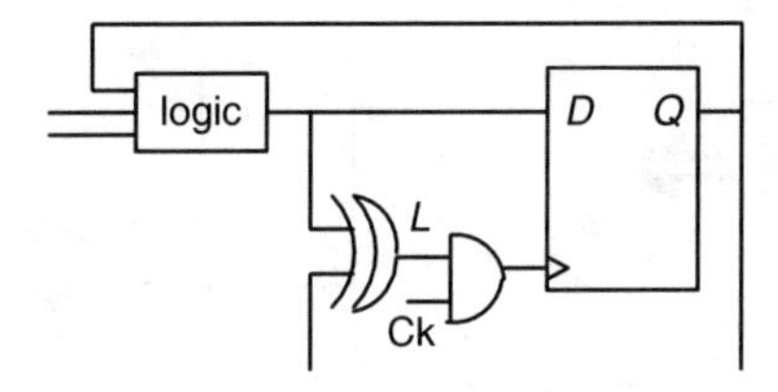

Figure 18.4 XOR-based approach.

- If the flip-flop is quite active, the XOR will consume a significant amount of power, which will compensate the power decrease that is due to the decreased activity of the clock circuitry inside the flip-flop (cf. the simple example in Section 18.4.2).

Exploiting Conditional Assignments. In our opinion, the most promising approach to obtain L is based on the following observation. The major part of a hardware description language (HDL) design description is typically formed by conditional assignments (if/else statements):

```
if cond1 then
              if cond2 then a <=...
              if cond3 then a <=...
/* else a <= a */
```

In many designs, there is an implicit "else : a < = a" assignment, which results in the structure in the left-hand side of Figure 18.5.

The discussion in the previous section highlighted the main problem in generating L: There is considerably less time available compared to the generation of D. Therefore, the expression for L should be simple. Fortunately, it is allowed to have L equal to 1 even when the flip-flop does not strictly need to be clocked: The expression for L may be simplified as long as L is a cover of $D \oplus Q$:

```
on-set(D⊕Q)⊂ on_set(L)
```

that is, L is 1 at least when $D \oplus Q$ is 1.

Our proposal is now to use for L a simplified cover of the union of the conditions under which the considered variable is explicitly assigned. Thus L is simplified cover of the multiplexer control signal in Figure 18.5.

In particular one could construct a simple expression for L by retaining the first level of if statements in the HDL code, that is, condition 1 in the above example.

Due to the implementation of L, the logic to generate the D input signal can be simplified: The original D-logic is split into D-logic and L-logic. In this way, logic synthesis might limit the area overhead that is due to the logic for L. This task appears to be more complicated that simply constructing a suitable L. Note, however, that this D-logic reduction is an additional optimization and is not essential for the proposed approach.

Figure 18.6 schematically shows the impact of the complexity of L on the power of the clock net internally in the flip-flop, on the area, and on the power needed to

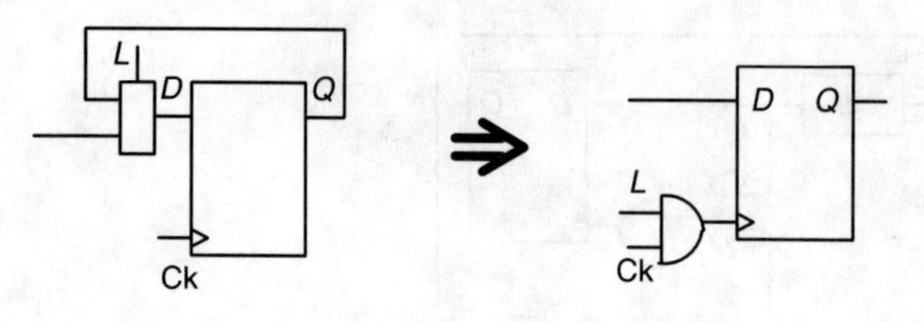

Figure 18.5 Feedback loop around flip-flop.

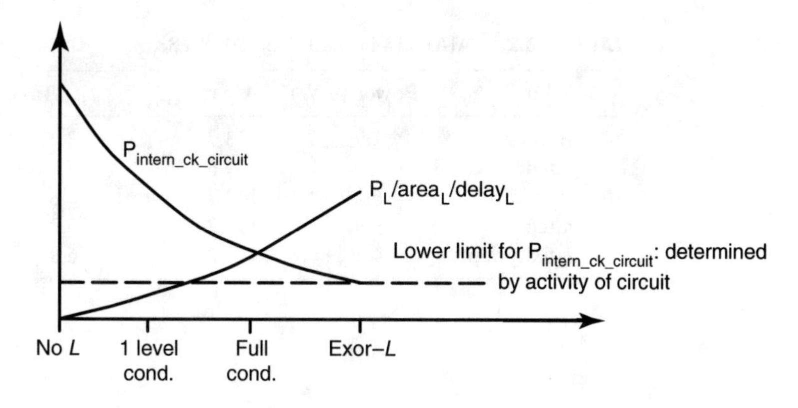

Figure 18.6 Impact of complexity of *L*.

generate *L* and its delay. Note that the impact on the total area is more difficult to predict because of the potential D-logic reduction.

18.4. RESULTS

In the sequel, the power is reported for a clock of 40 MHz with a duty cycle of 50%. Here, T_{min} is the minimum clock period at which the design could be clocked using a symmetric clock. The area is in NAND2 units.

18.4.1. Gated Clock Inside Library Modules

During library development, the synchronous gated-clock methodology can be applied in a more intelligent way than is possible by the automated approaches of Section 18.3.2.

For example, consider a simple counter:

```
a <= a+1;
```

A suitable expression for *L*[*i*] is

```
and(L[i-1], A[i])
```

where *L*[*i*] denotes the load signal for bit *i*.

Results are shown in Table 18.2. For the 4-bit counter, the activity of the flip-flops is such that the power gain due to the gated-clock strategy is offset by the overhead to generate the load signals. For the 16-bit counter, there is a power gain of 38%. Unfortunately, this gain comes at the expense of a large increase in T_{min}.

Therefore, the AND tree is cut in the last experiment. For the 4th, 8th, and 12th bitslice (i.e., $l = 4, 8, 12$), the following *L*[*k*] is tried:

```
L[k]=and(A[k-1], A[k])
```

TABLE 18.2 DATA FOR 4-BIT AND 16-BIT COUNTERS

	N-bit	Power (mW)	Area	T_{min} (ns)
4	Standard	2.9	33.7	2.7
	gatedc	2.9	39.6	5.8
16	Standard	6.3	147.7	6.8
	gated	3.9	175.2	16.2
	gatedc_depth4	4.1	175.2	6.8

instead of the above expression.

The result, called gatedc_depth4 in Table 18.2, is a power gain of 35% without increase in T_{min} and with an area increase of 19% with respect to the standard circuit.

18.4.2. Exploiting Conditional Expressions

Simple Example. To illustrate tne strategy proposed in Section 18.3.2 under Exploring Conditional Assignments, consider the following simple example:

```
if C then A <= B;
```

Results are shown in Table 18.3, assuming arrival times of 10 ns for *B* and *C*. Probabilities of 10 and 25% for the rising transistion of *A* and 5 and 50% for the high probability of *C* are considered.

The first row is the standard solution of logic synthesis; in the second row *C* is used as the load signal. The last row describes the result obtained by using commercial logic synthesis to generate *L*: A flip-flop with an incorporated mux, as shown in Figure 18.5, was provided as the only memory element. Traditional logic synthesis then leads to the Exor-based solution for *L* that was discussed in Section 18.3.2.

TABLE 18.3 DATA FOR SIMPLE EXAMPLE

	Power (mW)			Area (no. of gates)	T_{min} (ns)
	A:10%r		A:25%r,		
	C:5%h	C:50%h	C:50%h		
Standard	0.31	0.37	0.49	8.7	10.7
gMan	0.11	0.26	0.35	6.5	10.3
gAuto	0.24	0.21	0.46	10.1	22.6

As expected, the power reduction due to the gated clock is most apparent for the case when there is a small probability that A is updated (C has a probability of 5% to be high). The area reduction in the second row is due to the fact that the D-logic (i.e., a multiplexer) could be eliminated in that case. However, we should not generalize this conclusion; in general, we expect a small area overhead when the gated-clock strategy is introduced.

The case where A has a 25% probability to be rising and B has a 50% probability to be high presents the "high-activity" case. In that case, the power gain for reduced-clock toggling is offset by the power loss in the Exor.

Real-Life Example. To investigate the real-life usefulness of the proposed technique, we considered the ASIC mentioned in Section 18.2. It appeared that only about 25% of the chip is continuously processing data that is close to random. An estimated 75% of the chip has a very low activity, either because it is only activated for a few cycles per data frame of 2400 bytes or because it is part of an interface with an on-board controller.

Since the implementation of the load signals had to be done manually, we limited ourselves to a medium-size low-activity block. The PT_proc module contains only 10,000 gates but is representative of the major part of the chip that has a very low activity.

As described in the second part of Section 18.3.2, the HDL code was considered and the load signals were generated based on the outer level of if tests around the assignments. Attention was paid not to consider conditions containing variables that were modified by direct assignments (':=" in VHDL) before their use in the condition. As a result, simple logic functions were derived for the load signals of 74 of the 94 flip-flops. For the remaining 20 flip-flops, V_{dd} was used as load signal.

The results are shown in the second row of Table 18.4. The power of the block is reduced by 57% for an area overhead of 5% and a timing overhead of 1%. Note that no simplification of the D-logic was performed: The logic to generate the L signals is directly reported into overhead for the total area of the module.

The internal contribution of the clock circuitry in the flip-flop is reduced from 20.3 to 4.3 mW, and thus almost disappears. It should be stressed that the small gain of 16 mW compared to the 3.4 W of the complete ASIC is due to the fact that, awaiting automation of the described technique, the technique was tried out as an experiment on only 1% of the ASIC. Extrapolating these results to the complete chip, where about 75% of the logic has similar activity as the PT_proc module and the overall internal contribution of the clock circuitry in the flip-flop is about 40%, the overall potential gain is estimated at about 25%.

TABLE 18.4 DATA FOR PT–PROC MODULE

	Power (mW)	Area (no. of gates)	T_{min} (ns)
Standard	28.2	1398	8.5
gatedCond1level	12.2	1472	8.6

18.5. CONCLUSIONS

We have identified that in many telecom ASICs the most important power contribution is due to the clock circuitry inside the flip-flops and explained the reason. A synchronous gated-clock strategy has been described that enables an impressive power reduction while retaining the ease and safety of synchronous design.

References

[1] A. Chandrakasan and R. Brodersen, *Low Power Digital CMOS Design*, Kluwer Academic, Boston, 1996.

[2] J. Monteiro, S. Devadas, and A. Ghosh, "Returning Sequential Circuits for Low Power," *International Conference on Computer-Aided Design*, pp. 398–402, 1993.

[3] P. Vanoostende, P. Six, J. Vandewalle, and H. De Man, "Estimation of Typical Power of Synchronous Circuits Using a Hierarchy of Simulators," *IEEE J. Solid-State Circuits*, vol. 28, no. 1, pp. 26–39, 1993.

[4] *Proceedings of the 1994 International Workshop on Low Power Design*, Napa, CA, April 1994.

Index

A

B

C

D

Q

R

S

About the Editors

Edgar Sánchez-Sinencio was born in Mexico City on October 27, 1944. He received the degree in communications and electrical engineering (professional degree) from the National Polythecnic Institute of Mexico in 1966. He received the M.S.E.E. degree from Stanford University, Stanford, CA, and the Ph.D. degree from the University of Illinois at Champaign-Urbana, in 1970 and 1973, respectively.

He did an industrial postdoctoral with Nippon Electric Company, Kawasaki, Japan, in 1973–1974. Currently, he is with the Department of Electrical Engineering, Texas A&M University, College Station, as a professor. He is the co-author of *Switched-Capacitor Circuits* (New York: Van Nostrand-Reinhold, 1984) and co-editor of *Artificial Neural Networks: Paradigms, Applications, and Hardware Implementation* (Piscataway, NJ: IEEE Press, 1992). He has been the guest editor or co-editor of three special issues on neural network hardware (*IEEE Transactions on Neural Networks*, March 1991, May 1992, May 1993) and one special issue on low-voltage, low-power analog and mixed-signal circuits and systems (*IEEE Transactions Circuit and Systems-I*, November 1995). His present interests are in the area of solid-state processing circuits, including BiCMOS, CMOS RF communication circuits, data converters, and testing. He was the IEEE/CAS Technical Committee Chairman on Analog Signal Processing (1994–1995). He has been associate editor for the different IEEE magazines and transactions from 1982 until the present. He was the IEEE Video Editor for the *IEEE Transactions on Neural Networks*; IEEE Neural Network Council Fellow Committee Chairman, 1994 and 1995; a member of the IEEE Circuits and Systems Society Board of Governors (1990–1992), and the 1993–1994 IEEE Circuits and Systems Vice-President-Publications and a member of the IEEE Press Editorial Board. Currently, he is the editor for the *IEEE Transactions Circuits and Systems-II*. In 1992 he was elected as a fellow of the IEEE for contributions to monolithic analog filter design. In 1995, he received an Honoris Causa Doctorate from the National Institute for Astrophysics, Optics and Electronics, Puebla, Mexico.

Andreas G. Andreou was born in Nicosia, Cyprus in 1957. He received his Ph.D. in electrical engineering and computer science in 1986 from Johns Hopkins University. Between 1986 and 1989 he held postdoctoral fellow and associate research scientist positions in the electrical and computer engineering department, and he was also a member of the professional staff at the Applied Physics Laboratory. Dr. Andreou became an assistant professor of electrical and computer engineering in 1989, associate professor in 1993, and professor in 1997. In 1995 he was a visiting associate professor in the computation and neural systems program at the California Institute of Technology.

He is a recipient of a National Science Foundation Research Initiation Award and is the co-founder of the Center for Language and Speech Processing at Johns Hopkins University. In 1989 and 1991 he was awarded the Applied Physics Laboratory R.W. Hart Prize for his work on integrated circuits for space applications. He is also the recipient of the 1995 and 1997 Myril B. Reed Best Paper Award for his work on the silicon retina and low-power integrated circuits. Dr. Andreou's research interests include electron devices, very large scale integrated circuits, integrated sensory microsystems, and physics of computation.